VOLUME NINETY EIGHT

VITAMINS AND HORMONES

Hormones and Transport Systems

VOLUME NINETY EIGHT

VITAMINS AND HORMONES

Hormones and Transport Systems

Series Editor

GERALD LITWACK, PhD

Toluca Lake, North Hollywood, California

ELSEVIER

AMSTERDAM • BOSTON • HEIDELBERG • LONDON
NEW YORK • OXFORD • PARIS • SAN DIEGO
SAN FRANCISCO • SINGAPORE • SYDNEY • TOKYO

Academic Press is an imprint of Elsevier

Cover photo credit:
Nicola, J.P., Carrasco, N., Masini-Repiso, A.M.
Dietary I^- Absorption: Expression and Regulation of the Na^+/I^- Symporter in the Intestine
Vitamins and Hormones (2015) **98**, pp. 1–32

Academic Press is an imprint of Elsevier
225 Wyman Street, Waltham, MA 02451, USA
525 B Street, Suite 1800, San Diego, CA 92101-4495, USA
125 London Wall, London, EC2Y 5AS, UK
The Boulevard, Langford Lane, Kidlington, Oxford OX5 1GB, UK

First edition 2015

Notices

Knowledge and best practice in this field are constantly changing. As new research and experience broaden our understanding, changes in research methods, professional practices, or medical treatment may become necessary.

Practitioners and researchers must always rely on their own experience and knowledge in evaluating and using any information, methods, compounds, or experiments described herein. In using such information or methods they should be mindful of their own safety and the safety of others, including parties for whom they have a professional responsibility.

To the fullest extent of the law, neither the Publisher nor the authors, contributors, or editors, assume any liability for any injury and/or damage to persons or property as a matter of products liability, negligence or otherwise, or from any use or operation of any methods, products, instructions, or ideas contained in the material herein.

ISBN: 978-0-12-803008-0
ISSN: 0083-6729

CONTENTS

CONTRIBUTORS

Yasaman Aghazadeh
The Research Institute of the McGill University Health Centre, and Department of Medicine, McGill University, Montreal, Quebec, Canada

Denovan P. Begg
School of Psychology, University of New South Wales (UNSW, Australia), Sydney, New South Wales, Australia

Liliana G. Bianciotti
Cátedra de Fisiopatología, Facultad de Farmacia y Bioquímica, Universidad de Buenos Aires, Instituto de Inmunología, Genética y Metabolismo (INIGEM-CONICET), Buenos Aires, Argentina

Nabila Boukelmoune
Department of Integrative Biology and Pharmacology, The University of Texas Health Science Center at Houston, Houston, Texas, USA

Rafael Brito
Program of Neurosciences, Fluminense Federal University, Niterói, Rio de Janeiro, Brazil

David A. Buckley
Department of Pharmacy, School of Applied Sciences, University of Huddersfield, Huddersfield, United Kingdom

Nancy Carrasco
Department of Cellular and Molecular Physiology, Yale University School of Medicine, New Haven, Connecticut, USA

Narattaphol Charoenphandhu
Center of Calcium and Bone Research (COCAB), and Department of Physiology, Faculty of Science, Mahidol University, Bangkok, Thailand

Lihe Chen
Graduate School of Biomedical Sciences, The University of Texas Health Science Center at Houston, and Division of Renal Diseases and Hypertension, Department of Internal Medicine, University of Texas Medical School at Houston, Houston, Texas, USA

Nádia A. de Oliveira
Program of Neurosciences, Fluminense Federal University, Niterói, Rio de Janeiro, Brazil

Alexandre dos Santos-Rodrigues
Program of Neurosciences, Fluminense Federal University, Niterói, Rio de Janeiro, Brazil

Peying Fong
Department of Anatomy and Physiology, Kansas State University College of Veterinary Medicine, Manhattan, Kansas, USA

Peter A. Friedman
Department of Pharmacology & Chemical Biology, University of Pittsburgh School of Medicine, Pittsburgh, Pennsylvania, USA

Jyothsna Gattineni
Department of Pediatrics, University of Texas Southwestern Medical Center, Dallas, Texas, USA

Jürg Gertsch
Institute of Biochemistry and Molecular Medicine, NCCR TransCure, University of Bern, Bern, Switzerland

María J. Guil
Cátedra de Fisiología e Instituto de la Química y Metabolismo del Fármaco (IQUIMEFA-CONICET), Facultad de Farmacia y Bioquímica, Universidad de Buenos Aires, Buenos Aires, Argentina

Sandra I. Hope
Cátedra de Fisiología e Instituto de la Química y Metabolismo del Fármaco (IQUIMEFA-CONICET), Facultad de Farmacia y Bioquímica, Universidad de Buenos Aires, Buenos Aires, Argentina

Masahiro Ikeda
Department of Veterinary Pharmacology, University of Miyazaki, Miyazaki, Japan

Eric Madden
Department of Integrative Biology and Pharmacology, The University of Texas Health Science Center at Houston, Houston, Texas, USA

Mykola Mamenko
Department of Integrative Biology and Pharmacology, The University of Texas Health Science Center at Houston, Houston, Texas, USA

Ana María Masini-Repiso
Departamento de Bioquímica Clínica, Facultad de Ciencias Químicas, Universidad Nacional de Córdoba, Córdoba, Argentina

Toshiyuki Matsuzaki
Department of Anatomy and Cell Biology, Gunma University Graduate School of Medicine, Maebashi, Japan

Patrick C. McHugh
Department of Pharmacy, School of Applied Sciences, University of Huddersfield, Huddersfield, United Kingdom

Juan Pablo Nicola
Departamento de Bioquímica Clínica, Facultad de Ciencias Químicas, Universidad Nacional de Córdoba, Córdoba, Argentina

Simon Nicolussi
Institute of Biochemistry and Molecular Medicine, NCCR TransCure, University of Bern, Bern, Switzerland

Roberto Paes-de-Carvalho
Program of Neurosciences, Fluminense Federal University, Niterói, Rio de Janeiro, Brazil

Vassilios Papadopoulos
The Research Institute of the McGill University Health Centre; Department of Medicine; Department of Biochemistry, and Department of Pharmacology & Therapeutics, McGill University, Montreal, Quebec, Canada

Mariana R. Pereira
Program of Neurosciences, Fluminense Federal University, Niterói, Rio de Janeiro, Brazil

Oleh Pochynyuk
Department of Integrative Biology and Pharmacology, The University of Texas Health Science Center at Houston, Houston, Texas, USA

Matthias Quick
Department of Psychiatry, Division of Molecular Therapeutics, Columbia University College of Physicians and Surgeons, New York State Psychiatric Institute, New York, USA

Lei Shi
Department of Physiology and Biophysics, Institute for Computational Biomedicine, Weill Medical College of Cornell University, New York, USA

Andrey Sorokin
Division of Nephrology, Department of Medicine, Medical College of Wisconsin, Milwaukee, Wisconsin, USA

Alexander Staruschenko
Department of Physiology, Medical College of Wisconsin, Milwaukee, Wisconsin, USA

Marcelo S. Vatta
Cátedra de Fisiología e Instituto de la Química y Metabolismo del Fármaco (IQUIMEFA-CONICET), Facultad de Farmacia y Bioquímica, Universidad de Buenos Aires, Buenos Aires, Argentina

Kannikar Wongdee
Office of Academic Management, Faculty of Allied Health Sciences, Burapha University, Chonburi, and Center of Calcium and Bone Research (COCAB), Faculty of Science, Mahidol University, Bangkok, Thailand

Oleg Zaika
Department of Integrative Biology and Pharmacology, The University of Texas Health Science Center at Houston, Houston, Texas, USA

Wenzheng Zhang
Graduate School of Biomedical Sciences, The University of Texas Health Science Center at Houston, and Division of Renal Diseases and Hypertension, Department of Internal Medicine, University of Texas Medical School at Houston, Houston, Texas, USA

Xi Zhang
Division of Renal Diseases and Hypertension, Department of Internal Medicine, University of Texas Medical School at Houston, Houston, Texas, USA

Barry R. Zirkin
Department of Biochemistry and Molecular Biology, Johns Hopkins University Bloomberg School of Public Health, Baltimore, Maryland, USA

PREFACE

Movements of hormones and ions through intracellular membranes and through the plasma membrane to the cell exterior and movement of these substances from the bloodstream into other cells require the agency of molecular transporters. The functionality of these transporters is essential to the actions of hormones, such as insulin, norepinephrine, and dopamine, or to the actions of ions, such as sodium, calcium, phosphate, and iodide, or to the actions of other substances, such as cholesterol, vitamins, adenosine, endogenous cannabinoids (one is anandamide), and even water molecules. If a transporter is not functioning properly, a disease condition may follow. If there is an excess of a substance being transported and the availability of that substance needs to be reduced, a transporter can become a target for chemotherapy. Of the many steps in the mechanisms of all of these critical molecules or atoms, the transporters themselves become vital regulators. In this volume, the latest research is reviewed on these many topics.

To open this area, the transporters involved in the formation and action of thyroid hormones are considered. The first topic is that of J.P. Nicola, N. Carrasco, and A.M. Masini-Repiso on "Dietary I^- Absorption: Expression and Regulation of the Na^+/I^- Symporter in the Intestine." "Apical Iodide Efflux in Thyroid" is reviewed by P. Fong. D. Braun and U. Schweitzer contribute "Thyroid Hormone Transport and Transporters."

A discussion of the movement of sodium ion and the comovement of other molecules, in some cases, occurs through the following reviews. M. Quick and L. Shi offer "The Sodium/Multivitamin Transporter: A Multipotent System with Therapeutic Implications." "Regulation of αENaC Transcription" is authored by L. Chen, X. Zhang, and W. Zhang. M. Mamenko, O. Zaika, M. Boukelmoune, E. Madden, and O. Pochynyuk write on "Control of ENaC-Mediated Sodium Reabsorption in the Distal Nephron by Bradykinin." This topic is concluded with "Inhibition of ENaC by Endothelin-1," a report by A. Sorokin and A. Staruschenko.

There are many other systems to be considered. Of these, Y. Aghazadeh, B.R. Zirkin, and V. Papadopoulos describe "Pharmacological Regulation of the Cholesterol Transport Machinery in Steroidogenic Cells of the Testis." D.P. Begg has written on "Insulin Transport into the Brain and Cerebrospinal Fluid." "Regulation of Hormone-Sensitive Renal Phosphate Transport" is the focus of J. Gattineni and P.A. Friedman. M. Ikeda and

T. Matsuzaki review "Regulation of Aquaporins by Vasopressin in the Kidney." D.A. Buckley and P.C. McHugh contribute "The Structure and Function of the Dopamine Transporter and Its Role in CNS Diseases." M.S. Vatta, L.G. Bianciotti, M.J. Guil, and S.I. Hope are the authors of "Regulation of the Norepinephrine Transporter by Endothelins: A Potential Therapeutic Target." K. Wongdee and N. Charoenphandhu cover "Vitamin D-Enhanced Duodenal Calcium Transport." "Endocannabinoid Transport Revisited" is the subject of S. Nicolussi and J. Gertsch. The final contribution is that of A. dos Santos-Rodrigues, M.R. Pereira, R. Brito, N.A. de Oliveira, and R. Paes-de-Carvalho who describe "Adenosine Transporters and Receptors: Key Elements for Retinal Function and Neuroprotection."

As always, Helene Kabes of Elsevier (Oxford, UK) and Vignesh Tamilselvvan of Elsevier (Chennai, India) have expedited the final preparations for the publication of this volume.

The cover illustration is taken from Fig. 1 of chapter entitled "Dietary I^- Absorption: Expression and Regulation of the Na^+/I^- Symporter in the Intestine" by J.P. Nicola, N. Carrasco, and A.M. Masini-Repiso.

Gerald Litwack
North Hollywood, California
October 23, 2014

CHAPTER ONE

Dietary I^- Absorption: Expression and Regulation of the Na^+/I^- Symporter in the Intestine

Juan Pablo Nicola*, Nancy Carrasco†,1, Ana María Masini-Repiso*,1

*Departamento de Bioquímica Clínica, Facultad de Ciencias Químicas, Universidad Nacional de Córdoba, Córdoba, Argentina

†Department of Cellular and Molecular Physiology, Yale University School of Medicine, New Haven, Connecticut, USA

[1]Corresponding authors: e-mail address: nancy.carrasco@yale.edu; amasini@fcq.unc.edu.ar

Contents

Abstract

Thyroid hormones are critical for the normal development, growth, and functional maturation of several tissues, including the central nervous system. Iodine is an essential constituent of the thyroid hormones, the only iodine-containing molecules in vertebrates. Dietary iodide (I^-) absorption in the gastrointestinal tract is the first step in I^- metabolism, as the diet is the only source of I^- for land-dwelling vertebrates. The Na^+/I^- symporter (NIS), an integral plasma membrane glycoprotein located in the brush border of enterocytes, constitutes a central component of the I^- absorption system in the small intestine. In this chapter, we review the most recent research on structure/function relations in NIS and the protein's I^- transport mechanism and stoichiometry, with a special focus on the tissue distribution and hormonal regulation of NIS, as well as the role of NIS in mediating I^- homeostasis. We further discuss recent findings concerning the autoregulatory effect of I^- on I^- metabolism in enterocytes: high

Vitamins and Hormones, Volume 98
ISSN 0083-6729
http://dx.doi.org/10.1016/bs.vh.2014.12.002

intracellular I^- concentrations in enterocytes decrease NIS-mediated uptake of I^- through a complex array of posttranscriptional mechanisms, e.g., downregulation of NIS expression at the plasma membrane, increased NIS protein degradation, and reduction of NIS mRNA stability leading to decreased NIS mRNA levels. Since the molecular identification of NIS, great progress has been made not only in understanding the role of NIS in I^- homeostasis but also in developing protocols for NIS-mediated imaging and treatment of various diseases.

1. THE IMPORTANCE OF IODIDE IN HUMAN HEALTH

Iodide (I^-) uptake in the thyroid gland is the first step in the biosynthesis of thyroid hormones—triiodothyronine (T_3) and thyroxine (T_4) (Portulano, Paroder-Belenitsky, & Carrasco, 2014). Thyroid hormones are the only iodine-containing hormones in vertebrates and are required for the development and maturation of the central nervous system, skeletal muscle, and lungs in the fetus and the newborn. They are also primary regulators of intermediate metabolism and effect pleiotropic modulation in virtually all organs and tissues throughout life (Yen, 2001).

Iodine is an extremely scarce element in the environment and is supplied to the body exclusively through the diet. Insufficient dietary I^- intake may cause mild to severe hypothyroidism and subsequently goiter, stunted growth, retarded psychomotor development, and even cretinism (impairment of physical growth and irreversible mental retardation due to severe thyroid hormone deficiency during childhood) (Zimmermann, 2009). I^- deficiency-associated diseases are the most common preventable cause of mental retardation in the world and were slated for global eradication by iodination of table salt by the year 1990 by the World Health Organization. Although significant progress has been made, there were still an estimated 1.88 billion people suffering from insufficient I^- intake in 2011 (Andersson, Karumbunathan, & Zimmermann, 2012).

As iodine is an irreplaceable component of thyroid hormones, normal thyroid physiology relies on adequate dietary I^- intake, gastrointestinal I^- absorption, and proper I^- accumulation in thyrocytes. Therefore, the evolution of a highly efficient system to avidly accumulate I^- appears to be a physiological adaptation to compensate for the environmental scarcity of iodine.

2. THE Na^+/I^- SYMPORTER

The thyroid gland has developed a remarkably efficient system to ensure an adequate supply of I^- for thyroid hormone biosynthesis. Under physiological conditions, the thyroid concentrates I^- approximately 40-fold with respect to the plasma concentration (Wolff & Maurey, 1961). Moreover, the ability of the thyroid to concentrate I^- has provided the molecular basis for the use of radioiodide in the diagnosis, treatment, and follow-up of thyroid pathology (Bonnema & Hegedus, 2012; Reiners, Hanscheid, Luster, Lassmann, & Verburg, 2011). A major breakthrough in the field—as important as the introduction of radioactive I^- isotopes into the study of thyroid physiology near the middle of the twentieth century (Hertz, Roberts, Means, & Evans, 1940)—was the identification of the complementary DNA (cDNA) encoding the Na^+/I^- symporter (NIS), the protein that mediates I^- transport in the thyroid (Dai, Levy, & Carrasco, 1996). The identification of NIS started a new era of intensive I^- research.

2.1 Molecular identification of NIS

The journey toward the identification of NIS began with the isolation of poly(A^+) RNA from FRTL-5 cells, a line of highly differentiated rat thyroid-derived cells which, microinjected into *Xenopus laevis* oocytes, produced Na^+-dependent I^- transport (Vilijn & Carrasco, 1989). Thereafter, the cDNA encoding NIS was isolated by expression cloning in *X. laevis* oocytes using cDNA libraries generated from FRTL-5 cells (Dai et al., 1996). The full nucleotide sequence revealed an open reading frame of 1,854 nucleotides encoding a protein of 618 amino acids. Shortly thereafter, the screening of a human thyroid cDNA library with rat NIS probes enabled the identification of human NIS (Smanik et al., 1996), which exhibits 84% identity and 93% similarity to rat NIS. The human NIS gene was mapped to chromosome 19p13.11 and comprises 15 exons with an open reading frame of 1,929 nucleotides, giving rise to a protein of 643 amino acids (Smanik, Ryu, Theil, Mazzaferri, & Jhiang, 1997).

NIS is an intrinsic plasma membrane glycoprotein. The current, experimentally tested NIS secondary structure model shows a hydrophobic protein with 13 transmembrane segments (TMSs), an extracellular amino terminus and an intracellular carboxy terminus (Levy et al., 1997, 1998; Fig. 1A). Moreover, NIS is a highly *N*-glycosylated protein, although *N*-glycosylation is not essential for I^- transport or NIS trafficking to the plasma membrane (Levy et al., 1998).

NIS-driven active transport of I^- into the thyroid is electrogenic and relies on the driving force of the Na^+ gradient generated by the Na^+/K^+ ATPase and the electrical potential across the plasma membrane. By

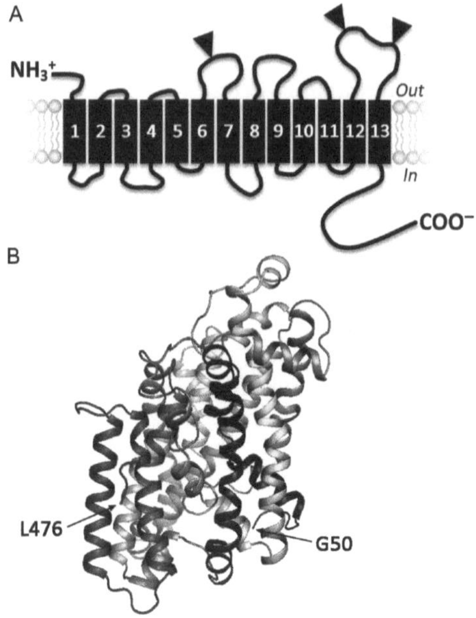

Figure 1 NIS secondary and tertiary structure. (A) Secondary structure. NIS secondary structure model showing the 13 transmembrane segments from the extracellular amino terminus to the intracellular carboxy terminus. Black triangles mark N-linked glycosylation sites at N225, N485, and N497. (B) Tertiary structure. Membrane plane of the NIS homology model built using the rat NIS sequence including residues G50 through L476 (Paroder-Belenitsky et al., 2011), based on the X-ray structure of vSGLT. The NIS homology model is shown as a ribbon representation and rainbow colored by sequence, from the amino terminus (blue) to the carboxy terminus (red). (See the color plate.)

coupling the inward transport of Na^+ down its electrochemical gradient to the translocation of I^- against its electrochemical gradient across the plasma membrane, NIS avidly concentrates I^- into the cells (Dai et al., 1996; Eskandari et al., 1997).

Like all membrane transporters, NIS belongs to the solute-carrier gene (SLC) superfamily. In particular, NIS is a member of solute-carrier family 5A (SLC5A) and has been designated SLC5A5 according to the Human Genome Organization (HUGO) Gene Nomenclature Committee. To date, the only crystal structure of a member of SLC5A is that of the *Vibrio parahaemolyticus* Na^+/galactose transporter (vSGLT), a bacterial homologue of the human SGLT1 (SLC5A1) (Faham et al., 2008). Despite the lack of sequence homology, as predicted by De la Vieja, Reed, Ginter, and Carrasco (2007), the structure of vSGLT revealed the same fold—an inverted topology repeat and unwound helices in regions critical for substrate binding—and a Na^+ coordination similar to that observed in the high-resolution (1.65 Å) crystal structure of the leucine transporter (LeuT) from *Aquifex aeolicus* (LeuT) (Yamashita, Singh, Kawate, Jin, & Gouaux, 2005). Remarkably, NIS shares significant identity (27%) and homology (58%) with vSGLT—almost as much as SGLT1 does (31% identity, 62% homology). Therefore, Paroder-Belenitsky et al. (2011) generated a structural homology model for rat NIS, comprising residues 50–476, using as template the crystal structure of vSGLT (Fig. 1B). Importantly, the development of the 3D homology model helped bridge the gap between the secondary and tertiary structures and further contributed to our understanding of the relation between NIS structure and function. Using our NIS homology model, we uncovered the interaction between the δ-amino group of Arg-124 with the thiol group of Cys-440, concluding that the interaction between intracellular loop (IL)-2 and IL-6 is critical for the local folding required for NIS maturation and targeting to the plasma membrane (Paroder, Nicola, Ginter, & Carrasco, 2013). Moreover, we proposed that the side chain of Asn-441 interacts with the main chain amino group of Gly-444, capping the α-helix of TMS XII and thus stabilizing NIS structure (Li, Nicola, Amzel, & Carrasco, 2013).

2.2 NIS-mediated transport: Substrates and stoichiometry

Using electrophysiological techniques, Eskandari et al. (1997) demonstrated NIS-elicited inward currents when Na^+-dependent I^- accumulation occurs in NIS-expressing *X. laevis* oocytes. Simultaneous flux experiments with radioactive tracers and electrophysiological data established that NIS-mediated I^- transport is electrogenic, with a 2 Na^+/1 I^- stoichiometry (Eskandari et al., 1997). Similar inward currents were observed with different NIS-transported anions. However, surprisingly, the environmental pollutant and well-known inhibitor of thyroidal I^- uptake perchlorate (ClO_4^-) did not elicit currents and, further, abolished I^--induced inward currents (Eskandari et al., 1997). The blockage of I^- transport by ClO_4^- has been used in the treatment of hyperthyroidism and is currently used in the detection of I^- organification defects (ClO_4^- discharge test) (Hilditch, Horton, McCruden, Young, & Alexander, 1982). As radioactive $^{36}ClO_4^-$ was not available for flux experiments, the most likely interpretation was that ClO_4^- blocked NIS activity. A decade later, Dohan et al. (2007) conclusively demonstrated that ClO_4^- is actively transported by NIS. The kinetic parameters of NIS-mediated ClO_4^- transport were determined using the structurally related anion perrhenate (ReO_4^-). Flux experiments using ^{186}Re revealed active accumulation of $^{186}ReO_4^-$, and Na^+-dependent initial rates of ReO_4^- transport indicated an electroneutral stoichiometry (1 Na^+/1 ReO_4^- or ClO_4^-). Therefore, these results demonstrated that NIS translocates different substrates with different stoichiometries (Dohan et al., 2007). NIS-mediated ClO_4^- accumulation has been reported using chromatography-electrospray ionization-tandem mass spectrometry (Tran et al., 2008) and yellow fluorescent protein-based genetic biosensors (Cianchetta, di Bernardo, Romeo, & Rhoden, 2010).

2.3 The role of physiological Na^+ concentrations in NIS affinity for I^-

Na^+-driven symporters such as NIS are expected to exist in at least two conformations, an open-out conformation in which they are open to the extracellular milieu and bind the substrates to be transported, and an open-in state where they are open to the cytoplasm and release the substrates (Krishnamurthy & Gouaux, 2012; Yamashita et al., 2005). The current model of coupled transport is the alternating access model, according to which structural changes occur between the two conformations, allowing the transport of substrates across biological membranes. During the transition, uncoupled flux is prevented by intermediate states that close off access to the binding sites, and after the substrates are released into cytoplasm, the transporter reverts to the open-out state with the binding sites empty.

In Na^+-driven symporters, the coupling mechanism requires that the conformational changes occur when the transporter has bound both Na^+ and substrate. To fulfill this requirement, the transporter must use binding site occupancy to control conformational transitions. Experimental evidence suggests that Na^+ triggers a conformational change, as Na^+ stabilizes the open-out state until the substrate binds (Zhao et al., 2010). Moreover, substrate binding to the open-out conformation was proposed to initiate the conformational change by overcoming the stabilizing effect of Na^+ binding (Zhao et al., 2010). Nevertheless, the coupling mechanism remains poorly understood at the molecular level.

Very recently, Nicola, Carrasco, and Amzel (2014) addressed a fundamental mechanistic question: how NIS binds and releases its substrates. Taking advantage of the fact that NIS translocates I^- and ReO_4^- with different stoichiometries, the authors analyzed initial rates of transport measured at different concentrations of substrates using statistical thermodynamics and determined the affinity of NIS for the transported ions as well as the relative populations of the different NIS species present during the transport cycle. They showed that empty NIS has a very low intrinsic affinity for I^- ($K_d = 224$ μM), but it increases 10 times ($K_d = 22.4$ μM) when two Na^+ ions are bound to the transporter. Moreover, at physiological Na^+ concentrations, approximately 79% of NIS molecules are occupied by two Na^+ ions, and hence poised to bind and transport I^-, even though the physiological concentration of I^- in the blood is in the submicromolar range, well below the affinity of NIS for I^- (Nicola et al., 2014). Ultimately, understanding the conformational changes that NIS undergoes during the transport cycle and the changes in Na^+/anion stoichiometry will require us to obtain structural information on NIS with different substrates bound and in different conformations.

3. NIS EXPRESSION BEYOND THE THYROID

In addition to the thyroid, I^- uptake has been demonstrated in other tissues, including the lacrimal drainage system, choroid plexus, salivary glands, stomach, and lactating breast. Indeed, radioiodide accumulation outside the thyroid is routinely observed in whole-body radioiodide scintiscans (Bruno et al., 2004). Interestingly, patients with congenital hypothyroidism due to NIS mutations display no I^- transport in the thyroid or any extrathyroidal tissue, highlighting the role of NIS in mediating I^- transport in all these tissues (Spitzweg & Morris, 2010). NIS was initially thought to be a thyroid-specific protein, but since NIS was cloned and NIS-specific antibodies generated, various groups have detected NIS protein expression in extrathyroidal locations previously known to actively accumulate I^-, such as salivary glands, stomach, and lactating breast (Altorjay et al., 2007; La Perle et al., 2013; Spitzweg, Joba, Schriever, et al., 1999; Tazebay et al., 2000; Vayre et al., 1999; Wapnir et al., 2003). In addition, NIS expression was demonstrated in the lacrimal sac and nasolacrimal duct, kidney, placenta, and ovary (Di Cosmo et al., 2006; Donowitz et al., 2007; Mitchell et al., 2001; Morgenstern et al., 2005; Riesco-Eizaguirre et al., 2014; Spitzweg et al., 2001).

The functional significance of NIS expression is clear in some extrathyroidal tissues but in others remains largely unknown. The placenta allows I^- to pass from the maternal to the fetal circulation for normal fetal thyroid function. The observation that NIS is mainly expressed at the apical membrane of cytotrophoblasts is consistent with this (Di Cosmo et al., 2006; Mitchell et al., 2001). In the lactating breast, NIS is expressed at the basolateral membrane of ductal epithelial cells (Tazebay et al., 2000). NIS translocates I^- from the bloodstream to the maternal milk, where it reaches a concentration of approximately 150 μg/L, thus providing the nursing newborn with a supply of I^- adequate for thyroid hormone biosynthesis. Although basolateral NIS expression has been demonstrated in the mucus-secreting and parietal cells of the stomach and ductal epithelial cells in the salivary glands (Altorjay et al., 2007; La Perle et al., 2013; Spitzweg, Joba, Schriever, et al., 1999; Vayre et al., 1999; Wapnir et al., 2003), the physiological role of I^- accumulation in the saliva and gastric juice is a matter of debate. Given the scarcity of I^-, some authors have speculated that the secretion of I^- into the gastrointestinal tract may serve as an I^- recycling mechanism (Venturi & Venturi, 2009), as I^- that is not accumulated in the thyroid or released by the action of iodothyronine deiodinases in peripheral tissues is secreted into the saliva and gastric juice and likely reabsorbed further down the gastrointestinal tract along with newly ingested I^-, thus preventing excessive renal excretion. Moreover, I^- has been proposed to serve antioxidant and antimicrobial functions in these tissues (El Hassani et al., 2005; Geiszt, Witta, Baffi, Lekstrom, & Leto, 2003). It is worth emphasizing that NIS in the stomach is not involved absorbing dietary I^- from the stomach lumen into the bloodstream, as previously suggested (Kotani et al., 1998). Importantly, I^- accumulation in the saliva has long served as a key diagnostic tool in the detection of genetic defects in I^- transport (patients with NIS-inactivating mutations do not accumulate I^- in the saliva; Portulano et al., 2014).

The excretion of I^- occurs primarily through glomerular filtration in the kidney. Measurement of urinary I^- is the simplest method to assess I^- intake, as under I^- sufficiency almost all ingested I^- is excreted in the urine (Vejbjerg et al., 2009). I^- clearance involves glomerular filtration and partial tubular reabsorption as well as secretion from the plasma. However, the events that regulate tubular I^- handling remain poorly understood. Immunohistochemical analysis has revealed NIS expression in the tubular system of the human kidney. Using a monoclonal anti-human NIS antibody, Spitzweg et al. (2001) observed predominant intracellular immunostaining throughout the entire tubular system, without evidence of plasma membrane localization. Later, Wapnir et al. (2003) showed NIS expression in six out of six tissue microarray cores derived from normal human kidney samples. The protein was localized at the apical surface of principal and intercalated cells of renal distal and collecting tubules, suggesting a potential role for NIS in mediating I^- reabsorption. Other immunohistochemical studies did not reveal NIS staining in kidney tissue (Lacroix et al., 2001; Vayre et al., 1999). However, none of these studies measured NIS-mediated renal I^- transport, either absorption or secretion. So, it is still an open question whether NIS is functionally expressed and regulated in the kidney.

Recently, Riesco-Eizaguirre et al. (2014) reported NIS expression at the basolateral membrane of ovarian surface epithelial cells and in secretory cells of the epithelium of the fallopian fimbriae, but not in ovarian stromal cells, in 14 out of 14 healthy women. NIS expression in the ovary was functionally evaluated using a gamma camera; 49 out of 345 women (15%) accumulated $^{99m}TcO_4^-$ in the ovary region, suggesting that NIS mediates physiological I^- accumulation in the reproductive tract.

Elucidating the mechanisms of NIS expression and regulation in extrathyroidal tissues may help us not only to understand I^- metabolism and prevent or minimize side effects of radioiodide therapy but also to better handle patients under treatment (Bonnema & Hegedus, 2012; Reiners & Luster, 2012). The most common side effects are swelling; nausea and vomiting; gastritis; dry mouth, taste changes, and sialadenitis; dry eyes and conjunctivitis; disturbances of female reproductive function; and decreased testicular function. Unsurprisingly, these side effects may be related to NIS-mediated radioiodide accumulation in the relevant tissues. Therefore, understanding tissue-specific NIS regulation may help us selectively downregulate NIS expression to minimize side effects as well as enhance NIS expression in particular tissues to increase the efficiency of radioiodide therapy.

Immunohistochemical analysis of frozen or paraffin-embedded tissue sections offers the advantage of revealing not only the expression of NIS but also its subcellular localization. However, as mentioned, conflicting results have been reported regarding NIS expression in several extrathyroidal tissues. This may be related to the quality of the different anti-NIS antibodies used, in terms of specificity and affinity for the relevant epitope and the procedures used to obtain and preserve the tissue. Moreover, studies in tissue microarrays are optimal for high-throughput screening but intrinsically limited because of the size of the samples and uncertainty about tissue preservation conditions. Detecting NIS mRNA or protein expression by real-time PCR or immunoblot analysis, respectively, may not be trivial. The samples would have to be highly enriched for a specific cell type before preparing tissue lysates to ensure that NIS expression is not diluted out to undetectable levels (see NIS expression in the small intestine).

4. TARGETING OF NIS TO THE PLASMA MEMBRANE

Epithelial tissues are composed of polarized cells with an apical membrane facing the external or internal surface of the body (as in the skin or small intestine), or the lumen of a gland (i.e., as in thyroid), and a basolateral membrane facing the connective tissue. Apical-to-basolateral polarity defines different domains in terms of membrane protein expression and determines normal cell function. Therefore, differential sorting of membrane proteins to specific membrane domains is necessary for the generation and maintenance of biochemical polarity.

As previously mentioned, NIS displays different polarized localizations in different tissues. NIS is expressed at the basolateral membrane in the thyroid, stomach, salivary gland, and lactating breast. In contrast, NIS is targeted to the apical surface of placental cytotrophoblasts and the collecting tubules of the kidney. Although one may interpret this as indicating that different polarized NIS targetings arise from different tissue-specific I^--handling requirements, the mechanisms responsible for this behavior remain largely uncharacterized. Thus, these findings have raised new and intriguing biological questions about the posttranslational regulation of NIS in different tissues and about how different epithelia selectively interpret NIS sorting signals.

Little has been reported on the signals and molecular regions involved in the polarized targeting of NIS in the thyroid or other tissues. NIS sequencing in different tissues yielded the same protein identity (Spitzweg, Joba, Eisenmenger, & Heufelder, 1998), suggesting that factors other than the NIS sequence may regulate the polarized targeting of NIS. Analysis of the NIS intracellular carboxy terminus revealed the presence of conserved sorting sequences known to participate in retention, endocytosis, and targeting to the plasma membrane of proteins. In particular, the last four amino acids of the carboxy terminus of NIS constitute a putative class I PDZ-binding motif potentially involved in basolateral targeting. In addition, L556 and L557 constitute a potential di-leucine motif which may interact with the clathrin-coated system involved in protein endocytosis.

A major limitation in the study of NIS polarized targeting has been the nonexistence of highly functional polarized thyroid cell lines for *in vitro* studies. However, the Madin–Darby canine kidney cell line has been shown to recapitulate the native polarity of several thyroid proteins (Paroder et al., 2006; Zhang, Riedel, Carrasco, & Arvan, 2002) and is therefore an interesting cell system in which to study NIS polarization signals.

NIS-mediated radioiodide therapy used to ablate thyroid cancer metastases and remnants after thyroidectomy has been the most successful targeted internal radiation anticancer therapy ever designed (Bonnema & Hegedus, 2012; Reiners et al., 2011). Radioiodide therapy depends on the ability of thyroid tumors to accumulate radioiodide, which is ultimately dependent on functional NIS expression at the plasma membrane (Schlumberger, Lacroix, Russo, Filetti, & Bidart, 2007). However, thyroid tumors often exhibit less I^- transport than normal thyroid tissue (or even no detectable transport) and are diagnosed as cold nodules on thyroid scintigraphy. Several reports have demonstrated that 70–80% of thyroid tumors in fact overexpress NIS when compared to surrounding normal tissue, suggesting the presence of trafficking abnormalities (Dohan, Baloch, Banrevi, Livolsi, & Carrasco, 2001; Kollecker et al., 2012; Tonacchera et al., 2002; Wapnir et al., 2003). No NIS mutations have been identified in thyroid tumors (Neumann et al., 2004; Russo et al., 2001), so it cannot be structural defects that impair targeting of NIS in these tumors; this stands in contrast to the situation in some patients with congenital I^- transport deficiency (Li et al., 2013; Paroder et al., 2013). Therefore, it is crucial that we understand the mechanisms that regulate the trafficking of NIS to the cell surface in normal and diseased tissue.

To date, only one NIS-interacting protein has been reported that may be involved in NIS plasma membrane targeting: the pituitary tumor-transforming gene binding factor (PBF). PBF expression is frequently upregulated in thyroid tumors. Smith et al. (2009) reported that ectopic PBF overexpression resulted in the redistribution of NIS from the plasma membrane into CD63-positive intracellular vesicles associated with clathrin-dependent endocytosis. Therefore, improving NIS-mediated radioiodide therapy for thyroid cancer may require that greater priority be given to developing strategies aimed at enhancing NIS plasma membrane expression, as opposed to just stimulating NIS transcription.

5. HORMONAL REGULATION OF NIS EXPRESSION

Hormonal regulation of NIS expression seems to be tissue specific. Thyrotropin (TSH) has long been known to be a key regulator of NIS expression and activity in the thyroid. Transgenic mice that do not express the TSH receptor do not show detectable thyroidal NIS expression (Marians et al., 2002). Similarly, hypophysectomized rats show the same phenotype, but NIS expression can be restored in these animals by TSH administration (Levy et al., 1997). TSH regulates several steps in the biogenesis of NIS, including NIS expression at both the transcriptional and the posttranscriptional level (Kogai et al., 1997; Ohno, Zannini, Levy, Carrasco, & di Lauro, 1999; Riedel, Levy, & Carrasco, 2001). Detailed functional analysis of the NIS promoter has revealed that the transcription factor Pax8 plays a critical role in NIS transcription (Ohno et al., 1999).

The role of TSH in regulating NIS expression in the thyroid has been well established, but TSH does not regulate NIS expression in any extrathyroidal tissue. Importantly, withdrawal of thyroid hormone to increase endogenous TSH concentrations and administration of recombinant TSH are routinely used to stimulate I^- uptake in differentiated thyroid cancer to prepare patients receiving radioiodide for diagnostic scintigraphy and radioiodide therapy (Schlumberger et al., 2007). Tissue-specific NIS regulation makes it possible to improve the therapeutic outcome of stimulating radioiodide accumulation in the tumor cells and to simultaneously reduce the therapeutic dose of radioiodide, thereby decreasing its side effects.

NIS expression seems to be constitutive in the stomach and salivary glands and no hormonal regulation has yet been reported in these tissues. The mechanisms behind the differential regulation of NIS in different tissues remain largely unknown; clearly, the elucidation of these mechanisms will be a valuable contribution to basic science and likely to clinical medicine as well. For example, the development of novel strategies for allowing selective inhibition of NIS expression in salivary glands and stomach, thereby reducing tissue damage in thyroid cancer patients undergoing radiotherapy and decreasing radioiodide clearance, may permit a reduction of the therapeutic dose of radioiodide.

Although NIS is not expressed in healthy nonlactating breast tissue, NIS expression becomes evident toward the end of gestation and persists throughout lactation (Cho et al., 2000; Tazebay et al., 2000). In the lactating breast, NIS expression is stimulated by a combination of various hormones, including estrogen, prolactin, and oxytocin (Cho et al., 2000; Tazebay et al., 2000), and suckling is essential for maintaining NIS expression in the lactating breast after delivery (Tazebay et al., 2000). The combined administration of 17-β-estradiol and oxytocin in ovariectomized mice resulted in NIS expression, indicating that the effect of oxytocin on NIS expression in the mammary gland requires the presence of estrogen (Tazebay et al., 2000).

Placental NIS expression is regulated by pregnancy-related hormones such as human chorionic gonadotropin (hCG), prolactin, and oxytocin. These hormones increase I^- uptake in primary cultures of human placental cytotrophoblasts and human placental choriocarcinoma cell lines (Arturi et al., 2002; Burns, O'Herlihy, & Smyth, 2013). However, although neither 17-β-estradiol nor progesterone itself had any significant effect on NIS expression levels, the two hormones appear to work synergistically by increasing the effect of prolactin and oxytocin on NIS expression in the placenta (Burns et al., 2013). Pax8 expression has been described in placental tissue and placental cell lines. hCG increased cAMP-dependent Pax8 expression and DNA-binding activity. However, placental cells transfected with a Pax8-specific small interfering RNA did not show changes in NIS mRNA expression in response to hCG stimulation (Ferretti et al., 2005). These findings indicate that NIS expression in trophoblasts is modulated by transcription factors other than Pax8.

Physiological I^- accumulation in the rat female reproductive tract correlates with the reproductive cycle: NIS-mediated I^- accumulation coincides with the rise of estrogens during the follicular phase (Riesco-Eizaguirre et al., 2014). Interestingly, unligated estrogen receptor α cooperates with Pax8 to upregulate NIS transcriptional expression in transiently transfected HeLa cells. On the basis of these findings, Riesco-Eizaguirre et al. (2014) suggested that attention should be paid to when in their menstrual cycle women are given radioiodide.

6. DIETARY I^- ABSORPTION

I^- is supplied to the body exclusively through the diet; therefore, I^- absorption in the gastrointestinal tract constitutes the first step in I^- metabolism. Given the physiological importance of I^-, it has long been of major interest where and how dietary I^- is absorbed in the gastrointestinal tract.

To our knowledge, I^- absorption in the gastrointestinal tract was first reported by Hanzlik in 1912 (Hanzlik, 1912). This author showed that the most I^- absorption took place between the pylorus and the colon, and that the duodenum, jejunum, and ileum maintain this absorption rate. Later, Cohn (1932) measured I^- absorption in isolated canine small intestine, reporting that ingested inorganic iodine and iodate may be reduced to I^- in the gastrointestinal tract before being absorbed in the small intestine. Ingested I^- appears to be absorbed almost entirely in the gastrointestinal tract. When euthyroid human patients were given a single oral dose of radioiodide, less than 1% of it was found in their feces, suggesting that ingested radioiodide is absorbed remarkably efficiently (Fisher, Oddie, & Epperson, 1965).

An important step toward the characterization of I^- absorption in the small intestine was the study by Josefsson, Grunditz, Ohlsson, and Ekblad (2002) involving ligation of the gastrointestinal tract. The authors demonstrated that pyloric ligation virtually abolished I^- accumulation in the thyroid after oral administration of radioiodide, but did not modify thyroid I^- accumulation after parenteral administration (Fig. 2). The reduction in I^- accumulation in the thyroid after oral administration in pylorus-ligated animals was accompanied by lower levels of I^- in the blood, indicating deficient I^- absorption (Fig. 2). Furthermore, animals receiving I^- intravenously showed substantial accumulation of I^- in the stomach, confirming that I^- is secreted into the lumen of the stomach rather than absorbed from it (Josefsson et al., 2002).

Although I^- absorption was restricted to the small intestine, it was initially not known whether there was a dedicated intestinal I^- transporter. Shortly after the cloning of NIS, several studies investigated NIS expression in the small intestine to test the hypothesis that NIS participates in dietary I^- accumulation; conflicting results were obtained. Using semiquantitative RT-PCR, Perron, Rodriguez, Leblanc, and Pourcher (2001) detected low

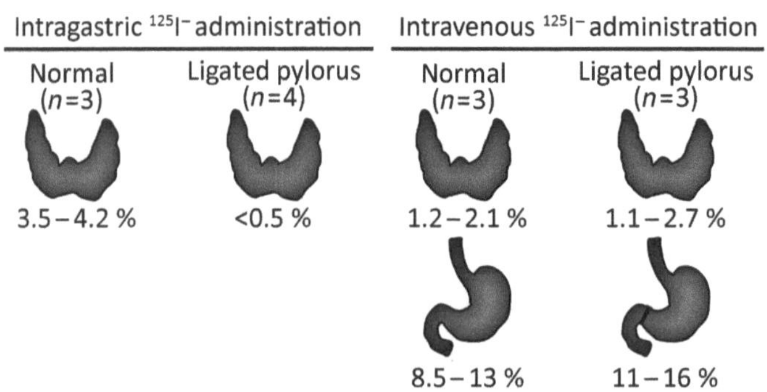

Figure 2 Gastrointestinal absorption and secretion of I^-. Untreated or pylorus-ligated rats received a bolus dose of radioiodide by intragastric or intravenous administration. After 60 min, radioactivity of thyroid glands and gastric washouts were determined and expressed as percentage of the total administered radioiodide dose. Values are indicated as ranges for each group and *n* indicates the number of animals per group (Josefsson et al., 2002). *Adapted from Josefsson (2009). Reproduced with permission.*

levels of NIS mRNA in the mouse small intestine. In contrast, previous reports did not detect NIS mRNA expression in whole human small intestine extracts by Northern blot analysis using a radiolabeled human NIS-specific probe (Spitzweg et al., 1998) or real-time PCR analysis (Lacroix et al., 2001).

The presence in a cell or tissue of NIS mRNA does not in itself show that the NIS protein is biosynthesized, targeted to the plasma membrane, or functional. Indeed, several independent studies using immunohistochemical procedures with unrelated anti-NIS antibodies failed to detect NIS protein expression in frozen and paraffin-embedded normal or pathological human small intestine specimens (Altorjay et al., 2007; Lacroix et al., 2001; Vayre et al., 1999). In contrast, Wapnir et al. (2003) investigated NIS protein expression in tissue microarrays containing cores from normal human small intestine by immunohistochemical analysis and found weak expression in two out of three samples. In agreement with this result, Donowitz et al. (2007) performed a detailed proteomic analysis of mouse jejunal brush-border enterocytes and demonstrated NIS protein expression by immunoblot and immunofluorescence staining, which validated NIS as a brush-border protein. These reports furnished more persuasive evidence that NIS may be involved in I^- absorption in the small intestine, but this evidence was not quite conclusive, as no functional data were provided.

Consistent with these findings, we have characterized I^- absorption in the small intestine and concluded that NIS may play a key role in dietary I^- absorption, since we demonstrated that NIS is functionally expressed on the apical surface of the absorptive epithelium (Nicola et al., 2009). We analyzed paraffin-embedded sections of small intestine from rats and mice by immunohistochemistry using an affinity-purified polyclonal anti-NIS antibody (Levy et al., 1997). In an immunohistochemistry study in which tissue samples were collected with special care, we consistently observed NIS protein expression along all three sections of the small intestine (duodenum, jejunum, and ileum), but exclusively on the brush border or microvilli, the finger-like projections that protrude from the apical membrane of absorptive enterocytes into the intestinal lumen (Fig. 3A–C; Nicola et al., 2009). This observation is compatible with the notion that NIS may translocate I^- from the intestinal lumen into absorptive enterocytes.

To demonstrate NIS protein expression by immunoblot, we followed a protocol described by Weiser (1973) to isolate villus-tip epithelial cells and further purify brush-border apical membranes because NIS expression is restricted to the most villus-tip-differentiated enterocytes. As expected, the procedure resulted in a pronounced enrichment (~20-fold) of the

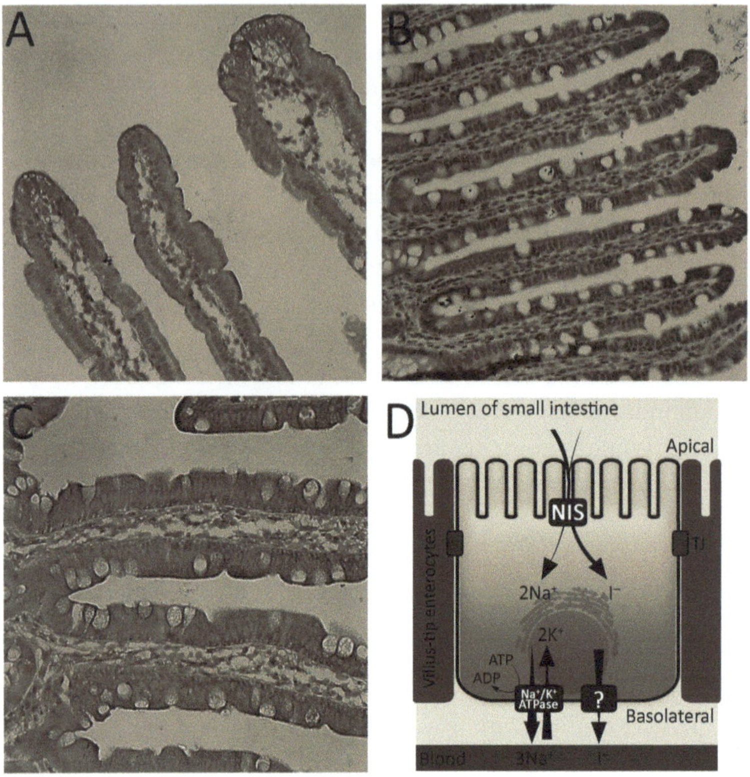

Figure 3 NIS expression in rat small intestine. Immunohistochemistry analysis demonstrating NIS expression in the three sections of the rat small intestine. NIS-specific immunostaining is evident in the apical membrane of the absorptive enterocytes (Nicola et al., 2009). (A) Duodenum, (B) jejunum, and (C) ileum. All pictures are presented at 40 × magnification. (D) Schematic representation of NIS-mediated I^- absorption in villus-tip small intestine enterocytes. NIS mediates transcellular apical transport of dietary I^- against its concentration gradient by coupling it to Na^+ transport. The Na^+ gradient is generated by the Na^+/K^+ ATPase in the basolateral membrane, maintaining the Na^+ electrochemical gradient that favors the Na^+-dependent accumulation of substrates. Transport of I^- into the blood may be facilitated by still uncharacterized channels or transporters. It is not known whether I^- is translocated across the intestinal epithelium through the intercellular space between the enterocytes (paracellular transport). TJ, tight junctions. (See the color plate.)

activity of alkaline phosphatase, a villus-tip marker, in membranes over cell homogenates. Immunoblot analysis of enriched brush-border membranes from villus-tip enterocytes revealed a 90-kDa polypeptide corresponding to intestinal NIS, whose electrophoretic mobility was identical to that of NIS from the thyroid cell line FRTL-5 (Nicola et al., 2009). It is worth noting that NIS expression was undetectable in intestinal cell homogenates, but became evident upon enrichment.

To determine more precisely the functional significance of NIS expression in the enterocyte, we prepared sealed brush-border membrane vesicles (BBMVs) from isolated villus-tip enterocytes and performed steady-state I^- transport assays. I^- uptake in BBMVs was both Na^+-dependent and ClO_4^- – sensitive, two hallmarks of NIS activity. Moreover, kinetic analysis of I^- transport in BBMVs showed an affinity for I^- of 13.4 ± 2.0 μM, a value comparable to those reported for NIS-mediated I^- transport in thyroid membrane vesicles (O'Neill, Magnolato, & Semenza, 1987). We also investigated whether NIS mediates I^- absorption *in vivo*. We administered pertechnetate ($^{99m}TcO_4^-$), a widely used radioactive NIS substrate with a short half-life, alone or together with the NIS inhibitor ClO_4^- via duodenal catheterization, and collected blood samples through a jugular catheter placed in the right atrium. Interestingly, rats simultaneously treated with ClO_4^- absorbed 27–48% less $^{99m}TcO_4^-$ than rats treated with $^{99m}TcO_4^-$ alone (Nicola et al., 2009). These data provide strong evidence that NIS is a significant and possibly central component of the I^- absorption system in the small intestine (Fig. 3D).

However, our data do not rule out the possibility that channels or transporters other than NIS, such as chloride channels or anion exchangers participate, or the possibility that passive paracellular transport is involved in the absorption of I^- from the intestinal lumen—both of which are consistent with the partial inhibition of $^{99m}TcO_4^-$ absorption by ClO_4^-. de Carvalho and Quick (2011) have reported that the Na^+/multivitamin transporter (SLC5A6), the protein with the highest sequence homology with NIS, actively mediates Na^+-dependent but ClO_4^- – insensitive I^- transport, albeit with a lower affinity than NIS. On the basis of the intestinal expression of the Na^+/multivitamin transporter, this protein has been proposed to provide a complementary pathway for I^- absorption in the small intestine. This hypothesis can now be fruitfully tested in enterocyte-specific Na^+/multivitamin transporter knockout mice, a recently developed system (Ghosal, Lambrecht, Subramanya, Kapadia, & Said, 2013).

One important finding in the study of intestinal NIS has been the detection of functional NIS expression in late-passage IEC-6 cells (IEC-6 cells are a line of rat small intestine-derived cells). Performing flux experiments under steady-state conditions, we showed Na^+-dependent, ClO_4^- – sensitive I^- accumulation in IEC-6 cells (Nicola et al., 2009). Active I^- accumulation levels were higher in IEC-6 cells than in FRTL-5 cells, a result consistent with the higher NIS protein expression levels observed by immunoblot in IEC-6 than in FRTL-5 cells. We analyzed the kinetic properties of NIS in IEC-6 cells and FRTL-5 thyroid cells. Intestinal NIS exhibited an affinity for I^- (K_m $I^- = 20.3 \pm 3.9$ μM) similar to that of thyroid NIS (K_m $I^- = 23.2 \pm 3.7$ μM) (Nicola et al., 2009). Therefore, IEC-6 cells may constitute a good *in vitro* model in which to study NIS regulation in intestinal cells.

Crohn's disease is an inflammatory bowel disease that may affect any part of the gastrointestinal tract from mouth to anus, most commonly the terminal ileum of the small intestine. The disease causes a wide variety of symptoms including diarrhea and malabsorption syndrome. Jarnerot (1975) investigated I^- metabolism in patients with chronic inflammatory bowel disease, including Crohn's disease. His results demonstrated that 10 out of 50 patients with chronic inflammatory bowel disease excreted less than 40 μg I^- in the urine over a 24-h period, compared with 5 out of 102 healthy controls. Moreover, 16 out of 38 patients showed a 24-h thyroid radioiodide uptake higher than 50% of the administered dose, compared with 4 out of 36 controls. Although these results suggested an increased occurrence of I^- deficiency in patients with chronic inflammatory bowel diseases, no evidence was found of impaired absorption of inorganic iodide from the gut as the amount of orally administered radioiodide they absorbed was not significantly different from the corresponding amount for control patients (Jarnerot, 1975). However, an accurate classification of patients with Crohn's disease according to the location of the inflammation (ileum, colon, or both) will shed light on the role of the small intestine in dietary I^- absorption.

Navarro, Suen, Souza, De Oliveira, and Marchini (2005) investigated the possible influence of intestinal malabsorption on I^- status in patients with severe bowel malabsorption due to chronic pancreatitis or short bowel syndrome who were fed exclusively parenterally and in control subjects. The study demonstrated that severe bowel malabsorption does not significantly affect I^- status, as patients and control subjects receiving equal dietary intakes over a period of 24 h did not show significant changes in daily urinary I^- excretion. However, a major limitation of the study was the small size of the population analyzed—only nine patients per group. Follow-up studies with more patients are needed to obtain conclusive data.

Recently, Michalaki et al. (2014) reported that dietary I^- absorption is not influenced by malabsorptive bariatric surgery. Urinary excretion of I^- was not reduced in obese patients following malabsorptive bariatric surgery, although the stomach, the duodenum, and a substantial part of the jejunum were bypassed. This indicates that sufficient I^- is absorbed along the remainder of the gastrointestinal tract, as expected, given that NIS is expressed all along the small intestine (Nicola et al., 2009).

7. REGULATION OF INTESTINAL NIS EXPRESSION

I^- plays a key role in thyroid physiology, not only as an irreplaceable constituent of the thyroid hormones but also as a regulator of thyroid physiology and NIS expression and function (Portulano et al., 2014). Saturating concentrations of I^- downregulate thyroid function by inhibiting thyroid hormone biosynthesis, a phenomenon, ill understood at the molecular level, known as the Wolff–Chaikoff effect (Wolff & Chaikoff, 1948). This effect is followed by downregulation of I^- uptake leading to an "escape" from the effect, which restores thyroid hormone biosynthesis (Braverman & Ingbar, 1963). High I^- concentration-reduced I^- transport has been associated with a decrease in NIS expression. In thyroid cells, the regulation of NIS mRNA levels by I^- excess has mainly been attributed to a transcriptional effect (Eng et al., 1999; Spitzweg, Joba, Morris, & Heufelder, 1999; Uyttersprot et al., 1997). However, more recent data suggest that NIS regulation by I^- takes place at the posttranscriptional and posttranslational levels (Dohan, De la Vieja, & Carrasco, 2006; Leoni, Kimura, Santisteban, & De la Vieja, 2011; Serrano-Nascimento, Calil-Silveira, & Nunes, 2010). The "escape" from the Wolff–Chaikoff effect seems to be an adaptive response that serves to reduce intracellular I^- levels, thus protecting thyrocytes from the oxidative effects of I^- excess. Very recent evidence suggests a link between thyroid oxidative state and the Wolff–Chaikoff effect. Serrano-Nascimento et al. (2014) reported the involvement of phosphatidylinositol-4,5-bisphosphate 3-kinase (PI3K) signaling activation in I^- excess-downregulated NIS function in thyroid cells. Interestingly, I^- excess led to increased generation of mitochondrial reactive oxygen species that trigger activation of PI3K signaling (Serrano-Nascimento et al., 2014).

We investigated the regulatory effect of dietary I^- intake on NIS expression and function in rat small intestine. Animals were divided into four groups, three of which received a high concentration of I^- (0.05%) in their drinking water for 12–48 h. The effect of I^- administration was compared with that in a fourth group of animals, which received regular water. After treatment, villus-tip enterocytes were isolated and sealed BBMVs prepared from treated and nontreated rats to perform steady-state I^- transport studies.

High concentrations of I^- significantly reduced NIS-mediated I^- uptake by 55% at 24 h, and the decrease in I^- transport became more pronounced with longer exposure times. Moreover, BBMVs were subjected to immunoblot analysis to assess NIS protein expression. Consistent with the reduced I^- uptake, high dietary I^- levels decreased NIS protein levels by 49% at 12 h, relative to the levels of the control group. The same treatment given for 48 h decreased NIS protein expression by as much as 83% (Nicola et al., 2009). Importantly, no significant changes were observed in the expression of the differentiation marker alkaline phosphatase. Thus, our data demonstrate that high I^- concentrations inhibit NIS-mediated I^- uptake *in vivo* in the small intestine, just as in the thyroid.

To further our understanding of the molecular mechanism involved in the high I^- concentration-regulated expression of intestinal NIS, we investigated the effect of I^- *in vitro* using IEC-6 cells (Nicola, Reyna-Neyra, Carrasco, & Masini-Repiso, 2012). When these cells were incubated with a high concentration of I^- (100 μM), there was a significant reduction in I^- transport 3 h after I^- treatment, which became more pronounced with longer incubation times. This is due to a decrease in I^- influx rather than an increase in I^- efflux. Analysis of the kinetic parameters of I^- transport demonstrated that I^- excess did not affect the apparent affinity of NIS for its substrates. Together, these findings suggested that high concentrations of I^- may decrease the number of functional NIS molecules at the plasma membrane of enterocytes.

Indeed, surface biotinylation experiments revealed a significant time-dependent reduction in NIS expression at the plasma membrane by 6 h after I^- treatment. However, total NIS protein levels were only reduced 24 h after I^- treatment. Moreover, immunofluorescence studies showed time-dependent decreased colocalization of NIS and the Na^+/K^+ ATPase (a plasma membrane marker) and increased NIS intracellular staining in response to I^- excess by 6 h. Complementarily, we determined intracellular NIS protein levels by assaying the supernatant remaining after streptavidin-purified biotin-labeled surface proteins. Immunoblot analysis showed that NIS intracellular expression increased from 6 to 12 h after I^- treatment, but the amount of intracellular NIS decreased after 24 h, consistent with the observed reduction in total lysates (Nicola et al., 2012).

Given that only NIS molecules at the plasma membrane take up I^-, we established a correlation between the reduction in I^- accumulation and the corresponding lowering of NIS expression at the cell surface induced by I^- excess (Nicola et al., 2012). The prompt recruitment of NIS from the

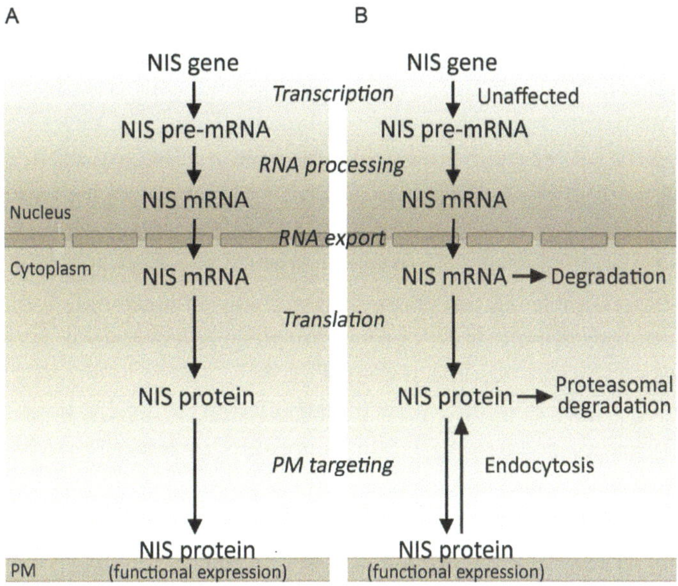

Figure 4 Different levels of intestinal NIS regulation induced by high concentrations of I^-. (A) Under normal conditions, the NIS gene is transcribed into NIS pre-mRNA. Then introns are removed and exons reconnected to generate mature NIS mRNA, which is exported to the cytoplasm and translated into protein. NIS protein is glycosylated and targeted to the plasma membrane of small intestine absorptive enterocytes where it mediates I^- transport. (B) Under I^- excess, although the NIS gene is normally transcribed, mature NIS mRNA levels are decreased due to a reduction in its stability. NIS protein expression levels are diminished in response to increased proteasomal degradation, and NIS expression at the plasma membrane is downregulated due to increased amiloride-sensitive internalization. As a result, I^- excess decreased NIS-mediated accumulation of I^- in enterocytes (Nicola et al., 2012). (See the color plate.)

plasma membrane to intracellular compartments upon I^- treatment suggests the existence of posttranslational mechanisms for reducing the number of NIS molecules at the plasma membrane. Interestingly, we observed that the physiological control of NIS expression at the cell surface of enterocytes seems to involve constitutive macropinocytosis-dependent endocytosis, as amiloride treatment increased I^- transport in IEC-6 cells and abolished I^- excess-induced NIS internalization (Nicola et al., 2012) (Fig. 4).

In addition to NIS endocytosis, we observed a significant reduction in NIS protein expression in I^--treated IEC-6 cell lysates after 24 h. Therefore, we investigated the effect of I^- on NIS levels after cycloheximide treatment in IEC-6 cells to study any changes there might be in NIS protein stability (Nicola et al., 2012). We found that the half-life of NIS in cells treated with excess I^- was 36% lower than that of NIS in control cells, suggesting that increased NIS protein degradation is partially responsible for the lower levels of I^--induced NIS protein expression in intestinal cells. To determine the proteolytic pathways involved in I^--induced NIS protein degradation, we incubated IEC-6 cells with lysosomal or proteasome inhibitors (Nicola et al., 2012). The lysosomal inhibitor cocktail chloroquine plus ammonium chloride did not have an effect on NIS levels, independently of the presence of I^-. In contrast, the proteasome inhibitor MG132 markedly increased NIS protein expression levels and prevented excess I^--induced reduction of NIS expression, indicating that NIS protein turnover in enterocytes is regulated via the ubiquitin–proteasome system (Fig. 4).

Although increased NIS protein degradation could by itself account for the observed reduction in NIS expression in response to excess I^-, a reduction of NIS mRNA levels is also compatible with a decreased protein translation, which in turn may lead to reduced protein biosynthesis. Indeed, real-time PCR analysis demonstrated an I^--induced time-dependent reduction in NIS mRNA levels in IEC-6 cells, without a significant change in the mRNA levels of alkaline phosphatase (Nicola et al., 2012) (Fig. 4). We further determined NIS mRNA expression *in vivo* in response to an I^--rich diet. Rats received 0.05% I^--supplemented drinking water for different periods of time, whereas control rats received regular water. Villus-tip small intestine epithelial cells were isolated and further processed for total RNA extraction. Quantification of NIS mRNA levels showed substantial reduction in enterocytes subjected to the I^--supplemented diet after 24 h, while I^- had no effect on alkaline phosphatase mRNA expression. In a complementary experiment, we observed a significant increase in intestinal NIS mRNA levels in enterocytes of animals fed with an I^--deficient diet for 2 and 4 weeks, but no effect on alkaline phosphatase mRNA expression (Nicola et al., 2012).

Messenger RNA expression levels are the result of gene transcription and mRNA stability and degradation. Therefore, we investigated a potential negative transcriptional effect exerted by excess I^- on NIS transcriptional activity (Nicola et al., 2012). We transiently transfected IEC-6 cells with a luciferase reporter construct containing a 2867-bp DNA fragment from the rat NIS promoter (−2854 to +13, +1 being the adenosine of the start codon) and determined whether excess I^- had an effect on transcriptional promoter activity. Although the NIS regulatory region showed strong transcriptional activity in intestinal cells, this activity was not modified by I^- excess; however, the possibility cannot be entirely ruled out that transcriptional regulation occurs outside the region tested (Fig. 4). In agreement with our data, Leoni et al. (2011) reported that high concentrations of I^- do not modulate NIS promoter activity in thyroid cells.

As excess I^- reduced NIS mRNA levels in the absence of a transcriptional effect, we investigated whether excess I^- regulates NIS mRNA stability in IEC-6 cells (Nicola et al., 2012). We evaluated the half-life of NIS mRNA in cells treated (or not) with I^- in the presence of the mRNA synthesis inhibitor actinomycin D for several periods of time. Whereas I^- significantly shortened the half-life of NIS mRNA by almost 75%, I^- did not have an effect on the half-life of alkaline phosphatase mRNA. Our data taken together are consistent with the notion that there is in enterocytes a previously unknown mode of posttranscriptional regulation of NIS by its own substrate. Consistent with this, several findings have suggested that thyroid NIS mRNA stability or translation efficiency may decrease owing to shortening of the molecule's poly-A tail in response to I^- administration (Serrano-Nascimento et al., 2010).

Untranslated regions (UTRs) of mRNAs play crucial roles in the posttranscriptional regulation of gene expression. In particular, 3′-UTRs harbor determinants that control mRNA stability and translation efficiency (Jackson, Hellen, & Pestova, 2010; Sonenberg & Hinnebusch, 2009). Therefore, to study the involvement of NIS UTRs in the I^--triggered regulation of NIS mRNA, we generated heterologous green fluorescent protein (GFP) reporters containing both the NIS mRNA 5′-UTR sequence (−92 to −1) upstream of the GFP open reading frame (5′-UTR-GFP) and the 3′-UTR sequence (+1858 to +2761) downstream of the GFP coding sequence (GFP-3′-UTR) (Nicola et al., 2012). IEC-6 cells were transiently transfected with the aforementioned reporters or a control vector expressing GFP and treated with excess I^-. Interestingly, I^- excess had no effect on GFP mRNA expression in GFP- and 5′-UTR-GFP-transfected cells but markedly reduced GFP mRNA expression in GFP-3′-UTR-transfected cells. Given that all the GFP-based reporters are controlled by the cytomegalovirus promoter, these results suggest that the NIS 3′-UTR sequence regulated GFP mRNA stability, rather than its transcription in the presence of I^-.

Interestingly, other trace elements also regulate at the posttranscriptional level the expression of genes coding for proteins involved in their own transport or metabolism. For example, iron, selenium, zinc, and calcium regulate the mRNA abundance of transferrin receptor, glutathione peroxidase, the zinc transporter ZnT5, and parathyroid hormone, respectively. The mechanisms that underlie this regulation involve mRNA UTRs, particularly 3′-UTRs (Bermano, Arthur, & Hesketh, 1996; Erlitzki, Long, & Theil, 2002; Jackson et al., 2007; Moallem, Kilav, Silver, & Naveh-Many, 1998; Nechama, Uchida, Mor Yosef-Levi, Silver, & Naveh-Many, 2009; Owen & Kuhn, 1987). Although the mechanism related to I^--induced NIS mRNA decay remains unknown, our findings suggest the presence of functional *cis*-acting elements in the NIS mRNA 3′-UTR sequence related to the I^- regulatory effect. We identified a homozygous mutation -54C>T in the NIS 5′-UTR as responsible for dyshormonogenic congenital hypothyroidism due to reduced NIS mRNA translation efficiency (Nicola et al., 2011), highlighting the importance of UTR sequences in gene expression regulation. These results taken together demonstrate that dietary I^- plays an essential role in enterocyte physiology by controlling its own NIS-mediated absorption and thus regulating the supply of I^- to the body (Nicola et al., 2009, 2012).

There are currently no data regarding hormonal regulation of NIS expression in the small intestine. However, given the great importance of I^- in thyroid hormone biosynthesis, the thyroid hormones themselves might regulate intestinal I^- absorption and NIS expression. Thyroid hormones have been shown to modulate the developmental processes responsible for intestinal maturation, such as the onset of digestive enzyme expression and the regulation of intestinal homeostasis. Thyroid hormone receptor (TR)-α is mainly involved in postnatal small intestine development, as TR-α-deficient mice display alterations in bone and small intestine development, in contrast to TR-β-deficient mice, which do not display retarded intestinal development (Plateroti et al., 1999). Importantly, strong nuclear expression of TR-α was observed in the differentiated epithelial cells of the intestinal villi, indicating that thyroid hormones may regulate gene expression in enterocytes (Gauthier et al., 2001). Therefore, we hypothesize that thyroid hormone levels decrease under chronic I^- deficiency, increasing intestinal NIS expression to maximize dietary I^- absorption.

8. CONCLUSIONS AND FUTURE DIRECTIONS

In light of the findings here reviewed, we propose that NIS is a key molecule in I^- metabolism: it mediates dietary I^- absorption in the small intestine; I^- uptake in the thyroid; and I^- accumulation in breast milk, fetal blood, saliva, and gastric juice. In these last two secretions, I^- accumulation causes the anion to return to the gastrointestinal lumen, where it is again reabsorbed by NIS in the small intestine as part of the I^- conservation system.

NIS expression at the apical surface of the absorptive epithelium of the small intestine is crucial for I^- absorption, the first step in I^- metabolism. These findings underscore the physiological and regulatory significance of the apical localization of intestinal NIS, which contrasts with the basolateral localization observed in virtually all other tissues that express NIS, including the thyroid.

Regulation of intestinal NIS may be elucidated by experiments on the NIS promoter aimed at understanding the molecular mechanisms involved in the expression and regulation of intestinal NIS, as well as the identification of transcription factors required for intestinal NIS expression.

A topic of great significance closely related to that of NIS expression in the small intestine is the fact that NIS mediates active transport of the environmental pollutant ClO_4^-. As ClO_4^- is a frequent contaminant in drinking water sources, intestinal NIS may be a conduit through which ClO_4^- enters the bloodstream. Because NIS actively transports ClO_4^-, the effects of ClO_4^- exposure on public health are more detrimental than previously supposed, particularly for pregnant and nursing women with partial I^- deficiency and, even more worryingly, for their children, the exposure of whom to high ClO_4^- levels puts them at risk of impaired development, not only physically but also intellectually.

There is no pathological condition currently known to result in gastrointestinal I^- malabsorption. Although we have mentioned a few studies investigating I^- absorption in bowel malabsorption syndromes, most of them suffer from several limitations, mainly having to do with sample size. In our view, studies involving more patients may shed light on whether or not patients suffering malabsorption syndromes need I^- supplementation. It will be of great interest to investigate frequent small intestine-restricted malabsorption syndromes such as celiac disease.

We expect that generating small intestine-specific NIS knockout mice will unequivocally establish the role of NIS expression in dietary I^- absorption. This aim will also be well served by a systematic study of patients with NIS mutations that cause congenital I^- transport defect.

ACKNOWLEDGMENTS

We would like to thank all the members of our laboratories who contributed to the research described in this chapter at various stages. This work was supported by awards from the Latin American Thyroid Society and Brown-Coxe postdoctoral fellowship (to J. P. N.) and grants from Fondo Nacional de Ciencia y Tecnología, Secretaría de Ciencia y Tecnología de la Universidad Nacional de Córdoba, Agencia Córdoba Ciencia (to A. M. M.-R.), and the National Institutes of Health grant DK-41544 (to N. C.).

REFERENCES

Altorjay, A., Dohan, O., Szilagyi, A., Paroder, M., Wapnir, I. L., & Carrasco, N. (2007). Expression of the Na^+/I^- symporter (NIS) is markedly decreased or absent in gastric cancer and intestinal metaplastic mucosa of Barrett esophagus. *BMC Cancer*, *7*, 5.

Andersson, M., Karumbunathan, V., & Zimmermann, M. B. (2012). Global iodine status in 2011 and trends over the past decade. *The Journal of Nutrition*, *142*, 744–750.

Arturi, F., Lacroix, L., Presta, I., Scarpelli, D., Caillou, B., Schlumberger, M., et al. (2002). Regulation by human chorionic gonadotropin of sodium/iodide symporter gene expression in the JAr human choriocarcinoma cell line. *Endocrinology*, *143*, 2216–2220.

Bermano, G., Arthur, J. R., & Hesketh, J. E. (1996). Role of the 3′ untranslated region in the regulation of cytosolic glutathione peroxidase and phospholipid-hydroperoxide glutathione peroxidase gene expression by selenium supply. *The Biochemical Journal*, *320*, 891–895.

Bonnema, S. J., & Hegedus, L. (2012). Radioiodine therapy in benign thyroid diseases: Effects, side effects, and factors affecting therapeutic outcome. *Endocrine Reviews*, *33*, 920–980.

Braverman, L. E., & Ingbar, S. H. (1963). Changes in thyroidal function during adaptation to large doses of iodide. *The Journal of Clinical Investigation*, *42*, 1216–1231.

Bruno, R., Giannasio, P., Ronga, G., Baudin, E., Travagli, J. P., Russo, D., et al. (2004). Sodium iodide symporter expression and radioiodine distribution in extrathyroidal tissues. *Journal of Endocrinological Investigation*, *27*, 1010–1014.

Burns, R., O'Herlihy, C., & Smyth, P. P. (2013). Regulation of iodide uptake in placental primary cultures. *European Thyroid Journal*, *2*, 243–251.

Cho, J. Y., Leveille, R., Kao, R., Rousset, B., Parlow, A. F., Burak, W. E., Jr., et al. (2000). Hormonal regulation of radioiodide uptake activity and Na^+/I^- symporter expression in mammary glands. *The Journal of Clinical Endocrinology and Metabolism*, *85*, 2936–2943.

Cianchetta, S., di Bernardo, J., Romeo, G., & Rhoden, K. J. (2010). Perchlorate transport and inhibition of the sodium iodide symporter measured with the yellow fluorescent protein variant YFP-H148Q/I152L. *Toxicology and Applied Pharmacology*, *243*, 372–380.

Cohn, B. N. E. (1932). Absorption of compound solution of iodine from the gastro-intestinal tract. *Archives of Internal Medicine*, *49*, 950–956.

Dai, G., Levy, O., & Carrasco, N. (1996). Cloning and characterization of the thyroid iodide transporter. *Nature*, *379*, 458–460.

de Carvalho, F. D., & Quick, M. (2011). Surprising substrate versatility in SLC5A6: Na^+-coupled I^- transport by the human Na^+/multivitamin transporter (hSMVT). *The Journal of Biological Chemistry*, *286*, 131–137.

De la Vieja, A., Reed, M. D., Ginter, C. S., & Carrasco, N. (2007). Amino acid residues in transmembrane segment IX of the Na^+/I^- symporter play a role in its Na^+ dependence and are critical for transport activity. *The Journal of Biological Chemistry*, *282*, 25290–25298.

Di Cosmo, C., Fanelli, G., Tonacchera, M., Ferrarini, E., Dimida, A., Agretti, P., et al. (2006). The sodium-iodide symporter expression in placental tissue at different gestational age: An immunohistochemical study. *Clinical Endocrinology*, *65*, 544–548.

Dohan, O., Baloch, Z., Banrevi, Z., Livolsi, V., & Carrasco, N. (2001). Predominant intracellular overexpression of the Na(+)/I(−) symporter (NIS) in a large sampling of thyroid cancer cases. *The Journal of Clinical Endocrinology and Metabolism*, *86*, 2697–2700.

Dohan, O., De la Vieja, A., & Carrasco, N. (2006). Hydrocortisone and purinergic signaling stimulate sodium/iodide symporter (NIS)-mediated iodide transport in breast cancer cells. *Molecular Endocrinology, 20*, 1121–1137.
Dohan, O., Portulano, C., Basquin, C., Reyna-Neyra, A., Amzel, L. M., & Carrasco, N. (2007). The Na^+/I symporter (NIS) mediates electroneutral active transport of the environmental pollutant perchlorate. *Proceedings of the National Academy of Sciences of the United States of America, 104*, 20250–20255.
Donowitz, M., Singh, S., Salahuddin, F. F., Hogema, B. M., Chen, Y., Gucek, M., et al. (2007). Proteome of murine jejunal brush border membrane vesicles. *Journal of Proteome Research, 6*, 4068–4079.
El Hassani, R. A., Benfares, N., Caillou, B., Talbot, M., Sabourin, J. C., Belotte, V., et al. (2005). Dual oxidase2 is expressed all along the digestive tract. *American Journal of Physiology Gastrointestinal and Liver Physiology, 288*, G933–G942.
Eng, P. H., Cardona, G. R., Fang, S. L., Previti, M., Alex, S., Carrasco, N., et al. (1999). Escape from the acute Wolff-Chaikoff effect is associated with a decrease in thyroid sodium/iodide symporter messenger ribonucleic acid and protein. *Endocrinology, 140*, 3404–3410.
Erlitzki, R., Long, J. C., & Theil, E. C. (2002). Multiple, conserved iron-responsive elements in the 3′-untranslated region of transferrin receptor mRNA enhance binding of iron regulatory protein 2. *The Journal of Biological Chemistry, 277*, 42579–42587.
Eskandari, S., Loo, D. D., Dai, G., Levy, O., Wright, E. M., & Carrasco, N. (1997). Thyroid Na^+/I^- symporter. Mechanism, stoichiometry, and specificity. *The Journal of Biological Chemistry, 272*, 27230–27238.
Faham, S., Watanabe, A., Besserer, G. M., Cascio, D., Specht, A., Hirayama, B. A., et al. (2008). The crystal structure of a sodium galactose transporter reveals mechanistic insights into Na^+/sugar symport. *Science, 321*, 810–814.
Ferretti, E., Arturi, F., Mattei, T., Scipioni, A., Tell, G., Tosi, E., et al. (2005). Expression, regulation, and function of paired-box gene 8 in the human placenta and placental cancer cell lines. *Endocrinology, 146*, 4009–4015.
Fisher, D. A., Oddie, T. H., & Epperson, D. (1965). Effect of increased dietary iodide on thyroid accumulation and secretion in euthyroid Arkansas subjects. *The Journal of Clinical Endocrinology and Metabolism, 25*, 1580–1590.
Gauthier, K., Plateroti, M., Harvey, C. B., Williams, G. R., Weiss, R. E., Refetoff, S., et al. (2001). Genetic analysis reveals different functions for the products of the thyroid hormone receptor alpha locus. *Molecular and Cellular Biology, 21*, 4748–4760.
Geiszt, M., Witta, J., Baffi, J., Lekstrom, K., & Leto, T. L. (2003). Dual oxidases represent novel hydrogen peroxide sources supporting mucosal surface host defense. *The FASEB Journal, 17*, 1502–1504.
Ghosal, A., Lambrecht, N., Subramanya, S. B., Kapadia, R., & Said, H. M. (2013). Conditional knockout of the Slc5a6 gene in mouse intestine impairs biotin absorption. *American Journal of Physiology. Gastrointestinal and Liver Physiology, 304*, G64–G71.
Hanzlik, P. J. (1912). Quantitative studies on the gastro-intestinal absorption of drugs: II. The absorption of sodium iodide. *The Journal of Pharmacology and Experimental Therapeutics, 3*, 387–421.
Hertz, S., Roberts, A., Means, J. H., & Evans, R. D. (1940). Radioactive iodine as an indicator in thyroid physiology. *The American Journal of Physiology, 128*, 565–576.
Hilditch, T. E., Horton, P. W., McCruden, D. C., Young, R. E., & Alexander, W. D. (1982). Defects in intrathyroid binding of iodine and the perchlorate discharge test. *Acta Endocrinologica, 100*, 237–244.
Jackson, R. J., Hellen, C. U., & Pestova, T. V. (2010). The mechanism of eukaryotic translation initiation and principles of its regulation. *Nature Reviews. Molecular Cell Biology, 11*, 113–127.

Jackson, K. A., Helston, R. M., McKay, J. A., O'Neill, E. D., Mathers, J. C., & Ford, D. (2007). Splice variants of the human zinc transporter ZnT5 (SLC30A5) are differentially localized and regulated by zinc through transcription and mRNA stability. *The Journal of Biological Chemistry, 282*, 10423–10431.

Jarnerot, G. (1975). The thyroid in ulcerative colitis and Crohn's disease. I. Thyroid radioiodide uptake and urinary iodine excretion. *Acta Medica Scandinavica, 197*, 77–81.

Josefsson, M. (2009). *Sodium/iodide symporter (NIS) - Abundant and important in gastric mucosa* (Doctoral dissertation). Sweeden: Faculty of medicine, Lund University.

Josefsson, M., Grunditz, T., Ohlsson, T., & Ekblad, E. (2002). Sodium/iodide-symporter: Distribution in different mammals and role in entero-thyroid circulation of iodide. *Acta Physiologica Scandinavica, 175*, 129–137.

Kogai, T., Endo, T., Saito, T., Miyazaki, A., Kawaguchi, A., & Onaya, T. (1997). Regulation by thyroid-stimulating hormone of sodium/iodide symporter gene expression and protein levels in FRTL-5 cells. *Endocrinology, 138*, 2227–2232.

Kollecker, I., von Wasielewski, R., Langner, C., Muller, J. A., Spitzweg, C., Kreipe, H., et al. (2012). Subcellular distribution of the sodium iodide symporter in benign and malignant thyroid tissues. *Thyroid, 22*, 529–535.

Kotani, T., Ogata, Y., Yamamoto, I., Aratake, Y., Kawano, J. I., Suganuma, T., et al. (1998). Characterization of gastric Na^{+}/I^{-} symporter of the rat. *Clinical Immunology and Immunopathology, 89*, 271–278.

Krishnamurthy, H., & Gouaux, E. (2012). X-ray structures of LeuT in substrate-free outward-open and apo inward-open states. *Nature, 481*, 469–474.

La Perle, K. M., Kim, D. C., Hall, N. C., Bobbey, A., Shen, D. H., Nagy, R. S., et al. (2013). Modulation of sodium/iodide symporter expression in the salivary gland. *Thyroid, 23*, 1029–1036.

Lacroix, L., Mian, C., Caillou, B., Talbot, M., Filetti, S., Schlumberger, M., et al. (2001). Na(+)/I(−) symporter and Pendred syndrome gene and protein expressions in human extra-thyroidal tissues. *European Journal of Endocrinology, 144*, 297–302.

Leoni, S. G., Kimura, E. T., Santisteban, P., & De la Vieja, A. (2011). Regulation of thyroid oxidative state by thioredoxin reductase has a crucial role in thyroid responses to iodide excess. *Molecular Endocrinology, 25*, 1924–1935.

Levy, O., Dai, G., Riedel, C., Ginter, C. S., Paul, E. M., Lebowitz, A. N., et al. (1997). Characterization of the thyroid Na^{+}/I^{-} symporter with an anti-COOH terminus antibody. *Proceedings of the National Academy of Sciences of the United States of America, 94*, 5568–5573.

Levy, O., De la Vieja, A., Ginter, C. S., Riedel, C., Dai, G., & Carrasco, N. (1998). N-linked glycosylation of the thyroid Na^{+}/I^{-} symporter (NIS). Implications for its secondary structure model. *The Journal of Biological Chemistry, 273*, 22657–22663.

Li, W., Nicola, J. P., Amzel, L. M., & Carrasco, N. (2013). Asn441 plays a key role in folding and function of the Na^{+}/I^{-} symporter (NIS). *The FASEB Journal, 27*, 3229–3238.

Marians, R. C., Ng, L., Blair, H. C., Unger, P., Graves, P. N., & Davies, T. F. (2002). Defining thyrotropin-dependent and -independent steps of thyroid hormone synthesis by using thyrotropin receptor-null mice. *Proceedings of the National Academy of Sciences of the United States of America, 99*, 15776–15781.

Michalaki, M., Volonakis, S., Mamali, I., Kalfarentzos, F., Vagenakis, A. G., & Markou, K. B. (2014). Dietary iodine absorption is not influenced by malabsorptive bariatric surgery. *Obesity Surgery, 24*, 1921–1925.

Mitchell, A. M., Manley, S. W., Morris, J. C., Powell, K. A., Bergert, E. R., & Mortimer, R. H. (2001). Sodium iodide symporter (NIS) gene expression in human placenta. *Placenta, 22*, 256–258.

Moallem, E., Kilav, R., Silver, J., & Naveh-Many, T. (1998). RNA-protein binding and post-transcriptional regulation of parathyroid hormone gene expression by calcium and phosphate. *The Journal of Biological Chemistry, 273*, 5253–5259.

Morgenstern, K. E., Vadysirisack, D. D., Zhang, Z., Cahill, K. V., Foster, J. A., Burns, J. A., et al. (2005). Expression of sodium iodide symporter in the lacrimal drainage system: Implication for the mechanism underlying nasolacrimal duct obstruction in I(131)-treated patients. *Ophthalmic Plastic and Reconstructive Surgery, 21*, 337–344.

Navarro, A. M., Suen, V. M., Souza, I. M., De Oliveira, J. E., & Marchini, J. S. (2005). Patients with severe bowel malabsorption do not have changes in iodine status. *Nutrition, 21*, 895–900.

Nechama, M., Uchida, T., Mor Yosef-Levi, I., Silver, J., & Naveh-Many, T. (2009). The peptidyl-prolyl isomerase Pin1 determines parathyroid hormone mRNA levels and stability in rat models of secondary hyperparathyroidism. *The Journal of Clinical Investigation, 119*, 3102–3114.

Neumann, S., Schuchardt, K., Reske, A., Reske, A., Emmrich, P., & Paschke, R. (2004). Lack of correlation for sodium iodide symporter mRNA and protein expression and analysis of sodium iodide symporter promoter methylation in benign cold thyroid nodules. *Thyroid, 14*, 99–111.

Nicola, J. P., Basquin, C., Portulano, C., Reyna-Neyra, A., Paroder, M., & Carrasco, N. (2009). The Na^+/I^- symporter mediates active iodide uptake in the intestine. *American Journal of Physiology. Cell Physiology, 296*, C654–C662.

Nicola, J. P., Carrasco, N., & Mario Amzel, L. (2014). Physiological sodium concentrations enhance the iodide affinity of the Na(+)/I(−) symporter. *Nature Communications, 5*, 3948.

Nicola, J. P., Nazar, M., Serrano-Nascimento, C., Goulart-Silva, F., Sobrero, G., Testa, G., et al. (2011). Iodide transport defect: Functional characterization of a novel mutation in the Na^+/I^- symporter 5′-untranslated region in a patient with congenital hypothyroidism. *The Journal of Clinical Endocrinology and Metabolism, 96*, E1100–E1107.

Nicola, J. P., Reyna-Neyra, A., Carrasco, N., & Masini-Repiso, A. M. (2012). Dietary iodide controls its own absorption through post-transcriptional regulation of the intestinal Na^+/I^- symporter. *The Journal of Physiology, 590*, 6013–6026.

Ohno, M., Zannini, M., Levy, O., Carrasco, N., & di Lauro, R. (1999). The paired-domain transcription factor Pax8 binds to the upstream enhancer of the rat sodium/iodide symporter gene and participates in both thyroid-specific and cyclic-AMP-dependent transcription. *Molecular and Cellular Biology, 19*, 2051–2060.

O'Neill, B., Magnolato, D., & Semenza, G. (1987). The electrogenic, Na^+-dependent I^- transport system in plasma membrane vesicles from thyroid glands. *Biochimica et Biophysica Acta, 896*, 263–274.

Owen, D., & Kuhn, L. C. (1987). Noncoding 3′ sequences of the transferrin receptor gene are required for mRNA regulation by iron. *The EMBO Journal, 6*, 1287–1293.

Paroder, V., Nicola, J. P., Ginter, C. S., & Carrasco, N. (2013). The iodide-transport-defect-causing mutation R124H: A delta-amino group at position 124 is critical for maturation and trafficking of the Na^+/I^- symporter. *Journal of Cell Science, 126*, 3305–3313.

Paroder, V., Spencer, S. R., Paroder, M., Arango, D., Schwartz, S., Jr., Mariadason, J. M., et al. (2006). Na(+)/monocarboxylate transport (SMCT) protein expression correlates with survival in colon cancer: Molecular characterization of SMCT. *Proceedings of the National Academy of Sciences of the United States of America, 103*, 7270–7275.

Paroder-Belenitsky, M., Maestas, M. J., Dohan, O., Nicola, J. P., Reyna-Neyra, A., Follenzi, A., et al. (2011). Mechanism of anion selectivity and stoichiometry of the Na^+/I^- symporter (NIS). *Proceedings of the National Academy of Sciences of the United States of America, 108*, 17933–17938.

Perron, B., Rodriguez, A. M., Leblanc, G., & Pourcher, T. (2001). Cloning of the mouse sodium iodide symporter and its expression in the mammary gland and other tissues. *The Journal of Endocrinology, 170*, 185–196.

Plateroti, M., Chassande, O., Fraichard, A., Gauthier, K., Freund, J. N., Samarut, J., et al. (1999). Involvement of T3Ralpha- and beta-receptor subtypes in mediation of T3 functions during postnatal murine intestinal development. *Gastroenterology*, *116*, 1367–1378.

Portulano, C., Paroder-Belenitsky, M., & Carrasco, N. (2014). The Na^+/I^- symporter (NIS): Mechanism and medical impact. *Endocrine Reviews*, *35*, 106–149.

Reiners, C., Hanscheid, H., Luster, M., Lassmann, M., & Verburg, F. A. (2011). Radioiodine for remnant ablation and therapy of metastatic disease. *Nature Reviews. Endocrinology*, *7*, 589–595.

Reiners, C., & Luster, M. (2012). Radiotherapy: Radioiodine in thyroid cancer-how to minimize side effects. *Nature Reviews. Clinical Oncology*, *9*, 432–434.

Riedel, C., Levy, O., & Carrasco, N. (2001). Post-transcriptional regulation of the sodium/iodide symporter by thyrotropin. *The Journal of Biological Chemistry*, *276*, 21458–21463.

Riesco-Eizaguirre, G., Leoni, S. G., Mendiola, M., Estevez-Cebrero, M. A., Gallego, M. I., Redondo, A., et al. (2014). NIS mediates iodide uptake in the female reproductive tract and is a poor prognostic factor in ovarian cancer. *The Journal of Clinical Endocrinology and Metabolism*, *99*, E1199–E1208.

Russo, D., Manole, D., Arturi, F., Suarez, H. G., Schlumberger, M., Filetti, S., et al. (2001). Absence of sodium/iodide symporter gene mutations in differentiated human thyroid carcinomas. *Thyroid*, *11*, 37–39.

Schlumberger, M., Lacroix, L., Russo, D., Filetti, S., & Bidart, J. M. (2007). Defects in iodide metabolism in thyroid cancer and implications for the follow-up and treatment of patients. *Nature Reviews. Endocrinology*, *3*, 260–269.

Serrano-Nascimento, C., Calil-Silveira, J., & Nunes, M. T. (2010). Posttranscriptional regulation of sodium-iodide symporter mRNA expression in the rat thyroid gland by acute iodide administration. *American Journal of Physiology. Cell Physiology*, *298*, C893–C899.

Serrano-Nascimento, C., da Silva Teixeira, S., Nicola, J. P., Nachbar, R. T., Masini-Repiso, A. M., & Nunes, M. T. (2014). The acute inhibitory effect of iodide excess on sodium/iodide symporter expression and activity involves the PI3K/Akt signaling pathway. *Endocrinology*, *155*, 1145–1156.

Smanik, P. A., Liu, Q., Furminger, T. L., Ryu, K., Xing, S., Mazzaferri, E. L., et al. (1996). Cloning of the human sodium iodide symporter. *Biochemical and Biophysical Research Communications*, *226*, 339–345.

Smanik, P. A., Ryu, K. Y., Theil, K. S., Mazzaferri, E. L., & Jhiang, S. M. (1997). Expression, exon-intron organization, and chromosome mapping of the human sodium iodide symporter. *Endocrinology*, *138*, 3555–3558.

Smith, V. E., Read, M. L., Turnell, A. S., Watkins, R. J., Watkinson, J. C., Lewy, G. D., et al. (2009). A novel mechanism of sodium iodide symporter repression in differentiated thyroid cancer. *Journal of Cell Science*, *122*, 3393–3402.

Sonenberg, N., & Hinnebusch, A. G. (2009). Regulation of translation initiation in eukaryotes: Mechanisms and biological targets. *Cell*, *136*, 731–745.

Spitzweg, C., Dutton, C. M., Castro, M. R., Bergert, E. R., Goellner, J. R., Heufelder, A. E., et al. (2001). Expression of the sodium iodide symporter in human kidney. *Kidney International*, *59*, 1013–1023.

Spitzweg, C., Joba, W., Eisenmenger, W., & Heufelder, A. E. (1998). Analysis of human sodium iodide symporter gene expression in extrathyroidal tissues. *The Journal of Clinical Endocrinology and Metabolism*, *83*, 1746–1751.

Spitzweg, C., Joba, W., Morris, J. C., & Heufelder, A. E. (1999). Regulation of sodium iodide symporter gene expression in FRTL-5 rat thyroid cells. *Thyroid*, *9*, 821–830.

Spitzweg, C., Joba, W., Schriever, K., Goellner, J. R., Morris, J. C., & Heufelder, A. E. (1999). Analysis of human sodium iodide symporter immunoreactivity in human exocrine glands. *The Journal of Clinical Endocrinology and Metabolism*, *84*, 4178–4184.

Spitzweg, C., & Morris, J. C. (2010). Genetics and phenomics of hypothyroidism and goiter due to NIS mutations. *Molecular and Cellular Endocrinology*, *322*, 56–63.

Tazebay, U. H., Wapnir, I. L., Levy, O., Dohan, O., Zuckier, L. S., Zhao, Q. H., et al. (2000). The mammary gland iodide transporter is expressed during lactation and in breast cancer. *Nature Medicine*, *6*, 871–878.

Tonacchera, M., Viacava, P., Agretti, P., de Marco, G., Perri, A., di Cosmo, C., et al. (2002). Benign nonfunctioning thyroid adenomas are characterized by a defective targeting to cell membrane or a reduced expression of the sodium iodide symporter protein. *The Journal of Clinical Endocrinology and Metabolism*, *87*, 352–357.

Tran, N., Valentin-Blasini, L., Blount, B. C., McCuistion, C. G., Fenton, M. S., Gin, E., et al. (2008). Thyroid-stimulating hormone increases active transport of perchlorate into thyroid cells. *American Journal of Physiology. Endocrinology and Metabolism*, *294*, E802–E806.

Uyttersprot, N., Pelgrims, N., Carrasco, N., Gervy, C., Maenhaut, C., Dumont, J. E., et al. (1997). Moderate doses of iodide in vivo inhibit cell proliferation and the expression of thyroperoxidase and Na^+/I^- symporter mRNAs in dog thyroid. *Molecular and Cellular Endocrinology*, *131*, 195–203.

Vayre, L., Sabourin, J. C., Caillou, B., Ducreux, M., Schlumberger, M., & Bidart, J. M. (1999). Immunohistochemical analysis of Na^+/I^- symporter distribution in human extra-thyroidal tissues. *European Journal of Endocrinology*, *141*, 382–386.

Vejbjerg, P., Knudsen, N., Perrild, H., Laurberg, P., Andersen, S., Rasmussen, L. B., et al. (2009). Estimation of iodine intake from various urinary iodine measurements in population studies. *Thyroid*, *19*, 1281–1286.

Venturi, S., & Venturi, M. (2009). Iodine in evolution of salivary glands and in oral health. *Nutrition and Health*, *20*, 119–134.

Vilijn, F., & Carrasco, N. (1989). Expression of the thyroid sodium/iodide symporter in *Xenopus laevis* oocytes. *The Journal of Biological Chemistry*, *264*, 11901–11903.

Wapnir, I. L., van de Rijn, M., Nowels, K., Amenta, P. S., Walton, K., Montgomery, K., et al. (2003). Immunohistochemical profile of the sodium/iodide symporter in thyroid, breast, and other carcinomas using high density tissue microarrays and conventional sections. *The Journal of Clinical Endocrinology and Metabolism*, *88*, 1880–1888.

Weiser, M. M. (1973). Intestinal epithelial cell surface membrane glycoprotein synthesis. I. An indicator of cellular differentiation. *The Journal of Biological Chemistry*, *248*, 2536–2541.

Wolff, J., & Chaikoff, I. L. (1948). Plasma inorganic iodide as a homeostatic regulator of thyroid function. *The Journal of Biological Chemistry*, *174*, 555–564.

Wolff, J., & Maurey, J. R. (1961). Thyroidal iodide transport. II. Comparison with non-thyroid iodide-concentrating tissues. *Biochimica et Biophysica Acta*, *47*, 467–474.

Yamashita, A., Singh, S. K., Kawate, T., Jin, Y., & Gouaux, E. (2005). Crystal structure of a bacterial homologue of Na^+/Cl^--dependent neurotransmitter transporters. *Nature*, *437*, 215–223.

Yen, P. M. (2001). Physiological and molecular basis of thyroid hormone action. *Physiological Reviews*, *81*, 1097–1142.

Zhang, X., Riedel, C., Carrasco, N., & Arvan, P. (2002). Polarized trafficking of thyrocyte proteins in MDCK cells. *Molecular and Cellular Endocrinology*, *188*, 27–36.

Zhao, Y., Terry, D., Shi, L., Weinstein, H., Blanchard, S. C., & Javitch, J. A. (2010). Single-molecule dynamics of gating in a neurotransmitter transporter homologue. *Nature*, *465*, 188–193.

Zimmermann, M. B. (2009). Iodine deficiency. *Endocrine Reviews*, *30*, 376–408.

CHAPTER TWO

Apical Iodide Efflux in Thyroid

Peying Fong[1]
Department of Anatomy and Physiology, Kansas State University College of Veterinary Medicine, Manhattan, Kansas, USA
[1]Corresponding author: e-mail address: pfong@vet.k-state.edu

Contents

Abstract

Thyroid follicular epithelial cells produce thyroxine (T_4) and its physiologically active derivative, 3,3′,5-triiodothyronine (T_3), hormones that regulate critical developmental and metabolic functions. In order for the thyroid to form hormone precursor, iodide, the defining element in thyroid hormone, must cross both blood-facing and luminal sides of the follicular epithelium. The pathway for uptake from blood is well understood, but the mechanism(s) that enable iodide to cross the luminally facing apical membrane remain obscure. This chapter considers the physiological properties of several molecularly characterized anion transport proteins, all of which potentially contribute to the overall mechanism of apical iodide efflux.

Vitamins and Hormones, Volume 98
ISSN 0083-6729
http://dx.doi.org/10.1016/bs.vh.2014.12.015

1. INTRODUCTION

Thyroid hormones govern critical processes throughout the lifetime of every vertebrate. They play diverse roles that range from regulation of key developmental processes during fetal life to modulation of cardiac function and metabolism in adults. Overt disturbances in thyroid function, as well as subclinical thyroid dysfunction, affect practically all systems of the body.

The defining elemental component of thyroid hormones is iodide (I^-), which is ingested as dietary iodine (I_2) and carried to the thyroid gland in the ionic form. There, it is incorporated into thyroglobulin (Tg) and stored as iodinated hormone precursor. The rarity of I_2 in the environment—and hence, diet—dictates that thyroid I^- uptake requires highly specialized membrane transport capabilities, together with a robust means of chemical coupling to Tg. Sufficient dietary iodine levels are required for health and recommended daily intake levels for I^- per unit body weight accordingly adjust for the needs of actively growing individuals (infants, young children, and pregnant women). Much of the earth's I^- is found in the ocean, so populations inhabiting land-locked regions of the world are particularly susceptible to I^- deficiency disorders. These conditions, such as profoundly impaired cognitive and physical development, inflict a heavy collective societal burden. More sobering is the persistence of I^- deficiency-produced impairments, despite the fact that they can be addressed effectively at comparatively low cost (i.e., incorporation of iodized table salt into the diet). Worldwide public health initiatives promoting the use of iodized dietary salt represent a move toward preventing and eliminating the debilitating conditions resulting from I^- deficiency.

This chapter considers the emergent details of one essential step in thyroid uptake and utilization of I^-: transport into the follicular space, a process necessary for its coupling to Tg. Although the understanding of apical membrane I^- exit presently is in its infancy, the rapid advances in available technologies make this an especially exciting time. It is anticipated that the accelerated pace of investigations targeting this important topic will bring greater clarity and resolution of the present gaps in understanding.

What presently is understood about how I^- moves into the thyroid follicular lumen? Systematic treatment of this topic necessitates that several fundamental spatial and functional relationships are presented briefly. To this end, the organization of epithelial follicles comprising the thyroid gland is considered first, followed by a description of biochemical processes that combine key substrates and ultimately form thyroid hormones. A summary of general epithelial transport processes and how they pertain to thyroid I^- accumulation precedes the chapter's primary focus: the anion transport proteins expressed at thyrocyte apical membranes that potentially mediate I^- movement into the follicular lumen. Although the functional characteristics of many anion transport proteins inhabit the literature, this chapter is limited to those that are identified on both the functional and molecular levels (i.e., the cDNA has been isolated). The physiological signatures of these transport proteins are considered within the context of their putative roles in mediating follicular I^- accumulation.

2. IODIDE AND THYROID HORMONE SYNTHESIS

In vertebrates, the thyroid gland—a discrete, bi-lobular cluster of numerous epithelial follicles—assumes critical functions of thyroid hormone production: I^- uptake, synthesis, and secretion of Tg into the follicular lumen, iodination of Tg to form prohormone, and ultimately, cleavage of prohormone to the circulating thyroxine (T_4) and 3,3′,5-triiodothyronine (T_3). Storage of this large *extracellular* reservoir of prohormone is unique to the thyroid. Although Tg, iodotyrosines, and T_4 are elaborated by select cells within the endostyle of nonvertebrate chordates, such as ascidians, the follicular arrangement is unique to the vertebrate thyroid (Barrington & Thorpe, 1965; Thorndyke, 1978).

2.1 Thyroid organization

As described in the previous section, thyroid follicles, spheroidal collections of epithelial cells (thyrocytes), function in the production and storage of thyroid hormone and its precursor prohormone. Follicular epithelial cells are organized with their apices directed toward a luminal cavity, with their basolateral aspects adjacent to a highly vascularized interstitium (Fig. 1A). The luminal compartment is filled with colloid comprised of the thyrocytes' characteristic primary protein secretion: Tg. As for all epithelia, the apical (luminal) and basolateral (blood) membrane domains are demarcated and separated by tight junctional complexes. The tight junctions not only

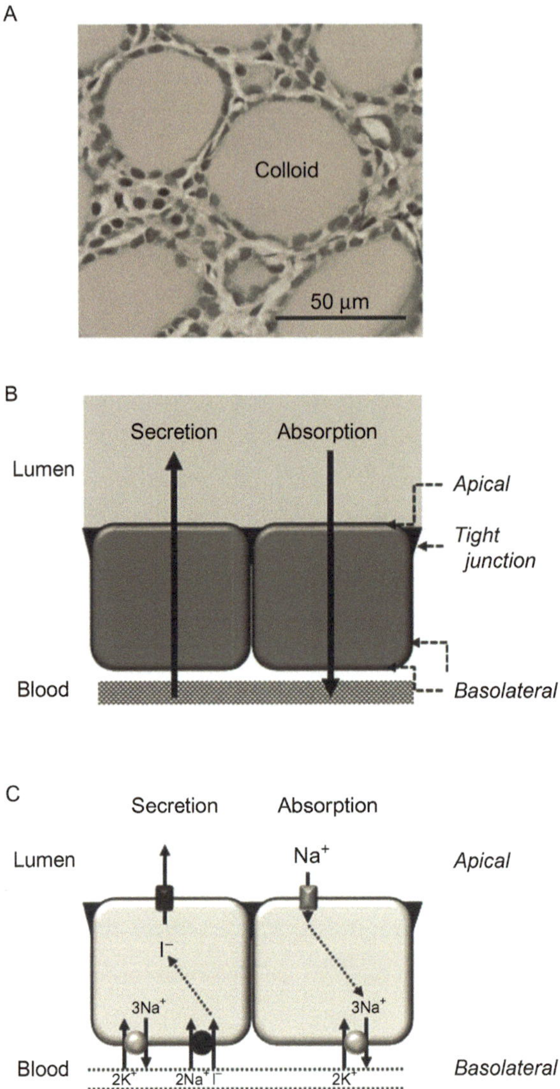

Figure 1 Organization of the thyroid epithelium and basic transport schema. (A) Micrograph shows a section of porcine thyroid showing the follicular arrangement of thyroid epithelial cells. Note the thyroglobulin colloid (C) present in the follicular lumen. (B) Schematic diagram of an epithelium shows epithelial polarity and basic vectorial transport mechanisms. Note the tight junctions, which separate the apical (lumen-facing) and basolateral (blood-facing) domains. (C) Diagram shows transport proteins known to be involved in two thyroid follicular secretion and absorption processes. Cell depicted at left shows a model for I^- secretion. Note Na^+–I^- symporter (NIS)-mediated, Na^+-coupled uptake of I^- at the basolateral membrane. The exit pathways for I^- exit into the luminal space, here represented as a lumped efflux pathway, may involve many different transport proteins, as discussed herein. Cell depicted at right shows a model for Na^+ absorption. Absorption depends on passive luminal entry through epithelial sodium channels (ENaC). Note that both processes depend on activity of the Na^+/K^+ ATPase.

introduce a physical barrier between the two compartments but also contribute to electrical resistivity of the follicular epithelium.

Interspersed among the follicles reside a minority population of calcitonin-secreting C cells. It is important to acknowledge their presence but, as they do not contribute to follicular I^- accumulation, they are not discussed further in this chapter.

2.2 Thyroid hormone synthesis

The Tg iodination process relies on enzymatic activities at the interface of the thyrocyte apical membrane and the follicular lumen. Of key importance are (1) dual oxidases (DUOX1/A1 and DUOX2/A2) that supply hydrogen peroxide (H_2O_2) (Carvalho & Dupuy, 2014) as well as (2) thyroperoxidase (TPO), the enzyme that catalyzes the reaction of I^-, H_2O_2, and Tg, thereby forming iodinated Tg (Tg-I).

Note that the follicular arrangement of thyrocytes endows the thyroid gland with capacity for extracellular Tg-I storage. From this reservoir, hormone can be deployed rapidly upon demand (i.e., stimulation by circulating thyrotropin (thyroid-stimulating hormone, TSH)). Localization of TSH receptors to the basolateral—that is, the vasculature-facing side—enables this process. Subsequent endocytosis and cleavage liberate T_4 and the physiologically active T_3, which then cross the basolateral membrane, partially via a monocarboxylate transporter (MCT8), and ultimately enter the blood. Because the thyroid primarily releases T_4, with lesser amounts of T_3, into the circulation, downstream conversion is required. This is accomplished through the expression in target tissues of the iodothyronine deiodinases. The actions of these important enzymes depend critically on the presence of a selenocysteine within their active centers (Gereben et al., 2008).

Figure 2 summarizes the actions of deiodinases 1, 2, and 3. The type 1 deiodinase is capable of removing I^- from either outer or inner rings of T_4. This generates either T_3 or [*reverse* T_3], respectively. However, little is understood about deiodinase 1's role in the healthy human thyroid; it is believed to contribute to the high serum T_3 levels in the hyperthyroid state. Instead, tight regulation of type 2 deiodinase, which removes I^- from the outer phenol ring, ensures an adequate supply of T_3. On the other hand, the type 3 deiodinase can blunt potential T_4 to T_3 conversion by deiodination of the inner tyrosyl ring, thereby forming inactive rT_3. Moreover, type 3 deiodinase disables T_3 function by catalyzing its deiodination to inactive T_2. This gentle tug-of-war between the type 2 and type 3 deiodinases thereby ensures fine control of T_3 levels in the healthy thyroid.

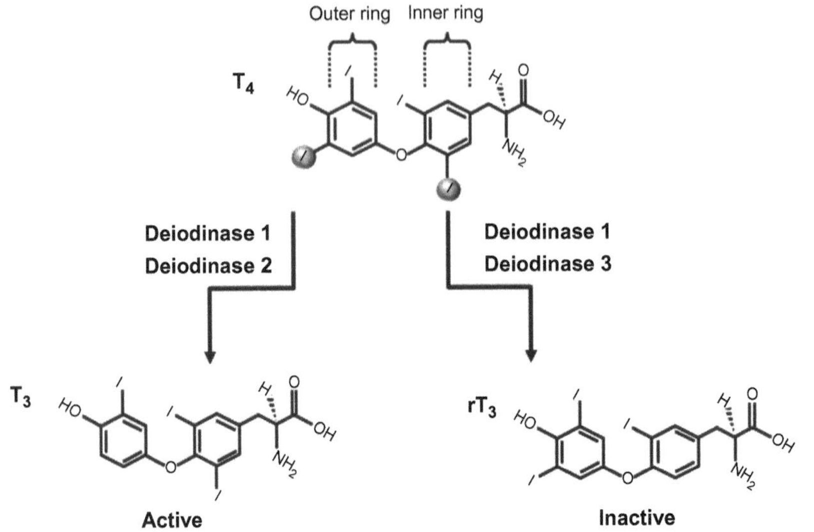

Figure 2 Deiodinases regulate thyroid hormone activity at target tissues. Deiodinases 2 and 3 primarily govern this process, with deiodinase 2 catalyzing removal of I^- from the outer ring of T_4, thereby generating T_3 (left branch). Action of deiodinase 3, on the other hand, catalyzes removal from the inner ring, forming the inactive rT_3 (right branch). Deiodinase 1 can perform removal from either inner or outer ring, but its function in humans is obscure (see text). The shaded I^- residues indicate those removed by action of respective deiodinases depicted. (See the color plate.)

3. VECTORIAL TRANSPORT PROCESSES IN EPITHELIA AND THYROID I^- ACCUMULATION

3.1 Brief overview of basic epithelial transport processes

Epithelia act as barriers and are organized to promote directional (*vectorial*) movement of substrates, either from the blood to the luminal side (*secretion*) or from the luminal to blood side (*absorption*) (refer to Fig. 1B). Vectorial transport processes rely on the polarized distribution of transport proteins within epithelial cells, which in turn is maintained by the integrity of tight junctional complexes.

The term "transport protein" refers to many functional species. It comprises active transporters, such as the Na^+/K^+ ATPase ("the pump"), which, fueled by ATP hydrolysis, move Na^+ and K^+ against their electrochemical gradients. It also includes coupled ("secondary active") transporters, membrane proteins that pair the movement of a substrate *against* its electrochemical gradient to that of a substance that moves *down* its gradient. The latter is often, but not always, Na^+. "Transport protein" also can refer to channels (and also uniporters) that move a substance passively, *down* its electrochemical gradient. Thus, transport is determined by substrate availability, driven by concentration gradients and, for charged substrates, by the membrane potential.

An important early description of flux coupling, that of glucose to the Na^+ gradient (Lyon & Crane, 1966a, 1966b), continues to guide our present understanding of how diverse substrates gain entry into the cell via a spectrum of specific, Na^+-coupled transporters. Intracellular Na^+ is maintained at low levels as a result of its active extrusion via the Na^+/K^+ ATPase, which—with the exception of the choroid and the retinal pigment epithelium—resides at the basolateral membrane of epithelia. This generates a concentration gradient favoring Na^+ entry. Moreover, the stoichiometry of the pump is $3Na^+:2K^+$. Thus, it is electrogenic and contributes to maintenance of the cytosol-negative membrane potential (V_m). The resultant *electrochemical* gradient for Na^+ that is created by Na^+/K^+ ATPase pump activity thus powers Na^+ into the cell, across *either* apical or basolateral membranes.

The inward gradient for Na^+ can be harnessed to accomplish *either* secretion or absorption. In many secreting epithelia, anion uptake from the blood is coupled to the basolateral entry of Na^+ via cotransport mechanisms. For example, airway submucosal gland serous cells take up Cl^- and HCO_3^- from the circulation via $Na^+/K^+/2Cl^-$ cotransport and Na^+/HCO_3^- cotransport (Devor et al., 1999). Thus, in the thyrocyte, one substrate that is taken up from blood is I^-, which also exploits a Na^+-coupled vehicle of entry (Fig. 1C, left side). Overall, secretion results as intracellular anion accumulates to levels sufficient to drive its passive efflux across the apical membrane.

Bear in mind that the thyroid follicular epithelium is bidirectionally transporting. Although this chapter focuses on the membrane transport proteins that mediate the second step of I^- secretion into the follicular lumen, well-documented Na^+-absorptive processes also govern overall thyroid fluid and electrolyte transport (Armstrong, Matainaho, et al., 1992; Penel, Gerard, Mauchamp, & Verrier, 1989). In many absorptive epithelia, Na^+ absorption from the luminal space results from passive entry via epithelial sodium channels, a process that also relies on the inwardly directed Na^+ gradient (Fig. 1C, right side). Indeed, Na^+ absorption likely accounts for the measured low Na^+ levels in thyroid follicular lumina (Yen-Chow, Chow, & Woodbury, 1984). These low luminal Na^+ levels are important to keep in mind, for they may in fact even regulate apical I^- release (see Section 4.3).

3.2 Basolateral iodide uptake

Many anion channels and transporters accept I^- as a substrate. For this reason, I^- (and particularly isotopic I^- (e.g., ^{125}I)) often is utilized as a surrogate for probing the function of Cl^--transporting proteins. In the physiological setting, however, serum I^- levels are dwarfed by those of Cl^-. Therefore, I^- entry via Cl^--transporting proteins cannot drive its thyroid accumulation. This necessitates that a high-affinity, specific mechanism for uptake of I^- be in place in the thyrocyte. Indeed, the transporter that fulfills this need has been identified. The Na^+–I^- symporter (NIS) was identified initially by functionally screening cDNAs isolated from a rat thyroid line (Dai, Levy, & Carrasco, 1996) and subsequently isolated by RT-PCR of human thyroid total RNA (Smanik, Ryu, Theil, Mazzaferri, & Jhiang, 1997). Rat and human NIS are encoded by the *Slc5a5* and *SLC5A5* genes, respectively, and are predicted to span the membrane 13 times, with the amino terminus situated extracellularly and the carboxyl terminus positioned intracellularly. Thyrocytes comprising the follicular epithelium actively accumulate circulating I^- through the basolaterally situated NIS, which like other Na^+-coupled cotransport mechanisms depends on activity of the Na^+/K^+ ATPase. NIS moves one I^- ion into the cell via coupled entry with two Na^+ ions; therefore, like the Na^+/K^+ ATPase, it is electrogenic. However, its overall effect depolarizes V_m and diminishes the inward Na^+ gradient created by the pump.

Thus, in the absence of additional mechanisms, NIS-mediated I^- uptake would be self-limiting. How is this limitation overcome by the thyrocyte?

Despite observed functional differences between rodent and human orthologs of NIS, I^- uptake by NIS is robust (reviewed by Darrouzet, Lindenthal, Marcellin, Pellequer, & Pourcher, 2014). Early studies utilizing FRTL-5 cells, a Fisher rat thyroid line that accumulates I^- in response to TSH, demonstrated that intracellular I^- levels can achieve up to 30-fold those found in the bathing medium (Weiss, Philp, & Grollman, 1984b). FRTL-5 cells display increased NIS expression in response to TSH stimulation, consistent with the notion that NIS functions as the critical I^- uptake transporter (Kogai et al., 1997). The observation that a point mutation in NIS (T354P) causes human congenital hypothyroidism further highlights its critical role in thyroid I^- uptake (Levy, Ginter, De la Vieja, Levy, & Carrasco, 1998).

This brings us back to the problem of the self-limiting nature of I^- uptake, as shown in the scheme depicted in Fig. 3A. NIS carries more

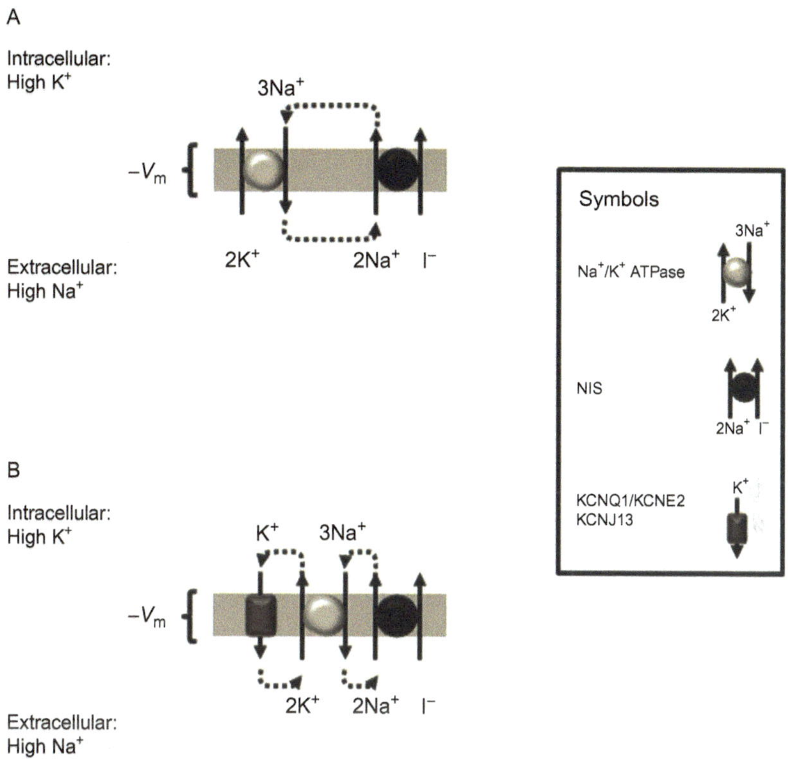

Figure 3 Schema depicting the functional coupling of NIS to Na^+/K^+ATPase in the absence (A) and presence (B) of a K^+ conductance. (A) Turnover of NIS, which has a 2:1 stoichiometry, acts to diminish the Na^+ gradient (and membrane potential, V_m) that is generated by Na^+/K^+ ATPase activity. This ultimately limits NIS-mediated I^- uptake. (B) Presence of a K^+ conductance helps maintain the inward Na^+ gradient and V_m, and thereby promotes further I^- uptake via NIS.

Na^+ ions than I^- in each transport cycle means, so the inward Na^+ gradient falls, thereby decreasing the driving force for Na^+-coupled I^- entry via NIS.

How, then, do thyrocytes sustain I^- accumulation? Consider the other consequence of Na^+/K^+ pump activity: maintenance of high intracellular K^+ concentration. Closer inspection of the mechanism depicted in Fig. 3B reveals how the presence of basolateral K^+ conductances can drive K^+ recycling and solve the problem. Indeed, channels that enable recycling of K^+ by the Na^+/K^+-ATPase also drive I^- entry via NIS. Of the many families of K^+ channels, representative members of at least two—KCNQ and KCNJ—likely function as K^+ recycling conductances. Roepke and colleagues demonstrated that *Kcne2*$^{-/-}$ mice display a phenotype characteristic of profound thyroid impairment, consistent with a mechanism in which *Kcne2* knockout disrupts function of TSH-stimulated Kcnq1/Kcne2 channels and thereby disables both the K^+ recycling conductance and I^- uptake (Roepke et al., 2009). In the case of KCNJ13 (also known as Kir7.1), expression on the RNA and protein levels, as well as basolateral localization data, suggests functional coupling with the Na^+/K^+-ATPase (Nakamura et al., 1999). The contribution of KCNJ13 to basolateral K^+ recycling and I^- uptake remains to be determined.

3.3 Apical iodide release

In order for intracellular I^- to be coupled to Tg, it must traverse the apical membrane and enter the follicular space. Pathways that potentially mediate this process must satisfy several criteria, foremost of which is permeability to I^-. However, in contrast to NIS, selectivity for I^- over other anions does not appear necessary. As noted by Yoshida and colleagues, organification lowers follicular I^- levels to the extent that a large electrochemical gradient for cytosolic-to-luminal transport is preserved (Yoshida et al., 1999). In the healthy thyroid, apical release should not be a limiting factor. Inspection of the literature offers a handful of candidate transport proteins that may function to mediate I^- exit across the apical membrane. Although these are primarily considered in most tissues to be Cl^- channels, their role in the context of the thyroid may be instead to function directly in the carriage of I^- into the follicular lumen, or alternatively, to sustain gradient-coupled I^- transport. In Section 4, structural and functional aspects of these four putative contributors—SLC26A4 (pendrin), cystic fibrosis transmembrane conductance regulator (CFTR), sodium-monocarboxylic acid transporter (SMCT1; SLC5A8), and the anoctamin, TMEM16A—are presented in further detail.

4. CHLORIDE TRANSPORT PROTEINS AND LUMINAL I^- TRANSLOCATION

Genetic and molecular biological methods, combined with dynamic functional readouts, have advanced our understanding of how anions traverse epithelial apical membranes. This is especially evident when considering how, in thyrocytes, two transporters that are linked to inherited human diseases, SLC26A4 and CFTR, contribute to trans-apical I^- exit. Much remains to be learned about the more recently identified SLC5A8 and TMEM16A. Definitive mechanisms explaining how all four candidate transport proteins contribute to the overarching goal of mediating cell-to-lumen I^- flux still remain incomplete. However, several attractive—and importantly, testable—models emerge from the studies to date and will be discussed.

4.1 SLC26A4 (Pendrin)

SLC26A4 is a member of the SLC26 family of sulfate transporters (Alper & Sharma, 2013). In humans, SLC26A4 comprises 780 amino acids. The predicted secondary structural elements of SLC26 transporters include intracellular amino and carboxyl termini which precede and follow, respectively, 12 transmembrane-spanning domains (Li, Xia, & Reithmeier, 2014) (See Fig. 4A).

The carboxyl terminus of many SLC26 transporters, including SLC26A4, harbors a region important for regulatory interactions, the *s*ulfate *t*ransporter and *a*nti-*s*igma factor antagonist (STAS) domain (Aravind & Koonin, 2000; Ko et al., 2004; Sharma, Rigby, & Alper, 2011). This moniker originates from the similarity of this domain to bacterial antagonists of anti-sigma factor such as SpoIIAA, a sporulation regulatory factor (Aravind & Koonin, 2000). Of significance is the interaction of the STAS domain with CFTR's R domain (see Section 4.2).

Pendred syndrome is an inherited human condition, first described by Vaughan Pendred (Pendred, 1896). Pendred syndrome is produced by *SLC26A4* gene mutations that encode a defective transporter protein, eponymously named Pendrin (Everett et al., 1997); it is referred to as SLC26A4 throughout this chapter. Most Pendred syndrome-producing mutations entail coding errors that incorporate point mutations in the SLC26A4 protein, thereby producing profound deafness through disruption of cochlear fluid homeostasis (Wangemann, 2006).

The sensorineural deafness associated with Pendred syndrome presents during childhood and accounts for a significant portion (7.5%) of all cases of prelingual deafness (Reardon & Trembath, 1996). Pendred syndrome is accompanied sporadically with goiter during the second decade of life. In addition, patients exhibit positive perchlorate discharge tests, an indication of defective I^- organification. The thyroid phenotypes distinguish Pendred syndrome from nonsyndromic hearing loss. These observations suggest the possibility that *SLC26A4* encodes a protein important for thyroid function. Using cells heterologously expressing SLC26A4, Scott and coworkers found no evidence for sulfate transport. Instead, SLC26A4 negotiates the exchange of Cl^- for I^- when overexpressed heterologously in a variety of systems, including insect cells, *Xenopus laevis* oocytes, and the

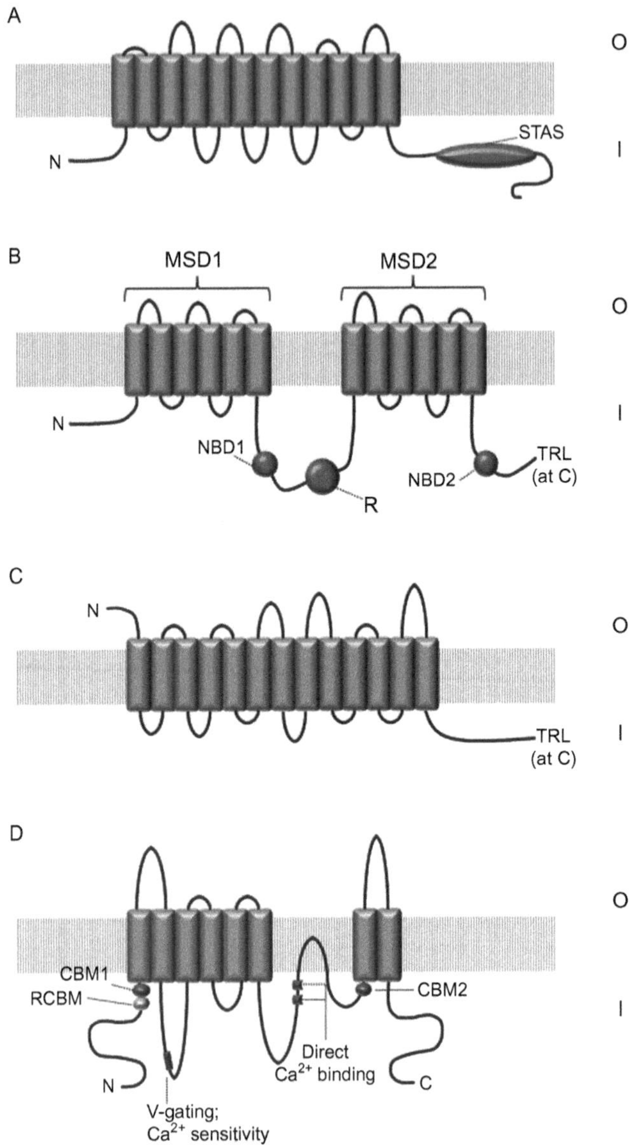

Figure 4 General topology and domain organization of SLC26A4, CFTR, SLC5A8, and TMEM16A are shown. (A) SLC26A4, with 12 transmembrane domains and intracellular amino (N) and carboxyl (C) termini, features the important STAS domain that is common to all SLC26 family members. (B) CFTR's predicted 12 transmembrane domains are organized in two clusters (membrane spanning domains, MSD1 and MSD2), each comprising 6 transmembrane domains. The N- and C-termini localize intracellularly. Nucleotide-binding domains 1 and 2 (NBD1, NBD2) follow MSD1 and MSD2, respectively. Unique to CFTR is a regulatory (R) domain, a region containing many regulatory consensus sites

mammalian fibroblast cell lines, COS-7 (African green monkey kidney) and CHO (Chinese hamster ovary) (Scott, Wang, Kreman, Sheffield, & Karniski, 1999; Yoshida et al., 2002). Moreover, mutations resulting in preservation of low I^- transport levels associated with nonsyndromic deafness, whereas those completely lacking transport function associated with Pendred syndrome (Scott et al., 2000). So great was the excitement that "the" thyrocyte apical I^- efflux pathway had been identified that subsequent work indicating that SLC26A4 transports formate (Scott & Karniski, 2000) and HCO_3^- (Royaux et al., 2001) became temporarily overshadowed.

The directness of the resultant working model became obscured by the finding that *Slc26a4* knockout mice presented no indication of thyroid insufficiency, exhibiting sensorineural deafness as the sole common feature with human Pendred syndrome (Everett, Morsli, Wu, & Green, 1999). Indeed, *Slc26a4*$^{-/-}$ mice did not develop goiter and showed no differences in thyroid hormone (total T_4 and T_3, as well as rT_3) and TSH levels. Moreover, restriction of dietary I^- did not reveal evidence of thyroid insufficiency in *Slc26a4*$^{-/-}$ mice (Calebiro et al., 2011). Still, evidence for at least partial involvement of SLC26A4 in thyroid I^- transport is clear. In the PCCL-3 Fisher rat thyroid cell line, immunoblot analysis of surface-biotinylated membrane fractions provides evidence consistent with increased surface expression of SLC26A4 upon stimulation with TSH. Confocal immunofluorescence data were consistent with these conclusions. Moreover, this is accompanied by augmented I^- efflux (Pesce et al., 2012). Collectively, these disparate findings therefore raise the possibility that additional apical I^- efflux pathways/mechanisms compensate for the lack of functional Slc26a4 in mice, and possibly also for SLC26A4 in the portion of Pendred syndrome-afflicted humans presenting without obvious thyroid phenotype.

for phosphorylation. The R domain follows NBD1 and precedes MSD2. The carboxyl terminus ends with a PDZ domain-binding tripeptide, TRL. (C) SLC5A8 contains a predicted 13 transmembrane domains, with an extracellular N-terminus and an intracellular C terminus that contains the same PDZ domain-binding motif as CFTR. (D) TMEM16A, an anoctamin, contains eight transmembrane domains, intracellular amino and carboxyl termini, and multiple sites for regulation. These include two residues mediating the direct binding of Ca^{2+} in an intracellular region between the sixth and seventh transmembrane domains, as well as a peptide sequence within the first intracellular loop that couples voltage gating and Ca^{2+} sensitivity. TMEM16A also contains calmodulin-binding motifs that include a regulatory calmodulin-binding motif (RCBM), as well as two that modulate HCO_3^- permeation (CBM1 and CBM2). See text for further details.

4.1.1 SLC26A4, HCO_3^-, luminal pH

The lack of measureable thyroid phenotype in the *Slc26a4* knockout mouse also made it important to establish whether Slc26a4 plays a role in other aspects of thyroid physiology. Using the same *Slc26a4*$^{-/-}$ mouse model, Wangemann and colleagues tested whether Slc26a4-mediated HCO_3^- transport could be detected in thyroid by performing measurements of intrafollicular (luminal) pH (Wangemann et al., 2009). Luminal pH measured in *Slc26a4*$^{-/-}$ mice was reduced, compared to heterozygous control animals. This suggests defective luminal alkalinization, possibly due to impaired HCO_3^- secretion. Thus, Slc26a4 also may mediate HCO_3^- transport in the thyroid, as it does in cochlea and renal intercalated cells (Wall, 2006; Wangemann et al., 2007). This observation may provide a clue toward unraveling how expression of CLC-5, an electrogenic Cl^-/H^+ antiporter (2:1), impacts that of SLC26A4; indeed, van den Hove and colleagues reported the presence of euthyroid goiter, partial I^- organification defect, and diminishment of both *Slc26a4* RNA and Slc26a4 protein expression in *Clcn5* knockout mice (Fong, 2011; van den Hove et al., 2006).

By transporting Cl^- into the follicular lumen and removing H^+ from this space, CLC-5 furnishes Cl^- that drives SLC26A4 turnover and at the same time alkalinizes the lumen. This presents an attractive scenario whereby SLC26A4 and CLC-5 synergize to promote alkalinization of the follicular space—the former by mediating luminal HCO_3^- entry and the latter by removing H^+ from the follicular space. Is, for instance, luminal pH critical for the overall process of organification? A detailed understanding of the contributions of Slc26a4/SLC26A4-mediated HCO_3^- transport to thyroid physiology will provide important clues toward elucidating the etiology of goiter and I^- organification defect in Pendred syndrome patients.

4.2 Cystic fibrosis transmembrane conductance regulator

Another human disease caused by mutations in a gene encoding an ion transport protein is cystic fibrosis (CF), caused by mutations in the *CFTR* gene (Riordan et al., 1989). *CFTR* encodes CFTR, a complex protein formed of 1480 amino acids. CFTR organizes as two sets of 6-helix clusters that traverse the membrane (transmembrane domains 1 and 2, or alternatively, TMD1 and TMD2), a signature pair of nucleotide-binding domains (NBD1 and NBD2) and an important and unique regulatory (R) domain. These features are highlighted in the topological model shown in Fig. 4B. The first two structural features identify CFTR as a member of the *A*TP-*b*inding *c*assette (ABC) family of transport proteins. In contrast, the regulatory (R) domain is not found in any other ABC transporters identified to date.

Many ABC transporters move substrate against electrochemical gradients in an ATP hydrolysis-dependent fashion, whereas CFTR operates through a channel mechanism, allowing "downhill" movement of Cl^-. Its gating not only relies on dimerization of the NBDs upon ATP binding, but also is regulated by cyclic AMP (cAMP)-dependent, protein kinase A-mediated phosphorylation of the R domain.

CF disease impairs multiple organ systems, notably the respiratory, gastrointestinal, and reproductive tracts. Nearly 2000 identified mutations (1974 as of this writing; www.genet.sickkids.on.ca) within *CFTR* are known to underlie disease. *CFTR* mutations can be classified based on how they ultimately disrupt protein synthesis, translation, trafficking, and function (both channel regulation and channel conductance) (Choo-Kang & Zeitlin, 2000). Class I ("null") mutations result from nonsense codons that result in premature termination, frameshifts within the coding region, or splicing defects. Class II mutations (deletion and missense) are caused by either omission of critical residues or incorporation of inappropriate amino acids. These mutations block CFTR membrane trafficking. For example, the most common *CFTR* mutation deletes phenylalanine at position 508 (ΔF508), causing misfolding and hence impaired membrane trafficking of the mutant protein. Class III mutations arise from incorporation of incorrect amino acids, again the result of missense codons. These mutations ultimately cause defective CFTR gating. Class IV mutations also involve missense codons, but these instead reduce CFTR channel conductance. Class V mutations comprise missense mutations or mutations that lead to altered gene splicing. They cause reduced levels of normal protein to be made.

The example of ΔF508 highlights how membrane trafficking events bear significantly on the pathophysiology of CFTR (Farinha, Matos, & Amaral, 2013). For this reason, many groups over the last two decades focused on aspects of this subcellular function. CFTR contains a protein interaction motif, formed by its terminal three amino acids: T–R–L. These comprise a consensus sequence for binding to PDZ domains (for postsynaptic density-*Discs large homologous*-zona occludens 1, three proteins in which the domains were first identified), which are important protein interaction modules (Sheng & Sala, 2001). Many membrane scaffolding proteins contain PDZ domains. Interactions between CFTR and such scaffolding proteins govern its membrane residency and stability (Cheng et al., 2002; Moyer et al., 1999; Short et al., 1998), as well as its activity (Raghuram, Mak, & Foskett, 2001; Wang, Yue, Derin, Guggino, & Li, 2000). Moreover, PDZ domain-containing scaffolding proteins also may play a central role in organizing macromolecular signaling complexes (Sun et al., 2000). CFTR's role also as a regulator of other membrane transporters and channels, while controversial, has accumulated steadily. Interactions with PDZ domain-containing proteins therefore may enable this function as well.

Inadequate anion transport caused by defective CFTR can explain many disease symptoms that present in affected tissues. Examples of such include airway obstruction and chronic infection, pancreatic insufficiency, intestinal malabsorption, and infertility. Not as well understood, however, is the impact of CF on the endocrine system, where primary defects may produce subtle and complex phenotypic outcomes. Thyroid status in CF patients may influence disease severity and outcome profoundly. In addition, because CF disease-causing mutations fall into distinct functional classes, knowledge of which most closely associate with thyroid insufficiency should inform models of how CFTR functions in the thyroid.

Importantly, a possible link between subclinical hypothyroidism and stunted growth in CF has been conjectured (Panesar, 1999). Growth retardation associated with CF more conventionally is ascribed to malnutrition resulting from impaired pancreatic and digestive system function. In principle, insufficiency of hormones important to growth and development, such as the thyroid hormones, might produce a similar outcome. More recently, Naehrlich and colleagues presented compelling evidence for subclinical hypothyroidism, as assessed by elevated levels of TSH as well as depressed free T_4, within a cohort of 129 CF patients (Naehrlich, Dorr, Bagheri-Behrouzi, & Rauh, 2013). Iodine deficiency, as evidenced by reduced urinary I^- excretion, was also found.

In evaluating whether CFTR plays a role in orchestrating—if not directly underlying—the thyroid apical I^- efflux pathway, cellular-level studies also must be considered. Evidence for TSH- and cAMP-stimulated anion secretion in primary porcine thyroid cultures accumulated in the early 1990s, with reports using both ^{125}I flux measurements and short-circuit current measurements of polarized primary porcine thyrocyte cultures (Armstrong, Cragoe, Bourke, Huxham, & Manley, 1992; Armstrong, Matainaho, et al., 1992; Nilsson, Bjorkman, Ekholm, & Ericson, 1990).

Moreover, single-channel recordings performed on confluent primary thyrocyte cultures (Champigny, Verrier, Gerard, Mauchamp, & Lazdunski, 1990) first revealed active Cl^- channels in membrane patches prestimulated with 8-chloro-cAMP and bearing the biophysical signature of CFTR. Additional in-depth studies independently confirmed the stimulatory action of cAMP agonists on a similar, CFTR-like channel having linear current–voltage relationships and a low unitary conductance (Bourke, Sand, Abel, Huxham, & Manley, 1995). With respect to thyroid transport, it is important to recognize that, in addition to Cl^-, CFTR is capable of conducting I^- and other small anions (Linsdell, 2014; Tabcharani, Chang, Riordan, & Hanrahan, 1992). It remains to be determined whether direct carriage of I^- by CFTR contributes substantially to overall follicular accumulation, and if so, under what physiological conditions.

Devuyst and colleagues demonstrated CFTR expression on both RNA and protein levels using bovine and human thyroid preparations (Devuyst et al., 1997). Also, primary monolayer cultures of neonatal $CFTR^{-/-}$ pig thyroids confirmed the requirement for CFTR in mediating a normal secretory response to cAMP-elevating agents (Li, Ganta, & Fong, 2010). Interestingly, cultures grown from $CFTR^{\Delta F508/\Delta F508}$ thyroids produced a near-normal (~2/3 of wild type) response, despite its drastically reduced cell surface expression (Li, Ganta, & Fong, 2012). These observations evoke the exciting possibility that the measured secretory response may correspond to CFTR-dependent regulation of an additional transporter important to thyroid function.

4.2.1 CFTR and SLC26A4 interplay

Might CFTR exert regulatory interactions with SLC26A4 in thyroid? Indeed, mounting evidence obtained by studies in other cell systems suggests that, in principle, this is highly feasible. Ko et al. postulated—and provided persuasive evidence for—a model whereby CFTR's R domain interacts with the STAS domain of SLC26 family members (Ko et al., 2002, 2004). Although their findings focus primarily on the related anion exchangers, SLC26A3 and SLC26A6, the authors elude to a similar mechanism involving SLC26A4. A role for SLC26A4 interactions with CFTR in the context of HCO_3^- secretion by the airway serous cell line, Calu-3, has been suggested (Garnett et al., 2011). In the thyroid, one might envisage a situation in which CFTR-mediated Cl^- exit supplies the counter-ion for I^- exit through SLC26A4 (see Fig. 5A).

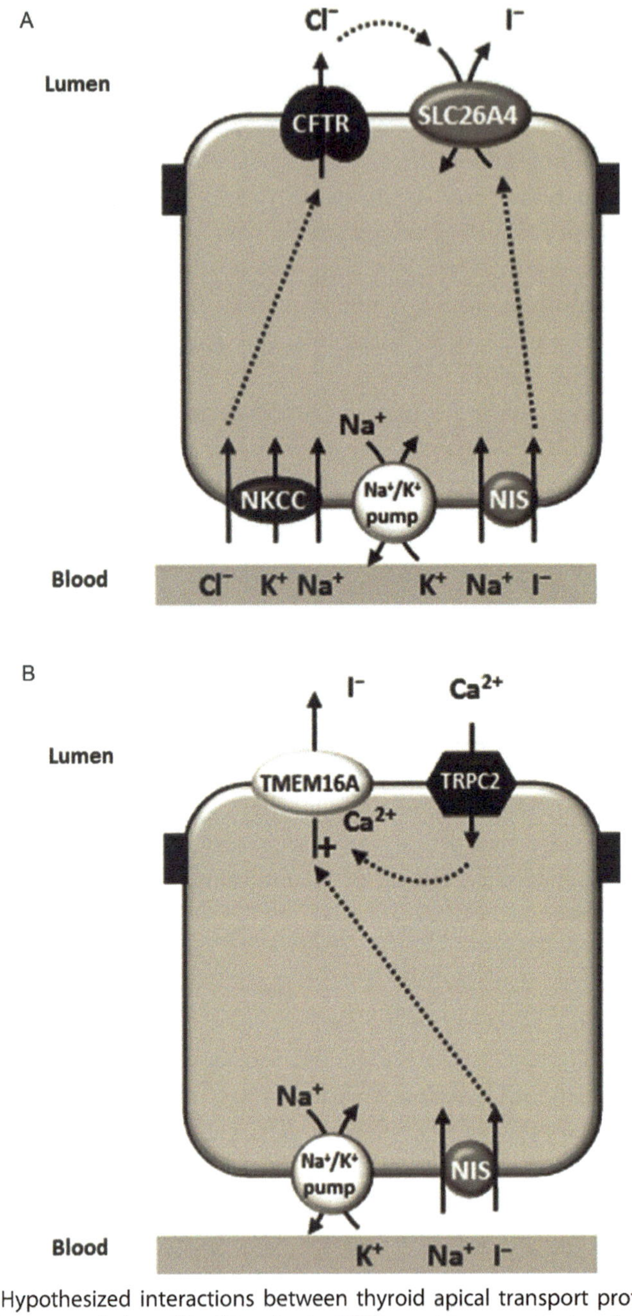

Figure 5 Hypothesized interactions between thyroid apical transport proteins implicated in I^- accumulation. (A) Cl^- exit via CFTR provides counter-ion for SLC26A4-mediated anion, thereby driving I^- efflux. Note that intracellular accumulation of Cl^- results from action of the $Na^+/K^+/2Cl^-$ cotransporter, NKCC. Likewise, I^- enters via NIS. (B) Entry of Ca^{2+} (here shown as the canonical transient receptor potential cation channel, TRPC2) activates TMEM16A by direct binding and other Ca^{2+}-regulated (e.g., calmodulin) pathways (see Fig. 4). Intracellular I^- then crosses the apical membrane via TMEM16A, entering into the follicular lumen.

However, it is important to bear in mind the fact that both SLC26A3 and SLC26A6 include carboxyl terminal PDZ domain-binding tripeptide motifs (TKF and TRL, respectively), whereas SLC26A4 lacks this feature. Therefore, the capacity for regulation is limited to interactions between its STAS domain and CFTR's R domain.

Recently, Cai et al. reported their observation of CFTR inhibition by thyroid hormones (Cai, Li, Chen, & Sheppard, 2013). Their findings provide an interesting twist: if CFTR-mediated anion conductance is important for optimal apical I^- efflux, then blocking its activity effectively would inhibit efflux and limit Tg iodination. Such a scenario suggests a potential means for feedback autoregulation of thyroid hormone production (Fong, 2013).

4.3 SLC5A8, a sodium-monocarboxylate transporter (hAIT; SLC5A8; SMCT1)

Another candidate apical I^- efflux transport protein is encoded by the *SLC5A8* gene. Despite important advances toward fully understanding SLC5A8 transport function, the role of this transporter in the thyroid remains enigmatic.

Rodriguez et al. used an RT-PCR-based approach to screen a human kidney cDNA library and identify a cDNA bearing significant homology to NIS (Rodriguez et al., 2002). The translated, primary amino acid sequence, consisting of 610 residues, shares ~46% identity and 70% similarity to NIS. Moreover, its secondary structural organization resembles that of NIS: 13 transmembrane helices, with the amino terminus placed outside the cell and the carboxyl terminus inside (see Fig. 4C). Interestingly, the terminal three amino acids are identical to those of CFTR, which leaves open the possibility that SLC5A8, too, may interact with PDZ domain-containing scaffolding proteins. Quantitative RT-PCR of human tissues revealed comparably stronger expression of the isolated cDNA in thyroid than in kidney, exceeding 10-fold. Indeed, immunohistochemical localization using antibodies targeted against a unique region of the predicted amino acid sequence revealed robust staining that localized to the apical aspect of human thyrocytes. The earliest functional study, using ^{125}I tracer uptake as the outcome measure, reported that, compared to COS-7 cells transiently expressing NIS alone, coexpression with this cDNA diminished I^- accumulation at equilibrium (Rodriguez et al., 2002). These findings align with the notion that the newly identified transporter mediates I^- efflux. Taken together with the expression pattern and cellular localization studies, these observations led to the hypothesis that this novel cDNA encoded an apical I^- transporter. Hence, it was christened *h*uman *A*pical *I*⁻ *T*ransporter, hAIT. The authors prudently left open the possibility that hAIT might perform other transport functions that reflect its molecular similarity to NIS.

Shortly thereafter, Li and colleagues independently identified the *SLC5A8* gene as a tumor suppressor: abnormal methylation of *SLC5A8* and consequent gene silencing associated with colon cancer (Li et al., 2003). The *SLC5A8* gene product is identical to that encoded by the cRNA isolated by Rodriguez and coworkers. In keeping with convention, the gene product from this point is referred to throughout this chapter as SLC5A8.

Importantly, Na^+-sensitive microelectrode measurements performed on *Xenopus* oocytes heterologously expressing SLC5A8 convincingly demonstrated that SLC5A8 functions in Na^+ accumulation. This observation confirms that SLC5A8, like other SLC5 family members (i.e., NIS), retains the ability to transport Na^+. However, distinction between function as a passive Na^+ leak or as a Na^+-coupled mechanism of substrate uptake cannot be made from the reported studies. Similarly, these studies do not rule out the possibility of an additional passive I^- transport mode.

Perhaps it was with this ambiguity and the need for further characterization in mind that studies designed to identify substrate dependence were undertaken by two groups independently (Coady et al., 2004; Miyauchi, Gopal, Fei, & Ganapathy, 2004). Taking as a clue the abundant colonic expression of SLC5A8, the hypothesis that SLC5A8 mediates Na^+-coupled uptake of bacterial fermentation products (i.e., short-chain fatty acids/ monocarboxylic acids such as propionate, pyruvate, and butyrate) was tested. Indeed, two-microelectrode voltage clamp recordings of *Xenopus* oocytes expressing SLC5A8 revealed substrate-dependent inward currents, presumably by moving more Na^+ ions than substrate anion per transport cycle. This functionality inspired the renaming of the hAIT/SLC5A8 as SMCT, for *s*odium-*m*ono*c*arboxylate *t*ransporter, introducing perhaps a further source of confusion. Thus, SLC5A8-mediated uptake of butyrate by colonic epithelia may exert tumor suppressive actions via butyrate's action in inhibiting histone deacetylase and promoting apoptosis. These observations launched a plethora of studies directed at SLC5A8's function as a tumor suppressor in many tissues, including thyroid. But how does this inform understanding of SLC5A8's potential role in thyroid hormone production? Does it mediate apical I^- efflux and if so, how?

In contrast to the findings of Rodriguez et al., no evidence for currents carried by I^- could be detected. Instead, Coady and colleagues initially noted reduction of Na^+-dependent cotransport upon substitution of extracellular Cl^- with cyclamate and detected an obscure anionic leak current as well (Coady et al., 2004). To confound the picture even more, a murine *Slc5a8* knockout model showed neither an increased propensity to develop colonic tumors nor did it display disrupted thyroid function (Frank et al., 2008). In addition, the double knockout of *Slc5a8* and *Slc26a4* (murine pendrin) failed to produce thyroid deficiency. This outcome further strengthens the growing suspicion that multiple transport proteins contribute to apical I^- efflux in thyroid.

Taken together, these observations raise several questions. If SLC5A8 does not transport I^-, then what substrate in the thyroid does it carry and what physiological function might this subserve? If SLC5A8 does transport I^-, what specific conditions permit this functional mode?

Subsequent studies by Coady et al. shed light on the latter question by demonstrating the capability of SLC5A8 to conduct anions, including I^-, albeit under very specific conditions (Coady, Wallendorff, Bourgeois, & Lapointe, 2010). In fact, anion leak currents associated with SLC5A8 showed lyotropic selectivity, favoring I^- over Cl^- and moreover not supporting HCO_3^- permeation. Significantly, the presence of extracellular Na^+ blocked the anion current; conversely, reduction of extracellular Na^+ revealed anion current. Given the low concentrations of Na^+ in the follicular space (see Section 3.1), this raises the possibility that SLC5A8 indeed may facilitate a component of apical I^- exit in the thyroid.

Thus, the contributions of SLC5A8 to overall movement of I^- into the thyroid follicular lumen are complex. Whether operation in the cotransport mode is of significance to thyroid function is also untested. It is clear that much remains to be understood about this enigmatic transport protein.

4.4 TMEM16A (anoctamin 1)

Of the candidate pathways that potentially mediate apical release of I^- from the thyrocytes, TMEM16A represents the only one known to be activated directly by intracellular Ca^{2+} (Caputo et al., 2008; Schroeder, Cheng, Jan, & Jan, 2008; Yang et al., 2008). TMEM16A is the latest entrant into the pool of channels that ostensibly activate in response to this important cellular signal, the others being CLCA channels and bestrophin (Gruber et al., 1998; Sun, Tsunenari, Yau, & Nathans, 2002). The TMEM16 family comprises, to date, 10 members, all membrane proteins. Several, but not all, function as anion channels. Topology of the prototypic member, TMEM16A, comprises eight transmembrane-spanning regions (see Fig. 4D). Hence,

TMEM16 proteins often are referred to as anoctamins (anion; *octa*-, eight), and TMEM16A also is called anoctamin 1. Inspection of the sequence reveals numerous regions important for Ca^{2+} activation, including a domain conferring voltage dependence and Ca^{2+} sensitivity resting within the first intracellular loop, multiple calmodulin-binding domains (two of which modulate HCO_3^- permeation), and two glutamate residues residing within the intracellular loop following transmembrane domain 6 that mediate direct Ca^{2+} binding (Yu, Duran, Qu, Cui, & Hartzell, 2012). Interestingly, although some TMEM family members, including TMEM16A, function as Ca^{2+}-activated Cl^- channels, several also act as membrane scramblases. The interested reader is directed to an excellent recent review for further details (Pedemonte & Galietta, 2014).

The intense, cumulative efforts devoted to understanding function of Ca^{2+}-activated Cl^- currents underscore its importance to diverse physiological systems. Three groups contemporaneously identified mouse (Yang et al., 2008), human (Caputo et al., 2008), and *X. laevis* (Schroeder et al., 2008) orthologs of TMEM16A, respectively. The routes by which each group independently arrived at the same channel protein attest to the creative flexibility afforded by modern molecular approaches. Important to recognize is the fact that—in the case of all three groups—rigorous validation of TMEM1A function relied critically on the enormous information base acquired from earlier physiological studies. Such functional validation criteria included assessment of (1) activation by diverse agonists and other stimuli that result ultimately in increased intracellular Ca^{2+}, (2) biophysical properties such as single-channel conductance, voltage-sensitivity of Ca^{2+} activation, and rectification, (3) currents produced by mutation of putative pore-lining residues, (4) pharmacology consistent with that of native Ca^{2+}-activated Cl^- currents, and (5) tissue distribution.

How did these groups independently discover TMEM16A? Searching databases for orphan proteins having multiple transmembrane regions, Yang and coworkers cast their attentions in the direction of a novel candidate which was predicted to encode a protein product having eight transmembrane-spanning regions, and comprising 10 homologs in humans (Yang et al., 2008). In this case again, functional interrogation through systematic heterologous expression studies ultimately characterized the murine *Tmem16a* cDNA. Mouse Tmem16a consists of 980 amino acids. Larger anions permeated Tmem16a more readily; experiments performed under bi-ionic conditions yielded the following halide permeability preference

$I^- > Br^- > Cl^- > F^-$, similar to previous measurements of native Ca^{2+}-activated Cl^- channels in lacrimal glands (Evans & Marty, 1986). Moreover, the expressed currents were blocked by compounds previously shown to block Ca^{2+}-activated Cl^- channels, such as niflumic acid. Tissue distribution, as determined by immunostaining, revealed the presence of Tmem16a in transporting epithelia, including pancreatic acini, renal proximal and distal tubules, and bronchioles. Staining also was found in extra-epithelial tissues, such as retina and dorsal root ganglia.

In the case of Caputo et al., their creative strategy was informed by earlier findings which showed dramatic upregulation of Ca^{2+}-activated Cl^- current in human airway cells that had been challenged with proinflammatory cytokines (Caputo et al., 2008). Taking a functional genomics screening approach, gene microarrays of human bronchial epithelial cells stimulated with interleukin 4 (IL-4) were compared with microarrays of control cells. Caputo et al. identified genes upregulated by IL-4, subsequently utilizing siRNA to screen these for functional knockdown of Ca^{2+}-activated anion transport function in human bronchial and pancreatic ductal cell lines, and thereby identifying the critical gene as the human *TMEM16A*. Human TMEM16A encodes a protein that consists of 986 amino acids. Further interrogation of TMEM16A's function was facilitated by the establishment of a stably expressing epithelial cell line having the reproducible stringency required for high-throughput pharmacological screens.

In contrast, Schroeder et al. employed classical expression cloning methods—with a clever twist (Schroeder et al., 2008). Recognizing that endogenous Ca^{2+}-activated Cl^- currents often obscure measurement of unrelated, exogenous Cl^- currents heterologously expressed in the *X. laevis* oocyte system, this group chose to use *Xenopus* oocytes instead as a *source* of channel-encoding mRNA.

They also recognized that oocytes of the axolotl, *Ambystoma mexicana*, lack this background current. Thus, evaluation of currents produced upon expression of polyA$^+$ RNA fractions in this system enabled isolation of an active fraction and, ultimately, the cDNA clone encoding *Xenopus* Tmem16a, a channel comprising 979 amino acid residues. Schroeder and colleagues moreover compared function of *Xenopus* Tmem16a with that of the mouse ortholog. Murine Tmem16a, as well as the closely related Tmem16b, also yielded functional currents when expressed in axolotl oocytes. Both produced similar, outwardly rectifying currents. Similar to the findings of Yang et al., as well as measurements performed previously on native *Xenopus* Ca^{2+}-activated Cl^- channels, TMEM16A preferred I^- over Cl^-, showing a permeability sequence of $SCN^- > I^- > Br^- > Cl^-$ (Qu & Hartzell, 2000). Also, drugs known to block Ca^{2+}-activated Cl^- channels (i.e., niflumic acid) diminished Tmem16a currents. *In situ* hybridization studies indicated expression in mammary and salivary gland epithelia, findings which align with the immunostaining data of Yang et al.

With respect to thyroid I^- efflux, the observation that TMEM16A permeates I^- more readily than Cl^- clearly bears important implications. In FRTL-5 cells, TSH not only increases cAMP and drives I^- uptake via NIS but also raises intracellular Ca^{2+}, leading to I^- release (Weiss, Philp, Ambesi-Impiombato, & Grollman, 1984; Weiss, Philp, & Grollman, 1984a). An important source of Ca^{2+} is the extracellular fluid. External Ca^{2+} might gain entry into thyrocytes through diverse classes of Ca^{2+}-conducting channels that populate the plasma membrane. These can comprise receptor-operated Ca^{2+} channels, as well as voltage-gated Ca^{2+} channels.

The preceding discussion provides a point of departure for highlighting a novel mechanism that is supported by recent studies exploring the role of a *t*ransient *r*eceptor *p*otential channel, Trpc2, in regulating Tmem16a channels in FRTL-5 cells (Viitanen, Sukumaran, Lof, & Tornquist, 2013). TRPC channels form the canonical branch of the large superfamily of TRP cation channels. Relevant to the present discussion is the ability of TRPCs to permeate Ca^{2+} (Owsianik, Talavera, Voets, & Nilius, 2006). Thus, in the emergent model of Viitanen et al., extracellular Ca^{2+} passively enters the cell through Trpc2, driven by its electrochemical gradient (see Fig. 5B). The subsequent elevation in intracellular Ca^{2+} thereby directly triggers TMEM16A activation and I^- efflux into the luminal space. Alternatively, the resultant membrane depolarization might also activate voltage-gated Ca^{2+} channels and potentiate the signal.

There is, however, one important caveat: in contrast to rodent Trpc2, which functions in pheromone detection, TRPC2 is a pseudogene in the human genome. Thus, although Viitanen and colleagues revealed in FRTL-5 cells the contribution of Trpc2-mediated Ca^{2+} entry in I^- efflux, in order for the general scheme to function in humans, alternative routes for raising intracellular Ca^{2+} must exist. These include both ionotropic and metabotropic classes of purinergic receptors, which have been identified in porcine and human thyroid (Bourke, Abel, Huxham, Cooper, & Manley, 1999; Burnstock, 2014; Schofl, Rossig, Potter, von zur Muhlen, & Brabant, 1995), and possibly also voltage-gated Ca^{2+} channels

(Takasu et al., 1987). In principle, activation of any of these routes ostensibly would raise intracellular Ca^{2+} to levels sufficient for gating TMEM16A directly, and thereby promote I^- exit from the thyroid epithelial cell.

5. CONCLUSIONS

The collective knowledge of thyroid function has benefited greatly from concerted efforts directed at understanding the basic structure and function of membrane transport proteins underlying a critical step in hormonogenesis: the uptake of I^-. All four transport proteins presented herein—SLC26A4, CFTR, SLC5A8, and TMEM16A—have been identified as key players in the process of thyroid I^- accumulation, and hence, hormone production. Technological advances have driven acquisition of detailed information regarding the molecular organization and functionality of all four candidates at an unprecedented rate. The data indicate the likelihood that these—and possibly additional transport proteins—work together to facilitate optimal I^- accumulation. Evidence points to the notion that this cannot be accomplished by any one candidate transport protein alone. Coherent mechanistic schemes have emerged swiftly from these nascent data and there are abundant questions to address. In responding to physiological stimuli, do these transport proteins functionally interact within spatially defined domains? How is dynamic organization within the membrane orchestrated and regulated? What are the roles of scaffolding proteins and how are interactions governed? The answers to these basic cell physiological questions will inform deeper understanding of the delicate balance of processes that must be regulated to ensure thyroid health.

ACKNOWLEDGMENT

Research in the Fong laboratory currently is supported by the National Institutes of Health/National Institute of General Medical Sciences (1R15GM101674).

REFERENCES

Alper, S. L., & Sharma, A. K. (2013). The SLC26 gene family of anion transporters and channels. *Molecular Aspects of Medicine, 34*(2–3), 494–515.

Aravind, L., & Koonin, E. V. (2000). The STAS domain—A link between anion transporters and antisigma-factor antagonists. *Current Biology, 10*(2), R53–R55.

Armstrong, J. W., Cragoe, E. J., Jr., Bourke, J. R., Huxham, G. J., & Manley, S. W. (1992). Chloride conductance of apical membrane in cultured porcine thyroid cells activated by cyclic AMP. *Molecular and Cellular Endocrinology, 88*(1–3), 105–110.

Armstrong, J., Matainaho, T., Cragoe, E. J., Jr., Huxham, G. J., Bourke, J. R., & Manley, S. W. (1992). Bidirectional ion transport in thyroid: Secretion of anions by monolayer cultures that absorb sodium. *The American Journal of Physiology*, *262*(1 Pt 1), E40–E45.

Barrington, E. J., & Thorpe, A. (1965). The identification of monoiodotyrosine, diiodotyrosine and thyroxine in extracts of the endostyle of the ascidian, *Ciona intestinalis* L. *Proceedings of the Royal Society of London—Series B: Biological Sciences*, *163*, 136–149.

Bourke, J., Abel, K., Huxham, G., Cooper, V., & Manley, S. (1999). UTP-preferring P2 receptor mediates inhibition of sodium transport in porcine thyroid epithelial cells. *British Journal of Pharmacology*, *127*(8), 1787–1792.

Bourke, J. R., Sand, O., Abel, K. C., Huxham, G. J., & Manley, S. W. (1995). Chloride channels in the apical membrane of thyroid epithelial cells are regulated by cyclic AMP. *The Journal of Endocrinology*, *147*(3), 441–448.

Burnstock, G. (2014). Purinergic signalling in endocrine organs. *Purinergic Signal*, *10*(1), 189–231.

Cai, Z., Li, H., Chen, J. H., & Sheppard, D. N. (2013). Acute inhibition of the cystic fibrosis transmembrane conductance regulator (CFTR) Cl^- channel by thyroid hormones involves multiple mechanisms. *American Journal of Physiology Cell Physiology*, *305*(8), C817–C828.

Calebiro, D., Porazzi, P., Bonomi, M., Lisi, S., Grindati, A., De Nittis, D., et al. (2011). Absence of primary hypothyroidism and goiter in *PDS* mice fed on a low iodine diet. *Journal of Endocrinological Investigation*, *34*(8), 593–598.

Caputo, A., Caci, E., Ferrera, L., Pedemonte, N., Barsanti, C., Sondo, E., et al. (2008). TMEM16A, a membrane protein associated with calcium-dependent chloride channel activity. *Science*, *322*(5901), 590–594.

Carvalho, D. P., & Dupuy, C. (2014). Role of the NADPH oxidases DUOX and NOX4 in thyroid oxidative stress. *European Thyroid Journal*, *2*(3), 160–167.

Champigny, G., Verrier, B., Gerard, C., Mauchamp, J., & Lazdunski, M. (1990). Small conductance chloride channels in the apical membrane of thyroid cells. *FEBS Letters*, *259*(2), 263–268.

Cheng, J., Moyer, B. D., Milewski, M., Loffing, J., Ikeda, M., Mickle, J. E., et al. (2002). A Golgi-associated PDZ domain protein modulates cystic fibrosis transmembrane regulator plasma membrane expression. *The Journal of Biological Chemistry*, *277*(5), 3520–3529.

Choo-Kang, L. R., & Zeitlin, P. L. (2000). Type I, II, III, IV, and V cystic fibrosis transmembrane conductance regulator defects and opportunities for therapy. *Current Opinion in Pulmonary Medicine*, *6*(6), 521–529.

Coady, M. J., Chang, M. H., Charron, F. M., Plata, C., Wallendorff, B., Sah, J. F., et al. (2004). The human tumour suppressor gene SLC5A8 expresses a Na^+-monocarboxylate cotransporter. *The Journal of Physiology*, *557*(Pt 3), 719–731.

Coady, M. J., Wallendorff, B., Bourgeois, F., & Lapointe, J. Y. (2010). Anionic leak currents through the Na^+/monocarboxylate cotransporter SMCT1. *American Journal of Physiology Cell Physiology*, *298*(1), C124–C131.

Dai, G., Levy, O., & Carrasco, N. (1996). Cloning and characterization of the thyroid iodide transporter. *Nature*, *379*(6564), 458–460.

Darrouzet, E., Lindenthal, S., Marcellin, D., Pellequer, J. L., & Pourcher, T. (2014). The sodium/iodide symporter: State of the art of its molecular characterization. *Biochimica et Biophysica Acta*, *1838*(1 Pt B), 244–253.

Devor, D. C., Singh, A. K., Lambert, L. C., DeLuca, A., Frizzell, R. A., & Bridges, R. J. (1999). Bicarbonate and chloride secretion in Calu-3 human airway epithelial cells. *The Journal of General Physiology*, *113*(5), 743–760.

Devuyst, O., Golstein, P. E., Sanches, M. V., Piontek, K., Wilson, P. D., Guggino, W. B., et al. (1997). Expression of CFTR in human and bovine thyroid epithelium. *The American Journal of Physiology*, *272*(4 Pt 1), C1299–C1308.

Evans, M. G., & Marty, A. (1986). Calcium-dependent chloride currents in isolated cells from rat lacrimal glands. *The Journal of Physiology, 378*, 437–460.

Everett, L. A., Glaser, B., Beck, J. C., Idol, J. R., Buchs, A., Heyman, M., et al. (1997). Pendred syndrome is caused by mutations in a putative sulphate transporter gene (PDS). *Nature Genetics, 17*(4), 411–422.

Everett, L. A., Morsli, H., Wu, D. K., & Green, E. D. (1999). Expression pattern of the mouse ortholog of the Pendred's syndrome gene (*Pds*) suggests a key role for pendrin in the inner ear. *Proceedings of the National Academy of Sciences of the United States of America, 96*(17), 9727–9732.

Farinha, C. M., Matos, P., & Amaral, M. D. (2013). Control of cystic fibrosis transmembrane conductance regulator membrane trafficking: Not just from the endoplasmic reticulum to the Golgi. *The FEBS Journal, 280*(18), 4396–4406.

Fong, P. (2011). Thyroid iodide efflux: A team effort? *The Journal of Physiology, 589*(24), 5929–5939.

Fong, P. (2013). Time for T. . .(T3, T4, rT3)? Focus on "Acute inhibition of the cystic fibrosis transmembrane conductance regulator (CFTR) Cl^- channel by thyroid hormones involves multiple mechanisms". *American Journal of Physiology. Cell Physiology, 305*(8), C800–C802.

Frank, H., Groger, N., Diener, M., Becker, C., Braun, T., & Boettger, T. (2008). Lactaturia and loss of sodium-dependent lactate uptake in the colon of SLC5A8-deficient mice. *The Journal of Biological Chemistry, 283*(36), 24729–24737.

Garnett, J. P., Hickman, E., Burrows, R., Hegyi, P., Tiszlavicz, L., Cuthbert, A. W., et al. (2011). Novel role for pendrin in orchestrating bicarbonate secretion in cystic fibrosis transmembrane conductance regulator (CFTR)-expressing airway serous cells. *The Journal of Biological Chemistry, 286*(47), 41069–41082.

Gereben, B., Zavacki, A. M., Ribich, S., Kim, B. W., Huang, S. A., Simonides, W. S., et al. (2008). Cellular and molecular basis of deiodinase-regulated thyroid hormone signaling. *Endocrine Reviews, 29*(7), 898–938.

Gruber, A. D., Elble, R. C., Ji, H. L., Schreur, K. D., Fuller, C. M., & Pauli, B. U. (1998). Genomic cloning, molecular characterization, and functional analysis of human CLCA1, the first human member of the family of Ca^{2+}-activated Cl^- channel proteins. *Genomics, 54*(2), 200–214.

Ko, S. B., Shcheynikov, N., Choi, J. Y., Luo, X., Ishibashi, K., Thomas, P. J., et al. (2002). A molecular mechanism for aberrant CFTR-dependent HCO_3^- transport in cystic fibrosis. *The EMBO Journal, 21*(21), 5662–5672.

Ko, S. B., Zeng, W., Dorwart, M. R., Luo, X., Kim, K. H., Millen, L., et al. (2004). Gating of CFTR by the STAS domain of SLC26 transporters. *Nature Cell Biology, 6*(4), 343–350.

Kogai, T., Endo, T., Saito, T., Miyazaki, A., Kawaguchi, A., & Onaya, T. (1997). Regulation by thyroid-stimulating hormone of sodium/iodide symporter gene expression and protein levels in FRTL-5 cells. *Endocrinology, 138*(6), 2227–2232.

Levy, O., Ginter, C. S., De la Vieja, A., Levy, D., & Carrasco, N. (1998). Identification of a structural requirement for thyroid Na^+/I^- symporter (NIS) function from analysis of a mutation that causes human congenital hypothyroidism. *FEBS Letters, 429*(1), 36–40.

Li, H., Ganta, S., & Fong, P. (2010). Altered ion transport by thyroid epithelia from $CFTR^{-/-}$ pigs suggests mechanisms for hypothyroidism in cystic fibrosis. *Experimental Physiology, 95*(12), 1132–1144.

Li, Y., Ganta, S., & Fong, P. (2012). Endogenous surface expression of ΔF508 CFTR mediates cAMP-stimulated Cl^- current in $CFTR^{\Delta F508/\Delta F508}$ pig thyroid epithelial cells. *Experimental Physiology, 97*(1), 115–124.

Li, H., Myeroff, L., Smiraglia, D., Romero, M. F., Pretlow, T. P., Kasturi, L., et al. (2003). SLC5A8, a sodium transporter, is a tumor suppressor gene silenced by methylation in human colon aberrant crypt foci and cancers. *Proceedings of the National Academy of Sciences of the United States of America, 100*(14), 8412–8417.

Li, J., Xia, F., & Reithmeier, R. A. (2014). N-glycosylation and topology of the human SLC26 family of anion transport membrane proteins. *American Journal of Physiology. Cell Physiology, 306*(10), C943–C960.

Linsdell, P. (2014). Functional architecture of the CFTR chloride channel. *Molecular Membrane Biology, 31*(1), 1–16.

Lyon, I., & Crane, R. K. (1966a). Studies on transmural potentials in vitro in relation to intestinal absorption. I. Apparent Michaelis constants for Na^+-dependent sugar transport. *Biochimica et Biophysica Acta, 112*(2), 278–291.

Lyon, I., & Crane, R. K. (1966b). Studies on transmural potentials in vitro in relation to intestinal absorption. II. An effect of ouabain on glucose-dependent increment of transmural potential of rat small intestine. *Biochimica et Biophysica Acta, 126*(1), 146–153.

Miyauchi, S., Gopal, E., Fei, Y. J., & Ganapathy, V. (2004). Functional identification of SLC5A8, a tumor suppressor down-regulated in colon cancer, as a Na^+-coupled transporter for short-chain fatty acids. *The Journal of Biological Chemistry, 279*(14), 13293–13296.

Moyer, B. D., Denton, J., Karlson, K. H., Reynolds, D., Wang, S., Mickle, J. E., et al. (1999). A PDZ-interacting domain in CFTR is an apical membrane polarization signal. *Journal of Clinical Investigation, 104*(10), 1353–1361.

Naehrlich, L., Dorr, H. G., Bagheri-Behrouzi, A., & Rauh, M. (2013). Iodine deficiency and subclinical hypothyroidism are common in cystic fibrosis patients. *Journal of Trace Elements in Medicine & Biology, 27*(2), 122–125.

Nakamura, N., Suzuki, Y., Sakuta, H., Ookata, K., Kawahara, K., & Hirose, S. (1999). Inwardly rectifying K^+ channel Kir7.1 is highly expressed in thyroid follicular cells, intestinal epithelial cells and choroid plexus epithelial cells: Implication for a functional coupling with Na^+/K^+ ATPase. *The Biochemical Journal, 342*(Pt 2), 329–336.

Nilsson, M., Bjorkman, U., Ekholm, R., & Ericson, L. E. (1990). Iodide transport in primary cultured thyroid follicle cells: Evidence of a TSH-regulated channel mediating iodide efflux selectively across the apical domain of the plasma membrane. *European Journal of Cell Biology, 52*(2), 270–281.

Owsianik, G., Talavera, K., Voets, T., & Nilius, B. (2006). Permeation and selectivity of TRP channels. *Annual Review of Physiology, 68*, 685–717.

Panesar, N. S. (1999). Could growth retardation in cystic fibrosis be partly due to deficient steroid and thyroid hormonogenesis? *Medical Hypotheses, 53*(6), 530–532.

Pedemonte, N., & Galietta, L. J. (2014). Structure and function of TMEM16 proteins (anoctamins). *Physiological Reviews, 94*(2), 419–459.

Pendred, V. (1896). Deaf-mutism and goitre. *Lancet, ii*, 532.

Penel, C., Gerard, C., Mauchamp, J., & Verrier, B. (1989). The thyroid cell monolayer in culture. A tight sodium absorbing epithelium. *Pflügers Archiv, 414*(5), 509–515.

Pesce, L., Bizhanova, A., Caraballo, J. C., Westphal, W., Butti, M. L., Comellas, A., et al. (2012). TSH regulates pendrin membrane abundance and enhances iodide efflux in thyroid cells. *Endocrinology, 153*(1), 512–521.

Qu, Z., & Hartzell, H. C. (2000). Anion permeation in Ca^{2+}-activated Cl^- channels. *The Journal of General Physiology, 116*(6), 825–844.

Raghuram, V., Mak, D. O., & Foskett, J. K. (2001). Regulation of cystic fibrosis transmembrane conductance regulator single-channel gating by bivalent PDZ-domain-mediated interaction. *Proceedings of the National Academy of Sciences of the United States of America, 98*(3), 1300–1305.

Reardon, W., & Trembath, R. C. (1996). Pendred syndrome. *Journal of Medical Genetics, 33*(12), 1037–1040.

Riordan, J. R., Rommens, J. M., Kerem, B., Alon, N., Rozmahel, R., Grzelczak, Z., et al. (1989). Identification of the cystic fibrosis gene: Cloning and characterization of complementary DNA. *Science*, *245*(4922), 1066–1073.

Rodriguez, A. M., Perron, B., Lacroix, L., Caillou, B., Leblanc, G., Schlumberger, M., et al. (2002). Identification and characterization of a putative human iodide transporter located at the apical membrane of thyrocytes. *Journal of Clinical Endocrinology and Metabolism*, *87*(7), 3500–3503.

Roepke, T. K., King, E. C., Reyna-Neyra, A., Paroder, M., Purtell, K., Koba, W., et al. (2009). Kcne2 deletion uncovers its crucial role in thyroid hormone biosynthesis. *Nature Medicine*, *15*(10), 1186–1194.

Royaux, I. E., Wall, S. M., Karniski, L. P., Everett, L. A., Suzuki, K., Knepper, M. A., et al. (2001). Pendrin, encoded by the Pendred syndrome gene, resides in the apical region of renal intercalated cells and mediates bicarbonate secretion. *Proceedings of the National Academy of Sciences of the United States of America*, *98*(7), 4221–4226.

Schofl, C., Rossig, L., Potter, E., von zur Muhlen, A., & Brabant, G. (1995). Extracellular ATP and UTP increase cytosolic free calcium by activating a common P2u-receptor in single human thyrocytes. *Biochemical and Biophysical Research Communications*, *213*(3), 928–934.

Schroeder, B. C., Cheng, T., Jan, Y. N., & Jan, L. Y. (2008). Expression cloning of TMEM16A as a calcium-activated chloride channel subunit. *Cell*, *134*(6), 1019–1029.

Scott, D. A., & Karniski, L. P. (2000). Human pendrin expressed in *Xenopus laevis* oocytes mediates chloride/formate exchange. *American Journal of Physiology. Cell Physiology*, *278*(1), C207–C211.

Scott, D. A., Wang, R., Kreman, T. M., Andrews, M., McDonald, J. M., Bishop, J. R., et al. (2000). Functional differences of the *PDS* gene product are associated with phenotypic variation in patients with Pendred syndrome and non-syndromic hearing loss (DFNB4). *Human Molecular Genetics*, *9*(11), 1709–1715.

Scott, D. A., Wang, R., Kreman, T. M., Sheffield, V. C., & Karniski, L. P. (1999). The Pendred syndrome gene encodes a chloride-iodide transport protein. *Nature Genetics*, *21*(4), 440–443.

Sharma, A. K., Rigby, A. C., & Alper, S. L. (2011). STAS domain structure and function. *Cellular Physiology and Biochemistry*, *28*(3), 407–422.

Sheng, M., & Sala, C. (2001). PDZ domains and the organization of supramolecular complexes. *Annual Review of Neuroscience*, *24*, 1–29.

Short, D. B., Trotter, K. W., Reczek, D., Kreda, S. M., Bretscher, A., Boucher, R. C., et al. (1998). An apical PDZ protein anchors the cystic fibrosis transmembrane conductance regulator to the cytoskeleton. *The Journal of Biological Chemistry*, *273*(31), 19797–19801.

Smanik, P. A., Ryu, K. Y., Theil, K. S., Mazzaferri, E. L., & Jhiang, S. M. (1997). Expression, exon-intron organization, and chromosome mapping of the human sodium iodide symporter. *Endocrinology*, *138*(8), 3555–3558.

Sun, F., Hug, M. J., Lewarchik, C. M., Yun, C. H., Bradbury, N. A., & Frizzell, R. A. (2000). E3KARP mediates the association of ezrin and protein kinase A with the cystic fibrosis transmembrane conductance regulator in airway cells. *The Journal of Biological Chemistry*, *275*(38), 29539–29546.

Sun, H., Tsunenari, T., Yau, K. W., & Nathans, J. (2002). The vitelliform macular dystrophy protein defines a new family of chloride channels. *Proceedings of the National Academy of Sciences of the United States of America*, *99*(6), 4008–4013.

Tabcharani, J. A., Chang, X. B., Riordan, J. R., & Hanrahan, J. W. (1992). The cystic fibrosis transmembrane conductance regulator chloride channel. Iodide block and permeation. *Biophysical Journal*, *62*(1), 1–4.

Takasu, N., Murakami, M., Nagasawa, Y., Yamada, T., Shimizu, Y., Kojima, I., et al. (1987). BAY-K-8644, a calcium channel agonist, induces a rise in cytoplasmic free calcium and iodide discharge in thyroid cells. *Biochemical and Biophysical Research Communications, 143*(3), 1107–1111.

Thorndyke, M. C. (1978). Evidence for a 'mammalian' thyroglobulin in endostyle of the ascidian Styela clava. *Nature, 271*(5640), 61–62.

van den Hove, M. F., Croizet-Berger, K., Jouret, F., Guggino, S. E., Guggino, W. B., Devuyst, O., et al. (2006). The loss of the chloride channel, ClC-5, delays apical iodide efflux and induces a euthyroid goiter in the mouse thyroid gland. *Endocrinology, 147*(3), 1287–1296.

Viitanen, T. M., Sukumaran, P., Lof, C., & Tornquist, K. (2013). Functional coupling of TRPC2 cation channels and the calcium-activated anion channels in rat thyroid cells: Implications for iodide homeostasis. *Journal of Cellular Physiology, 228*(4), 814–823.

Wall, S. M. (2006). The renal physiology of pendrin (SLC26A4) and its role in hypertension. *Novartis Foundation Symposium, 273*, 231–239, discussion 239–243, 261–264.

Wang, S., Yue, H., Derin, R. B., Guggino, W. B., & Li, M. (2000). Accessory protein facilitated CFTR-CFTR interaction, a molecular mechanism to potentiate the chloride channel activity. *Cell, 103*(1), 169–179.

Wangemann, P. (2006). Supporting sensory transduction: Cochlear fluid homeostasis and the endocochlear potential. *The Journal of Physiology, 576*(Pt 1), 11–21.

Wangemann, P., Kim, H. M., Billings, S., Nakaya, K., Li, X., Singh, R., et al. (2009). Developmental delays consistent with cochlear hypothyroidism contribute to failure to develop hearing in mice lacking Slc26a4/pendrin expression. *American Journal of Physiology Renal Physiology, 297*(5), F1435–F1447.

Wangemann, P., Nakaya, K., Wu, T., Maganti, R. J., Itza, E. M., Sanneman, J. D., et al. (2007). Loss of cochlear HCO_3^- secretion causes deafness via endolymphatic acidification and inhibition of Ca^{2+} reabsorption in a Pendred syndrome mouse model. *American Journal of Physiology. Renal Physiology, 292*(5), F1345–F1353.

Weiss, S. J., Philp, N. J., Ambesi-Impiombato, F. S., & Grollman, E. F. (1984). Thyrotropin-stimulated iodide transport mediated by adenosine 3′,5′-monophosphate and dependent on protein synthesis. *Endocrinology, 114*(4), 1099–1107.

Weiss, S. J., Philp, N. J., & Grollman, E. F. (1984a). Effect of thyrotropin on iodide efflux in FRTL-5 cells mediated by Ca^{2+}. *Endocrinology, 114*(4), 1108–1113.

Weiss, S. J., Philp, N. J., & Grollman, E. F. (1984b). Iodide transport in a continuous line of cultured cells from rat thyroid. *Endocrinology, 114*(4), 1090–1098.

Yang, Y. D., Cho, H., Koo, J. Y., Tak, M. H., Cho, Y., Shim, W. S., et al. (2008). TMEM16A confers receptor-activated calcium-dependent chloride conductance. *Nature, 455*(7217), 1210–1215.

Yen-Chow, Y. C., Chow, S. Y., & Woodbury, D. M. (1984). Distribution of sodium in cellular and luminal compartments of turtle thyroid glands. *The American Journal of Physiology, 246*(5 Pt 1), E436–E443.

Yoshida, A., Hattori, K., Hisatome, I., Taniguchi, S., Ueta, Y., Hukui, H., et al. (1999). A TSH/dibutyryl cAMP activated Cl^-/I^- channel in FRTL-5 cells. *Biochemical and Biophysical Research Communications, 259*(3), 631–635.

Yoshida, A., Taniguchi, S., Hisatome, I., Royaux, I. E., Green, E. D., Kohn, L. D., et al. (2002). Pendrin is an iodide-specific apical porter responsible for iodide efflux from thyroid cells. *Journal of Clinical Endocrinology and Metabolism, 87*(7), 3356–3361.

Yu, K., Duran, C., Qu, Z., Cui, Y. Y., & Hartzell, H. C. (2012). Explaining calcium-dependent gating of anoctamin-1 chloride channels requires a revised topology. *Circulation Research, 110*(7), 990–999.

CHAPTER THREE

The Sodium/Multivitamin Transporter: A Multipotent System with Therapeutic Implications

Matthias Quick*[,1], Lei Shi†

*Department of Psychiatry, Division of Molecular Therapeutics, Columbia University College of Physicians and Surgeons, New York State Psychiatric Institute, New York, USA
†Department of Physiology and Biophysics, Institute for Computational Biomedicine, Weill Medical College of Cornell University, New York, USA
[1]Corresponding author: e-mail address: mq2102@columbia.edu

Contents

Abstract

The Na^+/multivitamin transporter (SMVT) is a member of the solute:sodium symporter family that catalyzes the Na^+-dependent uptake of the structurally diverse water-soluble vitamins pantothenic acid (vitamin B5) and biotin (vitamin H), α-lipoic acid—a vitamin-like substance with strong antioxidant properties—and iodide. The organic substrates of SMVT play central roles in the cellular metabolism and are, therefore, essential for normal human health and development. For example, biotin deficiency leads to growth retardation, dermatological disorders, and neurological disorders. Animal studies have shown that biotin deficiency during pregnancy is directly correlated to embryonic

Vitamins and Hormones, Volume 98
ISSN 0083-6729
http://dx.doi.org/10.1016/bs.vh.2014.12.003

growth retardation, congenital malformation, and death of the embryo. This chapter focuses on the structural and functional features of the human isoform of SMVT (hSMVT); the discovery of which was greatly facilitated by the cloning and expression of hSMVT in tractable expression systems. Special emphasis will be given to mechanistic implications of the transport process of hSMVT that will inform our understanding of the molecular determinants of hSMVT-mediated transport in dynamic context to alleviate the development and optimization of hSMVT as a multipotent platform for drug delivery.

1. INTRODUCTION

The Na^+/multivitamin transporter (SMVT) is a member of the solute: sodium symporter (SSS) family, a gene family of membrane proteins that catalyze the coupled transport of their respective solute(s) to the flux of sodium ions (Na^+) across the plasma membrane of all living organisms (Jung, 2002; Saier, 2000; Wright, Loo, Hirayama, & Turk, 2004). The human isoform of SMVT (hSMVT) is one of 12 members of the solute carrier gene family 5 (SLC5A) (Hediger et al., 2004) which includes Na^+-coupled transporters for sugars, myo-inositol, iodide, short-chain fatty acids, and choline (Wright, Loo, & Hirayama, 2011; Wright & Turk, 2004). A breakthrough in the analysis of SMVT came in 1998 with the cloning of the rat placental cDNA of *SMVT* (*rSMVT*) by Prasad and colleagues (Prasad et al., 1998), shortly followed by the isolation of the cDNA of *hSMVT* from placenta (Wang et al., 1999). Following its cloning and expression in suitable expression systems, it was shown that SMVT mediates the Na^+-dependent uptake of the structurally diverse water-soluble vitamins pantothenic acid (vitamin B5) and biotin (vitamin H) (Prasad et al., 1998; Wang et al., 1999). In addition, confirming previous indirect studies (Wang et al., 1999), recent radiotracer uptake measurements have revealed the Na^+-coupled uptake of α-lipoic acid, a vitamin-like substance with strong antioxidant properties, as well as the uptake of iodide (I^-) (de Carvalho & Quick, 2011). The organic substrates of SMVT play central roles in the cellular metabolism and are, therefore, essential for normal human health and development. For example, biotin deficiency leads to growth retardation, dermatological disorders, and neurological disorders (Ahmad & Ahmad, 1991). Animal studies have shown that biotin deficiency during pregnancy is directly correlated to embryonic growth retardation, congenital malformation, and death (Mock, Quirk, & Mock, 2002; Watanabe, 1983, 1996; Zempleni & Mock, 2000). Whereas there are several excellent reviews with scope on the physiology of the substrates of SMVT, in particular, for biotin and pantothenic acid (Bender, 1999; Kelly, 2011; Said & Mohammed, 2006; Tahiliani & Beinlich, 1991; Trumbo, 2014; Zempleni & Kuroishi, 2012; Zempleni & Mock, 1999; Zempleni, Wijeratne, & Hassan, 2009), this chapter relates the advances in SMVT research in context with the recent progress in the field of Na^+-coupled transport processes at the mechanistic and structural level.

2. "ACTIVE" TRANSPORT

Active transport of solutes across biological membranes driven by electrochemical gradients (i.e., secondary active transport) plays a central role in fundamental cellular processes, such as nutrient uptake, excretion of toxic compounds, and signal transduction (DeFelice, 2004; Saier & Ren, 2006). Actively excreted by the action of the Na^+/K^+-ATPase located in membranes of animal cells, Na^+ is the predominant coupling cation in secondary active transport of a diverse range of molecules, such as ions, nutrients, vitamins, and osmolytes in higher organisms (Jung, 2002; Wright et al., 1994, 2004; Wright, Loo, Turk, & Hirayama, 1996). Formulated in the 1960s by Robert Crane (Crane, 1962, 1965), Na^+-driven substrate cotransport (symport) was hypothesized for the active transport of glucose across the brush border membrane of the enterocytes lining the intestine. The Na^+/glucose symport hypothesis has been tested and confirmed by work performed in the labs of Ernest Wright and others (see Wright et al., 2011, for a recent review), and in 1987, using expression cloning, the gene product that catalyzes Na^+/glucose symport was identified and isolated from a rabbit intestinal cDNA library (Hediger, Coady, Ikeda, & Wright, 1987). Two years later, the human isoform of the sodium/glucose transporter (hSGLT1) was cloned using a similar approach (Hediger, Turk, & Wright, 1989), thus making possible the functional characterization of substrate flux and substrate-induced electrical currents (Hediger et al., 1987; Ikeda et al., 1989; Parent, Supplisson, Loo, & Wright, 1992a). For SGLT1, this approach has led to the formulation of an ordered kinetic model of transport (Parent, Supplisson, Loo, & Wright, 1992b) (Fig. 1) that follows the key principles of the so-called "alternating access model," a concept that was introduced in the 1950s (Patlak, 1957). According to this model transporters alternate between two key conformations, the "inward-open" and "outward-open" state, thereby making accessible, in reciprocal fashion, the Na^+ and substrate sites to one side of the membrane but never to both

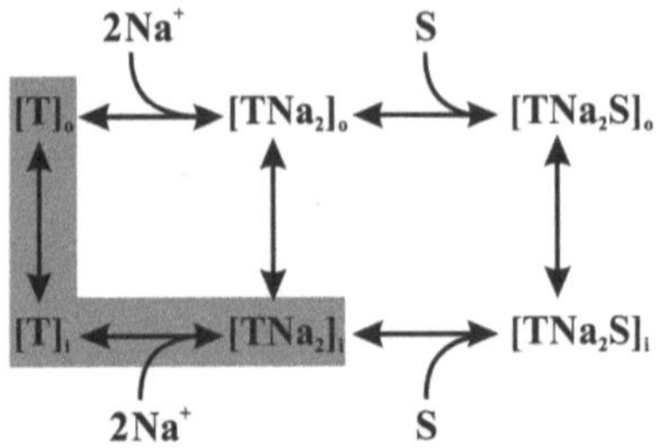

Figure 1 Kinetic model for the SSS member SGLT1. A six-state kinetic model of the hSGLT1 transport cycle was proposed (Parent et al., 1992b; Wright et al., 1994) according to which two external Na^+ bind to the empty, outward-facing transporter $[T]_O$. The Na^+-bound transporter $[TNa_2]_O$ is poised to bind sugar (S), yielding the Na^+ and sugar-bound intermediate $[TNa_2S]_O$. Under physiological conditions, this conformational state changes its accessibility to the inward-facing conformation $[TNa_2S]_i$ which is followed by the internal sequential release of sugar ($[TNa_2]_i$) and Na^+ ($[T]_i$). However, all state transitions are reversible. Since the transient charge relaxations are eliminated by transported sugars (compare Fig. 9B, left panel for hSGLT1), only the three shaded conformational states participate effectively in the observed charge transfer. Membrane voltage affects Na^+ binding and the transition between $[T]_O \leftrightarrow [T]_i$ (Loo, Hirayama, Cha, Bezanilla, & Wright, 2005). In the absence of substrate, the uncoupled flux of Na^+ (Na^+-leak) is observed, which is ~10% of the maximum sugar-induced currents (Loo, Hazama, Supplisson, Turk, & Wright, 1993). ^{14}C-sugar and $^{22}Na^+$ flux studies show the direct correlation between the fluxes of substrates and charge (Mackenzie, Harper, Taylor, & Rennie, 1994).

sides at the same time (Abramson & Wright, 2009; Jardetzky, 1966; Weyand et al., 2011). In recent years, a number of X-ray crystallographic structures of Na^+-coupled symporters have been reported and these static snapshots of proteins in a particular conformational state support this basic principle of the alternating access mechanism (Faham et al., 2008; Krishnamurthy, Piscitelli, & Gouaux, 2009; Ressl, Terwisscha van Scheltinga, Vonrhein, Ott, & Ziegler, 2009; Weyand et al., 2008; Yamashita, Singh, Kawate, Jin, & Gouaux, 2005; Zhou et al., 2014).

3. IDENTIFICATION OF THE MULTIVITAMIN TRANSPORTER

Similar to the initial characterization of Na^+/glucose transport in animal tissue prior to the cloning of the *SGLT1* gene, key elements of the transport of biotin, pantothenic acid, and lipoic acid have been described in studies involving animal and human tissue (Baur & Baumgartner, 1993, 2000; Beinlich, Naumovitz, Song, & Neely, 1990; Chatterjee, Kumar,

Ortiz, Rubin, & Said, 1999; Prasad, Ramamoorthy, Leibach, & Ganapathy, 1997; Said, Korchid, Horne, & Howard, 1990; Said et al., 1998; Said & Redah, 1988; Said, Redha, & Nylander, 1987, 1988a; Spector & Mock, 1987; Spencer & Brody, 1964). In particular, radiotracer uptake studies by Grassl (1992a, 1992b) in brush border membrane vesicles from human term placenta indicated that the transport of radiolabelled pantothenate into membrane vesicles was inhibited by biotin and lipoate; likewise, transport of radiolabelled biotin was inhibited by pantothenate and lipoate. The uptake of radiolabeled biotin was inhibited by nonradioactive pantothenate in concentration-dependent manner, and the concentration of pantothenate to yield half-maximum inhibition (K_i) of biotin uptake was comparable to the concentration of radiolabeled pantothenate to yield the half-maximum velocity of transport (K_m). Likewise, the K_i of biotin for radiolabeled pantothenate transport was similar to the K_m of radiolabeled biotin transport, thus, leading to the hypothesis of the involvement of a single transport system that mediates the uptake of these diverse substrates.

In 1998, Prasad et al. (Prasad et al., 1998) reported the isolation of the first mammalian gene that codes for a single polypeptide that, when expressed in HeLa cells, exhibits the key characteristics of this long-sought system—which was termed SMVT—from a rat placental cDNA library. This breakthrough at the molecular level was followed by the cloning and expression of *hSMVT*, which was isolated from a placental choriocarcinoma cell line (JAR) (Wang et al., 1999), and the isolation of a rabbit intestinal *SMVT* (*riSMVT*) (Prasad et al., 1999). In addition, the mouse isoform of *SMVT* was cloned in the Said lab (unpublished, GenBank® Accession number AY572835). The sequence and kinetic properties of the cloned SMVTs are very similar and, thus, for simplicity, the following description will focus on hSMVT.

4. THE *hSMVT* GENE

The *hSMVT* gene is about 14.3 kbp in length and contains 17 exons separated by 16 introns (GenBank® Accession number AF116241). All exon–intron boundaries are consistent with consensus donor–acceptor sequences (gt/ag) for RNA splicing (Wang et al., 1999). Using chromosomal assignment studies with somatic cell hybrid analysis and fluorescence *in situ* hybridization of metaphase chromosomes revealed that the *hSMVT* gene is located on the p23 band on the short arm of the human chromosome 2 (Wang et al., 1999).

4.1 Expression of *hSMVT* in various tissues

Analysis of the expression of *hSMVT* in human tissues was conducted using Northern blots with commercial mRNA samples at the time when *hSMVT* was cloned (Wang et al., 1999). According to this seminal study, hSMVT is ubiquitously distributed in the human body as positive signals were obtained in all eight tested tissue samples (heart, brain, placenta, lung, liver, skeletal muscle, kidney, and pancreas). Since the strongest signals were detected in absorptive tissues such as the liver, kidney, and placenta it seems justified to conclude that hSMVT plays a central role in the homeostasis of the essential vitamins H and B5 in the human body. Those studies involving cloned *hSMVT* (Wang et al., 1999) are consistent with observations regarding biotin and pantothenate uptake studies performed with human or animal tissues (Baur & Baumgartner, 1993, 2000; Beinlich et al., 1990; Said et al., 1987, 1988a, 1990, 1998; Said & Redah, 1988; Spector & Mock, 1987). Biotin uptake studies performed in brush border membrane vesicles isolated from different regions in the rabbit or rat intestine further demonstrated that uptake of this vitamin is higher in the duodenum and jejunum than in the ileum (Said & Redah, 1988). However, studies at the mRNA level reveal no significant difference in the expression of *hSMVT* along the rabbit intestinal tract (Prasad et al., 1999), thus suggesting that additional regulatory elements may influence protein expression, trafficking, and/or insertion of SMVT in the plasma membrane. Regulation of the *hSMVT* gene expression at the transcriptional level has been shown (Reidling, Nabokina, & Said, 2007; Reidling & Said, 2007) and was suggested to counteract conditions of biotin deficiencies (Sweetman & Nyhan, 1986; Wolf, Heard, McVoy, & Raetz, 1984). Guided by experiments that showed adaptive changes of biotin uptake in rats in response to the availability of extracellular biotin with respect to the maximal velocity (V_{max}) of biotin transport (Said, Mock, & Collins, 1989), the hSMVT activity was tested under biotin-deficient and biotin-sufficient states of the cells. The results of this study revealed that the upregulation of hSMVT-mediated biotin transport was accompanied by an increase of the mRNA and protein levels of hSMVT (Reidling et al., 2007). Further *in vitro* studies showed that *hSMVT* expression in human intestinal epithelial cells is regulated at the transcriptional level via the specific *cis*-regulatory elements, Kruppel-like factor 4 and activator protein-2, in the minimal promoter region of *hSMVT*. *In vivo* studies that utilized a transgenic mouse model supported the *in vitro* findings, and although corresponding data are not available for the transcriptional regulation of *hSMVT* expression in the human intestine, the authors suggest a similar regulation pattern in humans as the promoters for mouse and human *SMVT* share a high degree of sequence homology (Reidling & Said, 2007).

4.2 An additional high-affinity hSMVT-like uptake system?

Several studies involving animal and human tissues indicated the presence of a single saturable transport system that mediates the Na^+-dependent uptake of biotin and/or pantothenate in the intestine (Chatterjee, Kumar, Ortiz, Rubin, & Said, 1999; Said et al., 1987), kidney (Baur & Baumgartner, 1993), liver (Said et al., 1990), brain microvessel endothelial cells of the blood–brain barrier (Baur & Baumgartner, 2000; Spector & Mock, 1987), placenta (Grassl, 1992a, 1992b), heart (Beinlich et al., 1990), as well as in keratinocytes (Grafe, Wohlrab, Neubert, & Brandsch, 2003). Whereas the majority of these studies point to the involvement of SMVT in the transport process due to the kinetic properties and specificity/inhibition profile of the transport features, few studies suggest the presence of an additional uptake system with SMVT-like properties. For example, Grafe et al. (2003) reported that biotin uptake in keratinocytes shows features that are reminiscent of the kinetic properties of hSMVT (i.e., μM affinity for ^{3}H-biotin uptake that was inhibited by pantothenate and lipoate with K_is in the μM range); however, the authors also identified an additional component (or transport system) with preference for pantothenate and high-affinity biotin uptake kinetics (with a K_m in the nM range). Likewise, Zempleni and Mock (1998) described a high-affinity biotin transport in human peripheral blood mononuclear cells (PBMC) with a K_m for biotin transport of 2.6 nM. Furthermore, *RS*-lipoic acid failed to inhibit transport of biotin. A study that targeted the high-affinity biotin transport phenomenon in Caco-2 and HepG2 cells concluded that such a high-affinity system is not present/functional in human intestinal or liver epithelial cells (Balamurugan, Ortiz, & Said, 2003). Using siRNA-silencing techniques to repress the expression of hSMVT it was shown that under these conditions biotin uptake (tested at 2.6 nM ^{3}H-biotin concentration) was inhibited by more than 85%. The authors, thus, concluded that hSMVT is the major (and probably the only) biotin uptake system in intestinal and hepatic cells. It remains to be seen whether the reported high-affinity transport of biotin reflects a unprecedented component in the hSMVT transport reaction that was elicited by particular experimental conditions or whether it stems from an additional, hitherto unknown transport system present in keratinocytes or PBMCs.

5. FROM GENE TO PROTEIN

The *hSMVT* gene product is 635 amino acids in length and has a calculated molecular weight of 68696.97 in unmodified form. On the basis of hydrophobicity analysis (Kyte–Doolittle plot (Kyte & Doolittle, 1982)), Wang et al. (1999) proposed a secondary structure model according to which hSMVT consists of 12 hydrophobic transmembrane segments (TMs) connected by hydrophilic loops with the amino (N) and carboxy (C) termini located in the cytoplasm (Fig. 2). An elegant study in the Said lab involving the introduction of a green fluorescent protein-based assay concluded that the C terminus of hSMVT is indeed located in the cytoplasm and that several domains within the C-terminal region are important for the trafficking of hSMVT to the apical cell membrane (Subramanian, Marchant, Boulware, Ma, & Said, 2009).

On the basis of the 12-TM model, Wang et al. (1999) further predicted four *N*-linked glycosylation sites in hSMVT. *N*-linked glycosylation is a common protein modification of Asn residues in the sequence Asn-X-Ser/Thr-X (X defines any amino acid except Pro) and is catalyzed in the

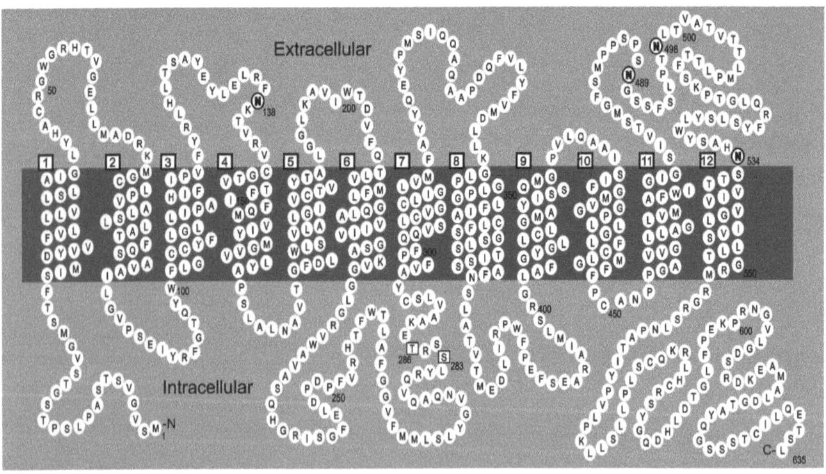

Figure 2 Deduced topology of hSMVT. The secondary structure of hSMVT is based on hydrophobicity calculations (Kyte & Doolittle, 1982) and contains 12 transmembrane domains connected by hydrophilic extracellular and intracellular loops. N and C termini are facing in the cytoplasm. Putative *N*-glycosylation sites are highlighted (bold) and the two putative PKC phosphorylation sites are shown as rectangles (Wang et al., 1999).

endoplasmatic reticulum of virtually all eukaryotic cells. The predicted *N*-glycosylation sites in hSMVT are Asn138, Asn489, Asn498, and Asn534 (compare Fig. 2). While the first putative site (Asp138) is located between TM 3 and 4, the other three putative sites are located between transmembrane domains 11 and 12. Furthermore, the deduced hSMVT amino acid sequence features two sites with consensus sequence for protein kinase C (PKC)-dependent phosphorylation. These two putative sites are Ser283 and Thr286 and both are located in the cytoplasmic loop between transmembrane domains 6 and 7. Supporting previous immunological studies that demonstrated that hSMVT has a higher molecular mass than 68.7 kDa predicted from its molecular mass (Nabokina, Subramanian, & Said, 2010; Subramanian, Marchant, Boulware, Ma, & Said, 2009), Ghosal, Subramanian, and Said (2011) showed that hSMVT is indeed subjected to posttranslational modifications. The posttranslational modification of hSMVT has also been confirmed by studies in one of the authors' laboratory (de Carvalho & Quick, 2011); however, in contrast to the aforementioned studies performed in the Said lab (Nabokina et al., 2010; Subramanian et al., 2009), hSMVT is detected as fully glycosylated protein at a molecular weight of about 70 kDa and as core-glycosylated/unglycosylated species at ~55 kDa (Fig. 3). Although the predicted molecular mass of hSMVT is 68.7 kDa, the unglycosylated polypeptide migrates faster (55 kDa) than its predicted molecular mass. Such an electrophoretic

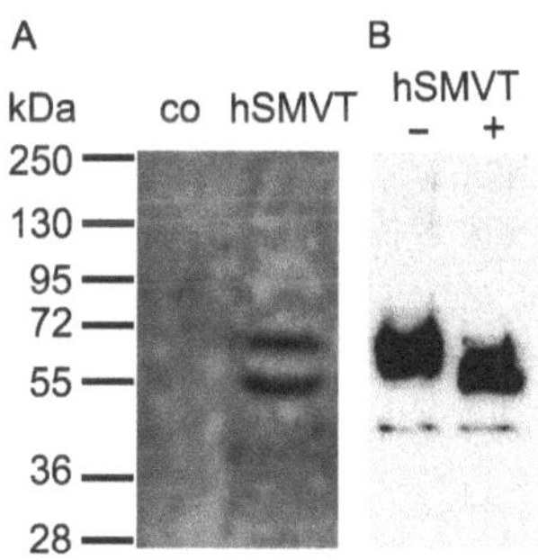

Figure 3 Characterization of the hSMVT protein. (A) Immunological detection of native hSMVT using a polyclonal antibody against SMVT. Noninjected oocytes (co) served as control. *This research was originally published in de Carvalho and Quick (2011). © The American Society for Biochemistry and Molecular Biology.* (B) hSMVT was subjected to peptidyl *N*-glycosidase F treatment (indicated by "+") to show the glycosylation of hSMVT (M. Q., unpublished).

behavior has been observed for many other membrane proteins (Jung, Tebbe, Schmid, & Jung, 1998; Levy et al., 1998; Turk, Kerner, Lostao, & Wright, 1996) and is probably due to differential (and incomplete) solvation of the highly hydrophobic membrane-embedded polytopic polypeptides by SDS (Rath, Glibowicka, Nadeau, Chen, & Deber, 2009). A similar observation with respect to the presence of 70 kDa and 55 kDa protein bands has been reported for the rabbit isoform of SMVT by Griffin, Stanley, and Zempleni (2002). This discrepancy in the size of the glycosylated form(s) is probably due to the variations in the glycosylation patterns in different expression host systems. Nevertheless, these separate studies emphasize consistently the modification of hSMVT by *N*-glycosylation. Using site-directed mutagenesis, Ghosal et al. targeted the individual putative glycosylation sites and found that positions Asn138 and Asn489 are important for the function of hSMVT as replacement of these sites significantly reduces the V_{max} of biotin uptake (Ghosal et al., 2011), thus highlighting the importance of glycosylation on the function of hSMVT. Since previous studies identified a significant effect on the level of intestinal biotin uptake by modulators of the PKC-mediated regulatory pathway without affecting the affinity of the uptake process (Said, 1999; Said et al., 1998), Ghosal and colleagues tested the effect of mutating the proposed PKC sites by mutating them in individual fashion. Whereas replacing Thr286 led to a significant decrease in the PKC-mediated inhibition of biotin uptake (note that this effect was not associated to changes in the level of expression or insertion of hSMVT in the cell membrane), the replacement of Ser283 did not show any significant effect when compared to the wild-type protein, thus emphasizing the modulation of the hSMVT-mediated transport process by protein regulatory elements.

6. FAMILY TIES

The amino acid sequences of hSMVT (635 amino acids) shares 84% identity and 89% similarity with its rat counterpart (634 amino acids) and 87% and 92% identity and similarity, respectively, with riSMVT (636 amino acids). With the mouse SMVT (634 amino acids) hSMVT shares 82% identity and 88% similarity. Based on homology studies, the identified SMVT proteins are encompassed in the SSS family (Saier, 1999). Together with other human members of the SSS family hSMVT is grouped in the solute carrier family series 5 (SLC5) (Hediger et al., 2004). hSMVT (SLC5A6) shares the highest sequence identity and similarity with the Na^+/iodide

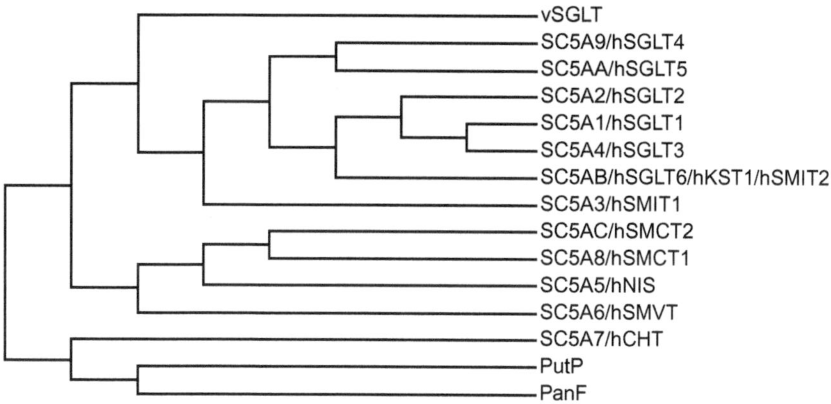

Figure 4 The phylogenetic tree of the SSS family. The tree was generated by the workflow integrated in the phylogeny.fr server (Dereeper et al., 2008). Briefly, the sequences of selected SSS members were aligned by the MUSCLE program (Edgar, 2004) and curated by the Gblock program (Castresana, 2000); the tree was constructed with the maximum likelihood method using PhyML (Guindon & Gascuel, 2003) and rendered by TreeDyn (Chevenet, Brun, Banuls, Jacq, & Christen, 2006). The SSS members are indicated by their Swissprot codes and commonly known abbreviations. For simplicity, the lengths of the branches were not indicated.

symporter (NIS; SLC5A5) (Dai, Levy, & Carrasco, 1996) (40% and 64%, respectively). Surprisingly, the functional homologous bacterial SSS member PanF (the pantothenate transporter) (Vallari & Rock, 1985) shares only 15% identity and 18% similarity with hSMVT, similar to values obtained for sequence homology comparison studies between hSMVT and PutP, the bacterial Na^+/proline transporter (Jung, 2002) (Fig. 4).

7. THE PREDICTED STRUCTURE OF hSMVT

Unlike soluble proteins, membrane proteins are generally notoriously resistant to the determination of their high-resolution structure by 3D crystallographic methods most likely due to their hydrophobic and dynamic nature—the key to their biological role as mediator of a "system in flux" that is situated in the membrane. To circumvent this bottleneck in their structural elucidation a new trend emerged in the late 1980s that encompassed the developed molecular biological techniques to engineer membrane proteins for site-directed biochemical and biophysical assays. For example, substituted cysteine scanning mutagenesis (Karlin & Akabas, 1998) in combination with chemical cross-linking and spectroscopic methods has become an important tool that has provided—and still provides—invaluable information about the structure and dynamics of membrane transport proteins (Kazmier et al., 2014; Raba et al., 2014; Zhao et al., 2011). The work performed in the lab of Ron Kaback and others on the lactose permease (LacY) of *Escherichia coli* is probably the best example for the extensive use of these techniques to collect structural information in lieu of a crystal structure of a particular membrane protein (Guan & Kaback, 2006; Kaback, Sahin-Toth, & Weinglass, 2001; Kaback, Smirnova, Kasho, Nie, & Zhou, 2011; Kaback & Wu, 1997). Moreover, this information provides a comprehensive framework for the interpretation of data collected from crystallographic X-ray analyses to amount to a coherent mechanistic picture (Abramson et al., 2003; Chaptal et al., 2011; Guan, Mirza, Verner, Iwata, & Kaback, 2007; Jiang, Driessen, Feringa, & Kaback, 2013; Kumar et al., 2014; Serdiuk et al., 2014; Smirnova et al., 2007).

While proteins of eukaryotic origin are well-suited to a variety of functional studies that allow the detailed collection of functional data (i.e., tracer flux and electrophysiology; see below), their low abundance in native tissue or mammalian expression systems represents a serious bottleneck for their large-scale purification—a prerequisite for their crystallization. On the other hand, to date only crystallization yields detailed high-resolution information about 3D protein structures. Whereas bacterial homologs of a particular gene family have been used successfully as model systems for crystallization (Abramson & Wright, 2009; Krishnamurthy et al., 2009), their expression in bacterial systems precludes the application of many, if not most of the commonly used functional characterization techniques applicable to membrane proteins expressed in eukaryotic systems.

In this context, direct (crystallographic) or indirect (biochemical/biophysical) structural approaches employing isolated (purified) hSMVT are still missing. On the basis of hydrophobicity analysis (Kyte–Doolittle plot, Kyte & Doolittle, 1982) a secondary structure model according to which hSMVT consists of 12 hydrophobic TMs was proposed (Wang et al., 1999; see Fig. 2). This secondary structure model was consistent with the common transmembrane protein model on the basis of the available structural information gathered for LacY (Kaback & Wu, 1997). However, at about the time when hSMVT was cloned this 12-TM model was implicitly challenged by several studies involving members of the SSS family. Most notably, the N termini of hSGLT1 (Turk et al., 1996) and NIS (Levy et al., 1998) were shown to be located on the extracellular face of the membrane, a result that had been confirmed in a study in which PutP was subjected to reporter-protein fusion analyses and site-directed cysteine-accessibility studies (Jung, Rubenhagen, et al., 1998). The study also showed that the C terminus of PutP is located in the cytoplasm, yielding a topological model of this SSS member of 13 TMs with an extracellularly located N terminus. The experimental elucidation of the extracellular location of the N terminus, and the consequential rearrangement of the N-terminal region in the PutP model, revealed that several residues that were identified to be critical for Na^+ and substrate binding and transport (Quick & Jung, 1997; Quick, Stolting, & Jung, 1999; Quick, Tebbe, & Jung, 1996) were, in fact, located in TM2 and not in extracellular loops as suggested by the 12-TM model (Nakao, Yamato, & Anraku, 1987). Wright and colleagues proposed a general structural motif for all members of the SSS family, featuring an extracellular N terminus and a common core of 13 TMs, but suggested that some SSS (including all SGLT1s and their bacterial homolog, vSGLT of *Vibrio parahaemolyticus*) have one additional TM appended to the C terminus (Turk & Wright, 1997).

This prediction has been confirmed by the available crystal structures of vSGLT, and it is indeed composed of 14 TMs (Faham et al., 2008; Watanabe et al., 2010). To date, vSGLT is the only member of the SSS family for which high-resolution structural information is available (Abramson & Wright, 2009). Surprisingly, however, although vSGLT and LeuT, a Na^+-coupled amino acid transporter and member of the neurotransmitter:sodium symporter (NSS) family, share only insignificant sequence homology (Wong et al., 2012), they have a common so-called "LeuT-like" structural fold. The LeuT fold refers to a core structure of 10 TMs arranged in two inverted repeats of five TMs (in vSGLT these include TMs 2–6 and 7–11, which are structurally homologous to TMs 1–5 and 6–10 in LeuT) (Abramson & Wright, 2009; Krishnamurthy et al., 2009). This common protein architecture was first identified in the crystal structure of LeuT (Yamashita et al., 2005) and is shared among several Na^+ or proton (H^+)-coupled transporter families including—in addition to vSGLT of the SSS family—the nucleobase:cation symporter-1 (Shimamura et al., 2010; Weyand et al., 2008), the betaine/carnitine/choline transporter (Ressl et al., 2009; Schulze, Koster, Geldmacher, Terwisscha van Scheltinga, & Kuhlbrandt, 2010; Tang, Bai, Wang, & Jiang, 2010), and the amino acid/polyamine/organocation transporter (Shaffer, Goehring, Shankaranarayanan, & Gouaux, 2009) families.

Despite these advances in structural understanding of related SSS members, the 12-TM model of hSMVT (Fig. 2) is still commonly used as a working model for the interpretation of functional and regulatory data for hSMVT (Ghosal & Said, 2011; Said, 2009; Subramanian et al., 2009). For comparison, we present here a model of hSMVT which contains 13 TMs, with the N terminus located on the extracellular side of the membrane and the C terminus situated in the cytoplasm. This proposed model emerged from our sequence alignment of an exhaustive collection of the relevant pro- and eukaryotic SSS sequences and the topology revealed by the vSGLT crystal structure (Fig. 5) (Mazier, Quick, & Shi, 2011). Note that Asn138 in the original TM-12 model of hSMVT is located on the extracellular side of the membrane, whereas the 13-TM model of hSMVT predicts it to be in the cytoplasm. Previous studies have shown that mutating this residue to Ala leads to a dramatically reduced V_{max} of biotin uptake; however, this phenomenon could be explained by the ~20-fold reduced expression/insertion of the mutated gene product in the cytoplasm membrane as detected with Western blot analysis (Ghosal et al., 2011). So far, the observed effects could be attributed solely to the reduced amount of protein in the membrane caused by the presence of Ala at position 138 but the study did not unequivocally reveal whether Asn138 is indeed an *N*-glycosylation site and/or that glycosylation of this particular residue is critical for the function of hSMVT.

8. THE (Co)SUBSTRATES OF hSMVT

As mentioned above, hSMVT transports the water-soluble vitamins pantothenic acid (vitamin B5) and biotin (vitamin H) (Wang et al., 1999), as well as α-lipoic acid, a vitamin-like substance with strong antioxidant properties (de Carvalho & Quick, 2011). In addition, the transport of I^- by hSMVT expressed in *Xenopus* oocytes has been reported (de Carvalho & Quick, 2011). Biotin and pantothenate are essential cofactors in key metabolic reactions that are vital for normal growth and development (Said, 1999; Trumbo, 2014; Zempleni & Kuroishi, 2012). Since they are not synthesized in the body, humans (and mammals in general) depend on external sources from which these vitamins can be ingested. In the human intestine, there are two sources of these vitamins: (a) the dietary source and (b) a bacterial source, the latter of which stems from the microflora predominantly found in the large intestine (Said & Mohammed, 2006; Said et al., 1998; Said, Redha, & Nylander, 1988b; Tahiliani & Beinlich, 1991). Found in virtually all dietary sources, these vitamins exist in protein-bound and free form. Only the free forms of the vitamins can be transported by hSMVT (Said, 2004). Prior to the

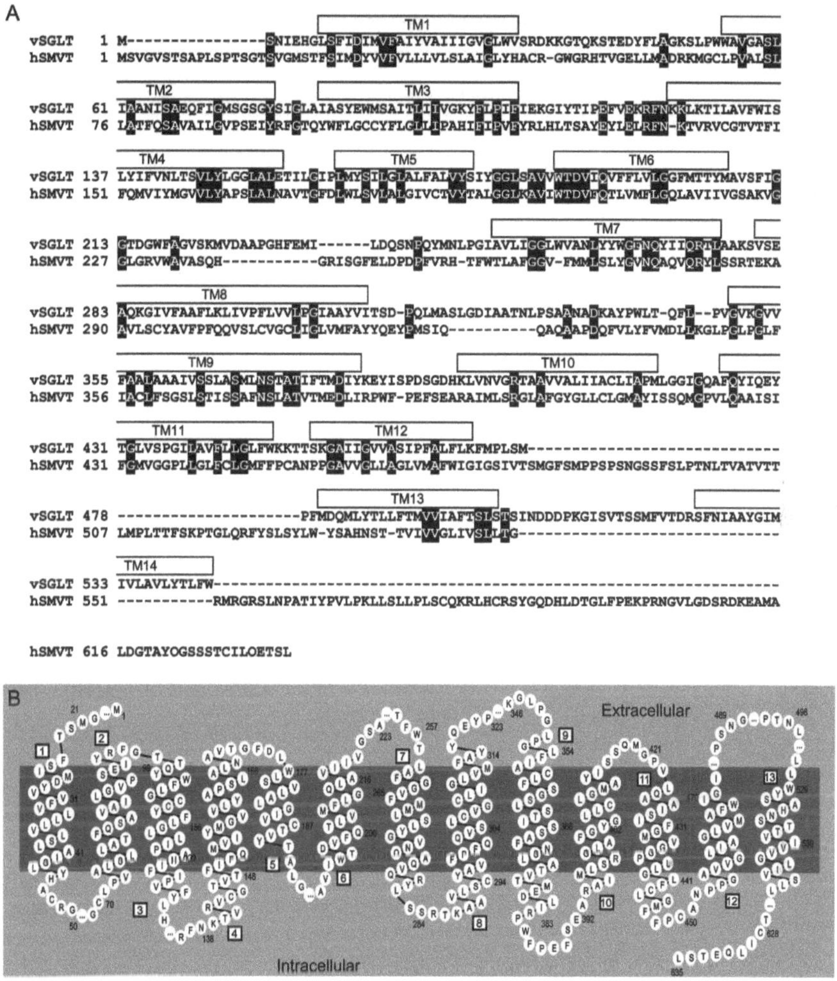

Figure 5 Predicted topology of hSMVT. (A) Pair-wise sequence alignment between hSMVT and vSGLT extracted from a structure-based and manually adjusted multiple sequence alignment of all the available nonredundant prokaryotic and eukaryotic SSS sequences from refseq database (>1000 sequences). Identical residues in vSGLT and hSMVT are highlighted and the transmembrane domains in vSGLT (Faham et al., 2008) are indicated. (B) The diagram of hSMVT secondary structure generated by the RbDe program (Skrabanek, Campagne, & Weinstein, 2003), based on the pair-wise sequence alignment in (A).

hSMVT-mediated absorption in the small intestine, protein-bound biotin and pantothenate are digested by gastrointestinal proteases and peptidases, leading to the generation of biocytin (and biotin-short peptides) and 4′-phosphopantetheine, respectively (Wolf et al., 1984). The enzyme biotinidase then liberates free biotin (Said, 2012), whereas pantetheinase dephosphorylates 4′-phosphopantetheine to pantothenate (Trumbo, 2014). Lipoic acid is readily absorbed from the diet and can be removed from protein complexes in the intestine by lipoamidase (Jiang & Cronan, 2005; Yoshikawa et al., 1996) (note that lipoamidase and biotinidase may be identical enzymes (Garganta & Wolf, 1990; Nilsson & Kagedal, 1993; Nilsson & Ronge, 1992)).

There is a compelling evidence that hSMVT couples the transport of its substrates exclusively to the electrochemical gradient of Na^+ across the plasma membrane. In contrast to other SSS members (Eskandari et al., 1997; Hirayama, Loo, & Wright, 1994, 1997; Quick & Jung, 1998; Quick, Loo, & Wright, 2001), hSMVT does not perform H^+ and/or Li^+-dependent substrate transport (de Carvalho & Quick, 2011; Grassl, 1992a, 1992b; Wang et al., 1999). In analogy to the SSS family member SGLT1(Mackenzie, Loo, & Wright, 1998), hSMVT is proposed to couple the flux of two Na^+ to the flux of one substrate molecule per transport cycle (Wang et al., 1999). Thus, substrate transport by hSMVT is rheogenic (see Section 9.1).

Among the members of the SSS family, hSMVT exhibits a relatively broad substrate specificity encompassing unrelated organic compounds and an inorganic anion. Whereas the SGLT group of SSS members features highly specific uptake of sugars (see Wright et al., 2011, for a recent review), NIS, on the other hand, transports a variety of inorganic anions such a I^-, ClO_3^-, SCN^-, $SeCN^-$, NO_3^-, Br^-, BF_4^-, IO_4^-, BrO_3^- (Eskandari et al., 1997), as well as perchlorate (ClO_4^-) (Dohan et al., 2007). In this regard, hSMVT seems to have an intermediate position in the SSS family as it transports the structurally diverse vitamin B5, vitamin H, and α-lipoic acid, as well as the anion I^-. Although hSMVT and NIS share the highest sequence identity/similarity (40% and 64%, respectively) (Dai et al., 1996) among the SSS family, with the exception of I^- hSMVT does not transport any of the other monovalent anions transported by NIS (M. Quick, unpublished). Chloride (Cl^-), an anion that is involved in the Na^+-coupled transport—either in a Cl^- cotransport (Keynan & Kanner, 1988) or Cl^-/Cl^- exchange mechanism (Loo, Eskandari, Boorer, Sarkar, & Wright, 2000)—of neurotransmitters by most, if not all mammalian NSS proteins (Rudnick, 2002; Sonders, Quick, & Javitch, 2005; Zomot et al., 2007), seems to be not involved in the transport reaction catalyzed by NIS or by hSMVT (de Carvalho & Quick, 2011; Eskandari et al., 1997). Interestingly, I^- transport by hSMVT has been adumbrated by Wang and colleagues

(Wang et al., 1999) when their team studied the effect of replacing Cl^- as the counter ion for Na^+. However, the role of I^- transport by hSMVT in human physiology is not entirely understood. The central role of NIS in the physiology of the I^--containing thyroid hormones thyroxine (T4) and triiodothyronine (T3) by mediating the Na^+-coupled transport of I^- from the blood across the basolateral membrane of thyroid cells (Dohan et al., 2003) as well as in the intestine (Nicola et al., 2009). However, early reports (Acland & Illman, 1959; Pastan, 1957) indicate a rather complex I^- homeostasis in the gastrointestinal tract that may involve other proteins in addition to NIS. This notion is supported by the fact that the accumulation of pertechnetate (TcO_4^-), a commonly used NIS substrate, in the blood of rats that were administered $^{99m}TcO_4^-$ via a duodenal catheter, was only partially inhibited by the potent NIS blocker/substrate ClO_4^- (Nicola et al., 2009). Since hSMVT is expressed in virtually every tissue in the body (Said, 1999; Wang et al., 1999), it may play a hitherto unknown role in the homeostasis of I^-.

The other substrates of hSVMT, pantothenic acid (vitamin B_5), biotin (vitamin H), and *R*-α-lipoic acid play vital roles in the energy metabolism of all cells. All three substrates contain a carboxyl group attached to a side chain of different lengths (Fig. 6). Biotin has an imidazole- and thiazole-fused ring structure and pantothenic acid has a pseudo-ring structure in which the amino group of the β-alanine moiety and the hydroxyl group of the pantoic acid moiety are connected via a hydrogen bond. The disulfite bond in lipoic acid (in its oxidized form) confers to the ring structure of the molecule, whereas the reduced form of lipoic acid (dihydrolipoic acid) does not exist as a ring structure.

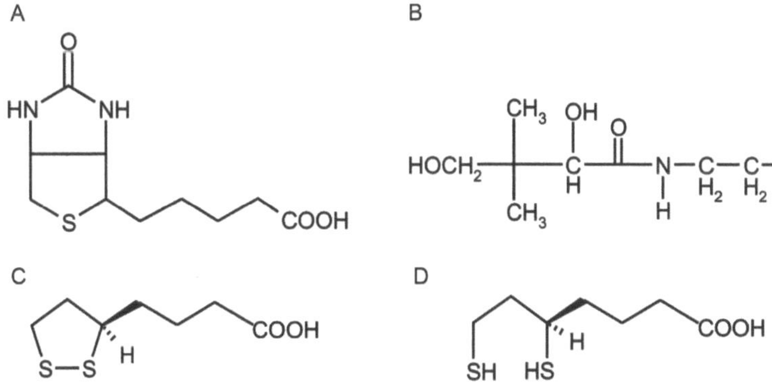

Figure 6 The structures of the organic substrates of hSMVT. (A) Biotin, (B) pantothenic acid, (C) (oxidized) *R*-α-lipoic acid, and (D) *R*-dihydrolipoic acid.

Pantothenic acid (3-[(2,4-dihydroxy-3,3-dimethylbutanoyl)amino] propanoic acid) is synthesized *de novo* by plants and microorganisms and it is an essential vitamin for humans (Smith & Song, 1996). Pantothenate is the precursor of the 4′-phosphopantetheine moiety of coenzyme A (CoA) and the acyl carrier protein (Alberts & Vagelos, 1966), two essential cellular enzymes involved in acyl-group transfer reactions (e.g., in the synthesis of fatty acids, cholesterol, acetylcholine and heme, and in the tricitric acid (TCA) cycle), and in the synthesis of lipids, respectively (Bender, 1999; Higdon, 2003; Leonardi, Zhang, Rock, & Jackowski, 2005; Tahiliani & Beinlich, 1991).

Biotin (5-[(3a*S*,4*S*,6a*R*)-2-oxohexahydro-1*H*-thieno[3,4-*d*]imidazol-4-yl]pentanoic acid) is an essential cofactor for several carboxylase-mediated CO_2 transfer reactions (Knowles, 1989). Synthesized only in bacteria, yeasts, molds, algae, and some plant species (Mock, 1999) biotin is a coenzyme for the following carboxylases in mammals: pyruvate carboxylase (EC 6.4.1.1), propionyl-CoA carboxylase (EC 6.4.1.3), methylcrotonyl-CoA carboxylase (EC 6.4.1.4), and acetyl-CoA carboxylase (EC 6.4.1.2; both isoforms). In addition, biotin was recently found to be central in epigenetic gene regulation and cell signaling via biotinylation of histones (Zempleni, 2005; Zempleni et al., 2009).

α-Lipoic acid ((*R*)-5-(1,2-dithiolan-3-yl)pentanoic acid; thioctic acid) is a naturally occurring substance that serves as a cofactor in dehydrogenase complexes (Milne et al., 2006), enzymes that catalyze central redox reactions of all cells (Bustamante et al., 1998). Lipoic acid is synthesized in small amounts in mitochondria of humans and other mammals and plants, as well as in bacteria (Carreau, 1979). The biological active form of α-lipoic acid, the *R*-enantiomer, is endogenously synthesized *de novo* from octanoic acid (Reed, 2001) being covalently bound to specific lipoate-dependent enzyme complexes (Zhao, Miller, Jiang, Marletta, & Cronan, 2003). Most notably, *R*-α-lipoic acid serves as a cofactor in the pyruvate dehydrogenase (PDG) and α-ketoglutarate dehydrogenase (KGDG) complexes (Milne et al., 2006), enzymes that catalyze central redox reactions related to energy production in the cell as well as in oxidative decarboxylations of α-keto acids and amino acids (Bustamante et al., 1998). For example, the PDG complex is responsible for the transition of pyruvate to acetyl-CoA, while the KGDG complex catalyzes the production of succinyl-CoA from α-ketoglutarate, both important intermediates in the TCA cycle. *R*-α-lipoic acid also serves as a cofactor in the catabolism of the branched-chain amino acids, leucine, isoleucine, and valine (Harris, Joshi, Jeoung, & Obayashi, 2005). Whereas the *R*-enantiomer is the biologically active form of lipoic acid, there is evidence that the *S*-enantiomer assists in the reduction of *R*-lipoic acid to dihydrolipoic acid (DHLA) (Biewenga et al., 1997). DHLA is the form of lipoic acid that is most commonly found within cells (Fig. 6D).

9. THE CHARACTERIZATION OF THE CLONED *hSMVT*

The cloning and expression of several mammalian, and in particular human membrane transport proteins, in the 1980s and 1990s (Dai et al., 1996; Guastella et al., 1990; Hediger et al., 1987; Kilty, Lorang, & Amara, 1991; Kopito & Lodish, 1985; Mueckler et al., 1985) have permitted rapid progress in the identification of the sites of gene expression, the generation of target-specific antibodies, and the biophysical/biochemical analysis of transporter function (such as substrate specificity, substrate and inhibitor kinetics, transport-associated electric currents, and currents that are not immediately associated with substrate transport). The cloning of *hSMVT* by Wang and colleagues (Wang et al., 1999) opened the door for the detailed analysis of the human isoform of the multivitamin transporter in heterologous expression systems. Expression of *hSMVT* in human retinal pigment epithelial (HRPE) cells was used by the authors to perform uptake studies with radiolabeled biotin and pantothenic acid and to test the effect of the ionic composition of the assay medium on hSMVT-mediated transport. Consistent with previous measurements in native tissue transport of pantothenic acid was strictly dependent on the presence of Na^+ but not on Li^+ or K^+. hSMVT-mediated transport did not require Cl^- as anion; however, replacement of Cl^- with I^- reduced the uptake of pantothenic acid by about 50%, an observation that led to the discovery of I^- transport by hSMVT (de Carvalho & Quick, 2011). Uptake of radiolabeled pantothenic acid was abolished by the three known physiological substrates at 100-fold concentration excess. The related vitamins thiamine and niacinamide had no effect on the transport of radiolabeled pantothenic acid, and neither had diaminobiotin or iminobiotin. Desthiobiotin, at 100-fold molar excess, showed a similar inhibitory efficacy like that observed with biotin. These data highlighted that the interaction of biotin with hSMVT requires the keto group at the second position of the imidazole ring in biotin (missing in diaminobiotin and iminobiotin), whereas the tetrahydrothiophene ring of biotin (present in desthiobiotin) is dispensable for substrate recognition by hSMVT. Likewise, recent studies indicate that the β-alanyl fragment of pantothenic acid is crucial for the recognition of this substrate by hSMVT

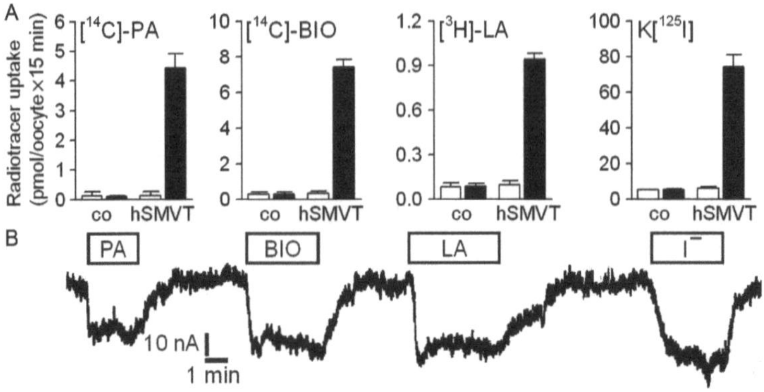

Figure 7 Substrate transport by hSMVT in oocytes. (A) Uptake of 1.4 μM ^{14}C-pantothenic acid (PA), 2.2 μM ^{14}C-biotin (BIO), 1.6 μM ^{3}H-*R*-α-lipoic acid (LA) or 500 μM K[$^{125}I^{-}$] was measured for 15 min in hSMVT-expressing or control (co) oocytes in 100 mM choline chloride (open) or NaCl (solid). (B) Substrate-elicited inward currents in hSMVT-expressing oocyte at a holding potential of −50 mV in 100 mM NaCl. Twenty-five micromolar of BIO, PA, LA, or NaI (I^{-}) were added as indicated. *This research was originally published in de Carvalho and Quick (2011).* *© The American Society for Biochemistry and Molecular Biology.*

(Chirapu et al., 2013). A closer analysis of the concentration dependence of the three organic substrates on the transport reaction revealed that uptake of radiolabeled biotin and pantothenic acid exhibits micromolar (μM) affinities (1.5 μM for pantothenate and 3.2 μM for biotin). Since radiolabeled lipoic acid was not commercially available until recently, the authors determined the apparent affinity for lipoic acid with competition assays in which the concentration of nonlabeled lipoic acid required to obtain 50% inhibition of radiolabeled pantothenic acid transport (termed K_i or IC_{50}) was found to be 2.7 μM for oxidized lipoate and 7.3 μM for reduced lipoate (i.e., DHLA). The unequivocal hSMVT-mediated uptake of lipoic acid was shown in hSMVT-expressing *Xenopus* oocytes using radiolabeled α-lipoic acid (as *RS*-racemic mixture) (de Carvalho & Quick, 2011) (the uptake of *R*-α-lipoic acid is depicted in Fig. 7A) and the kinetic details and specificity of α-lipoic acid uptake by hSMVT are currently under investigation in the Quick lab (Zehnpfennig et al., in preparation).

9.1 Electrogenicity of hSMVT-mediated transport

Whereas previous reports claimed the uptake system for biotin and pantothenic acid in human and animal tissue to be Na^{+}-dependent, the Hill coefficient—an indirect measure of the Na^+-to-substrate coupling stoichiometry (Segel, 1975)—was calculated to be 1 for Na^+-dependent biotin transport (Grassl, 1992a; Ma, Dyer, & Said, 1994) and 2 for Na^+-coupled pantothenic transport (Prasad et al., 1997). Since biotin and pantothenic acid are negatively charged at physiological pH (the pK_a values of biotin and pantothenic acid are 4.51 (Said et al., 1988b) and 4.4 (Ball, 2005), respectively), this discrepancy indicates that biotin transport by hSMVT would be electroneutral whereas pantothenic acid would be electrogenic (i.e., 2 Na^+:1 pantothenic $acid^-$).

In contrast to these diverging findings in native tissue and based on the Na^+-dependence of radiolabeled biotin or pantothenic acid transport in HRPE cells, Wang and colleagues determined an apparent Na^+ affinity for hSMVT of about 30 mM with a Hill coefficient of nearly 2 for the transport of biotin and pantothenic acid. Although the Hill coefficient does not provide an absolute measure for the Na^+-to-substrate flux stoichiometry as the individual events may not affect each other in cooperative fashion, the results indicate that hSMVT might mediate electrogenic Na^+-coupled substrate transport. Like the Na^+-coupled transport of sugars by SGLT1 (Hirayama et al., 1997) or I^- by NIS (Eskandari et al., 1997), the addition of the physiological substrates (biotin, pantothenic acid, and lipoic acid) to the Na^+-containing bath solution of hSMVT-expressing oocytes elicited electrical currents (Wang et al., 1999), thus confirming the proposed rheogenic nature of hSMVT (Fig. 7B). These inward currents increased in magnitude with increasing concentrations of the physiological substrates and with hyperpolarization of the applied membrane potential following similar features observed for other SLC5 family members when expressed in oocytes (see Wright et al., 2011, for review).

9.2 Mechanistic implications

Since hSMVT mediates the electrogenic transport of its physiological substrates in the presence of Na^+, the expression of the protein in *Xenopus* oocytes in combination with the two-electrode voltage-clamp technique provided a unique opportunity to assess kinetic features of hSMVT with a combination of flux and electrophysiological experiments. Similar experiments have been successfully used for the analysis of the well-studied SLC5 members SGLT1 (Parent et al., 1992a; Wright et al., 1994) and NIS (Eskandari et al., 1997), and led to the formulation of a kinetic model (Parent et al., 1992b) (Fig. 1). Figure 7B depicts the effect of the addition of the physiological substrates to oocytes injected with hSMVT-mRNA superfused in 100 mM NaCl-containing buffer under voltage-clamped conditions at a holding membrane voltage (V_h) of −50 mV. The addition of pantothenic acid, biotin, lipoic acid, and I^- (at the same concentration) elicited the same magnitude of inward charge fluxes, which were reversible upon the removal of each compound from the bath solution. Analysis of the ion species that could contribute to the charge movement indicated that electrical currents were only observed in the presence of Na^+, thus supporting previous observations that hSMVT-mediated transport is exclusively dependent on the presence of Na^+ and that Na^+ is most likely also the cotransported (charge-carrying) cation. A closer examination of the ionic conditions in one of the authors' laboratory showed that in striking contrast to several members of the SLC5 family (Wright et al., 2004), hSMVT exhibited no detectable Na^+ leak currents under the selected experimental conditions (M. Quick, unpublished; Fig. 8A) but elicited substrate-induced currents only in the presence of Na^+ (Fig. 8B–D). For example, these leak currents have been observed for NIS (Eskandari et al., 1997) and SGLT1

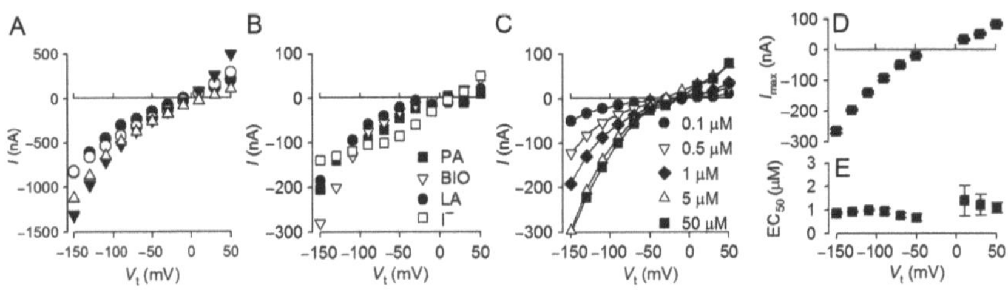

Figure 8 hSMVT-mediated ionic currents. (A) Current–voltage *(I/V)* relationship of total steady-state currents in hSMVT-expressing oocytes upon stepping the holding membrane potential of −50 mV to a series of test voltages (V_t) from +50 to −150 mV in the presence of 100 mM choline chloride, pH 7.4 (●) or pH 5.5 (▼), 100 mM NaCl= Na-gluconate (○) or 100 mM LiCl (△), at pH 7.4. hSMVT exhibited no detectable Na^+ leak currents under the experimental conditions (de Carvalho & Quick, 2011). Also note that substrate-elicited currents were only observed in the presence of Na^+. (B) Current–voltage *(I/V)* relationship of substrate-induced steady-state currents elicited in hSMVT-expressing oocytes upon stepping the holding membrane potential of −50 mV to a series of test voltages (V_t) from +50 to −150 mV. Twenty-five micromolar of PA, BIO, LA, or KI (same abbreviations as in Fig. 7) were added to buffer containing 100 mM NaCl. (C) *I/V* relationship of currents induced by varying concentrations of PA in 100 mM NaCl. PA-induced currents were plotted as a function of [PA] for each V_t (except −10 and −30 mV), yielding (D) the maximum current (I_{max}) and (E) the apparent affinity (EC_{50}) of PA-induced electrical currents. *This research was originally published in de Carvalho and Quick (2011). © The American Society for Biochemistry and Molecular Biology.*

(Hirayama et al., 1997) when 100 mM choline chloride was replaced with 100 mM NaCl in assay buffer at pH 7.4. For NIS and SGLT1, the Na^+ leak currents reached ~35% and ~10% of the maximum Na^+/substrate cotransport currents, respectively, and they are attributed to the uncoupled flux (uniport) of Na^+ in the absence of substrate as described in the kinetic model of SGLT1 from $TNa_o \rightarrow TNa_i$ in Fig. 1 (Panayotova-Heiermann, Loo, & Wright, 1995; Quick et al., 2001; Vandenberg, Arriza, Amara, & Kavanaugh, 1995; Wright et al., 1996). However, a recent publication from the Wright, Abramson, and Grabe labs (Watanabe et al., 2010) suggests, based on new structural information and molecular dynamics simulations, that—in contrast to the original kinetic model for SGLT1 shown in Fig. 1—the sequence of Na^+ and substrate release on the cytoplasmic side of the membrane follows that proposed for the NSS member LeuT (Shi, Quick, Zhao, Weinstein, & Javitch, 2008). This modification of the "classic" ordered binding model for SGLT1 inevitably affects the proposed mechanism of Na^+ uniport as the conformation $[TNa_2]_i$ can only exist after the release of substrate. Recent evidence, originating from studies on mouse SGLT3b, suggests that Na^+ is indeed released into the cytoplasm before sugar (Diez-Sampedro & Barcelona, 2011).

A key feature of SGLT1 that led in major part to the formulation of the kinetic model depicted in Fig. 1 is the appearance of relaxation currents upon stepping the test voltage (V_t) from a holding potential (V_h). Figure 9A shows a side-by-side comparison of current traces recorded in oocytes injected with hSMVT- or hSGLT1-mRNA in the presence of 100 mM NaCl upon stepping the voltage from −50 mV for 100 ms to V_t from +50 to −150 mV in 20 mV increments. As described (Quick et al., 2001), hSGLT1-expressing oocytes exhibited fast membrane capacitive currents (that are also observed in water-injected oocytes (Hazama, Loo, & Wright, 1997; Loo et al., 1993)) followed by transient currents that relaxed to steady-state currents. Integration of the transporter-mediated relaxations (termed presteady-state currents) revealed the hSGLT1-specific charge movement within the membrane electric field. Plotting the total charge as a function of membrane potential (V_t) yielded a sigmoidal charge–voltage (Q–V) curve that followed a Boltzmann equation,

$$\frac{Q - Q_{hyp}}{Q_{max}} = (1 + \exp[z(V_t - V_{0.5})F/RT])^{-1}$$

where Q is the displaced charge, $Q_{max} = Q_{dep} - Q_{hyp}$ (with Q_{hyp} and Q_{dep} for Q at hyperpolarizing and depolarizing limits, respectively); $V_{0.5}$ is the V_t

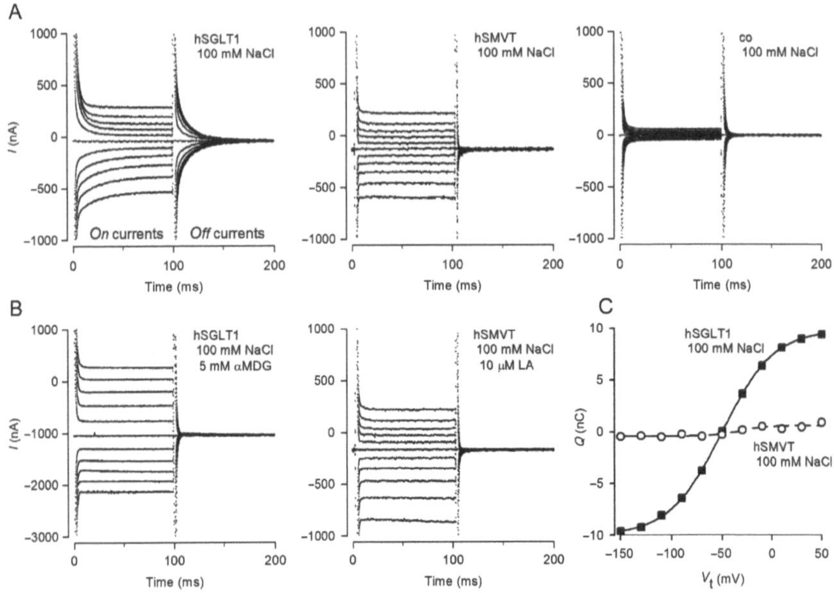

Figure 9 Current recordings in *Xenopus laevis* ooctes injected with hSGLT1- or hSMVT-cRNA in response to step changes in the test voltage. The test membrane potential (V_t) was stepped from a holding potential (V_h) of −50 mV to V_t of +50 to −150 mV in 20 mV decrements for 100 ms (*on* currents) before returning to V_h (*off* currents). (A) Currents were recorded in the presence of 100 mM external NaCl in individual hSGLT-, hSMVT-, or water-injected control (co) oocytes. (B) Currents were recorded the same oocytes shown in (A) after the addition of 5 mM α-methyl-D-glucopyranoside (αMDG; hSGLT1, left panel) or 10 μM LA (hSMVT, right panel). The addition of 5 mM αMDG or 10 μM LA did not affect the current response in control oocytes (not shown). (C) Charge transfer. The charge–voltage (*Q*–*V*) relationship for hSGLT1 (top panel) and hSMVT (lower panel) in the presence of 100 mM NaCl (corresponding to the left and center panel in A) were calculated according to the Boltzmann equation.

at which 50% of the charge has moved in the membrane field, z is the apparent valence of the moveable charge (1), F is the Faraday constant, R is the gas constant, and T is the temperature in K. The maximum charge (Q_{max}) and $V_{0.5}$ were calculated to be ~20 nC and −50.6 mV, respectively, for hSGLT1 (Fig. 9C). In contrast, hSMVT-elicited currents revealed only marginal relaxations that were at the border of resolution (Fig. 9C).

Several other secondary active transporters exhibit a comparable electrophysiological profile (Coady, Wallendorff, Bourgeois, & Lapointe, 2010; Eskandari et al., 1997; Forster, Hernando, Biber, & Murer, 1998) and it was assumed that their kinetic model was similar to that proposed for SGLT1 (Parent et al., 1992a, 1992b) (Fig. 1). However, given the limitation in the classical characterization of hSMVT—either due to the low expression level of hSMVT in the oocyte membrane or due to its different kinetic properties—using electrophysiological techniques that were so helpful for other members of the SLC5 family, the formulation of a mechanistic model for hSMVT-mediated transport requires alternative approaches that will hopefully lead to a coherent picture of the molecular determinants that govern the transport of its substrates. A few studies employing site-directed mutagenesis have been performed to identify functionally important amino acid residues in hSMVT (Ghosal & Said, 2011, 2012) but the interpretation of the data in functional context has been hampered by the effect caused by the mutations on the level of surface expression.

10. MEDICAL IMPLICATIONS

As outlined above, the organic substrates of hSMVT play central roles in key reactions in the body. For example, biotin is essential for normal growth and development (Bender, 1999; Sweetman & Nyhan, 1986) and severe biotin deficiency is directly associated with clinical symptoms (Whitehead, 1985). In particular, biotin deficiency during pregnancy has been shown to cause growth retardation of the embryo, congenital malformation, and even death (Watanabe, 1983, 1996). Deficiency of biotin has been observed in patients with inborn errors of biotin metabolism (Sweetman & Nyhan, 1986; Wolf, 2001) and in patients on long-term therapy with anticonvulsant drugs (Krause et al., 1988), or in those on long-term parenteral nutrition (Forbes & Forbes, 1997), as well as in patients with seboric dermatitis and Leiner's disease (Nisenson, 1957). It has been shown that biotin therapy rapidly improves skin and other congenital abnormalities (Miller, 1989; Mock, 1991; Tsao & Kien, 2002). To date, most cases of biotin deficiency can be attributed to two major genetic disorders in the biotin metabolism of humans involving either (a) the holocarboxylase synthetase (HCLS) deficiency (leading to a defect in the biotinylation) or (b) the biotinidase deficiency (preventing the release of free biotin from conjugated forms) (McMahon, 2002; Sweetman & Nyhan, 1986). There are also a few reports that suggest that hSMVT may be defective in patients in whom the HCLS and biotinidase function normal (Baumgartner, Suormala, Wick, Geisert, & Lehnert, 1982; Munnich et al., 1981; Thoene, Lemons, & Baker, 1983). However, to date there is no confirmed evidence that mutations in the hSMVT gene are directly correlated with biotin deficiency.

Like biotin, pantothenic acid is ubiquitously available in foods and, thus, dietary deficiency is rare. Pantothenic acid deficiency symptoms include mood changes, sleep, neurological, cardiac and gastrointestinal disturbances, and skin abnormalities (Smith & Song, 1996; Trumbo, 2014). Also, malnourished prisoners of World War II in Asia reported numbness and burning sensation in their feet, and these symptoms were remedied by the administration of pantothenic acid supplementation (Davenport, Spaide, & Hodges, 1971). Also, the beneficial effect of pantothenic acid has been shown in the maintenance of healthy skin and in wound healing processes (Ebner, Heller, Rippke, & Tausch, 2002; Kelly, 2011; Shibata et al., 2013; Weimann & Hermann, 1999; Yang et al., 2014). Again, the direct involvement of hSMVT in pantothenate deficiency is not confirmed.

In contrast to the two essential vitamins that are transported by hSMVT, α-lipoic acid is synthesized in the body but is also readily absorbed in free form from the diet via the action of hSMVT (de Carvalho & Quick, 2011; Wang et al., 1999) after enzymatic cleavage of protein complexes by lipoamidase (Jiang & Cronan, 2005). Besides its role in the energy metabolism of the cell, there is emerging evidence that free lipoic acid is a potent biological antioxidant and anti-inflammatory agent that has been used in the treatment of several human diseases (for recent reviews, see Goraca et al., 2011; Maczurek et al., 2008). The therapeutic effects of lipoic acid have been attributed to the free radical scavenger properties of the lipoic acid (oxidized)/DHLA redox couple, which—due to the high redox potential of −320 mV—is also a very potent reducing agent. Both the oxidized and reduced forms of lipoic acid are antioxidants that possess anti-inflammatory and neuroprotective properties (Packer, Witt, & Tritschler, 1995; Smith, Shenvi, Widlansky, Suh, & Hagen, 2004). Several chemical features have been identified that make this redox couple an outstanding "metabolic antioxidant" (Roy & Packer, 1998): it can (i) scavenge reactive oxygen and nitrogen oxide species (ROS and RNOS, respectively), (ii) chelate toxic metals, (iii) regenerate endogenous antioxidants, (iv) is prone to destruction when quenching free radicals, (v) can pass the blood–brain barrier, and (vi) is amphipathic and may thus exert its antioxidant properties in lipophilic as well as in hydrophobic environments. ROS and RNOS are byproducts of oxidative metabolism reactions as partially reduced molecular oxygen, such as superoxide anion and hydrogen peroxide, or nitric oxide (NO), e.g., peroxynitrite, nitroxyl, and nitrogen dioxide. Whereas ROS and/or RNOS may function as signaling molecules in various signal transduction pathways (Lander, 1997), high levels of ROS and

RNOS have been considered the key players in oxidative stress to potentially damage macromolecules, such as lipids, DNA, and proteins (Finkel & Holbrook, 2000). Thus, ROS and RNOS have been implicated in the pathogenesis and progression of chronic inflammatory diseases, atherosclerosis, diabetes, cancer, as well as in aging (Halliwell, Gutteridge, & Cross, 1992). The harmful effect of free radicals in the cell can be counteracted by the action of the antioxidant defense system, including enzymatic scavengers, e.g., superoxide dismutase, catalase, and glutathione peroxidase, as well as nonenzymatic antioxidants, such as glutathione (GSH), thioredoxin, ascorbic acid (vitamin C), and lipoic acid (Nordberg & Arner, 2001). The unique chemical properties of the lipoic acid/DHLA redox couple make lipoic acid a potent and versatile "natural antioxidant drug" in the nonenzymatic antioxidant defense system as it can also regenerate other antioxidants, e.g., the glutathione/glutathione disulfide pair. Being able to cross the blood–brain barrier (Moini, Packer, & Saris, 2002; Packer, Tritschler, & Wessel, 1997), it is suggested to be used for detoxification of heavy metals (e.g., Hg^{2+}) localized in brain cells. There, it can mobilize bound heavy metals into the blood stream where another chelator, such as dimercaptosuccinic acid or methylsulfonylmethane, both of which cannot cross the blood–brain barrier in humans, is used to transfer mercury safely into the urine for excretion (Gregus, Stein, Varga, & Klaassen, 1992).

These diverse actions highlight that lipoic acid acts by multiple mechanisms, both physiologically and pharmacologically as safe and multipotent "natural drug" with strong antioxidant and anti-inflammatory properties useful in the treatment and prevention of human diseases attributed to free radical damage, and, thus, highlight the potential of hSMVT for the delivery of therapeutic compounds. In particular, the elucidation of key mechanistic features of the cloned hSMVT, together with its ubiquitous distribution in the human body, presents an excellent opportunity for its use as molecular target for drug delivery. While the therapeutic benefits of lipoic acid have been addressed above, new branches of research focus on the development of conjugates of the physiological substrates of hSMVT to increase the efficacy and bioavailability of drugs. For example, work in the Mitra lab has focused on the delivery of (pro)drugs conjugated to biotin (Vadlapudi, Vadlapatla, Earla, et al., 2013; Vadlapudi, Vadlapatla, Kwatra, et al., 2012; Vadlapudi, Vadlapatla, & Mitra, 2012), e.g., in the treatment against breast cancer (Vadlapudi, Vadlapatla, Pal, & Mitra, 2013).

11. CONCLUSION AND FUTURE DIRECTIONS

The cloning of hSMVT has led to a dramatic improvement in our understanding of the biology associated with the uptake of the two essential vitamins, vitamin B5 and vitamin H, α-lipoic acid, and I^- by this SLC5 member. Connecting the knowledge gathered in studies involving human and animal tissues prior to the cloning of the hSMVT gene with the detailed characterization of hSMVT expressed in suitable expression systems presents an unique opportunity to draw a coherent picture of the action of this protein in health and disease. Of course, further work is required to elucidate the interaction of Na^+ and the multiple substrates with the protein as this information will shine light on mechanistic details associated with the molecular determinants of hSMVT-mediated transport. Together with an improved understanding of the structure of hSMVT this knowledge will inform us about the action of this promising drug delivery tool in determining key functional characteristics in a structure-dynamic context.

ACKNOWLEDGMENTS

This work was supported in part by the National Science Foundation Grant MCB-1330730 (to M. Q.) and the National Institutes of Health Grant DA023694 (to L. S.). The authors thank Dr. S. Mazier for her assistance with the generation of the vSGLT-hSMVT alignment, and Ms. Estacio for critical reading of the chapter.

REFERENCES

Abramson, J., Smirnova, I., Kasho, V., Verner, G., Kaback, H. R., & Iwata, S. (2003). Structure and mechanism of the lactose permease of *Escherichia coli*. *Science*, *301*(5633), 610–615.

Abramson, J., & Wright, E. M. (2009). Structure and function of Na^+-symporters with inverted repeats. *Current Opinion in Structural Biology*, *19*, 425–432.

Acland, J. D., & Illman, O. (1959). Studies on iodide transport against a concentration gradient by the small intestine of the rat in vitro. *The Journal of Physiology*, *147*, 260–268.

Ahmad, P. M., & Ahmad, F. (1991). Mammalian pyruvate carboxylase: Effect of biotin on the synthesis and translocation of apo-enzyme into 3T3-L adipocyte mitochondria. *FASEB Journal*, *5*, 2482–2485.

Alberts, A. W., & Vagelos, P. R. (1966). Acyl carrier protein. 8. Studies of acyl carrier protein and coenzyme A in *Escherichia coli* pantothenate or betaalanine auxotrophs. *Journal of Biological Chemistry*, *241*, 5201–5204.

Balamurugan, K., Ortiz, A., & Said, H. M. (2003). Biotin uptake by human intestinal and liver epithelial cells: Role of the SMVT system. *American Journal of Physiology Gastrointestinal and Liver Physiology*, *285*, G73–G77.

Ball, G. F. M. (2005). *Vitamins in foods: Analysis, bioavailability, and stability*. Boca Raton: Saint Lucie Press/Taylor & Francis G.

Baumgartner, R., Suormala, T., Wick, H., Geisert, J., & Lehnert, W. (1982). Infantile multiple carboxylase deficiency: Evidence for normal intestinal absorption but renal loss of biotin. *Helvetica Paediatrica Acta*, *37*, 499–502.

Baur, B., & Baumgartner, E. R. (1993). Na^{+}-dependent biotin transport into brush-border membrane vesicles from human kidney cortex. *Pflügers Archiv*, *422*, 499–505.

Baur, B., & Baumgartner, E. R. (2000). Biotin and biocytin uptake into cultured primary calf brain microvessel endothelial cells of the blood-brain barrier. *Brain Research*, *858*, 348–355.

Beinlich, C. J., Naumovitz, R. D., Song, W. O., & Neely, J. R. (1990). Myocardial metabolism of pantothenic acid in chronically diabetic rats. *Journal of Molecular and Cellular Cardiology*, *22*, 323–332.

Bender, D. A. (1999). Optimum nutrition: Thiamin, biotin and pantothenate. *The Proceedings of the Nutrition Society*, *58*, 427–433.

Biewenga, G. P., Haenen, G. R. M. M., Groen, B. H., Biewenga, J. E., van Grodelle, R., & Bast, A. (1997). Combined non-enzymatic and enzymatic reduction favors bioactivation of racemic lipoic acid: An advantage of a racemic drug? *Chirality*, *9*, 362–366.

Bustamante, J., Lodge, J. K., Marcocci, L., Tritschler, H. J., Packer, L., & Rihn, B. H. (1998). Alpha-lipoic acid in liver metabolism and disease. *Free Radical Biology & Medicine*, *24*, 1023–1039.

Carreau, J. P. (1979). Biosynthesis of lipoic acid via unsaturated fatty acids. *Methods in Enzymology*, *62*, 152–158.

Castresana, J. (2000). Selection of conserved blocks from multiple alignments for their use in phylogenetic analysis. *Molecular Biology and Evolution*, *17*, 540–552.

Chaptal, V., Kwon, S., Sawaya, M. R., Guan, L., Kaback, H. R., & Abramson, J. (2011). Crystal structure of lactose permease in complex with an affinity inactivator yields unique insight into sugar recognition. *Proceedings of the National Academy of Sciences of the United States of America*, *108*, 9361–9366.

Chatterjee, N. S., Kumar, C. K., Ortiz, A., Rubin, S. A., & Said, H. M. (1999). Molecular mechanism of the intestinal biotin transport process. *The American Journal of Physiology*, *277*(4 Pt. 1), C605–C613.

Chevenet, F., Brun, C., Banuls, A. L., Jacq, B., & Christen, R. (2006). TreeDyn: Towards dynamic graphics and annotations for analyses of trees. *BMC Bioinformatics*, 7, 439.

Chirapu, S. R., Rotter, C. J., Miller, E. L., Varma, M. V., Dow, R. L., & Finn, M. G. (2013). High specificity in response of the sodium-dependent multivitamin transporter to derivatives of pantothenic acid. *Current Topics in Medicinal Chemistry*, *13*, 837–842.

Coady, M. J., Wallendorff, B., Bourgeois, F., & Lapointe, J. Y. (2010). Anionic leak currents through the Na^{+}/monocarboxylate cotransporter SMCT1. *American Journal of Physiology. Cell Physiology*, *298*, C124–C131.

Crane, R. K. (1962). Hypothesis for mechanism of intestinal active transport of sugars. *Federation Proceedings*, *21*, 891–895.

Crane, R. K. (1965). Na^{+} -dependent transport in the intestine and other animal tissues. *Federation Proceedings*, *24*(5), 1000–1006.

Dai, G., Levy, O., & Carrasco, N. (1996). Cloning and characterization of the thyroid iodide transporter. *Nature*, *379*(6564), 458–460.

Davenport, R. E., Spaide, J., & Hodges, R. E. (1971). An evaluation of various survival rations. *The American Journal of Clinical Nutrition*, *24*, 513–523.

de Carvalho, F. D., & Quick, M. (2011). Surprising substrate versatility in SLC5A6: Na^{+}-coupled I^{-} transport by the human Na^{+}/multivitamin transporter (hSMVT). *The Journal of Biological Chemistry*, *286*, 131–137.

DeFelice, L. J. (2004). Transporter structure and mechanism. *Trends in Neurosciences*, *27*, 352–359.

Dereeper, A., Guignon, V., Blanc, G., Audic, S., Buffet, S., Chevenet, F., et al. (2008). Phylogeny.fr: Robust phylogenetic analysis for the non-specialist. *Nucleic Acids Research*, *36*(Web Server issue), W465–W469.

Diez-Sampedro, A., & Barcelona, S. (2011). Sugar binding residue affects apparent Na^+ affinity and transport stoichiometry in mouse sodium/glucose cotransporter type 3B. *The Journal of Biological Chemistry*, *286*, 7975–7982.

Dohan, O., De la Vieja, A., Paroder, V., Riedel, C., Artani, M., Reed, M., et al. (2003). The sodium/iodide symporter (NIS): Characterization, regulation, and medical significance. *Endocrine Reviews*, *24*, 48–77.

Dohan, O., Portulano, C., Basquin, C., Reyna-Neyra, A., Amzel, L. M., & Carrasco, N. (2007). The Na^+/I^- symporter (NIS) mediates electroneutral active transport of the environmental pollutant perchlorate. *Proceedings of the National Academy of Sciences of the United States of America*, *104*, 20250–20255.

Ebner, F., Heller, A., Rippke, F., & Tausch, I. (2002). Topical use of dexpanthenol in skin disorders. *American Journal of Clinical Dermatology*, *3*, 427–433.

Edgar, R. C. (2004). MUSCLE: Multiple sequence alignment with high accuracy and high throughput. *Nucleic Acids Research*, *32*, 1792–1797.

Eskandari, S., Loo, D. D., Dai, G., Levy, O., Wright, E. M., & Carrasco, N. (1997). Thyroid Na^+/I^- symporter. Mechanism, stoichiometry, and specificity. *Journal of Biological Chemistry*, *272*, 27230–27238.

Faham, S., Watanabe, A., Besserer, G. M., Cascio, D., Specht, A., Hirayama, B. A., et al. (2008). The crystal structure of a sodium galactose transporter reveals mechanistic insights into Na^+/sugar symport. *Science*, *321*, 810–814.

Finkel, T., & Holbrook, N. J. (2000). Oxidants, oxidative stress and the biology of ageing. *Nature*, *408*, 239–247.

Forbes, G. M., & Forbes, A. (1997). Micronutrient status in patients receiving home parenteral nutrition. *Nutrition*, *13*, 941–944.

Forster, I., Hernando, N., Biber, J., & Murer, H. (1998). The voltage dependence of a cloned mammalian renal type II Na^+/P_i cotransporter (NaPi-2). *The Journal of General Physiology*, *112*, 1–18.

Garganta, C. L., & Wolf, B. (1990). Lipoamidase activity in human serum is due to biotinidase. *Clinica Chimica Acta*, *189*, 313–325.

Ghosal, A., & Said, H. M. (2011). Structure-function activity of the human sodium-dependent multivitamin transporter: Role of His115 and His254. *American Journal of Physiology Cell Physiology*, *300*, C97–C104.

Ghosal, A., & Said, H. M. (2012). Cys_{294} is essential for the function of the human sodium-dependent multivitamin transporter. *Biochimica et Biophysica Acta*, *1818*, 97–102.

Ghosal, A., Subramanian, V. S., & Said, H. M. (2011). Role of the putative *N*-glycosylation and PKC-phosphorylation sites of the human sodium-dependent multivitamin transporter (hSMVT) in function and regulation. *Biochimica et Biophysica Acta*, *1808*, 2073–2080.

Goraca, A., Huk-Kolega, H., Piechota, A., Kleniewska, P., Ciejka, E., & Skibska, B. (2011). Lipoic acid—Biological activity and therapeutic potential. *Pharmacological Reports*, *63*, 849–858.

Grafe, F., Wohlrab, W., Neubert, R. H., & Brandsch, M. (2003). Transport of biotin in human keratinocytes. *The Journal of Investigative Dermatology*, *120*, 428–433.

Grassl, S. M. (1992a). Human placental brush-border membrane Na^+-biotin cotransport. *The Journal of Biological Chemistry*, *267*, 17760–17765.

Grassl, S. M. (1992b). Human placental brush-border membrane Na^+-pantothenate cotransport. *The Journal of Biological Chemistry*, *267*, 22902–22906.

Gregus, Z., Stein, A. F., Varga, F., & Klaassen, C. D. (1992). Effect of lipoic acid on biliary excretion of glutathione and metals. *Toxicology and Applied Pharmacology*, *114*, 88–96.

Griffin, J. B., Stanley, J. S., & Zempleni, J. (2002). Synthesis of a rabbit polyclonal antibody to the human sodium-dependent multivitamin transporter. *International Journal for Vitamin and Nutrition Research*, *72*, 195–198.

Guan, L., & Kaback, H. R. (2006). Lessons from lactose permease. *Annual Review of Biophysics and Biomolecular Structure*, *35*, 67–91.

Guan, L., Mirza, O., Verner, G., Iwata, S., & Kaback, H. R. (2007). Structural determination of wild-type lactose permease. *Proceedings of the National Academy of Sciences of the United States of America*, *104*, 15294–15298.

Guastella, J., Nelson, N., Nelson, H., Czyzyk, L., Keynan, S., Miedel, M. C., et al. (1990). Cloning and expression of a rat brain GABA transporter. *Science*, *249*, 1303–1306.

Guindon, S., & Gascuel, O. (2003). A simple, fast, and accurate algorithm to estimate large phylogenies by maximum likelihood. *Systematic Biology*, *52*, 696–704.

Halliwell, B., Gutteridge, J. M., & Cross, C. E. (1992). Free radicals, antioxidants, and human disease: Where are we now? *The Journal of Laboratory and Clinical Medicine*, *119*, 598–620.

Harris, R. A., Joshi, M., Jeoung, N. H., & Obayashi, M. (2005). Overview of the molecular and biochemical basis of branched-chain amino acid catabolism. *Journal of Nutrition*, *135*, 1527S–1530S.

Hazama, A., Loo, D. D., & Wright, E. M. (1997). Presteady-state currents of the rabbit Na^+/glucose cotransporter (SGLT1). *The Journal of Membrane Biology*, *155*, 175–186.

Hediger, M. A., Coady, M. J., Ikeda, T. S., & Wright, E. M. (1987). Expression cloning and cDNA sequencing of the Na^+/glucose co-transporter. *Nature*, *330*, 379–381.

Hediger, M. A., Romero, M. F., Peng, J. B., Rolfs, A., Takanaga, H., & Bruford, E. A. (2004). The ABCs of solute carriers: Physiological, pathological and therapeutic implications of human membrane transport proteins. *Pflügers Archiv*, *447*, 465–468.

Hediger, M. A., Turk, E., & Wright, E. M. (1989). Homology of the human intestinal Na^+/glucose and *Escherichia coli* Na^+/proline cotransporters. *Proceedings of the National Academy of Sciences of the United States of America*, *86*, 5748–5752.

Higdon, J. (2003). *An evidence-based approach to vitamins and minerals: Health benefits and intake recommendations*. New York: Thieme Medical Publishers.

Hirayama, B. A., Loo, D. D., & Wright, E. M. (1994). Protons drive sugar transport through the Na^+/glucose cotransporter (SGLT1). *The Journal of Biological Chemistry*, *269*, 21407–21410.

Hirayama, B. A., Loo, D. D., & Wright, E. M. (1997). Cation effects on protein conformation and transport in the Na^+/glucose cotransporter. *The Journal of Biological Chemistry*, *272*, 2110–2115.

Ikeda, T. S., Hwang, E. S., Coady, M. J., Hirayama, B. A., Hediger, M. A., & Wright, E. M. (1989). Characterization of a Na^+/glucose cotransporter cloned from rabbit small intestine. *The Journal of Membrane Biology*, *110*, 87–95.

Jardetzky, O. (1966). Simple allosteric model for membrane pumps. *Nature*, *211*, 969–970.

Jiang, Y., & Cronan, J. E. (2005). Expression cloning and demonstration of *Enterococcus faecalis* lipoamidase (pyruvate dehydrogenase inactivase) as a Ser-Ser-Lys triad amidohydrolase. *The Journal of Biological Chemistry*, *280*, 2244–2256.

Jiang, X., Driessen, A. J., Feringa, B. L., & Kaback, H. R. (2013). The periplasmic cavity of LacY mutant Cys154→Gly: How open is open? *Biochemistry*, *52*, 6568–6574.

Jung, H. (2002). The sodium/substrate symporter family: Structural and functional features. *FEBS Letters*, *529*(1), 73–77.

Jung, H., Rubenhagen, R., Tebbe, S., Leifker, K., Tholema, N., Quick, M., et al. (1998). Topology of the Na^+/proline transporter of *Escherichia coli*. *The Journal of Biological Chemistry*, *273*(41), 26400–26407.

Jung, H., Tebbe, S., Schmid, R., & Jung, K. (1998). Unidirectional reconstitution and characterization of purified Na^+/proline transporter of *Escherichia coli*. *Biochemistry*, *37*, 11083–11088.

Kaback, H. R., Sahin-Toth, M., & Weinglass, A. B. (2001). The kamikaze approach to membrane transport. *Nature Reviews Molecular Cell Biology*, *2*, 610–620.

Kaback, H. R., Smirnova, I., Kasho, V., Nie, Y., & Zhou, Y. (2011). The alternating access transport mechanism in LacY. *The Journal of Membrane Biology*, *239*, 85–93.

Kaback, H. R., & Wu, J. (1997). From membrane to molecule to the third amino acid from the left with a membrane transport protein. *Quarterly Reviews of Biophysics*, *30*, 333–364.

Karlin, A., & Akabas, M. H. (1998). Substituted-cysteine accessibility method. *Methods in Enzymology*, *293*, 123–145.

Kazmier, K., Sharma, S., Quick, M., Islam, S. M., Roux, B., Weinstein, H., et al. (2014). Conformational dynamics of ligand-dependent alternating access in LeuT. *Nature Structural & Molecular Biology*, *21*, 472–479.

Kelly, G. S. (2011). Pantothenic acid. Monograph. *Alternative Medicine Review*, *16*, 263–274.

Keynan, S., & Kanner, B. I. (1988). gamma-Aminobutyric acid transport in reconstituted preparations from rat brain: Coupled sodium and chloride fluxes. *Biochemistry*, *27*, 12–17.

Kilty, J. E., Lorang, D., & Amara, S. G. (1991). Cloning and expression of a cocaine-sensitive rat dopamine transporter. *Science*, *254*, 578–579.

Knowles, J. R. (1989). The mechanism of biotin-dependent enzymes. *Annual Review of Biochemistry*, *58*, 195–221.

Kopito, R. R., & Lodish, H. F. (1985). Primary structure and transmembrane orientation of the murine anion exchange protein. *Nature*, *316*, 234–238.

Krause, K. H., Bonjour, J. P., Berlit, P., Kynast, G., Schmidt-Gayk, H., & Schellenberg, B. (1988). Effect of long-term treatment with antiepileptic drugs on the vitamin status. *Drug-Nutrient Interactions*, *5*, 317–343.

Krishnamurthy, H., Piscitelli, C. L., & Gouaux, E. (2009). Unlocking the molecular secrets of sodium-coupled transporters. *Nature*, *459*, 347–355.

Kumar, H., Kasho, V., Smirnova, I., Finer-Moore, J. S., Kaback, H. R., & Stroud, R. M. (2014). Structure of sugar-bound LacY. *Proceedings of the National Academy of Sciences of the United States of America*, *111*, 1784–1788.

Kyte, J., & Doolittle, R. F. (1982). A simple method for displaying the hydropathic character of a protein. *Journal of Molecular Biology*, *157*, 105–132.

Lander, H. M. (1997). An essential role for free radicals and derived species in signal transduction. *FASEB Journal*, *11*, 118–124.

Leonardi, R., Zhang, Y. M., Rock, C. O., & Jackowski, S. (2005). Coenzyme A: Back in action. *Progress in Lipid Research*, *44*, 125–153.

Levy, O., De la Vieja, A., Ginter, C. S., Riedel, C., Dai, G., & Carrasco, N. (1998). *N*-linked glycosylation of the thyroid Na^+/I^- symporter (NIS). Implications for its secondary structure model. *Journal of Biological Chemistry*, *273*, 22657–22663.

Loo, D. D., Eskandari, S., Boorer, K. J., Sarkar, H. K., & Wright, E. M. (2000). Role of Cl^- in electrogenic Na^+-coupled cotransporters GAT1 and SGLT1. *The Journal of Biological Chemistry*, *275*, 37414–37422.

Loo, D. D., Hazama, A., Supplisson, S., Turk, E., & Wright, E. M. (1993). Relaxation kinetics of the Na^+/glucose cotransporter. *Proceedings of the National Academy of Sciences of the United States of America*, *90*, 5767–5771.

Loo, D. D., Hirayama, B. A., Cha, A., Bezanilla, F., & Wright, E. M. (2005). Perturbation analysis of the voltage-sensitive conformational changes of the Na^+/glucose cotransporter. *The Journal of General Physiology*, *125*, 13–36.

Ma, T. Y., Dyer, D. L., & Said, H. M. (1994). Human intestinal cell line Caco-2: A useful model for studying cellular and molecular regulation of biotin uptake. *Biochimica et Biophysica Acta*, *1189*, 81–88.

Mackenzie, B., Harper, A. A., Taylor, P. M., & Rennie, M. J. (1994). Na^+/amino acid coupling stoichiometry of rheogenic system $B^{0,+}$ transport in *Xenopus* oocytes is variable. *Pflügers Archiv*, *426*, 121–128.

Mackenzie, B., Loo, D. D., & Wright, E. M. (1998). Relationships between Na^+/glucose cotransporter (SGLT1) currents and fluxes. *The Journal of Membrane Biology, 162*, 101–106.

Maczurek, A., Hager, K., Kenklies, M., Sharman, M., Martins, R., Engel, J., et al. (2008). Lipoic acid as an anti-inflammatory and neuroprotective treatment for Alzheimer's disease. *Advanced Drug Delivery Reviews, 60*, 1463–1470.

Mazier, S., Quick, M., & Shi, L. (2011). Conserved tyrosine in the first transmembrane segment of solute:sodium symporters is involved in Na^+-coupled substrate co-transport. *The Journal of Biological Chemistry, 286*, 29347–29355.

McMahon, R. J. (2002). Biotin in metabolism and molecular biology. *Annual Review of Nutrition, 22*, 221–239.

Miller, S. J. (1989). Nutritional deficiency and the skin. *Journal of the American Academy of Dermatology, 21*, 1–30.

Milne, J. L., Wu, X., Borgnia, M. J., Lengyel, J. S., Brooks, B. R., Shi, D., et al. (2006). Molecular structure of a 9-MDa icosahedral pyruvate dehydrogenase subcomplex containing the E2 and E3 enzymes using cryoelectron microscopy. *The Journal of Biological Chemistry, 281*, 4364–4370.

Mock, D. M. (1991). Skin manifestations of biotin deficiency. *Seminars in Dermatology, 10*, 296–302.

Mock, D. M. (1999). *Biotin* (9th ed.). Baltimore: Lippincott Williams & Wilkins.

Mock, D. M., Quirk, J. G., & Mock, N. I. (2002). Marginal biotin deficiency during normal pregnancy. *The American Journal of Clinical Nutrition, 75*, 295–299.

Moini, H., Packer, L., & Saris, N.-E. L. (2002). Antioxidant and prooxidant activities of [alpha]-lipoic acid and dihydrolipoic acid. *Toxicology and Applied Pharmacology, 182*, 84–90.

Mueckler, M., Caruso, C., Baldwin, S. A., Panico, M., Blench, I., Morris, H. R., et al. (1985). Sequence and structure of a human glucose transporter. *Science, 229*, 941–945.

Munnich, A., Saudubray, J. M., Carre, G., Coude, F. X., Ogier, H., Charpentier, C., et al. (1981). Defective biotin absorption in multiple carboxylase deficiency. *Lancet, 2*, 263.

Nabokina, S. M., Subramanian, V. S., & Said, H. M. (2010). Association of PDZ-containing protein PDZD11 with the human sodium-dependent multivitamin transporter, hSMVT. *American Journal of Physiology Gastrointestinal and Liver Physiology, 300*, G561–G567.

Nakao, T., Yamato, I., & Anraku, Y. (1987). Nucleotide sequence of *putP*, the proline carrier gene of *Escherichia coli* K12. *Molecular & General Genetics, 208*, 70–75.

Nicola, J. P., Basquin, C., Portulano, C., Reyna-Neyra, A., Paroder, M., & Carrasco, N. (2009). The Na^+/I^- symporter mediates active iodide uptake in the intestine. *American Journal of Physiology Cell Physiology, 296*, C654–C662.

Nilsson, L., & Kagedal, B. (1993). Co-purification of human serum lipoamidase and biotinidase: Evidence that the two enzyme activities are due to the same enzyme protein. *The Biochemical Journal, 291*, 545–551.

Nilsson, L., & Ronge, E. (1992). Lipoamidase and biotinidase deficiency: Evidence that lipoamidase and biotinidase are the same enzyme in human serum. *European Journal of Clinical Chemistry and Clinical Biochemistry, 30*, 119–126.

Nisenson, A. (1957). Seborrheic dermatitis of infants and Leiners's disease: A biotin deficiency. *The Journal of Pediatrics, 51*, 537–548.

Nordberg, J., & Arner, E. S. (2001). Reactive oxygen species, antioxidants, and the mammalian thioredoxin system. *Free Radical Biology & Medicine, 31*, 1287–1312.

Packer, L., Tritschler, H. J., & Wessel, K. (1997). Neuroprotection by the metabolic antioxidant alpha-lipoic acid. *Free Radical Biology & Medicine, 22*, 359–378.

Packer, L., Witt, E. H., & Tritschler, H. J. (1995). alpha-Lipoic acid as a biological antioxidant. *Free Radical Biology & Medicine, 19*, 227–250.

Panayotova-Heiermann, M., Loo, D. D., & Wright, E. M. (1995). Kinetics of steady-state currents and charge movements associated with the rat Na^+/glucose cotransporter. *The Journal of Biological Chemistry*, *270*, 27099–27105.

Parent, L., Supplisson, S., Loo, D. D., & Wright, E. M. (1992a). Electrogenic properties of the cloned Na^+/glucose cotransporter: I. Voltage-clamp studies. *Journal of Membrane Biology*, *125*, 49–62.

Parent, L., Supplisson, S., Loo, D. D., & Wright, E. M. (1992b). Electrogenic properties of the cloned Na^+/glucose cotransporter: II. A transport model under nonrapid equilibrium conditions. *Journal of Membrane Biology*, *125*, 63–79.

Pastan, I. (1957). Absorption and secretion of iodide by the intestine of the rat. *Endocrinology*, *61*, 93–97.

Patlak, C. S. (1957). Contributions of the theory of active transport: II The gate type non-carrier mechanism and generalizations concerning tracer flow, efficiency, and measurement of energy expenditure. *The Bulletin of Mathematical Biophysics*, *19*, 209–235.

Prasad, P. D., Ramamoorthy, S., Leibach, F. H., & Ganapathy, V. (1997). Characterization of a sodium-dependent vitamin transporter mediating the uptake of pantothenate, biotin and lipoate in human placental choriocarcinoma cells. *Placenta*, *18*, 527–533.

Prasad, P. D., Wang, H., Huang, W., Fei, Y. J., Leibach, F. H., Devoe, L. D., et al. (1999). Molecular and functional characterization of the intestinal Na^+-dependent multivitamin transporter. *Archives of Biochemistry and Biophysics*, *366*, 95–106.

Prasad, P. D., Wang, H., Kekuda, R., Fujita, T., Fei, Y. J., Devoe, L. D., et al. (1998). Cloning and functional expression of a cDNA encoding a mammalian sodium-dependent vitamin transporter mediating the uptake of pantothenate, biotin, and lipoate. *The Journal of Biological Chemistry*, *273*, 7501–7506.

Quick, M., & Jung, H. (1997). Aspartate 55 in the Na^+/proline permease of *Escherichia coli* is essential for Na^+-coupled proline uptake. *Biochemistry*, *36*, 4631–4636.

Quick, M., & Jung, H. (1998). A conserved aspartate residue, Asp187, is important for Na^+-dependent proline binding and transport by the Na^+/proline transporter of *Escherichia coli*. *Biochemistry*, *37*, 13800–13806.

Quick, M., Loo, D. D., & Wright, E. M. (2001). Neutralization of a conserved amino acid residue in the human Na^+/glucose transporter (hSGLT1) generates a glucose-gated H^+ channel. *The Journal of Biological Chemistry*, *276*, 1728–1734.

Quick, M., Stolting, S., & Jung, H. (1999). Role of conserved Arg40 and Arg117 in the Na^+/proline transporter of *Escherichia coli*. *Biochemistry*, *38*, 13523–13529.

Quick, M., Tebbe, S., & Jung, H. (1996). Ser57 in the Na^+/proline permease of *Escherichia coli* is critical for high-affinity proline uptake. *European Journal of Biochemistry*, *239*, 732–736.

Raba, M., Dunkel, S., Hilger, D., Lipiszko, K., Polyhach, Y., Jeschke, G., et al. (2014). Extracellular loop 4 of the proline transporter PutP controls the periplasmic entrance to ligand binding sites. *Structure*, *22*, 769–780.

Rath, A., Glibowicka, M., Nadeau, V. G., Chen, G., & Deber, C. M. (2009). Detergent binding explains anomalous SDS-PAGE migration of membrane proteins. *Proceedings of the National Academy of Sciences of the United States of America*, *106*, 1760–1765.

Reed, L. J. (2001). A trail of research from lipoic acid to alpha-keto acid dehydrogenase complexes. *The Journal of Biological Chemistry*, *276*, 38329–38336.

Reidling, J. C., Nabokina, S. M., & Said, H. M. (2007). Molecular mechanisms involved in the adaptive regulation of human intestinal biotin uptake: A study of the hSMVT system. *American Journal of Physiology Gastrointestinal and Liver Physiology*, *292*, G275–G281.

Reidling, J. C., & Said, H. M. (2007). Regulation of the human biotin transporter hSMVT promoter by KLF-4 and AP-2: Confirmation of promoter activity *in vivo*. *American Journal of Physiology. Cell Physiology*, *292*, C1305–C1312.

Ressl, S., Terwisscha van Scheltinga, A. C., Vonrhein, C., Ott, V., & Ziegler, C. (2009). Molecular basis of transport and regulation in the Na^+/betaine symporter BetP. *Nature, 458*, 47–52.

Roy, S., & Packer, L. (1998). Redox regulation of cell functions by alpha-lipoate: Biochemical and molecular aspects. *Biofactors, 8*, 17–21.

Rudnick, G. (Ed.), (2002). *Mechanisms of biogenic amine neurotransmitter transporters* (2nd ed.). Totowa, New Jersey: Humana Press Inc.

Said, H. M. (1999). Cellular uptake of biotin: Mechanisms and regulation. *The Journal of Nutrition, 129*, 490S–493S.

Said, H. M. (2004). Recent advances in carrier-mediated intestinal absorption of water-soluble vitamins. *Annual Review of Physiology, 66*, 419–446.

Said, H. M. (2009). Cell and molecular aspects of human intestinal biotin absorption. *The Journal of Nutrition, 139*, 158–162.

Said, H. M. (2012). Biotin: Biochemical, physiological and clinical aspects. *Subcellular Biochemistry, 56*, 1–19.

Said, H. M., Korchid, S., Horne, D. W., & Howard, M. (1990). Transport of biotin in basolateral membrane vesicles of rat liver. *The American Journal of Physiology, 259*, G865–G872.

Said, H. M., Mock, D. M., & Collins, J. C. (1989). Regulation of intestinal biotin transport in the rat: Effect of biotin deficiency and supplementation. *The American Journal of Physiology, 256*, G306–G311.

Said, H. M., & Mohammed, Z. M. (2006). Intestinal absorption of water-soluble vitamins: An update. *Current Opinion in Gastroenterology, 22*, 140–146.

Said, H. M., Ortiz, A., McCloud, E., Dyer, D., Moyer, M. P., & Rubin, S. (1998). Biotin uptake by human colonic epithelial NCM460 cells: A carrier-mediated process shared with pantothenic acid. *The American Journal of Physiology, 275*, C1365–C1371.

Said, H. M., & Redah, R. (1988). Ontogenesis of the intestinal transport of biotin in the rat. *Gastroenterology, 94*, 68–72.

Said, H. M., Redha, R., & Nylander, W. (1987). A carrier-mediated, Na^+ gradient-dependent transport for biotin in human intestinal brush-border membrane vesicles. *The American Journal of Physiology, 253*, G631–G636.

Said, H. M., Redha, R., & Nylander, W. (1988a). Biotin transport in basolateral membrane vesicles of human intestine. *Gastroenterology, 94*, 1157–1163.

Said, H. M., Redha, R., & Nylander, W. (1988b). Biotin transport in the human intestine: Site of maximum transport and effect of pH. *Gastroenterology, 95*, 1312–1317.

Saier, M. H., Jr. (1999). A functional-phylogenetic system for the classification of transport proteins. *Journal of Cellular Biochemistry, 32–33*, 84–94.

Saier, M. H., Jr. (2000). A functional-phylogenetic classification system for transmembrane solute transporters. *Microbiology and Molecular Biology Reviews, 64*, 354–411.

Saier, M. H., Jr., & Ren, Q. (2006). The bioinformatic study of transmembrane molecular transport. *Journal of Molecular Microbiology and Biotechnology, 11*, 289–290.

Schulze, S., Koster, S., Geldmacher, U., Terwisscha van Scheltinga, A. C., & Kuhlbrandt, W. (2010). Structural basis of Na^+-independent and cooperative substrate/product antiport in CaiT. *Nature, 467*, 233–236.

Segel, I. H. (1975). *Enzyme kinetics: Behavior and analysis of rapid equilibrium and steady-state enzyme systems.* New York: John Wiley & Sons.

Serdiuk, T., Madej, M. G., Sugihara, J., Kawamura, S., Mari, S. A., Kaback, H. R., et al. (2014). Substrate-induced changes in the structural properties of LacY. *Proceedings of the National Academy of Sciences of the United States of America, 111*, E1571–E1580.

Shaffer, P. L., Goehring, A., Shankaranarayanan, A., & Gouaux, E. (2009). Structure and mechanism of a Na^+-independent amino acid transporter. *Science, 325*, 1010–1014.

Shi, L., Quick, M., Zhao, Y., Weinstein, H., & Javitch, J. A. (2008). The mechanism of a neurotransmitter:sodium symporter—Inward release of Na^+ and substrate is triggered by substrate in a second binding site. *Molecular Cell, 30*, 667–677.

Shibata, K., Fukuwatari, T., Higashiyama, S., Sugita, C., Azumano, I., & Onda, M. (2013). Pantothenic acid refeeding diminishes the liver, perinephrical fats, and plasma fats accumulated by pantothenic acid deficiency and/or ethanol consumption. *Nutrition, 29*, 796–801.

Shimamura, T., Weyand, S., Beckstein, O., Rutherford, N. G., Hadden, J. M., Sharples, D., et al. (2010). Molecular basis of alternating access membrane transport by the sodium-hydantoin transporter Mhp1. *Science, 328*, 470–473.

Skrabanek, L., Campagne, F., & Weinstein, H. (2003). Building protein diagrams on the web with the residue-based diagram editor RbDe. *Nucleic Acids Research, 31*, 3856–3858.

Smirnova, I., Kasho, V., Choe, J. Y., Altenbach, C., Hubbell, W. L., & Kaback, H. R. (2007). Sugar binding induces an outward facing conformation of LacY. *Proceedings of the National Academy of Sciences of the United States of America, 104*, 16504–16509.

Smith, A. R., Shenvi, S. V., Widlansky, M., Suh, J. H., & Hagen, T. M. (2004). Lipoic acid as a potential therapy for chronic diseases associated with oxidative stress. *Current Medicinal Chemistry, 11*, 1135–1146.

Smith, C. M., & Song, W. O. (1996). Comparative nutrition of pantothenic acid. *Journal of Nutritional Biochemistry*, 7, 312–321.

Sonders, M. S., Quick, M., & Javitch, J. A. (2005). How did the neurotransmitter cross the bilayer? A closer view. *Current Opinion in Neurobiology, 15*, 296–304.

Spector, R., & Mock, D. (1987). Biotin transport through the blood-brain barrier. *Journal of Neurochemistry, 48*, 400–404.

Spencer, R. P., & Brody, K. R. (1964). Biotin transport by small intestine of rat, hamster, and other species. *The American Journal of Physiology, 206*, 653–657.

Subramanian, V. S., Marchant, J. S., Boulware, M. J., Ma, T. Y., & Said, H. M. (2009). Membrane targeting and intracellular trafficking of the human sodium-dependent multivitamin transporter in polarized epithelial cells. *American Journal of Physiology Cell Physiology, 296*, C663–C671.

Sweetman, L., & Nyhan, W. L. (1986). Inheritable biotin-treatable disorders and associated phenomena. *Annual Review of Nutrition, 6*, 317–343.

Tahiliani, A. G., & Beinlich, C. J. (1991). Pantothenic acid in health and disease. *Vitamins and Hormones, 46*, 165–228.

Tang, L., Bai, L., Wang, W. H., & Jiang, T. (2010). Crystal structure of the carnitine transporter and insights into the antiport mechanism. *Nature Structural & Molecular Biology, 17*, 492–496.

Thoene, J. G., Lemons, R., & Baker, H. (1983). Impaired intestinal absorption of biotin in juvenile multiple carboxylase deficiency. *The New England Journal of Medicine, 308*, 639–642.

Trumbo, P. R. (2014). Pantothenic acid. In A. C. Ross, B. Caballero, R. J. Cousins, K. L. Tucker, & T. R. Ziegler (Eds.), *Modern nutrition in health and disease* (11th ed., pp. 351–357). Philadelphia, PA: Lippincott Williams & Wilkins.

Tsao, C. Y., & Kien, C. L. (2002). Complete biotinidase deficiency presenting as reversible progressive ataxia and sensorineural deafness. *Journal of Child Neurology, 17*, 146.

Turk, E., Kerner, C. J., Lostao, M. P., & Wright, E. M. (1996). Membrane topology of the human Na^+/glucose cotransporter SGLT1. *The Journal of Biological Chemistry, 271*, 1925–1934.

Turk, E., & Wright, E. M. (1997). Membrane topology motifs in the SGLT cotransporter family. *The Journal of Membrane Biology, 159*, 1–20.

Vadlapudi, A. D., Vadlapatla, R. K., Earla, R., Sirimulla, S., Bailey, J. B., Pal, D., et al. (2013). Novel biotinylated lipid prodrugs of acyclovir for the treatment of herpetic

keratitis (HK): Transporter recognition, tissue stability and antiviral activity. *Pharmaceutical Research*, *30*, 2063–2076.
Vadlapudi, A. D., Vadlapatla, R. K., Kwatra, D., Earla, R., Samanta, S. K., Pal, D., et al. (2012). Targeted lipid based drug conjugates: A novel strategy for drug delivery. *International Journal of Pharmaceutics*, *434*, 315–324.
Vadlapudi, A. D., Vadlapatla, R. K., & Mitra, A. K. (2012). Sodium dependent multivitamin transporter (SMVT): A potential target for drug delivery. *Current Drug Targets*, *13*, 994–1003.
Vadlapudi, A. D., Vadlapatla, R. K., Pal, D., & Mitra, A. K. (2013). Biotin uptake by T47D breast cancer cells: Functional and molecular evidence of sodium-dependent multivitamin transporter (SMVT). *International Journal of Pharmaceutics*, *441*, 535–543.
Vallari, D. S., & Rock, C. O. (1985). Pantothenate transport in *Escherichia coli*. *Journal of Bacteriology*, *162*, 1156–1161.
Vandenberg, R. J., Arriza, J. L., Amara, S. G., & Kavanaugh, M. P. (1995). Constitutive ion fluxes and substrate binding domains of human glutamate transporters. *The Journal of Biological Chemistry*, *270*, 17668–17671.
Wang, H., Huang, W., Fei, Y. J., Xia, H., Yang-Feng, T. L., Leibach, F. H., et al. (1999). Human placental Na^+-dependent multivitamin transporter. Cloning, functional expression, gene structure, and chromosomal localization. *Journal of Biological Chemistry*, *274*, 14875–14883.
Watanabe, T. (1983). Teratogenic effects of biotin deficiency in mice. *The Journal of Nutrition*, *113*, 574–581.
Watanabe, T. (1996). Morphological and biochemical effects of excessive amounts of biotin on embryonic development in mice. *Experientia*, *52*, 149–154.
Watanabe, A., Choe, S., Chaptal, V., Rosenberg, J. M., Wright, E. M., Grabe, M., et al. (2010). The mechanism of sodium and substrate release from the binding pocket of vSGLT. *Nature*, *468*, 988–991.
Weimann, B. I., & Hermann, D. (1999). Studies on wound healing: Effects of calcium D-pantothenate on the migration, proliferation and protein synthesis of human dermal fibroblasts in culture. *International Journal for Vitamin and Nutrition Research*, *69*, 113–119.
Weyand, S., Shimamura, T., Beckstein, O., Sansom, M. S., Iwata, S., Henderson, P. J., et al. (2011). The alternating access mechanism of transport as observed in the sodium-hydantoin transporter Mhp1. *Journal of Synchrotron Radiation*, *18*, 20–23.
Weyand, S., Shimamura, T., Yajima, S., Suzuki, S., Mirza, O., Krusong, K., et al. (2008). Structure and molecular mechanism of a nucleobase-cation-symport-1 family transporter. *Science*, *322*, 709–713.
Whitehead, C. C. (1985). Assessment of biotin deficiency in animals. *Annals of the New York Academy of Sciences*, *447*, 86–96.
Wolf, B. (2001). Disorders of biotin metabolism. In C. R. Scriver, A. L. Beaudet, W. S. Sly, D. Valle, B. Childs, K. W. Kinzler, & B. Vogelstein (Eds.), *The metabolic and molecular bases of inherited diseases: Vol. 8*. (pp. 3935–3962). New York: McGraw-Hill Medical Publishing Division.
Wolf, B., Heard, G. S., McVoy, J. R., & Raetz, H. M. (1984). Biotinidase deficiency: The possible role of biotinidase in the processing of dietary protein-bound biotin. *Journal of Inherited Metabolic Disease*, 7(Suppl. 2), 121–122.
Wong, F. H., Chen, J. S., Reddy, V., Day, J. L., Shlykov, M. A., Wakabayashi, S. T., et al. (2012). The amino acid-polyamine-organocation superfamily. *Journal of Molecular Microbiology and Biotechnology*, *22*, 105–113.
Wright, E. M., Loo, D. D., & Hirayama, B. A. (2011). Biology of human sodium glucose transporters. *Physiological Reviews*, *91*, 733–794.
Wright, E. M., Loo, D. D., Hirayama, B. A., & Turk, E. (2004). Surprising versatility of Na^+-glucose cotransporters: SLC5. *Physiology (Bethesda)*, *19*, 370–376.

Wright, E. M., Loo, D. D., Panayotova-Heiermann, M., Lostao, M. P., Hirayama, B. H., Mackenzie, B., et al. (1994). 'Active' sugar transport in eukaryotes. *The Journal of Experimental Biology*, *196*, 197–212.

Wright, E. M., Loo, D. D., Turk, E., & Hirayama, B. A. (1996). Sodium cotransporters. *Current Opinion in Cell Biology*, *8*, 468–473.

Wright, E. M., & Turk, E. (2004). The sodium/glucose cotransport family SLC5. *Pflügers Archiv*, *447*, 510–518.

Yamashita, A., Singh, S. K., Kawate, T., Jin, Y., & Gouaux, E. (2005). Crystal structure of a bacterial homologue of Na^+/Cl-dependent neurotransmitter transporters. *Nature*, *437*, 215–223.

Yang, M., Moclair, B., Hatcher, V., Kaminetsky, J., Mekas, M., Chapas, A., et al. (2014). A randomized, double-blind, placebo-controlled study of a novel pantothenic acid-based dietary supplement in subjects with mild to moderate facial acne. *Dermatology and Therapy*, *4*, 93–101.

Yoshikawa, K., Hayakawa, K., Katsumata, N., Tanaka, T., Kimura, T., & Yamauchi, K. (1996). High-performance liquid chromatographic determination of lipoamidase (lipoyl-X hydrolase) activity with a novel substrate, lipoyl-6-aminoquinoline. *Journal of Chromatography B, Biomedical Applications*, *679*, 41–47.

Zempleni, J. (2005). Uptake, localization, and noncarboxylase roles of biotin. *Annual Review of Nutrition*, *25*, 175–196.

Zempleni, J., & Kuroishi, T. (2012). Biotin. *Advances in Nutrition*, *3*, 213–214.

Zempleni, J., & Mock, D. M. (1998). Uptake and metabolism of biotin by human peripheral blood mononuclear cells. *The American Journal of Physiology*, *275*, C382–C388.

Zempleni, J., & Mock, D. M. (1999). Biotin biochemistry and human requirements. *The Journal of Nutritional Biochemistry*, *10*, 128–138.

Zempleni, J., & Mock, D. M. (2000). Marginal biotin deficiency is teratogenic. *Proceedings of the Society for Experimental Biology and Medicine*, *223*, 14–21.

Zempleni, J., Wijeratne, S. S., & Hassan, Y. I. (2009). Biotin. *Biofactors*, *35*, 36–46.

Zhao, X., Miller, J. R., Jiang, Y., Marletta, M. A., & Cronan, J. E. (2003). Assembly of the covalent linkage between lipoic acid and its cognate enzymes. *Chemistry & Biology*, *10*, 1293–1302.

Zhao, Y., Terry, D. S., Shi, L., Quick, M., Weinstein, H., Blanchard, S. C., et al. (2011). Substrate-modulated gating dynamics in a Na^+-coupled neurotransmitter transporter homologue. *Nature*, *474*, 109–113.

Zhou, X., Levin, E. J., Pan, Y., McCoy, J. G., Sharma, R., Kloss, B., et al. (2014). Structural basis of the alternating-access mechanism in a bile acid transporter. *Nature*, *505*, 569–573.

Zomot, E., Bendahan, A., Quick, M., Zhao, Y., Javitch, J. A., & Kanner, B. I. (2007). Mechanism of chloride interaction with neurotransmitter:sodium symporters. *Nature*, *449*, 726–730.

CHAPTER FOUR

Regulation of αENaC Transcription

Lihe Chen*[,†], Xi Zhang[†], Wenzheng Zhang*[,†,1]

*Graduate School of Biomedical Sciences, The University of Texas Health Science Center at Houston, Houston, Texas, USA

[†]Division of Renal Diseases and Hypertension, Department of Internal Medicine, University of Texas Medical School at Houston, Houston, Texas, USA

[1]Corresponding author: e-mail address: wenzheng.zhang@uth.tmc.edu

Contents

Abstract

Aldosterone is a major regulator of Na^+ absorption and acts primarily by controlling the epithelial Na^+ channel (ENaC) function at multiple levels including transcription. ENaC consists of α, β, and γ subunits. In the classical model, aldosterone enhances

Vitamins and Hormones, Volume 98
ISSN 0083-6729
http://dx.doi.org/10.1016/bs.vh.2014.12.004

transcription primarily by activating mineralocorticoid receptor (MR). However, how aldosterone induces chromatin alternation and thus leads to gene activation or repression remains largely unknown. Emerging evidence suggests that Dot1a–Af9 complex plays an important role in repression of α*ENaC* by directly binding and modulating targeted histone H3 K79 hypermethylation at the specific subregions of *αENaC* promoter. Aldosterone impairs Dot1a–Af9 formation by decreasing expression of Dot1a and Af9 and by inducing Sgk1, which, in turn, phosphorylates Af9 at S435 to weaken Dot1a–Af9 interaction. MR counterbalances Dot1a–Af9 action by competing with Dot1a for binding Af9. Af17 derepresses *αENaC* by competitively interacting with Dot1a and facilitating Dot1a nuclear export. Consistently, $MR^{-/-}$ mice have impaired ENaC expression at day 5 after birth, which may contribute to progressive development of pseudohypoaldosteronism type 1 in a later stage. $Af17^{-/-}$ mice have decreased ENaC expression, renal Na^+ retention, and blood pressure. In contrast, $Dot1l^{AC}$ mice have increased αENaC expression, despite a 20% reduction of the principal cells. This chapter reviews these findings linking aldosterone action to ENaC transcription through chromatin modification. Future direction toward the understanding the role of Dot1a–Af9 complex beyond ENaC regulation, in particular, in renal fibrosis is also briefly discussed.

1. INTRODUCTION

1.1 Aldosterone is a ligand for the mineralocorticoid receptor and glucocorticoid receptor

Members of steroid hormone family include aldosterone, glucocorticoid, estrogen, progesterone, and androgen. The hormones induce genomic and nongenomic effects. In the classical mode of action, steroid hormones diffuse into cytoplasm, where they bind their intracellular receptors. The hormone–receptor complexes then move into the nucleus, bind the hormone response elements (HREs), and stimulate or inhibit transcription of their target genes. This ligand-dependent modulation of transcription by the ligand–receptor complex has been termed a "genomic" effect and is sensitive to inhibitors of transcription (actinomycin D) and translation (cycloheximide). In addition to the classical mode of receptor action, steroid hormones might also act through alternative, "nongenomic" (also called nontranscriptional or nonclassical) pathways, particularly in cardiovascular systems. These actions are rapid (<1 min), independent of transcription and translation, and presumably mediated by distinct plasma membrane/cytosolic receptors that have not been identified yet (Losel, Feuring, & Wehling, 2002). Hence, this mode of hormone action is referred as nongenomic (Alangari, 2010).

Aldosterone is produced in the *zona glomerulosa* of the adrenal cortex. The hormone serves as a ligand for two distinct but similar types of nuclear receptors: mineralocorticoid receptor (MR) and glucocorticoid receptor (GR). Both MR and GR, like other members of the steroid hormone receptor family, act as ligand-inducible transcription factors. Upon binding their ligand aldosterone, the receptors activate or repress the transcription of target genes by directly binding as monomers, homodimers, or heterodimers to HREs and produce long-lasting physiological effects of aldosterone stimulation. HREs are normally presented in the 5′-flanking region of target genes.

Unique receptor complexes may target distinct *cis*-acting elements (Stockand, 2002). These modes of steroid hormone action involve the *trans*-activation and *trans*-repression via interaction with cognate DNA-binding sites, such as HRE. In contrast to direct *trans*-activation and repression, the steroid receptor may regulate gene expression by protein–protein interaction between the receptor and other *trans*-acting factors. In these cases, the receptor does not need to physically bind DNA (Stockand, 2002). Obviously, before interaction of the nuclear receptors with the cognate HRE, the ligand–receptor complex must have the accessibility to the DNA, which is compacted into the chromatin.

Interestingly, many coregulators of steroid hormone receptors are enzymes that have either histone-modifying activities or ATP-dependent chromatin remodeling activities to promote the accessibility of transcription factors to HRE (Narlikar, Fan, & Kingston, 2002; Westin, Rosenfeld, & Glass, 2000). For example, aldosterone induces the binding of RNA helicase A (RHA) to MR, then recruits a complex with histone acetyltransferase (HAT) activity that contains cAMP-response element-binding protein (CREB)-binding protein (CBP), leading to the cooperative potentiation of MR transcriptional activity by RHA and the HAT complex (Kitagawa et al., 2002).

Although many aldosterone up- or downregulated genes have been identified in different systems including the renal collecting duct (CD) (Kellner et al., 2003; Robert-Nicoud et al., 2001; Spindler & Verrey, 1999) and IMCD3 cells (Gumz, Popp, Wingo, & Cain, 2003), our understanding about the underlying molecular mechanisms by which aldosterone induces chromatin alternation and thus leads to gene activation or repression is hindered. No aldosterone-regulated genes had been identified to have a known function in histone modifications or chromatin remodeling until our finding of histone H3 K79 methyltransferase Dot1a as an integrate component of aldosterone-signaling network regulating *αENaC* transcription.

1.2 Epithelial sodium channel (ENaC) is a major target of aldosterone action and a key ion channel in regulating Na^+ balance

The renin–angiotensin–aldosterone system plays a major role in the control of blood pressure, extracellular fluid volume, and electrolyte balance, largely through the regulation of urinary Na^+ excretion. Perturbations in the normal or adaptive mechanisms controlling this system, which are frequently seen in clinical practice, can result in organ system dysfunction and even death. In the kidney, aldosterone-dependent regulation of ENaC takes place in the aldosterone-sensitive distal nephron (ASDN). ASDN comprises the late distal convoluted tubule (DCT2), connecting tubule (CNT), and CD. The latter includes the cortical CD (CCD) and the outer and inner medullary CD (OMCD and IMCD, respectively).

In the ASDN, transepithelial Na^+ absorption occurs by apical Na^+ entry via ENaC and basolateral Na^+ exit via the Na^+, K^+ ATPase. ENaC, composed of three subunits α, β, and γ, constitutes the rate-limiting step in this process, and changes in the synthesis of the subunits, its open probability and/or plasma membrane abundance constitute key regulatory steps, and determine the rate of Na^+ entry. Defects in ENaC subunits cause two human genetic diseases, Liddle's syndrome and the autosomal recessive form of pseudohypoaldosteronism type 1 (PHA-1). Liddle's syndrome is characterized by early onset of hypertension associated with normal plasma aldosterone levels and is caused by gain-of-function mutations in βENaC or γENaC that produce inappropriately large Na^+ absorption by CD. In contrast, PHA-1 is a salt-wasting syndrome with hypotension and high plasma aldosterone levels, resulting from loss-of-function mutations in any of the three subunits. Classical gene inactivation of all three genes encoding the ENaC subunits leads to perinatal-lethal phenotype, characterized by lung fluid-clearance failure, and/or by an acute PHA-1 with severe hyperkalemia and metabolic acidosis (reviewed in Hummler & Vallon, 2005).

2. DOT1A–AF9 COMPLEX MEDIATES REPRESSION OF *α*ENaC

2.1 Histone H3 K79 methyltransferase Dot1a

2.1.1 Dot1 proteins are a unique class of histone methyltransferases

Eukaryotic cells coil DNA around the histone octamer to form the basic unit, nucleosome, of the higher-order chromatin structures. Dynamic alternations in the chromatin structure control accessibility to regulatory factors and, thus, function critically in all DNA-based biological processes including replication, repair, recombination, and transcription. Among many mechanisms by which cells modulate their chromatin structure is the posttranslational covalent modifications of histones N-terminal tails, such as acetylation, methylation, phosphorylation, and ubiquitination. According to the histone code hypothesis (Strahl & Allis, 2000), many regulatory factors gain access to these exposed and unstructured histone tails and write/read the various modifications of specific amino acids or their combinations to control particular cellular processes. The effects of histone methylation on gene expression depend on the chromosomal location, the context of the targeted lysines and arginines, and the methyltransferase involved (Jenuwein & Allis, 2001).

Histone arginine methylation typically activates genes. In this scenario, histone methyltransferases (HMTs) are recruited as coactivators to the target promoters. One of such examples is the coactivator-associated arginine methyltransferase, protein arginine methyltransferase (CARM1/PRMT1) family, which primarily targets histones H3 or H4 (Wang et al., 2001). In contrast, gene silencing has been attributed to the suppressor of variegation, enhancer of zeste, and trithorax (SET) domain-containing lysine HMT family. The Suvar39 enzyme represents the first identified member of this family. It methylates histone H3 K9 and serves as a corepressor (Rea et al., 2000). A third class of HMTs comprises disruptor of telomeric silencing (Dot1) family members (van Leeuwen, Gafken, & Gottschling, 2002). They do not contain a SET domain. Unlike the other two types of HMTs, which target the N-terminal tails of histones, Dot1 proteins methylate lysine residue (K79) in the globular domain of histone H3. This methylation *in vivo* strongly depends on the intact nucleosomal structure and Rad6-dependent ubiquitination of histone H2B at K123 and Paf1 protein complex (Krogan et al., 2003), which is associated with the elongating RNA polymerase II.

2.1.2 Dot1 proteins and H3 K79 methylation have diverse functions

Dot1 was originally identified as a gene affecting telomeric silencing in *Saccharomyces cerevisiae* and is highly conserved from yeast to human (Feng et al., 2002; Singer et al., 1998). In *S. cerevisiae*, deletion of *Dot1* or alteration of histone H3 K79 significantly impedes telomeric and HM silencing and Sir protein association at these loci (van Leeuwen et al., 2002). Surprisingly, overexpression of Dot1 also impairs silencing and Sir protein association at these loci (Singer et al., 1998). To accommodate these paradoxical results, two models have been proposed in which Sir proteins associate preferentially with histone H3 harboring nonmethylated (van Leeuwen et al., 2002) or methylated (Ng et al., 2002) K79. It is generally held that Dot1 HMTs specifically methylate histone H3 K79 in a nontarget manner in spite of the nonrandom methylation pattern observed (van Leeuwen et al., 2002). For instance, about 90% of all H3 K79 in yeast is methylated (van Leeuwen et al., 2002) and hDot1L aa 1–416 exhibited apparently nonsequence-specific binding with DNA *in vitro* (Min, Feng, Li, Zhang, & Xu, 2003). However, it has been reported that the level of H3 K79 methylation is low at silenced loci in both yeast and mammalian cells (Ng, Ciccone, Morshead, Oettinger, & Struhl, 2003) and is concentrated at the recombination-active rather than at the inactive loci in mammalian cells. These observations raise the possibility that Dot1-mediated H3 K79 methylation occurs in a targeted manner, which is now clearly demonstrated by our work on Dot1a–Af9-mediated repression of *αENaC* (Zhang, Yu, et al., 2013). Since the identification of Dot1 encoding a HMT, various roles of H3 K79 methylation have been proposed. The modification is considered as a conserved hallmark of active chromatin regions (Ng et al., 2003) and involved in multiple biological processes including telomeric and HM silencing, cell cycle regulation, cell proliferation, meiotic checkpoint, DNA replication, apoptosis, and leukemogenesis (reviewed in Nguyen & Zhang, 2011). The functional diversity of Dot1 and H3 K79 methylation may in part comes from the ability of the enzyme to add one, two, and three methyl groups to K79, leading to mono-, di-, and trimethylated K79, respectively (van Leeuwen et al., 2002). These methylation events are referred as H3 m1, m2, and m3K79.

2.1.3 Dot1a is the first aldosterone-regulated target with a known function in epigenetics

We cloned mouse Dot1-like *(Dot1l)* gene, characterized its mRNA expression in the mouse kidney and IMCD3 cells (derived from the mouse inner medullary CD), identified five alternative splicing variants (Dot1a–e), and defined its methyltransferase activity specific for histone H3 K79 (Zhang, Hayashizaki, & Kone, 2004). Among these isoforms, Dot1a is the most extensively studied isoform. It contains a methyltransferase domain, an Af9/Af17-interacting domain, three potential nuclear localization signals, and a putative leucine zipper (Fig. 1). Dot1a shares 84% identity in the full length of amino acid sequence and 97% identity in the methyltransferase domain with its closest human counterpart hDOT1L.

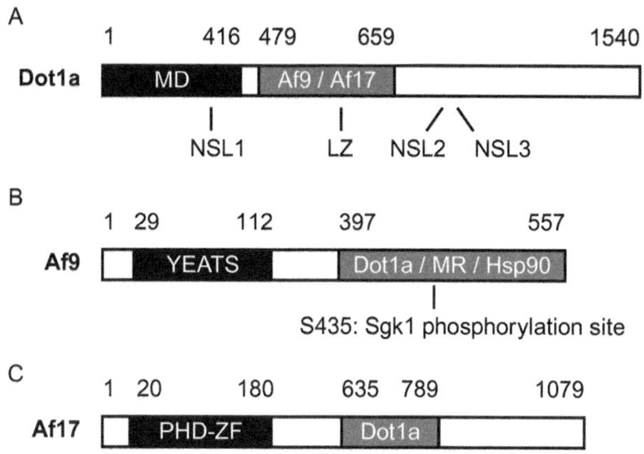

Figure 1 Diagrams of Dot1a, Af9, and Af17. Structural features in mouse Dot1a (NP_955354.1) (A), mouse Af9 (AAH21420.1) (B), and mouse Af17 (NP_647472.2) (C). MD, methyltransferase domain (Min et al., 2003); NSL, nuclear localization signal (Reisenauer, Wang, Xia, & Zhang, 2010); LZ, leucine zipper (Reisenauer et al., 2009; Zhang, Xia, Reisenauer, Hemenway, & Kone, 2006); Af9/Af17, Af9/Af17-interacting domain (Reisenauer et al., 2009; Zhang, Xia, Reisenauer, et al., 2006); YEATS, YEATS domain; Dot1a/MR/Hsp90, Dot1a/MR/Hsp90-interacting domain; PHD-ZF, PHD-zinc-finger-like domain. *Domain search: http://www.ncbi.nlm.nih.gov/Structure/cdd/wrpsb.cgi*

Our subsequent analyses identified Dot1a as the first aldosterone-regulated target that has a known function in histone modifications. Aldosterone dynamically regulates Dot1a mRNA levels, leading to a delayed overall reduction of H3 K79 methylation in IMCD3 cells (Zhang, Xia, Jalal, et al., 2006). Aldosterone-treated IMCD3 cells displayed Dot1a mRNA levels that were ~30%, ~65%, and ~110% of vehicle-treated controls at the 2-, 4-, and 7-h time points, respectively. However, changes in H3m2K79 in bulk histones were not obvious until the 7-h time point, at which H3m2K79 was significantly reduced, followed by progressive recovery till 72-h time point. The effect of aldosterone on H3m2K79 seems specific, since H3 acetylated K9 as well as the total histone H3 remained constant at all time points examined (Zhang, Xia, Jalal, et al., 2006). Aldosterone also downregulated hDOT1L mRNA expression in 293T cells (Reisenauer et al., 2009), suggesting that the mechanism is likely conserved between mouse and human cells. Consistently, there are multiple putative glucocorticoid response elements (GRE) throughout the 2-kb 5′-flanking region of mouse *Dot1l* (-283, -359, -649, -716, -748, -900, -1274, -1553, -1613, -1755, -1805, and -1846; relative to the putative transcription start site 238 bp upstream of the translation initiation codon of Dot1a), as revealed by the analyses using the Transcription Element Search System (http://www.cbil.upenn.edu/tess/). Nevertheless, whether aldosterone decreases Dot1a mRNA levels by repressing its transcription through these GRE-like sites or by impacting its stability/degradation remains unexplored. Whether aldosterone downregulates Dot1a mRNA abundance *in vivo* in kidney and distal colon, two major physiological targets of aldosterone action, still lacks direct evidence.

2.1.4 Dot1a modulates targeted H3 K79 methylation at the αENaC promoter and represses αENaC in a methyltransferase-dependent manner

The physiological relevance and significance of aldosterone-induced downregulation of Dot1a are established by the identification of Dot1a as a negative regulator of *αENaC* transcription. There are several lines of evidence to support this notion. First, *trans*-activation of *αENaC* after aldosterone administration is associated with significantly impaired association of methylated histone H3 K79 at specific areas of the *αENaC* 5′-flanking region (Fig. 2). The impairment varies to different degrees, depending on the subregions (Ra and R0–R3) examined (Zhang, Xia, Jalal, et al., 2006). Second, RNAi-mediated knockdown of Dot1a mimics aldosterone treatment, producing similar effects including increased activity of a stably integrated

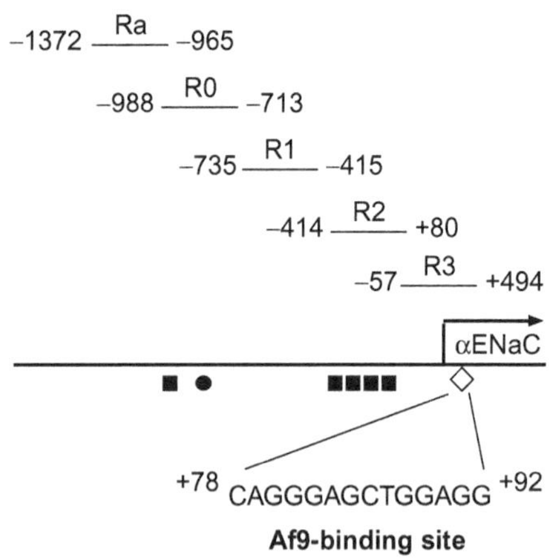

Figure 2 Diagram of the 5′-flanking region of *αENaC*. Fragments designated Ra–R3 are shown along with their relative positions to the major transcription start site (+1) of *αENaC*. • and ■ represent the putative GRE site (-811) and GRE half sites (-983, -416, -325, -241, and -234), respectively. ◇ indicates the Af9-binding site (+78) (Zhang, Xia, Jalal, et al., 2006; Zhang, Xia, Reisenauer, et al., 2006; Zhang, Yu, et al., 2013).

αENaC promoter-luciferase construct and expression of endogenous αENaC mRNA. Third, overexpression of WT Dot1a, but not the methyltransferase-dead mutant, causes increased association of H3m2K79 at the specific subregions of the *αENaC* 5′-flanking regions, repression of αENaC mRNA expression, and reduced activity of the *αENaC* promoter-luciferase construct (Zhang, Xia, Jalal, et al., 2006). In aggregate, our data identify Dot1a as a new aldosterone-downregulated histone modification enzyme, and a negative regulator of *αENaC* transcription, most likely by directly or indirectly binding the *αENaC* promoter and hypermethylating histone H3 K79 associated with the *αENaC* promoter. These initial findings from IMCD3 cells have been replicated in 293T cells and mouse cortical CD M1 cells and extended to other aldosterone-regulated genes including *βENaC*, *γENaC*, *connecting tissue growth factor* (*CTGF*), *period* homolog, *preproendothelin*, and *Sgk1* (Reisenauer et al., 2009; Reisenauer, Wang, Xia, & Zhang, 2010; Wu, Chen, Zhou, & Zhang, 2011). Generation and characterization of mice lacking function of Dot1a and its regulators further solidify these findings *in vivo* in mouse kidneys (see below).

2.1.5 Dot1a-mediated repression apparently requires its nuclear expression as well as its methyltransferase activity and targeted H3 K79 methylation

Transiently expressed GFP-Dot1a fusion in 293T cells was primarily located in the cytoplasm, nucleus, or both. The cellular distribution of Dot1a apparently regulated histone H3 K79 (Reisenauer et al., 2009). Sequence analysis revealed three potential NLSs spanning aa 394–416 (NLS1), 1089–1112 (NLS2), and 1165–1172 (NLS3) (Fig. 1). These NLSs are also highly conserved in rat and human with identical NLS1 sequences, one mismatch in human NLS2, rat NLS2, and rat NLS3, respectively. The biological significance of these putative NLSs in regulating Dot1a cellular distribution was evaluated by mutagenesis. Analyses of GFP-Dot1a fusions harboring various mutations in the three NSLs revealed that NLS1, NLS2, and NLS3 most likely play a role in directing Dot1a nuclear expression, with NLS1 and NLS2 being more important than NLS3. Moreover, Dot1a-mediated repression of the *αENaC* promoter-luciferase construct was largely relieved by mutations that impede Dot1a nuclear expression, remove the Af9/Af17-interacting domain, or remove the methyltransferase domain. A Dot1a mutant harboring deletions of all three NLSs was almost exclusively cytoplasmic and failed to inhibit *αENaC* promoter activity (Reisenauer et al., 2010). These results suggest that nuclear expression is essential for Dot1a to repress *αENaC* promoter. Manipulation of Dot1a nuclear/cytoplasmic distribution offers an additional strategy to remove Dot1a-mediated repression. Indeed, Af17 upregulates *αENaC* promoter by stimulating Dot1a nuclear export.

2.1.6 Dot1a-mediated repression of αENaC raised new questions

Dot1a-mediated repression of *αENaC* raised three major questions regarding the function and regulation of Dot1a and H3 K79 methylation. The first question comes from Dot1a independency of aldosterone. Although it is clear that aldosterone stimulates *αENaC* transcription, it appears that Dot1a-mediated effects on *αENaC* transcription are at least partially independent of aldosterone–MR interactions with the *αENaC* promoter. Manipulation of Dot1a expression or its methyltransferase activity without aldosterone treatment is sufficient to induce changes in histone H3 K79 methylation and *αENaC* transcription.

The second question is if H3K79 methylation is active or repressive. In the broader context of other targets impacted by histone H3 K79 methylation, our work suggests that the modification may have bimodal effects on transcription. Studies in yeast of Sir-dependent silencing loci and in mammalian cells of recombination-active versus recombination-inactive genomic loci suggest that histone H3 K79 hypermethylation functions as a marker of active chromatin regions (Ng et al., 2003). On the contrary, our data indicate that increased H3 K79 methylation is correlated with repression. Several explanations exist for these inconsistent findings. (1) Even in yeast, the role of histone H3 K79 methylation in association with the Sir proteins and thus in gene silencing is still unclear. As mentioned above, deletion or overexpression of Dot1 shares a similar phenotype: impaired Sir protein binding and disrupted silencing at telomeric and HM loci. (2) The cells contain a mixture of non-, mono-, di-, or trimethylated H3 K79 (van Leeuwen et al., 2002). A mechanistic insight of differential methylation states is mysterious. While the specificity of the anti-dimethyl H3 K79 antibody (from Upstate Biotechnology) to methylated K79 in histones has been established (Ng et al., 2002; Zhang et al., 2004), the cross-reactivity of this antibody toward mono- and trimethylated H3 K79 has not been rigorously examined. In case of *αENaC*, it remains to be determined whether hypomethylation induced by aldosterone or hypermethylation resulted from Dot1a overexpression represents a change in the relative ratio of each K79-methylated form to the non-modified one. Since the same antibody was utilized in the other studies

(Ng et al., 2002, 2003), it is possible that a similar increase or decrease in histone H3 K79 methylation observed by chromatin immunoprecipitation (ChIP) with the antibody in different systems may represent a different shift of the spectrum of differential methylation status and thus point to different functions. Finally, it is also possible that histone H3 K79 methylation behaves similarly as H3 S10 phosphorylation (Cheung et al., 2000). Both of them may lead to gene activation and repression. Their ultimate effects on gene expression are dependent on the patterns of other histone modifications leading to either the association or the dissociation of distinct effector proteins.

The third question lies in the mode of Dot1a action. Dot1 family members are generally considered as nontargeted methyltransferases, although H3m2K79 is not uniformly detected throughout the genome in yeast or mammalian cells (Ng et al., 2002, 2003; van Leeuwen et al., 2002). Our data raise the additional possibility that targeted interactions of Dot1a with the *αENaC* 5′-flanking region not only occur but also are regulated in response to changes in Dot1a expression levels achieved by aldosterone treatment, overexpression, or knockdown of Dot1a. These results argue against nonsequence-specific DNA-binding activity of Dot1a in this promoter context and favor the hypothesis that Dot1a is recruited to these specific regions, presumably by Dot1a-interacting protein(s) that exerts DNA-binding activity to the implicated regions of the *αENaC* promoter. Driven by this hypothesis, we have identified Af9 as a Dot1a binding partner and a specific DNA-binding protein that also regulates *αENaC* promoter in an aldosterone-sensitive manner.

2.2 Putative transcription factor Af9

2.2.1 Af9 interacts with Dot1a

ALL1-fused gene from chromosome 9 (*hAF9*) was initially cloned from a leukemia patient containing t(9,11)(p22;q23) translocation (Iida et al., 1993). The translocation results in in-frame fusion of hAF9 aa 376–568 (corresponding to mouse Af9 aa 365–557) to mixed lineage leukemia (MLL) gene. To date, there are over 40 MLL fusion partners. These fusion events are considered as the cause of leukemogenesis. MLL-hAF9 is one of the most common forms. hAF9 is highly homologous to ENL and interacts with AF4. Like hAF9, both AF4 and ENL are also MLL fusion partners (Erfurth, Hemenway, De Erkenez, & Domer, 2004). Other hAF9 interacting proteins include specific isoforms of BCL-6 corepressor (BCoR) (Srinivasan, de Erkenez, & Hemenway, 2003), a component of histone deacetylase complexes, and MPc3, a member of the Polycomb group multiprotein complexes involved in gene silencing by modifying chromatin structure (Hemenway, de Erkenez, & Gould, 2001). MPc3 can simultaneously interact with hAF9 and RING1, and in turn RING1 may bind other Polycomb group proteins (Hemenway et al., 2001). Thus hAF9 may acquire a dominant-negative function by recruitment of, or to, corepressors or chromatin-modifying proteins. hAF9 and its mouse homolog Af9 belong to YEATS family, which is named after "YNK7," "ENL," "AF9," and "TFIIF small subunit." This family is characterized by the possession of a YEATS domain and thought to have a transcription stimulatory activity. Apart from these observations, little is known about the hAF9 function in normal cells.

To identify Dot1a-interacting protein that may modulate targeted H3 K79 methylation, we screened a mouse kidney cDNA library ligated to the GAL4 activation domain using Dot1a as bait in a yeast two-hybrid assay. This effort led to identification of Af9 as a specific binding partner of Dot1a. The specificity of the interaction was subsequently confirmed *in vitro* and *in vivo* by multiple assays (GST pulldown, Co-IP, colocalization, and mammalian two-hybrid). Further analyses revealed that aa 479–659 in Dot1a and the very C-terminal part (aa 397–557) of Af9 are capable of and sufficient for mediating the interaction (Fig. 1). Since the domain responsible for interacting with Af9 in Dot1a also binds Af17 (see below), it is referred as Af9/Af17-interacting domain. This domain is not overlapped with the methyltransferase domain (aa 1–416) and does not exist in yeast Dot1. Since the Af9 very C-terminal part also mediates its interaction with MR and Hsp90, it is named as Dot1a/MR/Hsp90-interacting domain. This fragment is retained in the leukemia patient used to identify hAF9 from the t(9,11)(p22;q23) translocation. Hence, our study also provided additional evidence in favor of the hypothesis that mistargeting hDOT1L is a critical event for leukemogenesis (Okada et al., 2005).

2.2.2 Af9 directly binds αENaC promoter and represses its transcription

The interaction of Af9 with Dot1a implies that Af9 may be a component of aldosterone-signaling network and regulate *αENaC* transcription. Indeed, like Dot1a, Af9 expression is also downregulated by aldosterone at both mRNA and protein levels in IMCD3 cells. Af9 overexpression causes hypermethylation of histone H3 K79 at R0–R3, but not at Ra subregions of the *αENaC* promoter. It also decreases expression of the endogenous αENaC mRNA expression and acts synergistically with Dot1a to inhibit *αENaC* promoter-luciferase constructs. RNAi-mediated knockdown of Af9 induces the opposite effects. ChIP assays reveal that endogenous Af9 and Dot1a as well as overexpressed FLAG-Af9 are each associated with R0–R3, but not Ra subregions of *αENaC* promoter (Zhang, Xia, Reisenauer, et al., 2006; Zhang et al., 2007). Gel shift and antibody competition assays revealed that Af9 has specific DNA-binding activity and binds a functional Af9 *cis*-element (+78/+92) in the R3 subregion of *αENaC* promoter, the principal site for Dot1a–Af9 interaction, in IMCD3 cells (Zhang, Yu, et al., 2013). Mutation of the Af9 *cis*-element resulted in greatly reduced association of Af9 and Dot1a with the *αENaC* promoter-reporter constructs, higher basal *αENaC* promoter activity, and impaired Dot1a-mediated inhibition in *trans*-repression assays (Zhang, Yu, et al., 2013). Thus, it can be concluded that Af9 recruits Dot1a, primarily through directly binding +78/+92 of *αENaC*, to mediate targeted H3 K79 methylation and to repress basal and aldosterone-sensitive *αENaC* transcription. In addition, NAD^+-dependent deacetylase sirtuin 1 (Sirt1) functionally and physically interacts with Dot1 to enhance the distributive activity of Dot1a on H3K79 methylation and thereby represses *αENaC* transcription in IMCD3 cells (Zhang, Li, Cruz, & Kone, 2009). Dot1a–Af9-mediated repression may be more broadly applied to at least some other aldosterone-regulated genes, including *βENaC*, *γENaC*, *CTGF*, *period* homolog, *preproendothelin*, and *Sgk1* (Wu et al., 2011; Zhang, Xia, Reisenauer, et al., 2006; Zhang et al., 2007). However, compared to *αENaC*, these genes are much less characterized in terms of Dot1a–Af9-mediated targeted H3 K79 methylation associated with their promoters.

3. DOT1A–AF9-MEDIATED *αENaC* REPRESSION IS RELIEVED BY MULTIPLE MECHANISMS

3.1 Sgk1 relieves Dot1a–Af9-mediated repression by phosphorylating Af9

3.1.1 Sgk1 is an early target of aldosterone and regulates ENaC expression and activity

Prevailing evidence suggests that aldosterone increases ENaC function in two phases. The early phase involves the upregulation of preexisting transport machinery and of aldosterone-induced regulatory proteins, notably Sgk1. The delayed phase of aldosterone action involves *de novo* synthesis of ENaC, either through the liganded MR directly binding HREs in the *αENaC* promoter to activate transcription (Thomas & Itani, 2004) or through indirect mechanisms involving other proteins (Asher, Wald, Rossier, & Garty, 1996; Sayegh et al., 1999).

Studies on acute transcriptional effects of aldosterone led to identification of several transcripts, including Sgk1, induced in the early phase of aldosterone stimulation (Chen et al., 1999). Sgk1 expression can be induced as soon as 30 min after hormone administration (Chen et al., 1999). Rapid induction of Sgk1 mRNA by aldosterone is coupled to increased ENaC cell surface expression (Loffing et al., 2000, 2001), which can be achieved by enhancing ENaC trafficking from the cytosol to the apical membrane or by decreasing its removal from the membrane. How Sgk1 regulates these processes is still not fully defined. A widely accepted theme is that Sgk1 regulates ENaC abundance at the cell surface, in part through phosphorylation of the ubiquitin ligase Nedd4-2 (Debonneville et al., 2001). Nedd4-2 phosphorylation reduces its affinity with and hence binding ENaC, and induces its interaction with 14-3-3. The concerted actions of Sgk1 and 14-3-3 appear to disrupt Nedd4-2-mediated ubiquitination and subsequent degradation of ENaC, leading to accumulation of ENaC channels at the cell surface (Debonneville et al., 2001; Ichimura et al., 2005; Lee, Campbell, Cook, & Dinudom, 2008; Snyder, Olson, & Thomas, 2002). Aldosterone also induces ubiquitin-specific protease Usp2-45, which deubiquitylates ENaC and increases its surface expression and activity (Fakitsas et al., 2007; Verrey, Fakitsas, Adam, & Staub, 2008). In this regard, part of the early effect of aldosterone is mediated by inhibiting ENaC removal from the membrane (Kamynina & Staub, 2002; Snyder et al., 2002).

However, following observations support additional mechanisms for Sgk1-mediated ENaC regulation. (1) While the PY motif of ENaC is required for interaction with Nedd4-2, its requirement for Sgk1-stimulated Na^+ transport is controversial (Alvarez de la Rosa, Zhang, Naray-Fejes-Toth, Fejes-Toth, & Canessa, 1999; Debonneville et al., 2001). For example, the isolated collecting tubules and primary cultures of CCD from a Liddle mouse model expressing a PY-missing βENaC subunit still displayed aldosterone-stimulated Na^+ transport (Dahlmann et al., 2003; Pradervand et al., 2003). Some ion-transporters such as renal outer medullary K^+ channel ROMK1 (Yoo et al., 2003; Yun et al., 2002) and Na^+, K^+ ATPase (Setiawan et al., 2002) that have neither a PY motif nor a known interaction with Nedd4-2 can also be regulated by Sgk1. (2) Sgk1 also apparently plays a role at the transcriptional level. Sgk1 has been shown to regulate the transcription of *CTGF* (Vallon et al., 2006) and α- and β*ENaC* (Boyd & Naray-Fejes-Toth, 2005; Zhang et al., 2007). Examination of αENaC mRNA expression in the presence or absence of corticosteroids in mouse CCD cells stably expressing either full-length Sgk1 or a kinase-dead, dominant-negative (K127M)-Sgk1 mutant revealed that Sgk1 upregulates *αENaC* expression in a corticosteroid-independent, but kinase-dependent manner (Boyd & Naray-Fejes-Toth, 2005). Accordingly, Sgk1 is proposed to activate unidentified transcription factors or inactivate an inhibitor of *αENaC* transcription (Boyd & Naray-Fejes-Toth, 2005). Our finding supports the latter. That is, Sgk1 impairs Dot1a–Af9-mediated repression of *αENaC* through phosphorylating Af9 to impair Dot1a–Af9 interaction.

In addition to aldosterone, Sgk1 is also rapidly induced in response to other stimuli in a tissue-specific manner (Cooper et al., 2001; Cowling & Birnboim, 2000; Waldegger, Barth, Raber, & Lang, 1997; Waldegger, Gabrysch, Barth, Fillon, & Lang, 2000). Sgk1 can be shuttled between nucleus and cytoplasm in a cell cycle- and stimulus-dependent manner (Firestone, Giampaolo, & O'Keeffe, 2003). The activation of Sgk1 is dependent upon phosphoinositide 3-kinase activity and requires the phosphorylation of the two regulatory sites (T256 and S422) (Park et al., 1999). Thus, Sgk1 serves as a functional convergence point between several types of signaling pathways and cellular phosphorylation cascades. The Sgk1 gene is highly conserved from yeast to human and expressed in a variety of tissues and cell lines in mammals. In addition to Nedd4-2, Sgk1 also phosphorylates Af9, forkhead-like transcription factor FOXO3a and several other substrates (Sahin, McCaig, Jeevahan, Murray, & Hainsworth, 2013; Zhang, Xia, Reisenauer, et al., 2006). All of these proteins contain the Sgk1 consensus phosphorylation sites (RXRXXS/T).

3.1.2 Sgk1 phosphorylates Af9 at S435 in vitro *and* in vivo

As mentioned above, aldosterone decreases expression of Dot1a and Af9 and thus the formation of Dot1a–Af9 complex. In addition, aldosterone rapidly induces Sgk1, which inactivates the Dot1a–Af9 complex by targeting Af9 for phosphorylation. Af9 possesses a highly conserved consensus Sgk1 phosphorylation site S435. *In vitro*, purified Sgk1 can phosphorylate recombinant GST-Af9, but not GST-Af9 S435A. *In vivo* in IMCD3 cells, the overall level of FLAG-Af9 S435A phosphorylation was about 50% of the wild-type FLAG-Af9 phosphorylation, suggesting that S435 is a major phosphorylation site. Aldosterone increased Sgk1 protein level slightly at 1-h and significantly at 1.5- and 2-h time points, which was accompanied with a significant enhancement of the endogenous Af9 S435 phosphorylation at 1 h. The phosphorylation was peaked at 1.5 h of aldosterone treatment.

Hence, Sgk1 may contribute to aldosterone-stimulated Af9 S435 phosphorylation in IMCD3 cells. RNAi-mediated knockdown of Sgk1 by ~50% resulted in ~40% reduction in Af9 S435 phosphorylation. In reciprocal experiments, transient overexpression of Sgk1, but not its kinase-dead mutant, increased by ~160% Af9 S435 phosphorylation in IMCD3 cells (Zhang et al., 2007). *Sgk1*$^{-/-}$ versus *Sgk1*$^{+/+}$ mice had significantly reduced Af9 S435 phosphorylation in the kidney (Zhang et al., 2007). Taken together, Sgk1 phosphorylates Af9 at S435 *in vitro* and *in vivo* in multiple experimental settings.

3.1.3 Sgk1-mediated Af9 phosphorylation decreases the Dot1a–Af9 interaction in vitro and at the αENaC promoter in IMCD3 cells

The location of S435 in the Dot1a/MR/Hsp90-interacting domain (aa 397–557) provides the structural basis for its role in regulating the Dot1a–Af9 interaction through Sgk1-mediated phosphorylation. Consistently, the ability of Sgk1-phosphorylated GST-Af9 397–557 to retain GFP-Dot1-a 479–659 was decreased to 24% of the nonphosphorylated GST-Af9 fusion. Such ability to retain the GFP-Dot1a fusion and sensitivity to Sgk1 were significantly impaired, but not abolished by S435A mutation (Zhang et al., 2007). These results suggest that Sgk1-mediated Af9 S435 phosphorylation decreases but does not completely abolish the Dot1a–Af9 interaction, and that S435 plays a critical role in binding Dot1a. Analogous to our findings, protein kinase A phosphorylation of GST-neurabin I significantly reduced, but did not completely eliminate, its interaction with protein phosphatase 1 (McAvoy et al., 1999). Similarly, the non-phosphorylated form of PC4 displayed a lower, but not eliminated, affinity for Rep proteins compared to phosphorylated PC4 (Weger, Wendland, Kleinschmidt, & Heilbronn, 1999), and phosphorylated merlin had lower, but not abolished, interactions with CD44 and hepatocyte growth factor-regulated tyrosine kinase substrate in GST-pulldown assays (Rong, Surace, Haipek, Gutmann, & Ye, 2004). In addition, S435A and S435D to mimic nonphosphorylated and phosphorylated forms, respectively, yielded a similar phenotype, indicating that phosphorylation induces changes in structure, charge, or both. Observations of identical behavior of Ala and Asp mutations of putative serine phosphorylation sites also have been reported before. For example, changing the putative phosphorylation site S33 in the intermediate filament protein keratin 18 into Ala or Asp disrupted the keratin 18 interaction with 14-3-3 proteins (Ku, Liao, & Omary, 1998). Similarly, substitution of the putative phosphorylation site

S444 in the K^+ channel KCNQ4 with Ala or Asp abolished Sgk1 modulation of KCNQ4 (Seebohm et al., 2005). In addition, the kinase activity of Sgk1, while necessary for lessening the Dot1a–Af9 interaction, does not appear to be required for its association with the R0–R3 subregions of the *αENaC* promoter in IMCD3 cells (Zhang et al., 2007).

Since ChIP revealed that endogenous Af9 and FLAG-Af9 associated with specific subregions R0–R3, but not with Ra of the *αENaC* promoter (Zhang, Xia, Reisenauer, et al., 2006; Zhang et al., 2007), it is predicted that Sgk1 would associate with the *αENaC* promoter, decrease the association of the Dot1a–Af9 complex, and thus limit histone H3 K79 methylation at the *αENaC* promoter. These predictions were validated through ChIP and Re-ChIP in IMCD3 cells transfected with FLAG-tagged Sgk1 or the kinase-dead mutant Sgk1. In particular, Sgk1, like Af9, specifically associated with the R0–R3, but not Ra, subregions of the *αENaC* promoter in IMCD3 cells. Sgk1 impairs the ability of Af9 to interact with Dot1a at these subregions without impacting Af9 DNA-binding activity, leading to targeted histone H3 K79 hypomethylation. In all subregions except R1, Sgk1 induced these effects in a kinase-dependent manner and presumably by phosphorylating Af9 (Zhang et al., 2007).

Mice with high salt intake versus normal salt intake exhibited comparable low levels of Af9 S435 phosphorylation and mRNA expression of Sgk1 and αENaC in their kidneys. Salt restriction coordinately induced Sgk1 expression and Af9 S435 phosphorylation, which was associated with increased αENaC mRNA levels in mouse kidney (Zhang et al., 2007). These data provide first evidence that Af9 is a bona fide substrate of Sgk1 for phosphorylation at S435 *in vivo* in mouse kidney. These findings are also consistent with previous studies showing that low-salt diet-induced renal αENaC mRNA expression (Hou, Speirs, Seckl, & Brown, 2002; Ono, Kusano, Muto, Ando, & Asano, 1997).

More direct evidence showing that Sgk1 phosphorylates Af9 at S435 and regulates *αENaC* transcription *in vivo* in the kidney was acquired by analysis of *Sgk1*$^{-/-}$ mice (Huang et al., 2006; Vallon, Huang, et al., 2005; Vallon, Wulff, et al., 2005; Wulff et al., 2002). *Sgk1*$^{-/-}$ mice exhibited significantly lower levels of renal Af9 S435 phosphorylation than WT controls under a normal- or low-salt diet, indicating that Af9 can be phosphorylated in a Sgk1-dependent or Sgk1-independent manner in mouse kidney. The reduced Af9 phosphorylation was accompanied with impaired *αENaC* expression at both mRNA and protein levels (Zhang et al., 2007).

Mechanistically, Sgk1-mediated phosphorylation of Af9 impairs but does not fully abolish the ability of Af9 to interact with Dot1a. Therefore, other factors controlling the Dot1a–Af9 complex are operative. Among these factors are MR and Af17 as described below.

3.2 MR counterbalances Dot1a–Af9 by interacting with Af9

A central question is how aldosterone coordinately activates *αENaC* through MR-dependent mechanism and removes Dot1a–Af9-mediated repression. In the unliganded state, MR is localized in the cytosol as a part of a multiprotein complex containing heat-shock protein 90 (Hsp90) (Faresse, Ruffieux-Daidie, Salamin, Gomez-Sanchez, & Staub, 2010). MR has been known to interact with GR (Liu, Wang, Sauter, & Pearce, 1995; Savory et al., 2001) and multiple coactivators and corepressors (reviewed in Yang & Young, 2009). Af9 binds Dot1a (Zhang, Xia, Reisenauer, et al., 2006), Sgk1 (Zhang et al., 2007), Af4 (Erfurth et al., 2004), Aff4 (Biswas et al., 2011), Hsp90 (Lin & Hemenway, 2010), CBX8 (Hemenway et al., 2001), and BCoR (Srinivasan et al., 2003). Because they have opposite roles in *αENaC* transcription, MR and Af9 may interact to mutually antagonize their opponent effect. Yeast two-hybrid assays revealed a specific interaction between GAL4-AD-MR and GAL4-BD-Af9 fusions (Zhang, Zhou, et al., 2013). The interaction was subsequently confirmed by Co-IP and colocalization experiments. MR coimmunoprecipitated with Af9 in the whole-kidney lysates from the WT, but not from the $MR^{-/-}$ mice. This demonstrates not only the interaction but also the specificity of the MR–Af9 interaction. Further yeast two-hybrid assays revealed that MR, like Dot1a, also binds the Dot1a/MR/Hsp90-interacting domain (aa 397–557), but not the N-terminal part (aa 2–406) of Af9 (Zhang, Zhou, et al., 2013). This suggests that MR may interfere with Dot1a–Af9 interaction by competing for Af9 binding. Such hypothesis was subsequently validated by GST-pulldown assays to investigate Dot1a–Af9 interaction with MR as a competitor. Increasing the amount of lysate containing overexpressed MR gradually reduced the amount of the GFP-Dot1a fusion retained by a fixed amount of the GST-Af9 fusion, indicating that MR inhibited Dot1a–Af9 interaction. Consistently, the Dot1a–Af9-mediated repression of *α*-, *β*-, and *γENaC* was progressively abolished and eventually reversed by increasing amounts of overexpressed MR in the presence of aldosterone. Hence, MR/aldosterone is capable of antagonizing Dot1a–Af9-mediated repression of ENaC mRNA expression, at least in IMCD3 cells under the conditions tested (Zhang, Zhou, et al., 2013). In these experiments, the aldosterone concentration in the medium (100 nM) was higher than the physiological concentration of the hormone. However, the actual effective intracellular concentration may be much lower because steroid hormones may not diffuse across cell membranes freely (Gumz et al., 2003; Pietras, Nemere, & Szego, 2001). This may explain why there is a wide range of aldosterone concentrations including 10 nM (Dooley, Angibaud, Yusef, Thomas, & Harvey, 2013), 30 nM (Faresse, Debonneville, & Staub, 2013), 1000 nM (Gumz et al., 2003; Reisenauer et al., 2009, 2010; Soodvilai, Jia, Fongsupa, Chatsudthipong, & Yang, 2012; Wu et al., 2011; Zhang, Xia, Jalal, et al., 2006; Zhang, Xia, Reisenauer, et al., 2006; Zhang et al., 2007; Zhang, Yu, et al., 2013), and even 1500 nM (Helms et al., 2005) used in different cell culture systems. Aldosterone elicited the most dramatic effect in IMCD3 cells at 1000 nM, which can be completely blocked by MR and GR inhibitors, used alone or in combination (Gumz et al., 2003).

On the other hand, reciprocal experiments demonstrated that overexpression of Dot1a–Af9 suppressed MR-mediated transcriptional activation of ENaC genes in IMCD3 cells. Transfection with the MR construct alone significantly increased α-, β-, and γENaC mRNA levels, compared to the vector-transfected control. Increasing amounts of the Dot1a and Af9 constructs added to the transfection mixture progressively impaired the MR-mediated activation (Zhang, Zhou, et al., 2013). Taken together, Dot1a–Af9 and MR mutually impair their opponent's effect on ENaC transcription under the conditions tested.

3.3 Af17 impairs Dot1a–Af9-mediated repression by competitively binding Dot1a and facilitating Dot1a nuclear export

3.3.1 Af17 competes with Af9 for binding Dot1a

Aldosterone-independent impairment of Dot1a–Af9 interaction and relief of Dot1a–Af9-mediated repression were first revealed by identification of ALL1-fused gene from chromosome 17 (hAF17) as another Dot1a-interacting molecule (Reisenauer et al., 2009; Wu et al., 2011). *hAF17* gene was originally cloned as a less frequent fusion partner of the *MLL* gene in t(11;17)(q23;q21) translocations from some acute myeloid leukemias (Prasad et al., 1994; Suzukawa et al., 2005). hAF17 is proposed to be a transcriptional regulator and a downstream target of the β-catenin/T-cell factor pathway and regulates G2–M progression. hAF17 binds PC2 glutamine/q-rich-associated protein (PCQAP) and CCAAT/enhancer binding protein (C/EBP) in yeast two-hybrid assay (Lin et al., 2001). However, the biological relevance and significance of these protein–protein interactions have not been well characterized.

In search of additional Dot1a partners by yeast two-hybrid screen, we identified mouse Af17 aa 635–786 is capable of and sufficient for interacting with Dot1a. Interestingly, like Af9, mouse Af17 binds the same Af9/Af17-interacting domain of Dot1a, but not the methyltransferase domain (aa 1–416), the putative leucine zipper domain (aa 576–597), or the C-terminal portion (aa 1112–1540) of Dot1a. The Dot1a–Af17 interaction was mutually confirmed by multiple other approaches (GST pulldown, Co-IP, mammalian two-hybrid, and colocalization) (Reisenauer et al., 2009).

The fact that the same domain of Dot1a can interact with Af17 and Af9 suggests that these two proteins may compete for binding Dot1a. In the GST-pulldown assays to investigate Dot1a–Af17 interaction with FLAG-Af9 as a competitor, increasing the amount of FLAG-Af9 gradually reduced the ability of GST–Af17 fusion to retain GFP-Dot1a. In competitive mammalian two-hybrid assays, the luciferase reporter activity driven by Dot1a–Af17 interaction was inversely correlated with the amount of FLAG-Af9 added. Reciprocally, the Dot1a–Af9 interaction was progressively prohibited by elevating amounts of FLAG-Af17 635–786 as a competitor in GST-pulldown assays and in competitive mammalian two-hybrid assays (Reisenauer et al., 2009). Hence, Af17 and Af9 apparently inhibited their opponent to interact with Dot1a *in vitro* and *in vivo*.

3.3.2 Af17 facilitates Dot1a nuclear export and upregulates **ENaC** *genes*

RNAi-mediated depletion of the endogenous hAF17 induced a shift of GFP-Dot1a expression from cytoplasm to nucleus in 293T cells. In contrast, overexpression of RFP-hAF17 enhances cytoplasmic expression of GFP-Dot1a at the expense of its nuclear expression in 293T cells. In the absence of their partners, GFP-Dot1a and RFP-hAF17 were primarily, but not exclusively, located in the nucleus and cytoplasm, respectively. However, when these fusion proteins were coexpressed, Dot1a nuclear expression was significantly decreased, which was accompanied with a dramatic increase of Dot1a colocalization with hAF17 in the cytoplasm (Fig. 3). The redistribution of Dot1a from the nucleus to the cytoplasm was accompanied by a substantial reduction of H3m2K79 and H3m3K79, but not

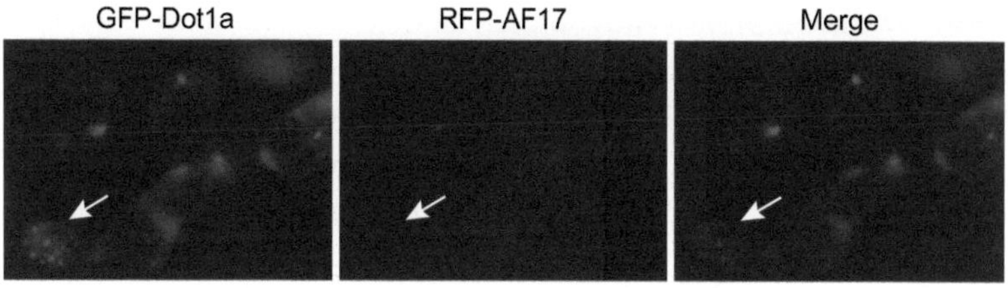

Figure 3 Representative images showing that AF17 facilitates Dot1a cytoplasmic distribution. 293T cells were transiently transfected with GFP-Dot1a and RFP-AF17, and examined by epifluorescence microscopy (20 ×). Arrow: a cell showing typical nuclear localization pattern of GFP-Dot1a in the absence of detectable RFP-AF17. *This figure was originally published in Reisenauer et al. (2009). © The American Society for Biochemistry and Molecular Biology.* (See the color plate.)

dimethylated H3 K9 (H3m2K9) in bulk histones. The effect of hAF17 overexpression on H3 K79 hypomethylation can be blocked by the nuclear export inhibitor leptomycin B. ChIP and Re-ChIP revealed that hAF17 overexpression significantly impaired the Dot1a–Af9 interaction at all four subregions (R0–R3) of *αENaC* promoter and upregulates *αENaC* transcription in 293T cells and in mouse cortical CD M1 cells (Reisenauer et al., 2009; Wu et al., 2011). Therefore, like Sgk1 and MR, hAF17 appears to inhibit the Dot1a–Af9 interaction at the promoter without measurably affecting the association of Af9 with the *αENaC* promoter. However, unlike Sgk1 that impairs Dot1a–Af9 interaction by phosphorylation of Af9, hAF17 and its mouse homolog Af17 do so presumably by competing with Af9 for binding Dot1a. hAF17 overexpression also led to upregulation of several other aldosterone-regulated genes examined, including *β*- and *γ-ENaC*, *Sgk1*, and *CTGF* (Reisenauer et al., 2009).

3.4 Hsp90 relieves Dot1a–Af9-mediated repression by directly modulating the spatial distribution of Af9

The heat-shock protein Hsp90 plays an important role in nuclear import of steroid receptors. It is also an effector of transcriptional activation through diverse mechanisms. Hsp90 was identified by mass spectrometry as one of the proteins bound a FLAG-AF9-ct column (amino acids 475–568 of hAF9, identical to mouse Af9 aa 464–557) in THP-1 whole-cell extract (Lin & Hemenway, 2010). The interaction was subsequently confirmed by Co-IP and colocalization assays and appeared to be a part of an Hsp90–Hsp70–p60/Hop chaperone complex. Both inhibition of Hsp90 function by the inhibitor novobiocin and knockdown of Hsp90 with siRNA resulted in redistribution of hAF9 from a primarily nuclear to cytoplasmic location, which was associated with a decrease in hAF9 occupancy at the *αENaC* promoter and increased *αENaC* expression (Lin & Hemenway, 2010). Interestingly, the Af9-ct fragment mediating the interaction with Hsp90 resides within the domain capable of binding Dot1a and MR, we name the later as Dot1a/MR/Hsp90-intercating domain. It remains to determine if Hsp90 can compete with Dot1a and/or MR for binding Af9. However, it is very unlikely that Sgk1-mediated Af9 S435 phosphorylation is important for regulation of Af9–Hsp90 interaction since the Af9-ct fragment lacks S435.

4. TRANSCRIPTIONAL CHANGES IN ENaC GENES ARE TRANSLATED INTO CHANGES IN ENaC ACTIVITY

The physiological significance of Dot1a–Af9-mediated repression is indicated by translation of the changes in the mRNA levels of ENaC genes into the corresponding changes in ENaC activity. ENaC activity can be measured with different approaches using ENaC inhibitors such as benzamil. Single-cell fluorescence imagining with SBFI-AM is often used to measure benzamil-sensitive intracellular sodium concentration ($[Na^+]_i$) as an index of ENaC activity. With this approach, we found that aldosterone-induced ENaC mRNA upregulation led to elevated ENaC activity, as seen by an increase in the benzamil-sensitive $[Na^+]_I$ in IMCD3 cells (Wu et al., 2011). Both transcription inhibitor actinomycin D and translation inhibitor cycloheximide largely blocked this effect (Wu et al., 2011). These results indicate that new mRNA and protein synthesis is required for aldosterone-induced ENaC-mediated Na^+ uptake in IMCD3 cells. Furthermore, overexpression of Dot1a and Af9 attenuated the benzamil-sensitive $[Na^+]_i$, while knockdown of Dot1a and Af9 increased Na^+ uptake in IMCD3 cells (Wu et al., 2011).

The second approach is the measurement of the benzamil-sensitive equivalent short-circuit current (I_{sc}) of the confluent monolayer of cells on filter units. Af9 overexpression significantly lessened the benzamil-sensitive I_{sc} (ΔI_{sc}) in IMCD3 and M1 cells. While Dot1a-transfected cells were able to transiently overexpress Dot1a, they failed to maintain Dot1a overexpression during the formation of confluent monolayers on permeable support. This excluded the possibility to evaluate the effect of Dot1a overexpression on the benzamil-sensitive I_{sc} (Wu et al., 2011).

The third approach is whole-cell patch clamping. Nevertheless, this approach is proved inappropriate in IMCD3 and M1 cells due to their fragility. To circumvent this limitation, we performed whole-cell patch clamping in 293T cells, treated with vehicle as control or aldosterone. Alternatively, 293T cells were transiently transfected with an empty vector or a construct encoding Dot1a, Af9, or Af17. A transcriptional increase in ENaC genes by either aldosterone treatment or overexpressing Af17 resulted in an increase in benzamil-sensitive Na^+ current density. Oppositely, overexpression of Dot1a and Af9 decreased the ENaC-mediated Na^+ current (Reisenauer et al., 2009). Thus, multiple and independent lines of data indicate that transcriptional changes in ENaC genes by modulating Dot1a, Af9, and Af17 expression through aldosterone treatment or transient/stable transfection are translated into changes in ENaC activity in all cell model systems (293T, IMCD3, and M1 cells) examined.

5. MOUSE MODELS WITH GENETIC DEFECTS IN ENaC REGULATORS

5.1 *Sgk1*$^{-/-}$ mice

On a standard NaCl diet, *Sgk*$^{-/-}$ mice showed apparently normal renal water and electrolyte excretion, which is consistent with no significant changes in αENaC mRNA and protein abundance; despite of a significant lower level of Af9 phosphorylation, compared with WT littermates. Under Na^+ restriction, however, *Sgk1*$^{-/-}$ mice failed to upregulate renal tubular Na^+ reabsorption, leading to Na^+ depletion despite the presence of increased plasma aldosterone levels (Wulff et al., 2002). This phenotype is associated with impaired Af9 phosphorylation and reduced αENaC expression at both mRNA and protein levels. Taken together, the results indicate that Sgk1 plays an important role in the adaptational response to altered physiological or pathological conditions.

5.2 *MR*$^{-/-}$ mice

The biological significance of aldosterone classical action via MR is clearly demonstrated by the generation and characterization of *MR*$^{-/-}$ mice. *MR*$^{-/-}$ mice die in the second week after birth, showing at day 8 PHA-1 phenotype with hyponatremia, hyperkalemia, high renal salt wasting, and a strongly activated RAAS. Unexpectedly RNase protection analysis of the 8-day-old $MR^{-/-}$ mice did not reveal a significant decrease in the mRNA abundance of the subunits of ENaC and Na^+,K^+ ATPase (Berger et al., 1998). Our recent study revealed that the transcriptional defect in ENaC occurs at an earlier developmental stage and is masked by a compensatory mechanism. $MR^{-/-}$ mice have decreased mRNA levels of ENaC genes at day 5 after birth (Zhang, Zhou, et al., 2013), possibly due to less activated RAAS. The plasma aldosterone concentration in these mutant neonates may be insufficient to attenuate Dot1a–Af9-mediated repression and to activate substantial amount of GR. The impaired ENaC expression may be at least partially responsible for the progressively developed PHA-1 phenotype.

5.3 *Af17*$^{-/-}$ mice

Af17$^{-/-}$ versus *Af17*$^{+/+}$ mice fed a normal Na^+ diet have impaired renal water and electrolyte retention and decreased BP (Chen et al., 2011). This phenotype is associated with increased H3 K79 methylation in whole-kidney lysate and at the *αENaC* promoter; impaired ENaC expression at both mRNA and protein levels, and significantly lower active channel number, open probability, and effective activity of ENaC. In contrast, inducing high levels of plasma aldosterone by dietary Na^+ restriction, high dietary potassium loading, or aldosterone infusion completely compensated for *Af17* deficiency with respect to sodium handling and blood pressure (BP) (Chen et al., 2011).

5.4 *Dot1l*^{AC} mice

Despite having 20% fewer principal cells in which ENaC is expressed (Wu, Chen, Zhou, et al., 2013), CNT/CD-specific *Dot1l* conditional knockout mice (*Dot1l*^{AC}) expressed 36% greater levels of *αENaC* mRNA compared with *Dot1l*$^{f/f}$ mice as measured by real-time RT-qPCR (Zhang, Yu, et al., 2013). The effect of CNT/CD-specific *Dot1l* deletion on *αENaC* mRNA expression in the *Dot1l*^{AC} mice would likely have been even greater if the mice expressed normal numbers of principal cells. Taken together, these data support the conclusion that genetic inactivation of *Dot1l* removes Dot1a-mediated repression of *αENaC* transcription and leads to upregulation of *αENaC* expression, and they reinforce the notion that Dot1a functions as a negative regulator of basal *αENaC* transcription.

6. REGULATION OF ENaC ACTIVITY BY OTHER REGULATORY PROTEINS

Other aldosterone-stimulated ENaC regulatory factors include the *S*-adenosyl-L-homocysteine hydrolase (SAHHase) (Stockand, Al-Baldawi, Al-Khalili, Worrell, & Eaton, 1999; Stockand et al., 2001), glucocorticoid-induced leucine zipper protein (GILZ) (Bhalla, Soundararajan, Pao, Li, & Pearce, 2006; Muller et al., 2003; Soundararajan, Wang, Melters, & Pearce, 2007; Soundararajan, Zhang, Wang, Vandewalle, & Pearce, 2005), the small G protein K-Ras2A (Brennan & Fuller, 2006; Mastroberardino et al., 1998; Verrey, 1999; Verrey et al., 2000), and the channel-activating protease CAP1 (Narikiyo et al., 2002). SAHHase activity, but not expression was enhanced by aldosterone, which in part, increases ENaC open probability (P_o). GILZ markedly increased ENaC membrane expression and ENaC-mediated Na^+ transport, possibly by inhibiting extracellular signal-regulated kinase (ERK) signaling. K-Ras2A also increased the mean activity of ENaC at the cell surface. The complexity of ENaC regulation is also supported by the recent findings that maturation and activation of ENaC subunits involve α- and γENaC cleavage mediated by CAP1 (Caldwell, Boucher, & Stutts, 2004; Vallet, Chraibi, Gaeggeler, Horisberger, & Rossier, 1997; Vuagniaux, Vallet, Jaeger, Hummler, & Rossier, 2002; Vuagniaux et al., 2000) or furin (Bruns et al., 2007; Harris, Garcia-Caballero, Stutts, Firsov, & Rossier, 2008; Hughey et al., 2004; Sheng, Carattino, Bruns, Hughey, & Kleyman, 2006). CAP1 stimulates ENaC activity probably by cleaving γENaC (Bruns et al., 2007; Thomas & Itani, 2004) and/or by increasing ENaC P_o (Caldwell et al., 2004; Vallet et al., 1997; Vuagniaux et al., 2000, 2002). Furin-mediated cleavage of α- and γENaC activates the channels by relieving Na^+ self-inhibition (Sheng et al., 2006). The vesicle traffic regulatory protein syntaxin 1A contains distinct inhibitory and stimulatory domains that interact with ENaC subunits and determine the overall ENaC functionality/regulation under distinct physiological conditions (Condliffe, Zhang, & Frizzell, 2004; Qi et al., 1999; Saxena, Singh, Kaur, & George, 2007). However, the aldosterone effect on furin or syntaxin 1A expression has not been documented yet.

7. CONCLUSION AND FUTURE DIRECTIONS

We propose a model summarizing our understanding how Dot1a–Af9-mediated repression of *αENaC* transcription is achieved or relieved in the absence or presence of aldosterone (Fig. 4). It can be speculated that

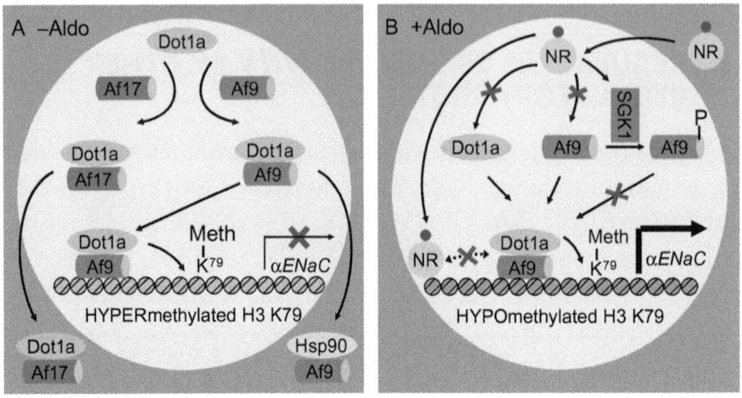

Figure 4 Model for Dot1a–Af9-mediated repression of *αENaC* transcription. Under basal conditions (−Aldo), Af9 recruits Dot1a to form a nuclear complex, which indirectly or directly through Af9 DNA-binding activity binds specific sites of the *αENaC* promoter, leading to hypermethylation of histone H3 K79 and repression of *αENaC* transcription (Zhang, Xia, Jalal, et al., 2006; Zhang, Xia, Reisenauer, et al., 2006; Zhang, Yu, et al., 2013). Af17 relieves the repression by competing with Af9 for binding Dot1a and promoting Dot1a redistribution from the nucleus to cytoplasm (Reisenauer et al., 2009; Wu et al., 2011). Hsp90 also interacts with Af9 and apparently promotes its cytoplasmic distribution to derepress the *αENaC* promoter (Lin & Hemenway, 2010). Aldosterone (+Aldo) stimulates *αENaC* transcription by multiple mechanisms. Through the classical action, aldosterone binds and activates the nuclear hormone receptors (NR) that are either glucocorticoid receptor or mineralocorticoid receptor homo- or heterodimers to bind the glucocorticoid response element in the *αENaC* promoter and transactivate *αENaC*. In parallel, aldosterone releases Dot1a–Af9-mediated repression by limiting the formation of the complex through (A) downregulating Dot1a (Zhang, Xia, Jalal, et al., 2006) and Af9 expression (Zhang, Xia, Reisenauer, et al., 2006), presumably via nuclear receptor-dependent or -independent (not shown) mechanisms, (B) decreasing the Dot1a–Af9 interaction via Sgk1-mediated phosphorylation of Af9 at S435 (Zhang et al., 2007), and (C) counterbalancing Dot1a–Af9 complex by activating MR to compete for binding Af9 (Zhang, Zhou, et al., 2013). These concerted actions lead to histone H3 K79 hypomethylation at specific subregions of the *αENaC* promoter. Sgk1 might join and downregulate the Dot1a–Af9 complex associated with the *αENaC* promoter (not shown for simplicity). In all cases, Af9-free Dot1a binds DNA nonspecifically and catalyzes histone H3 K79 methylation throughout the genome under basal conditions (not shown). Meth, methylation. (See the color plate.)

deregulation of the Dot1a–Af9 complex may result in defects in renal Na^+ excretion, and, thus, potentially abnormal blood pressure in humans. Phenotypically, such deregulation may be reminiscent of the monogenic forms of Na^+ retention and hypertension (Liddle's syndrome) or renal Na^+ wasting and hypotension (pseudohypoaldosteronism type 1), two types of disorders due to gain-of-function and loss-of-function mutations in ENaC genes, respectively. The evidence showing the physiological relevance of Dot1a–Af9 in regulating blood pressure is emerging. In a recent clinical study, a polymorphism in *hDOT1L* (rs2269879) is associated with blood pressure response to hydrochlorothiazide in Caucasians, and a polymorphism in *hAF9* (rs12350051) is associated with untreated blood pressure in African-Americans (Duarte et al., 2012).

In addition to regulation of Na^+ balance, aldosterone may contribute to glomerulosclerosis and interstitial fibrosis (Hostetter & Ibrahim, 2003). Inactivation of *Dot1l* in CNT/CD results in upregulation of many genes including *endothelin 1* and *Lcn2* (Wu, Chen, Zhang, et al., 2013). These genes have profibrotic properties. Endothelin 1 has been previously shown as a target of aldosterone action (Zhang, Xia, Reisenauer, et al., 2006; Zhou et al., 2012). It is directly regulated by Dot1a–Af9 complex (Zhou et al., 2012). Hence, future research to identify if some patients with chronic kidney disease have genetic defects in Dot1a–Af9 pathways and to determine if $Dot1l^{AC}$ mice develop kidney fibrosis with aging may extend our knowledge of Dot1–Af9-mediated repression far beyond ENaC and may lead to new diagnostic, prognostic, and therapeutic strategy for management of kidney fibrosis.

ACKNOWLEDGMENT

This work was supported by National Institutes of Health Grant R01 DK080236 (to W. Z. Z.).

REFERENCES

Alangari, A. A. (2010). Genomic and non-genomic actions of glucocorticoids in asthma. *Annals of Thoracic Medicine*, *5*(3), 133–139. http://dx.doi.org/10.4103/1817-1737.65040.

Alvarez de la Rosa, D., Zhang, P., Naray-Fejes-Toth, A., Fejes-Toth, G., & Canessa, C. M. (1999). The serum and glucocorticoid kinase sgk increases the abundance of epithelial sodium channels in the plasma membrane of Xenopus oocytes. *The Journal of Biological Chemistry*, *274*(53), 37834–37839.

Asher, C., Wald, H., Rossier, B. C., & Garty, H. (1996). Aldosterone-induced increase in the abundance of Na+ channel subunits. *The American Journal of Physiology*, *271*(2 Pt. 1), C605–C611.

Berger, S., Bleich, M., Schmid, W., Cole, T. J., Peters, J., Watanabe, H., et al. (1998). Mineralocorticoid receptor knockout mice: Pathophysiology of Na+ metabolism. *Proceedings of the National Academy of Sciences of the United States of America*, *95*(16), 9424–9429.

Bhalla, V., Soundararajan, R., Pao, A. C., Li, H., & Pearce, D. (2006). Disinhibitory pathways for control of sodium transport: Regulation of ENaC by SGK1 and GILZ. *American Journal of Physiology. Renal Physiology*, *291*(4), F714–F721.

Biswas, D., Milne, T. A., Basrur, V., Kim, J., Elenitoba-Johnson, K. S., Allis, C. D., et al. (2011). Function of leukemogenic mixed lineage leukemia 1 (MLL) fusion proteins through distinct partner protein complexes. *Proceedings of the National Academy of Sciences of the United States of America, 108*(38), 15751–15756. http://dx.doi.org/10.1073/pnas.1111498108.

Boyd, C., & Naray-Fejes-Toth, A. (2005). Gene regulation of ENaC subunits by serum- and glucocorticoid-inducible kinase-1. *American Journal of Physiology. Renal Physiology, 288*(3), F505–F512.

Brennan, F. E., & Fuller, P. J. (2006). Mammalian K-ras2 is a corticosteroid-induced gene in vivo. *Endocrinology, 147*(6), 2809–2816.

Bruns, J. B., Carattino, M. D., Sheng, S., Maarouf, A. B., Weisz, O. A., Pilewski, J. M., et al. (2007). Epithelial Na+ channels are fully activated by furin- and prostasin-dependent release of an inhibitory peptide from the gamma-subunit. *The Journal of Biological Chemistry, 282*(9), 6153–6160.

Caldwell, R. A., Boucher, R. C., & Stutts, M. J. (2004). Serine protease activation of near-silent epithelial Na+ channels. *American Journal of Physiology. Cell Physiology, 286*(1), C190–C194.

Chen, S. Y., Bhargava, A., Mastroberardino, L., Meijer, O. C., Wang, J., Buse, P., et al. (1999). Epithelial sodium channel regulated by aldosterone-induced protein sgk. *Proceedings of the National Academy of Sciences of the United States of America, 96*(5), 2514–2519.

Chen, L., Wu, H., Pochynyuk, O. M., Reisenauer, M. R., Zhang, Z., Huang, L., et al. (2011). Af17 deficiency increases sodium excretion and decreases blood pressure. *Journal of the American Society of Nephrology, 22*(6), 1076–1086. http://dx.doi.org/10.1681/ASN.2010121270 (selected by Faculty of 1000. Invited author interview by Hemodialysis.com, pii: ASN.2010121270).

Cheung, P., Tanner, K. G., Cheung, W. L., Sassone-Corsi, P., Denu, J. M., & Allis, C. D. (2000). Synergistic coupling of histone H3 phosphorylation and acetylation in response to epidermal growth factor stimulation. *Molecular Cell, 5*(6), 905–915.

Condliffe, S. B., Zhang, H., & Frizzell, R. A. (2004). Syntaxin 1A regulates ENaC channel activity. *The Journal of Biological Chemistry, 279*(11), 10085–10092.

Cooper, M. S., Bujalska, I., Rabbitt, E., Walker, E. A., Bland, R., Sheppard, M. C., et al. (2001). Modulation of 11beta-hydroxysteroid dehydrogenase isozymes by proinflammatory cytokines in osteoblasts: An autocrine switch from glucocorticoid inactivation to activation. *Journal of Bone and Mineral Research, 16*(6), 1037–1044.

Cowling, R. T., & Birnboim, H. C. (2000). Expression of serum- and glucocorticoid-regulated kinase (sgk) mRNA is up-regulated by GM-CSF and other proinflammatory mediators in human granulocytes. *Journal of Leukocyte Biology, 67*(2), 240–248.

Dahlmann, A., Pradervand, S., Hummler, E., Rossier, B. C., Frindt, G., & Palmer, L. G. (2003). Mineralocorticoid regulation of epithelial Na+ channels is maintained in a mouse model of Liddle's syndrome. *American Journal of Physiology. Renal Physiology, 285*(2), F310–F318.

Debonneville, C., Flores, S. Y., Kamynina, E., Plant, P. J., Tauxe, C., Thomas, M. A., et al. (2001). Phosphorylation of Nedd4-2 by Sgk1 regulates epithelial Na(+) channel cell surface expression. *The EMBO Journal, 20*(24), 7052–7059.

Dooley, R., Angibaud, E., Yusef, Y. R., Thomas, W., & Harvey, B. J. (2013). Aldosterone-induced ENaC and basal Na+/K+-ATPase trafficking via protein kinase D1-phosphatidylinositol 4-kinaseIIIbeta trans Golgi signalling in M1 cortical collecting duct cells. *Molecular and Cellular Endocrinology, 372*(1–2), 86–95. http://dx.doi.org/10.1016/j.mce.2013.03.011.

Duarte, J. D., Zineh, I., Burkley, B., Gong, Y., Langaee, T. Y., Turner, S. T., et al. (2012). Effects of genetic variation in H3K79 methylation regulatory genes on clinical blood pressure and blood pressure response to hydrochlorothiazide. *Journal of Translational Medicine, 10*, 56. http://dx.doi.org/10.1186/1479-5876-10-56.

Erfurth, F., Hemenway, C. S., De Erkenez, A. C., & Domer, P. H. (2004). MLL fusion partners AF4 and AF9 interact at subnuclear foci. *Leukemia, 18*(1), 92–102.

Fakitsas, P., Adam, G., Daidie, D., van Bemmelen, M. X., Fouladkou, F., Patrignani, A., et al. (2007). Early aldosterone-induced gene product regulates the epithelial sodium channel by deubiquitylation. *Journal of the American Society of Nephrology, 18*(4), 1084–1092.

Faresse, N., Debonneville, A., & Staub, O. (2013). USP2-45 represses aldosterone mediated responses by decreasing mineralocorticoid receptor availability. *Cellular Physiology and Biochemistry, 31*(2–3), 462–472. http://dx.doi.org/10.1159/000343382.

Faresse, N., Ruffieux-Daidie, D., Salamin, M., Gomez-Sanchez, C. E., & Staub, O. (2010). Mineralocorticoid receptor degradation is promoted by Hsp90 inhibition and the ubiquitin-protein ligase CHIP. *American Journal of Physiology. Renal Physiology, 299*(6), F1462–F1472. http://dx.doi.org/10.1152/ajprenal.00285.2010.

Feng, Q., Wang, H., Ng, H. H., Erdjument-Bromage, H., Tempst, P., Struhl, K., et al. (2002). Methylation of H3-lysine 79 is mediated by a new family of HMTases without a SET domain. *Current Biology, 12*(12), 1052–1058.

Firestone, G. L., Giampaolo, J. R., & O'Keeffe, B. A. (2003). Stimulus-dependent regulation of serum and glucocorticoid inducible protein kinase (SGK) transcription, subcellular localization and enzymatic activity. *Cellular Physiology and Biochemistry, 13*(1), 1–12.

Gumz, M. L., Popp, M. P., Wingo, C. S., & Cain, B. D. (2003). Early transcriptional effects of aldosterone in a mouse inner medullary collecting duct cell line. *American Journal of Physiology. Renal Physiology, 285*(4), F664–F673.

Harris, M., Garcia-Caballero, A., Stutts, M. J., Firsov, D., & Rossier, B. C. (2008). Preferential assembly of epithelial sodium channel (ENaC) subunits in Xenopus oocytes: Role of furin-mediated endogenous proteolysis. *The Journal of Biological Chemistry, 283*(12), 7455–7463. http://dx.doi.org/10.1074/jbc.M707399200.

Helms, M. N., Yu, L., Malik, B., Kleinhenz, D. J., Hart, C. M., & Eaton, D. C. (2005). Role of SGK1 in nitric oxide inhibition of ENaC in Na+-transporting epithelia. *American Journal of Physiology. Cell Physiology, 289*(3), C717–C726.

Hemenway, C. S., de Erkenez, A. C., & Gould, G. C. (2001). The polycomb protein MPc3 interacts with AF9, an MLL fusion partner in t(9;11)(p22;q23) acute leukemias. *Oncogene, 20*(29), 3798–3805.

Hostetter, T. H., & Ibrahim, H. N. (2003). Aldosterone in chronic kidney and cardiac disease. *Journal of the American Society of Nephrology, 14*(9), 2395–2401.

Hou, J., Speirs, H. J., Seckl, J. R., & Brown, R. W. (2002). Sgk1 gene expression in kidney and its regulation by aldosterone: Spatio-temporal heterogeneity and quantitative analysis. *Journal of the American Society of Nephrology, 13*(5), 1190–1198.

Huang, D. Y., Boini, K. M., Friedrich, B., Metzger, M., Just, L., Osswald, H., et al. (2006). Blunted hypertensive effect of combined fructose and high-salt diet in gene-targeted mice lacking functional serum- and glucocorticoid-inducible kinase SGK1. *American Journal of Physiology. Regulatory, Integrative and Comparative Physiology, 290*(4), R935–R944.

Hughey, R. P., Bruns, J. B., Kinlough, C. L., Harkleroad, K. L., Tong, Q., Carattino, M. D., et al. (2004). Epithelial sodium channels are activated by furin-dependent proteolysis. *The Journal of Biological Chemistry, 279*(18), 18111–18114.

Hummler, E., & Vallon, V. (2005). Lessons from mouse mutants of epithelial sodium channel and its regulatory proteins. *Journal of the American Society of Nephrology, 16*(11), 3160–3166.

Ichimura, T., Yamamura, H., Sasamoto, K., Tominaga, Y., Taoka, M., Kakiuchi, K., et al. (2005). 14-3-3 Proteins modulate the expression of epithelial Na+ channels by phosphorylation-dependent interaction with Nedd4-2 ubiquitin ligase. *The Journal of Biological Chemistry, 280*(13), 13187–13194.

Iida, S., Seto, M., Yamamoto, K., Komatsu, H., Tojo, A., Asano, S., et al. (1993). MLLT3 gene on 9p22 involved in t(9;11) leukemia encodes a serine/proline rich protein homologous to MLLT1 on 19p13. *Oncogene, 8*(11), 3085–3092.

Jenuwein, T., & Allis, C. D. (2001). Translating the histone code. *Science, 293*(5532), 1074–1080.

Kamynina, E., & Staub, O. (2002). Concerted action of ENaC, Nedd4-2, and Sgk1 in transepithelial Na(+) transport. *American Journal of Physiology. Renal Physiology, 283*(3), F377–F387.

Kellner, M., Peiter, A., Hafner, M., Feuring, M., Christ, M., Wehling, M., et al. (2003). Early aldosterone up-regulated genes: New pathways for renal disease? *Kidney International, 64*(4), 1199–1207.

Kitagawa, H., Yanagisawa, J., Fuse, H., Ogawa, S., Yogiashi, Y., Okuno, A., et al. (2002). Ligand-selective potentiation of rat mineralocorticoid receptor activation function 1 by a CBP-containing histone acetyltransferase complex. *Molecular and Cellular Biology, 22*(11), 3698–3706.

Krogan, N. J., Dover, J., Wood, A., Schneider, J., Heidt, J., Boateng, M. A., et al. (2003). The Paf1 complex is required for histone H3 methylation by COMPASS and Dot1p: Linking transcriptional elongation to histone methylation. *Molecular Cell, 11*(3), 721–729.

Ku, N. O., Liao, J., & Omary, M. B. (1998). Phosphorylation of human keratin 18 serine 33 regulates binding to 14-3-3 proteins. *The EMBO Journal, 17*(7), 1892–1906.

Lee, I. H., Campbell, C. R., Cook, D. I., & Dinudom, A. (2008). Regulation of epithelial Na+ channels by aldosterone: Role of Sgk1. *Clinical and Experimental Pharmacology & Physiology, 35*(2), 235–241.

Lin, J. J., & Hemenway, C. S. (2010). Hsp90 directly modulates the spatial distribution of AF9/MLLT3 and affects target gene expression. *The Journal of Biological Chemistry, 285*(16), 11966–11973. http://dx.doi.org/10.1074/jbc.M110.101642 (pii: M110.101642).

Lin, Y. M., Ono, K., Satoh, S., Ishiguro, H., Fujita, M., Miwa, N., et al. (2001). Identification of AF17 as a downstream gene of the beta-catenin/T-cell factor pathway and its involvement in colorectal carcinogenesis. *Cancer Research, 61*(17), 6345–6349.

Liu, W., Wang, J., Sauter, N. K., & Pearce, D. (1995). Steroid receptor heterodimerization demonstrated in vitro and in vivo. *Proceedings of the National Academy of Sciences of the United States of America, 92*(26), 12480–12484.

Loffing, J., Pietri, L., Aregger, F., Bloch-Faure, M., Ziegler, U., Meneton, P., et al. (2000). Differential subcellular localization of ENaC subunits in mouse kidney in response to high- and low-Na diets. *American Journal of Physiology. Renal Physiology, 279*(2), F252–F258.

Loffing, J., Zecevic, M., Feraille, E., Kaissling, B., Asher, C., Rossier, B. C., et al. (2001). Aldosterone induces rapid apical translocation of ENaC in early portion of renal collecting system: Possible role of SGK. *American Journal of Physiology. Renal Physiology, 280*(4), F675–F682.

Losel, R., Feuring, M., & Wehling, M. (2002). Non-genomic aldosterone action: From the cell membrane to human physiology. *The Journal of Steroid Biochemistry and Molecular Biology, 83*(1–5), 167–171.

Mastroberardino, L., Spindler, B., Forster, I., Loffing, J., Assandri, R., May, A., et al. (1998). Ras pathway activates epithelial Na+ channel and decreases its surface expression in Xenopus oocytes. *Molecular Biology of the Cell, 9*(12), 3417–3427.

McAvoy, T., Allen, P. B., Obaishi, H., Nakanishi, H., Takai, Y., Greengard, P., et al. (1999). Regulation of neurabin I interaction with protein phosphatase 1 by phosphorylation. *Biochemistry, 38*(39), 12943–12949.

Min, J., Feng, Q., Li, Z., Zhang, Y., & Xu, R. M. (2003). Structure of the catalytic domain of human DOT1L, a non-SET domain nucleosomal histone methyltransferase. *Cell, 112*(5), 711–723.

Muller, O. G., Parnova, R. G., Centeno, G., Rossier, B. C., Firsov, D., & Horisberger, J. D. (2003). Mineralocorticoid effects in the kidney: Correlation between alphaENaC,

GILZ, and Sgk-1 mRNA expression and urinary excretion of Na+ and K+. *Journal of the American Society of Nephrology, 14*(5), 1107–1115.

Narikiyo, T., Kitamura, K., Adachi, M., Miyoshi, T., Iwashita, K., Shiraishi, N., et al. (2002). Regulation of prostasin by aldosterone in the kidney. *The Journal of Clinical Investigation, 109*(3), 401–408.

Narlikar, G. J., Fan, H. Y., & Kingston, R. E. (2002). Cooperation between complexes that regulate chromatin structure and transcription. *Cell, 108*(4), 475–487.

Ng, H. H., Ciccone, D. N., Morshead, K. B., Oettinger, M. A., & Struhl, K. (2003). Lysine-79 of histone H3 is hypomethylated at silenced loci in yeast and mammalian cells: A potential mechanism for position-effect variegation. *Proceedings of the National Academy of Sciences of the United States of America, 100*(4), 1820–1825.

Ng, H. H., Feng, Q., Wang, H., Erdjument-Bromage, H., Tempst, P., Zhang, Y., et al. (2002). Lysine methylation within the globular domain of histone H3 by Dot1 is important for telomeric silencing and Sir protein association. *Genes & Development, 16*(12), 1518–1527. http://dx.doi.org/10.1101/gad.1001502.

Nguyen, A. T., & Zhang, Y. (2011). The diverse functions of Dot1 and H3K79 methylation. *Genes & Development, 25*(13), 1345–1358. http://dx.doi.org/10.1101/gad.2057811.

Okada, Y., Feng, Q., Lin, Y., Jiang, Q., Li, Y., Coffield, V. M., et al. (2005). hDOT1L links histone methylation to leukemogenesis. *Cell, 121*(2), 167–178.

Ono, S., Kusano, E., Muto, S., Ando, Y., & Asano, Y. (1997). A low-Na+ diet enhances expression of mRNA for epithelial Na+ channel in rat renal inner medulla. *Pflügers Archiv, 434*(6), 756–763.

Park, J., Leong, M. L., Buse, P., Maiyar, A. C., Firestone, G. L., & Hemmings, B. A. (1999). Serum and glucocorticoid-inducible kinase (SGK) is a target of the PI 3-kinase-stimulated signaling pathway. *The EMBO Journal, 18*(11), 3024–3033.

Pietras, R. J., Nemere, I., & Szego, C. M. (2001). Steroid hormone receptors in target cell membranes. *Endocrine, 14*(3), 417–427. http://dx.doi.org/10.1385/ENDO:14:3:417 (pii: ENDO:14:3:417).

Pradervand, S., Vandewalle, A., Bens, M., Gautschi, I., Loffing, J., Hummler, E., et al. (2003). Dysfunction of the epithelial sodium channel expressed in the kidney of a mouse model for Liddle syndrome. *Journal of the American Society of Nephrology, 14*(9), 2219–2228.

Prasad, R., Leshkowitz, D., Gu, Y., Alder, H., Nakamura, T., Saito, H., et al. (1994). Leucine-zipper dimerization motif encoded by the AF17 gene fused to ALL-1 (MLL) in acute leukemia. *Proceedings of the National Academy of Sciences of the United States of America, 91*(17), 8107–8111.

Qi, J., Peters, K. W., Liu, C., Wang, J. M., Edinger, R. S., Johnson, J. P., et al. (1999). Regulation of the amiloride-sensitive epithelial sodium channel by syntaxin 1A. *The Journal of Biological Chemistry, 274*(43), 30345–30348.

Rea, S., Eisenhaber, F., O'Carroll, D., Strahl, B. D., Sun, Z. W., Schmid, M., et al. (2000). Regulation of chromatin structure by site-specific histone H3 methyltransferases. *Nature, 406*(6796), 593–599.

Reisenauer, M. R., Anderson, M., Huang, L., Zhang, Z., Zhou, Q., Kone, B. C., et al. (2009). AF17 competes with AF9 for binding to Dot1a to up-regulate transcription of epithelial Na+ channel alpha. *The Journal of Biological Chemistry, 284*(51), 35659–35669. http://dx.doi.org/10.1074/jbc.M109.038448.

Reisenauer, M. R., Wang, S. W., Xia, Y., & Zhang, W. (2010). Dot1a contains three nuclear localization signals and regulates the epithelial Na+ channel (ENaC) at multiple levels. *American Journal of Physiology. Renal Physiology, 299*(1), F63–F76. http://dx.doi.org/10.1152/ajprenal.00105.2010.

Robert-Nicoud, M., Flahaut, M., Elalouf, J. M., Nicod, M., Salinas, M., Bens, M., et al. (2001). Transcriptome of a mouse kidney cortical collecting duct cell line: Effects of aldosterone and vasopressin. *Proceedings of the National Academy of Sciences of the United States of America, 98*(5), 2712–2716.

Rong, R., Surace, E. I., Haipek, C. A., Gutmann, D. H., & Ye, K. (2004). Serine 518 phosphorylation modulates merlin intramolecular association and binding to critical effectors important for NF2 growth suppression. *Oncogene, 23*(52), 8447–8454.

Sahin, P., McCaig, C., Jeevahan, J., Murray, J. T., & Hainsworth, A. H. (2013). The cell survival kinase SGK1 and its targets FOXO3a and NDRG1 in aged human brain. *Neuropathology and Applied Neurobiology, 39*(6), 623–633. http://dx.doi.org/10.1111/nan.12023.

Savory, J. G., Prefontaine, G. G., Lamprecht, C., Liao, M., Walther, R. F., Lefebvre, Y. A., et al. (2001). Glucocorticoid receptor homodimers and glucocorticoid-mineralocorticoid receptor heterodimers form in the cytoplasm through alternative dimerization interfaces. *Molecular and Cellular Biology, 21*(3), 781–793. http://dx.doi.org/10.1128/MCB.21.3.781-793.2001.

Saxena, S. K., Singh, M., Kaur, S., & George, C. (2007). Distinct domain-dependent effect of syntaxin1A on amiloride-sensitive sodium channel (ENaC) currents in HT-29 colonic epithelial cells. *International Journal of Biological Sciences, 3*(1), 47–56.

Sayegh, R., Auerbach, S. D., Li, X., Loftus, R. W., Husted, R. F., Stokes, J. B., et al. (1999). Glucocorticoid induction of epithelial sodium channel expression in lung and renal epithelia occurs via trans-activation of a hormone response element in the 5′-flanking region of the human epithelial sodium channel alpha subunit gene. *The Journal of Biological Chemistry, 274*(18), 12431–12437.

Seebohm, G., Strutz-Seebohm, N., Baltaev, R., Korniychuk, G., Knirsch, M., Engel, J., et al. (2005). Regulation of KCNQ4 potassium channel prepulse dependence and current amplitude by SGK1 in Xenopus oocytes. *Cellular Physiology and Biochemistry, 16*(4–6), 255–262.

Setiawan, I., Henke, G., Feng, Y., Bohmer, C., Vasilets, L. A., Schwarz, W., et al. (2002). Stimulation of Xenopus oocyte Na(+), K(+)ATPase by the serum and glucocorticoid-dependent kinase sgk1. *Pflügers Archiv, 444*(3), 426–431.

Sheng, S., Carattino, M. D., Bruns, J. B., Hughey, R. P., & Kleyman, T. R. (2006). Furin cleavage activates the epithelial Na+ channel by relieving Na+ self-inhibition. *American Journal of Physiology. Renal Physiology, 290*(6), F1488–F1496.

Singer, M. S., Kahana, A., Wolf, A. J., Meisinger, L. L., Peterson, S. E., Goggin, C., et al. (1998). Identification of high-copy disruptors of telomeric silencing in *Saccharomyces cerevisiae. Genetics, 150*(2), 613–632.

Snyder, P. M., Olson, D. R., & Thomas, B. C. (2002). Serum and glucocorticoid-regulated kinase modulates Nedd4-2-mediated inhibition of the epithelial Na+ channel. *The Journal of Biological Chemistry, 277*(1), 5–8.

Soodvilai, S., Jia, Z., Fongsupa, S., Chatsudthipong, V., & Yang, T. (2012). Liver X receptor agonists decrease ENaC-mediated sodium transport in collecting duct cells. *American Journal of Physiology. Renal Physiology, 303*(12), F1610–F1616. http://dx.doi.org/10.1152/ajprenal.00283.2012.

Soundararajan, R., Wang, J., Melters, D., & Pearce, D. (2007). Differential activities of glucocorticoid-induced leucine zipper protein isoforms. *The Journal of Biological Chemistry, 282*(50), 36303–36313.

Soundararajan, R., Zhang, T. T., Wang, J., Vandewalle, A., & Pearce, D. (2005). A novel role for glucocorticoid-induced leucine zipper protein in epithelial sodium channel-mediated sodium transport. *The Journal of Biological Chemistry, 280*(48), 39970–39981.

Spindler, B., & Verrey, F. (1999). Aldosterone action: Induction of p21(ras) and fra-2 and transcription-independent decrease in myc, jun, and fos. *The American Journal of Physiology, 276*(5 Pt 1), C1154–C1161.

Srinivasan, R. S., de Erkenez, A. C., & Hemenway, C. S. (2003). The mixed lineage leukemia fusion partner AF9 binds specific isoforms of the BCL-6 corepressor. *Oncogene, 22*(22), 3395–3406.

Stockand, J. D. (2002). New ideas about aldosterone signaling in epithelia. *American Journal of Physiology. Renal Physiology, 282*(4), F559–F576.

Stockand, J. D., Al-Baldawi, N. F., Al-Khalili, O. K., Worrell, R. T., & Eaton, D. C. (1999). *S*-Adenosyl-L-homocysteine hydrolase regulates aldosterone-induced Na+ transport. *The Journal of Biological Chemistry, 274*(6), 3842–3850.

Stockand, J. D., Zeltwanger, S., Bao, H. F., Becchetti, A., Worrell, R. T., & Eaton, D. C. (2001). *S*-Adenosyl-L-homocysteine hydrolase is necessary for aldosterone-induced activity of epithelial Na(+) channels. *American Journal of Physiology. Cell Physiology, 281*(3), C773–C785.

Strahl, B. D., & Allis, C. D. (2000). The language of covalent histone modifications. *Nature, 403*(6765), 41–45.

Suzukawa, K., Shimizu, S., Nemoto, N., Takei, N., Taki, T., & Nagasawa, T. (2005). Identification of a chromosomal breakpoint and detection of a novel form of an MLL-AF17 fusion transcript in acute monocytic leukemia with t(11;17)(q23;q21). *International Journal of Hematology, 82*(1), 38–41.

Thomas, C. P., & Itani, O. A. (2004). New insights into epithelial sodium channel function in the kidney: Site of action, regulation by ubiquitin ligases, serum- and glucocorticoid-inducible kinase and proteolysis. *Current Opinion in Nephrology and Hypertension, 13*(5), 541–548.

Vallet, V., Chraibi, A., Gaeggeler, H. P., Horisberger, J. D., & Rossier, B. C. (1997). An epithelial serine protease activates the amiloride-sensitive sodium channel. *Nature, 389*(6651), 607–610.

Vallon, V., Huang, D. Y., Grahammer, F., Wyatt, A. W., Osswald, H., Wulff, P., et al. (2005). SGK1 as a determinant of kidney function and salt intake in response to mineralocorticoid excess. *American Journal of Physiology. Regulatory, Integrative and Comparative Physiology, 289*(2), R395–R401.

Vallon, V., Wulff, P., Huang, D. Y., Loffing, J., Volkl, H., Kuhl, D., et al. (2005). Role of Sgk1 in salt and potassium homeostasis. *American Journal of Physiology. Regulatory, Integrative and Comparative Physiology, 288*(1), R4–R10.

Vallon, V., Wyatt, A. W., Klingel, K., Huang, D. Y., Hussain, A., Berchtold, S., et al. (2006). SGK1-dependent cardiac CTGF formation and fibrosis following DOCA treatment. *Journal of Molecular Medicine, 84*(5), 396–404.

van Leeuwen, F., Gafken, P. R., & Gottschling, D. E. (2002). Dot1p modulates silencing in yeast by methylation of the nucleosome core. *Cell, 109*(6), 745–756.

Verrey, F. (1999). Early aldosterone action: Toward filling the gap between transcription and transport. *The American Journal of Physiology, 277*(3 Pt. 2), F319–F327.

Verrey, F., Fakitsas, P., Adam, G., & Staub, O. (2008). Early transcriptional control of ENaC (de)ubiquitylation by aldosterone. *Kidney International, 73*(6), 691–696. http://dx.doi.org/10.1038/sj.ki.5002737.

Verrey, F., Pearce, D., Pfeiffer, R., Spindler, B., Mastroberardino, L., Summa, V., et al. (2000). Pleiotropic action of aldosterone in epithelia mediated by transcription and post-transcription mechanisms. *Kidney International, 57*(4), 1277–1282.

Vuagniaux, G., Vallet, V., Jaeger, N. F., Hummler, E., & Rossier, B. C. (2002). Synergistic activation of ENaC by three membrane-bound channel-activating serine proteases (mCAP1, mCAP2, and mCAP3) and serum- and glucocorticoid-regulated kinase (Sgk1) in Xenopus oocytes. *The Journal of General Physiology, 120*(2), 191–201.

Vuagniaux, G., Vallet, V., Jaeger, N. F., Pfister, C., Bens, M., Farman, N., et al. (2000). Activation of the amiloride-sensitive epithelial sodium channel by the serine protease mCAP1 expressed in a mouse cortical collecting duct cell line. *Journal of the American Society of Nephrology, 11*(5), 828–834.

Waldegger, S., Barth, P., Raber, G., & Lang, F. (1997). Cloning and characterization of a putative human serine/threonine protein kinase transcriptionally modified during anisotonic and isotonic alterations of cell volume. *Proceedings of the National Academy of Sciences of the United States of America, 94*(9), 4440–4445.

Waldegger, S., Gabrysch, S., Barth, P., Fillon, S., & Lang, F. (2000). h-sgk serine-threonine protein kinase as transcriptional target of p38/MAP kinase pathway in HepG2 human hepatoma cells. *Cellular Physiology and Biochemistry, 10*(4), 203–208.

Wang, H., Huang, Z. Q., Xia, L., Feng, Q., Erdjument-Bromage, H., Strahl, B. D., et al. (2001). Methylation of histone H4 at arginine 3 facilitating transcriptional activation by nuclear hormone receptor. *Science, 293*(5531), 853–857.

Weger, S., Wendland, M., Kleinschmidt, J. A., & Heilbronn, R. (1999). The adeno-associated virus type 2 regulatory proteins rep78 and rep68 interact with the transcriptional coactivator PC4. *Journal of Virology, 73*(1), 260–269.

Westin, S., Rosenfeld, M. G., & Glass, C. K. (2000). Nuclear receptor coactivators. *Advances in Pharmacology, 47*, 89–112.

Wu, H., Chen, L., Zhang, X., Zhou, Q., Li, J. M., Berger, S., et al. (2013). Aqp5 is a new transcriptional target of dot1a and a regulator of aqp2. *PLoS One, 8*(1), e53342. http://dx.doi.org/10.1371/journal.pone.0053342.

Wu, H., Chen, L., Zhou, Q., & Zhang, W. (2011). AF17 facilitates Dot1a nuclear export and upregulates ENaC-mediated Na+ transport in renal collecting duct cells. *PLoS One, 6*(11), e27429. http://dx.doi.org/10.1371/journal.pone.0027429.

Wu, H., Chen, L., Zhou, Q., Zhang, X., Berger, S., Bi, J., et al. (2013). Aqp2-expressing cells give rise to renal intercalated cells. *Journal of the American Society of Nephrology, 24*(2), 243–252. http://dx.doi.org/10.1681/ASN.2012080866 (Editorial in the same issue of J Am Soc Nephrol pp. 163–165, Selection by Faculty of 1000).

Wulff, P., Vallon, V., Huang, D. Y., Volkl, H., Yu, F., Richter, K., et al. (2002). Impaired renal Na(+) retention in the sgk1-knockout mouse. *The Journal of Clinical Investigation, 110*(9), 1263–1268.

Yang, J., & Young, M. J. (2009). The mineralocorticoid receptor and its coregulators. *Journal of Molecular Endocrinology, 43*(2), 53–64. http://dx.doi.org/10.1677/JME-09-0031.

Yoo, D., Kim, B. Y., Campo, C., Nance, L., King, A., Maouyo, D., et al. (2003). Cell surface expression of the ROMK (Kir 1.1) channel is regulated by the aldosterone-induced kinase, SGK-1, and protein kinase A. *The Journal of Biological Chemistry, 278*(25), 23066–23075.

Yun, C. C., Palmada, M., Embark, H. M., Fedorenko, O., Feng, Y., Henke, G., et al. (2002). The serum and glucocorticoid-inducible kinase SGK1 and the Na+/H+ exchange regulating factor NHERF2 synergize to stimulate the renal outer medullary K+ channel ROMK1. *Journal of the American Society of Nephrology, 13*(12), 2823–2830.

Zhang, W., Hayashizaki, Y., & Kone, B. C. (2004). Structure and regulation of the mDot1 gene, a mouse histone H3 methyltransferase. *The Biochemical Journal, 377*(Pt. 3), 641–651.

Zhang, D., Li, S., Cruz, P., & Kone, B. C. (2009). Sirtuin 1 functionally and physically interacts with disruptor of telomeric silencing-1 to regulate alpha-ENaC transcription in collecting duct. *The Journal of Biological Chemistry, 284*(31), 20917–20926. http://dx.doi.org/10.1074/jbc.M109.020073.

Zhang, W., Xia, X., Jalal, D. I., Kuncewicz, T., Xu, W., Lesage, G. D., et al. (2006). Aldosterone-sensitive repression of ENaCalpha transcription by a histone H3 lysine-79 methyltransferase. *American Journal of Physiology. Cell Physiology, 290*(3), C936–C946.

Zhang, W., Xia, X., Reisenauer, M. R., Hemenway, C. S., & Kone, B. C. (2006). Dot1a-AF9 complex mediates histone H3 Lys-79 hypermethylation and repression of ENaC {alpha} in an aldosterone-sensitive manner. *The Journal of Biological Chemistry*, *281*(26), 18059–18068.

Zhang, W., Xia, X., Reisenauer, M. R., Rieg, T., Lang, F., Kuhl, D., et al. (2007). Aldosterone-induced Sgk1 relieves Dot1a-Af9-mediated transcriptional repression of epithelial Na+ channel alpha. *The Journal of Clinical Investigation*, *117*(3), 773–783 (Comment in the same issue of J Clin Invest pp. 592–595, Selection by Faculty of 1000).

Zhang, W., Yu, Z., Wu, H., Chen, L., Kong, Q., & Kone, B. C. (2013). An Af9 cis-element directly targets Dot1a to mediate transcriptional repression of the alphaENaC gene. *American Journal of Physiology. Renal Physiology*, *304*(4), F367–F375. http://dx.doi.org/10.1152/ajprenal.00537.2011.

Zhang, X., Zhou, Q., Chen, L., Berger, S., Wu, H., Xiao, Z., et al. (2013). Mineralocorticoid receptor antagonizes Dot1a-Af9 complex to increase alphaENaC transcription. *American Journal of Physiology. Renal Physiology*, *305*(10), F1436–F1444. http://dx.doi.org/10.1152/ajprenal.00202.2013.

Zhou, Q., Liu, K., Wu, H., Chen, L., Pouranan, V., Yuan, M., et al. (2012). Spironolactone rescues Dot1a-Af9-mediated repression of endothelin-1 and improves kidney injury in streptozotocin-induced diabetic rats. *PLoS One*, 7(10), e47360. http://dx.doi.org/10.1371/journal.pone.0047360.

CHAPTER FIVE

Control of ENaC-Mediated Sodium Reabsorption in the Distal Nephron by Bradykinin

Mykola Mamenko, Oleg Zaika, Nabila Boukelmoune, Eric Madden, Oleh Pochynyuk[1]

Department of Integrative Biology and Pharmacology, The University of Texas Health Science Center at Houston, Houston, Texas, USA

[1]Corresponding author: e-mail address: Oleh.M.Pochynyuk@uth.tmc.edu

Contents

Abstract

Kinins, such as Bradykinin (BK), are peptide hormones of the kallikrein–kinin system. Apart from being a vasodilator, BK also increases urinary sodium excretion to reduce systemic blood pressure. It is becoming appreciated that BK modulates function of the epithelial Na^+ channel in the distal part of the renal nephron to affect tubular sodium reabsorption. In this chapter, we outline the molecular details, as well as discuss the physiological relevance of this regulation for the whole organism sodium homeostasis and setting chronic blood pressure.

ABBREVIATIONS

ACE angiotensin converting enzyme
Ang II angiotensin II
ASDN aldosterone-sensitive distal nephron
B1R Bradykinin receptor type 1
B2R Bradykinin receptor type 2

Vitamins and Hormones, Volume 98
ISSN 0083-6729
http://dx.doi.org/10.1016/bs.vh.2014.12.005

BK Bradykinin
BP blood pressure
CCD cortical collecting duct
CD collecting duct
CNT connecting tubule
CYP cytochrome P450
ENaC epithelial sodium channel
GPCR G-protein coupled receptor
$G_{q/11}$ heterotrimeric G protein Q/11 subtype
KKS kallikrein-kinin system
KLD kallidin
NO nitric oxide
PIP_2 or $PI(4,5)P_2$ phosphatidylinositol 4,5-bisphosphate
PKC protein kinase C
PLA_2 phospholipase A2
PLC phospholipase C
PLD protein kinase D
RAAS renin–angiotensin–aldosterone system
SS Dahl rats salt-sensitive Dahl rats

1. INTRODUCTION

The kallikrein–kinin system (KKS) has been proposed to regulate a plethora of physiological processes, including inflammation, coagulation, pain, cell proliferation, vasodilation, and blood pressure (BP) (Kakoki & Smithies, 2009; Kayashima, Smithies, & Kakoki, 2012; Manolis, Marketou, Gavras, & Gavras, 2010; Regoli, Plante, & Gobeil, 2012). Furthermore, beneficial roles of KKS activation were documented in diabetic nephropathy (Kakoki & Smithies, 2009; Kakoki, Takahashi, Jennette, & Smithies, 2004), glomerulosclerosis (Hirawa et al., 1999), ischemia–reperfusion injury (Kakoki, McGarrah, Kim, & Smithies, 2007), chronic renal failure (Wolf, Yoshida, Agata, Chao, & Chao, 2000), and possibly in polycystic kidney disease (Braun et al., 2002). In general, KKS is viewed as a natural counterbalance to the renin–angiotensin–aldosterone system (RAAS) (Shen & El-Dahr, 2006). KKS acts to reduce systemic BP and generation of reactive oxygen species, thus playing a protective role against organ damage in the heart and kidney (Kayashima et al., 2012). The balance between RAAS and KKS affects BP salt sensitivity, circulating volume, and vascular tone (Schmaier, 2003; Tschope, Schultheiss, & Walther, 2002). In the kidney, KKS favors natriuresis by reducing tubular Na^+ reabsorption and this process depends on dietary sodium intake (Mamenko, Zaika,

Doris, & Pochynyuk, 2012; Sivritas, Ploth, & Fitzgibbon, 2008). Recent experimental effort indicates that these natriuretic actions are largely mediated via regulation of the activity of the epithelial Na^+ channel (ENaC) by Bradykinin (BK) (Mamenko, Zaika, Doris, et al., 2012; Zaika, Mamenko, O'Neil, & Pochynyuk, 2011).

2. KKS COMPONENTS

The nonapeptide BK and decapeptide kallidin/kallidin-like peptide (Lys-BK in humans and Arg-BK in rodents, respectively) are major effectors of the KKS (Kakoki & Smithies, 2009; Tang, Leung, & Lai, 2011). Kinins are generated from an inactive precursor kininogen mainly by the action of the serine protease kallikrein (Fig. 1A). Targeted disruption of the gene encoding kallikrein drastically reduces kinin production in many tissues, including the kidney, suggesting that kinin synthesis is a virtually kallikrein-dependent process (Tang et al., 2011). In addition, plasma aminopeptidases can cleave kallidin to produce BK. Both kinins interact with G protein-coupled B2 receptors (B2R) and, to a lesser extent, with B1 receptors (B1R) (Hall, 1992, 1997). The physiological actions of the KKS are mainly mediated by BK and have both beneficial (vasodilation, natriuresis, diuresis, decrease in insulin resistance, etc.) and detrimental (pain, inflammation, allergy, and edema) consequences (Regoli et al., 2012). B2R are constitutively expressed in many tissues, such as smooth muscles, neurons, vascular endothelium, and kidney epithelial cells (Ma, Wang, Chao, & Chao, 1994). In contrast, expression of B1R is low during normal physiology but is markedly upregulated by tissue injury and inflammation (Marceau, Larrivee, Saint-Jacques, & Bachvarov, 1997).

BK is a local (or paracrine) hormone. This property arises from very rapid BK degradation. Thus, plasma half-life of BK is only around 30 s (Cyr et al., 2001). Kininase II (also known as angiotensin I-converting enzyme, ACE) is a carboxydipeptidase, which cleaves physiologically active BK (1–9) and kallidin (1–10) into inactive BK (1–7) and kallidin (1–8), respectively (Regoli et al., 2012; Tang et al., 2011) (see also Fig. 1A). Interestingly, *in vitro* studies suggest that ACE is approximately 30 times more potent in cleaving kinins than in producing Ang II from Ang I (Shen & El-Dahr, 2006; Tschope et al., 2002). It is proposed that beneficial effects of ACE inhibition in treating hypertension and related cardiovascular diseases extend beyond prevention of Ang II synthesis and involve accumulation of BK. Thus, ACE inhibition also results in KKS activation further contributing to

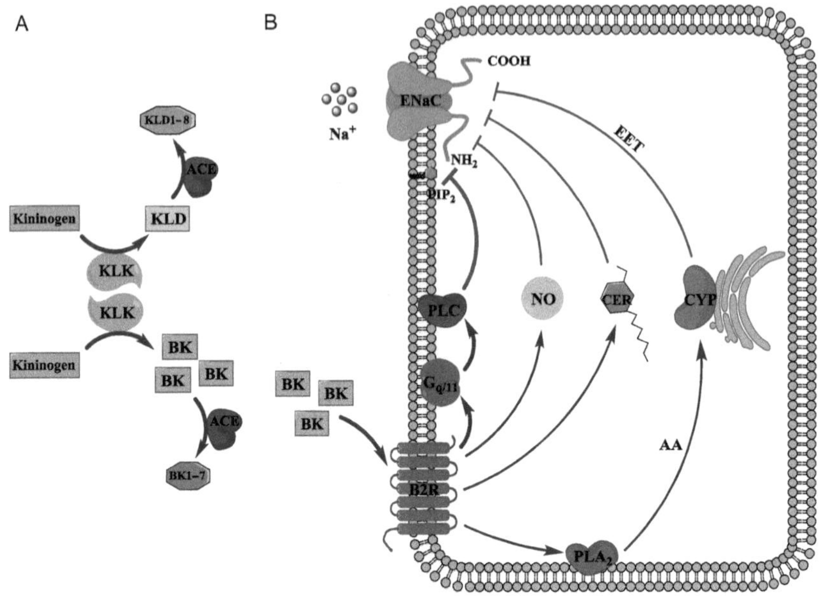

Figure 1 Kallikrein–kinin system in regulation of ENaC activity in the distal nephron. (A) Principal scheme of Bradykinin synthesis and degradation. KLK, kallikrein; KLD, kallidin; KLD1–8, kallidin (1–8); BK, bradykinin; BK1–7, bradykinin (1–7); ACE, angiotensin converting enzyme. (B) Intracellular signaling pathways coupling BK with changes in ENaC activity. Bold arrows represent signaling pathways supported by direct experimental evidence; and thin arrows indicate additional possible mechanisms of regulation. BK, bradykinin; B2R, bradykinin type 2 receptor; ENaC, epithelial Na^+ channel; $G_{q/11}$, heterotrimeric G protein Q/11 subtype; PLC, phospholipase C; PLA_2, phospholipase A2; NO, nitric oxide; CYP, cytochrome P450; AA, arachidonic acid; CER, ceramide; EET, epoxyeicosatrienoic acids; PIP_2, phosphatidylinositol 4,5-bisphosphate. NH_2 and COOH demonstrate amino- and carboxyl-termini of ENaC subunits, respectively. Green and red arrows represent stimulatory and inhibitory actions, respectively. (See the color plate.)

lowering of BP (Regoli et al., 2012). Interestingly, inhibition of B2R consistently and significantly attenuates the hypotensive effect of ACE inhibition in both normotensive and hypertensive individuals (Gainer, Morrow, Loveland, King, & Brown, 1998).

3. KKS EXPRESSION IN THE KIDNEY

Over the years, substantial experimental evidence was accumulated suggesting that all functional units of the KKS are present in the kidney tissue, forming the so-called renal KKS. It is generally recognized that the major site of KKS presence is the distal part of renal tubule, namely, the connecting tubule (CNT) and collecting duct (CD). Indeed, kininogen is actively synthesized and stored in principal cells of these tubular segments (El-Dahr, 2004; Song, Wang, Harley, Chao, & Chao, 1996). Furthermore, expression of KKS components is intimately coordinated with the terminal differentiation of the distal nephron during nephrogenesis (El-Dahr, 2004). Immunostaining and *in situ* hybridization studies demonstrate that kallikrein is chiefly produced in CNT (El-Dahr, 2004; Tang et al., 2011). Thus, it is viewed that intraluminally generated kinins exert autocrine and paracrine actions on distal tubule cells down to the place of synthesis, favoring natriuresis and diuresis, as discussed further below. In addition, interstitially released kinins are likely participating in regulation of cortex and medullary blood flow via nitric oxide generation (El-Dahr, 2004; Regoli et al., 2012). Sampling of interstitial fluid in the rat kidney was able to detect 10–100 nM of BK levels, and the concentration of BK is higher in the cortex than that in the medulla (Siragy, Ibrahim, Jaffa, Mayfield, & Margolius, 1994). These values are sufficient to activate both B1R and B2R (Hall, 1997). In contrast, no evidence of kinin formation was obtained in the proximal tubules of dogs (Yamaguchi et al., 1986). Interestingly, prominent kallikrein mRNA and protein expression (while still lower than that in the CNT) was detected in proximal tubule cells from patients with diabetic nephropathy (Tang et al., 2011). It is possible that such upregulation of KKS in proximal tubule is beneficial during the pathology to reduce albuminuria and glomerulosclerosis (Kakoki & Smithies, 2009; Tang et al., 2011). Indeed, renoprotective and cardioprotective effects of KKS activation during diabetic nephropathy are described in the literature (Kakoki & Smithies, 2009; Kakoki et al., 2010).

In the kidney, B2R are detected in the CD and CNT (El-Dahr, 2004; Figueroa et al., 1995; Rohrwasser et al., 2003; Shen & El-Dahr, 2006; Song et al., 1996; Tang et al., 2011), while lower levels of the protein and mRNA can be also found in more proximal tubular segments (El-Dahr, 2004). Expression of B2R was also demonstrated during nephrogenesis in the developing uretic bud branches and the upper limb of S-shaped nephrons (El-Dahr, 2004). Interestingly, B2R are localized to the apical membrane in the developing kidney, whereas well-differentiated nephron segments and cortical collecting ducts display B2R on both apical and basolateral membranes (El-Dahr, 2004). This argues for an important role of the KKS in the nephrogenesis. Indeed, B2R-knockout mice develop renal dysgenesis and hypertension when stressed with elevated salt intake during embryogenesis. In contrast, renal B1R expression is virtually absent and it is unlikely that B1R participate in early nephrogenesis (El-Dahr, 2004). However, a potential role of B1R in terminal differentiation of proximal nephron was also proposed (Bulut, Dipp, & El-Dahr, 2009).

In summary, the existing evidence suggests that the renal KKS is perfectly positioned to control development and transport function of the distal portions of nephron.

4. MOLECULAR MECHANISMS OF THE DISTAL NEPHRON SODIUM REABSORPTION

It is generally recognized that kidneys play the central role in controlling circulating volume and consequently BP by regulating the total amount of extracellular sodium via the pressure-natriuresis mechanism (Guyton, 1991). Inability of the kidneys to adequately control tubular Na^+ reabsorption leads to disturbances in sodium balance and BP abnormalities (Khawaja & Wilcox, 2011; Lifton, Gharavi, & Geller, 2001). Sodium reabsorption in the distal nephron (also termed aldosterone-sensitive distal nephron), which includes the CNT and CD, determines final concentration of Na^+ excreted in urine (Staruschenko, 2012). The ENaC, which is localized to the apical plasma membrane of principal cells in the CNT and CD, mediates electrogenic Na^+ reabsorption at these sites (Bhalla & Hallows, 2008; Frindt, Ergonul, & Palmer, 2008; Hummler, 2003; Schild, 2004; Schild & Kellenberger, 2001; Staruschenko, 2012). ENaC is a highly Na^+-selective channel expressed in several epithelial tissues, including distal nephron, colon, salivary and sweat glands, and lung (Staruschenko, 2012). In the kidney, ENaC consists of three homologous subunits (α, β, and γ ENaC), which share ~30–40% identity in their amino acid sequences (Canessa et al., 1994; Fyfe, Quinn, & Canessa, 1998). Each subunit has two transmembrane domains with short cytoplasmic NH_2- and COOH termini and a large extracellular loop (Benos & Stanton, 1999; Fyfe et al., 1998; Garty & Palmer, 1997).

Proper ENaC functioning is critical for sodium handling by the kidney (Hummler, 1999, 2003; Schild, 1996, 2004). Despite the fact that distal nephron reabsorbs only approximately 5% of total filtered Na^+ load, the absolute amount of Na^+ which is reabsorbed daily via ENaC-dependent mechanisms exceeds that present in the circulation. A critical role of ENaC in development of elevated BP is now becoming recognized in humans (Pratt, 2005) and in many salt-sensitive (SS) hypertensive animal models, such as SS Dahl rats (Kakizoe et al., 2009). The GenSalt study suggests that single nucleotide polymorphisms in genes encoding ENaC subunits contribute to BP changes over time (Yang et al., 2014). Finally, gain-of-function/loss-of-function mutations in ENaC result in monogenic forms of hypertension (Liddle's syndrome)/hypotension (pseudohypoaldosteronism type II), respectively (Schild, 1996, 2004).

ENaC activity is inversely related to the dietary sodium intake (Bhalla & Hallows, 2008; Frindt & Palmer, 2009; Hummler, 2003). This results from ENaC being an important end-effector of the RAAS. Activation of the RAAS in response to volume contraction/low sodium intake stimulates the ENaC-mediated sodium reabsorption in the distal nephron. Both aldosterone and Ang II were reported to synergistically increase functional ENaC expression in native distal nephrons, as well as in cultured cells of distal nephron origin (Mamenko, Zaika, Ilatovskaya, Staruschenko, & Pochynyuk, 2012; Mamenko et al., 2013). While the RAAS-mediated endocrine control of ENaC activity provides a mechanistic explanation of augmented Na^+ reabsorption in the distal nephron during dietary sodium restriction, the concepts of ENaC inhibition during volume-expanded states are just beginning to emerge. Effective suppression of ENaC activity during high sodium intake is viewed as an important regulatory mechanism that allows avoiding excessive sodium retention and SS hypertension. Recent experimental evidence strongly suggest that KKS, and specifically BK, is essential for proper regulation of ENaC activity by dietary salt intake particularly during volume-expanded states (Mamenko, Zaika, Doris, et al., 2012; Zaika et al., 2011).

5. REGULATION OF DISTAL NEPHRON SODIUM REABSORPTION BY BK: A ROLE FOR ENaC

Multiple reports are consistent with the notion that renal KKS promotes diuresis and natriuresis largely due to inhibition of tubular reabsorption of water and sodium (de Souza et al., 2013; Mukai, Fitzgibbon, Bozeman, Margolius, & Ploth, 1996; Rhaleb, Yang, & Carretero, 2011; Siragy, 1993; Sivritas et al., 2008; Suzuki, Katori, Fujita, Kumagai, & Majima, 2000). It appears that these effects occur in the absence of changes in glomerular filtration rate and renal blood flow (Siragy, 1993; Sivritas et al., 2008). However, the evidence about a role of KKS in promoting vasodilation of renal vasculature also exists (Rhaleb et al., 2011). Taking into consideration the topology of renal KKS expression, it is suggested that inhibition of distal nephron water–electrolyte transport underlies the effects of kinins on renal function (Kauker & Nasjletti, 1986; Mukai et al., 1996; Zeidel, Jabs, Kikeri, & Silva, 1990). Indeed, direct application of BK inhibits amiloride-sensitive component of conductive Na^+ uptake at the apical membrane in rabbit CD cells (Zeidel et al., 1990). Amiloride is a fairly specific ENaC blocker. Consistently, BK decreases short-circuit current (I_{sc}) in polarized monolayers of cultured CCD M1 cells (Stoos, Carretero, & Garvin, 1992). Interestingly, BK had no net effect on Na^+ reabsorption in perfused CCDs from Dahl SS rats (Hawk & Schafer, 1993). In contrast, it was demonstrated that kinins exert tonic inhibitory actions on tubular Na^+ transport in normal Sprague-Dawley rats (Sivritas et al., 2008). Thus, it is tempting to speculate that inability of distal nephron cells to reduce Na^+ reabsorption in response to BK is likely indicating disruption of this inhibitory tone possibly contributing to the pathology of SS hypertension in Dahl SS rats.

Using a physiologically relevant object, freshly isolated split-opened distal nephrons (Fig. 2A), our group recently demonstrated a direct inhibitory effect of BK on ENaC activity (Fig. 2B; see also Zaika et al., 2011; Mamenko, Zaika, Doris, et al., 2012). Specifically, nanomolar concentrations of BK alter ENaC function by drastically reducing open probability (P_o) of the channel (Fig. 2B). In contrast, the number of functional ENaC residing on the apical plasma membrane is unaffected after application of BK. Consistently, mice lacking both receptors for BK have unchanged ENaC levels but inappropriately augmented ENaC P_o in the distal nephron cells further supporting the concept that KKS exerts an inhibitory effect on distal nephron sodium reabsorption (Mamenko, Zaika, Doris, et al., 2012).

Apart from its actions on ENaC, BK may also affect sodium reabsorption by regulating activity of Na^+/H^+ exchanger isoform 1 in IMCD-3 cells (Mukhin et al., 2001). Furthermore, BK decreases net Cl^- absorption in the perfused CCDs (Tomita, Pisano, Burg, & Knepper, 1986) possibly activating Ca^{2+}-dependent Cl^- conductance (possibly ClCA1/2) to promote Cl^- secretion (Kose et al., 2000). BK signaling decreases osmotic water uptake by interfering with apical translocation of AQP2 in CD8 cells (Tamma, Carmosino, Svelto, & Valenti, 2005). Consistently, B2R blockade enhances Cl^- and water absorption in the medullary CDs (Mukai et al., 1996). The complexity of BK-initiated signaling network involving multiple transporting proteins can potentially explain why BK alters *trans*-epithelial voltage in some studies (Stoos et al., 1992; Zeidel et al., 1990) but fails to do it in others (Tomita, Pisano, & Knepper, 1985).

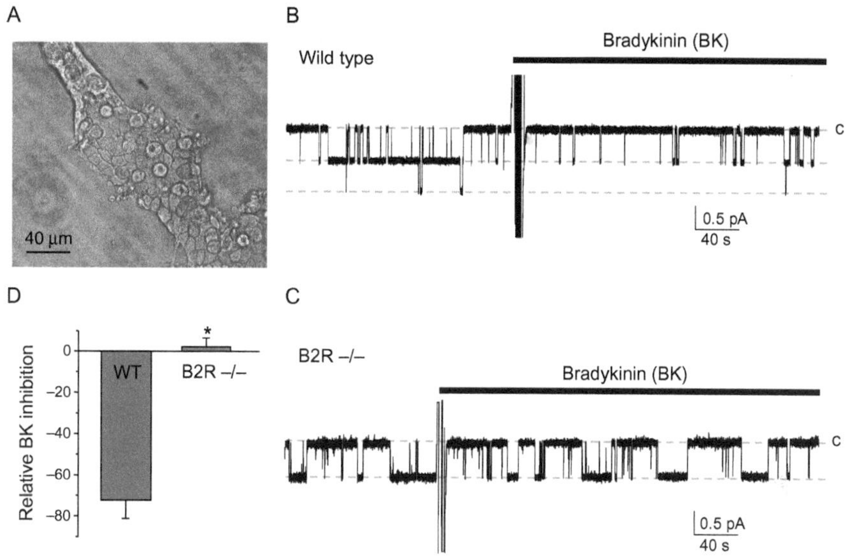

Figure 2 Bradykinin (BK) acutely inhibits ENaC open probability via activation of B2 receptors. (A) A representative micrograph of freshly isolated and split-opened murine cortical collecting duct suitable for patch clamp evaluation of ENaC activity. Scale bar is shown. (B) A representative continuous current trace from a cell-attached patch containing two ENaCs in the control condition and during application of 500 nM BK. The patch was formed on the apical plasma membrane of a principal cell within a split-opened area of a cortical collecting duct of a wild-type (WT, C57BL/6) mouse, similar to shown in (A). Dashed lines indicate the respective current state with "c" denoting the closed state. (C) A representative continuous current trace from a cell-attached patch monitoring ENaC activity in a cortical collecting duct isolated from a mouse lacking B2 receptors (B2R −/−) in the control condition and after addition of 500 nM BK. (D) Summary graph of relative BK inhibition of ENaC activity for WT and B2R −/− mice, respectively. *Significant difference versus WT.

6. SIGNALING PATHWAYS MEDIATING BK ACTIONS ON ENaC

Two distinct pharmacologically classified B1R and B2R serve as sites recognizing kinins at the plasma membrane (Blaukat, 2003; Hall, 1997; Marceau & Bachvarov, 1998). BK and kallidin preferentially bind to B2R, whereas B1R recognize carboxy-terminally truncated peptides desArg^{9}-BK and desArg^{10}-kallidin (Blaukat, 2003). As discussed above, renal epithelial and vascular cells express mainly B2R suggesting that this receptor subtype is essential for mediating BK actions in the kidney.

Indeed, pharmacological inhibition of B2R after intrarenal infusion of HOE-140 (icatibant) precludes BK-induced natriuresis and diuresis (Sivritas et al., 2008). Furthermore, genetic ablation of B2R abolishes inhibitory effect of BK on ENaC activity in split-opened murine distal nephrons (Fig. 2C and D).

With the aid of pharmacological and genetic techniques, our group recently demonstrated that stimulation of B2R-$G_{q/11}$-PLC signaling pathway is central for BK-mediated inhibition of ENaC in native principal cells of CNT and CD (Mamenko, Zaika, Doris, et al., 2012; Zaika et al., 2011). Interestingly, BK actions on ENaC activity do not seem to be dependent on elevations in $[Ca^{2+}]_i$ but rather determined by the levels of phospholipid, $PI(4,5)P_2$ in the apical plasma membrane, which is rapidly degraded in response to PLC activation (Zaika et al., 2011). Depletion of $PI(4,5)P_2$ levels upon stimulation of GPCR or receptor tyrosine kinase signaling rapidly decreases ENaC activity in expression systems (Pochynyuk et al., 2007; Pochynyuk, Tong, Staruschenko, Ma, & Stockand, 2006; Tong & Stockand, 2005), cultured epithelial cells (Kunzelmann et al., 2005; Ma, Saxena, & Warnock, 2002; Pochynyuk, Bugaj, Vandewalle, & Stockand, 2008), and freshly isolated split-opened ASDNs (Pochynyuk, Bugaj, Rieg, et al., 2008; Zaika et al., 2011). Moreover, putative $PI(4,5)P_2$ binding sites in the intracellular N-termini of β- and γ-ENaC subunits have been recently reported (Kunzelmann et al., 2005; Pochynyuk et al., 2007). In addition, activation of BK signaling is known to stimulate NO production via eNOS-dependent pathways in cortex kidney slices (Adler et al., 2001), augment production of cytochrome P450 derivatives in perfused rat kidney (Bagate et al., 2001), and activate PKC-PLA_2 (Lal, Kennedy, Proulx, & Hebert, 1997) and PLD-ceramide/phosphatidic acid (Liu, Kleine, Nasrallah, & Hebert, 1999) signaling cascades in rabbit CCD cells. All these pathways can also exert inhibitory actions on ENaC (Bao et al., 2007; Guo, Alli, Eaton, & Bao, 2013; Pavlov et al., 2011). Figure 1B summarizes possible signaling cascades that couple activation of B2R with changes in ENaC activity. However, relative contribution of these different mechanisms in regulation of the channel function by BK is currently unclear.

7. SALT SENSITIVITY OF BK ACTIONS ON DISTAL NEPHRON SODIUM REABSORPTION

Genetic animal models with dysfunctional KKS provide valuable insights about contribution of the system into systemic BP control.

Disruption of all KKS components: B2R (Alfie, Sigmon, Pomposiello, & Carretero, 1997; Alfie, Yang, Hess, & Carretero, 1996), kininogen (Majima et al., 1993), and kallikrein (Madeddu et al., 1997) leads to hypertensive phenotype during elevated Na^+ intake. Similar results were obtained in the presence of pharmacological inhibition of B2R and kallikrein in wild-type animals stressed with high-salt diet (Majima et al., 1993). Furthermore, urinary levels of BK and kallikrein positively correlate with amount of dietary Na^+ (Hettinger, Lukasova, Lewicka, & Hilgenfeldt, 2002). On the other side, it is well known that ENaC-mediated sodium reabsorption in the distal part of the renal nephron is regulated by dietary sodium intake (Staruschenko, 2012). Pharmacological inhibition (Madeddu et al., 1993) or genetic ablation (Emanueli et al., 1998) of B2R induces hypertension in animals with saturated mineralocorticoid status after deoxycorticosterone treatment. Altogether, this argues for a physiologically relevant role of KKS in promoting natriuresis due to inhibition of the distal nephron sodium reabsorption, particularly during volume expanded states, such as salt overloading. Furthermore, tissue kallikrein deficiency also leads to augmented aldosterone-salt-induced hypertension, which can be at least partially explained by renal sodium retention due to enhanced distal nephron sodium reabsorption (Waeckel et al., 2012). Consistently, our group recently demonstrated that genetic deletion of both BK receptors (B1R and B2R) results in increased ENaC activity under conditions of normal and elevated salt intake when compared to control group (Mamenko, Zaika, Doris, et al., 2012). In contrast, dietary sodium restriction abolishes differences in ENaC activity between control and mutant animals (Mamenko, Zaika, Doris, et al., 2012). This resonates with the observation that inhibition of B2R with HOE-140 in perfused kidney increases tubular Na^+ reabsorption in rats fed with normal, but not sodium-deficient diet (Sivritas et al., 2008). However, the precise contribution of ENaC-dependent and -independent mechanisms in development of SS hypertension in animals with dysfunctional KKS remains to be determined.

8. INHIBITION OF ENaC BY BK PROMOTES ACE-DEPENDENT NATRIURESIS

As was mentioned above, degradation of kinins is under control of ACE, which efficiently cleaves two amino acids at the carboxyl terminus of BK, converting it into inactive BK (1–7) (Regoli et al., 2012). In the kidney, sufficient ACE activity is present at the apical plasma membrane of principal cells in the distal nephron (Casarini et al., 1997; Redublo Quinto et al., 2008). Thus, it is proposed that ACE blockade will lead to accumulation of BK in the distal renal tubule, thus leading to a greater ENaC inhibition (Mamenko, Zaika, & Pochynyuk, 2014). Consistently, treatment with captopril (ACE inhibitor) greatly augments elevations of $[Ca^{2+}]_i$ in murine distal nephron cells in response to BK, indicating a higher level of B2R activation (Mamenko, Zaika, Doris, et al., 2012). Moreover, this also sensitizes inhibitory actions of BK on ENaC in split-opened distal nephrons of mice (Mamenko, Zaika, Doris, et al., 2012). Such augmentation of BK signaling could be possibly explained by close association of ACE with B2R, as was reported for different cell types (Sabatini et al., 2008). This may lead to a decreased actual concentration of the agonist in the vicinity of the receptors due to tonic kininase activity of ACE. Alternatively, ACE inhibitors can directly interact with B2R to enhance function of the receptor in an allosteric manner (Erdos, Tan, & Skidgel, 2010).

The observation, that BK-mediated inhibition of ENaC is controlled by ACE, suggests that augmented BK signaling in the distal nephron can contribute to the natriuresis and decreased BP promoted by ACE inhibition. Consistent with this, B2R inhibition significantly attenuates the hypotensive effect of the ACE inhibition in both normotensive and hypertensive individuals (Gainer et al., 1998). Chronic ACE blockade with captopril leads to a markedly diminished basal ENaC activity in the distal nephron cells of mice kept on normal sodium intake (Mamenko, Zaika, Doris, et al., 2012). Importantly, genetic ablation of BK receptors abolishes this inhibition pointing to a dominant role of BK signaling. This was also associated with blunted natriuresis in response to captopril in mutant mice (Mamenko, Zaika, Doris, et al., 2012).

9. CONCLUSIONS

Unraveling discrete systems controlling sodium homeostasis is fundamental for understanding physiology and treating cardiovascular pathologies, such as hypertension. ENaC-mediated Na^+ reabsorption in the distal nephron fine-tunes renal sodium excretion to match dietary sodium intake and maintain sodium balance on a daily basis. In this chapter, we summarized current knowledge about molecular details and physiological relevance of the renal KKS and specifically BK signaling cascade in regulation of ENaC-mediated sodium reabsorption in the distal nephron. BK via stimulation of B2R-PLC cascade and following depletion of $PI(4,5)P_2$ exerts tonic inhibitory actions on ENaC activity by reducing single channel capacity to conduct Na^+. Disruption of this regulation leads to overactive ENaC, which is detrimental under sodium loaded conditions contributing to SS hypertension in animal models with dysfunctional KKS. BK-induced inhibition of ENaC in the distal nephron also appears to be a critical component of the natriuresis promoted by ACE inhibition. From the clinical prospective, it is possible that genetic polymorphism in genes encoding B2R and other functional components of KKS may contribute to differences in susceptibility of patients to the antihypertensive actions of ACE blockade.

ACKNOWLEDGMENTS

This research was supported by NIH-NIDDK DK095029 (to O. P.), AHA-GIA-13GRNT16220002 (to O. P.), and 14POST20380979 (to M. M.).

REFERENCES

Adler, S., Huang, H., Loke, K. E., Xu, X., Tada, H., Laumas, A., et al. (2001). Endothelial nitric oxide synthase plays an essential role in regulation of renal oxygen consumption by NO. *American Journal of Physiology. Renal Physiology, 280*, F838–F843.

Alfie, M. E., Sigmon, D. H., Pomposiello, S. I., & Carretero, O. A. (1997). Effect of high salt intake in mutant mice lacking bradykinin-B2 receptors. *Hypertension, 29*, 483–487.

Alfie, M. E., Yang, X. P., Hess, F., & Carretero, O. A. (1996). Salt-sensitive hypertension in bradykinin B2 receptor knockout mice. *Biochemical and Biophysical Research Communications, 224*, 625–630.

Bagate, K., Grima, M., Imbs, J. L., Jong, W. D., Helwig, J. J., & Barthelmebs, M. (2001). Signal transduction pathways involved in kinin B2 receptor-mediated vasodilation in the rat isolated perfused kidney. *British Journal of Pharmacology, 132*, 1735–1742.

Bao, H. F., Zhang, Z. R., Liang, Y. Y., Ma, J. J., Eaton, D. C., & Ma, H. P. (2007). Ceramide mediates inhibition of the renal epithelial sodium channel by tumor necrosis factor-alpha through protein kinase C. *American Journal of Physiology. Renal Physiology, 293*, F1178–F1186.

Benos, D. J., & Stanton, B. A. (1999). Functional domains within the degenerin/epithelial sodium channel (Deg/ENaC) superfamily of ion channels. *The Journal of Physiology, 520*(Pt. 3), 631–644.

Bhalla, V., & Hallows, K. R. (2008). Mechanisms of ENaC regulation and clinical implications. *Journal of the American Society of Nephrology, 19*, 1845–1854.

Blaukat, A. (2003). Structure and signalling pathways of kinin receptors. *Andrologia, 35*, 17–23.

Braun, C., Kleemann, T., Hilgenfeldt, U., Riester, U., Rohmeiss, P., & van der Woude, F. J. (2002). Activity and functional significance of the renal kallikrein-kinin-system in polycystic kidney disease of the rat. *Kidney International, 61*, 2149–2156.

Bulut, O. P., Dipp, S., & El-Dahr, S. (2009). Ontogeny of bradykinin B1 receptors in the mouse kidney. *Pediatric Research, 66*, 519–523.

Canessa, C. M., Schild, L., Buell, G., Thorens, B., Gautschi, I., Horisberger, J. D., et al. (1994). Amiloride-sensitive epithelial Na^+ channel is made of three homologous subunits. *Nature, 367*, 463–467.

Casarini, D. E., Boim, M. A., Stella, R. C., Krieger-Azzolini, M. H., Krieger, J. E., & Schor, N. (1997). Angiotensin I-converting enzyme activity in tubular fluid along the rat nephron. *American Journal of Physiology*, *272*, F405–F409.

Cyr, M., Lepage, Y., Blais, C., Jr., Gervais, N., Cugno, M., Rouleau, J. L., et al. (2001). Bradykinin and des-Arg(9)-bradykinin metabolic pathways and kinetics of activation of human plasma. *American Journal of Physiology. Heart and Circulatory Physiology*, *281*, H275–H283.

de Souza, P., Crestani, S., da Silva Rde, C., Gasparotto, F., Kassuya, C. A., da Silva-Santos, J. E., et al. (2013). Involvement of bradykinin and prostaglandins in the diuretic effects of *Achillea millefolium* L. (Asteraceae). *Journal of Ethnopharmacology*, *149*, 157–161.

El-Dahr, S. S. (2004). Spatial expression of the kallikrein–kinin system during nephrogenesis. *Histology and Histopathology*, *19*, 1301–1310.

Emanueli, C., Fink, E., Milia, A. F., Salis, M. B., Conti, M., Demontis, M. P., et al. (1998). Enhanced blood pressure sensitivity to deoxycorticosterone in mice with disruption of bradykinin B2 receptor gene. *Hypertension*, *31*, 1278–1283.

Erdos, E. G., Tan, F., & Skidgel, R. A. (2010). Angiotensin I-converting enzyme inhibitors are allosteric enhancers of kinin B1 and B2 receptor function. *Hypertension*, *55*, 214–220.

Figueroa, C. D., Gonzalez, C. B., Grigoriev, S., Abd Alla, S. A., Haasemann, M., Jarnagin, K., et al. (1995). Probing for the bradykinin B2 receptor in rat kidney by anti-peptide and anti-ligand antibodies. *The Journal of Histochemistry and Cytochemistry*, *43*, 137–148.

Frindt, G., Ergonul, Z., & Palmer, L. G. (2008). Surface expression of epithelial Na channel protein in rat kidney. *The Journal of General Physiology*, *131*, 617–627.

Frindt, G., & Palmer, L. G. (2009). Surface expression of sodium channels and transporters in rat kidney: Efects of dietary sodium. *American Journal of Physiology. Renal Physiology*, *297*, F1249–F1255.

Fyfe, G. K., Quinn, A., & Canessa, C. M. (1998). Structure and function of the Mec-ENaC family of ion channels. *Seminars in Nephrology*, *18*, 138–151.

Gainer, J. V., Morrow, J. D., Loveland, A., King, D. J., & Brown, N. J. (1998). Effect of bradykinin-receptor blockade on the response to angiotensin-converting-enzyme inhibitor in normotensive and hypertensive subjects. *The New England Journal of Medicine*, *339*, 1285–1292.

Garty, H., & Palmer, L. G. (1997). Epithelial sodium channels: Function, structure, and regulation. *Physiological Reviews*, 77, 359–396.

Guo, L. J., Alli, A. A., Eaton, D. C., & Bao, H. F. (2013). ENaC is regulated by natriuretic peptide receptor-dependent cGMP signaling. *American Journal of Physiology. Renal Physiology*, *304*, F930–F937.

Guyton, A. C. (1991). Blood pressure control-special role of the kidneys and body fluids. *Science*, *252*, 1813–1816.

Hall, J. M. (1992). Bradykinin receptors: Pharmacological properties and biological roles. *Pharmacology & Therapeutics*, *56*, 131–190.

Hall, J. M. (1997). Bradykinin receptors. *General Pharmacology*, *28*, 1–6.

Hawk, C. T., & Schafer, J. A. (1993). Clonidine, but not bradykinin or ANP, inhibits Na^+ and water transport in Dahl SS rat CCD. *Kidney International*, *44*, 30–35.

Hettinger, U., Lukasova, M., Lewicka, S., & Hilgenfeldt, U. (2002). Regulatory effects of salt diet on renal renin-angiotensin-aldosterone, and kallikrein-kinin systems. *International Immunopharmacology*, *2*, 1975–1980.

Hirawa, N., Uehara, Y., Suzuki, T., Kawabata, Y., Numabe, A., Gomi, T., et al. (1999). Regression of glomerular injury by kallikrein infusion in Dahl salt-sensitive rats is a bradykinin B2-receptor-mediated event. *Nephron*, *81*, 183–193.

Hummler, E. (1999). Implication of ENaC in salt-sensitive hypertension. *The Journal of Steroid Biochemistry and Molecular Biology, 69*, 385–390.

Hummler, E. (2003). Epithelial sodium channel, salt intake, and hypertension. *Current Hypertension Reports, 5*, 11–18.

Kakizoe, Y., Kitamura, K., Ko, T., Wakida, N., Maekawa, A., Miyoshi, T., et al. (2009). Aberrant ENaC activation in Dahl salt-sensitive rats. *Journal of Hypertension, 27*, 1679–1689.

Kakoki, M., McGarrah, R. W., Kim, H. S., & Smithies, O. (2007). Bradykinin B1 and B2 receptors both have protective roles in renal ischemia/reperfusion injury. *Proceedings of the National Academy of Sciences of the United States of America, 104*, 7576–7581.

Kakoki, M., & Smithies, O. (2009). The kallikrein-kinin system in health and in diseases of the kidney. *Kidney International, 75*, 1019–1030.

Kakoki, M., Sullivan, K. A., Backus, C., Hayes, J. M., Oh, S. S., Hua, K., et al. (2010). Lack of both bradykinin B1 and B2 receptors enhances nephropathy, neuropathy, and bone mineral loss in Akita diabetic mice. *Proceedings of the National Academy of Sciences of the United States of America, 107*, 10190–10195.

Kakoki, M., Takahashi, N., Jennette, J. C., & Smithies, O. (2004). Diabetic nephropathy is markedly enhanced in mice lacking the bradykinin B2 receptor. *Proceedings of the National Academy of Sciences of the United States of America, 101*, 13302–13305.

Kauker, M. L., & Nasjletti, A. (1986). Segmental nephron function in rats treated with aprotinin, an inhibitor of kallikrein. *The Journal of Pharmacology and Experimental Therapeutics, 238*, 155–158.

Kayashima, Y., Smithies, O., & Kakoki, M. (2012). The kallikrein-kinin system and oxidative stress. *Current Opinion in Nephrology and Hypertension, 21*, 92–96.

Khawaja, Z., & Wilcox, C. S. (2011). Role of the kidneys in resistant hypertension. *International Journal of Hypertension, 2011*, 143471.

Kose, H., Boese, S. H., Glanville, M., Gray, M. A., Brown, C. D., & Simmons, N. L. (2000). Bradykinin regulation of salt transport across mouse inner medullary collecting duct epithelium involves activation of a Ca^{2+}-dependent Cl^{-} conductance. *British Journal of Pharmacology, 131*, 1689–1699.

Kunzelmann, K., Bachhuber, T., Regeer, R., Markovich, D., Sun, J., & Schreiber, R. (2005). Purinergic inhibition of the epithelial Na^{+} transport via hydrolysis of PIP_2. *The FASEB Journal, 19*, 142–143.

Lal, M. A., Kennedy, C. R., Proulx, P. R., & Hebert, R. L. (1997). Bradykinin-stimulated $cPLA_2$ phosphorylation is protein kinase C dependent in rabbit CCD cells. *The American Journal of Physiology, 273*, F907–F915.

Lifton, R. P., Gharavi, A. G., & Geller, D. S. (2001). Molecular mechanisms of human hypertension. *Cell, 104*, 545–556.

Liu, G., Kleine, L., Nasrallah, R., & Hébert, R. L. (1999). Bradykinin inhibits ceramide production and activates phospholipase D in rabbit cortical collecting duct cells. *The American Journal of Physiology, 276*, F589–F598.

Ma, H. P., Saxena, S., & Warnock, D. G. (2002). Anionic phospholipids regulate native and expressed epithelial sodium channel (ENaC). *The Journal of Biological Chemistry, 277*, 7641–7644.

Ma, J. X., Wang, D. Z., Chao, L., & Chao, J. (1994). Cloning, sequence analysis and expression of the gene encoding the mouse bradykinin B2 receptor. *Gene, 149*, 283–288.

Madeddu, P., Anania, V., Parpaglia, P. P., Demontis, M. P., Varoni, M. V., Fattaccio, M. C., et al. (1993). Chronic kinin receptor blockade induces hypertension in deoxycorticosterone-treated rats. *British Journal of Pharmacology, 108*, 651–657.

Madeddu, P., Varoni, M. V., Demontis, M. P., Chao, J., Simson, J. A., Glorioso, N., et al. (1997). Kallikrein-kinin system and blood pressure sensitivity to salt. *Hypertension, 29*, 471–477.

Majima, M., Yoshida, O., Mihara, H., Muto, T., Mizogami, S., Kuribayashi, Y., et al. (1993). High sensitivity to salt in kininogen-deficient brown Norway Katholiek rats. *Hypertension*, *22*, 705–714.

Mamenko, M., Zaika, O., Doris, P. A., & Pochynyuk, O. (2012). Salt-dependent inhibition of epithelial Na^+ channel-mediated sodium reabsorption in the aldosterone-sensitive distal nephron by bradykinin. *Hypertension*, *60*, 1234–1241.

Mamenko, M., Zaika, O., Ilatovskaya, D. V., Staruschenko, A., & Pochynyuk, O. (2012). Angiotensin II increases activity of the epithelial Na^+ channel (ENaC) in distal nephron additively to aldosterone. *The Journal of Biological Chemistry*, *287*, 660–671.

Mamenko, M., Zaika, O., & Pochynyuk, O. (2014). Direct regulation of ENaC by bradykinin in the distal nephron. Implications for renal sodium handling. *Current Opinion in Nephrology and Hypertension*, *23*, 122–129.

Mamenko, M., Zaika, O., Prieto, M. C., Jensen, V. B., Doris, P. A., Navar, L. G., et al. (2013). Chronic angiotensin II infusion drives extensive aldosterone-independent epithelial Na^+ channel activation. *Hypertension*, *62*, 1111–1122.

Manolis, A. J., Marketou, M. E., Gavras, I., & Gavras, H. (2010). Cardioprotective properties of bradykinin: Role of the B2 receptor. *Hypertension Research*, *33*, 772–777.

Marceau, F., & Bachvarov, D. R. (1998). Kinin receptors. *Clinical Reviews in Allergy & Immunology*, *16*, 385–401.

Marceau, F., Larrivee, J. F., Saint-Jacques, E., & Bachvarov, D. R. (1997). The kinin B1 receptor: An inducible G protein coupled receptor. *Canadian Journal of Physiology and Pharmacology*, *75*, 725–730.

Mukai, H., Fitzgibbon, W. R., Bozeman, G., Margolius, H. S., & Ploth, D. W. (1996). Bradykinin B2 receptor antagonist increases chloride and water absorption in rat medullary collecting duct. *The American Journal of Physiology*, *271*, R352–R360.

Mukhin, Y. V., Vlasova, T., Jaffa, A. A., Collinsworth, G., Bell, J. L., Tholanikunnel, B. G., et al. (2001). Bradykinin B2 receptors activate Na^+/H^+ exchange in mIMCD-3 cells via Janus kinase 2 and Ca^{2+}/calmodulin. *The Journal of Biological Chemistry*, *276*, 17339–17346.

Pavlov, T. S., Ilatovskaya, D. V., Levchenko, V., Mattson, D. L., Roman, R. J., & Staruschenko, A. (2011). Effects of cytochrome P-450 metabolites of arachidonic acid on the epithelial sodium channel (ENaC). *American Journal of Physiology. Renal Physiology*, *301*, F672–F681.

Pochynyuk, O., Bugaj, V., Rieg, T., Insel, P. A., Mironova, E., Vallon, V., et al. (2008). Paracrine regulation of the epithelial Na^+ channel in the mammalian collecting duct by purinergic P2Y2 receptor tone. *The Journal of Biological Chemistry*, *283*, 36599–36607.

Pochynyuk, O., Bugaj, V., Vandewalle, A., & Stockand, J. D. (2008). Purinergic control of apical plasma membrane $PI(4,5)P_2$ levels sets ENaC activity in principal cells. *American Journal of Physiology. Renal Physiology*, *294*, F38–F46.

Pochynyuk, O., Tong, Q., Medina, J., Vandewalle, A., Staruschenko, A., Bugaj, V., et al. (2007). Molecular determinants of $PI(4,5)P_2$ and $PI(3,4,5)P_3$ regulation of the epithelial Na^+ channel. *The Journal of General Physiology*, *130*, 399–413.

Pochynyuk, O., Tong, Q., Staruschenko, A., Ma, H. P., & Stockand, J. D. (2006). Regulation of the epithelial Na^+ channel (ENaC) by phosphatidylinositides. *American Journal of Physiology. Renal Physiology*, *290*, F949–F957.

Pratt, J. H. (2005). Central role for ENaC in development of hypertension. *Journal of the American Society of Nephrology*, *16*, 3154–3159.

Redublo Quinto, B. M., Camargo de Andrade, M. C., Ronchi, F. A., Santos, E. L., Alves Correa, S. A., Shimuta, S. I., et al. (2008). Expression of angiotensin I-converting enzymes and bradykinin B2 receptors in mouse inner medullary-collecting duct cells. *International Immunopharmacology*, *8*, 254–260.

Regoli, D., Plante, G. E., & Gobeil, F., Jr. (2012). Impact of kinins in the treatment of cardiovascular diseases. *Pharmacology & Therapeutics*, *135*, 94–111.

Rhaleb, N. E., Yang, X. P., & Carretero, O. A. (2011). The kallikrein-kinin system as a regulator of cardiovascular and renal function. *Comprehensive Physiology, 1*, 971–993.

Rohrwasser, A., Ishigami, T., Gociman, B., Lantelme, P., Morgan, T., Cheng, T., et al. (2003). Renin and kallikrein in connecting tubule of mouse. *Kidney International, 64*, 2155–2162.

Sabatini, R. A., Guimaraes, P. B., Fernandes, L., Reis, F. C., Bersanetti, P. A., Mori, M. A., et al. (2008). ACE activity is modulated by kinin B2 receptor. *Hypertension, 51*, 689–695.

Schild, L. (1996). The ENaC channel as the primary determinant of two human diseases: Liddle syndrome and pseudohypoaldosteronism. *Néphrologie, 17*, 395–400.

Schild, L. (2004). The epithelial sodium channel: From molecule to disease. *Reviews of Physiology Biochemistry and Pharmacology, 151*, 93–107.

Schild, L., & Kellenberger, S. (2001). Structure function relationships of ENaC and its role in sodium handling. *Advances in Experimental Medicine and Biology, 502*, 305–314.

Schmaier, A. H. (2003). The kallikrein-kinin and the renin-angiotensin systems have a multilayered interaction. *American Journal of Physiology. Regulatory, Integrative and Comparative Physiology, 285*, R1–R13.

Shen, B., & El-Dahr, S. S. (2006). Cross-talk of the renin-angiotensin and kallikrei-kinin systems. *Biological Chemistry, 387*, 145–150.

Siragy, H. M. (1993). Evidence that intrarenal bradykinin plays a role in regulation of renal function. *The American Journal of Physiology, 265*, E648–E654.

Siragy, H. M., Ibrahim, M. M., Jaffa, A. A., Mayfield, R., & Margolius, H. S. (1994). Rat renal interstitial bradykinin, prostaglandin E2, and cyclic guanosine 3′,5′-monophosphate. Effects of altered sodium intake. *Hypertension, 23*, 1068–1070.

Sivritas, S. H., Ploth, D. W., & Fitzgibbon, W. R. (2008). Blockade of renal medullary bradykinin B2 receptors increases tubular sodium reabsorption in rats fed a normal-salt diet. *American Journal of Physiology. Renal Physiology, 295*, F811–F817.

Song, Q., Wang, D. Z., Harley, R. A., Chao, L., & Chao, J. (1996). Cellular localization of low-molecular-weight kininogen and bradykinin B2 receptor mRNAs in human kidney. *The American Journal of Physiology, 270*, F919–F926.

Staruschenko, A. (2012). Regulation of transport in the connecting tubule and cortical collecting duct. *Comprehensive Physiology, 2*, 1541–1584.

Stoos, B. A., Carretero, O. A., & Garvin, J. L. (1992). ANF and bradykinin synergistically inhibit transport in M-1 cortical collecting duct cell line. *The American Journal of Physiology, 263*, F1–F6.

Suzuki, T., Katori, M., Fujita, T., Kumagai, Y., & Majima, M. (2000). Involvement of the renal kallikrein-kinin system in K^+-induced diuresis and natriuresis in anesthetized rats. *European Journal of Pharmacology, 399*, 223–227.

Tamma, G., Carmosino, M., Svelto, M., & Valenti, G. (2005). Bradykinin signaling counteracts cAMP-elicited aquaporin 2 translocation in renal cells. *Journal of the American Society of Nephrology, 16*, 2881–2889.

Tang, S. C., Leung, J. C., & Lai, K. N. (2011). The kallikrein-kinin system. *Contributions to Nephrology, 170*, 145–155.

Tomita, K., Pisano, J. J., Burg, M. B., & Knepper, M. A. (1986). Effects of vasopressin and bradykinin on anion transport by the rat cortical collecting duct. Evidence for an electroneutral sodium chloride transport pathway. *The Journal of Clinical Investigation, 77*, 136–141.

Tomita, K., Pisano, J. J., & Knepper, M. A. (1985). Control of sodium and potassium transport in the cortical collecting duct of the rat. Effects of bradykinin, vasopressin, and deoxycorticosterone. *The Journal of Clinical Investigation, 76*, 132–136.

Tong, Q., & Stockand, J. D. (2005). Receptor tyrosine kinases mediate epithelial Na^+ channel inhibition by epidermal growth factor. *American Journal of Physiology. Renal Physiology, 288*, F150–F161.

Tschope, C., Schultheiss, H. P., & Walther, T. (2002). Multiple interactions between the renin-angiotensin and the kallikrein-kinin systems: Role of ACE inhibition and AT1 receptor blockade. *Journal of Cardiovascular Pharmacology*, *39*, 478–487.

Waeckel, L., Potier, L., Chollet, C., Taveau, C., Bruneval, P., Roussel, R., et al. (2012). Antihypertensive role of tissue kallikrein in hyperaldosteronism in the mouse. *Endocrinology*, *153*, 3886–3896.

Wolf, W. C., Yoshida, H., Agata, J., Chao, L., & Chao, J. (2000). Human tissue kallikrein gene delivery attenuates hypertension, renal injury, and cardiac remodeling in chronic renal failure. *Kidney International*, *58*, 730–739.

Yamaguchi, Y., Shimamoto, K., Ura, N., Nishimiya, T., Nakagawa, M., Masuda, A., et al. (1986). Localization of renal kallikrein-kinin system components in the kidney. *Japanese Journal of Medicine*, *25*, 9–12.

Yang, X., He, J., Gu, D., Hixson, J. E., Huang, J., Rao, D. C., et al. (2014). Associations of epithelial sodium channel genes with blood pressure changes and hypertension incidence: The GenSalt study. *American Journal of Hypertension*, *27*, 1370–1376.

Zaika, O., Mamenko, M., O'Neil, R. G., & Pochynyuk, O. (2011). Bradykinin acutely inhibits activity of the epithelial Na^+ channel in mammalian aldosterone-sensitive distal nephron. *American Journal of Physiology. Renal Physiology*, *300*, F1105–F1115.

Zeidel, M. L., Jabs, K., Kikeri, D., & Silva, P. (1990). Kinins inhibit conductive Na^+ uptake by rabbit inner medullary collecting duct cells. *American Journal of Physiology*, *258*, F1584–F1591.

CHAPTER SIX

Inhibition of ENaC by Endothelin-1

Andrey Sorokin*[,1], Alexander Staruschenko†[,1]
*Division of Nephrology, Department of Medicine, Medical College of Wisconsin, Milwaukee, Wisconsin, USA
†Department of Physiology, Medical College of Wisconsin, Milwaukee, Wisconsin, USA
[1]Corresponding authors: e-mail address: sorokin@mcw.edu; staruschenko@mcw.edu

Contents

Abstract

The amiloride-sensitive epithelial Na^+ channel (ENaC) is a key player in the regulation of Na^+ homeostasis. Its functional activity is under continuous control by a variety of signaling molecules, including bioactive peptides of endothelin family. Since ENaC dysfunction is causative for disturbances in total body Na^+ levels associated with the abnormal regulation of blood volume, blood pressure, and lung fluid balance, uncovering the molecular mechanisms of inhibitory modulation or inappropriate activation of ENaC is crucial for the successful treatment of a variety of human diseases including hypertension. The precise regulation of ENaC is particularly important for normal Na^+ and fluid homeostasis in organs where endothelins are known to act: the kidneys, lung, and colon. Inhibition of ENaC by endothelin-1 (ET-1) has been established in renal cells, and several molecular mechanisms of inhibition of ENaC by ET-1 are proposed and will be reviewed in this chapter.

Vitamins and Hormones, Volume 98
ISSN 0083-6729
http://dx.doi.org/10.1016/bs.vh.2015.01.001

ABBREVIATIONS

AMPK 5′ adenosine monophosphate-activated protein kinase
Ang II angiotensin II
ANP atrial natriuretic peptide
ASDN aldosterone-sensitive distal nephron
AVP arginine vasopressin
BK large conductance calcium-activated Maxi-K channel
βPix p21-activated kinase-interacting exchange factor β
$[Ca^{2+}]_i$ cytosolic free Ca^{2+} concentration
CCD cortical collecting duct
CD collecting duct
CNT connecting tubules
CK2 casein kinase 2
DCT distal convoluted tubule
ECFV extracellular fluid volume
ENaC epithelial Na^+ channel
ERK extracellular-regulated kinase
ET-1 endothelin-1
ET-2 endothelin-2
ET-3 endothelin-3
ETRA endothelin receptor A
ETRB endothelin receptor B
GEF guanine nucleotide exchange factor
GRK2 G protein-coupled receptor kinase 2
ICT initial collecting tubule
IGF-1 insulin-like growth factor 1
IKKβ I kappa B kinase β
IMCD inner medullary collecting duct
IP_3 inositol trisphosphate
MAPK mitogen-activated protein kinase
NEDD4-2 neural precursor cell expressed, developmentally downregulated 4-2
NO nitric oxide
NOS nitric oxide synthase
OMCD outer medullary collecting duct
P_o channel open probability
PAH pulmonary arterial hypertension
PDGF platelet-derived growth factor
PE preeclampsia
Per1 circadian clock protein Period 1
PHAI pseudohypoaldosteronism type I
PI-3 kinase phosphoinositide 3 kinase
PKA protein kinase A
PKC protein kinase C
PKD1 polycystic kidney disease 1
PLC phospholipase C
PP2 a selective inhibitor of Src-family tyrosine kinases
Pyk2 proline-rich tyrosine kinase 2

RAAS renin–angiotensin–aldosterone system
RAGE receptor for advanced glycation end-products
ROMK renal outer medullary K channel
sFlt-1 soluble fms-like tyrosine kinase
SGK serum- and glucocorticoid-induced protein kinase
TAL thick ascending limb of Henle's loop
tALH thin ascending limb of Henle's loop
tDLH thin descending limb of Henle's loop
TNFα tumor necrosis factor alpha
TRPC transient receptor potential canonical protein

1. INTRODUCTION

The amiloride-sensitive epithelial Na^+ channel (ENaC) is expressed in the apical membrane of polarized epithelial cells in the distal part of the kidney tubule, the lung, and the distal colon. These cells are in constant contact with other cells, which is achieved by specific reception and secretion of a number of regulatory molecules, including potent bioactive peptides endothelins. The family of endothelin proteins consists of three members, endothelin-1, endothelin-2, and endothelin-3 (ET-1, ET-2, and ET-3, respectively), known to play important roles in renal physiology and pathophysiology due to their involvement into the physiological control of blood pressure and salt homeostasis (Kohan, Inscho, Wesson, & Pollock, 2011; Kohan, Rossi, Inscho, & Pollock, 2011; Simonson & Dunn, 1993; Sorokin & Kohan, 2003). ET-1 is the main isoform that is secreted by renal cells and that also operates on a number of renal targets, including its effects in different parts of nephron on top of action as multifunctional bioactive peptides in renal mesangium (Sorokin, 2011). Its production is largely controlled at the level of gene expression (Foschi et al., 2001; Sorokin & Kohan, 2003). ET-1 appears to be an important negative regulator of ENaC *in vivo*, and endothelin-dependent regulation of ENaC activity, critical aspect of the functional role of endothelins, will be the subject of this review.

2. REGULATION OF SODIUM REABSORPTION: THE ROLE OF ENaC

Hypertension represents one of the most prevalent chronic health problems of Western nations, but the mechanisms responsible for this disease remain poorly understood. Long-term control of blood pressure involves Na^+ homeostasis through the precise regulation of the ENaCs in the aldosterone-sensitive distal nephron (ASDN). Although only a small percent of the glomerular filtrate (<5–10%) reaches the connecting tubules (CNTs) and collecting ducts (CDs), these segments are critical for water and electrolyte homeostasis since fine-tuning of electrolyte and fluid balance is mediated in these nephron segments through reabsorption and secretion and these processes are under tight control of hormones and nonhormonal factors (Staruschenko, 2012). Na^+ reabsorption in the CNT and CDs is transcellular and is mediated by the CNT cells and the principal cells of CDs. Shown in Fig. 1 is a structure of the nephron with segments expressing ENaC. At the basolateral membrane of the principal cells, Na^+ extrusion is mediated by the Na^+/K^+-ATPase, which also provides the electrochemical driving force for the apical entry of Na^+ (Feraille & Doucet, 2001). ENaC, which appears to be a trimeric channel composed of one α-, one β-, and one γ-subunits (Jasti, Furukawa, Gonzales, & Gouaux, 2007; Staruschenko, Adams, Booth, &

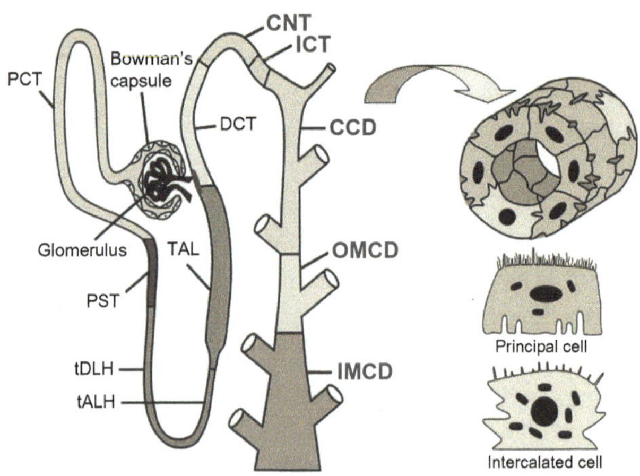

Figure 1 Structure of the nephron and specific segments involved in effects of ET-1 on ENaC. Every nephron contains a renal corpuscle (glomerulus and Bowman's capsule), a proximal tubule (proximal convoluted and straight tubules; PCT and PST, respectively), a loop of Henle (thin descending limb of Henle's loop (tDLH), thin ascending limb of Henle's loop (tALH), and thick ascending limb of Henle's loop (TAL)), a distal convoluted tubule (DCT), connecting tubule (CNT), and the collecting duct system, which includes the initial collecting tubule (ICT), the cortical collecting duct (CCD), the outer medullary collecting duct (OMCD), and the inner medullary collecting duct (IMCD). Structure of the CCD as a cross section and schematic presentation of principal and intercalated cells that comprise these segments are also shown. *Modified from Staruschenko (2012).* (See the color plate.)

Stockand, 2005), is responsible for sodium reabsorption in these segments. ENaC dysfunction is causative for disturbances in total body Na^+ levels associated with the abnormal regulation of blood volume and blood pressure, as well as alterations in lung fluid balance (Alvarez de la Rosa, Navarro-Gonzalez, & Giraldez, 2013; Bhalla & Hallows, 2008; Pearce et al., 2015; Rossier, 2014; Soundararajan, Pearce, Hughey, & Kleyman, 2010; Warnock et al., 2014). Mutations in genes encoding the ENaC subunits corroborate critical role of this channel in the control of blood pressure. For instance, Liddle syndrome is an autosomal dominant form of hypertension that results from the C-terminal truncation mutations in the β- or γ-ENaC subunits, which prevents the channel's retrieval from the apical membrane and subsequent degradation, thus leading to increased basal ENaC expression and activity at the apical membrane (Hansson, Nelson-Williams, et al., 1995; Hansson, Schild, et al., 1995; Lifton, Gharavi, & Geller, 2001; Shimkets et al., 1994). Loss-of-function mutations in any of the three different ENaC subunits also cause the autosomal recessive form of pseudohypoaldosteronism type I (PHAI) (Chang et al., 1996; Grunder et al., 1997; Lifton et al., 2001). ENaC-mediated electrogenic sodium entry also provides the driving force for luminal potassium exit via renal outer medullary K channel (ROMK) and large conductance calcium-activated Maxi-K (BK) potassium channels (Staruschenko, 2012; Welling, 2013; Wen, Cornelius, & Sansom, 2014). Mutations in ROMK channel result in type II Bartter syndrome (Simon et al., 1996; Srivastava et al., 2013; Welling & Ho, 2009).

The expression and activity of ENaCs are regulated by specific hormones and different extra- and intracellular regulatory mechanisms. Considering that ENaC is responsible for the fine-tuning of sodium reabsorption in the last nephron segment, the role of this channel in sodium reabsorption in the kidney is critical and unique. The tight regulation of transcellular Na^+ concentrations is so important that multiple mechanisms work in concert to control them. One of the main mechanisms controlling ENaC activity is the activation of the renin–angiotensin–aldosterone system (RAAS) (Alvarez de la Rosa et al. 2013; Pearce et al., 2015; Quinn, Harvey, & Thomas, 2014; Rossier, 2014). The activation of the RAAS is well known to enhance activity of ENaC. Aldosterone, as well as other corticosteroid hormones, binds to the mineralocorticoid receptor and triggers coordinated changes in the expression of multiple aldosterone-induced genes in a ligand-dependent manner (Kone, 2013; Stockand, 2002). Hyperaldosteronism results in enhanced sodium reabsorption in the distal nephron, whereas aldosterone insufficiency leads to reduced ENaC activity and urinary sodium wasting. However, substantial experimental data suggest a role for aldosterone-independent mechanisms in the regulation of ENaC (Zaika, Mamenko, Staruschenko, & Pochynyuk, 2013). For instance, mice lacking mineralocorticoid receptor still possess ENaC-mediated Na^+ reabsorption in the ASDN (Berger et al., 1998). Besides, it was reported that mice with surgically removed adrenal glands have no detectable circulating aldosterone, but possess robust ENaC activity in the distal nephron (Mironova, Bugaj, Roos, Kohan, & Stockand, 2012). Recent studies further revealed that ENaC activity remained above control levels during maximal mineralocorticoid receptor inhibition with spironolactone (Mamenko et al., 2013). Therefore, other signaling pathways, which modulate ENaC either acutely or at transcription level, work in parallel with the RAAS.

This chapter is focused on the control of ENaC by ET-1. However, as noticed above, multiple mechanisms are involved in the tight control of ENaC expression and activity in the ASDN. Hormones are key regulators of sodium transport in the kidney and in ASDN specifically. In addition to the RAAS, it was shown that arginine vasopressin (AVP) (Bankir, Bouby, & Ritz, 2013; Ecelbarger et al., 2000; Mironova et al., 2012; Sanghi et al., 2014; Stockand, 2012), atrial natriuretic peptide (ANP) (Guo, Alli, Eaton, & Bao, 2013; Kudo & Baird, 1984; Wang et al., 2006), and insulin (alongside with insulin-like growth factor 1, IGF-1) (Blazer-Yost, Liu, & Helman, 1998; Ilatovskaya, Pavlov, Levchenko, & Staruschenko, 2013; Li et al., 2013; Pavlov, Ilatovskaya, et al., 2013) are critical modulators of ENaC activity. The kallikrein–kinin systems through their peptide bradykinin also play a specific role in blunting ENaC activity, particularly under conditions of elevated sodium intake (Mamenko, Zaika, Doris, & Pochynyuk, 2012; Mamenko, Zaika, & Pochynyuk, 2014; Zaika, Mamenko, O'Neil, & Pochynyuk, 2011). In addition to the hormonal regulation of ENaC-mediated sodium transport in the kidney, a number of local autocrine and paracrine factors play critical roles in the modulation of ENaC. For instance, a recent review article by Stockand and colleagues highlights the regulation of ENaC-mediated sodium excretion and blood pressure by purinergic signaling (Mironova, Boiko, Bugaj, Kucher, & Stockand, 2015). Multiple evidences reveals that a robust inhibitory purinergic signaling system intrinsic to the ASDN dynamically regulates ENaC through paracrine ATP signaling via the metabotropic $P2Y_2$ purinergic receptor to properly match urinary Na^+ excretion to dietary Na^+ intake (Birch, Schwiebert, Peppiatt-Wildman, & Wildman, 2013; Pochynyuk et al., 2008, 2010; Rieg et al., 2007; Rieg, Gerasimova, Boyer, Insel, & Vallon, 2011; Stockand et al., 2010). This enables blood pressure to be maintained within a normal range despite broad changes in dietary Na^+ consumption. We and others also identified that members of the epidermal growth factors (EGF) family are involved in the control of ENaC (reviewed in Staruschenko, Palygin, Ilatovskaya, & Pavlov, 2013). Using the Dahl salt-sensitive rat model, we evaluated the role of EGF and identified that deficiency of renal cortical EGF increases ENaC activity and contributes to salt-sensitive hypertension (Pavlov, Levchenko, et al., 2013). Prostaglandins, cytochrome P450 metabolites, nitric oxide (NO), peroxisome proliferator-activated receptor agonists, and other molecules are also among important paracrine and autocrine factors modulating ENaC activity.

3. ENDOTHELIN SIGNALING AND CONTROL OF BLOOD PRESSURE

3.1 Endothelin receptors

Endothelins are multifunctional 21 amino-acid vasoactive peptides secreted by numerous cell types (Simonson & Dunn, 1993). All of endothelin's effects are elicited by binding to specific G-protein-coupled receptors endothelin receptor A (ETRA) (Arai, Hori, Aramori, Ohkubo, & Nakanishi, 1990) and endothelin receptor B (ETRB) (Sakurai et al., 1990). ETRA and ETRB demonstrate considerable variability in the distribution between different tissues and among species (Pollock, 1998). Human and rat renal tubular cells, vascular smooth muscle cells, glomerular mesangial cells, and many other cell types express substantial levels of both ETRA and ETRB (Bouallegue, Daou, & Srivastava, 2007; Kohan, Inscho, et al., 2011; Kohan, Rossi, et al., 2011; Orth et al., 2000; Sorokin, 2011). A progressive renal injury like inflammation or fibrosis is mediated via both ETRA and ETRB receptors, while constrictor effects are primarily transduced by ETRA receptors (Neuhofer & Pittrow, 2006). Thus, the actions of ET-1 depend on the cell type and physiological condition (Vignon-Zellweger, Heiden, & Emoto, 2011).

3.2 Endothelin signaling

ET-1 receptors couple to members of the $G\alpha_i$, $G\alpha_q$, $G\alpha_s$, and $G\alpha_{12/13}$ G-protein families (Sorokin & Kohan, 2003). Accordingly, when ET-1 binds to its receptor, the multiple intracellular pathways are triggered (Sorokin, 2011; Sorokin, Foschi, & Dunn, 2002). Among most important cellular events are the activation of phospholipases A, C, and D; the activation of serine–threonine kinases; the increased protein tyrosine phosphorylation;

and a stimulated influx of Ca^{2+} from outside the cell and mobilization of intracellular Ca^{2+} from internal sources.

ET-1 acts via ETRA to induce GTP loading of $G\alpha_q$ and $G\alpha_s$ to stimulate phospholipase C (PLC) and adenylate cyclase, respectively. Correspondingly, ETRB stimulates PLC via $G\alpha_q$ and causes the inhibition of adenylate cyclase through $G\alpha_i$. The ET receptors colocalize and interact with caveolin-1 in caveolae, and since caveolae integrate different receptor and signaling proteins, they add to the complexity of ET-1 signaling. It is of note that the activation of endothelin receptors leads to the compartmentalization and the binding of $G\alpha_q$ to guanine exchange factor βPix (p21-activated kinase-interacting exchange factor β) in caveolae (Chahdi & Sorokin, 2010b). Another important feature of ET-1 signaling is the transactivation of growth factor receptors, such as EGF receptors (Chahdi & Sorokin, 2010a; Hua, Munk, & Whiteside, 2003; Iwasaki, Eguchi, Ueno, Marumo, & Hirata, 1999; Kodama et al., 2002; Vacca, Bagnato, Catt, & Tecce, 2000) and PDGF receptors (Harada et al., 2014). ET-1 shares this ability with some other ligands of G-protein-coupled receptors (Daub, Weiss, Wallasch, & Ullrich, 1996; Little, 2013).

The contraction of smooth muscle and glomerular mesangial cells induced by ET-1 depends upon the stimulation of inositol triphosphate (IP_3) generation and elevation of cytosolic concentration of free Ca^{2+} ($[Ca^{2+}]_i$) (Badr et al., 1989; Bouallegue et al., 2007). The calcium influx and mobilization pathways activated by ET-1 vary immensely and are cell type-specific with regard to which ET receptor subtypes are involved (Tykocki & Watts, 2010). Initial transient increase in $[Ca^{2+}]_i$ after ET-1 treatment is followed by a sustained lesser increase. It has been shown that early transient peak reflects IP_3-mediated release of Ca^{2+} from intracellular sources, which is associated with cell alkalinization through augmented Na^+/H^+ exchange (Simonson & Dunn, 1991, 1993; Simonson et al., 1989). The continued plateau of $[Ca^{2+}]_i$ that over time decreases toward the initial level is due to the increased influx of Ca^{2+} across the plasma membrane. This pattern of initial increase followed by sustained influx is induced by ET-1 in both vascular and nonvascular cells (Tykocki & Watts, 2010). Vascular smooth muscle cells and glomerular mesangial cells express L-type Ca^{2+} channels (McDermott, Hurst, & Whiteside, 1993; Sonkusare et al., 2006) as well as transient receptor potential canonical (TRPC) proteins, a family of nonselective Ca^{2+}-permeable cation channels (Du et al., 2007; Gonzalez-Cobos & Trebak, 2010). Members of TRPC family that were shown to be activated by the stimulation of endothelin receptors in various cells or tissues include TRPC1, TRPC3, TRPC6, and TRPC7 (Horinouchi, Terada,

Higashi, & Miwa, 2013). Low concentrations of ET-1 added to mesangial cells induced slow sustained increase in $[Ca^{2+}]_i$ through a voltage channel-independent mechanism, whereas higher than 100 pM concentrations caused a transient increase in $[Ca^{2+}]_i$, which was dependent on Ca^{2+} release from intracellular stores and required the activation of PLC and protein kinase C (PKC) (Simonson et al., 1989).

Endothelin-mediated increase of $[Ca^{2+}]_i$ causes the activation of calcium-dependent intracellular tyrosine kinase Pyk2 (Kawanabe, Hashimoto, & Masaki, 2003; Park, Bose, Leszyk, & Czech, 2001; Sorokin, Kozlowski, Graves, & Philip, 2001), providing therefore the link between G-protein-coupled endothelin receptors and the induction of tyrosine phosphorylation via the mobilization of intracellular calcium. The activation of Pyk2 in ET-1-treated cells requires not only Ca^{2+} mobilization but also an increased activity of Src kinases (Sorokin et al., 2001). ET-1 was shown to activate Src kinases (Simonson & Herman, 1993), and it appears that at least some members of Src family are recruited to the complex of β-arrestin 1 with ETRA in an agonist-dependent manner (Imamura et al., 2001). Pyk2/p130Cas/BCAR3/Rap1 signaling cascade activated via ETRA receptors is involved in the regulation of adhesion and spreading of mesangial cells (Rufanova, Alexanian, Wakatsuki, Lerner, & Sorokin, 2009). Pyk2 also mediates ET-1 signaling to GLUT4 glucose transporters in 3T3-L1 adipocytes (Park et al., 2001).

The ability of ET-1 to act as a mitogen is well established (Bouallegue et al., 2007; Fukuda et al., 1996; Sorokin, 2011). Since GTP loading of the small GTPase Ras, which is followed by the activation of the Raf–MEK–ERK signaling pathway, is indispensable for cell proliferation, it is not surprising that ET-1 is not only a potent inducer of Ras activation but also an effective stimulator of ERK signaling cascade (Foschi, Chari, Dunn, & Sorokin, 1997; Wang, Simonson, Pouyssegur, & Dunn, 1992). Remarkably, other mitogen-activated protein kinase (MAPK)–JNK/SAPK cascades (Araki, Haneda, Togawa, & Kikkawa, 1997), p38 MAP kinase (Sorokin et al., 2001), and ERK5 (Dorado et al., 2008) are also induced by ET-1. The significance of the activation of these MAPKs by ET-1 has been investigated in different cellular systems and is partially responsible for the multifunctional activity of endothelins.

3.3 Targeting endothelins

The increase of ET-1 production accompanies a wide variety of renal diseases, and since the most efficient way to interfere with excessive ET-1 action is to block its specific reception by ETR receptors, the pharmacological blockade of ETR received unprecedented attention from multiple pharmaceutical companies. Quite a few nonselective (targeting both ETRA and ETRB) and selective (targeting either ETRA or ETRB) antagonists were created and evaluated for their efficiency (Motte, McEntee, & Naeije, 2005). This class of compounds showed noteworthy promise in different animal models of renal and cardiovascular diseases, and sufficient rationale for testing them for the treatment of human pathologies was obtained (Benigni et al., 1993; Boffa, Tharaux, Dussaule, & Chatziantoniou, 2001; Dhein et al., 2000; Fukuda et al., 1996; Gomez-Garre et al., 1996; Kohan & Pollock, 2013; Neuhofer & Pittrow, 2006). Numerous clinical trials addressing the benefits of blockade of ETR receptors in the treatment of cardiovascular and renal diseases have been initiated. It was reported that ETR antagonists can reverse proteinuric renal disease and glomerulosclerosis, and their remarkable antiproteinuric effects are additive to those of standard antiproteinuric therapy (Barton, 2008). There are two ETRA antagonists approved in the USA for the treatment of pulmonary hypertension: ambrisentan and bosentan. Sitaxsentan, the most selective ETRA antagonist clinically available, has been approved for use in Europe, Canada, and Australia but not in the USA. It was, however, recently voluntarily withdrawn by Pfizer driven by a review of evolving safety information from clinical trials regarding concerns about idiosyncratic, fatal hepatic failure (Raja & Raja, 2011). Nevertheless, experimental data and recent clinical studies suggest that despite limitations connected with side effects and relatively poor antagonist selectivity, the antagonists of ETRA hold promise in the treatment of hypertension and diabetic nephropathy (Meyers & Sethna, 2013). Antagonists of the ETRB are not usually beneficial since they might cause vasoconstriction along with salt and water retention. The blockade of ET-1 actions reduces the progression of chronic kidney disease and might even be beneficial in dialysis patients, even though the potential therapeutic effect of ETR antagonists for the treatment of end-stage kidney disease still requires the verification by clinical trials (Kohan & Pollock, 2013).

3.4 Endothelin-mediated control of blood pressure

The regulation of blood pressure by endothelin is a result of complex interweaving of endothelin's effects upon the vasculature, humoral systems, nervous system, heart function, and specialized kidney cells (Kohan, Inscho, et al., 2011; Kohan, Rossi, et al., 2011). It is generally accepted that the effects of endothelin upon blood pressure and Na^+ homeostasis are the consequences of vasodilation caused by stimulated NO production by endothelial cells, attenuation of water reabsorption in the CD, decrease of Na^+ transport in several segments of nephron, impairment of afferent nerve inputs, and inhibition of renin release (Kohan, Inscho, et al., 2011; Kohan, Rossi, et al., 2011).

ET-1 is among the most potent vasoconstrictors. In terms of contractility, the net effect of ETRA activation is contraction, whereas the net effect of activation of ETRB tends to be vasorelaxation (Sorokin & Kohan, 2003). Renal microvascular smooth muscle cells that express both ETRA and ETRB react to ET-1 by vasoconstriction, whereas ETRB microvascular endothelial cells promote vasodilator responses (Guan & Inscho, 2011; Pollock, Keith, & Highsmith, 1995). The vasorelaxant effect of ET-1 is due to stimulated secretion of endothelium-derived relaxing factor NO (Levin, 1996). The ability of ETRB to induce NO release is linked to nitric oxide synthase (NOS) activation, which is markedly increased when intracellular Ca^{2+} rises (Forstermann & Sessa, 2012). In turn, NO can modulate ET-1 activity via inhibiting ET-1 formation/secretion (Bourque, Davidge, & Adams, 2011). Since NO and ET-1 continuously interfere with each others actions, the pressure regulation by ET-1 and NO is more complex than simply their individual pressor and depressor action (Rapoport, 2014). Nonetheless, it is widely recognized that endothelial NOS keeps all types of blood vessels dilated, making NO/ET-1 interference an important factor of endothelin-mediated control of blood pressure (Forstermann & Sessa, 2012).

The role of ET-1 in the regulation of water reabsorption became evident from studies utilizing either *in vivo* administration of ET-1 or analysis of mice with CD-specific knockouts of ET-1. The administration of ET-1 inhibits vasopressin-stimulated osmotic water permeability in the CDs (Nadler, Zimpelmann, & Hebert, 1992; Oishi, Nonoguchi, Tomita, & Marumo, 1991). Mice lacking ET-1 production in the CD, the major renal site of ET-1 production, have impaired the excretion of an acute, but not a chronic, water load, suggesting that ET-1 functions as a physiological autocrine regulator of vasopressin-mediated water permeability in the CDs (Ge, Ahn, et al., 2005; Kohan, 2013). These effects seem to be mediated by ETRB, since blockade of ETRA was inefficient, while agonists of ETRB had inhibitory effects (Kohan, 2013). The data obtained with ETRB CD-specific knockout indicate however that ETRB of the CD only partially mediates the antihypertensive and natriuretic effects of ET-1 (Ge et al., 2006). It is likely that ET-1 effects are associated with the inhibition of vasopressin-stimulated cAMP accumulation, which is achieved by ET-1-triggered elevation of intracellular Ca^{2+} and PLC-mediated activation of PKC (Kohan, 2013; Teitelbaum & Berl, 1994).

The overall action of ET-1 to increase blood pressure and vascular tone is partially due to ET-1-mediated stimulation of the sympathetic nervous system (Agapitov & Haynes, 2002; Kohan, Inscho, et al., 2011; Kohan, Rossi, et al., 2011). The renal tissues are densely innervated by the sympathetic nerves, and increase in renal sympathetic activity, to which ET-1 contributes importantly, decreases urinary sodium excretion (Kopp, 2011). Endothelins are significantly involved in the modulation of the activation of the afferent renal nerves and can induce the release of neurotransmitters, promote the development of postganglionic neurons, and influence the generation of action potentials (Kohan, Inscho, et al., 2011; Kohan, Rossi, et al., 2011; Kopp, 2011). Even though the mechanisms responsible for the effects of ET-1 in the ganglia, peripheral nerves, and central nervous system resulting in blood regulation are not sufficiently studied, ET-1 and its receptors have been implicated in the enhanced sympathetic excitability associated with salt-sensitive hypertension, regulation of baroreflex activity, and systemic hemodynamics (Kohan, Inscho, et al., 2011; Kohan, Rossi, et al., 2011).

ET-1 was shown to inhibit renin release from rat kidney cortical slices and isolated juxtaglomerular cells (Kramer et al., 1994; Moe, Tejedor, Campbell, Alpern, & Henrich, 1991). Studies utilizing the administration of ET-1 intravenously suggest that endothelin inhibits renin release *in vivo* (Otsuka et al., 1989). Even though the inhibition of renin release by ET-1 *in vitro* and *in vivo* has been documented a long time ago, there is no consensus on what is the physiological significance of this effect. It has been hypothesized that high salt diet induces ET-1 production, which promotes Na excretion by attenuating the renal baroreceptor responsible for renin release (Kohan, Inscho, et al., 2011; Kohan, Rossi, et al., 2011). It has been reported that ET-1 generation is increased by Ang II and that vasoconstrictor effects of Ang II are partially mediated by ET-1 (Imai et al., 1992). Furthermore, the synergistic effects of ET-1 and Ang II on blood pressure have been observed in rats and the enhancement of ET-1-induced vasoconstriction by Ang II was shown through the upregulation of ETRA (Lin et al., 2014). The cross talk between the renin–angiotensin–aldosterone system and endothelin system is complex and is observed in a number of cellular systems, including cerebrovascular cells, hepatic stellate cells, and renal mesangial and endothelial cells (He, Miao, Li, & Qi, 2013; Konczalla et al., 2013; Lopez-Ongil, Diez-Marques, Griera, Rodriguez-Puyol, & Rodriguez-Puyol, 2005).

4. COLLECTING DUCT: ET-1 AND ENaC

4.1 ET receptor expression and endothelin production in CDs

Renal ET-1 production is facilitated by shear stress, inflammation, oxidative stress, the RAAS, and other systems and mechanisms (for review, see Kohan, Inscho, et al., 2011; Kohan, Rossi, et al., 2011). Changes in extracellular fluid volume also increase ET-1 levels. Importantly, most of these mechanisms result in the activation of ENaC in ASDN. Thus, the endothelin might represent a feedback mechanism, which plays protective role in the maintenance of ENaC-mediated sodium absorption in the kidney and blood pressure. As summarized in excellent reviews by Kohan and colleagues (Kohan, Inscho, et al., 2011; Kohan, Rossi, et al., 2011), the CDs are the major source of ET-1 in the kidney and may produce and release more ET-1 than any other cell type in the body. Consistent with this conclusion is that CD-specific ET-1 knockout mice have markedly reduced urinary ET-1 excretion in response to salt loading (Ahn et al., 2004). Numerous *in vitro* and *ex vivo* studies suggested that ET-1 inhibits Na^+ and water reabsorption in the CDs (Edwards, Stack, Pullen, & Nambi, 1993; Kohan, Padilla, & Hughes, 1993; Nadler et al., 1992; Takemoto, Uchida, Ogata, & Kurokawa, 1993; Tomita, Nonoguchi, Terada, & Marumo, 1993). However, central evidence of ET-1 in CDs was published only 10 years ago when Ahn et al. demonstrated that the absence of CD principal cell ET-1 causes hypertension and sodium retention (Ahn et al., 2004). Mice with CD-specific knockout of ET-1 (under the control of the AQP2 promoter) were hypertensive and had reduced sodium excretion in response to sodium loading. Systolic blood pressure was increased by approximately 15 and 35 mmHg during normal and high Na^+ intake, respectively (Ahn et al., 2004).

As discussed above, the effects of endothelins are mediated by the activation of ETRA and ETRB receptors. Both receptors are expressed in the CNT and entire CDs, where as ENaC is involved in sodium transport (see Fig. 1). Importantly, ET receptors are localized at the basolateral membranes with no evidence of their expression at the luminal side. Numerous studies identified the expression of the ETRA receptor in these nephron segments (Kohan, Hughes, & Perkins, 1992; Wendel, Knels, Kummer, & Koch, 2006; Yamamoto, Suzuki, Kubo, Matsumoto, & Uemura, 2008; Yukimura et al., 1996). While some studies suggest that ETRA receptor is expressed in the distal tubules, majority of work identified that the greatest levels of these receptors are localized in the inner medullary collecting ducts (IMCDs). Functional studies using genetic deletion of ETRA receptor in the principal cell of CDs (Ge, Stricklett, Hughes, Yanagisawa, & Kohan, 2005) or entire nephron (Stuart et al., 2012) revealed that renal ETRA receptor, at least under normal physiological conditions, does not play a major role in the regulation of blood pressure, sodium excretion, or systemic hemodynamics. In the following study using tissue-specific promoters to target ETRA receptors in the heart, smooth muscles, nephron, and CD, the authors demonstrated that fluid retention is due to a direct effect on the CDs and partially to a vascular effect but not related to cardiac function (Stuart, Chapman, Rees, Woodward, & Kohan, 2013). Recent studies tested the effects of ET-1 on ENaC-mediated (benzamil-sensitive) Na^+ transport by measuring lumen-to-bath isotopic ^{22}Na reabsorption (J_{Na}) using *in vitro* microperfusion and identified that ET-1 regulation of Na^+ reabsorption in the CDs involves both ETRA and ETRB receptors (Lynch, Welch, Kohan, Cain, & Wingo, 2013). However, despite the data demonstrating the expression of ETRA receptor and some evidence of its functional role, it appears that the ETRB receptor is the predominant isoform in the CDs and in the renal medulla and most effects on ENaC involved the activation of the ETRB receptor.

The expression of the ETRB receptor in the CDs was shown by multiple techniques in different species. While mice with CD-specific (or entire nephron) knockouts of ETRA receptor did not have altered blood pressure or sodium excretion Ge, Ahn, et al., 2005; Ge, Stricklett, et al., 2005; Stuart et al., 2012), CD-specific knockout of the ETRB receptor causes hypertension and sodium retention. However, CD-specific knockout of the ETRB receptor increases blood pressure to a lesser extent than CD ET-1 knockout (Ge et al., 2006). Combined knockout of both ETA and ETB receptors in CDs had increased blood pressure, which increased further with high salt intake. The degree of blood pressure elevation during normal Na^+ intake was similar to CD-specific ET-1 knockout mice and higher than in CD ETRB knockout animals.

4.2 ET-1 regulation of ENaC in CDs

Kurokawa et al. reported more than 20 years ago that in rabbit CCDs perfused *in vitro*, low concentrations of ET-1 gradually increased total cellular membrane resistance because of the inhibition of the luminal Na^+ channel (Kurokawa et al., 1993). Another evidence of the key role of ENaC in

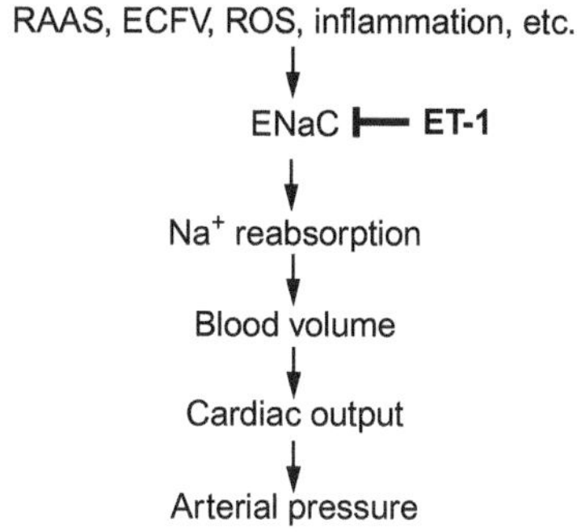

Figure 2 Simplified hypothesis to explain mechanisms of ENaC regulation by ET-1 in the setting of the development of hypertension. Changes in extracellular fluid volume (ECFV), inflammation, oxidative stress, the renin–angiotensin–aldosterone system (RAAS), and other systems and mechanisms positively modulate ENaC activity. Endothelin represents a feedback mechanism, which plays protective role in the maintenance of ENaC-mediated sodium absorption in the kidney and blood pressure.

ET-1-mediated processes was provided by Gariepy et al. who demonstrated that ETRB knockout rats develop hypertension on a high salt diet and that normal blood pressure in these salt-sensitive, hypertensive rats is restored after amiloride treatment (Gariepy, Ohuchi, Williams, Richardson, & Yanagisawa, 2000). Moreover, electrophysiological study showed that picomolar concentrations of ET-1 attenuate ENaC open probability (P_o) in an amphibian distal nephron cell line and this effect is prevented by ETRB antagonist but not ETRA antagonist (Gallego & Ling, 1996). Similar results were demonstrated in mammalian fibroblast cells by stably expressing genes for the three ENaC subunits (Gilmore, Stutts, & Milgram, 2001). Patch clamp studies of ENaC activity in isolated split-opened tubules provided compelling support for ET-1 acutely regulating ENaC activity and helped to identify signaling pathways and time line of these effects. Bugaj et al. demonstrated in native rat CD principal cells that ET-1 dynamically decreases ENaC P_o via ETRB receptors by about threefold within 5 min. Authors demonstrated that subsequent intracellular pathways involved Src tyrosine kinase activity and MAPK1/2 signaling (Bugaj et al., 2008). Following electrophysiological studies in CDs isolated from CD-specific ET-1, ETRA, ETRB, and ETRA/ETRB knockout mice further confirmed that ENaC regulation by ET-1 via ETB receptors contributes to the antihypertensive and natriuretic effects of the local endothelin system in the mammalian CDs (Bugaj, Mironova, Kohan, & Stockand, 2012). Shown in Fig. 2 is a basic scheme of ENaC regulation by ET-1.

5. LUNG, SMOOTH MUSCLE, AND DISTAL COLON: ET-1 AND ENaC

Pulmonary arterial hypertension (PAH) is one of human pathologies for which ET-1 is clearly a key mediator (Shao, Park, & Wort, 2011). The pathobiology of PAH is associated with increased pulmonary vascular resistance resulting from hyperplasia of vascular cells within vascular wall. ET-1 contributes to vascular remodeling by having direct effects on all three layers of the vessel wall: endothelial cells, smooth muscle cells, and fibroblasts. Excessive local pulmonary ET-1 production in patients with pulmonary hypertension has been established (Cacoub et al., 1997) and increased circulative levels of ET-1 are in correlation with increased mortality in PAH patients (Cacoub, Dorent, Nataf, & Carayon, 1993). Remarkably, PAH is one of a very short list of human diseases for which treatment with ET receptor antagonists has been approved. Endothelin-receptor antagonism with bosentan, which is a dual ETRA and ETRB antagonist, was beneficial for the therapy for PAH (Rubin et al., 2002; Sitbon et al., 2003). Currently, two orally active antagonists, dual antagonist bosentan and ETRA-selective antagonist ambrisentan, are approved as treatments for PAH (Rubin, 2012).

The lung is the site in the human body where fluid content is tightly controlled within certain limits and, accordingly, treatment of pulmonary edema by ENaC activators/stimulators has been the focus of scientific interest of multiple laboratories (Fronius, 2013). TGF-β, a key mediator of acute kidney injury, was shown to drive the internalization of the ENaC from the alveolar epithelial cell surface, leading to the persistence of alveolar flooding (Peters et al., 2014). In addition, these channels are potential contributors to cystic fibrosis lung disease (Donaldson & Galietta, 2013; Hobbs, Da, & Tarran, 2013). Thus, multiple lung pathologies are associated with ET-1 signaling and regulation of ENaC, but it appears that there is little overlap between the diseases linked with either the endothelin system or ENaC complex. ENaC abnormal function is described in the context of disintegrating fluid homeostasis control, whereas the endothelin system is primarily involved in lung pathologies accompanying vascular remodeling. Hence, the interaction between increased pulmonary production of ET-1 and activity of ENaC has not been sufficiently investigated yet. It is tempting to suggest that since ENaC-mediated pulmonary pathologies are generally explicated by increased ENaC activity, the excessive ET-1 production is supposed to attenuate ENaC action and correspondently should contribute to the prevention of pulmonary channelopathies. Nonetheless, multiple mechanisms have been described to be relevant for the regulation of ENaC in pulmonary epithelial cells (Eaton, Helms, Koval, Bao, & Jain, 2009). ENaC in the lung is under the control of β-adrenergic agents, purinergic agonists, steroids, inflammatory chemokines, reactive oxygen and nitrogen species, and low oxygen tension (Eaton et al., 2009). Epithelial pulmonary cells of the alveolar surface are exposed to air-filled compartment, and ENaCs were reported to be directly activated by shear stress (Fronius, Bogdan, Althaus, Morty, & Clauss, 2010). The receptor for advanced glycation end products regulated lung fluid balance via PKC–gp91^{phox} signaling to ENaC (Downs, Kreiner, Johnson, Brown, & Helms, 2015). SGK (serum- and glucocorticoid-regulated kinases), PKC, and PKA (protein kinase A) upregulate ENaC activity in a variety of epithelial cell systems (Baines, 2013; Eaton, Yue, Eaton, & Bao, 2014). Some of these signaling molecules and pathways are well-known components of signal transduction by ET-1.

ET-1 system contributes to the pathogenesis of vascular diseases and is known to activate at least three major signaling pathways in the vascular smooth muscle cells: phosphoinositide cascade, MAPK signaling cascades, and phosphoinositide 3-kinase (PI-3 kinase) pathway (Bouallegue et al., 2007). It is of note that β-ENaC subunit is expressed in the smooth muscle cells where they are supposed to act as mechanotransducers to regulate myogenic response and hypertension (Drummond, 2012).

Abnormal activation of the endothelin system is widely accepted as a common mechanism that is engaged in the progression of solid tumors (Rosano, Spinella, & Bagnato, 2013). Cancers where ET-1 signaling was shown to be an active player include colon cancer. Accordingly, the majority of the information with regard to ET-1 action in the colon is related to its cancer-promoting effect. The signaling via ETRA is involved in colon cancer progression and metastasis (Nie et al., 2014; Sorby, Kleiveland, Andersen, Bukholm, & Jacobsen, 2011). On the contrary, not only has the connection between ENaC activity and colon cancer progression not been revealed, but also even the concept of ENaC system playing a significant role in cancer progression is unproved.

The involvement of distal colon ENaC in the regulation of sodium reabsorption and the regulation of blood pressure has been proposed (Rossier, 2014). The ENaC-β and ENaC-γ expression was shown to be diminished by high salt intake in colonic epithelial cells (Lienhard, Lauterburg, Escher, Frey, & Frey, 2012). Whether ET-1-mediated inhibition of ENaC underlies the ability of colonic epithelial cells to contribute to the protection of the mammalian body against salt overload remains to be determined. It must be taken into consideration that whereas ET-1 and ETRA, but not ETRB, were expressed at a high level in primary and cultured colon carcinoma cells, in normal colon tissues, ET-1 level was very low or undetectable (Liakou et al., 2012). Whether high salt diets trigger increased production of ET-1 in colon tissues, as it does in the kidney, has not been established, and, accordingly, the connection between ET-1 signaling and the inhibition of ENaC function in colon cells, although probable, cannot be postulated.

Preeclampsia (PE), a systemic disorder that is one of the leading causes of maternal mortality worldwide, is characterized by hypertension, proteinuria, and edema (Al-Jameil, Aziz, Fareed, & Tabassum, 2014). It is instigated by abnormal insufficient placenta that releases in the bloodstream tumor necrosis factor-alpha (TNFα) and soluble fms-like tyrosine kinase (sFlt-1), causing the subsequent overexpression of vascular and renal ET-1 (Speed & Pollock, 2013). Multiple experimental models of PE are associated with elevated tissue levels of ET-1, and it appears that ET-1 serves as a final common pathway linking factor produced during placental ischemia to cause elevation of blood pressure (Palei, Spradley, Warrington, George, & Granger, 2013).

There is sufficient evidence indicating the involvement of ENaC with the progression of PE. One of the causes of PE is an inadequate cytotrophoblast migration, and since β-ENaC mediates cytotrophoblast migration, an altered expression of ENaC contributes to placental ischemia and hypertension (Warrington et al., 2014). A genetic variant of the β-subunit of ENaC was shown to be associated with the pathogenesis of PE and hypertension (Dhanjal, Owen, Anthony, Davidson, & Rayner, 2006; Jones, Owen, & Rayner, 2012).

Therefore, ET-1 is considered to be a key pathological factor in PE (Jain, 2012), and uncovering ET-1-mediated signaling pathways that contribute to hypertension in PE is of particular importance. The ET-1-mediated inhibition of ENaC could be one of the underappreciated mechanisms by which ET-1 is implicated in the pathobiology of PE. ETRA antagonists showed some promise for the treatment of pregnancy-induced hypertension, but the potential problem of teratogenicity cannot be ignored (Speed & Pollock, 2013). Similar to colon cancer, the connection between the overexpression of ET-1 and ENaC activity in PE has not been sufficiently investigated so far. It seems that the initial studies testing effects of ET-1 antagonists upon ENaC activity in cell cultures derived from placental and animal models of PE are due.

6. MOLECULAR MECHANISMS OF INHIBITION OF ENaC BY ET-1

There are several proposed molecular mechanisms of ET-1 effects on ENaC in the CDs. Both Src kinases and MAPK1/2 signaling were shown to mediate ET-1-dependent decreases in ENaC P_o in NIH 3T3 cells stably expressing genes for all three (α, β, and γ) ENaC subunits (Gilmore et al., 2001) and the split-open CD (Bugaj et al., 2008). The inhibitory effect of ET-1 on ENaC could be completely blocked when cells were pretreated with the selective Src family kinase inhibitor, PP2. Further studies revealed that basal Src family kinase activity strongly regulates ENaC inhibition. The inhibition of MAPK also abolishes ET-1 effects on ENaC (Bugaj et al., 2008). No roles for PLC or PKC in the rapid response of ENaC to ET-1 were identified.

The potential role for NO in the control of ENaC by ET-1 was proposed (Kohan, 2013) since it was reported that ET-1 increases NOS expression and NO production in the CDs (Hyndman & Pollock, 2013; Schneider, Ge, Pollock, Pollock, & Kohan, 2008; Stricklett, Hughes, & Kohan, 2006; Sullivan, Goodchild, Cai, Pollock, & Pollock, 2007). Inhibition of ENaC by NO in cultured cells was previously reported (Helms et al., 2005; Yu, Bao, Self, Eaton, & Helms, 2007). Furthermore, it was shown that NO reduces Cl^- absorption in the mouse CCDs through an ENaC-dependent mechanism (Pech et al., 2013). However, the role of NO in ET-1-mediated effects on ENaC was not directly studied and remains to be investigated.

Among the signaling molecules that contribute significantly to ET-1-mediated inhibition of ENaC are PKA and guanine nucleotide exchange factor (GEF) βPix. This mechanism of the inhibition of ENaC by ET-1 is likely to be coupled to the retention ability of NEDD4-2 to modulate ENaC despite SGK-mediated phosphorylation. SGK and PKA are important signaling molecules that have been shown to upregulate ENaC activity (Baines, 2013). Other kinases shown to upregulate ENaC include CK2, GRK2, IKKβ, and PKD1, while PKC, ERK1/2, and AMPK are inhibitory (Baines, 2013). Aldosterone-induced SGK1 is widely considered as one of the principal regulators of ENaC (Pao, 2012; Soundararajan, Lu, & Pearce, 2012). SGK1 not only associates with the apical membrane to regulate open probability of ENaC but also is recruited by scaffold proteins to multiunit signaling complexes to modulate ENaC expression (Pao, 2012). For SGK1 to be functionally active, it must be phosphorylated within its carboxy-terminal domain by serine–threonine kinase mammalian target of rapamycin (Lu et al., 2010). ENaC is known to be regulated by ubiquitylation following the binding of ubiquitin ligase Nedd4-2 protein–protein interaction domain WW to a PY motif in the ENaC subunits (Ronzaud & Staub, 2014; Rotin & Staub, 2012; Snyder, 2009). It appears that the phosphorylation of Nedd4-2 and two ENaC subunits (β and γ) are the main convergence targets for the action of kinases regulating ENaC. The phosphorylation of Nedd4-2 by SGK1 reduces Nedd4-2 ability to bind to ENaC, due to the interaction of phosphorylated Nedd4-2 with 14-3-3 proteins, and hence increases ENaC activity (Wiemuth et al., 2010). 14-3-3 proteins associate with SGK1-phosphorylated Nedd4-2 to maintain its phosphorylated/inactive state and thereby obstruct its physical association with ENaC (Bhalla et al., 2005; Ichimura et al., 2005; Liang, Butterworth, Peters, Walker, & Frizzell, 2008; Liang, Peters, Butterworth, & Frizzell, 2006; Nagaki et al., 2006). PKA also phosphorylates Nedd4-2 and, accordingly, regulates ENaC by decreasing Nedd4-2 binding to ENaC and congruently interfering with Nedd4-2 E3 ubiquitin-protein ligase activity that targets ENaC for degradation (Snyder, Olson, Kabra, Zhou, & Steines, 2004). The proposed mechanism of the negative regulation of ENaC by ET-1 suggests that ET-1-mediated formation of the complex comprising βPix, which acts as a scaffold in this case, and adaptor protein 14-3-3, is sufficient to prevent ubiquitin ligase Nedd4-2 from association with 14-3-3. Accordingly, phosphorylated Nedd4-2 retains ability to modify ENaC, which results in the decrease of ENaC function. It has been shown that SGK1 and Nedd4-2 physically associate with each other and with ENaC (Soundararajan et al., 2012). We have shown that ET-1-mediated recruitment of 14-3-3β by β_1Pix impairs the interaction of 14-3-3β with the ubiquitin ligase Nedd4-2, thereby promoting ubiquitination and degradation of ENaC (Pavlov et al., 2010). It is of note that βPix-mediated ET-1-triggered inhibition of ENaC activity is dispensable of βPix action as GEF (Pavlov et al., 2010) but depends on its ability to act as a scaffold protein (Staruschenko & Sorokin, 2012). The cellular model shown in Fig. 3 reveals a proposed signal transduction pathway coupling the ET receptors to ENaC.

In addition to the mechanisms described above, it was reported that gene encoding ET-1 is negatively regulated by the circadian clock protein Period 1 (Per1) (Richards et al., 2014). Interestingly, the transcription of α-ENaC subunit is also regulated by Per1 (Gumz et al., 2010, 2009). Besides, Per1 knockout mice have a hypotensive phenotype with a significant increase in cortical and medullary ET-1 production (Stow et al., 2012). These data

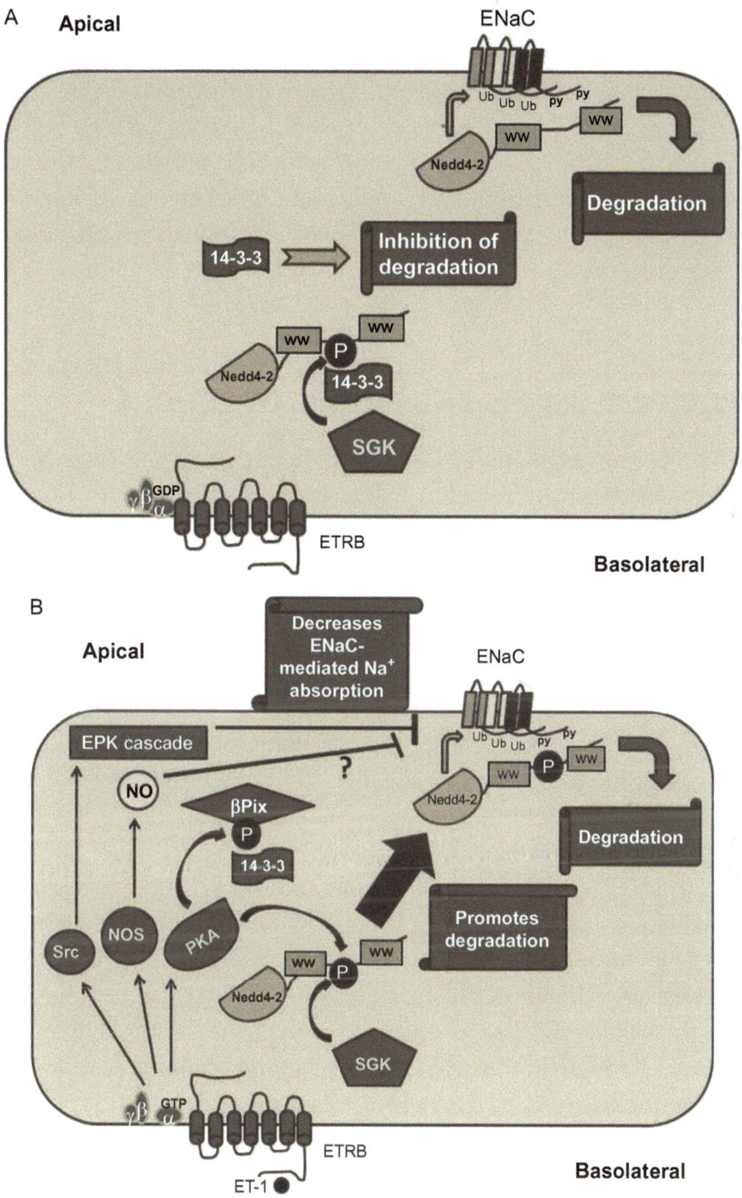

Figure 3 Proposed molecular mechanisms of ET-1-mediated inhibition of ENaC in CDs. (A) Quiescent cells with unoccupied ETRB. (B) Cells after the activation of ET-1 signaling cascade. (See the color plate.)

suggest that the lower blood pressure in Per1 knockout mice may be linked to increased ET-1-dependent natriuresis/diuresis and ENaC is involved in this molecular mechanism. Epigenetic factors altering endothelin signaling (Stow, Jacobs, Wingo, & Cain, 2011; Welch, Jacobs, Wingo, & Cain, 2013) might be also involved in the control of ENaC activity since this channel is under the control of several epigenetic mechanisms (Kone, 2013). However, these pathways and their connections remain to be studied in more details to make any definite conclusions.

7. CONCLUSIONS AND FUTURE DIRECTIONS

The precise regulation of ENaC is important for normal Na^+ and fluid homeostasis in the kidneys, lungs, and colon. These organs are also well-established bridgeheads for ET-1 actions and signaling. Even though the inhibition of ENaC by ET-1 has been studied and reliably demonstrated in the context of renal cells, there is a high probability that the modulation of ENaC by ET-1 also takes place in pulmonary channelopathies, PE, and distal colon. Whereas antagonists of ETRA represent probable candidates for being able to serve as modulators of ENaC function, no systematic analysis of their efficiency to contribute to the resolution of renal, lung, and colonic channelopathies caused by ENaC dysfunction has been carried out so far.

Molecular mechanisms of inhibition of ENaC by ET-1 are likely to be coupled to its effects through Src kinases and MAPK1/2 and retention of ability of NEDD4-2 to modulate ENaC despite SGK1-mediated phosphorylation. The current understanding of ET-1 effect upon ENaC function is based mostly upon cultured cell-based techniques. The major challenge is to confirm the validity of these mechanisms in the context of whole organism. Future breakthroughs in the field are likely to come from the analysis of renal function in models of channelopathies with animals in which precise modification of genes encoding proteins, involved in ET-1-mediated ENaC inhibition, has been achieved. The recent advances in targeted genome editing techniques suggest that this task could be achieved within next few years (Cheng & Alper, 2014).

ACKNOWLEDGMENTS

This work is supported by grants from National Institutes of Health to Andrey Sorokin (DK098159) and Alexander Staruschenko (HL108880).

REFERENCES

Agapitov, A. V., & Haynes, W. G. (2002). Role of endothelin in cardiovascular disease. *Journal of the Renin-Angiotensin-Aldosterone System, 3*, 1–15.

Ahn, D., Ge, Y., Stricklett, P. K., Gill, P., Taylor, D., Hughes, A. K., et al. (2004). Collecting duct-specific knockout of endothelin-1 causes hypertension and sodium retention. *Journal of Clinical Investigation, 114*, 504–511.

Al-Jameil, N., Aziz, K. F., Fareed, K. M., & Tabassum, H. (2014). A brief overview of preeclampsia. *Journal of Clinical Medicine & Research, 6*, 1–7.

Alvarez de la Rosa., Navarro-Gonzalez, J. F., & Giraldez, T. (2013). ENaC modulators and renal disease. *Current Molecular Pharmacology, 6*, 35–43.

Arai, H., Hori, S., Aramori, I., Ohkubo, H., & Nakanishi, S. (1990). Cloning and expression of a cDNA encoding an endothelin receptor. *Nature, 348*, 730–732.

Araki, S., Haneda, M., Togawa, M., & Kikkawa, R. (1997). Endothelin-1 activates c-Jun NH2-terminal kinase in mesangial cells. *Kidney International, 51*, 631–639.

Badr, K. F., Murray, J. J., Breyer, M. D., Takahashi, K., Inagami, T., & Harris, R. C. (1989). Mesangial cell, glomerular and renal vascular responses to endothelin in the rat kidney. Elucidation of signal transduction pathways. *Journal of Clinical Investigation, 83*, 336–342.

Baines, D. (2013). Kinases as targets for ENaC regulation. *Current Molecular Pharmacology, 6*, 50–64.

Bankir, L., Bouby, N., & Ritz, E. (2013). Vasopressin: A novel target for the prevention and retardation of kidney disease? *Nature Reviews Nephrology, 9*, 223–239.

Barton, M. (2008). Reversal of proteinuric renal disease and the emerging role of endothelin. *Nature Clinical Practice Nephrology, 4*, 490–501.

Benigni, A., Zoja, C., Corna, D., Orisio, S., Longaretti, L., Bertani, T., et al. (1993). A specific endothelin subtype A receptor antagonist protects against injury in renal disease progression. *Kidney International, 44*, 440–444.

Berger, S., Bleich, M., Schmid, W., Cole, T. J., Peters, J., Watanabe, H., et al. (1998). Mineralocorticoid receptor knockout mice: Pathophysiology of Na^+ metabolism. *Proceedings of the National Academy of Sciences of the United States of America, 95*, 9424–9429.

Bhalla, V., Daidie, D., Li, H., Pao, A. C., LaGrange, L. P., Wang, J., et al. (2005). Serum- and glucocorticoid-regulated kinase 1 regulates ubiquitin ligase neural precursor cell-expressed, developmentally down-regulated protein 4-2 by inducing interaction with 14-3-3. *Molecular Endocrinology, 19*, 3073–3084.

Bhalla, V., & Hallows, K. R. (2008). Mechanisms of ENaC regulation and clinical implications. *Journal of the American Society of Nephrology, 19*, 1845–1854.

Birch, R. E., Schwiebert, E. M., Peppiatt-Wildman, C. M., & Wildman, S. S. (2013). Emerging key roles for P2X receptors in the kidney. *Frontiers in Physiology, 4*, 262.

Blazer-Yost, B. L., Liu, X., & Helman, S. I. (1998). Hormonal regulation of ENaCs: Insulin and aldosterone. *American Journal of Physiology, 274*, C1373–C1379.

Boffa, J. J., Tharaux, P. L., Dussaule, J. C., & Chatziantoniou, C. (2001). Regression of renal vascular fibrosis by endothelin receptor antagonism. *Hypertension, 37*, 490–496.

Bouallegue, A., Daou, G. B., & Srivastava, A. K. (2007). Endothelin-1-induced signaling pathways in vascular smooth muscle cells. *Current Vascular Pharmacology, 5*, 45–52.

Bourque, S. L., Davidge, S. T., & Adams, M. A. (2011). The interaction between endothelin-1 and nitric oxide in the vasculature: New perspectives. *American Journal of Physiology Regulatory, Integrative and Comparative Physiology, 300*, R1288–R1295.

Bugaj, V., Mironova, E., Kohan, D. E., & Stockand, J. D. (2012). Collecting duct-specific endothelin B receptor knockout increases ENaC activity. *American Journal of Physiology. Cell Physiology, 302*, C188–C194.

Bugaj, V., Pochynyuk, O., Mironova, E., Vandewalle, A., Medina, J. L., & Stockand, J. D. (2008). Regulation of the epithelial Na^+ channel by endothelin-1 in rat collecting duct. *American Journal of Physiology. Renal Physiology, 295*, F1063–F1070.

Cacoub, P., Dorent, R., Nataf, P., & Carayon, A. (1993). Endothelin-1 in pulmonary hypertension. *The New England Journal of Medicine*, *329*, 1967–1968.

Cacoub, P., Dorent, R., Nataf, P., Carayon, A., Riquet, M., Noe, E., et al. (1997). Endothelin-1 in the lungs of patients with pulmonary hypertension. *Cardiovascular Research*, *33*, 196–200.

Chahdi, A., & Sorokin, A. (2010a). Endothelin-1 induces p66Shc activation through EGF receptor transactivation: Role of β_1Pix/Gα_{i3} interaction. *Cellular Signalling*, *22*, 325–329.

Chahdi, A., & Sorokin, A. (2010b). The role of β_1Pix/caveolin-1 interaction in endothelin signaling through Gα subunits. *Biochemical and Biophysical Research Communications*, *391*, 1330–1335.

Chang, S. S., Grunder, S., Hanukoglu, A., Rosler, A., Mathew, P. M., Hanukoglu, I., et al. (1996). Mutations in subunits of the epithelial sodium channel cause salt wasting with hyperkalaemic acidosis, pseudohypoaldosteronism type 1. *Nature Genetics*, *12*, 248–253.

Cheng, J. K., & Alper, H. S. (2014). The genome editing toolbox: A spectrum of approaches for targeted modification. *Current Opinion in Biotechnology*, *30C*, 87–94.

Daub, H., Weiss, F. U., Wallasch, C., & Ullrich, A. (1996). Role of transactivation of the EGF receptor in signalling by G-protein-coupled receptors. *Nature*, *379*, 557–560.

Dhanjal, M. K., Owen, E. P., Anthony, J. A., Davidson, J. S., & Rayner, B. L. (2006). Association of pre-eclampsia with the R563Q mutation of the beta-subunit of the epithelial sodium channel. *BJOG: An International Journal of Obstetrics and Gynaecology*, *113*, 595–598.

Dhein, S., Hochreuther, S., Aus Dem, S. C., Bollig, K., Hufnagel, C., & Raschack, M. (2000). Long-term effects of the endothelin(A) receptor antagonist LU 135252 and the angiotensin-converting enzyme inhibitor trandolapril on diabetic angiopathy and nephropathy in a chronic type I diabetes mellitus rat model. *Journal of Pharmacology and Experimental Therapeutics*, *293*, 351–359.

Donaldson, S. H., & Galietta, L. (2013). New pulmonary therapies directed at targets other than CFTR. *Cold Spring Harbor Perspectives in Medicine*, *3*(6), pii: a009787.

Dorado, F., Velasco, S., Esparis-Ogando, A., Pericacho, M., Pandiella, A., Silva, J., et al. (2008). The mitogen-activated protein kinase Erk5 mediates human mesangial cell activation. *Nephrology, Dialysis, Transplantation*, *23*, 3403–3411.

Downs, C. A., Kreiner, L. H., Johnson, N. M., Brown, L. A., & Helms, M. N. (2015). RAGE regulates lung fluid balance via PKC-gp91^{phox} signaling to epithelial sodium channels (ENaC). *American Journal of Respiratory Cell and Molecular Biology*, *52*, 75–87.

Drummond, H. A. (2012). βENaC is a molecular component of a VSMC mechanotransducer that contributes to renal blood flow regulation, protection from renal injury, and hypertension. *Frontiers in Physiology*, *3*, 341.

Du, J., Sours-Brothers, S., Coleman, R., Ding, M., Graham, S., Kong, D. H., et al. (2007). Canonical transient receptor potential 1 channel is involved in contractile function of glomerular mesangial cells. *Journal of the American Society of Nephrology*, *18*, 1437–1445.

Eaton, D. C., Helms, M. N., Koval, M., Bao, H. F., & Jain, L. (2009). The contribution of epithelial sodium channels to alveolar function in health and disease. *Annual Review of Physiology*, *71*, 403–423.

Eaton, A. F., Yue, Q., Eaton, D. C., & Bao, H. F. (2014). ENaC activity and expression is decreased in the lungs of protein kinase C-α knockout mice. *American Journal of Physiology. Lung Cellular and Molecular Physiology*, *307*, L374–L385.

Ecelbarger, C. A., Kim, G. H., Terris, J., Masilamani, S., Mitchell, C., Reyes, I., et al. (2000). Vasopressin-mediated regulation of epithelial sodium channel abundance in rat kidney. *American Journal of Physiology. Renal Physiology*, *279*, F46–F53.

Edwards, R. M., Stack, E. J., Pullen, M., & Nambi, P. (1993). Endothelin inhibits vasopressin action in rat inner medullary collecting duct via the ETB receptor. *Journal of Pharmacology and Experimental Therapeutics, 267*, 1028–1033.

Feraille, E., & Doucet, A. (2001). Sodium-potassium-adenosinetriphosphatase-dependent sodium transport in the kidney: Hormonal control. *Physiological Reviews, 81*, 345–418.

Forstermann, U., & Sessa, W. C. (2012). Nitric oxide synthases: Regulation and function. *European Heart Journal, 33*, 829–837.

Foschi, M., Chari, S., Dunn, M. J., & Sorokin, A. (1997). Biphasic activation of p21ras by endothelin-1 sequentially activates the ERK cascade and phosphatidylinositol 3-kinase. *The EMBO Journal, 16*, 6439–6451.

Foschi, M., Sorokin, A., Pratt, P., McGinty, A., La Villa, G., Franchi, F., et al. (2001). PreproEndothelin-1 expression in human mesangial cells: Evidence for a p38 mitogen-activated protein kinase/protein kinases-C-dependent mechanism. *Journal of the American Society of Nephrology, 12*, 1137–1150.

Fronius, M. (2013). Treatment of pulmonary edema by ENaC activators/stimulators. *Current Molecular Pharmacology, 6*, 13–27.

Fronius, M., Bogdan, R., Althaus, M., Morty, R. E., & Clauss, W. G. (2010). Epithelial Na^+ channels derived from human lung are activated by shear force. *Respiratory Physiology & Neurobiology, 170*, 113–119.

Fukuda, K., Yanagida, T., Okuda, S., Tamaki, K., Ando, T., & Fujishima, M. (1996). Role of endothelin as a mitogen in experimental glomerulonephritis in rats. *Kidney International, 49*, 1320–1329.

Gallego, M. S., & Ling, B. N. (1996). Regulation of amiloride-sensitive Na^+ channels by endothelin-1 in distal nephron cells. *American Journal of Physiology, 271*, F451–F460.

Gariepy, C. E., Ohuchi, T., Williams, S. C., Richardson, J. A., & Yanagisawa, M. (2000). Salt-sensitive hypertension in endothelin-B receptor-deficient rats. *Journal of Clinical Investigation, 105*, 925–933.

Ge, Y., Ahn, D., Stricklett, P. K., Hughes, A. K., Yanagisawa, M., Verbalis, J. G., et al. (2005). Collecting duct-specific knockout of endothelin-1 alters vasopressin regulation of urine osmolality. *American Journal of Physiology Renal Physiology, 288*, F912–F920.

Ge, Y., Bagnall, A., Stricklett, P. K., Strait, K., Webb, D. J., Kotelevtsev, Y., et al. (2006). Collecting duct specific knockout of the endothelin B receptor causes hypertension and sodium retention. *American Journal of Physiology Renal Physiology, 291*, F1274–F1280.

Ge, Y., Stricklett, P. K., Hughes, A. K., Yanagisawa, M., & Kohan, D. E. (2005). Collecting duct-specific knockout of the endothelin A receptor alters renal vasopressin responsiveness, but not sodium excretion or blood pressure. *American Journal of Physiology. Renal Physiology, 289*, F692–F698.

Gilmore, E. S., Stutts, M. J., & Milgram, S. L. (2001). SRC family kinases mediate epithelial Na^+ channel inhibition by endothelin. *The Journal of Biological Chemistry, 276*, 42610–42617.

Gomez-Garre, D., Largo, R., Liu, X. H., Gutierrez, S., Lopez-Armada, M. J., Palacios, I., et al. (1996). An orally active ETA/ETB receptor antagonist ameliorates proteinuria and glomerular lesions in rats with proliferative nephritis. *Kidney International, 50*, 962–972.

Gonzalez-Cobos, J. C., & Trebak, M. (2010). TRPC channels in smooth muscle cells. *Frontiers in Bioscience, 15*, 1023–1039.

Grunder, S., Firsov, D., Chang, S. S., Jaeger, N. F., Gautschi, I., Schild, L., et al. (1997). A mutation causing pseudohypoaldosteronism type 1 identifies a conserved glycine that is involved in the gating of the epithelial sodium channel. *The EMBO Journal, 16*, 899–907.

Guan, Z., & Inscho, E. W. (2011). Endothelin and the renal vasculature. *Contributions to Nephrology, 172*, 35–49.

Gumz, M. L., Cheng, K. Y., Lynch, I. J., Stow, L. R., Greenlee, M. M., Cain, B. D., et al. (2010). Regulation of αENaC expression by the circadian clock protein Period 1 in mpkCCD$_{c14}$ cells. *Biochimica et Biophysica Acta Gene Regulatory Mechanisms, 1799*, 622–629.

Gumz, M. L., Stow, L. R., Lynch, I. J., Greenlee, M. M., Rudin, A., Cain, B. D., et al. (2009). The circadian clock protein Period 1 regulates expression of the renal epithelial sodium channel in mice. *Journal of Clinical Investigation, 119*, 2423–2434.

Guo, L. J., Alli, A. A., Eaton, D. C., & Bao, H. F. (2013). ENaC is regulated by natriuretic peptide receptor-dependent cGMP signaling. *American Journal of Physiology. Renal Physiology, 304*, F930–F937.

Hansson, J. H., Nelson-Williams, C., Suzuki, H., Schild, L., Shimkets, R., Lu, Y., et al. (1995). Hypertension caused by a truncated epithelial sodium channel gamma subunit: Genetic heterogeneity of Liddle syndrome. *Nature Genetics, 11*, 76–82.

Hansson, J. H., Schild, L., Lu, Y., Wilson, T. A., Gautschi, I., Shimkets, R., et al. (1995). A de novo missense mutation of the beta subunit of the epithelial sodium channel causes hypertension and Liddle syndrome, identifying a proline-rich segment critical for regulation of channel activity. *Proceedings of the National Academy of Sciences of the United States of America, 92*, 11495–11499.

Harada, T., Horinouchi, T., Higa, T., Hoshi, A., Higashi, T., Terada, K., et al. (2014). Endothelin-1 activates extracellular signal-regulated kinases 1/2 via transactivation of platelet-derived growth factor receptor in rat L6 myoblasts. *Life Sciences, 104*, 24–31.

He, C., Miao, X., Li, J., & Qi, H. (2013). Angiotensin II induces endothelin-1 expression in human hepatic stellate cells. *Digestive Diseases and Sciences, 58*, 2542–2549.

Helms, M. N., Yu, L., Malik, B., Kleinhenz, D. J., Hart, C. M., & Eaton, D. C. (2005). Role of SGK1 in nitric oxide inhibition of ENaC in Na^+-transporting epithelia. *American Journal of Physiology. Cell Physiology, 289*, C717–C726.

Hobbs, C. A., Da, T. C., & Tarran, R. (2013). Does epithelial sodium channel hyperactivity contribute to cystic fibrosis lung disease? *The Journal of Physiology, 591*, 4377–4387.

Horinouchi, T., Terada, K., Higashi, T., & Miwa, S. (2013). Endothelin receptor signaling: New insight into its regulatory mechanisms. *Journal of Pharmacological Sciences, 123*, 85–101.

Hua, H., Munk, S., & Whiteside, C. I. (2003). Endothelin-1 activates mesangial cell ERK1/2 via EGF-receptor transactivation and caveolin-1 interaction. *American Journal of Physiology. Renal Physiology, 284*, F303–F312.

Hyndman, K. A., & Pollock, J. S. (2013). Nitric oxide and the A and B of endothelin of sodium homeostasis. *Current Opinion in Nephrology and Hypertension, 22*, 26–31.

Ichimura, T., Yamamura, H., Sasamoto, K., Tominaga, Y., Taoka, M., Kakiuchi, K., et al. (2005). 14-3-3 proteins modulate the expression of epithelial Na^+ channels by phosphorylation-dependent interaction with Nedd4-2 ubiquitin ligase. *The Journal of Biological Chemistry, 280*, 13187–13194.

Ilatovskaya, D. V., Pavlov, T. S., Levchenko, V., & Staruschenko, A. (2013). ROS production as a common mechanism of ENaC regulation by EGF, insulin, and IGF-1. *American Journal of Physiology. Cell Physiology, 304*, C102–C111.

Imai, T., Hirata, Y., Emori, T., Yanagisawa, M., Masaki, T., & Marumo, F. (1992). Induction of endothelin-1 gene by angiotensin and vasopressin in endothelial cells. *Hypertension, 19*, 753–757.

Imamura, T., Huang, J., Dalle, S., Ugi, S., Usui, I., Luttrell, L. M., et al. (2001). β-Arrestin-mediated recruitment of the Src family kinase Yes mediates endothelin-1-stimulated glucose transport. *The Journal of Biological Chemistry, 276*, 43663–43667.

Iwasaki, H., Eguchi, S., Ueno, H., Marumo, F., & Hirata, Y. (1999). Endothelin-mediated vascular growth requires p42/p44 mitogen-activated protein kinase and p70 S6 kinase cascades via transactivation of epidermal growth factor receptor. *Endocrinology, 140*, 4659–4668.

Jain, A. (2012). Endothelin-1: A key pathological factor in pre-eclampsia? *Reproductive Bio-Medicine Online, 25*, 443–449.

Jasti, J., Furukawa, H., Gonzales, E. B., & Gouaux, E. (2007). Structure of acid-sensing ion channel 1 at 1.9 Å resolution and low pH. *Nature, 449*, 316–323.

Jones, E. S., Owen, E. P., & Rayner, B. L. (2012). The association of the R563Q genotype of the ENaC with phenotypic variation in Southern Africa. *American Journal of Hypertension, 25*, 1286–1291.

Kawanabe, Y., Hashimoto, N., & Masaki, T. (2003). Involvements of voltage-independent Ca^{2+} channels and phosphoinositide 3-kinase in endothelin-1-induced PYK2 tyrosine phosphorylation. *Molecular Pharmacology, 63*, 808–813.

Kodama, H., Fukuda, K., Takahashi, T., Sano, M., Kato, T., Tahara, S., et al. (2002). Role of EGF receptor and Pyk2 in endothelin-1-induced ERK activation in rat cardiomyocytes. *Journal of Molecular and Cellular Cardiology, 34*, 139–150.

Kohan, D. E. (2013). Role of collecting duct endothelin in control of renal function and blood pressure. *American Journal of Physiology. Regulatory, Integrative and Comparative Physiology, 305*, R659–R668.

Kohan, D. E., Hughes, A. K., & Perkins, S. L. (1992). Characterization of endothelin receptors in the inner medullary collecting duct of the rat. *The Journal of Biological Chemistry, 267*, 12336–12340.

Kohan, D. E., Inscho, E. W., Wesson, D., & Pollock, D. M. (2011). Physiology of endothelin and the kidney. *Comprehensive Physiology, 1*, 883–919.

Kohan, D. E., Padilla, E., & Hughes, A. K. (1993). Endothelin B receptor mediates ET-1 effects on cAMP and PGE2 accumulation in rat IMCD. *American Journal of Physiology, 265*, F670–F676.

Kohan, D. E., & Pollock, D. M. (2013). Endothelin antagonists for diabetic and non-diabetic chronic kidney disease. *British Journal of Clinical Pharmacology, 76*, 573–579.

Kohan, D. E., Rossi, N. F., Inscho, E. W., & Pollock, D. M. (2011). Regulation of blood pressure and salt homeostasis by endothelin. *Physiological Reviews, 91*, 1–77.

Konczalla, J., Wanderer, S., Mrosek, J., Schuss, P., Platz, J., Guresir, E., et al. (2013). Crosstalk between the angiotensin and endothelin-system in the cerebrovasculature. *Current Neurovascular Research, 10*, 335–345.

Kone, B. C. (2013). Epigenetics and the control of the collecting duct epithelial sodium channel. *Seminars in Nephrology, 33*, 383–391.

Kopp, U. C. (2011). Endothelin in the control of renal sympathetic nerve activity. *Contributions to Nephrology, 172*, 107–119.

Kramer, B. K., Ritthaler, T., Ackermann, M., Holmer, S., Schricker, K., Riegger, G. A., et al. (1994). Endothelium-mediated regulation of renin secretion. *Kidney International, 46*, 1577–1579.

Kudo, T., & Baird, A. (1984). Inhibition of aldosterone production in the adrenal glomerulosa by atrial natriuretic factor. *Nature, 312*, 756–757.

Kurokawa, K., Yoshitomi, K., Ikeda, M., Uchida, S., Naruse, M., & Imai, M. (1993). Regulation of cortical collecting duct function: Effect of endothelin. *American Heart Journal, 125*, 582–588.

Levin, E. R. (1996). Endothelins as cardiovascular peptides. *American Journal of Nephrology, 16*, 246–251.

Li, L., Garikepati, R. M., Tsukerman, S., Kohan, D., Wade, J. B., Tiwari, S., et al. (2013). Reduced ENaC activity and blood pressure in mice with genetic knockout of the insulin receptor in the renal collecting duct. *American Journal of Physiology. Renal Physiology, 304*, F279–F288.

Liakou, P., Tepetes, K., Germenis, A., Leventaki, V., Atsaves, V., Patsouris, E., et al. (2012). Expression patterns of endothelin-1 and its receptors in colorectal cancer. *Journal of Surgical Oncology, 105*, 643–649.

Liang, X., Butterworth, M. B., Peters, K. W., Walker, W. H., & Frizzell, R. A. (2008). An obligatory heterodimer of 14-3-3β and 14-3-3ε is required for aldosterone regulation of the epithelial sodium channel. *The Journal of Biological Chemistry*, *283*, 27418–27425.

Liang, X., Peters, K. W., Butterworth, M. B., & Frizzell, R. A. (2006). 14-3-3 isoforms are induced by aldosterone and participate in its regulation of epithelial sodium channels. *The Journal of Biological Chemistry*, *281*, 16323–16332.

Lienhard, D., Lauterburg, M., Escher, G., Frey, F. J., & Frey, B. M. (2012). High salt intake down-regulates colonic mineralocorticoid receptors, epithelial sodium channels and 11β-hydroxysteroid dehydrogenase type 2. *PLoS One*, *7*, e37898.

Lifton, R. P., Gharavi, A. G., & Geller, D. S. (2001). Molecular mechanisms of human hypertension. *Cell*, *104*, 545–556.

Lin, Y. J., Kwok, C. F., Juan, C. C., Hsu, Y. P., Shih, K. C., Chen, C. C., et al. (2014). Angiotensin II enhances endothelin-1-induced vasoconstriction through upregulating endothelin type A receptor. *Biochemical and Biophysical Research Communications*, *451*, 263–269.

Little, P. J. (2013). GPCR responses in vascular smooth muscle can occur predominantly through dual transactivation of kinase receptors and not classical $G_{\alpha q}$ protein signalling pathways. *Life Sciences*, *92*, 951–956.

Lopez-Ongil, S., Diez-Marques, M. L., Griera, M., Rodriguez-Puyol, M., & Rodriguez-Puyol, D. (2005). Crosstalk between mesangial and endothelial cells: Angiotensin II down-regulates endothelin-converting enzyme 1. *Cellular Physiology and Biochemistry*, *15*, 135–144.

Lu, M., Wang, J., Jones, K. T., Ives, H. E., Feldman, M. E., Yao, L. J., et al. (2010). mTOR complex-2 activates ENaC by phosphorylating SGK1. *Journal of the American Society of Nephrology*, *21*, 811–818.

Lynch, I. J., Welch, A. K., Kohan, D. E., Cain, B. D., & Wingo, C. S. (2013). Endothelin-1 inhibits sodium reabsorption by ET_A and ET_B receptors in the mouse cortical collecting duct. *American Journal of Physiology Renal Physiology*, *305*, F568–F573.

Mamenko, M., Zaika, O., Doris, P. A., & Pochynyuk, O. (2012). Salt-dependent inhibition of epithelial Na^+ channel-mediated sodium reabsorption in the aldosterone-sensitive distal nephron by bradykinin. *Hypertension*, *60*, 1234–1241.

Mamenko, M., Zaika, O., & Pochynyuk, O. (2014). Direct regulation of ENaC by bradykinin in the distal nephron. Implications for renal sodium handling. *Current Opinion in Nephrology and Hypertension*, *23*, 122–129.

Mamenko, M., Zaika, O., Prieto, M. C., Jensen, V. B., Doris, P. A., Navar, L. G., et al. (2013). Chronic angiotensin II infusion drives extensive aldosterone-independent epithelial Na^+ channel activation. *Hypertension*, *62*, 1111–1122.

McDermott, G. F., Hurst, R. D., & Whiteside, C. I. (1993). Isolated rat glomerular cells demonstrate L-type Ca^{2+}-channel activity. *Cell Calcium*, *14*, 387–396.

Meyers, K. E., & Sethna, C. (2013). Endothelin antagonists in hypertension and kidney disease. *Pediatric Nephrology*, *28*, 711–720.

Mironova, E., Boiko, N., Bugaj, V., Kucher, V., & Stockand, J. D. (2015). Regulation of Na excretion and arterial blood pressure by purinergic signaling intrinsic to the distal nephron: Consequences and mechanisms. *Acta Physiologica*, *213*, 213–221.

Mironova, E., Bugaj, V., Roos, K. P., Kohan, D. E., & Stockand, J. D. (2012). Aldosterone-independent regulation of the epithelial Na^+ channel (ENaC) by vasopressin in adrenalectomized mice. *Proceedings of the National Academy of Sciences of the United States of America*, *109*, 10095–10100.

Moe, O., Tejedor, A., Campbell, W. B., Alpern, R. J., & Henrich, W. L. (1991). Effects of endothelin on in vitro renin secretion. *American Journal of Physiology*, *260*, E521–E525.

Motte, S., McEntee, K., & Naeije, R. (2005). Endothelin receptor antagonists. *Pharmacology & Therapeutics*, *93*, 434–442.

Nadler, S. P., Zimpelmann, J. A., & Hebert, R. L. (1992). Endothelin inhibits vasopressin-stimulated water permeability in rat terminal inner medullary collecting duct. *Journal of Clinical Investigation, 90*, 1458–1466.

Nagaki, K., Yamamura, H., Shimada, S., Saito, T., Hisanaga, S., Taoka, M., et al. (2006). 14-3-3 mediates phosphorylation-dependent inhibition of the interaction between the ubiquitin E3 ligase Nedd4-2 and epithelial Na^+ channels. *Biochemistry, 45*, 6733–6740.

Neuhofer, W., & Pittrow, D. (2006). Role of endothelin and endothelin receptor antagonists in renal disease. *European Journal of Clinical Investigation, 36*, 78–88.

Nie, S., Zhou, J., Bai, F., Jiang, B., Chen, J., & Zhou, J. (2014). Role of endothelin A receptor in colon cancer metastasis: In vitro and in vivo evidence. *Molecular Carcinogenesis, 53*, E85–E91.

Oishi, R., Nonoguchi, H., Tomita, K., & Marumo, F. (1991). Endothelin-1 inhibits AVP-stimulated osmotic water permeability in rat inner medullary collecting duct. *American Journal of Physiology, 261*, F951–F956.

Orth, S. R., Amann, K., Gehlen, F., Unger, L., Wagner, J., Raschack, M., et al. (2000). Adult human mesangial cells (HMCs) express endothelin-B-receptors which mediate endothelin-1-induced cell growth. *Journal of Cardiovascular Pharmacology, 36*, S232–S237.

Otsuka, A., Mikami, H., Katahira, K., Tsunetoshi, T., Minamitani, K., & Ogihara, T. (1989). Changes in plasma renin activity and aldosterone concentration in response to endothelin injection in dogs. *Acta Endocrinologica, 121*, 361–364.

Palei, A. C., Spradley, F. T., Warrington, J. P., George, E. M., & Granger, J. P. (2013). Pathophysiology of hypertension in pre-eclampsia: A lesson in integrative physiology. *Acta Physiologica, 208*, 224–233.

Pao, A. C. (2012). SGK regulation of renal sodium transport. *Current Opinion in Nephrology and Hypertension, 21*, 534–540.

Park, J. G., Bose, A., Leszyk, J., & Czech, M. P. (2001). PYK2 as a mediator of endothelin-1/$G_{\alpha 11}$ signaling to GLUT4 glucose transporters. *The Journal of Biological Chemistry, 276*, 47751–47754.

Pavlov, T. S., Chahdi, A., Ilatovskaya, D. V., Levchenko, V., Vandewalle, A., Pochynyuk, O., et al. (2010). Endothelin-1 inhibits the epithelial Na^+ channel through βPix/14-3-3/Nedd4-2. *Journal of the American Society of Nephrology, 21*, 833–843.

Pavlov, T. S., Ilatovskaya, D. V., Levchenko, V., Li, L., Ecelbarger, C. M., & Staruschenko, A. (2013). Regulation of ENaC in mice lacking renal insulin receptors in the collecting duct. *The FASEB Journal, 27*, 2723–2732.

Pavlov, T. S., Levchenko, V., O'Connor, P. M., Ilatovskaya, D. V., Palygin, O., Mori, T., et al. (2013). Deficiency of renal cortical EGF increases ENaC activity and contributes to salt-sensitive hypertension. *Journal of the American Society of Nephrology, 24*, 1053–1062.

Pearce, D., Soundararajan, R., Trimpert, C., Kashlan, O. B., Deen, P. M., & Kohan, D. E. (2015). Collecting duct principal cell transport processes and their regulation. *Clinical Journal of the American Society of Nephrology, 10*, 135–146.

Pech, V., Thumova, M., Dikalov, S. I., Hummler, E., Rossier, B. C., Harrison, D. G., et al. (2013). Nitric oxide reduces Cl^- absorption in the mouse cortical collecting duct through an ENaC-dependent mechanism. *American Journal of Physiology. Renal Physiology, 304*, F1390–F1397.

Peters, D. M., Vadasz, I., Wujak, L., Wygrecka, M., Olschewski, A., Becker, C., et al. (2014). TGF-β directs trafficking of the epithelial sodium channel ENaC which has implications for ion and fluid transport in acute lung injury. *Proceedings of the National Academy of Sciences of the United States of America, 111*, E374–E383.

Pochynyuk, O., Bugaj, V., Rieg, T., Insel, P. A., Mironova, E., Vallon, V., et al. (2008). Paracrine regulation of the epithelial Na^+ channel in the mammalian collecting duct by purinergic $P2Y_2$ receptor tone. *The Journal of Biological Chemistry, 283*, 36599–36607.

Pochynyuk, O., Rieg, T., Bugaj, V., Schroth, J., Fridman, A., Boss, G. R., et al. (2010). Dietary Na^+ inhibits the open probability of the epithelial sodium channel in the kidney by enhancing apical $P2Y_2$-receptor tone. *The FASEB Journal, 24*, 2056–2065.
Pollock, D. M. (1998). Endothelin receptor subtypes and tissue distribution. In R. F. Highsmith (Ed.), *Endothelin: Molecular biology, physiology, and pathology* (pp. 1–29). Totowa, New Jersey: Humana Press.
Pollock, D. M., Keith, T. L., & Highsmith, R. F. (1995). Endothelin receptors and calcium signaling. *The FASEB Journal, 9*, 1196–1204.
Quinn, S., Harvey, B. J., & Thomas, W. (2014). Rapid aldosterone actions on epithelial sodium channel trafficking and cell proliferation. *Steroids, 81*, 43–48.
Raja, S. G., & Raja, S. M. (2011). Treating pulmonary arterial hypertension: Current treatments and future prospects. *Therapeutic Advances in Chronic Disease, 2*, 359–370.
Rapoport, R. M. (2014). Acute nitric oxide synthase inhibition and endothelin-1-dependent arterial pressure elevation. *Frontiers in Pharmacology, 5*, 57.
Richards, J., Welch, A. K., Barilovits, S. J., All, S., Cheng, K. Y., Wingo, C. S., et al. (2014). Tissue-specific and time-dependent regulation of the endothelin axis by the circadian clock protein Per1. *Life Sciences, 118*, 255–262.
Rieg, T., Bundey, R. A., Chen, Y., Deschenes, G., Junger, W., Insel, P. A., et al. (2007). Mice lacking $P2Y_2$ receptors have salt-resistant hypertension and facilitated renal Na^+ and water reabsorption. *The FASEB Journal, 21*, 3717–3726.
Rieg, T., Gerasimova, M., Boyer, J. L., Insel, P. A., & Vallon, V. (2011). $P2Y_2$ receptor activation decreases blood pressure and increases renal Na^+ excretion. *American Journal of Physiology. Regulatory, Integrative and Comparative Physiology, 301*, R510–R518.
Ronzaud, C., & Staub, O. (2014). Ubiquitylation and control of renal Na^+ balance and blood pressure. *Physiology, 29*, 16–26.
Rosano, L., Spinella, F., & Bagnato, A. (2013). Endothelin 1 in cancer: Biological implications and therapeutic opportunities. *Nature Reviews Cancer, 13*, 637–651.
Rossier, B. C. (2014). Epithelial sodium channel (ENaC) and the control of blood pressure. *Current Opinion in Pharmacology, 15*, 33–46.
Rotin, D., & Staub, O. (2012). Nedd4-2 and the regulation of epithelial sodium transport. *Frontiers in Physiology, 3*, 212.
Rubin, L. J. (2012). Endothelin receptor antagonists for the treatment of pulmonary artery hypertension. *Life Sciences, 91*, 517–521.
Rubin, L. J., Badesch, D. B., Barst, R. J., Galie, N., Black, C. M., Keogh, A., et al. (2002). Bosentan therapy for pulmonary arterial hypertension. *The New England Journal of Medicine, 346*, 896–903.
Rufanova, V. A., Alexanian, A., Wakatsuki, T., Lerner, A., & Sorokin, A. (2009). Pyk2 mediates endothelin-1 signaling via p130Cas/BCAR3 cascade and regulates human glomerular mesangial cell adhesion and spreading. *Journal of Cellular Physiology, 219*, 45–56.
Sakurai, T., Yanagisawa, M., Takuwa, Y., Miyazaki, H., Kimura, S., Goto, K., et al. (1990). Cloning of a cDNA encoding a non-isopeptide-selective subtype of the endothelin receptor. *Nature, 348*, 732–735.
Sanghi, A., Zaringhalam, M., Corcoran, C. C., Saeed, F., Hoffert, J. D., Sandoval, P. C., et al. (2014). A knowledge base of vasopressin actions in kidney. *American Journal of Physiology. Renal Physiology, 307*, F747–F755.
Schneider, M. P., Ge, Y., Pollock, D. M., Pollock, J. S., & Kohan, D. E. (2008). Collecting duct-derived endothelin regulates arterial pressure and Na excretion via nitric oxide. *Hypertension, 51*, 1605–1610.
Shao, D., Park, J. E., & Wort, S. J. (2011). The role of endothelin-1 in the pathogenesis of pulmonary arterial hypertension. *Pharmacological Research, 63*, 504–511.

Shimkets, R. A., Warnock, D. G., Bositis, C. M., Nelson-Williams, C., Hansson, J. H., Schambelan, M., et al. (1994). Liddle's syndrome: Heritable human hypertension caused by mutations in the beta subunit of the epithelial sodium channel. *Cell*, *79*, 407–414.

Simon, D. B., Karet, F. E., Rodriguez-Soriano, J., Hamdan, J. H., DiPietro, A., Trachtman, H., et al. (1996). Genetic heterogeneity of Bartter's syndrome revealed by mutations in the K+ channel, ROMK. *Nature Genetics*, *14*, 152–156.

Simonson, M. S., & Dunn, M. J. (1991). Ca^{2+} signaling by distinct endothelin peptides in glomerular mesangial cells. *Experimental Cell Research*, *192*, 148–156.

Simonson, M. S., & Dunn, M. J. (1993). Endothelin peptides and the kidney. *Annual Review of Physiology*, *55*, 249–265.

Simonson, M. S., & Herman, W. H. (1993). Protein kinase C and protein tyrosine kinase activity contribute to mitogenic signaling by endothelin-1. Cross-talk between G protein-coupled receptors and pp60c-src. *The Journal of Biological Chemistry*, *268*, 9347–9357.

Simonson, M. S., Wann, S., Mene, P., Dubyak, G. R., Kester, M., Nakazato, Y., et al. (1989). Endothelin stimulates phospholipase C, Na^+/H^+ exchange, c-fos expression, and mitogenesis in rat mesangial cells. *Journal of Clinical Investigation*, *83*, 708–712.

Sitbon, O., Badesch, D. B., Channick, R. N., Frost, A., Robbins, I. M., Simonneau, G., et al. (2003). Effects of the dual endothelin receptor antagonist bosentan in patients with pulmonary arterial hypertension: A 1-year follow-up study. *Chest*, *124*, 247–254.

Snyder, P. M. (2009). Down-regulating destruction: Phosphorylation regulates the E3 ubiquitin ligase Nedd4-2. *Science Signaling*, *2*, e41.

Snyder, P. M., Olson, D. R., Kabra, R., Zhou, R., & Steines, J. C. (2004). cAMP and serum and glucocorticoid-inducible kinase (SGK) regulate the epithelial Na^+ channel through convergent phosphorylation of Nedd4-2. *The Journal of Biological Chemistry*, *279*, 45753–45758.

Sonkusare, S., Palade, P. T., Marsh, J. D., Telemaque, S., Pesic, A., & Rusch, N. J. (2006). Vascular calcium channels and high blood pressure: Pathophysiology and therapeutic implications. *Vascular Pharmacology*, *44*, 131–142.

Sorby, L. A., Kleiveland, C. R., Andersen, S. N., Bukholm, I. R., & Jacobsen, M. B. (2011). The endothelin axis in the metastatic process of colon carcinoma. *Anticancer Research*, *31*, 861–869.

Sorokin, A. (2011). Endothelin signaling and actions in the renal mesangium. *Contributions to Nephrology*, *172*, 50–62.

Sorokin, A., Foschi, M., & Dunn, M. J. (2002). Endothelin signalling and regulation of protein kinases in glomerular mesangial cells. *Clinical Science*, *103*, 132S–136S.

Sorokin, A., & Kohan, D. E. (2003). Physiology and pathology of endothelin-1 in renal mesangium. *American Journal of Physiology. Renal Physiology*, *285*, F579–F589.

Sorokin, A., Kozlowski, P., Graves, L., & Philip, A. (2001). Protein-tyrosine kinase Pyk2 mediates endothelin-induced p38 MAPK activation in glomerular mesangial cells. *The Journal of Biological Chemistry*, *276*, 21521–21528.

Soundararajan, R., Lu, M., & Pearce, D. (2012). Organization of the ENaC-regulatory machinery. *Critical Reviews in Biochemistry and Molecular Biology*, *47*, 349–359.

Soundararajan, R., Pearce, D., Hughey, R. P., & Kleyman, T. R. (2010). Role of epithelial sodium channels and their regulators in hypertension. *The Journal of Biological Chemistry*, *285*, 30363–30369.

Speed, J. S., & Pollock, D. M. (2013). Endothelin, kidney disease, and hypertension. *Hypertension*, *61*, 1142–1145.

Srivastava, S., Li, D., Edwards, N., Hynes, A. M., Wood, K., Al-Hamed, M., et al. (2013). Identification of compound heterozygous KCNJ1 mutations (encoding ROMK) in a kindred with Bartter's syndrome and a functional analysis of their pathogenicity. *Physiological Reports*, *1*, e00160.

Staruschenko, A. (2012). Regulation of transport in the connecting tubule and cortical collecting duct. *Comprehensive Physiology, 2*, 1541–1584.

Staruschenko, A., Adams, E., Booth, R. E., & Stockand, J. D. (2005). Epithelial Na^+ channel subunit stoichiometry. *Biophysical Journal, 88*, 3966–3975.

Staruschenko, A., Palygin, O., Ilatovskaya, D. V., & Pavlov, T. S. (2013). Epidermal growth factors in the kidney and relationship to hypertension. *American Journal of Physiology. Renal Physiology, 305*, F12–F20.

Staruschenko, A., & Sorokin, A. (2012). Role of βPix in the kidney. *Frontiers in Physiology, 3*, 154.

Stockand, J. D. (2002). New ideas about aldosterone signaling in epithelia. *American Journal of Physiology. Renal Physiology, 282*, F559–F576.

Stockand, J. D. (2012). The role of the epithelial Na^+ channel (ENaC) in high AVP but low aldosterone states. *Frontiers in Physiology, 3*, 304.

Stockand, J. D., Mironova, E., Bugaj, V., Rieg, T., Insel, P. A., Vallon, V., et al. (2010). Purinergic inhibition of ENaC produces aldosterone escape. *Journal of the American Society of Nephrology, 21*, 1903–1911.

Stow, L. R., Jacobs, M. E., Wingo, C. S., & Cain, B. D. (2011). Endothelin-1 gene regulation. *The FASEB Journal, 25*, 16–28.

Stow, L. R., Richards, J., Cheng, K. Y., Lynch, I. J., Jeffers, L. A., Greenlee, M. M., et al. (2012). The circadian protein period 1 contributes to blood pressure control and coordinately regulates renal sodium transport genes. *Hypertension, 59*, 1151–1156.

Stricklett, P. K., Hughes, A. K., & Kohan, D. E. (2006). Endothelin-1 stimulates NO production and inhibits cAMP accumulation in rat inner medullary collecting duct through independent pathways. *American Journal of Physiology. Renal Physiology, 290*, F1315–F1319.

Stuart, D., Chapman, M., Rees, S., Woodward, S., & Kohan, D. E. (2013). Myocardial, smooth muscle, nephron, and collecting duct gene targeting reveals the organ sites of endothelin A receptor antagonist fluid retention. *The Journal of Pharmacology and Experimental Therapeutics, 346*, 182–189.

Stuart, D., Rees, S., Woodward, S. K., Koesters, R., Strait, K. A., & Kohan, D. E. (2012). Disruption of the endothelin A receptor in the nephron causes mild fluid volume expansion. *BMC Nephrology, 13*, 166.

Sullivan, J. C., Goodchild, T. T., Cai, Z., Pollock, D. M., & Pollock, J. S. (2007). Endothelin(A) (ET(A)) and ET(B) receptor-mediated regulation of nitric oxide synthase 1 (NOS1) and NOS3 isoforms in the renal inner medulla. *Acta Physiologica, 191*, 329–336.

Takemoto, F., Uchida, S., Ogata, E., & Kurokawa, K. (1993). Endothelin-1 and endothelin-3 binding to rat nephrons. *American Journal of Physiology, 264*, F827–F832.

Teitelbaum, I., & Berl, T. (1994). Increased cytosolic Ca^{2+} inhibits AVP-stimulated adenylyl cyclase activity in rat IMCT cells by activation of PKC. *American Journal of Physiology, 266*, F486–F490.

Tomita, K., Nonoguchi, H., Terada, Y., & Marumo, F. (1993). Effects of ET-1 on water and chloride transport in cortical collecting ducts of the rat. *American Journal of Physiology, 264*, F690–F696.

Tykocki, N. R., & Watts, S. W. (2010). The interdependence of endothelin-1 and calcium: A review. *Clinical Science, 119*, 361–372.

Vacca, F., Bagnato, A., Catt, K. J., & Tecce, R. (2000). Transactivation of the epidermal growth factor receptor in endothelin-1-induced mitogenic signaling in human ovarian carcinoma cells. *Cancer Research, 60*, 5310–5317.

Vignon-Zellweger, N., Heiden, S., & Emoto, N. (2011). Renal function and blood pressure: Molecular insights into the biology of endothelin-1. *Contributions to Nephrology, 172*, 18–34.

Wang, W., Li, C., Nejsum, L. N., Li, H., Kim, S. W., Kwon, T. H., et al. (2006). Biphasic effects of ANP infusion in conscious, euvolumic rats: Roles of AQP2 and ENaC trafficking. *American Journal of Physiology. Renal Physiology, 290*, F530–F541.

Wang, Y., Simonson, M. S., Pouyssegur, J., & Dunn, M. J. (1992). Endothelin rapidly stimulates mitogen-activated protein kinase activity in rat mesangial cells. *Biochemical Journal, 287*, 589–594.

Warnock, D. G., Kusche-Vihrog, K., Tarjus, A., Sheng, S., Oberleithner, H., Kleyman, T. R., et al. (2014). Blood pressure and amiloride-sensitive sodium channels in vascular and renal cells. *Nature Reviews Nephrology, 10*, 146–157.

Warrington, J. P., Coleman, K., Skaggs, C., Hosick, P. A., George, E. M., Stec, D. E., et al. (2014). Heme oxygenase-1 promotes migration and β-epithelial Na^+ channel expression in cytotrophoblasts and ischemic placentas. *American Journal of Physiology. Regulatory, integrative and Comparative Physiology, 306*, R641–R646.

Welch, A. K., Jacobs, M. E., Wingo, C. S., & Cain, B. D. (2013). Early progress in epigenetic regulation of endothelin pathway genes. *British Journal of Pharmacology, 168*, 327–334.

Welling, P. A. (2013). Regulation of renal potassium secretion: Molecular mechanisms. *Seminars in Nephrology, 33*, 215–228.

Welling, P. A., & Ho, K. (2009). A comprehensive guide to the ROMK potassium channel: Form and function in health and disease. *American Journal of Physiology. Renal Physiology, 297*, F849–F863.

Wen, D., Cornelius, R. J., & Sansom, S. C. (2014). Interacting influence of diuretics and diet on BK channel-regulated K homeostasis. *Current Opinion in Pharmacology, 15*, 28–32.

Wendel, M., Knels, L., Kummer, W., & Koch, T. (2006). Distribution of endothelin receptor subtypes ETA and ETB in the rat kidney. *The Journal of Histochemistry & Cytochemistry, 54*, 1193–1203.

Wiemuth, D., Lott, J. S., Ly, K., Ke, Y., Teesdale-Spittle, P., Snyder, P. M., et al. (2010). Interaction of serum- and glucocorticoid regulated kinase 1 (SGK1) with the WW-domains of Nedd4-2 is required for epithelial sodium channel regulation. *PLoS One, 5*, e12163.

Yamamoto, T., Suzuki, H., Kubo, Y., Matsumoto, A., & Uemura, H. (2008). Endothelin A receptor-like immunoreactivity on the basal infoldings of rat renal tubules and collecting ducts. *Archives of Histology and Cytology, 71*, 77–87.

Yu, L., Bao, H. F., Self, J. L., Eaton, D. C., & Helms, M. N. (2007). Aldosterone-induced increases in superoxide production counters nitric oxide inhibition of epithelial Na channel activity in A6 distal nephron cells. *American Journal of Physiology. Renal Physiology, 293*, F1666–F1677.

Yukimura, T., Notoya, M., Mizojiri, K., Mizuhira, V., Matsuura, T., Ebara, T., et al. (1996). High resolution localization of endothelin receptors in rat renal medulla. *Kidney International, 50*, 135–147.

Zaika, O., Mamenko, M., O'Neil, R. G., & Pochynyuk, O. (2011). Bradykinin acutely inhibits activity of the epithelial Na^+ channel in mammalian aldosterone-sensitive distal nephron. *American Journal of Physiology. Renal Physiology, 300*, F1105–F1115.

Zaika, O., Mamenko, M., Staruschenko, A., & Pochynyuk, O. (2013). Direct activation of ENaC by angiotensin II: Recent advances and new insights. *Current Hypertension Reports, 15*, 17–24.

CHAPTER SEVEN

Pharmacological Regulation of the Cholesterol Transport Machinery in Steroidogenic Cells of the Testis

Yasaman Aghazadeh[*,†], **Barry R. Zirkin**[‡], **Vassilios Papadopoulos**[*,†,§,¶,1]

[*]The Research Institute of the McGill University Health Centre, Montreal, Quebec, Canada
[†]Department of Medicine, McGill University, Montreal, Quebec, Canada
[‡]Department of Biochemistry and Molecular Biology, Johns Hopkins University Bloomberg School of Public Health, Baltimore, Maryland, USA
[§]Department of Biochemistry, McGill University, Montreal, Quebec, Canada
[¶]Department of Pharmacology & Therapeutics, McGill University, Montreal, Quebec, Canada
[1]Corresponding author: e-mail address: vassilios.papadopoulos@mcgill.ca

Contents

Abstract

Reduced serum testosterone (T), or hypogonadism, is estimated to affect about 5 million American men, including both aging and young men. Low serum T has been linked to mood changes, worsening cognition, fatigue, depression, decreased lean body mass and bone mineral density, increased visceral fat, metabolic syndrome, decreased libido, and sexual dysfunction. Administering exogenous T, known as T-replacement therapy (TRT), reverses many of the symptoms of low T levels. However, this treatment can result in luteinizing hormone suppression which, in turn, can lead to reduced sperm numbers

Vitamins and Hormones, Volume 98
ISSN 0083-6729
http://dx.doi.org/10.1016/bs.vh.2014.12.006

and infertility, making TRT inappropriate for men who wish to father children. Additionally, TRT may result in supraphysiologic T levels, skin irritation, and T transfer to others upon contact; and there may be increased risk of prostate cancer and cardiovascular disease, particularly in aging men. Therefore, the development of alternate therapies for treating hypogonadism would be highly desirable. To do so requires greater understanding of the series of steps leading to T formation and how they are regulated, and the identification of key steps that are amenable to pharmacological modulation so as to induce T production. We review herein our current understanding of mechanisms underlying the pharmacological induction of T formation in hypogonadal testis.

ABBREVIATIONS

3βHSD/HSD3B 3β-hydroxysteroid dehydrogenase
17β-HSD/HSD3 17β-hydroxysteroid dehydrogenase
19-Atriol 3,17,19-androsten-5-triol
ACBD1/ABCD3 acyl-CoA binding domain 1/3
AKAP a kinase anchoring protein
ANT adenine nucleotide transporter
ATAD3 AAA+ ATPase domain 3
ATP adenosine three phosphate
Bcl-2 B-cell lymphoma 2
CAH congenital adrenal hyperplasia
cAMP cyclic adenosine monophosphate
CBP CREB-binding protein
CD spectroscopy circular dichroism spectroscopy
CDK5 cyclin-dependent kinase 5
CRAC cholesterol recognition/interaction consensus
CREB cAMP-response element-binding protein
CYP11A1 cytochrome P450 side-chain cleavage
CYP17A1 cytochrome P450 steroid 17-α-monooxygenase
ER endoplasmic reticulum
ERK1/2 extracellular signal-regulating kinase 1/2
FdR ferredoxin reductase
Fdx ferredoxin
FGIN-1-27 *N*,*N*-di-*n*-hexyl 2-(4-fluorophenyl)indole-3-acetamide
GABA gamma-aminobutyric acid
hCG human chorionic gonadotropin
Hsc70 heat-shock cognate protein of 70 kDa chaperone
HSP90 heat-shock protein 90
IMM inner mitochondrial membrane
IMS intermembrane space
LH luteinizing hormone
MEK1/2 mitogen-activated kinase kinase 1/2
MPTP mitochondrial permeability transition pore
MSF mitochondrial import-stimulating factor
NMR nuclear magnetic resonance

OMM outer mitochondrial membrane
P4 progesterone
P5 pregnenolone
PadX FdX precursor preadrenodoxin
PAP7 PKA-RIα-associated protein 7
PKA protein kinase A
PKA-C catalytic subunit
PKA-R regulatory subunit
PSA prostate-specific antigen
PTM posttranslational modification
RNAi interfering RNA
SAM Sorting and Assembly Machinery
SHBG sex hormone-binding globulin
STAR steroidogenic acute regulatory protein
START domain STAR-related lipid transfer domain
T testosterone
TIM translocase of the inner membrane
TOM translocase of the outer membrane
TRT testosterone replacement therapy
TSPO translocator protein
VDAC1 voltage-dependent anion channel 1

1. INTRODUCTION

Reduced serum testosterone (T), or hypogonadism, is estimated to affect about 5 million American men, including both aging and young. The condition is common in aging men, with 20–50% of men over age 60 reported to have serum T levels significantly below those of young men (age 20–30 years) (Lutz, Sanderson, & Scherbov, 1997; Oeppen & Vaupel, 2002). Age-related decline in serum T levels in these men typically is not in response to reduced luteinizing hormone (LH), but rather a consequence of Leydig cells becoming less responsive to LH (Huhtaniemi & Forti, 2011). This condition, referred to as primary hypogonadism, also occurs in many infertile men. Approximately 15% of couples seek infertility-related medical appointments, with male factor effects contributing to 40–50% of these cases (Carruthers, 2009; Huhtaniemi & Forti, 2011; Mitchell, Hollis, Rothwell, & Robertson, 1995; Perheentupa & Huhtaniemi, 2009; Rohrmann et al., 2011). About 30% of infertile men are diagnosed as idiopathic, among whom about 50% have primary hypogonadism (Ishunina et al., 2005; Valenti, 2005; Veldhuis, 2008). Whether in aging or young men, reduced serum T is linked to a number of metabolic and quality-of-life changes, including decreased lean body mass and bone mineral density, decreased muscle mass and strength, obesity, cardiovascular disorders, decreased libido and sexual function, altered mood, and fatigue (Belchetz, Barth, & Kaufman, 2010; Cattabiani et al., 2012; Kaplan, Meehan, & Shah, 2006; Morris & Channer, 2012; Oeppen & Vaupel, 2002; Tuck & Francis, 2009; Valenti, 2005).

Administering exogenous T, known as T-replacement therapy (TRT), reverses many of the symptoms of low T levels. The primary objective of TRT is to raise serum T levels so as to reduce symptoms of hypogonadism. Ideally, serum T concentration within the physiological range of adult men is the desired outcome. The T preparations in use to achieve this are injections, scrotal and nonscrotal transdermal patches, and oral, buccal, and gel preparations (Lang, Samaras, & Samaras, 2012; Lutz et al., 1997; Morris & Channer, 2012; Muraleedharan & Jones, 2010; Oeppen & Vaupel, 2002; Somani, Khan, & Donat, 2010; Wu et al., 2010). With injections, serum T levels initially are supraphysiologic and then reduced (ASA Position Statement, 2006; de Ronde, 2009; Dobs et al., 1999). This commonly used mode of T delivery requires serum T levels to be measured and sometimes adjusted between injections. High T levels may pose a risk for aging males due to possible prostate (benign prostatic hyperplasia; prostate cancer) and cardiovascular consequences (Morris & Channer, 2012). T administered by gels and other transdermal methods produce more constant serum T concentrations, but with the potential for T transfer via skin contact and skin irritation (Busse & Maibach, 2011; Muram, Melby, & Alles, 2012; Stahlman et al., 2012). Moreover, the administration of exogenous T by any means can suppress LH and thus result in reduced Leydig cell T formation and suppression of spermatogenesis. Indeed, contraception in men can be achieved by administering LH-suppressive T. Thus, the exogenous administration of T to ameliorate hypogonadism is inappropriate for men wishing to father children (Carruthers, 2009; Huhtaniemi & Forti, 2011; Perheentupa & Huhtaniemi, 2009; Riggs et al., 1982). There are methods in use that increase serum T without T administration, including hCG treatment in the case of men with secondary hypogonadism. In individuals with primary hypogonadism, aromatase inhibitors can increase T-to-estradiol ratios, particularly in men with severe infertility, but often this approach is ineffective (Perheentupa & Huhtaniemi, 2009).

Increasing intratesticular and serum T by stimulating the Leydig cells themselves could have great advantages. Such an approach should not elicit significant fluctuations in T levels because T formation would be regulated at least in part by negative feedback of T on LH. Nor would there be T transfer to others via contact. Fertility should be preserved, not suppressed, because the hypothalamic–pituitary axis should not be shut down as with exogenous T. Indeed, the local stimulation of Leydig cell T production by the use of drug ligands that target proteins involved in cholesterol import into mitochondria, the rate-determining step in T formation, might actually support or enhance spermatogenesis because intratesticular T levels should increase, not decrease. This advantage could be of great benefit to the many young men with primary hypogonadism who wish to father children. If successful, T replacement through the administration of drugs targeting proteins involved in T formation could constitute a paradigm shift in the treatment of hypogonadal men.

In summary, the design of new therapies that increase intratesticular bioactive androgen levels without affecting the hypothalamic–pituitary axis would be of great benefit to numerous patients. This approach requires an understanding of the series of steps leading to T formation and how they are regulated, and the identification of key steps amenable to pharmacological modulation to induce T.

2. ROLE OF T IN HEALTH AND WELL-BEING

Recent studies on human longevity indicated that women have a 10% longer life span than men (Lutz et al., 1997; Oeppen & Vaupel, 2002), suggesting that hormonal differences might play a role in general health and the rate and age of mortality (Cattabiani et al., 2012). T production by men is required for virilization, development of secondary male sex characteristics, fertility, and maintaining bone density and muscle mass. Thus, maintaining Leydig cell androgen formation is essential for well-being. However, reduced serum T occurs commonly in aging men (20–50% of men over age 60), subfertile/infertile young men (>50% of those diagnosed as "idiopathic infertile"), and many others (Huhtaniemi & Forti, 2011; Perheentupa & Huhtaniemi, 2009). A progressive decline in serum T levels can be caused by age-related reduction in the capacity of Leydig cells to produce T, a condition called primary hypogonadism, by interruptions in the hypothalamus–pituitary–testis axis, referred to as secondary hypogonadism (Belchetz et al., 2010; Mitchell et al., 1995; Veldhuis, 2008), or a mixed primary and secondary hypogonadism.

Approximately 1–2% of serum T is free and 30–50% is bound to albumin with low affinity, while the rest is bound to sex hormone-binding globulin

(SHBG) with high affinity. Therefore, free and albumin-bound T are considered bioactive. The assessment of total T levels of patients is initially used for diagnosis of hypogonadism. However, other factors could also affect the circulating levels of T, including (i) alteration in the levels of SHBG, leading to decreased free serum T levels while the overall T production is unaltered (Kupelian et al., 2006); (ii) increase in levels and activity of aromatase, an enzyme that converts T to estrogen (Ishunina et al., 2005; Valenti, 2005); and (iii) the levels of 5α-reductase which converts T to the active dihydrotestosterone (Valenti, 2005). Moreover, genetic and environmental factors including stress, smoking, and alcohol consumption, as well as health conditions such as obesity and metabolic syndrome (Kaplan et al., 2006; Muraleedharan & Jones, 2010; Somani et al., 2010), also can affect T production. Androgen deficiency has been found to be associated with conditions that include infertility, cardiovascular diseases, adrenal or testicular hyperplasia, neurodegeneration, altered mood, fatigue, decreased lean body mass, reduced bone mineral density, increased visceral fat, metabolic syndrome, and decreased libido and sexual function (Carruthers, 2009; Lang et al., 2012; Morris & Channer, 2012). These observations highlight the significance of reduced T in male health and well-being and provide the rationale for the development of therapeutic techniques to treat this condition.

3. T-REPLACEMENT THERAPY

The administration of T to hypogonadal men, referred to as TRT, has increased markedly in recent years. It is estimated that in 2011, there were 5.3 million prescriptions and a market of $1.6 billion for TRT, which was five times higher than that in 2010 (http://www.fda.gov/AdvisoryCommittees). TRT can reverse several symptoms of androgen deficiency. However, a large body of clinical studies suggests health-threatening side effects. Therefore, the Endocrine Society offered precautionary guidelines for the use of TRT (Bhasin et al., 2010; Valero, Marquez, Campos, Puigvert, & Prieto, 2013). TRT is not recommended for patients with breast or prostate cancer, a palpable prostate nodule, induration, or a prostate-specific antigen (PSA) estimate higher than 4 ng/ml for most men or 3 ng/ml in high-risk men (e.g., African-American men, or men with a family history of prostate cancer or cardiac conditions; Bhasin et al., 2010; Saad, 2009; Twiddy, Leon, & Wasan, 2011; Wang et al., 2009). On the other hand, TRT is recommended for many men showing symptoms of androgen deficiency, with the objectives to maintain secondary sex characteristic, improve sexual function (Isidori, Giannetta, Gianfrilli, et al., 2005; Wang et al., 2000), increase a sense of well-being, bone mineral density and free-fat mass (Benito et al., 2005; Isidori, Giannetta, Greco, et al., 2005), improve lipid profiles and insulin resistance (Jones et al., 2011; Marin, 1995), increase muscle strength (Sih et al., 1997), and treat depression (Amore et al., 2012).

Although TRT can reverse symptoms associated with low T, there are disadvantages and side effects to its use. TRT can result in LH suppression and therefore contraception (Howell et al., 2001; Nakazawa et al., 2006; Winters & Atkinson, 1997), fluctuating T levels, skin irritation, and the potential for T transfer to others upon skin contact (de Ronde, 2009; Lawrentschuk & Fleshner, 2009). It is noteworthy that studies on a large sample population of men showed that moderate rather than high serum levels of bioactive T are required for the treatment of hypogonadal men (Ferrini & Barrett-Connor, 1998).

Clearly, the development of alternate therapies for treating hypogonadism would be highly desirable. One such approach could be to stimulate T production by the hypofunctional Leydig cells themselves rather than providing T exogenously. Such an approach requires an understanding of the series of steps leading to T formation by Leydig cells, and the identification of key steps that are amenable to pharmacological modulation so as to induce T production.

4. STEROID BIOSYNTHESIS

There are four steroidogenic tissues in humans: adrenal cortex, gonads, placenta, and brain. Leydig cells of the testis are the main site of androgen formation in men, regulated by LH from the pituitary. Cholesterol is an amphipathic lipid synthesized from acetyl-CoA. Most tissues are able to synthesize cholesterol through a chain of enzymatic reactions. More than half of the cholesterol storage of the body is from *de novo* synthesis, 10% of which is produced in liver and another 10% by intestine. The remainder is provided by the diet. Cholesterol is the precursor of steroids, principally stored in the plasma membrane and lipid droplets of steroidogenic cells either as free cholesterol or combined with long-chain fatty acids as cholesteryl esters (Ikonen, 2008; Mesmin & Maxfield, 2009). Steroidogenesis is a result of a series of enzymatic reactions that occur in steroidogenic tissues, with the steroid products dependent upon tissue-specific enzymes (Payne & Hales, 2004).

Steroidogenic cells are defined by their ability to convert the precursor cholesterol to pregnenolone in the mitochondrial matrix through the function of the cytochrome P450 side-chain cleavage enzyme (CYP11A1) (Jefcoate, 2002), which is a member of a large family of P450 enzymes. Seven members of this family are targeted to mitochondria and the remaining 50 members to the endoplasmic reticulum (ER). Six of the members of this family are involved in steroidogenesis. Mitochondrial CYP11A1 converts cholesterol to pregnenolone, with the latter then converted to progesterone by 3β-hydroxysteroid dehydrogenase (3βHSD; HSD3B) present in both the mitochondria and ER. CYP17A1, present in the ER, possesses both 17α-hydroxylase and 17,20-lyase activities. In the gonads of men, CYP17A1 converts progesterone to dehydroepiandrosterone which is further metabolized to T by 17-βHSD (Fig. 1).

The rate-limiting step in steroidogenesis is the import of cholesterol into mitochondria to become accessible to CYP11A1. By limiting access of the hydrophobic cholesterol molecule to CYP11A1, steroidogenic cells are able to control the amount of steroids they produce (Rone, Fan, & Papadopoulos, 2009). Steroid hormones are continuously synthesized. However, in response to increased circulating peptide hormones (LH in the case of Leydig cells), the rate of steroid hormone synthesis is greatly increased. LH binds its cognate receptors on the surface of Leydig cells

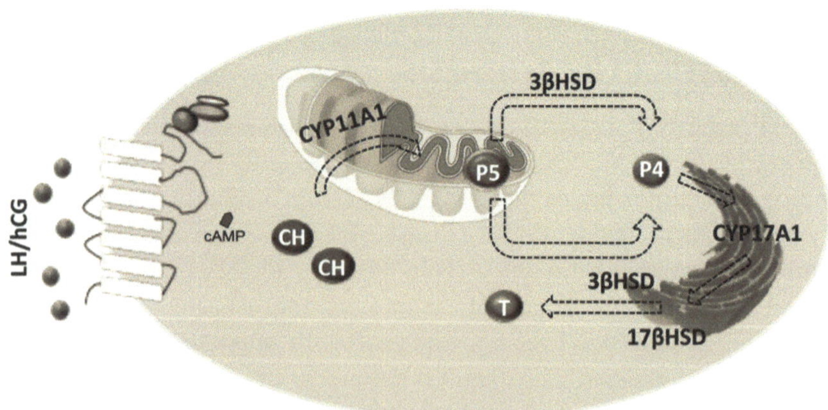

Figure 1 Summary of the steroidogenic pathway in the testis. The steroid hormone biosynthesis pathway in the Leydig cells of the testis is shown in this schematic representation. Cholesterol is the sole precursor of all steroids and it is through several enzymatic reactions in mitochondria, cytoplasm, and ER that different types of steroids are synthesized due to the presence and activity of specific enzymes. P5, pregnenolone; P4, progesterone; T, testosterone.

and stimulates intracellular signaling cascades, of which the cAMP pathway, through protein kinase A (PKA), is the most prominent. Acutely, these events stimulate the import of cholesterol to the OMM and then its transfer from OMM to IMM where it is converted to pregnenolone by CYP11A1 (Jefcoate, 2002; Papadopoulos, Liu, & Culty, 2007). In addition, hormonal and cAMP stimulation of steroidogenic cells are important for the chronic regulation of steroidogenesis, as continued stimulation is necessary to ensure proper expression levels of steroidogenic proteins and enzymes as well as steroidogenic metabolic flux (Simpson & Waterman, 1988).

The mitochondrion is a double-membrane organelle with an aqueous intermembrane space (IMS) between OMM and IMM, where CYP11A1 resides. Cholesterol molecules are hydrophobic. Thus, movement of cholesterol across the aqueous microenvironment of the mitochondria requires the involvement of intracellular machinery (Mesmin & Maxfield, 2009). Indeed, steroidogenic cells possess a multicomponent protein machine, the transduceosome, an ensemble of cytoplasmic and resident mitochondrial proteins that receives hormonal signals and is involved in the translocation of cholesterol across the IMS at contact sites between the OMM and IMM (Papadopoulos et al., 2007; Rone, Liu, et al., 2009) (Fig. 2). The transduceosome OMM proteins were identified as the 30-kDa voltage-dependent anion channel 1 (VDAC1) and the 18-kDa translocator protein TSPO, a high-affinity drug- and cholesterol-binding protein previously named the peripheral-type benzodiazepine receptor (McEnery, Snowman, Trifiletti, & Snyder, 1992; Papadopoulos et al., 2006). In addition to the mitochondrial proteins making up core structural and enzymatic components of the transduceosome, cytoplasmic proteins are instrumental in regulating its assembly and cholesterol transport activity. Hormonal stimulation was found to promote the clustering of TSPO, which was correlated with steroidogenesis and could be suppressed by the PKA inhibitor H-89 (Boujrad, Gaillard, Garnier, & Papadopoulos, 1994). This suggests an active role for PKA in the assembly of the transduceosome. Yeast two-hybrid screening for additional cellular partners of TSPO yielded the acyl-CoA-binding domain family protein ACBD3/PAP7 (Li, Degenhardt, et al., 2001), which serves to scaffold the cytosolic PKA-RI subunits to the transduceosome (Li, Degenhardt, et al., 2001; Liu, Rone, & Papadopoulos, 2006). An additional ACBD protein family member, ACBD1, participates in transduceosome function. Originally identified through its ability to displace benzodiazepine bound to GABA receptor sites in neurons (Costa & Guidotti, 1991), and hence originally named the diazepam-binding

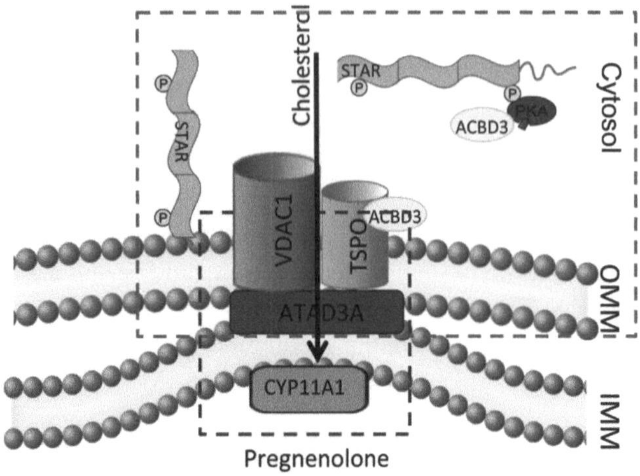

Figure 2 Transduceosome and metabolon complexes. The transduceosome consists of cytosolic and OMM proteins as indicated by gray margin. The metabolon, a protein complex that consists of OMM, IMS, and IMM proteins, is indicated in red (dark gray shade in the print version) margin. VDAC1 and TSPO are members of both the transduceosome and metabolon. CYP11A1 enzyme, located at the mitochondrial matrix side of IMM, and ATAD3, present at the contact sites of IMM and OMM, are members unique to the metabolon complex. The transduceosome mediates the import of cholesterol from cytosolic sources to mitochondria, while the metabolon serves to direct the cholesterol molecule to CYP11A1 where it can be converted to pregnenolone.

inhibitor, ACBD1 acts on TSPO and stimulates steroidogenesis (Boujrad et al., 1994; Papadopoulos, Berkovich, Krueger, Costa, & Guidotti, 1991).

While gene expression of the transduceosome components has been the focus of numerous studies, much less is known about mechanisms through which transduceosome components are targeted to their proper locations in the mitochondria, a factor critical to their function in steroidogenesis (Black, Harikrishna, Szklarz, & Miller, 1994).

4.1 Mitochondrial protein import and chaperones

Most proteins targeted to mitochondria are transcribed in the cytosol as preproteins containing mitochondrial signal sequences (Chacinska, Koehler, Milenkovic, Lithgow, & Pfanner, 2009). The import of such proteins occurs posttranslationally, shown by *in vitro* studies (Maccecchini, Rudin, Blobel, & Schatz, 1979) of cellular chaperones such as Hsc70 (heat-shock cognate protein of 70-kDa chaperone) and HSP90 (heat-shock protein 90). Upon interaction, Hsc70 and HSP90 stabilize the mitochondria-targeted, cytosolic preproteins and prevent protein malformation and/or aggregation by protecting the hydrophobic sequences (Hsc70) or high-affinity interactions with non-native forms of these proteins (HSP90) (Scheufler et al., 2000). Therefore, the preproteins are kept in mitochondrial import-competent state (Fan, Bhangoo, & Young, 2006; Voos & Rottgers, 2002; Young, Hoogenraad, & Hartl, 2003) until docking with the appropriate receptors at the OMM. ATP hydrolysis leads to dissociation of chaperones from preproteins and the mitochondrial import machinery continues the import process (Young & Hartl, 2000; Young et al., 2003). Import machinery includes the Translocase of the Outer Membrane (TOM) complex at the OMM, which assists in the import of α-helices proteins, Translocase of the Inner Membrane (TIM) complex at the IMM, which is involved in the import of β-sheet structures, and the Sorting and Assembly Machinery (SAM) complex at the IMS (Wiedemann et al., 2003).

4.2 Transduceosome

The transduceosome is composed of the cytosolic members STAR, PKA, and ACBD3, and OMM proteins including TSPO and VDAC1, which function together to mediate cholesterol transport through the OMM upon hormonal stimulation (Liu et al., 2006) (Fig. 2). Hormonal stimulation triggers the assembly of the transduceosome at the OMM. This protein complex mediates the import of lipophilic cholesterol from the cytosol through the aqueous IMS and therefore regulates the rate-limiting step in steroidogenesis. The transduceosome consists of cytosolic and mitochondrial proteins which are enriched at the contact site of OMM and IMM.

4.2.1 Translocator protein

TSPO, previously known as the peripheral-type benzodiazepine receptor, is an 18-kDa protein ubiquitously expressed in all mammalian cells and abundant in steroidogenic cells. TSPO localizes at the OMM where it spans across OMM through its 5α-helical domains. Therefore, it is widely considered to function in the cholesterol trail from intracellular stores to the inner mitochondrial membrane (Jaremko, Jaremko, Becker, & Zweckstetter, 2014; Jaremko, Jaremko, Giller, Becker, & Zweckstetter, 2014). Unlike most mitochondrial proteins, TSPO does not possess mitochondrial targeting presequences. However, it contains information on the C-terminus which is essential for its mitochondrial import (Rone, Liu, et al., 2009). Mutational analysis indicates that C-terminal amino acids 103–109 of TSPO form Shellman motif through which Hsc70 and HSP90 target TSPO to mitochondria. The insertion of TSPO is further carried out by Tom70 in an

ATP-dependent manner (Otera et al., 2007; Rone, Liu, et al., 2009). A more intricate mechanism of OMM insertion was proposed for TSPO when RNAi knockdown studies indicated that the classic TOM complex import mechanism involving Tom20, Tom22, and Tom40 was not required for TSPO import. Rather, a role for Tim8 and Metaxin 1, homolog of Sam37, was suggested (Otera et al., 2007; Rone, Liu, et al., 2009). After proper integration at the OMM and in response to hormonal stimulation, TSPO forms clusters at the interface of OMM and IMM in complex with VDAC1 through which cholesterol can enter mitochondria (Chung et al., 2013; Liu et al., 2006). Nuclear magnetic resonance (NMR) studies elucidated the presence of the cholesterol- and ligand-binding domains on TSPO (Jamin et al., 2005; Murail et al., 2008).

The TSPO cytosolic C-terminus has a cholesterol recognition/interaction domain (CRAC) which binds cholesterol with low nanomolar affinity (Ostuni et al., 2009; Papadopoulos et al., 2007). NMR spectroscopy of the TSPO CRAC motif demonstrated that the side-chain amino acids form a groove which accommodates the cholesterol molecule. The tyrosine residue in this groove was shown to play a critical role in cholesterol binding. Moreover, molecular modeling identified a wide range of androstenetriol derivatives that potently subdue steroid biosynthesis by targeting the TSPO CRAC domain (Midzak, Rammouz, & Papadopoulos, 2012). In particular, 3,17,19-androsten-5-triol (19-Atriol) pharmacologically blocks cholesterol binding to the CRAC domain, thereby suppressing steroid biosynthesis. 19-Atriol is potent at low micromolar concentrations and its effect is dose dependent in both hormone-responsive MA-10 mouse Leydig tumor cells and constitutive steroid-producing R2C rat Leydig tumor cells (Midzak, Akula, Lecanu, & Papadopoulos, 2011).

TSPO also contains a drug ligand-binding domain and therefore is a target of numerous exogenous drug ligands which induce steroidogenesis. Numerous previous studies have shown that TSPO drug ligands are able to induce the translocation of cholesterol from the inner to the outer mitochondrial membrane, leading to increased steroid formation by isolated mitochondria and steroidogenic cells (Chung et al., 2013; Romeo et al., 1992) (Fig. 3). These effects of the TSPO drugs will be further discussed in detail below.

4.2.2 Voltage-dependent anion channel 1

VDAC1 is an OMM β-barrel protein involved in the regulation of ions and small molecules across the OMM that influence multiple cellular processes,

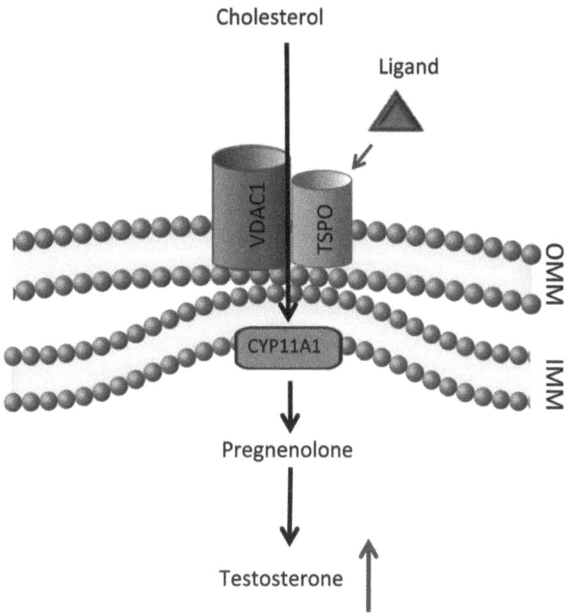

Figure 3 TSPO ligands induce endogenous T levels. TSPO is a high-affinity cholesterol- and drug-binding protein mediating cholesterol import into mitochondria. TSPO is closely interacting at the OMM with VDAC1. TSPO drug ligands (▲) stabilize the conformation of TSPO and induce the movement of cholesterol bound to TSPO into the mitochondria, leading to increased pregnenolone and then testosterone formation. Thus, these compounds can be used to induce endogenous T production.

including apoptosis and energy transduction (Vyssokikh & Brdiczka, 2003). The structure, but not amino acid sequence, of VDAC1 is similar to Tom40 as both proteins are β-barrel proteins that span the OMM 19 times (Zeth, 2010). shRNA knockdown of Tom40 in HeLa cells also reduced VDAC1 insertion into OMM (Kozjak-Pavlovic et al., 2007).

Inducing the loss of IMS proteins by using mitochondrial permeabilizing agents blocks VDAC1 import to this organelle (Krimmer et al., 2001), suggesting a role for Tim8–Tim13. As a B-barrel protein, VDAC1 is further transferred to SAM where Sam50 receives it and assists in its integration into mitochondria (Kutik et al., 2008). Knockdown of mammalian analogs of Sam37 (Metaxin 1) and Sam35 (Metaxin 2) in HeLa cells decreased VDAC1 assembly onto OMM by 50% and 90%, respectively (Armstrong, Komiya, Bergman, Mihara, & Bornstein, 1997; Armstrong, Saenz, & Bornstein, 1999; Kozjak-Pavlovic et al., 2007).

VDAC1 is a multifunctional protein that acts in cross talk between the cytosol and the mitochondria. VDAC1 interacts with TSPO (Liu et al., 2006; Papadopoulos & Miller, 2012), the Bcl-2 family of proteins (Brdiczka, Zorov, & Sheu, 2006), and adenine nucleotide transporter (ANT) (Liu et al., 2006). Upon hormonal stimulation, TSPO and VDAC1 form a cholesterol-binding complex through which the cytosolic cholesterol enters the mitochondria (Liu et al., 2006). VDAC1 also assists in anchoring the transduceosome to the OMM and in integration of STAR and OMM (Bose, Whittal, Ran, et al., 2008; Liu et al., 2006). The mechanisms and regulation of TSPO and VDAC1 interactions are not clear (Liu et al., 2006), but the presence of VDAC1 increases the ability of TSPO to bind to some of the TSPO drug ligands, such as benzodiazepines (Boujrad et al., 1994; Garnier et al., 1994). Through interactions with Bcl2, VDAC1 participates in apoptosis, and through its interactions with ANT, it forms the mitochondrial permeability transition pore (MPTP) through which small molecules and ions can enter mitochondria. In this way, it is involved in mitochondrial energy transduction (Sileikyte et al., 2011). Interestingly, TSPO ligands affect MPTP function (Sileikyte et al., 2011). It is noteworthy that VDAC1 knockout does not affect MPTP and mitochondrial function, opening a new area of investigation to identify the exact role of VDAC1 (Huang et al., 2013).

4.2.3 Steroidogenic acute regulatory protein

Of the cytoplasmic transduceosome components, the protein that has garnered the greatest research attention is the steroidogenic acute regulatory protein, STAR. STAR (STARD1) belongs to a family of proteins containing 15 members that are structurally related to STAR (Clark, 2012) as they contain the STAR-related lipid transfer (START) domain which binds a variety of lipids with low micromolar affinity. The members of this family STARD1–15, STARD1, STARD3 (MLN64), and STARD5 bind cholesterol. STARD5 also binds 25-hydroxycholesterol. STARD2 (phosphatidylcholine-transfer protein) binds phosphatidylcholine, and StarD10 binds phosphatidylcholine and phosphatidylethanolamine. Structural and knockdown studies suggested that among the cholesterol-binding members of this family, only STAR is involved in steroidogenesis (Akula, Midzak, Lecanu, & Papadopoulos, 2012; Midzak & Papadopoulos, 2014; Riegelhaupt, Waase, Garbarino, Cruz, & Breslow, 2010).

STAR was originally identified as a 30-kDa phosphoprotein located in the mitochondrial matrix and induced upon hormonal stimulation in adrenal and gonadal tissue (Pon, Hartigan, & Orme-Johnson, 1986; Pon &

Orme-Johnson, 1986). Later, it was shown to be synthesized as a 37-kDa cytoplasmic protein with a mitochondrial presequence tag that is imported into the mitochondrial matrix and processed to a 30-kDa form (Clark, Wells, King, & Stocco, 1994; King et al., 1995). STAR contains an N-terminal mitochondrial signal sequence that restricts its function to mitochondria (Clark et al., 1994), and a β-barrel START domain present at the C-terminus (Alpy & Tomasetto, 2005; Baker, Yaworsky, & Miller, 2005). Upon hormonal stimulation, STAR levels are induced in a manner parallel to steroidogenesis and STAR activation and action results to increased flow of cholesterol to mitochondria. STAR transcription in steroidogenic cells is rather rapid. In mouse Leydig cell lines, *Star* mRNA is induced 120 min following cAMP stimulation through the activity of cAMP-responsive element-binding protein 1 (CREB) and CREB-binding protein (CBP) (Clark, Combs, Hales, Hales, & Stocco, 1997; Lefrancois-Martinez et al., 2011). However, increase in STAR protein levels in these cells begins after 30 min of cAMP/hCG stimulation and reaches significantly higher levels after 60 min. Treatment of cells with actinomycin, a transcription inhibitor, does not block the induction in STAR protein levels, suggesting the presence of preexisting mRNA for STAR (Clark et al., 1997).

STAR activity is regulated by protein phosphorylation (Ascoli, Euffa, & Segaloff, 1987). Few protein kinases have been identified to date that regulate STAR function and activity. Extracellular signal-regulated kinase (ERK1/2) phosphorylates STAR on Ser232, which is a highly conserved ERK1/2-docking site. Such phosphorylation occurs at the mitochondria. However, the activity of ERK1/2 relies on cholesterol binding to STAR, whereas PKA phosphorylation occurs independently of cholesterol binding to this protein (Poderoso et al., 2008). Recent studies identified cyclin-dependent kinase 5 (CDK5) as interacting with and phosphorylating STAR. However, such phosphorylation is analogous to the role of PKA in regulating STAR phosphorylation (Lin, Chen, & Ku, 2009). Recently, the mitochondrial kinase MEK1/2 was also shown to phosphorylate STAR on S232, which contributes to STAR activation (Manna & Stocco, 2005).

PKA is the most prominent kinase that regulates STAR activity. Among the PKA phosphorylation sites, three residues are the most important, S554/56, S187, and S194. Mutational analysis suggests the presence of proteolytic cleavage sites at positions 39/40 and 55/56 of the bovine STAR protein, with similar predicted sites existing in other mammalian STAR homologs (Artemenko, Zhao, Hales, Hales, & Jefcoate, 2001; Yamazaki et al., 2006). In rodents, the STAR preprotein is targeted to mitochondria and proteolytically processed to yield a 32- and a final 30-kDa protein, the latter located in the mitochondrial matrix (Epstein & Orme-Johnson, 1991; Stocco & Sodeman, 1991). This suggests that phosphorylation might be required for such cleavage. PKA phosphorylation of S194 induces STAR activity by twofold, required for maximum steroidogenesis (Arakane et al., 1997). Therefore, mutation of this residue is considered to be one of the major causes of lipoid CAH, a severe adrenal and gonadal disorder in which steroidogenesis is severely impaired and fatty deposits accumulate in steroidogenic cells (Lin et al., 1995; Miller, 2007), leading to lipoid hyperplasia (Bose, Sugawara, Strauss, & Miller, 1996; Fluck et al., 2011; Jo, King, Khan, & Stocco, 2005). Similar results were shown in mutations of S187 phosphorylation sites. This observation led to the suggestion that STAR's role is as a sole importer of cholesterol. In this hypothesis, STAR was believed to bind cholesterol through its START domain and import this molecule from the cytosol to the mitochondrial matrix. However, further studies indicated that STAR is only functional at the OMM, and upon reaching the matrix, the 30-kDa STAR is degraded, with a half-life of 4–5 h (Granot et al., 2003). An insufficient stoichiometry of 1:1 cholesterol binding was assessed for STAR (Tsujishita & Hurley, 2000), suggesting that STAR function in steroidogenesis is distinct from its cholesterol-binding capacity (Baker, Epand, Epand, & Miller, 2007). STAR knockout mouse models were developed and found to require corticosteroid supplementation for their survival (Hasegawa et al., 2000). These mice had high levels of lipid deposition in the adrenal cortex, and much lower levels in the testes (Caron et al., 1997). Although Leydig cell steroidogenesis seems to not be affected in the neonatal STAR knockout mice, T levels and spermatogenesis were reduced in the adult compared to wild-type mice (Hasegawa et al., 2000). However, it is not clear whether the reduced testicular function over time is due to the lack of STAR or to the detrimental effect of the long-term corticosteroid treatment on Leydig cell development and function (Bernier, Gibb, Collu, & Ducharme, 1984; Dong et al., 2004; Welsh, Bambino, & Hsueh, 1982; Xiao, Huang, Hardy, Li, & Ge, 2010). In female mice, the gonads were not affected by STAR knockout (Hasegawa et al., 2000). Therefore, it is now clear that STAR functions in a multiprotein dynamic along with other members of the transduceosome to import cholesterol to mitochondria (Liu et al., 2006; Papadopoulos et al., 2007; Papadopoulos & Miller, 2012). These studies also suggested the possibility of differential roles for STAR in adrenal versus testicular steroidogenesis.

The mechanism underlying STAR's contribution to steroidogenesis has been addressed in a number of additional in-depth studies. CD spectroscopy suggested a "molten globule model" for START domain interactions with cholesterol. In this model, STAR was proposed to form Ω loops protecting the 259–282 residues from cholesterol docking. In the presence of OMM-resembling vesicles, STAR surface residues are protonated, leading to partial loosening of the tertiary structures. In this state, cholesterol can bind the START domain (Yaworsky et al., 2005). Furthermore, graphic modeling of the positioning and orientation of cholesterol in the START cavity led to high-throughput screening of steroid-structured compounds that can replace cholesterol. Two compounds that can dock onto the START domain with low energy decreased MA-10 steroidogenesis by 50% (Akula et al., 2012). As affixation of STAR at the cytosolic side of the OMM, but not on IMM or IMS, increased steroidogenesis (Bose, Lingappa, & Miller, 2002), it is now suggested that STAR binds cholesterol and is active at the OMM but does not shuttle cholesterol to CYP11A1 located at the IMM. Therefore, other candidate proteins such as TSPO and VDAC1 have been proposed to be integrally involved in cholesterol import.

Although observed at the IMS by crystallography studies, STAR import to mitochondria terminates its function, a mechanism proposed to be carried out by Lon protease with high turnover (Ondrovicova et al., 2005). However, the mitochondrial-binding partners of STAR are not yet identified. Interestingly, recent studies suggest the presence of an OMM acceptor for STAR. Knocking down TSPO or antagonizing its function in MA-10 cells suggested a role for TSPO in processing of the 37 kDa STAR preprotein to the mature 30 kDa form (Hauet et al., 2005; Midzak, Akula, et al., 2011), emphasizing the significance of protein–protein interactions in the transduceosome in driving steroidogenesis. This work was further complemented by a series of experiments in MA-10 cells in which mass spectrometry analysis identified STAR and VDAC1 in the same mitochondrial complexes, VDAC1 knockdown decreased STAR expression, and acute inhibition of VDAC1 led to accumulation of 32 kDa STAR preprotein in the cytosol (Bose, Whittal, Miller, & Bose, 2008). Despite the previous observations of STAR in complex with TSPO and VDAC1 (Bose, Whittal, Miller, et al., 2008; Liu et al., 2006), the lack of evidence of direct physical interaction between STAR with either VDAC1 or TSPO (Bose, Whittal, Gairola, & Bose, 2008; Bose, Whittal, Miller, et al., 2008; Liu et al., 2006) suggests the involvement of an adaptor or scaffold that mediates such physiological correlations.

4.2.4 Protein kinase A

Increase in cAMP levels leads to steroid production due to downstream activation of proteins such as PKA (Catt et al., 1980; Karlsson et al., 2010). PKA holoenzyme is a heterotetramer of two regulatory (R) and two catalytic (C) subunits. Upon binding of two cAMP molecules to each regulatory subunit, the two catalytic subunits are released to phosphorylate Ser/Thr residues on target proteins, among which are proteins involved in cholesterol import to mitochondria. Therefore, PKA activation enhances and transduces the hormonal signal (Hansson, Skalhegg, & Tasken, 1999). The R subunits of PKA compartmentalize at the target membranes in order to recruit the holoenzyme at the sites of cAMP action, thus amplifying the cAMP signal at specific subcellular compartments. PKA regulates steroidogenesis through multiple steps that include inducing STAR phosphorylation, Star gene expression, mRNA stability, and posttranslational modifications (PTMs) enhancing the protein activity.

In steroidogenic cells, both PKA and PKC phosphorylate CREB in a coordinated, time-sensitive manner, required for *star* transcription (Chrivia et al., 1993; Gonzalez & Montminy, 1989). Phosphorylation by PKA occurs on CREB Ser133, which is indispensable for its interactions with CBP to recruit the transcription factors on the proximal STAR promoter (Clem, Hudson, & Clark, 2005; Hiroi et al., 2004). Moreover, similar coordination between PKA and PKC was shown in order to phosphorylate c-Jun on S63 and c-Fos at Thr325. Such phosphorylations lead to dimerization and DNA binding of the two proteins which assist the recruitment of CBP to the STAR promoter (Manna, Wang, & Stocco, 2003). The interactions between phospho-CREB and CBP and the recruitment of CBP by phosphorylated c-Jun and c-Fos ultimately induce *star.* DNA expression and therefore PKA are the regulator of such transcription (Clem et al., 2005; Duan & Jefcoate, 2007; Manna et al., 2006; Manna & Stocco, 2005). The best-characterized contribution of PKA in steroidogenesis is the PTMs of STAR protein. Briefly, PKA phosphorylates STAR on several sites, critical for STAR cleavage, enhancement of activity, and prolongation of the duration of STAR activity at OMM before import and termination of its function.

4.2.5 Acetyl CoA-binding domain 3

ACBD3 is a 52-kDa protein formerly known as peripheral-type benzodiazepine receptor and PKA-RIα-associated protein 7 (PAP7). ACBD3 is expressed in all human tissues and is highly expressed in steroidogenic cells of gonads and adrenals as well as in glial and hippocampal neuronal cells in the brain (Morin et al., 2012). The primary localization of ACBD3 is at the Golgi apparatus, but this protein translocates to mitochondria during steroidogenesis (Li, Degenhardt, et al., 2001). ACBD3 contains a homology domain with AKAPs (A kinase anchoring proteins). These proteins are best known for their interactions with PKA enzymes and for enhancing the function of such kinases by increasing their proximity to their targets proteins and membranes (Esseltine & Scott, 2013). ACBD3 was shown to selectively interact with PKA-RIα *in vivo*, suggesting a role in transduction of PKA activity during steroidogenesis. ACBD3 has been shown to interact with TSPO at the OMM, suggesting that the translocation of this protein from the Golgi apparatus to mitochondria is through its interactions with TSPO (Fan, Liu, Culty, & Papadopoulos, 2010; Liu, Li, & Papadopoulos, 2003). Overexpression of ACBD3 induced hCG-stimulated steroidogenesis (Feliciello, Gottesman, & Avvedimento, 2001; Li, Degenhardt, et al., 2001), while overexpression of a sequence of ACBD3 that contained interaction sites with PKA-RIα and TSPO was shown to act as a dominant negative during steroidogenesis (Fan et al., 2010; Liu et al., 2003). Although the exact mechanism of ACBD3 action in steroidogenesis has not been clearly addressed as yet, it is likely that this protein tethers PKA-RIα to mitochondria in order to phosphorylate S194 on STAR and enhance the protein activity, and/or activates mitochondrial kinase MEK1/2 which phosphorylates STAR on S232 and contributes to STAR activation (Poderoso et al., 2008). Moreover, ACBD3 also anchors to mitochondria through interactions with TSPO, which serves to alter the topography of TSPO at the OMM so as to facilitate drug ligand binding during steroidogenesis (Fan et al., 2010; Liu et al., 2003).

4.3 Metabolon

The term "metabolon" refers to multiprotein complexes that perform a series of reactions on a particular substrate. The components of metabolons are tethered at membranes such as those of mitochondria. Their main function is to optimally increase substrate concentration and targeting to certain enzymes for further actions (Issop, Rone, & Papadopoulos, 2013). Recently, an 800-kDa metabolon complex was identified and characterized at the mitochondria consisting of OMM and IMM proteins (Moller, 2010). This protein complex links the transduceosome to CYP11A1 and therefore is an intermediate between cholesterol import into the mitochondria and its conversion to pregnenolone. The mitochondrial metabolon overlaps with the transduceosome at OMM as TSPO and VDAC1 are considered to be members of both complexes. The IMM enzyme, CYP11A1, is a part of the steroidogenic metabolon. Interactions between the IMM and OMM proteins are mediated through the IMM–OMM protein, AAA+ ATPase, ATAD3 (Fig. 2) (Rone et al., 2012).

4.3.1 Cholesterol side-chain cleavage cytochrome P450 (CYP11A1)

CY11A1 is a nucleus-encoded mitochondrial matrix protein. Cytosolic protein translation yields a 521-amino acid polypeptide CYP11A1 preprotein containing a 39-amino acid mitochondrial targeting signal sequences with three positive amino acids that are essential for protein import (DuBois, Simpson, Tuckey, Lambeth, & Waterman, 1981; Payne & Hales, 2004). Inhibition of CYP11A1 presequences using the metal chelator *o*-phenanthroline did not alter the association of CYP11A1 with the matrix or its enzymatic activity (Nabi, Kominami, Takemori, & Omura, 1983) as the presequences are cleaved upon import. The mechanism of CYP11A1 import suggests the involvement of Tim22 and Tim23 (Minenko, Novikova, Luzikov, & Kovaleva, 2008). Studies on yeast showed that CYP11A1 is imported completely into the matrix and then reassembled, leading to decreased enzymatic activity (Minenko et al., 2008). Once imported, 66 N-terminal amino acids of CYP11A1 form an F-G loop anchor to the IMM; Tim22 was suggested to mainly facilitate the anchorage process rather than be involved in mitochondrial import (Minenko et al., 2008). The N-terminal of CYP11A1 is therefore protected from trypsin proteolysis of isolated mitochondria (Ou, Ito, Morohashi, Fujii-Kuriyama, & Omura, 1986). Topology studies of the isolated mitochondrial membranes (Ou et al., 1986), and three-dimensional modeling of CYP11A1 structure, indicated that the IMM-anchoring residues are located on the opposite surface from the substrate-binding site (Nelson & Strobel, 1988). When properly localized, CYP11A1 catalyzes the conversion of cholesterol to pregnenolone by cleaving the cholesterol aliphatic side chain. The rate of CYP11A1 activity relies on the rate of electron transfer from the matrix proteins Fdx and FdxR (Minenko et al., 2008). The Fdx precursor preadrenodoxin (Padx) is imported through the function of Hsc70 or MSF. Upon import, a 58-amino acid signal sequence of Padx is cleaved, creating an interaction site with CYP11A1 as shown in *Escherichia coli* (Nabi & Omura, 1980). A more intricate model for Padx import was suggested by studies indicating that the ribosomes synthesizing Padx bind to the OMM (Hachiya et al., 1993, 1995), suggesting that this protein can be translated in the cytosol or at the level of mitochondria (Nabi & Omura, 1980, 1983). FdR also contains a mitochondrial signal sequence consisting of 32 amino acids that are cleaved after import (MacKenzie & Payne, 2004). Both proteins have matrix localizations and interact with CYP11A1 to provide electrons. The electrons are further utilized by CYP11A1 to activate an oxygen molecule during conversion of cholesterol to pregnenolone (Payne & Hales, 2004). The proteins are localized in the mitochondrial matrix where they interact with CYP11A1.

4.3.2 AAA+ ATPase, ATAD3

ATAD3 is an AAA+ ATPase protein that is present at the mitochondrial matrix and participates in maintenance of mitochondrial DNA integrity (He et al., 2007). This protein is enriched in OMM–IMM contact sites (Sagara, Takata, Miyata, Hara, & Horiuchi, 1987). The N-terminus of ATAD3 is anchored to OMM through which ATAD3 controls mitochondrial dynamics at the interface of the IMM and OMM (Gilquin et al., 2010). Recent studies reported that ATAD3 knockdown leads to reduced levels of Leydig cell steroidogenesis in response to hormonal stimulation (Rone et al., 2012). The mechanism through which ATAD3 contributes to steroidogenesis is under investigation.

5. CAN SERUM TESTOSTERONE LEVELS BE INCREASED BY STIMULATING THE LEYDIG CELLS THEMSELVES?

As indicated above, there are good reasons to increase serum T levels in hypogonadal men, but doing so by TRT is less than ideal for both young and aging men. There are methods to increase serum T without T administration, including administering LH (or hCG) for men with secondary hypogonadism. In men with primary hypogonadism, aromatase inhibitors can increase T-to-estradiol ratios, but often this approach is ineffective (Perheentupa & Huhtaniemi, 2009). Increasing intratesticular and serum T by stimulating the Leydig cells themselves could have great advantages in part because doing so would be expected to result in the physiological regulation of the T that is produced by its negative feedback on LH. This is in contrast to essentially flooding the system with T, as with exogenous T administration. Stimulating the Leydig cells should not elicit significant fluctuations in T levels, and nor there be T transfer to others via contact. Fertility should be preserved, not suppressed, because the hypothalamic–pituitary axis should not be shut down as with exogenous T. Indeed, the local stimulation of Leydig cell T production might actually enhance spermatogenesis because intratesticular

T levels should increase, not decrease. Moreover, Leydig cells produce more than just T, and so their stimulation might produce Leydig cell products in addition to T that might have effects on spermatogenesis. With an understanding of the series of steps leading to T formation derived from numerous studies conducted over the years, and the identification of key steps that have been shown to be amenable to pharmacological modulation so as to induce T production, it now might be possible to stimulate Leydig cells to produce additional T as a means by which to increase serum T levels.

5.1 TSPO drug ligands

These particular OMM proteins have attracted considerable interest in steroidogenesis research. Drugs binding to TSPO are able to stimulate steroid biosynthesis (Krueger & Papadopoulos, 1990; Papadopoulos, Berkovich, & Krueger, 1991), and the VDAC1–TSPO pair is the binding site of benzodiazepine and other drug ligands to the OMM (Garnier et al., 1994) (Fig. 2). TSPO itself is able to exclusively bind different classes of drugs with the capacity to stimulate steroidogenesis (Rupprecht et al., 2009). Moreover, blocking the interactions between TSPO and its ligands blocked steroidogenesis in primary cells (Chung et al., 2013; Lacapere & Papadopoulos, 2003; Midzak, Akula, et al., 2011).

Our recent *in vivo* studies showed that the pharmacologic activation of TSPO increased serum T in hypogonadal aged Brown Norway rats to the level of young rats (Chung et al., 2013). This exciting finding suggests that there is the potential to use of TSPO drug ligands to induce T formation in conditions of primary (or secondary) hypogonadism if their *in vivo* use proves to be specific for Leydig cells and safe. TSPO drug ligands have been shown to increase circulating corticosteroids in animals, though far less so in control than in hypophysectomized animals (Cavallaro, Korneyev, Guidotti, & Costa, 1992). These drugs also will induce neurosteroid formation in the brain *in vitro* and *in vivo* (Cavallaro et al., 1992; Costa & Guidotti, 1991; Romeo et al., 1992) and have been used to alleviate neurological and psychiatric disease symptoms (Romeo et al., 1992; Schule, Nothdurfter, & Rupprecht, 2014). These findings suggest that TSPO drug ligands may exert stimulatory effects on steroid formation by the adrenal and brain, with effects that are minor in normal *in vivo* settings but significant when TSPO levels are reduced due to conditions that include aging, hypophysectomy, depression, and anxiety. Compounds such as *N*,*N*-di-*n*-hexyl 2-(4-fluorophenyl) indole-3-acetamide (FGIN-1–27) are drug ligands of TSPO that can induce neurosteroidogenesis in rat glioma cells (Romeo et al., 1992), or androgen formation in rat testes (Aghazadeh, Martinez-Arguelles, Fan, Culty, & Papadopoulos, 2014). Taken together, these studies clearly indicate a potential role for TSPO drug ligands for possible therapeutic use in increasing serum T levels in hypogonadal men.

Knockdown of TSPO was shown in previous studies to ablate steroid biosynthesis in steroidogenic cells (Hauet et al., 2005; Papadopoulos et al., 1997). However, in a recent study, testis-specific TSPO knockout mice in which TSPO monomers were not present were reported to produce T in amounts comparable to those of control mice (Morohaku et al., 2014). This suggests a potential compensatory mechanism for maintenance of mitochondrial integrity and function (Papadopoulos, 2014). However, the identification and characterization of several pharmacological agents of distinct chemical structure that target TSPO and stimulate steroid formation, as well as structural studies of TSPO, indicate that TSPO is a valuable drug target for the induction of steroid and neurosteroid biosynthesis (Jaremko, Jaremko, Becker, et al., 2014; Jaremko, Jaremko, Giller, et al., 2014; Lacapere et al., 2001; Murail et al., 2008).

5.2 14-3-3γ and ε proteins

14-3-3 proteins are a family of small, acidic proteins that, through functioning adaptors, scaffolds, and chaperones, regulate a variety of intracellular signaling pathways. All species express at least two isoforms of 14-3-3 proteins; mammals express seven isoforms, named β, γ, ζ, ε, η, σ, and θ (α and δ are phosphorylated forms of β and ζ) (Ichimura et al., 1997; Toker et al., 1992). Each 14-3-3 monomer is composed of nine α-helices. There is a high level of homology among the 14-3-3 family members, and they are expressed ubiquitously in mammalian tissues (except the endothelial cell-specific 14-3-3σ and T cell-specific 14-3-3θ) (Aitken, 2006). Nonetheless, proteomic analysis identified unique, tissue-specific networks (Choudhary et al., 2009), and global microarray data suggest tissue-specific isoform expression (www.biogps.org). Therefore, we now understand that 14-3-3 family members also have isoform and tissue-specific roles in addition to their ubiquitous functions. Such unique roles rely in part on hetero-/homodimerization, which offer functional diversity to 14-3-3 proteins. The N-terminal helices contain highly varied amino acid sequences, and therefore, each isoform exerts different dimerization affinity for others (Liu et al., 1995; Xiao et al., 1995). The dimerization of 14-3-3 isoforms was previously believed to be essential for thermodynamic stability of the proteins of this family, but recent studies showed that dimer/monomer equilibrium for 14-3-3 proteins is influenced by PTMs, which also serve to stabilize 14-3-3 monomers. Moreover, a splicing variant was identified in 18 human tissues for 14-3-3ε, called sv-14-3-3ε, which lacks the 22 dimerization amino acids and is highly functional (Han et al., 2010). 14-3-3 dimers form a clamp-like shape where each monomer contains an amphipathic groove at the C-terminal through which it can bind to target proteins. Although the C-terminal of 14-3-3 proteins is highly homologous, the surrounding surface residues and tissue-specific PTMs dictate target specificity for each isoform. Thus, 14-3-3 proteins bind to more than 200 targets, and this list is expanding. The majority of target proteins contain 14-3-3-binding motif(s) with the following amino acid sequences: RSXpSXP (mode I) and RXY/FXpSXP (mode II) or -pS/pTX(1–2)–CO_2 H (mode III) (where X is not Pro) (Johnson et al., 2010). "Suboptimal" motifs are observed in majority of 14-3-3 targets, varying by 1–3 amino acids from the classical sequences. The presence of suboptimal motifs leads to transient association/dissociation of 14-3-3/target protein, required for adaptor characteristics of 14-3-3 family of proteins. The presence of more than one suboptimal motif induces 14-3-3-binding affinity by 30-fold (Yaffe et al., 1997). Upon interactions with targets, 14-3-3 proteins alter target protein modifications, activity, and cellular localization and therefore regulate five major cellular pathways: cell-cycle and apoptosis, signal transduction, metabolism, and intracellular protein trafficking.

14-3-3ε, also known as tyrosine 3/tryptophan 5-monooxygenase activation protein epsilon, has high affinity for heterodimerization with 14-3-3γ (Aitken et al., 2002). 14-3-3γ, also referred to as 3-monooxgenase/tryptophan 5-monooxygenase activation protein gamma, has high homodimerization affinity though its preferable heterodimerization partner is 14-3-3ε. A complex of the 14-3-3γ and ε is called the mitochondrial import-stimulating factor (MSF) (Alam et al., 1994; Hachiya et al., 1993, 1994). MSF is a cytosolic chaperone that delivers the mitochondrial-targeted β-barrel proteins and recognizes and binds to signal sequences on mitochondrial pre-proteins.

14-3-3ε is highly expressed in brain, lymphoblasts, adipocytes, and testes. The 14-3-3γ isoform is highly expressed in brain, skeletal muscle, heart, and embryonic stem cells (www.biogps.org). Recently, 14-3-3ε and γ were identified as negative regulators of androgen formation in the Leydig cells (Muslin, Tanner, Allen, & Shaw, 1996). *In silico* analysis predicted 14-3-3-binding motifs on transduceosome proteins; they were present on structural domains critical for their function (Aghazadeh et al., 2014). STAR contains three predicted 14-3-3 motifs, on sites of cleavage and activation (Aghazadeh et al., 2012); TSPO contains a predicted 14-3-3 motif with one amino acid distance from the CRAC domain (Li, Yao, Degenhardt, Teper, & Papadopoulos, 2001). VDAC1 contains two 14-3-3 motifs at the dimerization site and at the lateral surface accessible to other OMM proteins for interactions (Geula, Naveed, Liang, & Shoshan-Barmatz, 2012; Sorgato & Moran, 1993). 14-3-3γ is hormonally induced, similar to STAR, whereas the 14-3-3ε expression is not hormonally regulated. Rather, 14-3-3ε localizes to mitochondria during steroidogenesis. Low-throughput analysis suggests that the principal target of 14-3-3γ is STAR, while 14-3-3ε mainly targets VDAC1 in the transduceosome. Moreover, VDAC1 and START domains are classified as β-barrel protein and protein domain, and MSF was shown to mediate the transport of mitochondrial proteins belonging to this class (Geula et al., 2012; Sorgato & Moran, 1993; Tsujishita & Hurley, 2000). Different mechanisms of action were proposed for 14-3-3γ and ε, with their roles consecutive and complementary to each other.

14-3-3γ was shown to bind to 194 phosphorylation site on STAR at the initiation of steroidogenesis. PKA phosphorylation of S194 was shown previously to induce STAR activation by twofold (Arakane et al., 1997). Therefore, a model was proposed in which 14-3-3γ protein physically blocks the accessibility of STAR S194 for PKA phosphorylation and therefore maintains STAR at basal activity. Such negative regulation, however, is terminated by the dissociation of 14-3-3γ from STAR, thus allowing significant induction of steroidogenesis (Aghazadeh et al., 2012) (Fig. 4A). In light of the observations that the transduceosome assembly, TSPO polymerization, and increased STAR expression occur shortly after cAMP/hormone stimulation, it was not clear why there is a delay in reaching maximum steroidogenesis induction. The time-sensitive mechanism of action of 14-3-3γ suggests that this protein contributes, at least in part, to the delay in steroidogenesis induction. Why this lag period might be necessary for steroidogenic cells is yet to be elucidated. The termination of 14-3-3γ function was shown to coincide with increased homodimerization of this protein (Aghazadeh et al., 2012), suggesting a protein regulation mechanism. Indeed, 14-3-3 phosphorylation and/or high expression were shown to contribute to protein self-homodimerization which alters the protein function (Shen et al., 2003).

Once 14-3-3γ releases STAR, a second regulatory mechanism for cholesterol import to mitochondria is activated that involves 14-3-3ε. The regulatory role of 14-3-3ε has a later onset but is maintained long term. 14-3-3ε anchors to mitochondria through interacting with TSPO, thereby

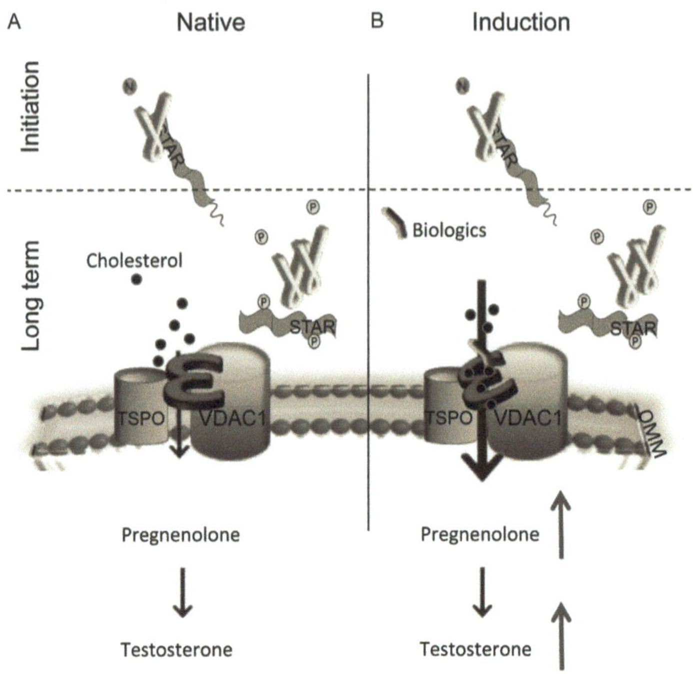

Figure 4 Induction of endogenous steroidogenesis via blockage of 14-3-3 protein negative regulation. 14-3-3γ and ε are negative regulators of steroidogenesis. Initially upon hormonal stimulation, acetylated 14-3-3γ monomers interact with STAR and block the PKA-dependent phosphorylation of this protein on S194, retaining STAR in its basal activity. Later on, 14-3-3γ phosphorylation and high levels of protein expression induce homodimerization of this protein and dissociation from STAR. (A) When steroidogenesis is highly increased, 14-3-3ε scaffold protein translocates to mitochondria where it regulates the TSPO microenvironment and, by intercalating between TSPO and VDAC1, negatively regulates the import of cholesterol into mitochondria, thus controlling the rates of T formation. (B) T production can be induced by using biologics that block the interactions between 14-3-3ε and VDAC1 on S167 (》), which allows the formation of more efficient TSPO–VDAC1 interactions. This prompts cholesterol import into mitochondria and increases T formation in the absence of gonadotropin.

regulating cholesterol and ligand binding to TSPO throughout steroidogenesis. When steroidogenesis is highly induced, 14-3-3ε interactions with VDAC1 are increased. The primary interaction sites were identified to be VDAC1 S167. This Ser residue is located on the lateral side of the protein accessible for interaction with OMM partners such as TSPO. Therefore, 14-3-3ε intercalation between TSPO and VDAC1 blocks their efficient interactions, thus affecting the rate of cholesterol entry into mitochondria (Aghazadeh et al., 2014, 2012) (Fig. 4A).

Additional studies were conducted using a cell penetrating sequence conjugated to a short sequence of VDAC1 containing S167, and of STAR containing S194. Such fusion peptides successfully compete out 14-3-3ε and 14-3-3γ interactions in MA-10 cells. As a result, the negative regulatory role of 14-3-3γ and ε is ablated and therefore cells produce more steroids acutely at the initiation of steroidogenesis, or long term, respectively. Furthermore, peptides containing VDAC1 S167, administered directly to the testes of adult male Sprague–Dawley rats, induced testicular and plasma T levels in a dose-dependent manner (Fig. 4B), independently of LH (Aghazadeh et al., 2014). Modeling studies suggest that these peptides can mimic the VDAC1 docking to 14-3-3ε in rats as well as in humans due to high homology of 14-3-3ε and VDAC1 across species and the 100% conservation of the 14-3-3ε motif. These biological fusion peptides are therefore potential tools to induce T levels in men diagnosed with androgen deficiency (Aghazadeh et al., 2014).

6. CONCLUSION

Cholesterol is the main precursor of T. Its import into mitochondria is the rate-limiting step in steroidogenesis. After a decade of intensive research, it is now clear that cholesterol import into the mitochondria in response to hormonal stimulation is carried out through a multiprotein complex called transduceosome, and not by individual proteins (Midzak, Rone, Aghazadeh, Culty, & Papadopoulos, 2011; Papadopoulos & Miller, 2012; Rone et al., 2012).

In search of mechanisms inducing T production in the Leydig cell as an alternative to TRT, components of the transduceosome were identified to be critical for T formation and "druggable" candidates. Indeed, TSPO and 14-3-3ε–VDAC1 interactions were shown to be viable targets which, when activated, result in increased T formation both *in vitro* and *in vivo*, thus providing new means for the treatment of androgen deficiency in hypogonadal men.

ACKNOWLEDGMENTS

This work was supported by grants from the Canadian Institutes of Health Research (MOP102647 and 125983), the National Institutes of Health (R37 AG21092) to B. Z., and a Canada Research Chair in Biochemical Pharmacology to V. P. Y. A. was supported in part by predoctoral fellowships from the Research Institute of MUHC and the McGill Center for the Study of Reproduction.

REFERENCES

Aghazadeh, Y., Martinez-Arguelles, D. B., Fan, J., Culty, M., & Papadopoulos, V. (2014). Induction of androgen formation in the male by a TAT-VDAC1 fusion peptide blocking 14-3-3 epsilon protein adaptor and mitochondrial VDAC1 interactions. *Molecular Therapy*, *22*, 1779–1791.

Aghazadeh, Y., Rone, M. B., Blonder, J., Ye, X., Veenstra, T. D., Hales, D. B., et al. (2012). Hormone-induced 14-3-3gamma adaptor protein regulates steroidogenic acute regulatory protein activity and steroid biosynthesis in MA-10 Leydig cells. *Journal of Biological Chemistry*, *287*, 15380–15394.

Aitken, A. (2006). 14-3-3 proteins: A historic overview. *Seminars in Cancer Biology*, *16*, 162–172.

Aitken, A., Baxter, H., Dubois, T., Clokie, S., Mackie, S., Mitchell, K., et al. (2002). Specificity of 14-3-3 isoform dimer interactions and phosphorylation. *Biochemical Society Transactions*, *30*, 351–360.

Akula, N., Midzak, A., Lecanu, L., & Papadopoulos, V. (2012). Identification of small-molecule inhibitors of the steroidogenic acute regulatory protein (STARD1) by structure-based design. *Bioorganic & Medicinal Chemistry Letters*, *22*, 4139–4143.

Alam, R., Hachiya, N., Sakaguchi, M., Kawabata, S., Iwanaga, S., Kitajima, M., et al. (1994). cDNA cloning and characterization of mitochondrial import stimulation factor (MSF) purified from rat liver cytosol. *Journal of Biochemistry*, *116*, 416–425.

Alpy, F., & Tomasetto, C. (2005). Give lipids a START: The StAR-related lipid transfer (START) domain in mammals. *Journal of Cell Science*, *118*, 2791–2801.

Amore, M., Innamorati, M., Costi, S., Sher, L., Girardi, P., & Pompili, M. (2012). Partial androgen deficiency, depression, and testosterone supplementation in aging men. *International Journal of Endocrinology*, *2012*, 280724.

Arakane, F., King, S. R., Du, Y., Kallen, C. B., Walsh, L. P., Watari, H., et al. (1997). Phosphorylation of steroidogenic acute regulatory protein (StAR) modulates its steroidogenic activity. *Journal of Biological Chemistry*, *272*, 32656–32662.

Armstrong, L. C., Komiya, T., Bergman, B. E., Mihara, K., & Bornstein, P. (1997). Metaxin is a component of a preprotein import complex in the outer membrane of the mammalian mitochondrion. *Journal of Biological Chemistry*, *272*, 6510–6518.

Armstrong, L. C., Saenz, A. J., & Bornstein, P. (1999). Metaxin 1 interacts with metaxin 2, a novel related protein associated with the mammalian mitochondrial outer membrane. *Journal of Cellular Biochemistry*, *74*, 11–22.

Artemenko, I. P., Zhao, D., Hales, D. B., Hales, K. H., & Jefcoate, C. R. (2001). Mitochondrial processing of newly synthesized steroidogenic acute regulatory protein (StAR), but not total StAR, mediates cholesterol transfer to cytochrome P450 side chain cleavage enzyme in adrenal cells. *Journal of Biological Chemistry*, *276*, 46583–46596.

ASA Position Statement. (2006). Testosterone replacement therapy for male aging. *Journal of Andrology*, *27*, 133–134.

Ascoli, M., Euffa, J., & Segaloff, D. L. (1987). Epidermal growth factor activates steroid biosynthesis in cultured Leydig tumor cells without affecting the levels of cAMP and potentiates the activation of steroid biosynthesis by choriogonadotropin and cAMP. *Journal of Biological Chemistry*, *262*, 9196–9203.

Baker, B. Y., Epand, R. F., Epand, R. M., & Miller, W. L. (2007). Cholesterol binding does not predict activity of the steroidogenic acute regulatory protein, StAR. *Journal of Biological Chemistry*, *282*, 10223–10232.

Baker, B. Y., Yaworsky, D. C., & Miller, W. L. (2005). A pH-dependent molten globule transition is required for activity of the steroidogenic acute regulatory protein, StAR. *Journal of Biological Chemistry*, *280*, 41753–41760.

Belchetz, P. E., Barth, J. H., & Kaufman, J. M. (2010). Biochemical endocrinology of the hypogonadal male. *Annals of Clinical Biochemistry*, *47*, 503–515.

Benito, M., Vasilic, B., Wehrli, F. W., Bunker, B., Wald, M., Gomberg, B., et al. (2005). Effect of testosterone replacement on trabecular architecture in hypogonadal men. *Journal of Bone and Mineral Research, 20*, 1785–1791.

Bernier, M., Gibb, W., Collu, R., & Ducharme, J. R. (1984). Effect of glucocorticoids on testosterone production by porcine Leydig cells in primary culture. *Canadian Journal of Physiology and Pharmacology, 62*, 1166–1169.

Bhasin, S., Cunningham, G. R., Hayes, F. J., Matsumoto, A. M., Snyder, P. J., Swerdloff, R. S., et al. (2010). Testosterone therapy in men with androgen deficiency syndromes: An Endocrine Society clinical practice guideline. *Journal of Clinical Endocrinology and Metabolism, 95*, 2536–2559.

Black, S. M., Harikrishna, J. A., Szklarz, G. D., & Miller, W. L. (1994). The mitochondrial environment is required for activity of the cholesterol side-chain cleavage enzyme, cytochrome P450scc. *Proceedings of the National Academy of Sciences of the United States of America, 91*, 7247–7251.

Bose, H. S., Lingappa, V. R., & Miller, W. L. (2002). Rapid regulation of steroidogenesis by mitochondrial protein import. *Nature, 417*, 87–91.

Bose, H. S., Sugawara, T., Strauss, J. F., III, & Miller, W. L. (1996). The pathophysiology and genetics of congenital lipoid adrenal hyperplasia. *The New England Journal of Medicine, 335*, 1870–1878.

Bose, M., Whittal, R. M., Gairola, C. G., & Bose, H. S. (2008). Molecular mechanism of reduction in pregnenolone synthesis by cigarette smoke. *Toxicology and Applied Pharmacology, 229*, 56–64.

Bose, M., Whittal, R. M., Miller, W. L., & Bose, H. S. (2008). Steroidogenic activity of StAR requires contact with mitochondrial VDAC1 and phosphate carrier protein. *Journal of Biological Chemistry, 283*, 8837–8845.

Bose, H. S., Whittal, R. M., Ran, Y., Bose, M., Baker, B. Y., & Miller, W. L. (2008). StAR-like activity and molten globule behavior of StARD6, a male germ-line protein. *Biochemistry, 47*, 2277–2288.

Boujrad, N., Gaillard, J. L., Garnier, M., & Papadopoulos, V. (1994). Acute action of choriogonadotropin on Leydig tumor cells: Induction of a higher affinity benzodiazepine-binding site related to steroid biosynthesis. *Endocrinology, 135*, 1576–1583.

Brdiczka, D. G., Zorov, D. B., & Sheu, S. S. (2006). Mitochondrial contact sites: Their role in energy metabolism and apoptosis. *Biochimica et Biophysica Acta, 1762*, 148–163.

Busse, K. L., & Maibach, H. I. (2011). Transdermal estradiol and testosterone transfer in man: Existence, models, and strategies for prevention. *Skin Pharmacology and Physiology, 24*, 57–66.

Caron, K. M., Soo, S. C., Wetsel, W. C., Stocco, D. M., Clark, B. J., & Parker, K. L. (1997). Targeted disruption of the mouse gene encoding steroidogenic acute regulatory protein provides insights into congenital lipoid adrenal hyperplasia. *Proceedings of the National Academy of Sciences of the United States of America, 94*, 11540–11545.

Carruthers, M. (2009). Time for international action on treating testosterone deficiency syndrome. *The Aging Male, 12*, 21–28.

Catt, K. J., Harwood, J. P., Clayton, R. N., Davies, T. F., Chan, V., Katikineni, M., et al. (1980). Regulation of peptide hormone receptors and gonadal steroidogenesis. *Recent Progress in Hormone Research, 36*, 557–662.

Cattabiani, C., Basaria, S., Ceda, G. P., Luci, M., Vignali, A., Lauretani, F., et al. (2012). Relationship between testosterone deficiency and cardiovascular risk and mortality in adult men. *Journal of Endocrinological Investigation, 35*, 104–120.

Cavallaro, S., Korneyev, A., Guidotti, A., & Costa, E. (1992). Diazepam-binding inhibitor (DBI)-processing products, acting at the mitochondrial DBI receptor, mediate adrenocorticotropic hormone-induced steroidogenesis in rat adrenal gland. *Proceedings of the National Academy of Sciences of the United States of America, 89*, 10598–10602.

Chacinska, A., Koehler, C. M., Milenkovic, D., Lithgow, T., & Pfanner, N. (2009). Importing mitochondrial proteins: Machineries and mechanisms. *Cell, 138*, 628–644.

Choudhary, C., Kumar, C., Gnad, F., Nielsen, M. L., Rehman, M., Walther, T. C., et al. (2009). Lysine acetylation targets protein complexes and co-regulates major cellular functions. *Science, 325*, 834–840.

Chrivia, J. C., Kwok, R. P., Lamb, N., Hagiwara, M., Montminy, M. R., & Goodman, R. H. (1993). Phosphorylated CREB binds specifically to the nuclear protein CBP. *Nature, 365*, 855–859.

Chung, J. Y., Chen, H., Midzak, A., Burnett, A. L., Papadopoulos, V., & Zirkin, B. R. (2013). Drug ligand-induced activation of translocator protein (TSPO) stimulates steroid production by aged brown Norway rat Leydig cells. *Endocrinology, 154*, 2156–2165.

Clark, B. J. (2012). The mammalian START domain protein family in lipid transport in health and disease. *Journal of Endocrinology, 212*, 257–275.

Clark, B. J., Combs, R., Hales, K. H., Hales, D. B., & Stocco, D. M. (1997). Inhibition of transcription affects synthesis of steroidogenic acute regulatory protein and steroidogenesis in MA-10 mouse Leydig tumor cells. *Endocrinology, 138*, 4893–4901.

Clark, B. J., Wells, J., King, S. R., & Stocco, D. M. (1994). The purification, cloning, and expression of a novel luteinizing hormone-induced mitochondrial protein in MA-10 mouse Leydig tumor cells. Characterization of the steroidogenic acute regulatory protein (StAR). *Journal of Biological Chemistry, 269*, 28314–28322.

Clem, B. F., Hudson, E. A., & Clark, B. J. (2005). Cyclic adenosine 3′,5′-monophosphate (cAMP) enhances cAMP-responsive element binding (CREB) protein phosphorylation and phospho-CREB interaction with the mouse steroidogenic acute regulatory protein gene promoter. *Endocrinology, 146*, 1348–1356.

Costa, E., & Guidotti, A. (1991). Diazepam binding inhibitor (DBI): A peptide with multiple biological actions. *Life Sciences, 49*, 325–344.

de Ronde, W. (2009). Hyperandrogenism after transfer of topical testosterone gel: Case report and review of published and unpublished studies. *Human Reproduction, 24*, 425–428.

Dobs, A. S., Meikle, A. W., Arver, S., Sanders, S. W., Caramelli, K. E., & Mazer, N. A. (1999). Pharmacokinetics, efficacy, and safety of a permeation-enhanced testosterone transdermal system in comparison with bi-weekly injections of testosterone enanthate for the treatment of hypogonadal men. *Journal of Clinical Endocrinology and Metabolism, 84*, 3469–3478.

Dong, Q., Salva, A., Sottas, C. M., Niu, E., Holmes, M., & Hardy, M. P. (2004). Rapid glucocorticoid mediation of suppressed testosterone biosynthesis in male mice subjected to immobilization stress. *Journal of Andrology, 25*, 973–981.

Duan, H., & Jefcoate, C. R. (2007). The predominant cAMP-stimulated 3×5 kb StAR mRNA contains specific sequence elements in the extended 3′UTR that confer high basal instability. *Journal of Molecular Endocrinology, 38*, 159–179.

DuBois, R. N., Simpson, E. R., Tuckey, J., Lambeth, J. D., & Waterman, M. R. (1981). Evidence for a higher molecular weight precursor of cholesterol side-chain-cleavage cytochrome P-450 and induction of mitochondrial and cytosolic proteins by corticotropin in adult bovine adrenal cells. *Proceedings of the National Academy of Sciences of the United States of America, 78*, 1028–1032.

Epstein, L. F., & Orme-Johnson, N. R. (1991). Regulation of steroid hormone biosynthesis. Identification of precursors of a phosphoprotein targeted to the mitochondrion in stimulated rat adrenal cortex cells. *Journal of Biological Chemistry, 266*, 19739–19745.

Esseltine, J. L., & Scott, J. D. (2013). AKAP signaling complexes: Pointing towards the next generation of therapeutic targets? *Trends in Pharmacological Sciences, 34*, 648–655.

Fan, A. C., Bhangoo, M. K., & Young, J. C. (2006). Hsp90 functions in the targeting and outer membrane translocation steps of Tom70-mediated mitochondrial import. *Journal of Biological Chemistry, 281*, 33313–33324.

Fan, J., Liu, J., Culty, M., & Papadopoulos, V. (2010). Acyl-coenzyme A binding domain containing 3 (ACBD3; PAP7; GCP60): An emerging signaling molecule. *Progress in Lipid Research*, *49*, 218–234.

Feliciello, A., Gottesman, M. E., & Avvedimento, E. V. (2001). The biological functions of A-kinase anchor proteins. *Journal of Molecular Biology*, *308*, 99–114.

Ferrini, R. L., & Barrett-Connor, E. (1998). Sex hormones and age: A cross-sectional study of testosterone and estradiol and their bioavailable fractions in community-dwelling men. *American Journal of Epidemiology*, *147*, 750–754.

Fluck, C. E., Pandey, A. V., Dick, B., Camats, N., Fernandez-Cancio, M., Clemente, M., et al. (2011). Characterization of novel StAR (steroidogenic acute regulatory protein) mutations causing non-classic lipoid adrenal hyperplasia. *PLoS One*, *6*, e20178.

Garnier, M., Dimchev, A. B., Boujrad, N., Price, J. M., Musto, N. A., & Papadopoulos, V. (1994). In vitro reconstitution of a functional peripheral-type benzodiazepine receptor from mouse Leydig tumor cells. *Molecular Pharmacology*, *45*, 201–211.

Geula, S., Naveed, H., Liang, J., & Shoshan-Barmatz, V. (2012). Structure-based analysis of VDAC1 protein: Defining oligomer contact sites. *Journal of Biological Chemistry*, *287*, 2179–2190.

Gilquin, B., Taillebourg, E., Cherradi, N., Hubstenberger, A., Gay, O., Merle, N., et al. (2010). The AAA+ ATPase ATAD3A controls mitochondrial dynamics at the interface of the inner and outer membranes. *Molecular and Cellular Biology*, *30*, 1984–1996.

Gonzalez, G. A., & Montminy, M. R. (1989). Cyclic AMP stimulates somatostatin gene transcription by phosphorylation of CREB at serine 133. *Cell*, *59*, 675–680.

Granot, Z., Geiss-Friedlander, R., Melamed-Book, N., Eimerl, S., Timberg, R., Weiss, A. M., et al. (2003). Proteolysis of normal and mutated steroidogenic acute regulatory proteins in the mitochondria: The fate of unwanted proteins. *Molecular Endocrinology*, *17*, 2461–2476.

Hachiya, N., Alam, R., Sakasegawa, Y., Sakaguchi, M., Mihara, K., & Omura, T. (1993). A mitochondrial import factor purified from rat liver cytosol is an ATP-dependent conformational modulator for precursor proteins. *The EMBO Journal*, *12*, 1579–1586.

Hachiya, N., Komiya, T., Alam, R., Iwahashi, J., Sakaguchi, M., Omura, T., et al. (1994). MSF, a novel cytoplasmic chaperone which functions in precursor targeting to mitochondria. *The EMBO Journal*, *13*, 5146–5154.

Hachiya, N., Mihara, K., Suda, K., Horst, M., Schatz, G., & Lithgow, T. (1995). Reconstitution of the initial steps of mitochondrial protein import. *Nature*, *376*, 705–709.

Han, D., Ye, G., Liu, T., Chen, C., Yang, X., Wan, B., et al. (2010). Functional identification of a novel 14-3-3 epsilon splicing variant suggests dimerization is not necessary for 14-3-3 epsilon to inhibit UV-induced apoptosis. *Biochemical and Biophysical Research Communications*, *396*, 401–406.

Hansson, V., Skalhegg, B. S., & Tasken, K. (1999). Cyclic-AMP-dependent protein kinase (PKA) in testicular cells. Cell specific expression, differential regulation and targeting of subunits of PKA. *Journal of Steroid Biochemistry and Molecular Biology*, *69*, 367–378.

Hasegawa, T., Zhao, L., Caron, K. M., Majdic, G., Suzuki, T., Shizawa, S., et al. (2000). Developmental roles of the steroidogenic acute regulatory protein (StAR) as revealed by StAR knockout mice. *Molecular Endocrinology*, *14*, 1462–1471.

Hauet, T., Yao, Z. X., Bose, H. S., Wall, C. T., Han, Z., Li, W., et al. (2005). Peripheral-type benzodiazepine receptor-mediated action of steroidogenic acute regulatory protein on cholesterol entry into Leydig cell mitochondria. *Molecular Endocrinology*, *19*, 540–554.

He, J., Mao, C. C., Reyes, A., Sembongi, H., Di, R. M., Granycome, C., et al. (2007). The AAA+ protein ATAD3 has displacement loop binding properties and is involved in mitochondrial nucleoid organization. *Journal of Cell Biology*, *176*, 141–146.

Hiroi, H., Christenson, L. K., Chang, L., Sammel, M. D., Berger, S. L., & Strauss, J. F., III. (2004). Temporal and spatial changes in transcription factor binding and histone modifications at the steroidogenic acute regulatory protein (stAR) locus associated with stAR transcription. *Molecular Endocrinology*, *18*, 791–806.

Howell, S. J., Radford, J. A., Adams, J. E., Smets, E. M., Warburton, R., & Shalet, S. M. (2001). Randomized placebo-controlled trial of testosterone replacement in men with mild Leydig cell insufficiency following cytotoxic chemotherapy. *Clinical Endocrinology*, *55*, 315–324.

Huang, H., Hu, X., Eno, C. O., Zhao, G., Li, C., & White, C. (2013). An interaction between Bcl-xL and the voltage-dependent anion channel (VDAC) promotes mitochondrial Ca^{2+} uptake. *Journal of Biological Chemistry*, *288*, 19870–19881.

Huhtaniemi, I., & Forti, G. (2011). Male late-onset hypogonadism: Pathogenesis, diagnosis and treatment. *Nature Reviews. Urology*, *8*, 335–344.

Ichimura, T., Ito, M., Itagaki, C., Takahashi, M., Horigome, T., Omata, S., et al. (1997). The 14-3-3 protein binds its target proteins with a common site located towards the C-terminus. *FEBS Letters*, *413*, 273–276.

Ikonen, E. (2008). Cellular cholesterol trafficking and compartmentalization. *Nature Reviews Molecular Cell Biology*, *9*, 125–138.

Ishunina, T. A., van, B. D., van der Meulen, G., Unmehopa, U. A., Hol, E. M., Huitinga, I., et al. (2005). Diminished aromatase immunoreactivity in the hypothalamus, but not in the basal forebrain nuclei in Alzheimer's disease. *Neurobiology of Aging*, *26*, 173–194.

Isidori, A. M., Giannetta, E., Gianfrilli, D., Greco, E. A., Bonifacio, V., Aversa, A., et al. (2005). Effects of testosterone on sexual function in men: Results of a meta-analysis. *Clinical Endocrinology*, *63*, 381–394.

Isidori, A. M., Giannetta, E., Greco, E. A., Gianfrilli, D., Bonifacio, V., Isidori, A., et al. (2005). Effects of testosterone on body composition, bone metabolism and serum lipid profile in middle-aged men: A meta-analysis. *Clinical Endocrinology*, *63*, 280–293.

Issop, L., Rone, M. B., & Papadopoulos, V. (2013). Organelle plasticity and interactions in cholesterol transport and steroid biosynthesis. *Molecular and Cellular Endocrinology*, *371*, 34–46.

Jamin, N., Neumann, J. M., Ostuni, M. A., Vu, T. K., Yao, Z. X., Murail, S., et al. (2005). Characterization of the cholesterol recognition amino acid consensus sequence of the peripheral-type benzodiazepine receptor. *Molecular Endocrinology*, *19*, 588–594.

Jaremko, L., Jaremko, M., Becker, S., & Zweckstetter, M. (2014). Toward the functional oligomerization state of tryptophan-rich sensory proteins. *Protein Science*, *23*, 1154–1160.

Jaremko, L., Jaremko, M., Giller, K., Becker, S., & Zweckstetter, M. (2014). Structure of the mitochondrial translocator protein in complex with a diagnostic ligand. *Science*, *343*, 1363–1366.

Jefcoate, C. (2002). High-flux mitochondrial cholesterol trafficking, a specialized function of the adrenal cortex. *Journal of Clinical Investigation*, *110*, 881–890.

Jo, Y., King, S. R., Khan, S. A., & Stocco, D. M. (2005). Involvement of protein kinase C and cyclic adenosine 3′,5′-monophosphate-dependent kinase in steroidogenic acute regulatory protein expression and steroid biosynthesis in Leydig cells. *Biology of Reproduction*, *73*, 244–255.

Johnson, C., Crowther, S., Stafford, M. J., Campbell, D. G., Toth, R., & Mackintosh, C. (2010). Bioinformatic and experimental survey of 14-3-3-binding sites. *The Biochemical Journal*, *427*, 69–78.

Jones, T. H., Arver, S., Behre, H. M., Buvat, J., Meuleman, E., Moncada, I., et al. (2011). Testosterone replacement in hypogonadal men with type 2 diabetes and/or metabolic syndrome (the TIMES2 study). *Diabetes Care*, *34*, 828–837.

Kaplan, S. A., Meehan, A. G., & Shah, A. (2006). The age related decrease in testosterone is significantly exacerbated in obese men with the metabolic syndrome. What are the implications for the relatively high incidence of erectile dysfunction observed in these men? *Journal of Urology*, *176*, 1524–1527.

Karlsson, A. B., Maizels, E. T., Flynn, M. P., Jones, J. C., Shelden, E. A., Bamburg, J. R., et al. (2010). Luteinizing hormone receptor-stimulated progesterone production by preovulatory granulosa cells requires protein kinase A-dependent activation/dephosphorylation of the actin dynamizing protein cofilin. *Molecular Endocrinology*, *24*, 1765–1781.

King, S. R., Ronen-Fuhrmann, T., Timberg, R., Clark, B. J., Orly, J., & Stocco, D. M. (1995). Steroid production after in vitro transcription, translation, and mitochondrial processing of protein products of complementary deoxyribonucleic acid for steroidogenic acute regulatory protein. *Endocrinology*, *136*, 5165–5176.

Kozjak-Pavlovic, V., Ross, K., Benlasfer, N., Kimmig, S., Karlas, A., & Rudel, T. (2007). Conserved roles of Sam50 and metaxins in VDAC biogenesis. *EMBO Reports*, *8*, 576–582.

Krimmer, T., Rapaport, D., Ryan, M. T., Meisinger, C., Kassenbrock, C. K., Blachly-Dyson, E., et al. (2001). Biogenesis of porin of the outer mitochondrial membrane involves an import pathway via receptors and the general import pore of the TOM complex. *Journal of Cell Biology*, *152*, 289–300.

Krueger, K. E., & Papadopoulos, V. (1990). Peripheral-type benzodiazepine receptors mediate translocation of cholesterol from outer to inner mitochondrial membranes in adrenocortical cells. *Journal of Biological Chemistry*, *265*, 15015–15022.

Kupelian, V., Page, S. T., Araujo, A. B., Travison, T. G., Bremner, W. J., & McKinlay, J. B. (2006). Low sex hormone-binding globulin, total testosterone, and symptomatic androgen deficiency are associated with development of the metabolic syndrome in nonobese men. *Journal of Clinical Endocrinology and Metabolism*, *91*, 843–850.

Kutik, S., Stojanovski, D., Becker, L., Becker, T., Meinecke, M., Kruger, V., et al. (2008). Dissecting membrane insertion of mitochondrial beta-barrel proteins. *Cell*, *132*, 1011–1024.

Lacapere, J. J., Delavoie, F., Li, H., Peranzi, G., Maccario, J., Papadopoulos, V., et al. (2001). Structural and functional study of reconstituted peripheral benzodiazepine receptor. *Biochemical and Biophysical Research Communications*, *284*, 536–541.

Lacapere, J. J., & Papadopoulos, V. (2003). Peripheral-type benzodiazepine receptor: Structure and function of a cholesterol-binding protein in steroid and bile acid biosynthesis. *Steroids*, *68*, 569–585.

Lang, P. O., Samaras, D., & Samaras, N. (2012). Testosterone replacement therapy in reversing "andropause": What is the proof-of-principle? *Rejuvenation Research*, *15*, 453–465.

Lawrentschuk, N., & Fleshner, N. (2009). Severe irritant contact dermatitis causing skin ulceration secondary to a testosterone patch. *ScientificWorldJournal*, *9*, 333–338.

Lefrancois-Martinez, A. M., Blondet-Trichard, A., Binart, N., Val, P., Chambon, C., Sahut-Barnola, I., et al. (2011). Transcriptional control of adrenal steroidogenesis: Novel connection between Janus kinase (JAK) 2 protein and protein kinase A (PKA) through stabilization of cAMP response element-binding protein (CREB) transcription factor. *Journal of Biological Chemistry*, *286*, 32976–32985.

Li, H., Degenhardt, B., Tobin, D., Yao, Z. X., Tasken, K., & Papadopoulos, V. (2001). Identification, localization, and function in steroidogenesis of PAP7: A peripheral-type benzodiazepine receptor- and PKA (RIalpha)-associated protein. *Molecular Endocrinology*, *15*, 2211–2228.

Li, H., Yao, Z., Degenhardt, B., Teper, G., & Papadopoulos, V. (2001). Cholesterol binding at the cholesterol recognition/interaction amino acid consensus (CRAC) of the peripheral-type benzodiazepine receptor and inhibition of steroidogenesis by an HIV TAT-CRAC peptide. *Proceedings of the National Academy of Sciences of the United States of America, 98*, 1267–1272.

Lin, H., Chen, M. C., & Ku, C. T. (2009). Cyclin-dependent kinase 5 regulates steroidogenic acute regulatory protein and androgen production in mouse Leydig cells. *Endocrinology, 150*, 396–403.

Lin, D., Sugawara, T., Strauss, J. F., III, Clark, B. J., Stocco, D. M., Saenger, P., et al. (1995). Role of steroidogenic acute regulatory protein in adrenal and gonadal steroidogenesis. *Science, 267*, 1828–1831.

Liu, D., Bienkowska, J., Petosa, C., Collier, R. J., Fu, H., & Liddington, R. (1995). Crystal structure of the zeta isoform of the 14-3-3 protein. *Nature, 376*, 191–194.

Liu, J., Li, H., & Papadopoulos, V. (2003). PAP7, a PBR/PKA-RIalpha-associated protein: A new element in the relay of the hormonal induction of steroidogenesis. *Journal of Steroid Biochemistry and Molecular Biology, 85*, 275–283.

Liu, J., Rone, M. B., & Papadopoulos, V. (2006). Protein-protein interactions mediate mitochondrial cholesterol transport and steroid biosynthesis. *Journal of Biological Chemistry, 281*, 38879–38893.

Lutz, W., Sanderson, W., & Scherbov, S. (1997). Doubling of world population unlikely. *Nature, 387*, 803–805.

Maccecchini, M. L., Rudin, Y., Blobel, G., & Schatz, G. (1979). Import of proteins into mitochondria: Precursor forms of the extramitochondrially made F1-ATPase subunits in yeast. *Proceedings of the National Academy of Sciences of the United States of America, 76*, 343–347.

MacKenzie, J. A., & Payne, R. M. (2004). Ribosomes specifically bind to mammalian mitochondria via protease-sensitive proteins on the outer membrane. *Journal of Biological Chemistry, 279*, 9803–9810.

Manna, P. R., Chandrala, S. P., King, S. R., Jo, Y., Counis, R., Huhtaniemi, I. T., et al. (2006). Molecular mechanisms of insulin-like growth factor-I mediated regulation of the steroidogenic acute regulatory protein in mouse Leydig cells. *Molecular Endocrinology, 20*, 362–378.

Manna, P. R., & Stocco, D. M. (2005). Regulation of the steroidogenic acute regulatory protein expression: Functional and physiological consequences. *Current Drug Targets Immune, Endocrine and Metabolic Disorders, 5*, 93–108.

Manna, P. R., Wang, X. J., & Stocco, D. M. (2003). Involvement of multiple transcription factors in the regulation of steroidogenic acute regulatory protein gene expression. *Steroids, 68*, 1125–1134.

Marin, P. (1995). Testosterone and regional fat distribution. *Obesity Research, 3*(Suppl. 4), 609S–612S.

McEnery, M. W., Snowman, A. M., Trifiletti, R. R., & Snyder, S. H. (1992). Isolation of the mitochondrial benzodiazepine receptor: Association with the voltage-dependent anion channel and the adenine nucleotide carrier. *Proceedings of the National Academy of Sciences of the United States of America, 89*, 3170–3174.

Mesmin, B., & Maxfield, F. R. (2009). Intracellular sterol dynamics. *Biochimica et Biophysica Acta, 1791*, 636–645.

Midzak, A., Akula, N., Lecanu, L., & Papadopoulos, V. (2011). Novel androstenetriol interacts with the mitochondrial translocator protein and controls steroidogenesis. *Journal of Biological Chemistry, 286*, 9875–9887.

Midzak, A., & Papadopoulos, V. (2014). Binding domain-driven intracellular trafficking of sterols for synthesis of steroid hormone, bile acid and oxysterols. *Traffic, 15*, 895–914.

Midzak, A., Rammouz, G., & Papadopoulos, V. (2012). Structure-activity relationship (SAR) analysis of a family of steroids acutely controlling steroidogenesis. *Steroids*, *77*, 1327–1334.

Midzak, A., Rone, M., Aghazadeh, Y., Culty, M., & Papadopoulos, V. (2011). Mitochondrial protein import and the genesis of steroidogenic mitochondria. *Molecular and Cellular Endocrinology*, *336*, 70–79.

Miller, W. L. (2007). StAR search—What we know about how the steroidogenic acute regulatory protein mediates mitochondrial cholesterol import. *Molecular Endocrinology*, *21*, 589–601.

Minenko, A. N., Novikova, L. A., Luzikov, V. N., & Kovaleva, I. E. (2008). Import of hybrid forms of CYP11A1 into yeast mitochondria. *Biochimica et Biophysica Acta*, *1780*, 1121–1130.

Mitchell, R., Hollis, S., Rothwell, C., & Robertson, W. R. (1995). Age related changes in the pituitary-testicular axis in normal men; lower serum testosterone results from decreased bioactive LH drive. *Clinical Endocrinology*, *42*, 501–507.

Moller, B. L. (2010). Plant science. Dynamic metabolons. *Science*, *330*, 1328–1329.

Morin, E., Mete, O., Wasserman, J. D., Joshua, A. M., Asa, S. L., & Ezzat, S. (2012). Carney complex with adrenal cortical carcinoma. *Journal of Clinical Endocrinology and Metabolism*, *97*, E202–E206.

Morohaku, K., Pelton, S. H., Daugherty, D. J., Butler, W. R., Deng, W., & Selvaraj, V. (2014). Translocator protein/peripheral benzodiazepine receptor is not required for steroid hormone biosynthesis. *Endocrinology*, *155*, 89–97.

Morris, P. D., & Channer, K. S. (2012). Testosterone and cardiovascular disease in men. *Asian Journal of Andrology*, *14*, 428–435.

Murail, S., Robert, J. C., Coic, Y. M., Neumann, J. M., Ostuni, M. A., Yao, Z. X., et al. (2008). Secondary and tertiary structures of the transmembrane domains of the translocator protein TSPO determined by NMR. Stabilization of the TSPO tertiary fold upon ligand binding. *Biochimica et Biophysica Acta*, *1778*, 1375–1381.

Muraleedharan, V., & Jones, T. H. (2010). Testosterone and the metabolic syndrome. *Therapeutic Advances in Endocrinology and Metabolism*, *1*, 207–223.

Muram, D., Melby, T., & Alles, K. E. (2012). Skin reactions in a phase 3 study of a testosterone topical solution applied to the axilla in hypogonadal men. *Current Medical Research and Opinion*, *28*, 761–766.

Muslin, A. J., Tanner, J. W., Allen, P. M., & Shaw, A. S. (1996). Interaction of 14-3-3 with signaling proteins is mediated by the recognition of phosphoserine. *Cell*, *84*, 889–897.

Nabi, N., Kominami, S., Takemori, S., & Omura, T. (1983). Contributions of cytoplasmic free and membrane-bound ribosomes to the synthesis of mitochondrial cytochrome P-450(SCC) and P-450(11 beta) and microsomal cytochrome P-450(C-21) in bovine adrenal cortex. *Journal of Biochemistry*, *94*, 1517–1527.

Nabi, N., & Omura, T. (1980). In vitro synthesis of adrenodoxin and adrenodoxin reductase: Existence of a putative large precursor form of adrenodoxin. *Biochemical and Biophysical Research Communications*, *97*, 680–686.

Nabi, N., & Omura, T. (1983). In vitro synthesis of NADPH-adrenodoxin reductase and adrenodoxin of bovine adrenal cortex, and the existence of a large precursor of adrenodoxin. *Journal of Biochemistry*, *94*, 1529–1538.

Nakazawa, R., Baba, K., Nakano, M., Katabami, T., Saito, N., Takahashi, T., et al. (2006). Hormone profiles after intramuscular injection of testosterone enanthate in patients with hypogonadism. *Endocrine Journal*, *53*, 305–310.

Nelson, D. R., & Strobel, H. W. (1988). On the membrane topology of vertebrate cytochrome P-450 proteins. *Journal of Biological Chemistry*, *263*, 6038–6050.

Oeppen, J., & Vaupel, J. W. (2002). Demography. Broken limits to life expectancy. *Science*, *296*, 1029–1031.
Ondrovicova, G., Liu, T., Singh, K., Tian, B., Li, H., Gakh, O., et al. (2005). Cleavage site selection within a folded substrate by the ATP-dependent ion protease. *Journal of Biological Chemistry*, *280*, 25103–25110.
Ostuni, M. A., Peranzi, G., Ducroc, R. A., Fasseu, M., Vidic, B., Dumont, J., et al. (2009). Distribution, pharmacological characterization and function of the 18 kDa translocator protein in rat small intestine. *Biology of the Cell*, *101*, 573–586.
Otera, H., Taira, Y., Horie, C., Suzuki, Y., Suzuki, H., Setoguchi, K., et al. (2007). A novel insertion pathway of mitochondrial outer membrane proteins with multiple transmembrane segments. *Journal of Cell Biology*, *179*, 1355–1363.
Ou, W. J., Ito, A., Morohashi, K., Fujii-Kuriyama, Y., & Omura, T. (1986). Processing-independent in vitro translocation of cytochrome P-450(SCC) precursor across mitochondrial membranes. *Journal of Biochemistry*, *100*, 1287–1296.
Papadopoulos, V. (2014). On the role of the translocator protein (18-kDa) TSPO in steroid hormone biosynthesis. *Endocrinology*, *155*, 15–20.
Papadopoulos, V., Amri, H., Li, H., Boujrad, N., Vidic, B., & Garnier, M. (1997). Targeted disruption of the peripheral-type benzodiazepine receptor gene inhibits steroidogenesis in the R2C Leydig tumor cell line. *Journal of Biological Chemistry*, *272*, 32129–32135.
Papadopoulos, V., Baraldi, M., Guilarte, T. R., Knudsen, T. B., Lacapere, J. J., Lindemann, P., et al. (2006). Translocator protein (18 kDa): New nomenclature for the peripheral-type benzodiazepine receptor based on its structure and molecular function. *Trends in Pharmacological Sciences*, *27*, 402–409.
Papadopoulos, V., Berkovich, A., & Krueger, K. E. (1991). The role of diazepam binding inhibitor and its processing products at mitochondrial benzodiazepine receptors: Regulation of steroid biosynthesis. *Neuropharmacology*, *30*, 1417–1423.
Papadopoulos, V., Berkovich, A., Krueger, K. E., Costa, E., & Guidotti, A. (1991). Diazepam binding inhibitor and its processing products stimulate mitochondrial steroid biosynthesis via an interaction with mitochondrial benzodiazepine receptors. *Endocrinology*, *129*, 1481–1488.
Papadopoulos, V., Liu, J., & Culty, M. (2007). Is there a mitochondrial signaling complex facilitating cholesterol import? *Molecular and Cellular Endocrinology*, *265–266*, 59–64.
Papadopoulos, V., & Miller, W. L. (2012). Role of mitochondria in steroidogenesis. *Best Practice & Research. Clinical Endocrinology & Metabolism*, *26*, 771–790.
Payne, A. H., & Hales, D. B. (2004). Overview of steroidogenic enzymes in the pathway from cholesterol to active steroid hormones. *Endocrine Reviews*, *25*, 947–970.
Perheentupa, A., & Huhtaniemi, I. (2009). Aging of the human ovary and testis. *Molecular and Cellular Endocrinology*, *299*, 2–13.
Poderoso, C., Converso, D. P., Maloberti, P., Duarte, A., Neuman, I., Galli, S., et al. (2008). A mitochondrial kinase complex is essential to mediate an ERK1/2-dependent phosphorylation of a key regulatory protein in steroid biosynthesis. *PLoS One*, *3*, e1443.
Pon, L. A., Hartigan, J. A., & Orme-Johnson, N. R. (1986). Acute ACTH regulation of adrenal corticosteroid biosynthesis. Rapid accumulation of a phosphoprotein. *Journal of Biological Chemistry*, *261*, 13309–13316.
Pon, L. A., & Orme-Johnson, N. R. (1986). Acute stimulation of steroidogenesis in corpus luteum and adrenal cortex by peptide hormones. Rapid induction of a similar protein in both tissues. *Journal of Biological Chemistry*, *261*, 6594–6599.
Riegelhaupt, J. J., Waase, M. P., Garbarino, J., Cruz, D. E., & Breslow, J. L. (2010). Targeted disruption of steroidogenic acute regulatory protein D4 leads to modest weight reduction and minor alterations in lipid metabolism. *Journal of Lipid Research*, *51*, 1134–1143.
Riggs, B. L., Wahner, H. W., Seeman, E., Offord, K. P., Dunn, W. L., Mazess, R. B., et al. (1982). Changes in bone mineral density of the proximal femur and spine with aging.

Differences between the postmenopausal and senile osteoporosis syndromes. *Journal of Clinical Investigation, 70,* 716–723.
Rohrmann, S., Platz, E. A., Selvin, E., Shiels, M. S., Joshu, C. E., Menke, A., et al. (2011). The prevalence of low sex steroid hormone concentrations in men in the Third National Health and Nutrition Examination Survey (NHANES III). *Clinical Endocrinology, 75,* 232–239.
Romeo, E., Auta, J., Kozikowski, A. P., Ma, D., Papadopoulos, V., Puia, G., et al. (1992). 2-Aryl-3-indoleacetamides (FGIN-1): A new class of potent and specific ligands for the mitochondrial DBI receptor (MDR). *Journal of Pharmacology and Experimental Therapeutics, 262,* 971–978.
Rone, M. B., Fan, J., & Papadopoulos, V. (2009). Cholesterol transport in steroid biosynthesis: Role of protein-protein interactions and implications in disease states. *Biochimica et Biophysica Acta, 1791,* 646–658.
Rone, M. B., Liu, J., Blonder, J., Ye, X., Veenstra, T. D., Young, J. C., et al. (2009). Targeting and insertion of the cholesterol-binding translocator protein into the outer mitochondrial membrane. *Biochemistry, 48,* 6909–6920.
Rone, M. B., Midzak, A. S., Issop, L., Rammouz, G., Jagannathan, S., Fan, J., et al. (2012). Identification of a dynamic mitochondrial protein complex driving cholesterol import, trafficking, and metabolism to steroid hormones. *Molecular Endocrinology, 26,* 1868–1882.
Rupprecht, R., Rammes, G., Eser, D., Baghai, T. C., Schule, C., Nothdurfter, C., et al. (2009). Translocator protein (18 kD) as target for anxiolytics without benzodiazepine-like side effects. *Science, 325,* 490–493.
Saad, F. (2009). The role of testosterone in type 2 diabetes and metabolic syndrome in men. *Arquivos Brasileiros de Endocrinologia e Metabologia, 53,* 901–907.
Sagara, Y., Takata, Y., Miyata, T., Hara, T., & Horiuchi, T. (1987). Cloning and sequence analysis of adrenodoxin reductase cDNA from bovine adrenal cortex. *Journal of Biochemistry, 102,* 1333–1336.
Scheufler, C., Brinker, A., Bourenkov, G., Pegoraro, S., Moroder, L., Bartunik, H., et al. (2000). Structure of TPR domain-peptide complexes: Critical elements in the assembly of the Hsp70-Hsp90 multichaperone machine. *Cell, 101,* 199–210.
Schule, C., Nothdurfter, C., & Rupprecht, R. (2014). The role of allopregnanolone in depression and anxiety. *Progress in Neurobiology, 113,* 79–87.
Shen, Y. H., Godlewski, J., Bronisz, A., Zhu, J., Comb, M. J., Avruch, J., et al. (2003). Significance of 14-3-3 self-dimerization for phosphorylation-dependent target binding. *Molecular Biology of the Cell, 14,* 4721–4733.
Sih, R., Morley, J. E., Kaiser, F. E., Perry, H. M., III, Patrick, P., & Ross, C. (1997). Testosterone replacement in older hypogonadal men: A 12-month randomized controlled trial. *Journal of Clinical Endocrinology and Metabolism, 82,* 1661–1667.
Sileikyte, J., Petronilli, V., Zulian, A., Dabbeni-Sala, F., Tognon, G., Nikolov, P., et al. (2011). Regulation of the inner membrane mitochondrial permeability transition by the outer membrane translocator protein (peripheral benzodiazepine receptor). *Journal of Biological Chemistry, 286,* 1046–1053.
Simpson, E. R., & Waterman, M. R. (1988). Regulation of the synthesis of steroidogenic enzymes in adrenal cortical cells by ACTH. *Annual Review of Physiology, 50,* 427–440.
Somani, B., Khan, S., & Donat, R. (2010). Screening for metabolic syndrome and testosterone deficiency in patients with erectile dysfunction: Results from the first UK prospective study. *BJU International, 106,* 688–690.
Sorgato, M. C., & Moran, O. (1993). Channels in mitochondrial membranes: Knowns, unknowns, and prospects for the future. *Critical Reviews in Biochemistry and Molecular Biology, 28,* 127–171.
Stahlman, J., Britto, M., Fitzpatrick, S., McWhirter, C., Testino, S. A., Brennan, J. J., et al. (2012). Effects of skin washing on systemic absorption of testosterone in hypogonadal

males after administration of 1.62% testosterone gel. *Current Medical Research and Opinion, 28*, 271–279.

Stocco, D. M., & Sodeman, T. C. (1991). The 30-kDa mitochondrial proteins induced by hormone stimulation in MA-10 mouse Leydig tumor cells are processed from larger precursors. *Journal of Biological Chemistry, 266*, 19731–19738.

Toker, A., Sellers, L. A., Amess, B., Patel, Y., Harris, A., & Aitken, A. (1992). Multiple isoforms of a protein kinase C inhibitor (KCIP-1/14-3-3) from sheep brain. Amino acid sequence of phosphorylated forms. *European Journal of Biochemistry, 206*, 453–461.

Tsujishita, Y., & Hurley, J. H. (2000). Structure and lipid transport mechanism of a StAR-related domain. *Nature Structural Biology, 7*, 408–414.

Tuck, S. P., & Francis, R. M. (2009). Testosterone, bone and osteoporosis. *Frontiers of Hormone Research, 37*, 123–132.

Twiddy, A. L., Leon, C. G., & Wasan, K. M. (2011). Cholesterol as a potential target for castration-resistant prostate cancer. *Pharmaceutical Research, 28*, 423–437.

Valenti, G. (2005). The pathway of partial androgen deficiency of aging male. *Journal of Endocrinological Investigation, 28*, 28–33.

Valero, R. J., Marquez, L. J., Campos, H. P., Puigvert, M. A., & Prieto, C. R. (2013). [Current recommendations about the diagnosis and treatment of testosterone deficit syndrome: Clinical guidelines]. *Archivos Españoles de Urología, 66*, 737–744.

Veldhuis, J. D. (2008). Aging and hormones of the hypothalamo-pituitary axis: Gonadotropic axis in men and somatotropic axes in men and women. *Ageing Research Reviews, 7*, 189–208.

Voos, W., & Rottgers, K. (2002). Molecular chaperones as essential mediators of mitochondrial biogenesis. *Biochimica et Biophysica Acta, 1592*, 51–62.

Vyssokikh, M. Y., & Brdiczka, D. (2003). The function of complexes between the outer mitochondrial membrane pore (VDAC) and the adenine nucleotide translocase in regulation of energy metabolism and apoptosis. *Acta Biochimica Polonica, 50*, 389–404.

Wang, C., Nieschlag, E., Swerdloff, R., Behre, H. M., Hellstrom, W. J., Gooren, L. J., et al. (2009). Investigation, treatment and monitoring of late-onset hypogonadism in males. *International Journal of Andrology, 32*, 1–10.

Wang, C., Swerdloff, R. S., Iranmanesh, A., Dobs, A., Snyder, P. J., Cunningham, G., et al. (2000). Transdermal testosterone gel improves sexual function, mood, muscle strength, and body composition parameters in hypogonadal men. *Journal of Clinical Endocrinology and Metabolism, 85*, 2839–2853.

Welsh, T. H., Jr., Bambino, T. H., & Hsueh, A. J. (1982). Mechanism of glucocorticoid-induced suppression of testicular androgen biosynthesis in vitro. *Biology of Reproduction, 27*, 1138–1146.

Wiedemann, N., Kozjak, V., Chacinska, A., Schonfisch, B., Rospert, S., Ryan, M. T., et al. (2003). Machinery for protein sorting and assembly in the mitochondrial outer membrane. *Nature, 424*, 565–571.

Winters, S. J., & Atkinson, L. (1997). Serum LH concentrations in hypogonadal men during transdermal testosterone replacement through scrotal skin: Further evidence that ageing enhances testosterone negative feedback. The Testoderm Study Group. *Clinical Endocrinology, 47*, 317–322.

Wu, F. C., Tajar, A., Beynon, J. M., Pye, S. R., Silman, A. J., Finn, J. D., et al. (2010). Identification of late-onset hypogonadism in middle-aged and elderly men. *The New England Journal of Medicine, 363*, 123–135.

Xiao, Y. C., Huang, Y. D., Hardy, D. O., Li, X. K., & Ge, R. S. (2010). Glucocorticoid suppresses steroidogenesis in rat progenitor Leydig cells. *Journal of Andrology, 31*, 365–371.

Xiao, B., Smerdon, S. J., Jones, D. H., Dodson, G. G., Soneji, Y., Aitken, A., et al. (1995). Structure of a 14-3-3 protein and implications for coordination of multiple signalling pathways. *Nature*, *376*, 188–191.

Yaffe, M. B., Rittinger, K., Volinia, S., Caron, P. R., Aitken, A., Leffers, H., et al. (1997). The structural basis for 14-3-3:phosphopeptide binding specificity. *Cell*, *91*, 961–971.

Yamazaki, T., Matsuoka, C., Gendou, M., Izumi, S., Zhao, D., Artemenko, I., et al. (2006). Mitochondrial processing of bovine adrenal steroidogenic acute regulatory protein. *Biochimica et Biophysica Acta*, *1764*, 1561–1567.

Yaworsky, D. C., Baker, B. Y., Bose, H. S., Best, K. B., Jensen, L. B., Bell, J. D., et al. (2005). pH-dependent interactions of the carboxyl-terminal helix of steroidogenic acute regulatory protein with synthetic membranes. *Journal of Biological Chemistry*, *280*, 2045–2054.

Young, J. C., & Hartl, F. U. (2000). Polypeptide release by Hsp90 involves ATP hydrolysis and is enhanced by the co-chaperone p23. *The EMBO Journal*, *19*, 5930–5940.

Young, J. C., Hoogenraad, N. J., & Hartl, F. U. (2003). Molecular chaperones Hsp90 and Hsp70 deliver preproteins to the mitochondrial import receptor Tom70. *Cell*, *112*, 41–50.

Zeth, K. (2010). Structure and evolution of mitochondrial outer membrane proteins of beta-barrel topology. *Biochimica et Biophysica Acta*, *1797*, 1292–1299.

CHAPTER EIGHT

Insulin Transport into the Brain and Cerebrospinal Fluid

Denovan P. Begg[1]
School of Psychology, University of New South Wales (UNSW, Australia), Sydney, New South Wales, Australia
[1]Corresponding author: e-mail address: d.begg@unsw.edu.au

Contents

Abstract

The pancreatic hormone insulin plays a well-described role in the periphery, based principally on its ability to lower circulating glucose levels via activation of glucose transporters. However, insulin also acts within the central nervous system (CNS) to alter a number of physiological outcomes ranging from energy balance and glucose homeostasis to cognitive performance. Insulin is transported into the CNS by a saturable receptor-mediated process that is proposed to be dependent on the insulin receptor. Transport of insulin into the brain is dependent on numerous factors including diet, glycemia, a diabetic state and notably, obesity. Obesity leads to a marked decrease in insulin transport from the periphery into the CNS and the biological basis of this reduction of transport remains unresolved. Despite decades of research into the effects of central insulin on a wide range of physiological functions and its transport from the periphery to the CNS, numerous questions remain unanswered including which receptor is responsible for transport and the precise mechanisms of action of insulin within the brain.

1. INTRODUCTION

Insulin, a 51-amino acid peptide hormone, is cleaved from preproinsulin and subsequently from proinsulin by proteases in the secretory vesicles of pancreatic β cells (Sanger & Tuppy, 1951a, 1951b). The primary function of insulin is reducing circulating glucose by activating glucose

Vitamins and Hormones, Volume 98
ISSN 0083-6729
http://dx.doi.org/10.1016/bs.vh.2014.12.007

transporters on cell membranes in peripheral tissues. The increase in glucose uptake can then be used as energy or stored as glycogen (Kitamura, Kahn, & Accili, 2003). While the role of insulin in the periphery was extensively examined following its discovery, there was little interest in (a) the effects of insulin in the central nervous system (CNS) and (b) the role of the CNS in the control of insulin secretion from the pancreas. This was due to two facts. First, in contrast to most other tissues, the brain can utilize glucose independent of the presence of glucose. Second, insulin is a large peptide hormone and it was simply too large to cross the blood–brain barrier (BBB; Park & Johnson, 1955). These are the reasons that for many decades it was deemed unlikely that insulin was relevant in the CNS.

From the late 1970s, it began to emerge that administration of exogenous insulin into the CNS produces multiple behavioral and biological effects including reductions in food intake (Woods, Lotter, McKay, & Porte, 1979), increased sympathetic output (Ward, Bardgett, Wolfgang, & Stocker, 2011), peripheral insulin secretion and glucose turnover (Obici, Zhang, Karkanias, & Rossetti, 2002), enhanced cognitive performance (Benedict et al., 2004), and increased energy expenditure (Muller et al., 1997). Given that almost all insulin is made in the pancreas, and there is no convincing evidence that insulin can be synthesized in the CNS of adult animals (Schwartz, Figlewicz, Baskin, Woods, & Porte, 1992), around four decades ago it was proposed that insulin passes from the blood to the cerebrospinal fluid (CSF) and then to the brain (Woods & Porte, 1976). It was suggested that insulin provides feedback to the brain as a hormonal indicator of peripheral adiposity.

The original hypothesis was that insulin first enters the CSF from plasma via the choroid plexus, then passes through the ependymal lining, and is free to act at insulin receptors on the adjacent neurons (Woods & Porte, 1976). This suggestion had merit, given that there are dense insulin-binding sites in the choroid plexus. Additionally, several nuclei in the ventral hypothalamus, which is located close to the wall of the third ventricle, contain high concentrations of insulin receptors (Baskin, Porte, Guest, & Dorsa, 1983; Havrankova, Roth, & Browstein, 1978; van Houten & Posner, 1979). Subsequent data examining the dynamics of insulin uptake into the CSF and brain determined that instead of entering the CNS via the choroid plexus and CSF, insulin is likely transported into the brain via a saturable, insulin receptor-mediated pathway (Banks, Jaspan, Huang, & Kastin, 1997; Baura et al., 1993, 1996). This chapter focuses on the studies of insulin transport into the brain/CSF and the effects of insulin within the CNS.

2. INSULIN SYNTHESIS, SECRETION, AND ACTION

The islets of Langerhans, in the pancreas, were first identified in the late-nineteenth century (Langerhans, 1869) and are richly innervated by sympathetic, parasympathetic, and sensory nerves (Ahren, 2000). While the islets account for less than 2% of pancreatic mass, they are highly vascularized and receive almost 20% of the blood-flow to the pancreas (Barbu et al., 2012; Jansson & Hellerstrom, 1983; Lifson, Kramlinger, Mayrand, & Lender, 1980). The vast majority of cells found in islets are insulin- and amylin-secreting β cells (Helmstaedter, Feurle, & Forssmann, 1976); however, there are also significant numbers of glucagon-secreting α and somatostatin-secreting δ cells. In addition, there are a small number of pancreatic polypeptide-secreting F-cells and ghrelin-secreting epsilon cells (Cooper et al., 1987; Helmstaedter et al., 1976; Kimmel, Hayden, & Pollock, 1975; Ku & Lee, 2005; Westermark et al., 1987). Despite a description of the islets as early as 1869, it would be a number of decades before their primary function was determined. Initially, von Mering and Minkowski (1889) reported that the removal of the pancreas produced a diabetic phenotype; it was subsequently found that crude pancreatic extracts produced moderate reductions in glycosuria (Piper, Allen, & Murlin, 1923). Then, in their ground-breaking work, Banting and Best (Banting, Best, Collip, Campbell, & Fletcher, 1922) demonstrated that insulin, isolated from pancreatic islets, could provide successful clinical treatment of type-1 diabetes. Insulin, purified from animal sources, rapidly became commercially available for the treatment of insulin-deficient (type-1) diabetes mellitus (Swann, 1986).

Preproinsulin, the polypeptide product of the *INS* gene, is processed in the β cells, a 24-amino acid signal peptide directs preproinsulin to the rough endoplasmic reticulum (ER). As preproinsulin is translocated to the lumen of the ER, the signal peptide is cleaved resulting in proinsulin. In the ER, three disulfide bonds are formed and the peptide is folded and undergoes Golgi transport and vesicle packaging. Active insulin is formed when endopeptidases cleave proinsulin at two positions, creating the C-peptide fragment, and leaving insulin, two peptide chains (A and B) that are connected via two disulfide bonds. Insulin is then packed into mature granules and awaits signals for release into the circulation.

Insulin release from secretory vesicles can be influenced by a number of factors; there is both a sustained low level release of insulin (basal insulin) that occurs throughout the day, vagally mediated cephalic releases of insulin that occur in anticipation of food intake (Ahren & Holst, 2001; Berthoud, Bereiter, Trimble, Siegel, & Jeanrenaud, 1981), and rapid releases that occur in response elevated metabolic signals, most commonly glucose but potentially also by amino acids (Floyd, Fajans, Conn, Knopf, & Rull, 1966). Thus, the exocytosis of insulin from the β-cell into the circulation is under both metabolic and neural control. At the β-cell, α-adrenergic stimulation inhibits secretion (Porte & Williams, 1966), whereas β-adrenergic agonists stimulate insulin release (Porte, 1967a, b), indicating that insulin secretion is under sympathetic control. Insulin secretion is also stimulated by acetylcholine (Campfield & Blocker, 1979; Campfield, Smith, & Eskinazi, 1984), demonstrating parasympathetic involvement. The cholinergic stimulation of insulin secretion is decreased in the presence of epinephrine (Campfield & Smith, 1983; Campfield, Smith, Settle, & Sohaey, 1986), indicating a complex neural parasympathetic and sympathetic control over β cells. In the case of metabolic stimulation of insulin secretion, glucose enters β cells through the GLUT2 glucose transporter, oxidation produces ATP which increases intracellular K^+ concentrations and a depolarization of the cell membrane. The cell depolarization opens voltage-gated Ca^{2+} channels, and the elevated intracellular Ca^{2+} levels then cause the release of insulin from the secretory vesicles (Kelly, McClenaghan, & Flatt, 2011; Wolf, Colca, Turk, Florholmen, & McDaniel, 1988).

In addition to the stimulation of insulin release metabolic signals, endocrine factors from the gut can potentiate the release of insulin from the pancreas. The most well-known example of this is glucagon-like peptide 1 (GLP-1), which promotes increased insulin secretion in the presence of glucose (Kreymann, Ghatei, Williams, & Bloom, 1987; Shima, Hirota, & Ohboshi, 1988). A number of other gastrointestinal hormones produce similar effects including cholecystokinin and glucose-dependent insulinotropic peptide (GIP; Andersen, Elahi, Brown, Tobin, & Andres, 1978; Begg & Woods, 2013; Szecowka, Lins, & Efendic, 1982). The incretins GLP-1 and GIP also act to inhibit glucagon release from α cells, furthering their glucose lowering capabilities (Ezcurra, Reimann, Gribble, & Emery, 2013; Gribble, 2012).

3. TRANSPORT OF INSULIN INTO THE CNS

For many decades, it was believed that insulin played no role in the CNS due to it being a large peptide and therefore unable to passively cross the BBB. In the mid-1950s, two studies described an inability of radioactively labeled insulin, injected in the periphery, to be detected in the brains of human or dogs (Elgee, Williams, & Lee, 1954; Haugaard, Vaughan, Haugaard, & Stadie, 1954). However, studies in the 1970s demonstrated that insulin was found in the CSF of numerous species (Stein et al., 1983; Woods & Porte, 1977). From these studies, important characteristics were identified including a strong relationship between the peripheral levels of insulin and levels in the CSF and a limit to the increase of CSF insulin when plasma insulin was at supraphysiological levels. Controversy over the source of CNS insulin continued into the late 1970s and early 1980s with the suggestion that insulin did not cross the BBB and any central insulin observed was produced within the CNS (Havrankova, Roth, & Brownstein, 1979). However, the evidence for production of insulin outside of the pancreas in mammals is weak; while there is evidence for insulin production in the olfactory epithelium of rats (Lacroix et al., 2008) and in a small subset of neurons from *Ins2* in the mouse (Mehran et al., 2012), these sources fail to account for CSF insulin levels and their close relationship to peripheral insulin.

Given that there is no significant source of brain-derived insulin and early data suggested that insulin could not cross the BBB, Woods and Porte (1976) hypothesized that insulin entered the CSF from plasma via the choroid plexus. The insulin could then pass through the ependymal lining and act at insulin receptors on nearby neurons. This was an understandable explanation given that there are dense insulin-binding sites in the choroid plexus and a number of nuclei in the ventral hypothalamus that contain significant numbers of insulin receptors are close to the wall of the CSF-containing third ventricle (Baskin, Porte, et al., 1983; Havrankova, Brownstein, & Roth, 1981; Havrankova et al., 1978; van Houten & Posner, 1979; van Houten, Posner, Kopriwa, & Brawer, 1980; Young, Kuhar, Roth, & Brownstein, 1980). However, when the transport of insulin was examined with more mechanistic studies throughout the 1990s, the dynamics of insulin uptake into the CSF and the brain were found to be different to the early theory. Rather than entering the CNS via the choroid plexus and CSF, insulin is transported into the brain via an insulin receptor-mediated, saturable pathway in brain capillary endothelial cells (Baura et al., 1993). Therefore, the normal transport of insulin through the CNS follows a three-compartment model (plasma to brain interstitial fluid to CSF; Banks, 2004; Banks, Jaspan, Huang, & Kastin, 1997; Schwartz et al., 1992). The transport of exogenous and radiolabeled insulin through brain capillary endothelial cells has been confirmed by numerous groups using a variety of methods (Banks, 2004; Banks, Jaspan, Huang, & Kastin, 1997). These findings demonstrated that insulin measured in the CSF has passed through the brain, not the choroid plexus. Insulin may be removed from the CNS with the reabsorption of CSF into the circulation or degraded by enzymes within the CNS such as insulin-degrading enzyme. Much of the research on the action of central insulin has focused on injection of insulin into the CSF; it is important to note that in order to engage neuronal insulin receptors, the insulin experimentally or therapeutically administered into the CSF must penetrate into the brain against the normal flow. The proposed mechanism of insulin transport into the brain is depicted in Fig. 1.

The saturable transport of insulin across the BBB has been demonstrated to be affected by many factors including obesity (Kaiyala, Prigeon, Kahn, Woods, & Schwartz, 2000), diet (Begg et al., 2013), glycemia and diabetes (Banks, Jaspan, & Kastin, 1997), nitric oxide and inflammation (Xaio, Banks, Niehoff, & Morley, 2001), and others (for review, see Banks, Owen, & Erickson, 2012), but unlike other hormones related to food intake (Kastin & Akerstrom, 2000; Schaeffer et al., 2013), insulin status is not altered by fasting (Strubbe, Porte, & Woods, 1988). Insulin transport into

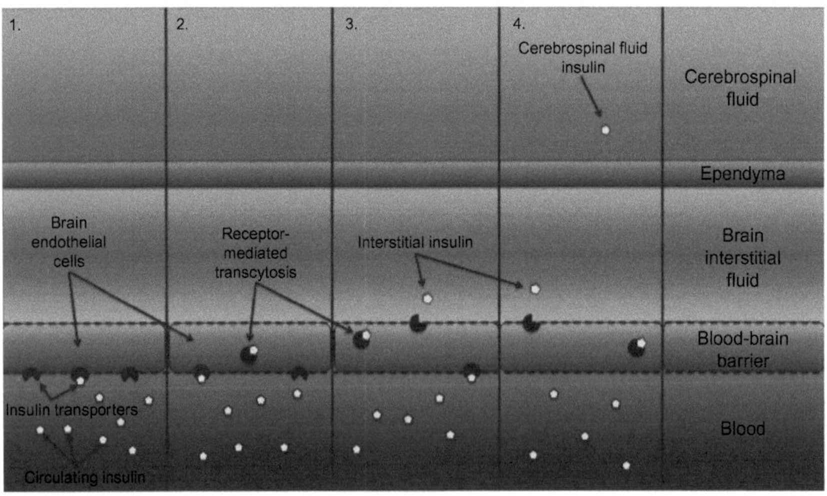

Figure 1 Transport of circulating insulin from the blood into the central nervous system. (1) Circulating insulin, released into the blood pancreatic β cells, binds to transporters on endothelial cells of the blood–brain barrier. (2) Insulin is transported through endothelial cells by receptor-mediated transcytosis. (3) Insulin is released into brain interstitial fluid, here it can act on neuronal insulin receptors. (4) Insulin then moves through the ependyma into the cerebrospinal fluid. (See the color plate.)

the CNS has been assumed to occur via the insulin receptor. This assumption has been made based on the circumstantial evidence, including that there are large numbers of insulin receptors on the endothelial cells of the BBB. Additionally, there is evidence that some analogous BBB transport occurs via the target receptor, i.e., when the receptor for tumor necrosis factor is knocked out, transport into the brain is eliminated (Pan & Kastin, 2002). However, it should be noted that evidence for this is mixed, for example, epidermal growth factor transport is not eliminated by inhibition of the target receptor (Pan & Kastin, 1999) and the data regarding leptin receptor knockout are also mixed (Banks, Niehoff, Martin, & Farrell, 2002; Kastin, Pan, Maness, Koletsky, & Ernsberger, 1999). Despite the continued interest in the CNS insulin system, and the fact that pancreatic insulin drives CNS insulin signaling, no studies have addressed the important issue of which protein enables insulin into the brain. There remains no direct evidence either in support or against involvement of the insulin receptor in the transport of insulin across the BBB.

The transport of insulin into the CNS appears to be somewhat dependent on the brain region of interest. Some regions of the brain having significantly greater accretion rates of insulin than others. For example, the area of the brain with the richest concentration of insulin receptors, the olfactory bulbs, also has a higher insulin concentration and greater insulin uptake (Hill, Lesniak, Pert, & Roth, 1986; Matsumoto & Rhoads, 1990). Some data also suggest that not all regions of the brain transport insulin (Banks & Kastin, 1998). However, given that significant differences in insulin uptake by the brain region have been observed between different strains of mice within the same lab, it is difficult to interpret how these data reflect the human condition.

4. THE EFFECTS OF INSULIN ON THE CNS

Insulin receptors are both abundant and widely distributed throughout the CNS (Hill et al., 1986). The distribution of insulin and the insulin receptor has been well characterized via immunohistochemistry (Baskin, Woods, et al., 1983), autoradiography (van Houten & Posner, 1979; van Houten, Posner, Kopriwa, & Brawer, 1979), and *in situ* hybridization of insulin receptor mRNA (Marks, Porte, Stahl, & Baskin, 1990). High levels of the insulin receptor are found in the olfactory bulbs, parts of the hypothalamus, the cerebral cortex, cerebellum, hippocampus, the choroid plexus, and regions of the lower brainstem (Baskin, Porte, et al., 1983; Baskin, Woods, et al., 1983; Marks & Eastman, 1990; Marks et al., 1990; van

Houten & Posner, 1979; van Houten et al., 1979). The widespread distribution of insulin receptors within the brain is evidence of the diverse central actions of insulin. Indeed, insulin effects within the CNS include roles in energy homeostasis (Benoit et al., 2002; Bruning et al., 2000; Woods et al., 1979), systemic glucose production (Fisher, Bruning, Lannon, & Kahn, 2005), reproductive development (Bruning et al., 2000), cognitive functioning (Stockhorst, de Fries, Steingrueber, & Scherbaum, 2004; Stockhorst, Huenig, Ziegler, & Scherbaum, 2011; van der Heide, Ramakers, & Smidt, 2006), and sympathetic nervous system activity (Rahmouni et al., 2004; Ward et al., 2011).

Central insulin administration has been demonstrated to reduce food intake in animal models when injected in to the third ventricle or the arcuate nucleus of the hypothalamus (Air, Benoit, Blake Smith, Clegg, & Woods, 2002; Begg et al., 2013; Clegg, Riedy, Smith, Benoit, & Woods, 2003) and also in humans following intranasal administration (Benedict, Kern, Schultes, Born, & Hallschmid, 2008; Hallschmid et al., 2004). These reductions in food intake are maintained as central insulin administration continues and lead to reduced body weight over time. Therefore, insulin within the brain has an opposing effect to peripheral insulin administration, which can lead to significant weight gain (Pontiroli, Miele, & Morabito, 2011). In addition, CNS insulin is involved in systemic glucose responses to meals and fluctuations of plasma glucose (Fisher et al., 2005; Levin, Magnan, Dunn-Meynell, & Le Foll, 2011; Sisley & Sandoval, 2011). The effects of CNS insulin on peripheral glucose may be due to insulin in the CNS providing a positive feedback loop to the pancreas, increasing peripheral insulin production. In addition, central injection of insulin reduces hepatic glucose production due to autonomic nervous system inhibition of gluconeogenesis (Obici et al., 2002).

Evidence for the role of brain insulin in cognitive function has been steadily growing. Early research demonstrated that levels of the insulin receptor are increased by memory consolidation in rat hippocampus (Zhao et al., 1999), and central insulin administration has been implicated both in place preference in the rat (Figlewicz, 2004) and object recognition in mice (Marks, Tucker, Cavallin, Mast, & Fadool, 2009). In humans, intranasal insulin improves verbal memory in both normal volunteers (Benedict et al., 2007) and patients with Alzheimer's disease (Craft et al., 1999, 2012). In addition to its role in cognitive function, downregulation of the insulin receptor in the hypothalamus leads to depressive-like behavior (Grillo et al., 2011), potentially implicating insulin in the pathophysiology of mood disorders.

5. CENTRAL INSULIN AND LEPTIN AS ADIPOSITY SIGNALS

The transport of insulin from the periphery to the CNS is mirrored by other peptides and growth factors, the most notable is leptin, which is produced in adipocytes and also crosses the BBB in a receptor-mediated manner. The insulin secreted from β cells and the leptin from adipocytes circulate in direct proportion to body fat levels (adiposity); leaner individuals have reduced insulin and leptin secretion relative to individuals with greater adiposity (Bagdade, 1968; Bagdade, Bierman, & Porte, 1967; Stephan, Reville, Thierry, & Schlienger, 1972; Woods, Decke, & Vasselli, 1974). Given that insulin and leptin levels in the blood increase when humans (Considine et al., 1996; Sims et al., 1973) or animals (Bernstein, Lotter, Kulkosky, Porte, & Woods, 1975; Weisinger et al., 2009) overeat and become fat, it was proposed that they enter into the CNS to act as a negative feedback controller of adiposity (Schwartz, Woods, Porte, Seeley, & Baskin, 2000). The concept is a simple form of homeostatic control; when an animal overeats, this leads to body weight gain and increased adiposity. The elevated levels of blood, and consequently brain, insulin and leptin provide a signal triggering corrective (negative feedback) responses (Begg & Woods, 2013). The increased insulin and leptin signaling cause the animals to eat less and return to a lower body weight. Alternatively, when an animal fasts, the reduced brain insulin and leptin signaling trigger a homeostatic reflex to increase food intake and regain the lost body weight. It is important to note that central insulin administration does not always promote reduced hypophagia and weight loss, rather brain insulin signaling acts to reduce the level of body adiposity that is maintained homeostatically; for example, central insulin infusion does not obligatorily reduce food intake but leads the animal to defend a lower body weight (Chavez, Kaiyala, Madden, Schwartz, & Woods, 1995). The data described have led to insulin and leptin commonly being referred to as adiposity signals and this is demonstrated in Fig. 2.

The concept of insulin as an adiposity signal was first tested experimentally in the previously noted experiment by Woods et al. (1979). This group infused insulin into the lateral cerebral ventricle of baboons and observed dose-dependent decreases in body weight associated with reduced food intake, over a several week period. The homeostatic nature of insulin was demonstrated by a reversal of these effects following cessation of insulin infusions. The general finding that central insulin administration reduces food

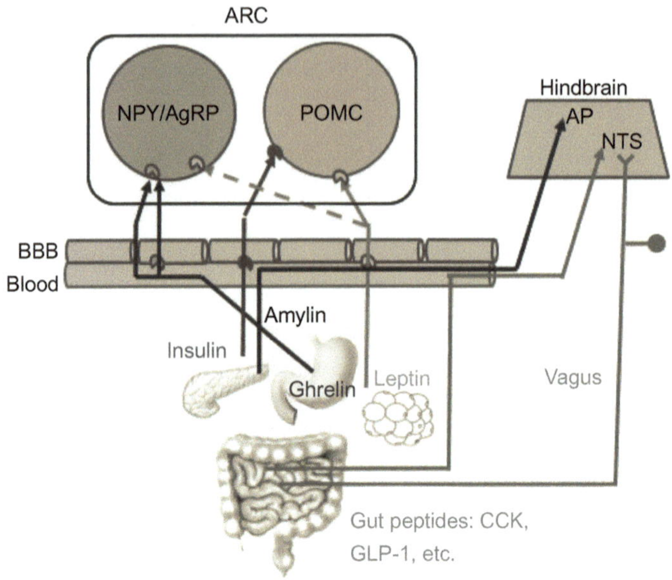

Figure 2 Numerous hormonal signals from the periphery influence different areas of the brain involved in food intake. The adiposity signal insulin, from the pancreas, interacts with a transporter on capillary endothelial cells comprising the blood–brain barrier (BBB), is transported into the brain and stimulates proopiomelanocortin (POMC) cells in the arcuate nucleus of the hypothalamus (ARC). The other adiposity signal leptin, produced by adipocytes, enters the brain via active transport and stimulates POMC cells while inhibiting neuropeptide-Y (NPY)/agouti-related peptide (AgRP) cells in the ARC. Ghrelin from the stomach has been reported to enter the brain not only via active transport, like insulin, but also via fenestrated capillaries during fasting, it stimulates NPY/AgRP cells in the ARC. Other factors act via the hindbrain to alter food intake, pancreatic amylin circulates directly to the area postrema (AP) in the hindbrain. Gut peptides such as glucagon-like peptide-1 (GLP-1) and cholecystokinin (CCK) either stimulate the nucleus of the solitary tract (NTS) directly or stimulate vagal afferent nerves whose axons end in the NTS. (See the color plate.)

intake and body weight has been replicated many times, most commonly in rats (Air, Benoit, Blake Smith, et al., 2002; Air, Benoit, Clegg, Seeley, & Woods, 2002; Ikeda et al., 1986; McGowan, Andrews, & Grossman, 1992) but also in mice (Brown, Clegg, Benoit, & Woods, 2006; Jaillard et al., 2009), chicks (Shiraishi et al., 2011), sheep (Foster, Ames, & Emery, 1991), and humans (Hallschmid et al., 2004). Similar weight-reducing effects have been observed with administration of the adiposity signal leptin (Bowen, Mitchell, & Harris, 2003; Seeley et al., 1997; Williams, Baskin, & Schwartz, 2006). An important factor is that the hypophagia and weight loss are not secondary outcomes to either aversion or decreased mobility following adiposity signal administration (Chavez, Kaiyala, et al., 1995; Chavez, Seeley, & Woods, 1995; Thiele et al., 1997).

Additional data consistent with the suggestion that insulin and leptin are adiposity signals come from studies of inhibition. For example, when insulin or leptin signaling in the CNS is reduced, by local administration of antibodies against each target, by genetically knocking out insulin or leptin receptors on neurons, or by interfering with the receptor mRNA message locally in the hypothalamus, animals have increased food intake and body weight (Bruning et al., 2000; Grillo et al., 2007; Leshan, Greenwald-Yarnell, Patterson, Gonzalez, & Myers, 2012; McGowan et al., 1992; Morton et al., 2003; Strubbe & Mein, 1977). This effect tends to be pronounced in leptin deficiency and more subtle with insulin inhibition (Bruning et al., 2000; Hill et al., 2010). In addition, insulin's crucial role in glucose homeostasis makes many insulin-receptor-deficient models unviable.

Administration of insulin into the CNS does not reduce food intake in genetically obese Zucker rats or animals made obese using diet-induced obesity models (Arase, Fisler, Shargill, York, & Bray, 1988; Chavez, Riedy, van Dijk, & Woods, 1996; Clegg et al., 2005, 2011; Ikeda et al., 1986). This is also true in obese humans receiving intranasal insulin (Hallschmid, Benedict, Schultes, Born, & Kern, 2008), demonstrating that obesity results in insulin resistance in the CNS as well as in peripheral tissues. The effects of central leptin administration are also suppressed in obesity (Seeley et al., 1996); however, this is arguably to a lesser degree. As previously mentioned, transport of insulin through the BBB is reduced by obesity (Israel et al., 1993); this also occurs for leptin (Banks, DiPalma, & Farrell, 1999).

Males are more sensitive to insulin's anorectic effect than females; this has been observed in both humans and animal models (Benedict et al., 2008; Clegg, Brown, Woods, & Benoit, 2006; Clegg et al., 2003). Interestingly, females are more sensitive to the anorectic effect of leptin administration, while females have higher circulating leptin and lower circulating insulin than comparably obese males (Clegg et al., 2006). These sex differences in hypophagia following central administration of insulin appear to be partially due to the actions of estrogen in the hypothalamus (Musatov et al., 2007).

Like insulin, hormones produced in the pancreas are highly neurally innervated, such that cholinergic stimulation of the pancreas can release not only insulin but also amylin, glucagon, and pancreatic polypeptide (Campfield & Blocker, 1979; Kaneto & Kosaka, 1974; Okita, Inui,

Baba, & Kasuga, 1997), whereas secretion of somatostatin by the pancreas is adrenergically controlled (Samols & Weir, 1979). Despite their often opposing effects on blood glucose, each of the islet hormones reduces food intake; however, the mechanisms controlling the reduced food intake differ greatly. While insulin, somatostatin, and pancreatic polypeptide act at the hypothalamus to decrease food intake (Aponte, Leung, Gross, & Yamada, 1984; Asakawa et al., 1999; Woods et al., 1979), glucagon acts on vagal afferents (Weatherford & Ritter, 1988) and amylin acts in the hindbrain at the area postrema (Potes et al., 2010). The neurally controlled release of pancreatic hormones, often referred to as cephalic release, generally occurs in anticipation of feeding. The cephalically released hormones appear to play an important role in preparation for food, preventing a glucose intolerant-like state when food is consumed. However, whether the cephalically driven increase act at the brain following release remains unresolved.

6. CONCLUSION AND FUTURE DIRECTIONS

Insulin acts throughout the periphery to reduce circulating glucose and to increase stored energy. However, within the brain, insulin is catabolic, reducing food intake, and increasing energy expenditure. Insulin is transported to the CNS by a saturable receptor-mediated process to act as an adiposity signal in a manner similar to the adipokine leptin. This transport can be affected by numerous factors including obesity, which results in impaired transport from the periphery to the brain. Despite decades of research into the role of central insulin and its transport, numerous questions remain including mechanisms by which transport occurs and the precise mechanisms of action.

REFERENCES

Ahren, B. (2000). Autonomic regulation of islet hormone secretion—Implications for health and disease. *Diabetelogia*, *43*(4), 393–410.

Ahren, B., & Holst, J. J. (2001). The cephalic insulin response to meal ingestion in humans is dependent on both cholinergic and noncholinergic mechanisms and is important for postprandial glycemia. *Diabetes*, *50*(5), 1030–1038.

Air, E. L., Benoit, S. C., Blake Smith, K. A., Clegg, D. J., & Woods, S. C. (2002). Acute third ventricular administration of insulin decreases food intake in two paradigms. *Pharmacology, Biochemistry & Behavior*, *72*(1–2), 423–429.

Air, E. L., Benoit, S. C., Clegg, D. J., Seeley, R. J., & Woods, S. C. (2002). Insulin and leptin combine additively to reduce food intake and body weight in rats. *Endocrinology*, *143*(6), 2449–2452.

Andersen, D. K., Elahi, D., Brown, J. C., Tobin, J. D., & Andres, R. (1978). Oral glucose augmentation of insulin secretion. Interactions of gastric inhibitory polypeptide with ambient glucose and insulin levels. *The Journal of Clinical Investigation, 62*(1), 152–161.

Aponte, G., Leung, P., Gross, D., & Yamada, T. (1984). Effects of somatostatin on food intake in rats. *Life Sciences, 35*(7), 741–746.

Arase, K., Fisler, J. S., Shargill, N. S., York, D. A., & Bray, G. A. (1988). Intracerebroventricular infusions of 3-OHB and insulin in a rat model of dietary obesity. *American Journal of Physiology, 255*, R974–R981.

Asakawa, A., Inui, A., Ueno, N., Fujimiya, M., Fujino, M. A., & Kasuga, M. (1999). Mouse pancreatic polypeptide modulates food intake, while not influencing anxiety in mice. *Peptides, 20*(12), 1445–1448.

Bagdade, J. D. (1968). Basal insulin and obesity. *Lancet, 2*, 630–631.

Bagdade, J. D., Bierman, E. L., & Porte, D., Jr. (1967). The significance of basal insulin levels in the evaluation of the insulin response to glucose in diabetic and nondiabetic subjects. *The Journal of Clinical Investigation, 46*, 1549–1557.

Banks, W. A. (2004). The source of cerebral insulin. *European Journal of Pharmacology, 490*, 5–12.

Banks, W. A., DiPalma, C. R., & Farrell, C. L. (1999). Impaired transport of leptin across the blood–brain barrier in obesity. *Peptides, 20*(11), 1341–1345.

Banks, W. A., Jaspan, J. B., Huang, W., & Kastin, A. J. (1997). Transport of insulin across the blood–brain barrier: Saturability at euglycemic doses of insulin. *Peptides, 18*, 1423–1429.

Banks, W. A., Jaspan, J. B., & Kastin, A. J. (1997). Effect of diabetes mellitus on the permeability of the blood-brain barrier to insulin. *Peptides, 18*(10), 1577–1584.

Banks, W. A., & Kastin, A. J. (1998). Differential permeability of the blood–brain barrier to two pancreatic peptides: Insulin and amylin. *Peptides, 19*(5), 883–889.

Banks, W. A., Niehoff, M. L., Martin, D., & Farrell, C. L. (2002). Leptin transport across the blood–brain barrier of the Koletsky rat is not mediated by a product of the leptin receptor gene. *Brain Research, 950*(1–2), 130–136.

Banks, W. A., Owen, J. B., & Erickson, M. A. (2012). Insulin in the brain: There and back again. *Pharmacology & Therapeutics, 136*(1), 82–93.

Banting, F. G., Best, C. H., Collip, J. B., Campbell, W. R., & Fletcher, A. A. (1922). Pancreatic extracts in the treatment of diabetes mellitus. *Canadian Medical Association Journal, 12*(3), 141–146.

Barbu, A., Johansson, A., Bodin, B., Kallskog, O., Carlsson, P. O., Sandberg, M., et al. (2012). Blood flow in endogenous and transplanted pancreatic islets in anesthetized rats: Effects of lactate and pyruvate. *Pancreas, 41*(8), 1263–1271.

Baskin, D. G., Porte, D., Jr., Guest, K., & Dorsa, D. M. (1983). Regional concentrations of insulin in the rat brain. *Endocrinology, 112*(3), 898–903.

Baskin, D. G., Woods, S. C., West, D. B., van, H. M., Posner, B. I., Dorsa, D. M., et al. (1983). Immunocytochemical detection of insulin in rat hypothalamus and its possible uptake from cerebrospinal fluid. *Endocrinology, 113*(5), 1818–1825.

Baura, G., Foster, D., Porte, D., Jr., Kahn, S. E., Bergman, R. N., Cobelli, C., et al. (1993). Saturable transport of insulin from plasma into the central nervous system of dogs in vivo: A mechanism for regulated insulin delivery to the brain. *The Journal of Clinical Investigation, 92*, 1824–1830.

Baura, G., Foster, D. M., Kaiyala, K., Porte, D., Jr., Kahn, S. E., & Schwartz, M. W. (1996). Insulin transport from plasma into the central nervous system is inhibited by dexamethasone in dogs. *Diabetes, 45*(1), 86–90.

Begg, D. P., Mul, J. D., Liu, M., Reedy, B. M., D'Alessio, D. A., Seeley, R. J., et al. (2013). Reversal of diet-induced obesity increases insulin transport into cerebrospinal fluid and restores sensitivity to the anorexic action of central insulin in male rats. *Endocrinology, 154*(3), 1047–1054.

Begg, D. P., & Woods, S. C. (2013). The endocrinology of food intake. *Nature Reviews. Endocrinology*, *9*(10), 584–597.
Benedict, C., Hallschmid, M., Hatke, A., Schultes, B., Fehm, H. L., Born, J., et al. (2004). Intranasal insulin improves memory in humans. *Psychoneuroendocrinology*, *29*(10), 1326–1334.
Benedict, C., Hallschmid, M., Schmitz, K., Schultes, B., Ratter, F., Fehm, H. L., et al. (2007). Intranasal insulin improves memory in humans: Superiority of insulin aspart. *Neuropsychopharmacology*, *32*(1), 239–243.
Benedict, C., Kern, W., Schultes, B., Born, J., & Hallschmid, M. (2008). Differential sensitivity of men and women to anorexigenic and memory-improving effects of intranasal insulin. *Journal of Clinical Endocrinology & Metabolism*, *93*(4), 1339–1344.
Benoit, S. C., Air, E. L., Coolen, L. M., Strauss, R., Jackman, A., Clegg, D. J., et al. (2002). The catabolic action of insulin in the brain is mediated by melanocortins. *Journal of Neuroscience*, *22*(20), 9048–9052.
Bernstein, I. L., Lotter, E. C., Kulkosky, P. J., Porte, D., Jr., & Woods, S. C. (1975). Effect of force-feeding upon basal insulin levels of rats. *Proceedings of the Society for Experimental Biology and Medicine. Society for Experimental Biology and Medicine (New York, N.Y.)*, *150*, 546–548.
Berthoud, H. R., Bereiter, D. A., Trimble, E. R., Siegel, E. G., & Jeanrenaud, B. (1981). Cephalic phase, reflex insulin secretion. Neuroanatomical and physiological characterization. *Diabetelogia*, *20*, 393–401.
Bowen, H., Mitchell, T. D., & Harris, R. B. (2003). Method of leptin dosing, strain, and group housing influence leptin sensitivity in high-fat-fed weanling mice. *American Journal of Physiology*, *284*(1), R87–R100.
Brown, L. M., Clegg, D. J., Benoit, S. C., & Woods, S. C. (2006). Intraventricular insulin and leptin reduce food intake and body weight in C57BL/6J mice. *Physiology & Behavior*, *30*, 687–691.
Bruning, J. C., Gautam, D., Burks, D. J., Gillette, J., Schubert, M., Orban, P. C., et al. (2000). Role of brain insulin receptor in control of body weight and reproduction. *Science*, *289*(5487), 2122–2125.
Campfield, L. A., & Blocker, D. C. (1979). Simulation of the autonomic neural control of insulin secretion. *Computers in Biology & Medicine*, *9*(3), 191–203.
Campfield, L. A., & Smith, F. J. (1983). Neural control of insulin secretion: Interaction of norepinephrine and acetylcholine. *American Journal of Physiology*, *244*(5), R629–R634.
Campfield, L. A., Smith, F. J., & Eskinazi, R. E. (1984). Glucose responsiveness and acetylcholine sensitivity of pancreatic beta-cells after vagotomy. *American Journal of Physiology*, *246*(6 Pt. 2), R985–R993.
Campfield, L. A., Smith, F. J., Settle, J. E., & Sohaey, R. (1986). Effect of the order of application of neural inputs on insulin secretion. *Advances in Experimental Medicine & Biology*, *211*, 343–349.
Chavez, M., Kaiyala, K., Madden, L. J., Schwartz, M. W., & Woods, S. C. (1995). Intraventricular insulin and the level of maintained body weight in rats. *Behavioral Neuroscience*, *109*, 528–531.
Chavez, M., Riedy, C. A., van Dijk, G., & Woods, S. C. (1996). Central insulin and macronutrient intake in the rat. *American Journal of Physiology*, *271*, R727–R731.
Chavez, M., Seeley, R. J., & Woods, S. C. (1995). A comparison between effects of intraventricular insulin and intraperitoneal lithium chloride on three measures sensitive to emetic agents. *Behavioral Neuroscience*, *109*(3), 547–550.
Clegg, D. J., Benoit, S. C., Reed, J. A., Woods, S. C., Dunn-Meynell, A., & Levin, B. E. (2005). Reduced anorexic effects of insulin in obesity-prone rats fed a moderate-fat diet. *American Journal of Physiology*, *288*(4), R981–R986.
Clegg, D. J., Brown, L. M., Woods, S. C., & Benoit, S. C. (2006). Gonadal hormones determine sensitivity to central leptin and insulin. *Diabetes*, *55*(4), 978–987.

Clegg, D. J., Gotoh, K., Kemp, C., Wortman, M. D., Benoit, S. C., Brown, L. M., et al. (2011). Consumption of a high-fat diet induces central insulin resistance independent of adiposity. *Physiology & Behavior, 103*(1), 10–16.
Clegg, D. J., Riedy, C. A., Smith, K. A., Benoit, S. C., & Woods, S. C. (2003). Differential sensitivity to central leptin and insulin in male and female rats. *Diabetes, 52*(3), 682–687.
Considine, R. V., Sinha, M. K., Heiman, M. L., Kriaucinas, A., Stephens, T. W., Nyce, M. R., et al. (1996). Serum immunoreactive-leptin concentrations in normal-weight and obese humans. *The New England Journal of Medicine, 334*, 292–295.
Cooper, G. J., Willis, A. C., Clark, A., Turner, R. C., Sim, R. B., & Reid, K. B. (1987). Purification and characterization of a peptide from amyloid-rich pancreases of type 2 diabetic patients. *Proceeding of the National Academy of Sciences of the United States of America, 84*(23), 8628–8632.
Craft, S., Asthana, S., Newcomer, J. W., Wilkinson, C. W., Matos, I. T., Baker, L. D., et al. (1999). Enhancement of memory in Alzheimer disease with insulin and somatostatin, but not glucose. *Archieves of General Psychiatry, 56*(12), 1135–1140.
Craft, S., Baker, L. D., Montine, T. J., Minoshima, S., Watson, G. S., Claxton, A., et al. (2012). Intranasal insulin therapy for Alzheimer disease and amnestic mild cognitive impairment: A pilot clinical trial. *Archieves of Neurology, 69*(1), 29–38.
Elgee, N. J., Williams, R. H., & Lee, N. D. (1954). Distribution and degradation studies with insulin I131. *The Journal of Clinical Investigation, 33*(9), 1252–1260.
Ezcurra, M., Reimann, F., Gribble, F. M., & Emery, E. (2013). Molecular mechanisms of incretin hormone secretion. *Current Opinion in Pharmacology, 13*(6), 922–927.
Figlewicz, D. P. (2004). Intraventricular insulin and leptin reverse place preference conditioned with high-fat diet in rats. *Behavioral Neuroscience, 118*(3), 479.
Fisher, S. J., Bruning, J. C., Lannon, S., & Kahn, C. R. (2005). Insulin signaling in the central nervous system is critical for the normal sympathoadrenal response to hypoglycemia. *Diabetes, 54*(5), 1447–1451.
Floyd, J. C., Jr., Fajans, S. S., Conn, J. W., Knopf, R. F., & Rull, J. (1966). Stimulation of insulin secretion by amino acids. *The Journal of Clinical Investigation, 45*(9), 1487–1502.
Foster, L. A., Ames, N. K., & Emery, R. S. (1991). Food intake and serum insulin responses to intraventricular infusions of insulin and IGF-I. *Physiology & Behavior, 50*(4), 745–749.
Gribble, F. M. (2012). The gut endocrine system as a coordinator of postprandial nutrient homoeostasis. *Proceedings of the Nutrition Society, 71*(4), 456–462.
Grillo, C. A., Piroli, G. G., Kaigler, K. F., Wilson, S. P., Wilson, M. A., & Reagan, L. P. (2011). Downregulation of hypothalamic insulin receptor expression elicits depressive-like behaviors in rats. *Behavioural Brain Research, 222*(1), 230–235.
Grillo, C. A., Tamashiro, K. L., Piroli, G. G., Melhorn, S., Gass, J. T., Newsom, R. J., et al. (2007). Lentivirus-mediated downregulation of hypothalamic insulin receptor expression. *Physiology & Behavior, 92*(4), 691–701.
Hallschmid, M., Benedict, C., Schultes, B., Born, J., & Kern, W. (2008). Obese men respond to cognitive but not to catabolic brain insulin signaling. *International Journal of Obesity, 32*(2), 275–282.
Hallschmid, M., Benedict, C., Schultes, B., Fehm, H. L., Born, J., & Kern, W. (2004). Intranasal insulin reduces body fat in men but not in women. *Diabetes, 53*, 3024–3029.
Haugaard, N., Vaughan, M., Haugaard, E. S., & Stadie, W. C. (1954). Studies of radioactive injected labeled insulin. *The Journal of Biological Chemistry, 208*(2), 549–563.
Havrankova, J., Brownstein, M., & Roth, J. (1981). Insulin and insulin receptors in the rodent brain. *Diabetologia, 20*, 268–273.
Havrankova, J., Roth, J., & Brownstein, M. J. (1979). Concentrations of insulin and insulin receptors in the brain are independent of peripheral insulin levels. Studies of obese and streptozotocin treated rodents. *The Journal of Clinical Investigation, 64*, 632–642.

Havrankova, J., Roth, J., & Browstein, M. (1978). Insulin receptors are widely distributed in the central nervous system of the rat. *Nature*, *272*, 827–829.

Helmstaedter, V., Feurle, G. E., & Forssmann, W. G. (1976). Insulin-, glucagon-, and somatostatin-immunoreactive endocrine cells in the equine pancreas. *Cell and Tissue Research*, *172*(4), 447–454.

Hill, J. M., Lesniak, M. A., Pert, C. B., & Roth, J. (1986). Autoradiographic localization of insulin receptors in rat brain: Prominence in olfactory and limbic areas. *Neuroscience*, *17*, 1127–1138.

Hill, J. W., Elias, C. F., Fukuda, M., Williams, K. W., Berglund, E. D., Holland, W. L., et al. (2010). Direct insulin and leptin action on pro-opiomelanocortin neurons is required for normal glucose homeostasis and fertility. *Cell Metabolism*, *11*(4), 286–297.

Ikeda, H., West, D. B., Pustek, J. J., Figlewicz, D. P., Greenwood, M. R., Porte, D., Jr., et al. (1986). Intraventricular insulin reduces food intake and body weight of lean but not obese Zucker rats. *Appetite*, *7*(4), 381–386.

Israel, P. A., Park, C. R., Schwartz, M. W., Green, P. K., Sipols, A. J., Woods, S. C., et al. (1993). Effect of diet-induced obesity and experimental hyperinsulinemia on insulin uptake into CSF of the rat. *Brain Research Bulletin*, *30*, 571–575.

Jaillard, T., Roger, M., Galinier, A., Guillou, P., Benani, A., Leloup, C., et al. (2009). Hypothalamic reactive oxygen species are required for insulin-induced food intake inhibition: An NADPH oxidase-dependent mechanism. *Diabetes*, *58*(7), 1544–1549.

Jansson, L., & Hellerstrom, C. (1983). Stimulation by glucose of the blood flow to the pancreatic islets of the rat. *Diabetologia*, *25*(1), 45–50.

Kaiyala, K., Prigeon, R. L., Kahn, S. E., Woods, S. C., & Schwartz, M. W. (2000). Obesity induced by a high-fat diet is associated with reduced brain insulin transport in dogs. *Diabetes*, *49*, 1525–1533.

Kaneto, A., & Kosaka, K. (1974). Stimulation of glucagon and insulin secretion by acetylcholine infused intrapancreatically. *Endocrinology*, *95*(3), 676–681.

Kastin, A. J., & Akerstrom, V. (2000). Fasting, but not adrenalectomy, reduces transport of leptin into the brain. *Peptides*, *21*(5), 679–682.

Kastin, A. J., Pan, W., Maness, L. M., Koletsky, R. J., & Ernsberger, P. (1999). Decreased transport of leptin across the blood–brain barrier in rats lacking the short form of the leptin receptor. *Peptides*, *20*, 1449–1453.

Kelly, C., McClenaghan, N. H., & Flatt, P. R. (2011). Role of islet structure and cellular interactions in the control of insulin secretion. *Islets*, *3*(2), 41–47.

Kimmel, J. R., Hayden, L. J., & Pollock, H. G. (1975). Isolation and characterization of a new pancreatic polypeptide hormone. *The Journal of Biological Chemistry*, *250*(24), 9369–9376.

Kitamura, T., Kahn, C. R., & Accili, D. (2003). Insulin receptor knockout mice. *Annual Review of Physiology*, *65*, 313–332.

Kreymann, B., Ghatei, M. A., Williams, G., & Bloom, S. R. (1987). Glucagon-like peptide-1 7-36: A physiological incretin in man. *Lancet*, *2*, 1300–1303.

Ku, S. K., & Lee, H. S. (2005). Distribution and frequency of endocrine cells in the pancreas of the ddY mouse: An immunohistochemical study. *European Journal of Histochemistry*, *49*(2), 125–130.

Lacroix, M. C., Badonnel, K., Meunier, N., Tan, F., Schlegel-Le Poupon, C., Durieux, D., et al. (2008). Expression of insulin system in the olfactory epithelium: First approaches to its role and regulation. *Journal of Neuroendocrinology*, *20*(10), 1176–1190.

Langerhans, P. (1869). *Beitrage zur mikroscopischen anatomie der bauchspeichel druse.* Inauguraldissertation, Berlin: Gustav Lange.

Leshan, R. L., Greenwald-Yarnell, M., Patterson, C. M., Gonzalez, I. E., & Myers, M. G., Jr. (2012). Leptin action through hypothalamic nitric oxide synthase-1-expressing neurons controls energy balance. *Nature Medicine*, *18*(5), 820–823.

Levin, B. E., Magnan, C., Dunn-Meynell, A., & Le Foll, C. (2011). Metabolic sensing and the brain: Who, what, where, and how? *Endocrinology, 152*(7), 2552–2557.
Lifson, N., Kramlinger, K. G., Mayrand, R. R., & Lender, E. J. (1980). Blood flow to the rabbit pancreas with special reference to the islets of Langerhans. *Gastroenterology, 79*(3), 466–473.
Marks, D. R., Tucker, K., Cavallin, M. A., Mast, T. G., & Fadool, D. A. (2009). Awake intranasal insulin delivery modifies protein complexes and alters memory, anxiety, and olfactory behaviors. *The Journal of Neuroscience, 29*(20), 6734–6751.
Marks, J. L., & Eastman, C. J. (1990). Ontogeny of insulin binding in different regions of the rat brain. *Developmental Neuroscience, 12*, 349–358.
Marks, J. L., Porte, D., Jr., Stahl, W. L., & Baskin, D. G. (1990). Localization of insulin receptor mRNA in rat brain by *in situ* hybridization. *Endocrinology, 127*, 3234–3236.
Matsumoto, H., & Rhoads, D. E. (1990). Specific binding of insulin to membranes from dendrodendritic synaptosomes of rat olfactory bulb. *Journal of Neurochemistry, 54*, 347–350.
McGowan, M. K., Andrews, K. M., & Grossman, S. P. (1992). Chronic intrahypothalamic infusions of insulin or insulin antibodies alter body weight and food intake in the rat. *Physiology & Behavior, 51*(4), 753–766.
Mehran, A. E., Templeman, N. M., Brigidi, G. S., Lim, G. E., Chu, K. Y., Hu, X., et al. (2012). Hyperinsulinemia drives diet-induced obesity independently of brain insulin production. *Cell Metabolism, 16*(6), 723–737.
Morton, G. J., Niswender, K. D., Rhodes, C. J., Myers, M. G., Jr., Blevins, J. E., Baskin, D. G., et al. (2003). Arcuate nucleus-specific leptin receptor gene therapy attenuates the obesity phenotype of Koletsky (fa(k)/fa(k)) rats. *Endocrinology, 144*(5), 2016–2024.
Muller, C., Voirol, M. J., Stefanoni, N., Surmely, J. F., Jequier, E., Gaillard, R. C., et al. (1997). Effect of chronic intracerebroventricular infusion of insulin on brown adipose tissue activity in fed and fasted rats. *International Journal of Obesity, 21*(7), 562–566.
Musatov, S., Chen, W., Pfaff, D. W., Mobbs, C. V., Yang, X. J., Clegg, D. J., et al. (2007). Silencing of estrogen receptor {alpha} in the ventromedial nucleus of hypothalamus leads to metabolic syndrome. *Proceeding of the National Academy of Sciences of the United States of America, 104*(7), 2501–2506.
Obici, S., Zhang, B. B., Karkanias, G., & Rossetti, L. (2002). Hypothalamic insulin signaling is required for inhibition of glucose production. *Nature Medicine, 8*(12), 1376–1382.
Okita, M., Inui, A., Baba, S., & Kasuga, M. (1997). Central cholinergic regulation of pancreatic polypeptide secretion in conscious dogs. *Journal of Endocrinology, 154*(2), 311–317.
Pan, W., & Kastin, A. J. (1999). Entry of EGF into brain is rapid and saturable. *Peptides, 20*(9), 1091–1098.
Pan, W., & Kastin, A. J. (2002). TNFalpha transport across the blood–brain barrier is abolished in receptor knockout mice. *Experimental Neurology, 174*(2), 193–200.
Park, C. R., & Johnson, L. H. (1955). Effect of insulin on transport of glucose and galactose into cells of rat muscle and brain. *American Journal of Physiology, 182*(1), 17–23.
Piper, H. A., Allen, R. S., & Murlin, J. R. (1923). Aqueous extracts of pancreas. II. Physical and chemical behavior of insulin. *The Journal of Biological Chemistry, 58*(1), 321.
Pontiroli, A. E., Miele, L., & Morabito, A. (2011). Increase of body weight during the first year of intensive insulin treatment in type 2 diabetes: Systematic review and meta-analysis. *Diabetes, Obesity and Metabolism, 13*(11), 1008–1019.
Porte, D., Jr. (1967a). Beta adrenergic stimulation of insulin release in man. *Diabetes, 16*(3), 150–155.
Porte, D., Jr. (1967b). A receptor mechanism for the inhibition of insulin release by epinephrine in man. *The Journal of Clinical Investigation, 46*(1), 86–94.

Porte, D., Jr., & Williams, R. H. (1966). Inhibition of insulin release by norepinephrine in man. *Science, 152*(3726), 1248–1250.
Potes, C. S., Turek, V. F., Cole, R. L., Vu, C., Roland, B. L., Roth, J. D., et al. (2010). Noradrenergic neurons of the area postrema mediate amylin's hypophagic action. *American Journal of Physiology, 299*(2), R623–R631.
Rahmouni, K., Morgan, D. A., Morgan, G. M., Liu, X., Sigmund, C. D., Mark, A. L., et al. (2004). Hypothalamic PI3K and MAPK differentially mediate regional sympathetic activation to insulin. *The Journal of Clinical Investigation, 114*(5), 652–658.
Samols, E., & Weir, G. C. (1979). Adrenergic modulation of pancreatic A, B, and D cells alpha-Adrenergic suppression and beta-adrenergic stimulation of somatostatin secretion, alpha-adrenergic stimulation of glucagon secretion in the perfused dog pancreas. *The Journal of Clinical Investigation, 63*(2), 230–238.
Sanger, F., & Tuppy, H. (1951a). The amino-acid sequence in the phenylalanyl chain of insulin. 2. The investigation of peptides from enzymic hydrolysates. *Biochemical Journal, 49*(4), 481–490.
Sanger, F., & Tuppy, H. (1951b). The amino-acid sequence in the phenylalanyl chain of insulin. I. The identification of lower peptides from partial hydrolysates. *Biochemical Journal, 49*(4), 463–481.
Schaeffer, M., Langlet, F., Lafont, C., Molino, F., Hodson, D. J., Roux, T., et al. (2013). Rapid sensing of circulating ghrelin by hypothalamic appetite-modifying neurons. *Proceeding of the National Academy of Sciences of the United States of America, 110*(4), 1512–1517.
Schwartz, M. W., Figlewicz, D. P., Baskin, D. G., Woods, S. C., & Porte, D., Jr. (1992). Insulin in the brain: A hormonal regulator of energy balance. *Endocrine Reviews, 13*, 387–414.
Schwartz, M. W., Woods, S. C., Porte, D., Jr., Seeley, R. J., & Baskin, D. G. (2000). Central nervous system control of food intake. *Nature, 404*, 661–671.
Seeley, R. J., van Dijk, G., Campfield, L. A., Smith, F. J., Burn, P., Nelligan, J. A., et al. (1996). Intraventricular leptin reduces food intake and body weight of lean rats but not obese Zucker rats. *Hormone and Metabolic Research, 28*(12), 664–668.
Seeley, R. J., Yagaloff, K. A., Fisher, S. L., Burn, P., Thiele, T. E., van Dijk, G., et al. (1997). Melanocortin receptors in leptin effects. *Nature, 390*(6658), 349.
Shima, K., Hirota, M., & Ohboshi, C. (1988). Effect of glucagon-like peptide-1 on insulin secretion. *Regulatory Peptides, 22*(3), 245–252.
Shiraishi, J., Yanagita, K., Fukumori, R., Sugino, T., Fujita, M., Kawakami, S., et al. (2011). Comparisons of insulin related parameters in commercial-type chicks: Evidence for insulin resistance in broiler chicks. *Physiology & Behavior, 103*(2), 233–239.
Sims, E. A., Danforth, E., Jr., Horton, E. S., Bray, G. A., Glennon, J. A., & Salans, L. B. (1973). Endocrine and metabolic effects of experimental obesity in man. *Recent Progress in Hormone Research, 29*, 457–496.
Sisley, S., & Sandoval, D. (2011). Hypothalamic control of energy and glucose metabolism. *Reviews in Endocrine and Metabolic Disorders, 12*(3), 219–233.
Stein, L. J., Dorsa, D. M., Baskin, D. G., Figlewicz, D. P., Ikeda, H., Frankmann, S. P., et al. (1983). Immunoreactive insulin levels are elevated in the cerebrospinal fluid of genetically obese Zucker rats. *Endocrinology, 113*(6), 2299–2301.
Stephan, F., Reville, P., Thierry, R., & Schlienger, J. L. (1972). Correlations between plasma insulin and body weight in obesity, anorexia nervosa and diabetes mellitus. *Diabetologia, 8*, 196–201.
Stockhorst, U., de Fries, D., Steingrueber, H. J., & Scherbaum, W. A. (2004). Insulin and the CNS: Effects on food intake, memory, and endocrine parameters and the role of intranasal insulin administration in humans. *Physiology & Behavior, 83*(1), 47–54.
Stockhorst, U., Huenig, A., Ziegler, D., & Scherbaum, W. (2011). Unconditioned and conditioned effects of intravenous insulin and glucose on heart rate variability in healthy men. *Physiology & Behavior, 103*, 31–38.

Strubbe, J. H., & Mein, C. G. (1977). Increased feeding in response to bilateral injection of insulin antibodies in the VMH. *Physiology & Behavior, 19*, 309–313.

Strubbe, J. H., Porte, D. J., & Woods, S. C. (1988). Insulin responses and glucose levels in plasma and cerebrospinal fluid during fasting and refeeding in the rat. *Physiology & Behavior, 44*, 205–208.

Swann, J. P. (1986). Insulin: A case study in the emergence of collaborative pharmacomedical research. *Pharmacy in History, 28*(2), 65–74.

Szecowka, J., Lins, P. E., & Efendic, S. (1982). Effects of cholecystokinin, gastric inhibitory polypeptide, and secretin on insulin and glucagon secretion in rats. *Endocrinology, 110*(4), 1268–1272.

Thiele, T. E., Van Dijk, G., Campfield, L. A., Smith, F. J., Burn, P., Woods, S. C., et al. (1997). Central infusion of GLP-1, but not leptin, produces conditioned taste aversions in rats. *American Journal of Physiology, 272*(2 Pt. 2), R726–R730.

van der Heide, L. P., Ramakers, G. M., & Smidt, M. P. (2006). Insulin signaling in the central nervous system: Learning to survive. *Progress in Neurobiology, 79*(4), 205–221.

van Houten, M., & Posner, B. I. (1979). Insulin binds to brain blood vessels in vivo. *Nature, 282*(5739), 623–625.

van Houten, M., Posner, B. I., Kopriwa, B. M., & Brawer, J. R. (1979). Insulin-binding sites in the rat brain: In vivo localization to the circumventricular organs by quantitative radioautography. *Endocrinology, 105*(3), 666–673.

van Houten, M., Posner, B. I., Kopriwa, B. M., & Brawer, J. R. (1980). Insulin binding sites localized to nerve terminals in rat median eminence and arcuate nucleus. *Science, 207*, 1081–1083.

von Mering, J. V., & Minkowski, O. (1889). Diabetes Mellitus nach pankreasextirpation. *Zbl Kiln Med, 10*, 393–394.

Ward, K. R., Bardgett, J. F., Wolfgang, L., & Stocker, S. D. (2011). Sympathetic response to insulin is mediated by melanocortin 3/4 receptors in the hypothalamic paraventricular nucleus. *Hypertension, 57*(3), 435–441.

Weatherford, S. C., & Ritter, S. (1988). Lesion of vagal afferent terminals impairs glucagon-induced suppression of food intake. *Physiology & Behavior, 43*(5), 645–650.

Weisinger, R. S., Stanley, T. K., Begg, D. P., Weisinger, H. S., Spark, K. J., & Jois, M. (2009). Angiotensin converting enzyme inhibition lowers body weight and improves glucose tolerance in C57BL/6J mice maintained on a high fat diet. *Physiology & Behavior, 98*, 192–197.

Westermark, P., Wernstedt, C., Wilander, E., Hayden, D. W., O'Brien, T. D., & Johnson, K. H. (1987). Amyloid fibrils in human insulinoma and islets of Langerhans of the diabetic cat are derived from a neuropeptide-like protein also present in normal islet cells. *Proceeding of the National Academy of Sciences of the United States of America, 84*(11), 3881–3885.

Williams, D. L., Baskin, D. G., & Schwartz, M. W. (2006). Leptin regulation of the anorexic response to glucagon-like peptide-1 receptor stimulation. *Diabetes, 55*(12), 3387–3393.

Wolf, B. A., Colca, J. R., Turk, J., Florholmen, J., & McDaniel, M. L. (1988). Regulation of Ca^{2+} homeostasis by islet endoplasmic reticulum and its role in insulin secretion. *American Journal of Physiology, 254*(2 Pt. 1), E121–E136.

Woods, S. C., Decke, E., & Vasselli, J. R. (1974). Metabolic hormones and regulation of body weight. *Psychological Review, 81*, 26–43.

Woods, S. C., Lotter, E. C., McKay, L. D., & Porte, D., Jr. (1979). Chronic intracerebroventricular infusion of insulin reduces food intake and body weight of baboons. *Nature, 282*, 503–505.

Woods, S. C., & Porte, D., Jr. (1976). Insulin and the set-point regulation of body weight. In D. Novin, G. A. Bray, & W. Wyrwichka (Eds.), *Hunger: Basic mechanisms and clinical implications* (pp. 273–280). New York, NY: Raven Press.

Woods, S. C., & Porte, D., Jr. (1977). Relationship between plasma and cerebrospinal fluid insulin levels of dogs. *American Journal of Physiology, 233*(4), E331–E334.

Xaio, H., Banks, W. A., Niehoff, M. L., & Morley, J. E. (2001). Effect of LPS on the permeability of the blood–brain barrier to insulin. *Brain Research, 896*, 36–42.

Young, W. S. I., Kuhar, M. J., Roth, J., & Brownstein, M. J. (1980). Radiohistochemical localization of insulin receptors in the adult and developing rat brain. *Neuropeptides, 1*, 15–22.

Zhao, W., Chen, H., Xu, H., Moore, E., Meiri, N., Quon, M., et al. (1999). Brain insulin receptors and spatial memory. Correlated changes in gene expression, tyrosine phosphorylation, and signaling molecules in the hippocampus of water maze trained rats. *The Journal of Biological Chemistry, 274*, 34893–34902.

CHAPTER NINE

Regulation of Hormone-Sensitive Renal Phosphate Transport

Jyothsna Gattineni*[,1], Peter A. Friedman†
*Department of Pediatrics, University of Texas Southwestern Medical Center, Dallas, Texas, USA
†Department of Pharmacology & Chemical Biology, University of Pittsburgh School of Medicine, Pittsburgh, Pennsylvania, USA
[1]Corresponding author: e-mail address: jyothsna.gattineni@utsouthwestern.edu

Contents

Abstract

Phosphate is essential for growth and maintenance of the skeleton and for generating high-energy phosphate compounds. Evolutionary adaptation to high dietary phosphorous in humans and other terrestrial vertebrates involves regulated mechanisms

Vitamins and Hormones, Volume 98
ISSN 0083-6729
http://dx.doi.org/10.1016/bs.vh.2015.01.002

assuring the efficient renal elimination of excess phosphate. These mechanisms prominently include PTH, FGF23, and Vitamin D, which directly and indirectly regulate phosphate transport. Disordered phosphate homeostasis is associated with pathologies ranging from kidney stones to kidney failure. Chronic kidney disease results in hyperphosphatemia, an elevated calcium × phosphate product with considerable morbidity and mortality, mostly associated with adverse cardiovascular events. This chapter highlights recent findings and insights regarding the hormonal regulation of renal phosphate transport along with imbalances of phosphate balance due to acquired or inherited diseases states.

1. INTRODUCTION

Phosphate plays a central and highly organized role in skeletal structure and cellular function in humans and other terrestrial vertebrates. Disordered phosphate metabolism, whether of genetic or disease origin, is associated with a variety of acute and chronic disabilities. Thus, the control of phosphate levels is essential to normal biological activity. The aim of this chapter is to present a current synopsis of the understanding of the mechanisms involved in regulating phosphate. Much of our understanding of renal phosphate transport has been reviewed frequently and comprehensively (Bergwitz & Juppner, 2010; Biber, Hernando, Forster, & Murer, 2009; Blaine, Weinman, & Cunningham, 2011; Murer & Biber, 2010; Virkki, Biber, Murer, & Forster, 2007). The present work is not intended to be an inclusive treatise on phosphate transport or homeostasis. Our intent principally lies on a translational approach that integrates cellular actions with clinical conditions. We begin this discussion with an appreciation of the biological chemistry of phosphate.

2. BIOLOGICAL FORMS OF PHOSPHATE

Phosphate, the oxidized form of phosphorous, is found in the body as both inorganic and organic compounds. Organic forms include phospholipids and various organic esters. Four moieties of inorganic orthophosphate can be present in biological solutions: H_3PO_4, $H_2PO_4^-$, HPO_4^{2-}, PO_4^{3-}. These different components are in a pH-dependent equilibrium. Only HPO_4^{2-} and $H_2PO_4^-$ are present at significant concentrations at physiological pH's. The Henderson relation determines the ratio of the two:

$$\frac{HPO_4^{2-}}{H_2PO_4^-} = 10^{pH - pK_a}$$

where pK_a, the dissociation constant, for phosphate is 6.8. Thus, at an extracellular pH of 7.4, the ratio approximates 4:1 with plasma phosphate having an intermediate valence of 1.8.[1] The abbreviation Pi is used here to refer to aggregate inorganic phosphate $\left[HPO_4^{2-} + H_2PO_4^-\right]$.

3. RENAL PHOSPHATE TRANSPORTERS

3.1 Physiology

Serum phosphate concentrations represent a balance between intestinal phosphate absorption and renal excretion, with bone serving as a storage depot as it does for calcium. Ninety percent of serum phosphate, whether ionized or complexed with Ca, Mg, or Na, is filtered at the glomerulus. The remaining 10% is bound to plasma proteins and is not filtered. Once filtered, only ionized phosphate is reabsorbed by tubular transport processes from the incipient urine; phosphate complexes are excreted. The cellular mechanisms responsible for renal phosphate transport are present primarily, if not exclusively, in proximal convoluted (S1, S2) and straight (S3) tubules and have been identified and highly characterized (Biber, Hernando, & Forster, 2013).

Phosphate reabsorption from the tubular fluid across the nephron is a two-step process, where Pi entry across apical brush border membranes is energetically coupled to the dissipative movement of Na. In this secondary active transport mechanism the favorable electrochemical driving force for Na entry energizes Pi uptake against its adverse electrochemical gradient. The driving force for Na^+ entry is provided by the continuous extrusion of Na^+ from the cytoplasm by the basolaterally located Na^+-K^+-ATPase and the maintenance of low intracellular Na^+ concentrations. Apical Na-coupled Pi entry is the rate-limiting step for absorption and is thus the site of endocrine regulatory action. This entry step has been well defined and its mechanisms extensively characterized and reviewed (Biber et al., 2013; Forster, Hernando, Biber, & Murer, 2006, 2013). Far less is known

[1] The net valence is calculated from the total number of negative charges, $2 \times 4 + 1$, divided by the number of molecules, $4 + 1$, as determined by the Henderson relation, viz., $9/5 = 1.8$. Thus, at pH 7.4, 1 mmol of phosphorous = 1.8 mEq.

about the diffusion of Pi within the cell or the mechanism of Pi efflux across basolateral membranes and back in to the circulation.

3.2 SLC34 transporters

NPT2A[2] (*SLC34A1*) and NPT2C (*SLC34A3*) are the principal proteins mediating apical phosphate uptake. Renal expression of Npt2a and Npt2c is limited to proximal tubules, where Npt2a is expressed in S1, S2, and S3 segments of the proximal nephron but Npt2c is found only in S1 segments (Nowik et al., 2008; Picard et al., 2010). Inactivation of murine Npt2a reduces serum Pi by 35% and is associated with threefold increases in Pi excretion (Beck et al., 1998) that is due to a 60–80% decrease of brush border Na-phosphate uptake (Hoag, Martel, Gauthier, & Tenenhouse, 1999; Segawa et al., 2009; Tenenhouse, Martel, Gauthier, Segawa, & Miyamoto, 2003). Thus, Npt2a mediates about 70% of apical Pi uptake, with Npt2c responsible for the bulk, but not all, of the remainder.

NPT2A consists of 12 membrane-spanning domains (Fig. 1). It transports 3 Na^+ ions per molecule of HPO_4^{2-} that carries two negative charges and hence is charge moving, or electrogenic. The residual 30% of Na-dependent Pi uptake following Npt2a ablation is due to Npt2c. Npt2c, in contrast to Npt2a, is electroneutral, moving 2 Na^+ ions for each molecule of Pi taken up. Despite the difference in stoichiometry between Npt2a and Npt2c, both preferentially transport divalent Pi; HPO_4^{2-}. Though Pi uptake by NPT2C is small in comparison with phosphate transport mediated by NPT2A, mutations of NPT2C are responsible for hereditary hypophosphatemic rickets with hypercalciuria (HHRH), an autosomal recessive form of hypophosphatemia characterized by profoundly reduced renal phosphate reabsorption and rickets accompanied by elevated serum levels of 1,25-dihydroxyvitamin D_3 (1,25$[OH]_2D_3$) leading to increased renal calcium excretion. This phenotype is not recapitulated in mice lacking *Slc34a3* (Lötscher et al., 1996) or animals genetically engineered to remove Npt2c from proximal tubules (Myakala et al., 2014). Constitutive deletion of *Slc34a3* causes neither hypophosphatemia nor hyperphosphaturia. Thus, NPT2C plays a more important role in phosphate economy in humans than in mice.

The mechanism regulating Na–Pi cotransporters differs from that of other Na-coupled transport proteins. Phosphate transport by SLC34 family

[2] By convention, human genes (italics) and gene products are designated by uppercase, whereas nonhuman forms are shown in lowercase.

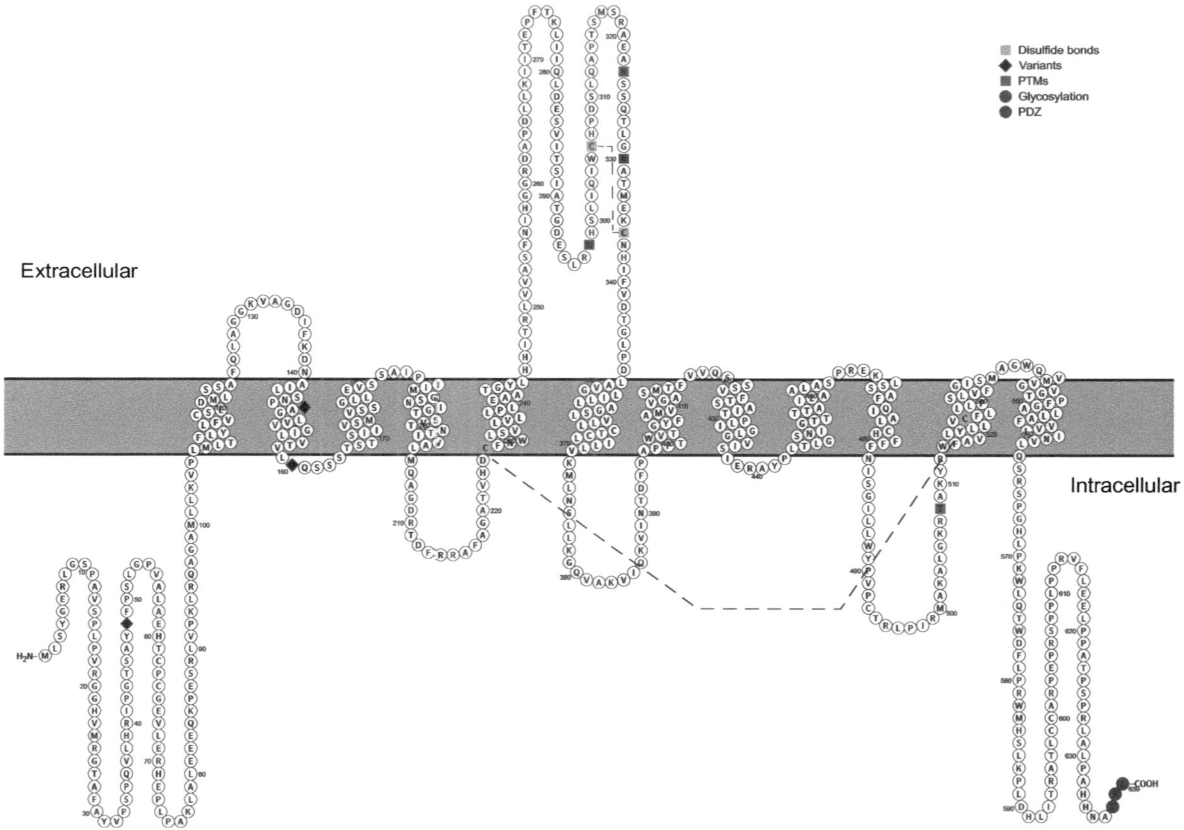

Figure 1 Structural organization of human NPT2A Na–Pi cotransporter. Described mutations (A48F, V147M), disulfide bridges, and the carboxy-terminal PDZ-binding domain are shown. Putative N-glycosylation sites are based on those described for the rat Npt2a (Hayes et al., 1994). The model was constructed using a hidden Markov model for topology prediction and visualized with Protter http://wlab.ethz.ch/protter/start/ (Omasits, Ahrens, Muller, & Wollscheid, 2014). (See the color plate.)

proteins is controlled by retrieval of the transporter from the cell membrane. That is, rather than altering affinity (K_m) or velocity (V_{max}) of the transporter, as with other Na-coupled uptake mechanisms, Pi transport by Npt2a and Npt2c is modulated by regulated endocytosis. Npt2a internalization proceeds through clathrin-coated pits in an actin-dependent fashion that involves myosin VI (Blaine et al., 2009; Mahon, 2008; Traebert, Roth, Biber, Murer, & Kaissling, 2000). Npt2a retrieval from brush border membranes begins within a few minutes of exposure to parathyroid hormone (PTH), with maximal internalization reaching some 60–75% by 2 h. Internalized Npt2a is trafficked from clathrin-coated pits to early endosomes (Bacic et al., 2006) and then degradation in lysozymes (Keusch et al., 1998; Murer, Hernando, Forster, & Biber, 2003; Pfister et al., 1998; Traebert, Roth, et al., 2000). Sorting to lysosomes may involve transit through the *trans*-Golgi network (Lanaspa et al., 2013). In contrast, Npt2c is not directed to lysozymes and seems to avoid degradation and may be recycled.

3.2.1 Npt2a-binding partners

Npt2a expresses a number of structural motifs that engage a variety of regulatory binding partners (Hernando et al., 2005). Other elements within Npt2a are subject to posttranscriptional modifications (Murer et al., 1999). Interestingly, though Npt2a harbors candidate intracellular serines, threonines, and lysines, neither phosphorylation nor ubiquitination, respectively, seem to play an important role in the behavior or fate of the transporter. Npt2a is constitutively phosphorylated (Jankowski, Hilfiker, Biber, & Murer, 2001), and PTH dephosphorylates the protein. Phosphatase inhibition failed to alter PTH-induced Npt2a downregulation leading to the conclusion that PTH-induced Npt2a dephosphorylation does not contribute to inhibition of Pi transport.

The most notable structural signature in NPT2A/Npt2a is a carboxy-terminal PDZ-recognition motif. The sequence signature consists of the form D/E-S/T-X-Φ, where X is any amino acid and Φ is a hydrophobic residue and is numbered from zero as the carboxy-terminal residue. Position −3 (D/E) is variably present. NPT2A/Npt2a harbor the sequence—TRL that mediates binding to NHERF1/EBP50 and other PDZ proteins importantly involved in regulation of Npt2a function. The mechanisms of interaction and function of PDZ proteins are discussed later in Section 7.

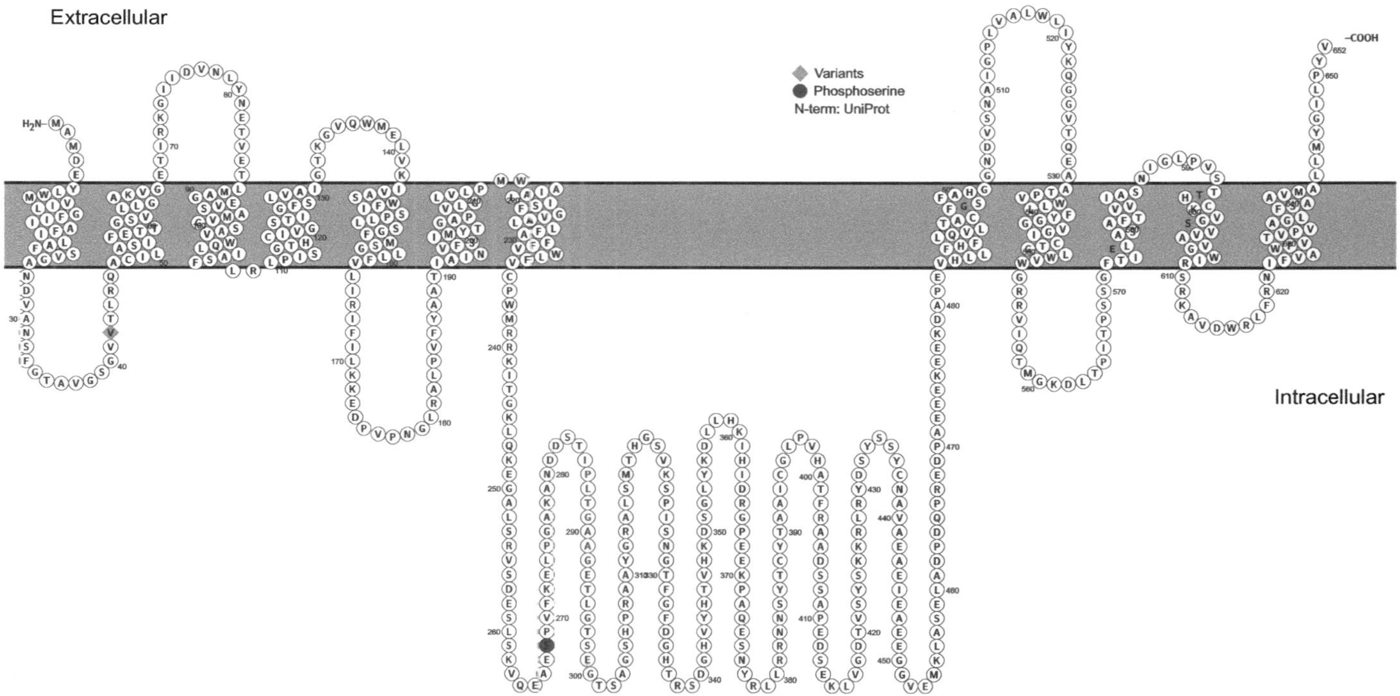

Figure 2 Human PIT-2 organization. The model was constructed as above and conforms to experimental findings (Bottger & Pedersen, 2011; Salaun, Rodrigues, & Heard, 2001). Naturally occurring variants are indicated in filled orange diamonds (Wang, Li, et al., 2012). (See the color plate.)

Table 1 Phosphate transporters

Type	Protein name	Gene symbol	Tissue localization	Size (AA)	Mass (Da)
Type 1	NPT1	*SLC17A1*	Kidney cortex, liver, and brain	467	51,132
Type 2	NPT2A	*SLC34A1*	Kidney	340	36,603
	NPT2B	*SLC34A2*	Intestine, lung, ovary, testis, prostate, and mammary gland	689	75,657
	NPT2C	*SLC34A3*	Kidney (spleen, brain, colon, breast, skin, ovary, and testis)[a]	599	63,550
Type 3	GLVR-1, PIT-1	*SLC20A1*	Ubiquitous	679	73,700
	RAM-1, PIT-2	*SLC20A2*	Ubiquitous	652	70,392

[a]GeneSeq.

3.3 SLC20 transporters

PIT-1 (Glvr-1, *SLC20A1*) and PIT-2 (Ram-1, *SLC20A2*) (Fig. 2) constitute the other family of phosphate transporters expressed in kidney, but elsewhere as well. Originally identified as retroviral receptors, they now form the type 3 Na–Pi cotransporters (Table 1). Transcripts for *Slc20a1* and *Slc20a2* represent just 1% of Na–Pi cotransporter mRNAs (Miyamoto et al., 1997; Tenenhouse, Gauthier, et al., 1998).

Slc20a1 and *Slc20a2* are both expressed in proximal and distal convoluted tubule cells, where they mediate Na-dependent Pi uptake (Tenenhouse, Roy, Martel, & Gauthier, 1998). Precise tubular localization remains uncertain for Pit-1. Pit-2 is expressed primarily in S1 proximal tubules (Picard et al., 2010), where it is found in brush border membranes (Breusegem et al., 2009). Pit-2, perhaps because of its more demonstrable regulatory modulation, is considered more physiologically significant. Pit-2 is importantly upregulated during phosphate deprivation (Villa-Bellosta et al., 2009). Pit-1 may serve a similar function (Nowik et al., 2008). A vital role for Pit-1 is underscored by the finding that eradication of *Slc20a1* is embryonically lethal due to hepatic hypoplasia, reduced hematopoiesis, and death from anemia (Beck et al., 2010).

PIT-1 and PIT-2 share certain structural features with NPT2A/C such as having 12 transmembrane-spanning domains but differ in others, such as both N- and C-termini being located at the extracellular cell surface in PIT1/2, whereas both are cytoplasmic in the type 2 transporters, NPT2A/B (Salaun et al., 2001). Structure–activity relations for PIT-2 have been characterized (Bottger & Pedersen, 2011). Heterozygous mutations in *SLC20A2* are associated with familial idiopathic basal ganglia calcification (Fahr's syndrome) that virtually eliminates Pi transport when assayed *in vitro* (Wang, Li, et al., 2012). Serum levels of calcium, phosphate, alkaline phosphatase, and PTH are normal.

Phosphoprotein analysis revealed the presence of specific phosphorylated serine residues in PIT-1 (268, 271, 273) and PIT-2 (256, 259, 316, 321, 385, 425, 436), as well as several tyrosine phosphorylation sites (PIT-1 421; PIT-2 386, 434) (Huttlin et al., 2010). The physiological role of these post-translational modifications has not been determined. It should be underscored that these are sites of constitutive phosphorylation and may be dephosphorylated upon hormonal challenge, and there are other residues that are not constitutively phosphorylated that may become so upon exposure to regulatory changes.

4. HORMONE REGULATION OF RENAL PHOSPHATE TRANSPORT

A variety of factors ranging from dietary influences, pH, hormones, and cytokines regulate renal Pi transport. By virtue of their diet, terrestrial vertebrates generally consume high levels of phosphorous and regulation is poised to eliminate this primarily through renal excretion by inhibiting phosphate reabsorption. Factors that diminish renal phosphate uptake do so by retrieving Npt2a or Npt2c from luminal brush border membranes. Influences that enhance phosphate reabsorption accomplish this by augmented transporter expression.

4.1 Parathyroid hormone

PTH is the principal endocrine regulator of renal phosphate excretion. PTH is synthesized as an 84-amino acid peptide. Decreases of extracellular calcium or increases of phosphate promote PTH release from parathyroid chief cells. While calcium causes rapid changes in PTH secretion, phosphate regulates secretion more slowly (Slatopolsky et al., 1996). Phosphate stimulates PTH release through direct and indirect mechanisms. Direct effects of phosphate on PTH release have been difficult to establish, perhaps because they seemingly require an intact parathyroid gland and cannot be demonstrated on dispersed parathyroid cells or primary cultures of parathyroid cells (Nielsen, Feldt-Rasmussen, & Olgaard, 1996; Slatopolsky et al., 1996). The fact that increases of PTH secretion are demonstrable only after several hours suggests that phosphate augments PTH release by enhancing the rate of PTH synthesis rather than by triggering secretion of pre-formed PTH. These effects of phosphate on PTH secretion are independent of calcium and Vitamin D (Kilav, Silver, & Naveh-Many, 1995). Phosphate also indirectly regulates PTH secretion through four counter-regulatory mechanisms, which are a compensatory adaption to other primary disturbances in mineral-ion homeostasis. First, hyperphosphatemia, by decreasing serum calcium, stimulates PTH secretion. Second, hyperphosphatemia inhibits renal $1,25[OH]_2D_3$ formation, which in turn decreases intestinal calcium absorption and leads to hypocalcemia and stimulation of PTH release. Third, $1,25[OH]_2D_3$ has direct inhibitory actions on PTH secretion. Hence, decreases of $1,25[OH]_2D_3$ in response to hyperphosphatemia promote PTH release. Finally, hyperphosphatemia stimulates parathyroid gland hyperplasia, which by increasing parathyroid gland mass, causes long-term increases in PTH secretion.

The liver and kidney rapidly metabolize PTH to an array of peptide fragments (Daugaard, 1996). Some of these peptide fragments retain biological activity, while others are inert or display partial agonism or are competitive antagonists at the PTH receptor (PTHR). Certain PTH fragments exhibit ligand-specific, or the so-called biased actions of, where some but not all the actions of PTH are retained (Sneddon et al., 2004). Some considerable attention has been paid to the effect of amino-truncated PTH fragments as they seem to accumulate in renal failure (Bringhurst, 2003; Brossard et al., 1996; D'Amour, 2012; Monier-Faugere et al., 2001).

PTHR activation by full agonists leads to stimulation of adenylyl cyclase, generation of cyclic AMP, and activation of protein kinase A (PKA), as well as phospholipase C and the protein kinase C (PKC) pathway. Semisynthetic PTH peptide fragments have been developed based on study of the structural determinants within PTH necessary for signaling through the PTHR. The findings established the framework for the generally accepted view that the first six residues of PTH are necessary for activating adenylyl cyclase and the attendant formation of cyclic AMP (cAMP) and downstream signaling through PKA or exchange protein directly activated by cAMP (EPAC). Peptide fragments such as (PTH[7-34]) lacking this amino-terminal sequence bind to the PTHR but do not activate adenylyl cyclase (Horiuchi, Holick, Potts, & Rosenblatt, 1983; Rosenblatt, 1986). Further analysis suggests that shorter amino-terminal deletions yielding PTH(3-34) are partial agonists, causing limited cAMP formation. Comparable studies of the PTH carboxy terminus indicates that the region 28–34 is needed for activation of PTHR signaling through the phospholipase C and PKC pathway (Friedman, Morley, Whitfield, Willick, & Gesek, 1996; Jouishomme et al., 1992, 1994). Application of these biased PTH peptides has proven useful in dissecting the mechanism of PTH action on renal Pi transport as described below.

4.1.1 PTH actions on renal Pi transport

PTH exerts its effects on the canonical type 1 PTHR. The PTHR is expressed at both apical and basolateral surfaces of kidney proximal tubules and at approximately equal density (Amizuka et al., 1997; Ba, Brown, & Friedman, 2003; Capuano et al., 2007; Guo, Song, Liu, & Mahon, 2012; Hilpert et al., 1999; Traebert, Völkl, Biber, Murer, & Kaissling, 2000). The relative action on Pi transport and the signaling pathways used by PTH from one cell surface or the other has been extensively explored but remains a subject of considerable controversy. The strategy generally employed to study this question involves growing opossum kidney cells to confluence and adding signaling-selective PTH ligands to one surface or the other, or using specific inhibitors of the various protein kinases implicated in the signaling mechanism, and measuring Pi uptake or transport across the epithelium (Pfister, Forgo, Ziegler, Biber, & Murer, 1999; Reshkin, Forgo, & Murer, 1990). Alternatively, intact proximal tubules have been studied with access and application of peptides to perfused luminal or serosal bath compartments (Traebert, Völkl, et al., 2000). The findings indicate that PTH(3-34), which largely lacks cAMP-activating capacity, internalizes Npt2a when applied to apical surfaces, whereas PTH(1-34) is effective from both epithelial membranes (Traebert, Völkl, et al., 2000). These peptide fragments exert comparable actions on Pi transport in OK cells (Cole, Forte, Eber, Thorne, & Poelling, 1988; Lederer, Sohi, Mathiesen, & Klein, 1998; Pfister et al., 1999). Taken together, this extensive and meticulously performed series of experiments suggest that at the luminal surface PTH inhibits Na–Pi cotransport primarily through a PKC-dependent pathway, whereas basolateral actions of PTH proceed principally through PKA. PTH exhibits about a 10-fold greater effect on Pi transport when applied to apical compared with the basolateral cell surface (Reshkin et al., 1990). This signaling bias may have its origins in the presence of the PDZ scaffold protein, NHERF1, as discussed later. Consistent with this conclusion is a battery of evidence that PKC inhibitors block PTH-dependent Pi transport (Cunningham et al., 2009). Perhaps needless to say, contrary results come to the opposite conclusion that inhibition of Pi transport by PTH follows the cAMP–PKA pathway as inhibitors of PLC and PKC did not suppress PTH action (K. J. Martin et al., 1994). Other studies using highly engineered PTH analogs that selectively activate adenylyl cyclase and PKA signaling point to a primary role of this pathway in regulating acute inhibition of renal phosphate transport (Nagai et al., 2011). More persistent inhibition of renal phosphate transport involves both PKA- and PKC-sensitive pathways (Guo et al., 2013; Nagai et al., 2011). Inherited disorders of pseudohypoparathyroidism types Ia or Ib, due to mutations in proximal tubule Gαs exhibit PTH resistance and hyperphosphatemia, thus pointing to the primacy of PKA signaling by the PTHR in regulating renal phosphate transport especially *in vivo* (Weinstein, Yu, Warner, & Liu, 2001). In isolated cell systems, considerable evidence points to the primacy of PKC signaling. Thus, PTH activates the PLC pathway at lower concentrations than are needed for stimulation of cyclic AMP (Murer & Biber, 1992; Quamme, Pfeilschifter, & Murer, 1989a). Of course, these studies were performed before the availability of biosensors for real-time and highly sensitive measurements of adenylyl cyclase activation. That said, it should be noted that many of these contradictory findings arise from experiments that were performed before bipolar membrane expression of the PTHR at apical and basolateral membranes of proximal tubule cells was appreciated or where PTH or PTH peptides were applied to nonpolarized OK cells. The target of PKA and PKC phosphorylation is not Npt2a, as once thought, but the scaffolding complex of NHERF1 and ezrin as discussed later.

As a corollary to analyzing the signaling pathways involved in PTH inhibition of phosphate transport mediated by Npt2a, PKA, and PKC pathways have been individually activated with forskolin and phorbol esters, respectively (Cole, Eber, Poelling, Thorne, & Forte, 1987; Malmstrom, Stange, & Murer, 1988). The results are consistent with the general view that both pathways are involved albeit to different degrees (Lederer et al., 1998; Pfister et al., 1999). A caveat of using OK cells for studies of PTH-sensitive phosphate transport is that a variety of sublines of these cells have been selected or generated and they exhibit distinct receptor and signaling characteristics (Cole, Forte, Krause, & Thorne, 1989; Silverstein, Spitzer, & Barac-Nieto, 2005). These considerable quantitative differences in action underscore a likely source of variability and for the contradictory findings and conclusions in the literature.

In addition to PKA and PKC, PTH stimulation leads to activation of extracellular signal-regulated kinases-1 and -2 (ERK1/2) (Cole, 1999; Lederer, Sohi, & McLeish, 2000; Sneddon, Gesek, & Friedman, 2000). Blockade of ERK1/2 suppresses PTH-inhibitable Npt2a endocytosis in mouse kidney, as well as internalization induced by direct activation of PKA, PKC, or cGMP pathways (Bacic, Schulz, et al., 2003). The results imply that that ERK1/2 lies downstream of these pathways and that all three pathways converge on ERK1/2.

A complementary strategy of using signaling-selective PTHR, where amino acids 317-320 (EKKY) in the second intracellular loop, were replaced with the sequence DSEL (Guo et al., 2010). This so-called DSEL receptor activates adenylyl cyclase and PKA normally but cannot activate the PLC–PKC pathway (Iida-Klein et al., 1997). DSEL mice exhibit an elevated serum phosphate concentration, suggesting a role for PKC signaling in the maintenance of phosphate homeostasis (Guo et al., 2010). PTH infusion resulted in a blunted phosphaturic response in the DSEL-mutant animals. A low-calcium diet exaggerated the hyperphosphatemia in DSEL mice compared to wild-type controls. Further, analysis of the *in vivo* role of PTHR signaling on phosphate balance was achieved by crossing DSEL knockin mice with animals lacking Npt2a or Npt2c (Guo et al., 2013). The results, which confirm the role of the PLC–PKC signaling pathway for long-term phosphate homeostasis, are not due to PTH resistance, *per se*, because PTH exerted normal regulatory effects on $1,25[OH]_2D_3$ (Guo et al., 2013). The conclusion that chronic regulation of Npt2a involves PKA and PKC pathways is consistent with the finding that both pathways are involved in PTH-dependent Npt2a mRNA downregulation (Murray, Khundmiri, Clark, & Lederer, 2014).

4.1.1.1 PKA/PKC inhibitors

Application of nominally selective pharmacological inhibitors of protein kinases has been a complementary approach to defining the signaling pathways employed by the PTHR to promote Npt2a endocytosis and thereby arrest Pi transport. It was appreciated at an early stage that PTH actions proceed through PKA-dependent and PKA-independent mechanisms (Cole et al., 1988). We recognize now that PKC and ERK1/2 represent the PKA-independent pathways utilized in PTHR signaling and inhibition of renal Pi transport. In some studies, inhibitors of PKC (calphostin C, chelerythrine, bisindolylmaleimide, GF109203X) blocked PTH-sensitive Pi transport (Cunningham et al., 2009), whereas other findings report preferential actions of PKA inhibitors. Notably, in OK cells where the PKC pathway was downregulated by phorbol ester treatment, PTH no longer inhibited Pi transport (Quamme, Pfeilschifter, & Murer, 1989), again emphasizing the dual signaling nature of PTH action on Npt2a.

PTH-related protein (PTHrP) exerts effects similar to the inhibitory action of PTH on renal Pi uptake by isolated brush border membrane vesicles (Muff, Caulfield, & Fischer, 1990; Saito, Hirano, Kubota, Tomino, & Koide, 1993). The biological importance of this action to the normal economy of phosphate is not known. Nor is it known if PTHrP exerts preferential or selective actions on apical or basolateral PTHR as does PTH.

4.2 Fibroblast growth factor 23

Fibroblast growth factor 23 (FGF23) belongs to the FGF19 subfamily of fibroblast growth factors (FGFs). FGF19 subfamily members (FGF15/19, FGF21, and FGF23) are classified as endocrine FGFs and unlike the other FGFs, which act in a paracrine or autocrine fashion. All FGFs have a heparin-binding site in their core sequence though with different affinities. The autocrine and paracrine FGFs require heparin sulfate (HS) proteoglycans to act as cofactors to bind to fibroblast growth factor receptors (FGFRs) in a 2:2:2 signaling complex stoichiometry of FGF:FGFR:HS. In contrast, FGF23 and FGF19 have decreased affinity for HS and thereby are able to circulate systemically and exert effects as endocrine hormones (Beenken & Mohammadi, 2012; Goetz et al., 2007). However, in the absence of a coreceptor, FGF19 family members have poor affinity for FGFRs. To circumvent this problem, Klotho evolved as the essential coreceptor for the endocrine FGF19 subfamily (Kurosu & Kuro, 2008; Kurosu & Kuro-o, 2008; Kurosu et al., 2006). There are four type-I high-affinity glycoprotein FGFRs characterized by an extracellular N-terminus and are encoded by four genes *Fgfr*1–*Fgfr*4. *Fgfr*1, 2, and 3 have two isoforms each, thereby making a total of seven FGFRs. Importantly, these different isoforms have conspicuous tissue specificity. A prototypical FGFR has a transmembrane domain, an immunoglobulin-like extracellular domain, and a tyrosine kinase intracellular domain (Fig. 3). Homodimerization of the FGFRs occurs in the presence of its ligand (FGFs) and coreceptor (heparin/Klotho), which in turn transphosphorylates the tyrosine kinase domains and initiates the cascade of intracellular signaling.

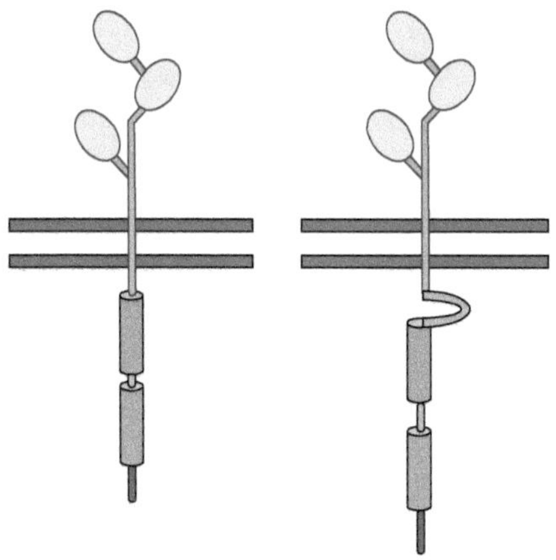

Figure 3 Features of FGFR. The extracellular domain contains 2 or 3 three immunoglobulin (IG)-like domains, a transmembrane domain, the absence or presence of a threonine–valine (TV) insert that defines a or b isoforms, and a split tyrosine kinase domain. Additionally, alternative splicing of the transcribed receptor genes results in a variety of receptor isoforms. Alternative splicing of the third IG-like domain occurs for FGFR1, FGFR2, and FGFR3. (See the color plate.)

Although FGFs can initiate signaling through many pathways including PLC-γ–PKC, CRKL, Ras-MAPK, and PI_3-Akt, the pathway most commonly used by FGF23 is the mitogen-activated protein kinase (MAPK) pathway. The MAPK pathway is activated by phosphorylation of FGF receptor substrate-2α (FRS2α), which is constitutively associated with the juxtamembrane domain of the FGFR (Beenken & Mohammadi, 2012).

FGF23 is a 32 kDa protein that is composed of 251 amino acids (aa) and is principally synthesized in the bone by osteocytes and osteoblasts. Low-FGF23 mRNA expression has been shown in other tissues including brain, thymus, small intestine, heart, and testis. However, the physiological role of FGF23 from these sites is unclear (Yoshiko et al., 2007). Secreted FGF23 is a potent phosphaturic hormone that causes renal phosphate wasting by decreasing the renal brush border expression of Npt2a and Npt2c in proximal tubules. Additionally, FGF23 decreases the circulating levels of 1,25 $[OH]_2D_3$ by decreasing the expression of 1α-hydroxylase and increasing the expression of 24-hydroxylase (Gattineni et al., 2009; Shimada, Hasegawa, et al., 2004). FGF23 has a signal sequence (24 aa), an N-terminal FGF core sequence (155 aa), and a unique C-terminal sequence (72 aa) (Goetz et al., 2007). Intact FGF23 is cleaved by yet poorly identified proteases at the site, $^{176}RXXR^{179}S^{180}$, where "R" is arginine and "S" is serine, to an 18 kDa N-terminal fragment and a 12 kDa C-terminal fragment

(Benet-Pagès et al., 2004; Shimada et al., 2002). Mature FGF23 protein is O-glycosylated by GALNT3 (UDP-*N*-acetyl-alpha-D-galactosamine: polypeptide *N*-acetylgalactosaminyl transferase 3) at many sites, especially at Thr^{178}, which confers stability to the FGF23 protein against proteolysis and aids in the secretion of intact FGF23. Recently, it was identified that Fam20C phosphorylates FGF23 at Ser^{180}, which inhibits FGF23 O-glycosylation, making FGF23 more prone to protease degradation (Tagliabracci et al., 2014). The N- and C-terminal FGF23 fragments are either biologically inactive (Benet-Pagès et al., 2004; Shimada et al., 2002) or can even act as antagonists. Isolated C-terminal FGF23(180-251), for instance, competes with intact FGF23 protein and inhibits FGF23 signaling. Administration of C-terminal FGF23 fragments to *Hyp* mice improved serum phosphate levels (Goetz et al., 2010). Using neutralizing antibodies to the N- and C-terminal fragments of FGF23, it was demonstrated that the N-terminal fragment of the intact FGF23 protein has high affinity for FGFR and the C-terminal for Klotho (Yamazaki et al., 2008). As discussed later, disorders of glycosylation, phosphorylation, and resistance to FGF23 degradation result in disturbances of phosphate homeostasis.

4.2.1 FGF23 and its interaction with FGFRs in regulating phosphate and 1,25 Vitamin D homeostasis

Humans and animal models of disorders of phosphate metabolism due to FGF23 dysregulation have been known for sometime but the receptors responsible for the actions of FGF23 eluded identification and have been the focus of intense research. *In vitro* studies identified FGF23 binding to FGFR1c, FGFR3c, and FGFR4 in the presence of Klotho (Kurosu et al., 2006). Using mouse models with deletions of various individual and multiple FGFRs, we and others narrowed down the receptors that interact with FGF23 *in vivo* to cause phosphaturia and low 1,25 Vitamin D (1,25 $[OH]_2D_3$) levels. Studying individual $Fgfr1^{-/-}$, $Fgfr3^{-/-}$, and $Fgfr4^{-/-}$ mice, it became apparent that Fgfr1 is the predominant receptor regulating the phosphaturic effects of FGF23, with a minor role for Fgfr4. However, baseline phosphate homeostasis was not disturbed in the individual $Fgfr^{-/-}$ mice, including the $Fgfr1^{-/-}$ mice (Gattineni et al., 2009; Liu, Vierthaler, Tang, Zhou, & Quarles, 2008). These data indicate that baseline compensatory mechanisms were in place when only a single FGF receptor was deleted. Compound deletion of $Fgfr3^{-/-}/Fgfr4^{-/-}$ resulted in resistance to the effects of FGF23 on $1{,}25[OH]_2D_3$ homeostasis implicating Fgfr3 and Fgfr4 in the regulation of $1{,}25[OH]_2D_3$ metabolism (Gattineni, Twombley, Goetz,

Mohammadi, & Baum, 2011; Li, Martin, David, & Quarles, 2011). Study of *Fgfr1*$^{-/-}$/*Fgfr4*$^{-/-}$ mice highlighted the complexity of the interaction of FGF23 with its receptors. *Fgfr1*$^{-/-}$/*Fgfr4*$^{-/-}$ is the first mouse model of Fgfr deletion with significantly elevated FGF23 levels (~50-fold), while the mice remain hyperphosphatemic. These data imply that Fgfr1 and Fgfr4 together are the principal receptors regulating the phosphaturic effects of FGF23. Notably, serum phosphate levels are not as elevated as *Fgf23*$^{-/-}$ or *Klotho*$^{-/-}$ mice indicating that even in *Fgfr1*$^{-/-}$/*Fgfr4*$^{-/-}$ mice there are compensatory mechanisms that prevent severe hyperphosphatemia. Interestingly, the 1,25 $[OH]_2D_3$ levels in *Fgfr1*$^{-/-}$/*Fgfr4*$^{-/-}$ mice are also elevated suggesting that Fgfr1, Fgfr3, and Fgfr4 together regulate the 1,25$[OH]_2D_3$ levels by FGF23 (Gattineni et al., 2014).

As discussed above, FGF23 in the presence of the FGFR and Klotho initiates intracellular signaling involving FRS2α and the extracellular signal-regulated kinases (ERK1/2). However, the precise mechanism of how this signaling pathway decreases the expression of Npt2a was unclear until recently when two groups independently showed that FGF23 signaling phosphorylates NHERF1 via activation of ERK1/2 and serum glucocorticoid-regulated kinase-1 (SGK1) (Andrukhova et al., 2012; Weinman, Steplock, Shenolikar, & Biswas, 2011). Phosphorylated NHERF1 causes disruption of its binding to Npt2a which results in endocytosis of Npt2a and decreased phosphate transport. Of note, Weinman et al. showed no difference in Npt2c protein expression in brush border membrane vesicles after treatment with FGF23 (10^{-9} M) and this could be due to the lower concentration of FGF23 used in the study (Weinman et al., 2011). Further studies are needed to elucidate the intracellular mechanisms responsible for the decreased expression of Npt2c, decreased expression of 1α-hydroxylase and increased expression of 24-hydroxylase by FGF23–FGFR–Klotho complex.

4.2.2 FGF23 and Klotho interaction

Deletion of FGF23 (*Fgf23*$^{-/-}$) or Klotho (*Klotho*$^{-/-}$) in mice results in an identical phenotype consisting of hyperphosphatemia, hypercalcemia, elevated 1,25$[OH]_2D_3$, vascular calcification, infertility, and shortened life span (Kuro-o et al., 1997; Shimada, Kakitani, et al., 2004). These murine models thus identified Klotho as an important coreceptor for the interaction of FGF23 with FGFRs. Although FGFRs are ubiquitously expressed, the actions of FGF23 are primarily limited to the kidney and the parathyroid glands; the limited expression of Klotho is responsible for this organ selectivity. The principal place of FGF23 action in the kidney is the proximal tubule, the site of phosphate transport and 1,25[OH]$_2$D$_3$ synthesis. However, Klotho is highly expressed in the distal tubule (Farrow, Davis, Summers, & White, 2009) with only limited expression in the proximal tubule (Andrukhova et al., 2012; Hu et al., 2010). Two possibilities have been advanced to explain the interaction of FGF23 with Klotho in the kidney; the first is that FGF23 interacts with Klotho in the distal tubule and the signal is somehow transmitted to the proximal tubule in a paracrine manner to decrease phosphate reabsorption and reduce the formation of 1,25 [OH]$_2$D$_3$. Alternatively, the second theory is that the limited Klotho expression is the proximal tubule is sufficient to augment the affinity of FGFRs to FGF23 and to initiate intracellular signaling. It is important to note that these two mechanisms are not mutually exclusive.

4.2.3 Regulation of FGF23

The regulation of FGF23 is quite complex due to the interplay of systemic factors that regulate mineral-ion homeostasis and local factors that are involved in bone mineralization. This is a focal point of intense research.

Although the role of FGF23 in regulating phosphate homeostasis is evident from dysregulation of FGF23 in humans and in animal models, the regulation of FGF23 by phosphate is poorly understood. Oral phosphate loading in healthy human subjects increased FGF23 levels and vice versa with phosphate restriction, though this is not a consistent finding (Antoniucci, Yamashita, & Portale, 2006; Burnett et al., 2006; Ferrari, Bonjour, & Rizzoli, 2005; Larsson, Nisbeth, Ljunggren, Juppner, & Jonsson, 2003; Vervloet et al., 2011). Recently, both intravenous and oral phosphate loading in healthy volunteers demonstrated elevated serum phosphate and FGF23 levels (Scanni, vonRotz, Jehle, Hulter, & Krapf, 2014). Animal studies complement the human data, showing that dietary phosphate restriction decreases serum FGF23 levels and phosphate loading increases FGF23 levels (Perwad et al., 2005; Yu, Sabbagh, Davis, Demay, & White, 2005). However, it is unclear how the osteocytes/osteoblasts sense phosphate load/serum phosphate to increase FGF23 production as a phosphate sensor is yet to be identified.

Although PTH and FGF23 both are phosphaturic hormones, they have contrasting actions on serum 1,25[OH]$_2$D$_3$ levels. FGF23 decreases and PTH increases 1,25[OH]$_2$D$_3$ levels. The role of FGF23 in regulating PTH seems to be uncertain. On the one hand, FGF23 decreases PTH mRNA and PTH secretion in both *in vitro* and *in vivo* studies (Ben-Dov et al., 2007; Krajisnik et al., 2007); on the other hand, transgenic mice-overexpressing FGF23 have elevated PTH levels (Bai, Miao, Li, Goltzman, & Karaplis, 2004; Larsson et al., 2004). These differences could be due to opposing effects of hypophosphatemia and low 1,25$[OH]_2D_3$ levels on PTH. The role of PTH in regulating FGF23 also remains unclear and controversial. FGF23 increases in animal models of primary hyperparathyroidism (Kawata et al., 2007) and in Jansen's metaphyseal chondrodysplasia (Brown et al., 2009), but FGF23 is not consistently elevated in humans with hyperparathyroidism (Kobayashi et al., 2006; Singh & Kumar, 2003; Tebben, Singh, Clarke, & Kumar, 2004). Parathyroidectomy in a mouse model of chronic kidney disease (CKD) prevents the increase in FGF23, and total parathyroidectomy in patients with secondary hyperparathyroidism due to CKD stage V decreased FGF23 levels (Lavi-Moshayoff, Wasserman, Meir, Silver, & Naveh-Many, 2010; Sato et al., 2004). Continuous infusion of PTH for 48 h results in elevated levels of FGF23 in healthy controls and in patients on dialysis and is not affected by bone histology (Wesseling-Perry et al., 2010). The contradictory results on the regulation of FGF23 by PTH could once again be secondary to modifying factors like serum phosphate and 1,25$[OH]_2D_3$ levels on PTH and FGF23. Additionally, this complex regulation may also depend on the bone turnover rate (Samadfam, Richard, Nguyen-Yamamoto, Bolivar, & Goltzman, 2009).

1,25$[OH]_2D_3$ regulation of FGF23 seems to be well understood and occurs in a Vitamin D receptor (VDR)-dependent and -independent fashion. Administration of 1,25$[OH]_2D_3$ to mice results in acute increases in *Fgf23* mRNA and FGF23 protein. Additionally, 1,25$[OH]_2D_3$ can directly stimulate FGF23 production in osteocytes through a Vitamin D responsive element (VDRE) on the *Fgf23* promoter (Liu et al., 2006). VDR-independent regulation of FGF23 was demonstrated in $VDR^{-/-}$ mice, where baseline FGF23 levels are low, as expected, due to elevated 1,25 $[OH]_2D_3$ levels. However, correction of hypocalcemia and hypophosphatemia via dietary manipulation results in elevated FGF23 levels (Yu et al., 2005). Thus, FGF23 decreases 1,25$[OH]_2D_3$, and 1,25 $[OH]_2D_3$ in turn increases FGF23 production creating a counter-regulatory feedback loop.

Klotho deficiency in humans and rodents results in extremely elevated levels of FGF23 (Ichikawa et al., 2007; Segawa et al., 2007). Possible explanations for the high FGF23 levels include hyperphosphatemia, elevated 1,25 $[OH]_2D_3$ levels, and/or a consequence of the absence of an essential coreceptor resulting in elevated levels of FGF23. The hyperphosphatemia and elevated 1,25[OH]$_2$D$_3$ levels improved significantly when *Klotho*$^{-/-}$ mice were placed on a Vitamin D-deficient diet (Tsujikawa, Kurotaki, Fujimori, Fukuda, & Nabeshima, 2003). FGF23 and phosphorus levels improved when *Klotho*$^{-/-}$ mice were placed on a low phosphate diet, supporting the hypothesis that hyperphosphatemia and hypervitaminosis D are responsible for elevated FGF23 levels in *Klotho*$^{-/-}$ mice (Segawa et al., 2007). Conversely, overexpression of circulating Klotho in rodents increases FGF23 levels (Smith et al., 2012). These data raise important unanswered questions in regards to the regulation of FGF23 by Klotho.

Calcium regulation of FGF23 is also complex with conflicting results. Addition of calcium to osteoblastic cell culture media fails to affect FGF23 promoter activity (Liu et al., 2006). However, when placed on a low-calcium, low Vitamin D diet, rats developed hypocalcemia and did not increase their FGF23 levels despite elevated PTH and 1,25[OH]$_2$D$_3$ levels (Rodriguez-Ortiz et al., 2012). Furthermore, baseline FGF23 levels were undetectable in *VDR*$^{-/-}$ mice despite elevated 1,25[OH]$_2$D$_3$ and PTH levels but with dietary supplementation of calcium and correction of calcium, FGF23 levels increased (Shimada et al., 2005). Further studies using *Cyp27b1*$^{-/-}$ and *Gcm2*$^{-/-}$ mice elegantly demonstrated that calcium plays a critical role in modulating the effects of 1,25[OH]$_2$D$_3$ and PTH on FGF23 regulation (David et al., 2013). It is appropriate that in the presence of hypocalcemia, elevated 1,25[OH]$_2$D$_3$ levels do not stimulate FGF23 production as elevated FGF23 levels will decrease 1,25[OH]$_2$D$_3$ levels resulting in worsening hypocalcemia, which would be detrimental to the organism.

Bone, the site of FGF23 synthesis, is increasingly recognized as an endocrine organ. Bone turnover releases calcium and phosphate to the systemic circulation, while bone mineralization requires calcium and phosphate to be deposited in the bone. Understandably, the changes in bone mineralization would necessitate alterations in the serum levels of calcium and phosphate and thus the need for modulating FGF23 levels. It seems logical that FGF23 production is linked with bone mineralization because FGF23 regulates phosphate and 1,25[OH]$_2$D$_3$ homeostasis, which directly and indirectly provide the building blocks for bone formation, thereby establishing the kidney–bone axis. Inactivating mutation of *Phex*, *Dmp1*, and *Enpp1* in rodents cause bone mineralization defects (Liu et al., 2009; Liu, Tang, Zhou, Vierthaler, & Quarles, 2007; Martin et al., 2008, 2011) and result in elevated FGF23 levels. Phex and DMP1 regulate *Fgf23* transcription in bone via Fgfr1 (Martin et al., 2011). Of interest is that defective mineralization seen in Vitamin D-deficient nutritional rickets is not associated with elevated FGF23 levels, and FGF23 levels decrease further when these patients are treated with Vitamin D supplements. Thus, defective bone mineralization alone is not a sufficient signal for increased FGF23 synthesis. The low-FGF23 levels during the treatment phase may be due to the fact that bone is being mineralized, which requires the need to be in positive phosphate and calcium balance (Quarles, 2012; Uzum et al., 2010).

The potential of iron regulating FGF23 is gaining clinical interest but again remains controversial and needs further study. Intact FGF23 (iFGF23) and c-terminal FGF23 (cFGF23) negatively correlate with iron levels in patients with autosomal-dominant hypophosphatemic rickets (ADHR) (Imel et al., 2011). Using an ADHR mouse model, in the presence of iron deficiency, wild-type mice demonstrate normal iFGF23 levels with increased cFGF23, whereas ADHR mice showed elevated iFGF23 and cFGF23 levels (Farrow et al., 2011). These findings indicate that iron possibly plays a role in FGF23 synthesis. Conversely, some forms of iron infusion in humans cause elevated FGF23 levels and hypophosphatemia but others have not (Gravesen, Hofman-Bang, Mace, Lewin, & Olgaard, 2013; Hryszko, Rydzewska-Rosolowska, Brzosko, Koc-Zorawska, & Mysliwiec, 2012; Schouten, Hunt, Livesey, Frampton, & Soule, 2009; Yamamoto, Okada, Mori, Fukumoto, & Tanaka, 2012).

4.3 1,25$[OH]_2$ Vitamin D_3

The sources of Vitamin D are dermal exposure to sunlight and dietary ingestion. Vitamin D is first 25-hydroxylated in the liver and then additionally hydroxylated primarily in the kidney at carbon-1 to its active form, 1,25 $[OH]_2D_3$, by 25-hydroxyvitamin D_3-1-α-hydroxylase (*CYP27B1*, 1α-hydroxylase). Under physiological conditions, the kidneys are the major source of circulating 1,25$[OH]_2D_3$, and even though other tissues make 1,25$[OH]_2D_3$, the extrarenal 1,25$[OH]_2D_3$ acts in a paracrine/autocrine fashion (Dusso, Brown, & Slatopolsky, 2005). 1α-Hydroxylase is highly regulated by calcium, PTH, phosphate, FGF23, and Klotho (Garabedian, Holick, DeLuca, & Boyle, 1972; Gattineni et al., 2011; Omdahl, Gray, Boyle, Knutson, & DeLuca, 1972; Shinki et al., 1997; Tanaka & DeLuca, 1973; Yoshida, Fujimori, & Nabeshima, 2002). Degradation of 1,25 $[OH]_2D_3$ is catalyzed by 1,25$[OH]_2D_3$-24-hydroxylase (*CYP24A1*), which is regulated by phosphate, PTH, FGF23, and 1,25$[OH]_2D_3$ itself (Gattineni et al., 2011; Henry & Norman, 1984; Ohyama et al., 1994; Wu et al., 1996). Vitamin D metabolites are primarily transported in the circulation by the

Vitamin D binding protein and only a small fraction of them circulate in their free form. 25(OH)Vitamin D bound to the Vitamin D binding protein is taken up by the brush border membrane of the proximal tubule via megalin/cubulin-mediated endocytosis (Nykjaer et al., 1999, 2001). Here, in the proximal tubule, $1,25[OH]_2D_3$ is synthesized and then released in to the circulation, where it is bound to the Vitamin D binding protein. Like other members of the steroid hormone family, $1,25[OH]_2D_3$ binds to VDR and then heterodimerizes with RXR (retinoid X receptor). This heterodimer binds to VDREs and initiates transcription (Dusso et al., 2005).

Regulation of phosphate homeostasis by direct effects of $1,25[OH]_2D_3$ on renal phosphate reabsorption is less well understood than the regulation of calcium reabsorption by the kidney. This is likely due to the complex interplay between $1,25[OH]_2D_3$, FGF23, and PTH on phosphate reabsorption. Even though a VDRE has been identified in *SLC34A1* (NPT2A) and *SLC34A3* (NPT2C) in humans, the clinical implications for the regulation of phosphate reabsorption by $1,25[OH]_2D_3$ is unclear (Kido, Kaneko, Tatsumi, Segawa, & Miyamoto, 2013; Taketani et al., 1998). Intestinal phosphate transport occurs via transcellular and paracellular pathways. The transcellular transport is mediated by NPT2B and is regulated by $1,25[OH]_2D_3$. $1,25[OH]_2D_3$ increases intestinal phosphate transport by enhancing the expression of Npt2b (Xu, Bai, Collins, & Ghishan, 2002). $1,25[OH]_2D_3$ is also important for bone formation and bone resorption (Dusso et al., 2005). In summary, in hypophosphatemic states, $1,25[OH]_2D_3$ regulates phosphate homeostasis by increasing gastrointestinal absorption of phosphate and release of phosphate from the bone; the effects on renal reabsorption remain unclear.

4.4 Klotho

Klotho was first identified in a mouse model of premature aging and has since has been defined as an important coreceptor for the endocrine family of FGFs regulating a multitude of functions including but not limited to mineral-ion homeostasis, fatty acid metabolism, and glucose homeostasis (Hu, Shiizaki, Kuro-o, & Moe, 2013; Kuro-o et al., 1997). For the purpose of this review, Klotho is used to refer to αKlotho. Klotho is a type-I membrane protein with a short intracellular domain and two tandem repeats of Kl (Kl1 and Kl2) forming an extracellular domain with a β-glucosidase-like domain. Alternative splicing of the Klotho gene can result in a shorter Klotho protein with only one Kl1 domain, which is one form of soluble Klotho (Matsumura et al., 1998). Another source of soluble Klotho results

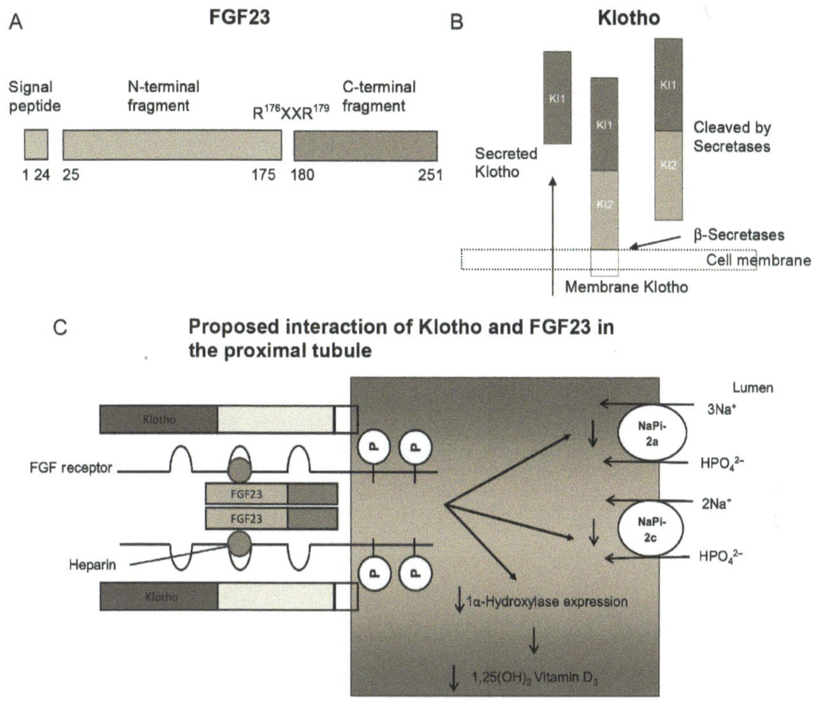

Figure 4 Structure of FGF23, Klotho, and the proposed interaction of FGF23, Klotho and FGFRs. (A) FGF23 structure depicting the signal sequence, N-terminal fragment that has homology to other FGFs, and the C-terminal which is divergent from other FGFs and has a high affinity to bind Klotho. (B) Structure of membrane Klotho with its two homologous Kl domains and short intracellular domain. The secreted form can be due to alternate splicing of the *Klotho* gene, alternatively, membrane Klotho can be cleaved by secretases. (C) The proposed interaction of 2FGF23:2FGFR:2Klotho with the site of heparin shown even though FGF23 has low affinity for heparin. Dimerization of the FGFRs with its ligands results in *trans*-phosphorylation of its tyrosine kinase domains which initiates intracellular signaling. Activation of the intracellular MAPK signaling ultimately decreases the expression of Npt2a and Npt2c with decreased phosphate reabsorption. Additionally, the expression of 1α-hydroxylase is decreased with decreased serum $1,25[OH]_2D_3$ levels. *Adapted with permission from Gattineni and Baum (2012).* (See the color plate.)

from cleavage of intact Klotho protein at the juxtamembrane area with the release of the entire extracellular domain into the circulation (Fig. 4). ADAMTS 10 and 7, members of the family of a disintegrin and metalloprotease with a thrombospodin type-1 motif, and β-site amyloid precursor protein-cleaving enzyme 1 (BACE1) cleave membrane Klotho (Bloch et al., 2009; Chen, Podvin, Gillespie, Leeman, & Abraham, 2007; Imura et al., 2004). Soluble Klotho has been detected in urine, cerebrospinal fluid, and plasma. Soluble Klotho has sialidase enzyme activity and has FGF23-independent effects on Npt2a. Soluble Klotho in the urine modifies the *N*-glycans on Npt2a, which results in its endocytosis, leading to phosphaturia (Hu et al., 2010).

5. OTHER HORMONES

5.1 Steroid hormones

5.1.1 Glucocorticoids

Glucocorticoids are well-known regulators of phosphate homeostasis. High levels of glucocorticoids promote phosphate excretion in humans and experimental animals (Biber et al., 2009; Blaine et al., 2011; Loffing et al., 1998). Little is known, however, about the mechanism of this effect though chronic treatment decreases Npt2a mRNA and protein (Levi et al., 1995; Loffing et al., 1998). Though adrenal insufficiency in rodents is associated with enhanced Npt2a expression, it is unclear that this affects serum phosphate levels (Loffing et al., 1998). Moreover, adrenal insufficiency or Addison's Disease in humans is not known to involve conspicuous changes in phosphate homeostasis.

Glucocorticoid action on Npt2a abundance is thought to be related to changes of lipid content of brush border membranes in proximal tubule cells (Levi et al., 1995). More recent studies suggest an alternate mechanism involving activation of SGK1 and phosphorylation of NHERF1 may be at play. According to the view that we advance here, glucocorticoids bind their cognate receptor in proximal tubule cells, thereby activating Sgk1 (Bhargava & Pearce, 2004). Sgk1 in pulmonary epithelia and the kidney phosphorylates NHERF1 (Andrukhova et al., 2012; Bomberger et al., 2014) that in turn disengages Npt2a, which is endocytosed and downregulated, resulting in diminished Pi reabsorption (Wang, Means, et al., 2012) as shown for FGF23 (Andrukhova et al., 2012). We propose that glucocorticoids may work in a similar manner to downregulate Npt2a, in addition to transcriptional effects and altered membrane composition.

5.1.2 Estradiol

Estradiol decreases serum phosphate levels in humans and in experimental animals (Blaine et al., 2011; Faroqui, Levi, Soleimani, & Amlal, 2008; Uemura et al., 2000). Further, blockade of estrogen receptors with raloxifene decreased phosphate excretion in men (Uebelhart, Herrmann, & Rizzoli, 2009). Increased phosphate excretion was associated with reduced

Npt2a in rats treated with 17β-estradiol (Faroqui et al., 2008). Moreover, estradiol treatment inhibits PTH-stimulated cAMP accumulation in OK cells (Stock et al., 1992). Estradiol itself had no effect on Pi uptake by OK cells and it attenuated the magnitude of PTH-inhibitable Pi transport, presumably by inhibiting adenylyl cyclase (Stock et al., 1992). This latter effect was of modest magnitude (25%) and was independent of the applied concentration of 17β-estradiol over the range 10^{-12}–10^{-8} M. It is now appreciated that inhibition of Pi transport involves phosphorylation of NHERF1, which tethers Npt2a at apical membranes. NHERF1 harbors an estrogen response element and estrogen treatment upregulates NHERF1 expression (Ediger, Kraus, Weinman, & Katzenellenbogen, 1999; Ediger, Park, & Katzenellenbogen, 2002). These findings suggest that there may be counterbalancing actions of estrogen on PTH-sensitive phosphate transport. Estradiol may reduce PTH-inhibitable phosphate transport by reducing cAMP formation, but may promote PTH action by increasing the amount of membrane-delimited Npt2a.

Estrogen treatment significantly increases serum FGF23 levels in an ovariectomized rat model of CKD (Carrillo-Lopez et al., 2009). Hence, the phosphaturic action of estradiol may be secondary to augmented FGF23 and its inhibitory action on Npt2a and phosphate transport.

5.2 Insulin-like growth factor-1

Insulin-like growth factor-1 (IGF1) stimulates Pi uptake by OK cells (Caverzasio & Bonjour, 1989; Jehle et al., 1998; Palmer, Bonjour, & Caverzasio, 1996; Silverstein, Spitzer, & Barac-Nieto, 2000). Whereas hormones that inhibit Npt2a do so by downregulating the protein, IGF1 enhances uptake by augmenting Npt2a stability (Jehle et al., 1998). The effects of IGF1 on Pi uptake are independent and unaffected by $1,25[OH]_2D_3$ or by the arachidonic acid metabolite 20-hydroxyeicosatetraneoic acid (20-HETE), which reportedly is increased by PTH (Omata, Abraham, & Schwartzman, 1992).

5.3 Serotonin

Serotonin (5-hydroxytryptamine) is a neurotransmitter, also synthesized in the kidney that like IGF1 stimulates renal phosphate transport. Acting through the 5-HT_{1B} receptor, serotonin stimulates Gαi and inhibits adenylyl cyclase (Murphy & Bylund, 1989). In this manner, serotonin blunts PTH (Hafdi et al., 1996) and dopamine (de Toledo et al., 1997) effects on Pi uptake. These actions demonstrated with OK cells are recapitulated upon

5-hydroxytryptophan infusion in rats, increasing renal phosphate reabsorption, an effect that was blocked by benserazide, an inhibitor of aromatic L-amino acid decarboxylase that catalyzes the formation of 5-hydroxytryptamine from tryptophan (and dopamine from L-dopa) (Hafdi et al., 1996). Unlike IGF1, the stimulatory action of serotonin does not augment Npt2a stability. The molecular basis for the stimulatory effect of serotonin on Pi transport has not been defined but could result from an enhanced driving force for apical Na^+ entry, secondary to stimulation of basolateral Na/K-ATPase activity or localization (Budu, Efendiev, Cinelli, Bertorello, & Pedemonte, 2002).

5.4 Adenosine

It has long been known that PTH action on proximal tubules is associated with the release and excretion of cAMP, which is referred to as nephrogenous cAMP (Chase & Aurbach, 1967). Infusion of cAMP, however, itself promotes phosphate excretion (Kuntziger, Amiel, Roinel, & Morel, 1974). This latter phenomenon occurs by virtue of the extracellular formation of adenosine from cAMP by ecto-5′-nucleotidase in the brush border membrane, with adenosine then acting in an autocrine manner on tubular adenosine receptors (Friedlander, Couette, Coureau, & Amiel, 1992). Although proximal tubules express A1, A2a, A2b, and A3 receptors (Vallon, Muhlbauer, & Osswald, 2006), the effects on phosphate transport are thought to be mediated by A_1 adenosine receptors. This conclusion is based on the action of selective A1R agonists (Coulson, Johnson, Olsson, Cooper, & Scheinman, 1991) and antagonists (Cai, Puschett, Guan, Batuman, & Puschett, 1995). The A1R can couple to Gi, Gs, and Gq in a ligand-biased fashion (Cordeaux, Ijzerman, & Hill, 2004). Further, at low concentrations (1 nM to 1 μM), adenosine analogs inhibit cAMP formation, while higher concentrations (0.1–100 μM) increase cAMP generation (Arend, Sonnenburg, Smith, & Spielman, 1987). Thus, reported actions of adenosine or adenosine analogs on renal Pi transport and the implied signaling mechanism should consider the agonist concentration that was used.

Adenosine stimulates cAMP accumulation in OK cells (Coulson et al., 1991) though the inhibitory action of adenosine on Pi uptake by OK cells is not mediated entirely by Gs and cAMP insofar as blockade of PKC with staurosporine abolished the inhibitory action of the stable adenosine analogue, R(−)N^6-phenylisopropyl-adenosine (PIA) (Coulson et al., 1991). The dual PKA + PKC mechanism is consistent with the nature of A1R

G protein coupling and reminiscent of the dual signaling pathways used by PTH in the regulation of Npt2a.

5.5 Stanniocalcin

Stanniocalcins are glycoproteins first isolated from the corpuscles of Stannius, an endocrine organ involved in calcium and phosphate metabolism in bony fish (Wagner, Hampong, Park, & Copp, 1986). Two stanniocalcins expressed in humans (STC1/2) and in mice (Stc1/2) exhibit 60% and 30% amino acid identity, respectively, with their fish homologs. Human STC1 and STC2 exhibit limited sequence identity (27%) and less similarity(16%). Both genes are expressed primarily in kidney. Stc1 is found in the circulation, whereas Stc2 is undetectable (Deol, Varghese, Wagner, & Dimattia, 2000).

Understanding of Stc1/2 actions on mineral-ion homeostasis comes from studies of overexpression or knockout experiments in mice. Overexpression of Stc1 was associated with increases of serum phosphate, implying that renal phosphate excretion may have been reduced; with no change of serum calcium (Varghese et al., 2002). In contrast, eliminating Stc1 was without effect on serum phosphate or calcium levels, and the mice responded equivalently to administered Vitamin D treatment (Chang, Cha, Koentgen, & Reddel, 2005). Moreover, the absence of an effect was not due to compensatory expression of Stc2, which was unchanged. Transgenic mice-overexpressing human STC2 likewise displayed no alterations in serum phosphate (Yang et al., 2014). A slight decline in serum calcium in female mice may be attributable to diminished bone turnover.

Although expression of Stc1 (Hung et al., 2012) and Stc2 (Takei et al., 2009) in OK cells is enhanced by 1,25$[OH]_2D_3$ and depressed by PTH, we are unaware of experiments showing that either Stc1 or Stc2 affects phosphate uptake in this default model or renal phosphate transport. Thus, no definitive statements can be made regarding possible actions of Stc1/2 on renal phosphate transport or a potential mechanism of action.

6. ADAPTER PROTEINS MODIFYING HORMONE-DEPENDENT PHOSPHATE TRANSPORT

6.1 PDZ proteins

PDZ proteins constitute a large family of cytoplasmic adapter proteins capable of assembling membrane-associated proteins with signaling and regulatory proteins in short-lived functional units. These proteins take their name from the common structural domain shared by the postsynaptic density protein (PSD95), Drosophila disc large tumor suppressor (DlgA), and zonula occludens 1 protein (ZO-1). PDZ proteins form partnerships with protein bearing a certain carboxy-terminal signature called a PDZ-recognition sequence (Romero, Von Zastrow, & Friedman, 2011). As noted earlier, Npt2a possesses such a motif (-TRL) that binds Na/H exchanger regulatory factors (NHERF). Four non-synonymous NHERF proteins have been described (Ardura & Friedman, 2011). All are capable of binding Npt2a but only NHERF1 importantly mediates hormone-sensitive renal phosphate transport.

6.1.1 NHERF1

In polarized epithelial cells, such as those forming the proximal tubule, NHERF1 localizes to apical plasma membranes (Weinman, Cunningham, Wade, & Shenolikar, 2005). Known also as the 50-kDa ezrin-binding protein (EBP50), NHERF1 possesses two tandem PDZ domains and an ezrin-binding domain (Fig. 5). Notably, NHERF1 itself harbors a carboxy-terminal PDZ-binding motif (-SNL). Npt2a colocalizes at apical brush border membranes with NHERF1 (Wade et al., 2003), where it binds principally to PDZ1 (Gisler et al., 2008, 2001; Hernando et al., 2002; Mamonova, Kurnikova, & Friedman, 2012; Wang, Means,

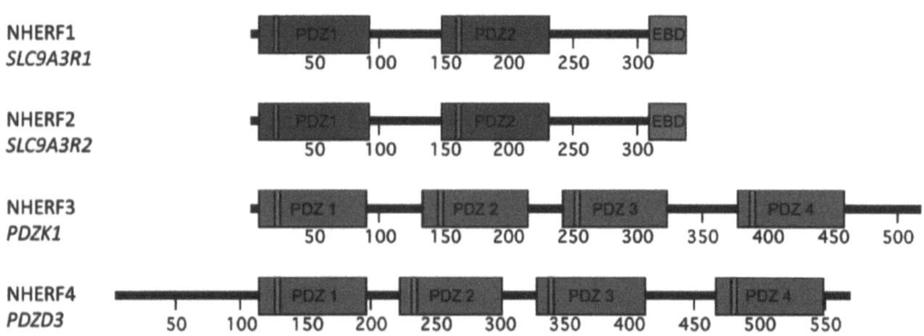

Figure 5 Schematic structural representation of the four described members of the human Na^+/H^+ exchanger (NHERF) family. Common and gene names are shown on the left. PDZ domains are shaded in blue and, where present, ezrin-binding domains are represented in red. Each PDZ domain possesses a single core-binding motif (pink band). The four-residue sequence may be the same (GYGF) as in the two PDZ domains of NHERF1 and NHERF2, or may be different as in the four PDZ domains of NHERF3 (GFFL, GFSL, GYGF, GYGF) or NHERF4 (GFHL, GFGF, GFGF, GFRL). Notably, NHERF1 and NHERF 2 terminate with a PDZ ligand (-SNL, -SNF, respectively) allowing them to form a closed confirmation (Wang, Means, et al., 2012) or functional heterodimers (Yang et al., 2014). (See the color plate.)

et al., 2012). A role for PDZ2 in Npt2a function has also been described (Khundmiri et al., 2008). Apical membrane localization of Npt2a in proximal tubules of NHERF1-null mice is markedly reduced with accumulation of the protein in the cytoplasm (Shenolikar, Voltz, Minkoff, Wade, & Weinman, 2002). NHERF1 is critically involved in renal Pi transport and in phosphate homeostasis. Targeted inactivation of Slc9a3r1 and the attendant absence of NHERF1 in mice is associated with profound hyperphosphaturia, hypophosphatemia, and associated skeletal abnormalities (Liu et al., 2012; Morales, Takahashi, Kreimann, & Georgescu, 2004; Shenolikar et al., 2002). Humans harboring mutations or polymorphisms of SLC9A3R1 exhibit a similar phenotype, including nephrocalcinosis, though haploinsufficient (Courbebaisse et al., 2012; Karim et al., 2008).

PTH-induced phosphorylation of Ser^{77} in PDZ1 releases Npt2a, promoting endocytosis and decreased phosphate absorption (Voltz et al., 2007; Weinman et al., 2007, 2010). Interestingly, purified PKC, but not PKA, phosphorylates Ser^{77} of recombinant PDZ1, whereas both PKC and PKA phosphorylate Ser^{77} in a cellular context (Weinman et al., 2007). These differences in kinase action may have several technical or biological origins, but also could speak to an indirect action of PKA in cells, where PKA phosphorylates an intermediary kinase that in turn phosphorylates NHERF1.

NHERF1 assembles a ternary complex consisting of Npt2a–NHERF1–ezrin. Ezrin functions here as an A kinase anchoring protein (AKAP) to engage PKA regulatory subunits that, in turn, phosphorylate NHERF1 leading to the disassembly of the complex and the attendant termination of phosphate transport (Wang, Means, et al., 2012). These experiments speak to an important role for ezrin and PKA in mediating PTH actions on Npt2a, and also explain the loss of hormone regulation of phosphate transport in the presence of defined NHERF1 mutations (Courbebaisse et al., 2012; Karim et al., 2008; Wang, Means, et al., 2012).

6.1.2 NHERF2

NHERF2 (Na^+/H^+ exchanger (NHE) type 3 kinase A regulatory protein, E3KARP) structurally resembles NHERF1 (Fig. 4) but NHERF2 localizes to the submicrovillar region of proximal tubule cells and is not expressed in microvilli (Wade et al., 2003). NHERF2 does not coimmunoprecipitate with Npt2a in the absence of NHERF1. NHERF2-null mutant mice have no described mineral-ion phenotype.

6.1.3 NHERF3

NHERF3 (PDZK1) harbors 4 PDZ domains but lacks an ezrin-binding domain (Fig. 4). NHERF3 was discovered in a search for genes differentially upregulated in animals on a restricted phosphate diet (Custer, Spindler, Verrey, Murer, & Biber, 1997). Npt2a binds NHERF3 through PDZ3 (Gisler et al., 2003, 2001). Hence, it was unexpected that gene ablation of *Pdzk1* had no demonstrable effect on Npt2a localization or on serum phosphate or calcium levels, or on phosphate excretion (Capuano et al., 2005; Kocher, Pal, Roberts, Cirovic, & Gilchrist, 2003). A role for NHERF3 in phosphate economy can be discerned from experiments showing that NHERF3-null mice placed on a high-phosphate diet exhibited reduced Npt2a abundance along with increased phosphate excretion, indicating that NHERF3 may participate in the trafficking or stability of Npt2a (Capuano et al., 2005).

6.1.4 NHERF4

NHERF4 (PDZD3), like NHERF3, possesses 4 PDZ domains (Fig. 4) and is expressed in kidney and intestine (Gisler et al., 2001). In contrast with NHERF3, which is localized in microvilli, NHERF4 is found in the subapical compartment and to a lower extent throughout the cytoplasm. Npt2a binds to PDZ3 of NHERF3 (Gisler et al., 2001).

6.1.5 PIST

The Golgi-associated PDZ and coiled-coil motif-containing protein (GOPC/Gopc, also called PIST [PDZ protein interacting specifically with TC10], CAL [CFTR-associated ligand], and FIG [Fused in glioblastoma]) is a PDZ protein with a single-PDZ domain. As described earlier, PTH-induced Npt2a endocytic retrieval proceeds through clathrin-coated pits and may traverse the *trans*-Golgi network before being sorted to lysosomes, where it is degraded. In the *trans*-Golgi network Npt2a colocalizes with GOPC (Lanaspa et al., 2013) though direct interaction of Npt2a through its PDZ ligand with the PDZ domain of GOPC was not demonstrated. Nonetheless, the findings are compatible with the idea that trafficking of proteins containing a PDZ-recognition motif may involve sequential hand offs from membrane-associated PDZ proteins such as NHERF1 to subcellular compartments harboring PDZ proteins such as GOPC or SNX27 (Belotti et al., 2013).

6.1.6 Megalin

Megalin is a large glycoprotein receptor that is found on apical membranes of proximal tubules and belongs to the low-density lipoprotein receptor family (Christensen & Birn, 2002; Saito, Sato, Iino, & Takeda, 2010). Megalin plays a central role in the biogenesis of 1,25[OH]$_2$D$_3$ (Nykjaer et al., 1999) and in PTH endocytosis (Hilpert et al., 1999). Although suppression of megalin expression did not affect Npt2a abundance or phosphate excretion, it delayed Npt2a internalization in response to PTH or direct application of second messengers (cAMP, cGMP, or an activator of PKC) (Bacic, Capuano, et al., 2003). Npt2a reportedly does bind megalin (Bacic et al., 2006; Biemesderfer, Nagy, DeGray, & Aronson, 1999). Megalin, however, terminates in a putative PDZ-recognition motif (-DSEV [human], -DSDV [mouse]) that permitted binding to NHERF1 in overlay assays (Wang & Friedman, unpublished observations) and to PDZ2 in coimmunoprecipitation experiments (Slattery et al., 2011). NHERF1 also binds to megalin through an internal megalin sequence (Slattery et al., 2011). Hence, the effect of megalin on Npt2a could arise from an indirect action negotiated by NHERF1. Indeed, Npt2a forms a multimolecular complex with NHERF1 and megalin (Tanimura et al., 2011).

Kidney-specific deletion of megalin further supports a role in retrieval and downregulation of Npt2a associated with reduced urinary phosphate excretion (Bachmann et al., 2004). Notably, Npt2a is internalized through receptor-mediated endocytosis, as described earlier (Bacic et al., 2006), whereas megalin mediates fluid-phase endocytosis, a constitutive and nonsaturable uptake process. This led Murer and his colleagues to deduce that altered Npt2a trafficking and function in the absence of megalin is due to structural alterations of the brush border membrane with disruption of nonspecific endocytic processes (Bachmann et al., 2004; Bacic et al., 2006).

7. GENETIC DISORDERS OF RENAL PHOSPHATE TRANSPORT DUE TO HORMONAL DYSREGULATION

Tight regulation of serum phosphate levels with in the physiological range in humans is critical for many biological functions, and therefore dysregulation of phosphate homeostasis has many undesired effects on multiple organs. We will discuss the clinical disturbances of phosphate regulation that are primarily due to genetic disorders of hormonal dysregulation. As discussed earlier, FGF23, Klotho, 1,25[OH]$_2$D$_3$, and PTH are the principal hormones regulating phosphate balance.

7.1 FGF23-mediated disorders

7.1.1 Autosomal-dominant hypophosphatemic rickets

ADHR is a rare disease that is inherited in an autosomal-dominant fashion and is due to activating mutations in the *FGF23* gene (ADHR Consortium, 2000). Around the time that the *FGF23* gene was identified as the cause of ADHR by the ADHR consortium, FGF23 protein was recognized as the causative factor for urinary phosphate wasting and low 1,25 $[OH]_2D_3$ levels in a patient with tumor-induced osteomalacia, where the tumors produced increased amounts of FGF23 (Shimada et al., 2001). In ADHR, the mutant FGF23 protein is resistant to cleavage by subtilisin/furin-like proprotein convertases due to missense mutations at the cleavage site, $R^{176}XXR^{179}$, resulting in increased circulating levels of FGF23 (Bai, Miao, Goltzman, & Karaplis, 2003; White et al., 2001).

The clinical presentation of patients with ADHR can be quite variable. Some patients present in childhood, whereas others do not manifest the disease until adolescence or as adults. Regardless of the age of presentation, the biochemical phenotype is similar with hypophosphatemia, phosphaturia, inappropriately low or normal levels of $1,25[OH]_2D_3$, and increased alkaline phosphatase. By studying a large kindred with ADHR, it became apparent that when ADHR was diagnosed in childhood, the presenting complaints included poor growth, lower extremity deformity, bone pain, delayed walking, short stature, and dental abnormalities. Two children exhibited spontaneous resolution of symptoms as adults in the kindred studied by Econs et al. and the etiology was unclear (Econs & McEnery, 1997). Patients with ADHR diagnosed in adulthood presented with bone pain, fractures, and muscle weakness but lacked bony deformities. All the patients that presented in adulthood were females, and pregnancy was a triggering factor in several of these patients. A recent study demonstrated that iron deficiency was associated with elevated intact and C-terminal FGF23 levels in patients with ADHR, whereas control subjects had only elevated C-terminal FGF23 levels during states of iron deficiency (Imel et al., 2011). These data indicate that iron deficiency stimulates FGF23 production in normal controls and in patients with ADHR, but since ADHR patients are unable to cleave intact FGF23, iron deficiency accentuates the levels of FGF23. This was confirmed by Farrow et al. elegantly using an *Fgf23* knockin mouse model with a known mutation seen in ADHR (Farrow et al., 2011). It is possible that pregnancy triggered the iron deficiency in patients harboring ADHR mutations and thus the delayed clinical presentation of the disease in adulthood during pregnancy. Treatment of ADHR is symptomatic with phosphate and 1,25 $[OH]_2D_3$ supplements.

7.1.2 X-linked-dominant hypophosphatemic rickets

XLH is the most common inherited form of rickets with an incidence of 1:20,000 and is primarily diagnosed in childhood, but there are reports of late diagnosis in adulthood. Children usually present before the age of 2 years due to delayed walking or bowing of legs when they start walking. Other symptoms include painful extremities, widened wrists, dental abscess, a waddling gait, poor growth, and short stature. Adults who were undiagnosed in childhood present with bone pain, fractures, bone deformities, dental abscess, sensorineural hearing loss, and features of osteomalacia on X-ray. Enthesopathies (painful calcification of joint capsule, ligaments, and tendons) is also a feature that is seen in adults with XLH (Winters, Graham, Williams, McFalls, & Burnett, 1991).

Using positional cloning techniques, mutations in the *PHEX* gene (phosphate regulating gene with homologies to endopeptidases on the X chromosome) were identified as the cause of XLH (HYP Consortium, 1995). XLH is inherited in an X-linked dominant fashion and the vast majority of patients have elevated FGF23 levels. Elevated FGF23 serum levels result in hypophosphatemia, phosphaturia, low or inappropriately normal levels of $1,25[OH]_2D_3$. These serum abnormalities are responsible for the bone phenotype of rickets and osteomalacia (Shimada, Hasegawa, et al., 2004). Serum calcium and PTH levels are usually normal in patients with XLH. The murine model of XLH, *Hyp* mice, also has elevated FGF23 levels and similar biochemical phenotypes as patients with XLH (Yamazaki et al., 2002). PHEX is primarily expressed in the bone and teeth (Du et al., 1996) but it is as yet unknown how mutations in the *PHEX* gene causes elevated FGF23 levels because FGF23 is not a substrate for PHEX (Liu et al., 2003). However, recent data demonstrate that Phex impairs bone mineralization in rodents which in turn stimulates Fgf23, which is probably to avoid phosphate overload in these animals (Liu et al., 2009, 2007). Deletion of *Fgf23* in the *Hyp* mice produced a phenotype identical to *Fgf23*-null mice with hyperphosphatemia and elevated $1,25[OH]_2D_3$ levels indicating that FGF23 is downstream of PHEX and that FGF23 is normally suppressed by PHEX. There appears to be a genotype–phenotype correlation in XLH patients with deletions, nonsense mutations, and splice mutations leading to premature stop codons, who present with a more severe biochemical phenotype than the missense mutations of the *PHEX* gene (Morey et al., 2011). Dosage effects of the *PHEX* gene were also examined by comparing heterozygous and homozygous female *Hyp* mice. It was demonstrated that the skeletal phenotype was worse in homozygous female *Hyp* mice, but both groups of mice had comparable serum FGF23 levels (Ichikawa, Gray, Bikorimana, & Econs, 2013).

Current standard of care for patients with XLH includes phosphate and $1,25[OH]_2D_3$ supplementation. The use of growth hormone to improve short stature is successful in some studies (Seikaly & Baum, 1995). The side effects of current treatment options include hyperparathyroidism and nephrocalcinosis with the risk of CKD associated with the latter (Alon, Donaldson, Hellerstein, Warady, & Harris, 1992). The use of FGF23 neutralizing antibodies in *Hyp* mice corrected the hypophosphatemia and low $1,25[OH]_2D_3$ levels, along with improvement of the bone phenotype, confirming the pathogenic role of FGF23 in XLH (Aono et al., 2011, 2009). The use of an anti-human monoclonal FGF23 antibody (KRN23) in patients with XLH has yielded promising results. KRN23 improved serum phosphate levels and $1,25[OH]_2D_3$ levels and had a favorable side effect profile when studied for a short duration (50 days) (Carpenter et al., 2014). Studies involving long-term use of KRN23 in patients with XLH are currently underway.

7.1.3 Autosomal recessive hypophosphatemic rickets

ARHR is also a rare disease that is inherited in a recessive fashion and is due to inactivating mutations of *DMP1* (dentin matrix protein 1), *ENPP1* (ectonucleotide pyrophosphatase/phosphodiesterase-1), or *FAM20c* (family with sequence similarity 20-member C) genes. The clinical and biochemical presentation is similar to patients with ADHR and XLH. Patients with ARHR can also develop osteosclerotic bone lesions and enthesopathies (Perry & Stamp, 1978). DMP1 is a protein that regulates mineralization process and thus plays an important role in the formation of teeth, bone, and cartilage. *Dmp1* null mice have elevated FGF23 levels, hypophosphatemia, and low $1,25[OH]_2D_3$ levels and have defects in their bones and teeth (Feng et al., 2006; Lorenz-Depiereux et al., 2006). Using a mouse model of transgenic expression of 57 kDa fragment of DMP1 rescued the phenotype in *Dmp1*-null mice (Lu et al., 2011) confirming the role of DMP1 in ARHR. ENPP1, on the other hand, is a mineralization inhibitor and does so by generating pyrophosphates. ENPP1 mutations were first described to cause generalized arterial calcification of infancy (Rutsch et al., 2003). Patients with ARHR due to ENPP1 mutations do not develop generalized calcifications likely due to hypophosphatemia because of elevated FGF23 levels (Levy-Litan et al., 2010; Lorenz-Depiereux, Schnabel, Tiosano, Hausler, & Strom, 2010).

Inactivating mutation of the FAM20c gene (family with sequence similarity 20-member C) with an autosomal recessive inheritance was recently identified in two siblings with hypophosphatemia, dental mineralization defects, intracerebral calcifications, osteosclerotic lesions, elevated FGF23 levels, and in the absence of rickets (Rafaelsen et al., 2013). Additionally, *Fam20c* null mice demonstrated elevated FGF23 levels along with hypophosphatemia and rickets in the absence of osteosclerotic lesions (Wang, Means, et al., 2012). The difference in the bone phenotype between murine models and humans is poorly understood at this time. FAM20c phosphorylates Ser^{180} at the $R^{176}XXR^{179}$ site, which inhibits the O-glycosylation of the FGF23 protein, and this in turn promotes proteolysis of FGF23. Thus, the absence of FAM20c disrupts the balance of phosphorylation and O-glycosylation of the FGF23 protein resulting in elevated FGF23 levels. Additionally, FAM20C has been shown to phosphorylate glycoproteins including DMP1, and therefore, FAM20C likely regulates FGF23 production via DMP1 also (Tagliabracci et al., 2012). Treatment of ARHR involves supplementation with phosphate 1,25 $[OH]_2D_3$.

Despite the advances in the field of FGF23, a lot remains unknown as to how different genetic mutations converge to increase FGF23 levels and cause an identical clinical phenotype of hypophosphatemia, low 1,25 $[OH]_2D_3$ levels, and skeletal mineralization defect.

7.1.4 Osteoglophonic dysplasia

OGD is due to activating mutations of FGFR1 and has been described only in a few patients. These patients have craniofacial abnormalities along with short stature, hypophosphatemia, and inappropriately normal levels of 1,25 $[OH]_2D_3$. Some of these patients have elevated FGF23 levels. Activation of FGFR1 using a monoclonal antibody in mice also demonstrated elevated FGF23 levels along with hypophosphatemia and decreased expression of 1α-hydroxylase. It is again unknown as to how activating mutations of FGFR1 cause elevated FGF23 levels (White et al., 2005).

7.1.5 Fibrous dysplasia

FD is due to an activating mutation of *GNAS1*, which is a gene that encodes for the alpha subunit of stimulatory G protein $G_s\alpha$. Some patients with FD develop hypophosphatemia, low $1,25[OH]_2D_3$ levels and elevated FGF23 levels, and the phenotype is related to the burden of the bone lesions. Normal bone is replaced by fibrous dysplastic tissue, which produces FGF23 in a cAMP-dependent fashion (Riminucci et al., 2003; Weinstein et al., 1991; Yamamoto et al., 2005). Recent work sheds light on the molecular mechanisms that are involved in FD. Patients with FD have an increased ratio of total/intact FGF23 levels when compared to patients with XLH and TIO. Furthermore, using *in vitro* studies, it was shown that increased cAMP decreased GALNT3 activity and increased furin activity, resulting in increased proteolysis of the FGF23 protein (Bhattacharyya et al., 2012).

7.1.6 Hyperphosphatemic familial tumoral calcinosis

HFTC is inherited in an autosomal recessive or autosomal-dominant fashion. The autosomal recessive forms of tumoral calcinosis can be due to mutations in the *FGF23* or *GALNT3* genes. GALNT3 encodes for UDP-*N*-acetyl-alpha-D-galactosamine:polypeptide *N*-acetylgalactosaminyl transferase 3, which is responsible for O-glycosylation of the threonine residues adjacent to the $R^{176}XXR^{179}$ site. This O-glycosylation stabilizes the FGF23 protein and decreases cleavage by furin-like proteases. Therefore, patients with GALNT3 mutations have elevated C-terminal FGF23 and decreased intact FGF23 levels. Similarly, elevated C-terminal FGF23 levels are seen in patients with inactivating mutations in the conserved region of the *FGF23* gene and the mutant protein is either prone to degradation by proteases or is not being secreted in its intact form (Benet-Pagès, Orlik, Strom, & Lorenz-Depiereux, 2005; Topaz et al., 2004). The precise etiology of the autosomal-dominant form of HFTC is yet unknown.

Patients with HFTC have hyperphosphatemia; elevated $1,25[OH]_2D_3$ levels; and painful deposition of calcium and phosphate in the skin, soft tissues, and joints. This phenotype is similar to that seen in *Fgf23* and *Klotho*-null mice. This is an extremely debilitating disease and the treatment options include phosphate binders and acetazolamide. Surgical options can provide temporary relief but in most instances, the lesions reoccur.

7.2 Klotho-mediated disorders

Translocation with a break point adjacent to the *α-Klotho* gene was identified in a patient who presented at the age of 13 months with hypophosphatemia, inappropriately low $1,25(OH)_2D_3$ levels, and rickets. The patient had elevated Klotho levels and hyperparathyroidism. This is the first description of Klotho overexpression in humans. Of interest, the patient also had elevated FGF23 levels and the mechanism is unknown at this time (Brownstein et al., 2008). Conversely, missense mutation in the *Klotho* gene in a 13-year-old resulted in a phenotype similar to tumoral calcinosis. The patient had hyperphosphatemia, hypercalcemia, hyperparathyroidism, and elevated levels of $1,25[OH]_2D_3$ levels. Furthermore, the patient was also noted to have elevated intact and C-terminal FGF23, and further work revealed a homozygous missense mutation in the *Klotho* gene. This patient demonstrated resistance to FGF23 due to the functional absence of its coreceptor, Klotho (Ichikawa et al., 2007).

7.3 PTH-mediated disorders

As discussed in this review earlier, PTH is a phosphaturic hormone but contrary to FGF23, PTH increases circulatory levels of $1,25(OH)_2$Vitamin D_3. As PTH is essential for calcium homeostasis, dysregulation of PTH causes disturbances in calcium metabolism in addition to phosphate metabolism.

Patients with inherited forms of hypoparathyroidism present with hypocalcemia, hypercalciuria, hyperphosphatemia, and low $1,25(OH)_2$Vitamin D_3 levels. Inheritance can be autosomal dominant or recessive and is due to defects either in the parathyroid gland development (*GCMB*) or in the *PTH* gene (Mannstadt et al., 2008; Parkinson & Thakker, 1992). Hypoparathyroidism can be isolated or be associated with syndromes including DiGeorge syndrome (hypoplasia of the pharyngeal arches impairing parathyroid gland development) (Lammer & Opitz, 1986), Kenny-Caffey syndrome (mutation in the microtubule assembly gene, TBCE), Kearns-Sayre syndrome (mitochondrial DNA defect), or part of an autoimmune polyglandular syndrome (Al-Azem & Khan, 2012).

Familial benign hypercalciuric hypocalcemia or autosomal dominant hypercalciuric hypocalcemia (FBHH/ADHH) is due to an activating mutation of the calcium-sensing receptor (CaSR), which causes the set point for PTH secretion to be shifted to the left. Patients with FBHH/ADHH are usually asymptomatic and are noted to have mild hypocalcemia, hypercalcuria, and inappropriately low levels of PTH (Pearce et al., 1996; Pollak et al., 1994).

Treatment options for patients with hypoparathyroidism include calcium and calcitriol ($1,25(OH)_2D_3$) supplements to normalize calcium. As calcitriol causes increased absorption of phosphate in addition to calcium, phosphate binders are also used to avoid hyperphosphatemia. Patients on therapy are at an increased risk of worsening hypercalciuria and the risk of developing nephrocalcinosis/nephrolithiasis which should be monitored. Treatment with PTH replacement therapy has been studied in a small number of patients with some success but future studies are needed to study the long-term effects of PTH especially in children due to their skeletal growth (Winer et al., 2003). Asymptomatic patients with ADHH should be monitored closely while avoiding calcium supplements as hypercalciuria worsens with treatment. The goal in ADHH would be to keep the patients asymptomatic and accept lower serum calcium levels.

Pseudohypoparathyroidism is characterized by hypocalcemia, hypercalciuria, and hyperphosphatemia. Contrary to patients with hypoparathyroidism that have low PTH levels, patients with pseudohypoparathyroidism have elevated PTH levels as they have resistance to the effects of PTH. Different forms of pseudohypoparathyroidism have been reported. Pseudohypoparathyroidism type 1a is also known as Albright's hereditary osteodystrophy (AHO) and patients with AHO have characteristic features of short stature, obesity, round face, brachydactyly, subcutaneous ossifications, and mental retardation. AHO is due to mutations in the *GNAS1* gene inherited from the mother which encodes for the $G_s\alpha$ subunit that stimulates cyclic AMP production. Patients with AHO usually have resistance to other hormones in addition to PTH (thyroid-stimulating hormone, growth follicle-stimulating hormone, and leuteinizing hormone). Pseudopseudohypoparathyroidism is due to paternal inheritance of the *GNAS* mutation and these patients have clinical features of AHO but lack any endocrine abnormalities. Of note, patients with both AHO and pseudopseudohypoparathyroidism can be found in large extended families due to the parental imprinting of the *GNAS* gene (Weinstein, Shenker, Gejman, & Merino, 1991; Weinstein, Yu, Warner, & Liu, 2001). Treatment options include calcium and calcitriol supplements while avoiding hypercalcemia and worsening hypercalciuria.

Hyperparathyroidism is commonly due to acquired parathyroid adenomas but can be seen in association with familial syndromes. The common biochemical abnormalities include hypercalcemia, hypophosphatemia, and elevated PTH levels. Multiple endocrine neoplasia type 1 (MEN 1) is inherited in an autosomal dominant fashion and patients with MEN 1 have tumors in the pituitary gland, pancreas and the parathyroid gland. MEN 1 is due to a germline inactivation of the *MEN1* gene. Treatment involves surgical removal of the gland with possible autotransplantation in the forearm (Gaztambide, Vazquez, & Castano, 2013).

FHH has an autosomal dominant inheritance pattern and is due to inactivating mutations of the CaSR where the set point for PTH release is shifted to the right. Patients with FHH have elevated PTH levels along with hypocalciuria, hypercalcemia, and hypophosphatemia but they usually remain asymptomatic. Due to persistent hypercalcemia, patients do have the increased risk for vascular calcification and pancreatitis (Egbuna & Brown, 2008; Giusti, Cavalli, Cavalli, & Brandi, 2013). Therapeutic considerations for symptomatic patients include thiazide diuretics to enhance urinary calcium excretion or calcimimetics. Neonatal severe primary hyperparathyroidism is an extremely rare disorder where patients present in the neonatal/infantile period with severe hypercalcemia, failure to thrive, hypophosphatemia, and elevated PTH levels. This disease is the homozygous version of FHH and is inherited in an autosomal recessive manner. Treatment includes aggressive hydration, calcimimetics, and surgical removal of the parathyroid gland is usually necessary. Postsurgical removal, these patients are at a high risk for hypocalcemia due to hungry bone syndrome (Davies & Shaw, 2012). Table 2 lists the hypophosphatemic and hyperphosphatemic disorders.

8. REGULATION OF PHOSPHATE IN CKD AND END STAGE RENAL DISEASE

Phosphate homeostasis is maintained with in a narrow range to avoid deleterious effects of hypophosphatemia or hyperphosphatemia in mammals. As described in this review, multiple hormones act in concert to maintain phosphate balance and the important hormones include PTH, FGF23, Klotho, and $1,25(OH)_2D_3$. Although these hormones originate from different tissues, act on different organs, and form a kidney–bone–intestine–parathyroid network to regulate phosphate homeostasis, they all have additional regulatory functions on one other. Diseases affecting one of the organs or hormones in this network result in some form of compensation from the others, but not usually sufficient to maintain normal serum phosphate levels.

In early CKD, as the functional nephron mass decreases, the initial abnormality is the increase in fractional phosphate excretion and this is likely due to increased FGF23 levels. FGF23 levels are elevated before any discernible changes in mineral metabolism markers and likely represent

Table 2 Disorders of phosphate due to hormone dysregulation

Disease	Genetic mutation	Pathogenesis of the disease
Hypophosphatemic disorders		
Autosomal-dominant hypophosphatemic rickets	*FGF23*	FGF23 protein is resistant to degradation
Autosomal recessive hypophosphatemic rickets	*DMP1* *ENPP1* *Fam20c*	Increased expression of FGF23 protein from the bone
X-linked hypophosphatemic rickets	*PHEX*	Increased expression of FGF23 protein from the bone
Fibrous dysplasia/McCune-Albright syndrome	*GNAS1*	Increased expression of FGF23 protein from the bone lesions
Hypophosphatemic rickets	*Klotho*	Overexpression of Klotho results in hypophosphatemia
Neonatal severe hyperparathyroidism	*CaSR*	Homozygous inactivating mutations of CaSR
Multiple endocrine neoplasia with hyperparathyroidism	*MEN1* and *MEN2*	Inherited tumors
Hyperphosphatemic disorders		
Tumoral Calcinosis	*FGF23* *GALNT3* *Klotho*	Decreased production or increased degradation of FGF23 or resistance to FGF23 due the absence of Klotho
Hypoparathyroidism	*GCMB* *PTH*	Parathyroid gland development defect or in the processing of PTH protein
Pseudohypoparathyroidism	*GNAS1*	Defective G protein α subunit leading to impaired cAMP production

a compensatory mechanism to avoid phosphate overload. Elevated FGF23 decreases circulating 1,25$(OH)_2D_3$ levels. It is generally accepted that an increased "phosphate load" is the stimulus for FGF23 release but a recent balance study did not demonstrate CKD patients to be in positive phosphate balance (Hill et al., 2013). The primary triggers, sensors, and sensing mechanisms that increase FGF23 levels in CKD remain to be identified but these data do suggest that elevated FGF23 seems to protect the patients from hyperphosphatemia, at least in the initial stages. With decreased

$1,25(OH)_2D_3$, PTH levels start to increase as the GFR is $\leq$50 ml/min/1.73 m^2. As CKD progresses further with a decline in nephron mass and function, serum phosphate levels begin to increase and complete spectrum of CKD-MBD (chronic kidney disease-mineral bone disease) is seen. Thus, phosphate seems to the prime initiator of this process. Patients with CKD/ESRD (end stage renal disease) exhibit increased mortality and morbidity not comparable to any chronic disease; a young adult on dialysis has life span of an 80-year-old adult ("Strategies for influencing outcomes in pre-ESRD and ESRD patients: A special conference," 1998). Even though elevated FGF23 initially is beneficial in preventing hyperphosphatemia, these FGF23 levels are not benign and are associated with increased cardiovascular mortality and morbidity, especially left ventricular hypertrophy (Faul et al., 2011; Gutierrez et al., 2008). Furthermore, attempts at neutralizing FGF23 in a rodent model of CKD lead to worsening hyperphosphatemia, increased levels of $1,25(OH)_2D_3$, hypercalcemia, low-PTH, vascular calcifications, and early mortality. These findings raise many questions regarding the role of elevated FGF23 as an adaptive protein versus a protein acting as a cardiotoxin, or somewhere in between. Phosphate also plays an important role in vascular calcification in many *in vivo* and *in vitro* studies (Kendrick & Chonchol, 2011).

In summary, regulation of phosphate homeostasis is critical for cellular function and survival. Alterations in serum phosphate levels can have detrimental consequences. Although regulation of phosphate balance is complex with the involvement of multiple hormones, it is crucial to understand the pathophysiological mechanisms of the disease processes with abnormal phosphate homeostasis. This will ultimately help treat patients with phosphate disorders adequately with minimal side effects.

REFERENCES

ADHR Consortium (2000). Autosomal dominant hypophosphataemic rickets is associated with mutations in FGF23. *Nature Genetics, 26*, 345–348.

Al-Azem, H., & Khan, A. A. (2012). Hypoparathyroidism. *Best Practice & Research. Clinical Endocrinology & Metabolism, 26*(4), 517–522.

Alon, U., Donaldson, D. L., Hellerstein, S., Warady, B. A., & Harris, D. J. (1992). Metabolic and histologic investigation of the nature of nephrocalcinosis in children with hypophosphatemic rickets and in the Hyp mouse. *Journal of Pediatrics, 120*, 899–905.

Amizuka, N., Lee, H. S., Kwan, M. Y., Arazani, A., Warshawsky, H., Hendy, G. N., et al. (1997). Cell-specific expression of the parathyroid hormone (PTH)/PTH-related peptide receptor gene in kidney from kidney-specific and ubiquitous promoters. *Endocrinology, 138*, 469–481.

Andrukhova, O., Zeitz, U., Goetz, R., Mohammadi, M., Lanske, B., & Erben, R. G. (2012). FGF23 acts directly on renal proximal tubules to induce phosphaturia through activation of the ERK1/2-SGK1 signaling pathway. *Bone, 51*, 621–628.

Anon. (1998). Strategies for influencing outcomes in pre-ESRD and ESRD patients: A special conference. *American Journal of Kidney Diseases, 32*, S1–S193.

Antoniucci, D. M., Yamashita, T., & Portale, A. A. (2006). Dietary phosphorus regulates serum fibroblast growth factor-23 concentrations in healthy men. *Journal of Clinical Endocrinology and Metabolism, 91*, 3144–3149.

Aono, Y., Hasegawa, H., Yamazaki, Y., Shimada, T., Fujita, T., Yamashita, T., et al. (2011). Anti-FGF-23 neutralizing antibodies ameliorate muscle weakness and decreased spontaneous movement of Hyp mice. *Journal of Bone and Mineral Research, 26*, 803–810.

Aono, Y., Yamazaki, Y., Yasutake, J., Kawata, T., Hasegawa, H., Urakawa, I., et al. (2009). Therapeutic effects of anti-FGF23 antibodies in hypophosphatemic rickets/osteomalacia. *Journal of Bone and Mineral Research, 24*, 1879–1888.

Ardura, J. A., & Friedman, P. A. (2011). Regulation of G protein-coupled receptor function by Na^+/H^+ exchange regulatory factors. *Pharmacological Reviews, 63*, 882–900.

Arend, L. J., Sonnenburg, W. K., Smith, W. L., & Spielman, W. S. (1987). A_1 and A_2 adenosine receptors in rabbit cortical collecting tubule cells. Modulation of hormone-stimulated cAMP. *Journal of Clinical Investigation, 79*, 710–714.

Ba, J., Brown, D., & Friedman, P. A. (2003). CaSR regulation of PTH-inhibitable proximal tubule phosphate transport. *American Journal of Physiology. Renal Physiology, 285*, F1233–F1243.

Bachmann, S., Schlichting, U., Geist, B., Mutig, K., Petsch, T., Bacic, D., et al. (2004). Kidney-specific inactivation of the megalin gene impairs trafficking of renal inorganic sodium phosphate cotransporter (NaPi-IIa). *Journal of the American Society of Nephrology, 15*, 892–900.

Bacic, D., Capuano, P., Gisler, S. M., Pribanic, S., Christensen, E. I., Biber, J., et al. (2003). Impaired PTH-induced endocytotic down-regulation of the renal type IIa Na+/Pi-cotransporter in RAP-deficient mice with reduced megalin expression. *Pflugers Archiv: European Journal of Physiology, 446*, 475–484.

Bacic, D., Lehir, M., Biber, J., Kaissling, B., Murer, H., & Wagner, C. A. (2006). The renal Na^+/phosphate cotransporter NaPi-IIa is internalized via the receptor-mediated endocytic route in response to parathyroid hormone. *Kidney International, 69*, 495–503.

Bacic, D., Schulz, N., Biber, J., Kaissling, B., Murer, H., & Wagner, C. A. (2003). Involvement of the MAPK-kinase pathway in the PTH-mediated regulation of the proximal tubule type IIa Na^+/P_i cotransporter in mouse kidney. *Pflugers Archiv: European Journal of Physiology, 446*, 52–60.

Bai, X. Y., Miao, D., Goltzman, D., & Karaplis, A. C. (2003). The autosomal dominant hypophosphatemic rickets R176Q mutation in fibroblast growth factor 23 resists proteolytic cleavage and enhances in vivo biological potency. *Journal of Biological Chemistry, 278*, 9843–9849.

Bai, X., Miao, D., Li, J., Goltzman, D., & Karaplis, A. C. (2004). Transgenic mice overexpressing human fibroblast growth factor 23 (R176Q) delineate a putative role for parathyroid hormone in renal phosphate wasting disorders. *Endocrinology, 145*, 5269–5279.

Beck, L., Karaplis, A. C., Amizuka, N., Hewson, A. S., Ozawa, H., & Tenenhouse, H. S. (1998). Targeted inactivation of *Npt2* in mice leads to severe renal phosphate wasting, hypercalciuria, and skeletal abnormalities. *Proceedings of the National Academy of Sciences of the United States of America, 95*, 5372–5377.

Beck, L., Leroy, C., Beck-Cormier, S., Forand, A., Salaun, C., Paris, N., et al. (2010). The phosphate transporter PiT1 (Slc20a1) revealed as a new essential gene for mouse liver development. *PLoS One, 5*, e9148.

Beenken, A., & Mohammadi, M. (2012). The structural biology of the FGF19 subfamily. *Advances in Experimental Medicine and Biology*, *728*, 1–24.

Belotti, E., Polanowska, J., Daulat, A. M., Audebert, S., Thome, V., Lissitzky, J. C., et al. (2013). The human PDZome: A gateway to PSD95-Disc Large-Zonula Occludens (PDZ)-mediated functions. *Molecular and Cellular Proteomics*, *12*, 2587–2603.

Ben-Dov, I. Z., Galitzer, H., Lavi-Moshayoff, V., Goetz, R., Kuro, O., Mohammadi, M., et al. (2007). The parathyroid is a target organ for FGF23 in rats. *Journal of Clinical Investigation*, *117*, 4003–4008.

Benet-Pagès, A., Lorenz-Depiereux, B., Zischka, H., White, K. E., Econs, M. J., & Strom, T. M. (2004). FGF23 is processed by proprotein convertases but not by PHEX. *Bone*, *35*, 455–462.

Benet-Pagès, A., Orlik, P., Strom, T. M., & Lorenz-Depiereux, B. (2005). An FGF23 missense mutation causes familial tumoral calcinosis with hyperphosphatemia. *Human Molecular Genetics*, *14*, 385–390.

Bergwitz, C., & Juppner, H. (2010). Regulation of phosphate homeostasis by PTH, vitamin D, and FGF23. *Annual Review of Medicine*, *61*, 91–104.

Bhargava, A., & Pearce, D. (2004). Mechanisms of mineralocorticoid action: Determinants of receptor specificity and actions of regulated gene products. *Trends in Endocrinology and Metabolism*, *15*, 147–153.

Bhattacharyya, N., Wiench, M., Dumitrescu, C., Connolly, B. M., Bugge, T. H., Patel, H. V., et al. (2012). Mechanism of FGF23 processing in fibrous dysplasia. *Journal of Bone and Mineral Research: The Official Journal of the American Society for Bone and Mineral Research*, *27*(5), 1132–1141. http://dx.doi.org/10.1002/jbmr.1546.

Biber, J., Hernando, N., & Forster, I. (2013). Phosphate transporters and their function. *Annual Review of Physiology*, *75*, 535–550.

Biber, J., Hernando, N., Forster, I., & Murer, H. (2009). Regulation of phosphate transport in proximal tubules. *Pflugers Archiv: European Journal of Physiology*, *458*, 39–52.

Biemesderfer, D., Nagy, T., DeGray, B., & Aronson, P. S. (1999). Specific association of megalin and the Na^+/H^+ exchanger isoform NHE3 in the proximal tubule. *Journal of Biological Chemistry*, *274*, 17518–17524.

Blaine, J., Okamura, K., Giral, H., Breusegem, S., Caldas, Y., Millard, A., et al. (2009). PTH-induced internalization of apical membrane NaPi2a: Role of actin and myosin VI. *American Journal of Physiology. Cell Physiology*, *297*, C1339–1346.

Blaine, J., Weinman, E. J., & Cunningham, R. (2011). The regulation of renal phosphate transport. *Advances in Chronic Kidney Disease*, *18*, 77–84.

Bloch, L., Sineshchekova, O., Reichenbach, D., Reiss, K., Saftig, P., Kuro-o, M., et al. (2009). Klotho is a substrate for alpha-, beta- and gamma-secretase. *FEBS Letters*, *583*, 3221–3224.

Bomberger, J. M., Coutermarsh, B. A., Barnaby, R. L., Sato, J. D., Chapline, M. C., & Stanton, B. A. (2014). Serum and glucocorticoid-inducible kinase1 increases plasma membrane wt-CFTR in human airway epithelial cells by inhibiting its endocytic retrieval. *PLoS One*, *9*, e89599.

Bottger, P., & Pedersen, L. (2011). Mapping of the minimal inorganic phosphate transporting unit of human PiT2 suggests a structure universal to PiT-related proteins from all kingdoms of life. *BMC Biochemistry*, *12*, 21.

Breusegem, S. Y., Takahashi, H., Giral-Arnal, H., Wang, X., Jiang, T., Verlander, J. W., et al. (2009). Differential regulation of the renal sodium-phosphate cotransporters NaPi-IIa, NaPi-IIc, and PiT-2 in dietary potassium deficiency. *American Journal of Physiology. Renal Physiology*, *297*, F350–361.

Bringhurst, F. R. (2003). Circulating forms of parathyroid hormone: Peeling back the onion. *Clinical Chemistry*, *49*, 1973–1975.

Brossard, J. H., Cloutier, M., Roy, L., Lepage, R., Gascon-Barré, N., & D'Amour, P. (1996). Accumulation of a non-(1-84) molecular form of parathyroid hormone (PTH) detected by intact PTH assay in renal failure: Importance in the interpretation of PTH values. *Journal of Clinical Endocrinology and Metabolism, 81*, 3923–3929.

Brown, W. W., Juppner, H., Langman, C. B., Price, H., Farrow, E. G., White, K. E., et al. (2009). Hypophosphatemia with elevations in serum fibroblast growth factor 23 in a child with Jansen's metaphyseal chondrodysplasia. *Journal of Clinical Endocrinology and Metabolism, 94*, 17–20.

Brownstein, C. A., Adler, F., Nelson-Williams, C., Iijima, J., Li, P., Imura, A., et al. (2008). A translocation causing increased alpha-klotho level results in hypophosphatemic rickets and hyperparathyroidism. *Proceedings of the National Academy of Sciences of the United States of America, 105*, 3455–3460.

Budu, C. E., Efendiev, R., Cinelli, A. M., Bertorello, A. M., & Pedemonte, C. H. (2002). Hormonal-dependent recruitment of Na+, K+-ATPase to the plasmalemma is mediated by PKCβ and modulated by [Na+]i. *British Journal of Pharmacology, 137*, 1380–1386.

Burnett, S. M., Gunawardene, S. C., Bringhurst, F. R., Juppner, H., Lee, H., & Finkelstein, J. S. (2006). Regulation of C-terminal and intact FGF-23 by dietary phosphate in men and women. *Journal of Bone and Mineral Research, 21*, 1187–1196.

Cai, H., Puschett, D. B., Guan, S., Batuman, V., & Puschett, J. B. (1995). Phosphate transport inhibition by KW-3902, an adenosine A1 receptor antagonist, is mediated by cyclic adenosine monophosphate. *American Journal of Kidney Diseases, 26*, 825–830.

Capuano, P., Bacic, D., Roos, M., Gisler, S. M., Stange, G., Biber, J., et al. (2007). Defective coupling of apical PTH receptors to phospholipase C prevents internalization of the Na^+-phosphate cotransporter NaPi-IIa in Nherf1-deficient mice. *American Journal of Physiology. Cell Physiology, 292*, C927–C934.

Capuano, P., Bacic, D., Stange, G., Hernando, N., Kaissling, B., Pal, R., et al. (2005). Expression and regulation of the renal Na/phosphate cotransporter NaPi-IIa in a mouse model deficient for the PDZ protein PDZK1. *Pflugers Archiv: European Journal of Physiology, 449*, 392–402.

Carpenter, T. O., Imel, E. A., Ruppe, M. D., Weber, T. J., Klausner, M. A., Wooddell, M. M., et al. (2014). Randomized trial of the anti-FGF23 antibody KRN23 in X-linked hypophosphatemia. *Journal of Clinical Investigation, 124*, 1587–1597.

Carrillo-Lopez, N., Roman-Garcia, P., Rodriguez-Rebollar, A., Fernandez-Martin, J. L., Naves-Diaz, M., & Cannata-Andia, J. B. (2009). Indirect regulation of PTH by estrogens may require FGF23. *Journal of the American Society of Nephrology, 20*, 2009–2017.

Caverzasio, J., & Bonjour, J. P. (1989). Insulin-like growth factor I stimulates Na-dependent Pi transport in cultured kidney cells. *American Journal of Physiology, 257*, F712–F717.

Chang, A. C., Cha, J., Koentgen, F., & Reddel, R. R. (2005). The murine stanniocalcin 1 gene is not essential for growth and development. *Molecular and Cellular Biology, 25*, 10604–10610.

Chase, L. R., & Aurbach, G. D. (1967). Parathyroid function and the renal excretion of 3'5'-adenylic acid. *Proceedings of the National Academy of Sciences of the United States of America, 58*, 518–525.

Chen, C. D., Podvin, S., Gillespie, E., Leeman, S. E., & Abraham, C. R. (2007). Insulin stimulates the cleavage and release of the extracellular domain of Klotho by ADAM10 and ADAM17. *Proceedings of the National Academy of Sciences of the United States of America, 104*, 19796–19801.

Christensen, E. I., & Birn, H. (2002). Megalin and cubilin: Multifunctional endocytic receptors. *Nature Reviews. Molecular Cell Biology, 3*, 256–266.

Cole, J. A. (1999). Parathyroid hormone activates mitogen-activated protein kinase in opossum kidney cells. *Endocrinology, 140*, 5771–5779.

Cole, J. A., Eber, S. L., Poelling, R. E., Thorne, P. K., & Forte, L. R. (1987). A dual mechanism for regulation of kidney phosphate transport by parathyroid hormone. *American Journal of Physiology, 253*, E221–E227.

Cole, J. A., Forte, L. R., Eber, S., Thorne, P. K., & Poelling, R. E. (1988). Regulation of sodium-dependent phosphate transport by parathyroid hormone in opossum kidney cells: Adenosine 3',5'-monophosphate- dependent and -independent mechanisms. *Endocrinology, 122*, 2981–2989.

Cole, J. A., Forte, L. R., Krause, W. J., & Thorne, P. K. (1989). Clonal sublines that are morphologically and functionally distinct from parental OK cells. *American Journal of Physiology, 256*, F672–F679.

Consortium, H. Y. P. (1995). A gene (PEX) with homologies to endopeptidases is mutated in patients with X-linked hypophosphatemic rickets. The HYP Consortium. *Nature Genetics, 11*, 130–136.

Cordeaux, Y., Ijzerman, A. P., & Hill, S. J. (2004). Coupling of the human A1 adenosine receptor to different heterotrimeric G proteins: Evidence for agonist-specific G protein activation. *British Journal of Pharmacology, 143*, 705–714.

Coulson, R., Johnson, R. A., Olsson, R. A., Cooper, D. R., & Scheinman, S. J. (1991). Adenosine stimulates phosphate and glucose transport in opossum kidney epithelial cells. *American Journal of Physiology, 260*, F921–928.

Courbebaisse, M., Leroy, C., Bakouh, N., Salaun, C., Beck, L., Grandchamp, B., et al. (2012). A new human NHERF1 mutation decreases renal phosphate transporter NPT2a expression by a PTH-independent mechanism. *PLoS One, 7*, e34764.

Cunningham, R., Biswas, R., Brazie, M., Steplock, D., Shenolikar, S., & Weinman, E. J. (2009). Signaling pathways utilized by PTH and dopamine to inhibit phosphate transport in mouse renal proximal tubule cells. *American Journal of Physiology. Renal Physiology, 296*, F355–F361.

Custer, M., Spindler, B., Verrey, F., Murer, H., & Biber, J. (1997). Identification of a new gene product (diphor-1) regulated by dietary phosphate. *American Journal of Physiology, 273*, F801–806.

D'Amour, P. (2012). Acute and chronic regulation of circulating PTH: Significance in health and in disease. *Clinical Biochemistry, 45*, 964–969.

Daugaard, H. (1996). Peripheral metabolism of parathyroid hormone—Studies with isolated perfused kidneys and livers from normal and uremic rats. *Danish Medical Bulletin, 43*, 203–215.

David, V., Dai, B., Martin, A., Huang, J., Han, X., & Quarles, L. D. (2013). Calcium regulates FGF-23 expression in bone. *Endocrinology, 154*, 4469–4482.

Davies, J. H., & Shaw, N. J. (2012). Investigation and management of hypercalcaemia in children. *Archives of Disease in Childhood, 97*(6), 533–538.

de Toledo, F. G., Beers, K. W., Berndt, T. J., Thompson, M. A., Tyce, G. M., Knox, F. G., et al. (1997). Opposite paracrine effects of 5-HT and dopamine on Na(+)-Pi cotransport in opossum kidney cells. *Kidney International, 52*, 152–156.

Deol, H. K., Varghese, R., Wagner, G. F., & Dimattia, G. E. (2000). Dynamic regulation of mouse ovarian stanniocalcin expression during gestation and lactation. *Endocrinology, 141*, 3412–3421.

Du, L., Desbarats, M., Viel, J., Glorieux, F. H., Cawthorn, C., & Ecarot, B. (1996). cDNA cloning of the murine Pex gene implicated in X-linked hypophosphatemia and evidence for expression in bone. *Genomics, 36*, 22–28.

Dusso, A. S., Brown, A. J., & Slatopolsky, E. (2005). Vitamin D. *American Journal of Physiology. Renal Physiology, 289*, F8–28.

Econs, M. J., & McEnery, P. T. (1997). Autosomal dominant hypophosphatemic rickets/osteomalacia: Clinical characterization of a novel renal phosphate-wasting disorder. *Journal of Clinical Endocrinology and Metabolism, 82*, 674–681.

Ediger, T. R., Kraus, W. L., Weinman, E. J., & Katzenellenbogen, B. S. (1999). Estrogen receptor regulation of the Na^+/H^+ exchange regulatory factor. *Endocrinology, 140*, 2976–2982.

Ediger, T. R., Park, S. E., & Katzenellenbogen, B. S. (2002). Estrogen receptor inducibility of the human Na^+/H^+ exchanger regulatory factor/ezrin-radixin-moesin binding protein 50 (NHE-RF/EBP50) gene involving multiple half-estrogen response elements. *Molecular Endocrinology, 16*, 1828–1839.

Egbuna, O. I., & Brown, E. M. (2008). Hypercalcaemic and hypocalcaemic conditions due to calcium-sensing receptor mutations. *Best Practice & Research. Clinical Rheumatology, 22*(1), 129–148.

Faroqui, S., Levi, M., Soleimani, M., & Amlal, H. (2008). Estrogen downregulates the proximal tubule type IIa sodium phosphate cotransporter causing phosphate wasting and hypophosphatemia. *Kidney International, 73*, 1141–1150.

Farrow, E. G., Davis, S. I., Summers, L. J., & White, K. E. (2009). Initial FGF23-mediated signaling occurs in the distal convoluted tubule. *Journal of the American Society of Nephrology, 20*, 955–960.

Farrow, E. G., Yu, X., Summers, L. J., Davis, S. I., Fleet, J. C., Allen, M. R., et al. (2011). Iron deficiency drives an autosomal dominant hypophosphatemic rickets (ADHR) phenotype in fibroblast growth factor-23 (Fgf23) knock-in mice. *Proceedings of the National Academy of Sciences of the United States of America, 108*(46), E1146–E1155. http://dx.doi.org/10.1073/pnas.1110905108. Epub 2011 Oct 17.

Faul, C., Amaral, A. P., Oskouei, B., Hu, M. C., Sloan, A., Isakova, T., et al. (2011). FGF23 induces left ventricular hypertrophy. *Journal of Clinical Investigation, 121*, 4393 4408.

Feng, J. Q., Ward, L. M., Liu, S., Lu, Y., Xie, Y., Yuan, B., et al. (2006). Loss of DMP1 causes rickets and osteomalacia and identifies a role for osteocytes in mineral metabolism. *Nature Genetics, 38*, 1310–1315.

Ferrari, S. L., Bonjour, J. P., & Rizzoli, R. (2005). Fibroblast growth factor-23 relationship to dietary phosphate and renal phosphate handling in healthy young men. *Journal of Clinical Endocrinology and Metabolism, 90*, 1519–1524.

Forster, I. C., Hernando, N., Biber, J., & Murer, H. (2006). Proximal tubular handling of phosphate: A molecular perspective. *Kidney International, 70*, 1548–1559.

Forster, I. C., Hernando, N., Biber, J., & Murer, H. (2013). Phosphate transporters of the SLC20 and SLC34 families. *Molecular Aspects of Medicine, 34*, 386–395.

Friedlander, G., Couette, S., Coureau, C., & Amiel, C. (1992). Mechanisms whereby extracellular adenosine 3',5'-monophosphate inhibits phosphate transport in cultured opossum kidney cells and in rat kidney. Physiological implication. *Journal of Clinical Investigation, 90*, 848–858.

Friedman, P. A., Morley, P., Whitfield, J. F., Willick, G., & Gesek, F. A. (1996). Structure-activity relations of amino-terminal PTH fragments on intracellular calcium in distal convoluted tubule cells. *Journal of the American Society of Nephrology*, 7, 1676.

Garabedian, M., Holick, M. F., DeLuca, H. F., & Boyle, I. T. (1972). Control of 25-hydroxycholecalciferol metabolism by parathyroid glands. *Proceedings of the National Academy of Sciences of the United States of America, 69*, 1673–1676.

Gattineni, J., Alphonse, P., Zhang, Q., Mathews, N., Bates, C. M., & Baum, M. (2014). Regulation of renal phosphate transport by FGF23 is mediated by FGFR1 and FGFR4. *American Journal of Physiology. Renal Physiology, 306*(3), F351–F358. http://dx.doi.org/10.1152/ajprenal.00232.2013. Epub 2013 Nov 20.

Gattineni, J., Bates, C., Twombley, K., Dwarakanath, V., Robinson, M. L., Goetz, R., et al. (2009). FGF23 decreases renal NaPi-2a and NaPi-2c expression and induces hypophosphatemia in vivo predominantly via FGF receptor 1. *American Journal of Physiology. Renal Physiology, 297*, F282–F291.

Gattineni, J., & Baum, M. (2012). Genetic disorders of phosphate regulation. *Pediatric Nephrology*, *27*, 1477–1487.
Gattineni, J., Twombley, K., Goetz, R., Mohammadi, M., & Baum, M. (2011). Regulation of serum 1,25(OH)2 vitamin D3 levels by fibroblast growth factor 23 is mediated by FGF receptors 3 and 4. *American Journal of Physiology. Renal Physiology*, *301*, F371–F377.
Gaztambide, S., Vazquez, F., & Castano, L. (2013). Diagnosis and treatment of multiple endocrine neoplasia type 1 (MEN1). *Minerva Endocrinologica*, *38*(1), 17–28.
Gisler, S. M., Kittanakom, S., Fuster, D., Wong, V., Bertic, M., Radanovic, T., et al. (2008). Monitoring protein-protein interactions between the mammalian integral membrane transporters and PDZ-interacting partners using a modified split-ubiquitin membrane yeast two-hybrid system. *Molecular and Cellular Proteomics*, 7, 1362–1377.
Gisler, S. M., Pribanic, S., Bacic, D., Forrer, P., Gantenbein, A., Sabourin, L. A., et al. (2003). PDZK1: I. A major scaffolder in brush borders of proximal tubular cells. *Kidney International*, *64*, 1733–1745.
Gisler, S. M., Stagljar, I., Traebert, M., Bacic, D., Biber, J., & Murer, H. (2001). Interaction of the type IIa Na/Pi cotransporter with PDZ proteins. *Journal of Biological Chemistry*, *276*, 9206–9213.
Giusti, F., Cavalli, L., Cavalli, T., & Brandi, M. L. (2013). Hereditary hyperparathyroidism syndromes. *Journal of Clinical Densitometry*, *16*(1), 69–74.
Goetz, R., Beenken, A., Ibrahimi, O. A., Kalinina, J., Olsen, S. K., Eliseenkova, A. V., et al. (2007). Molecular insights into the klotho-dependent, endocrine mode of action of fibroblast growth factor 19 subfamily members. *Molecular and Cellular Biology*, *27*, 3417–3428.
Goetz, R., Nakada, Y., Hu, M. C., Kurosu, H., Wang, L., Nakatani, T., et al. (2010). Isolated C-terminal tail of FGF23 alleviates hypophosphatemia by inhibiting FGF23-FGFR-Klotho complex formation. *Proceedings of the National Academy of Sciences of the United States of America*, *107*, 407–412.
Gravesen, E., Hofman-Bang, J., Mace, M. L., Lewin, E., & Olgaard, K. (2013). High dose intravenous iron, mineral homeostasis and intact FGF23 in normal and uremic rats. *BMC Nephrology*, *14*, 281.
Guo, J., Liu, M., Yang, D., Bouxsein, M. L., Thomas, C. C., Schipani, E., et al. (2010). Phospholipase C signaling via the parathyroid Hormone (PTH)/PTH-related peptide receptor Is essential for normal bone responses to PTH. *Endocrinology*, *151*, 3502–3513.
Guo, J., Song, L., Liu, M., & Mahon, M. J. (2012). Fluorescent ligand-directed co-localization of the parathyroid hormone 1 receptor with the brush-border scaffold complex of the proximal tubule reveals hormone-dependent changes in ezrin immunoreactivity consistent with inactivation. *Biochimica et Biophysica Acta*, *1823*, 2243–2253.
Guo, J., Song, L., Liu, M., Segawa, H., Miyamoto, K. I., Bringhurst, F. R., et al. (2013). Activation of a non-cAMP/PKA signaling pathway downstream of the PTH/PTHrP receptor Is essential for a sustained hypophosphatemic response to PTH infusion in male mice. *Endocrinology*, *154*, 1680–1689.
Gutierrez, O. M., Mannstadt, M., Isakova, T., Rauh-Hain, J. A., Tamez, H., Shah, A., et al. (2008). Fibroblast growth factor 23 and mortality among patients undergoing hemodialysis. *New England Journal of Medicine*, *359*, 584–592.
Hafdi, Z., Couette, S., Comoy, E., Prie, D., Amiel, C., & Friedlander, G. (1996). Locally formed 5-hydroxytryptamine stimulates phosphate transport in cultured opossum kidney cells and in rat kidney. *Biochemical Journal*, *320*(Pt 2), 615–621.
Hayes, G., Busch, A., Lötscher, M., Waldegger, S., Lang, F., Verrey, F., et al. (1994). Role of *N*-linked glycosylation in rat renal Na/P_i-cotransport. *Journal of Biological Chemistry*, *269*, 24143–24149.
Henry, H. L., & Norman, A. W. (1984) Vitamin D: Metabolism and biological actions. *Annual Review of Nutrition*, *4*, 493–520.

Hernando, N., Deliot, N., Gisler, S. M., Lederer, E., Weinman, E. J., Biber, J., et al. (2002). PDZ domain interactions and apical expression of type IIa Na/Pi cotransporters. *Proceedings of the National Academy of Sciences of the United States of America, 99*, 11957–11962.
Hernando, N., Gisler, S. M., Pribanic, S., Deliot, N., Capuano, P., Wagner, C. A., et al. (2005). NaPi-IIa and interacting partners. *Journal of Physiology, 567*, 21–26.
Hill, K. M., Martin, B. R., Wastney, M. E., McCabe, G. P., Moe, S. M., Weaver, C. M., et al. (2013). Oral calcium carbonate affects calcium but not phosphorus balance in stage 3-4 chronic kidney disease. *Kidney International, 83*, 959–966.
Hilpert, J., Nykjaer, A., Jacobsen, C., Wallukat, G., Nielsen, R., Moestrup, S. K., et al. (1999). Megalin antagonizes activation of the parathyroid hormone receptor. *Journal of Biological Chemistry, 274*, 5620–5625.
Hoag, H. M., Martel, J., Gauthier, C., & Tenenhouse, H. S. (1999). Effects of Npt2 gene ablation and low-phosphate diet on renal Na^+/phosphate cotransport and cotransporter gene expression. *Journal of Clinical Investigation, 104*, 679–686.
Horiuchi, N., Holick, M. F., Potts, J. T., Jr., & Rosenblatt, M. (1983). A parathyroid hormone inhibitor in vivo: Design and biological evaluation of a hormone analog. *Science, 220*, 1053–1055.
Hryszko, T., Rydzewska-Rosolowska, A., Brzosko, S., Koc-Zorawska, E., & Mysliwiec, M. (2012). Low molecular weight iron dextran increases fibroblast growth factor-23 concentration, together with parathyroid hormone decrease in hemodialyzed patients. *Therapeutic Apheresis and Dialysis, 16*, 146–151.
Hu, M. C., Shi, M., Zhang, J., Pastor, J., Nakatani, T., Lanske, B., et al. (2010). Klotho: A novel phosphaturic substance acting as an autocrine enzyme in the renal proximal tubule. *FASEB Journal, 24*, 3438–3450.
Hu, M. C., Shiizaki, K., Kuro-o, M., & Moe, O. W. (2013). Fibroblast growth factor 23 and Klotho: Physiology and pathophysiology of an endocrine network of mineral metabolism. *Annual Review of Physiology, 75*, 503–533.
Hung, N. T., Yamamoto, H., Takei, Y., Masuda, M., Otani, A., Kozai, M., et al. (2012). Up-regulation of stanniocalcin 1 expression by 1,25-dihydroxy vitamin D(3) and parathyroid hormone in renal proximal tubular cells. *Journal of Clinical Biochemistry and Nutrition, 50*, 227–233.
Huttlin, E. L., Jedrychowski, M. P., Elias, J. E., Goswami, T., Rad, R., Beausoleil, S. A., et al. (2010). A tissue-specific atlas of mouse protein phosphorylation and expression. *Cell, 143*, 1174–1189.
Ichikawa, S., Gray, A. K., Bikorimana, E., & Econs, M. J. (2013). Dosage effect of a Phex mutation in a murine model of X-linked hypophosphatemia. *Calcified Tissue International, 93*, 155–162.
Ichikawa, S., Imel, E. A., Kreiter, M. L., Yu, X., Mackenzie, D. S., Sorenson, A. H., et al. (2007). A homozygous missense mutation in human KLOTHO causes severe tumoral calcinosis. *Journal of Clinical Investigation, 117*, 2684–2691.
Iida-Klein, A., Guo, J., Takemura, M., Drake, M. T., Potts, J. T., Jr., Abou-Samra, A., et al. (1997). Mutations in the second cytoplasmic loop of the rat parathyroid hormone (PTH)/PTH-related protein receptor result in selective loss of PTH-stimulated phospholipase C activity. *Journal of Biological Chemistry, 272*, 6882–6889.
Imel, E. A., Peacock, M., Gray, A. K., Padgett, L. R., Hui, S. L., & Econs, M. J. (2011). Iron modifies plasma FGF23 differently in autosomal dominant hypophosphatemic rickets and healthy humans. *Journal of Clinical Endocrinology and Metabolism, 96*, 3541–3549.
Imura, A., Iwano, A., Tohyama, O., Tsuji, Y., Nozaki, K., Hashimoto, N., et al. (2004). Secreted Klotho protein in sera and CSF: Implication for post-translational cleavage in release of Klotho protein from cell membrane. *FEBS Letters, 565*, 143–147.
Jankowski, M., Hilfiker, H., Biber, J., & Murer, H. (2001). The opossum kidney cell type IIa Na/Pi cotransporter is a phosphoprotein. *Kidney and Blood Pressure Research, 24*, 1–4.

Jehle, A. W., Forgo, J., Biber, J., Lederer, E., Krapf, R., & Murer, H. (1998). IGF-I and vanadate stimulate Na/Pi-cotransport in OK cells by increasing type II Na/Pi-cotransporter protein stability. *Pflügers Archiv*, *437*, 149–154.

Jouishomme, H., Whitfield, J. F., Chakravarthy, B., Durkin, J. P., Gagnon, L., Isaacs, R. J., et al. (1992). The protein kinase-C activation domain of the parathyroid hormone. *Endocrinology*, *130*, 53–60.

Jouishomme, H., Whitfield, J. F., Gagnon, L., MacLean, S., Isaacs, R., Chakravarthy, B., et al. (1994). Further definition of the PKC activation domain of the parathyroid hormone. *Journal of Bone and Mineral Research*, *9*, 943–949.

Karim, Z., Gerard, B., Bakouh, N., Alili, R., Leroy, C., Beck, L., et al. (2008). NHERF1 mutations and responsiveness of renal parathyroid hormone. *New England Journal of Medicine*, *359*, 1128–1135.

Kawata, T., Imanishi, Y., Kobayashi, K., Miki, T., Arnold, A., Inaba, M., et al. (2007). Parathyroid hormone regulates fibroblast growth factor-23 in a mouse model of primary hyperparathyroidism. *Journal of the American Society of Nephrology*, *18*, 2683–2688.

Kendrick, J., & Chonchol, M. (2011). The role of phosphorus in the development and progression of vascular calcification. *American Journal of Kidney Diseases*, *58*, 826–834.

Keusch, I., Traebert, M., Lotscher, M., Kaissling, B., Murer, H., & Biber, J. (1998). Parathyroid hormone and dietary phosphate provoke a lysosomal routing of the proximal tubular Na/Pi-cotransporter type II. *Kidney International*, *54*, 1224–1232.

Khundmiri, S. J., Ahmad, A., Bennett, R. E., Weinman, E. J., Steplock, D., Cole, J., et al. (2008). Novel regulatory function for NHERF-1 in Npt2a transcription. *American Journal of Physiology. Renal Physiology*, *294*, F840–F849.

Kido, S., Kaneko, I., Tatsumi, S., Segawa, H., & Miyamoto, K. (2013). Vitamin D and type II sodium-dependent phosphate cotransporters. *Contributions to Nephrology*, *180*, 86–97.

Kilav, R., Silver, J., & Naveh-Many, T. (1995). Parathyroid hormone gene expression in hypophosphatemic rats. *Journal of Clinical Investigation*, *96*, 327–333.

Kobayashi, K., Imanishi, Y., Miyauchi, A., Onoda, N., Kawata, T., Tahara, H., et al. (2006). Regulation of plasma fibroblast growth factor 23 by calcium in primary hyperparathyroidism. *European Journal of Endocrinology*, *154*, 93–99.

Kocher, O., Pal, R., Roberts, M., Cirovic, C., & Gilchrist, A. (2003). Targeted disruption of the PDZK1 gene by homologous recombination. *Molecular and Cellular Biology*, *23*, 1175–1180.

Krajisnik, T., Bjorklund, P., Marsell, R., Ljunggren, O., Akerstrom, G., Jonsson, K. B., et al. (2007). Fibroblast growth factor-23 regulates parathyroid hormone and 1alpha-hydroxylase expression in cultured bovine parathyroid cells. *Journal of Endocrinology*, *195*, 125–131.

Kuntziger, H., Amiel, C., Roinel, N., & Morel, F. (1974). Effects of parathyroidectomy and cyclic AMP on renal transport of phosphate, calcium and magnesium. *American Journal of Physiology*, *227*, 905–911.

Kuro-o, M., Matsumura, Y., Aizawa, H., Kawaguchi, H., Suga, T., Utsugi, T., et al. (1997). Mutation of the mouse klotho gene leads to a syndrome resembling ageing. *Nature*, *390*, 45–51.

Kurosu, H., & Kuro, O. (2008). The Klotho gene family as a regulator of endocrine fibroblast growth factors. *Molecular and Cellular Endocrinology*, *299*, 72–78.

Kurosu, H., & Kuro-o, M. (2008). The Klotho gene family and the endocrine fibroblast growth factors. *Current Opinion in Nephrology and Hypertension*, *17*, 368–372.

Kurosu, H., Ogawa, Y., Miyoshi, M., Yamamoto, M., Nandi, A., Rosenblatt, K. P., et al. (2006). Regulation of fibroblast growth factor-23 signaling by klotho. *Journal of Biological Chemistry*, *281*, 6120–6123.

Lammer, E. J., & Opitz, J. M. (1986). The DiGeorge anomaly as a developmental field defect. *American Journal of Medical Genetics*, *2*(Suppl.), 113–127.

Lanaspa, M. A., Caldas, Y. A., Breusegem, S. Y., Andres-Hernando, A., Cicerchi, C., Levi, M., et al. (2013). Inorganic phosphate modulates the expression of the NaPi-2a transporter in the trans-Golgi network and the interaction with PIST in the proximal tubule. *BioMed Research International, 2013*, 513932.

Larsson, T., Marsell, R., Schipani, E., Ohlsson, C., Ljunggren, O., Tenenhouse, H. S., et al. (2004). Transgenic mice expressing fibroblast growth factor 23 under the control of the alpha1(I) collagen promoter exhibit growth retardation, osteomalacia, and disturbed phosphate homeostasis. *Endocrinology, 145*, 3087–3094.

Larsson, T., Nisbeth, U., Ljunggren, O., Juppner, H., & Jonsson, K. B. (2003). Circulating concentration of FGF-23 increases as renal function declines in patients with chronic kidney disease, but does not change in response to variation in phosphate intake in healthy volunteers. *Kidney International, 64*, 2272–2279.

Lavi-Moshayoff, V., Wasserman, G., Meir, T., Silver, J., & Naveh-Many, T. (2010). PTH increases FGF23 gene expression and mediates the high-FGF23 levels of experimental kidney failure: A bone parathyroid feedback loop. *American Journal of Physiology. Renal Physiology, 299*, F882–F889.

Lederer, E. D., Sohi, S. S., Mathiesen, J. M., & Klein, J. B. (1998). Regulation of expression of type II sodium-phosphate cotransporters by protein kinases A and C. *American Journal of Physiology. Renal Physiology, 275*, F270–F277.

Lederer, E. D., Sohi, S. S., & McLeish, K. R. (2000). Parathyroid hormone stimulates extracellular signal-regulated kinase (ERK) activity through two independent signal transduction pathways: Role of ERK in sodium-phosphate cotransport. *Journal of the American Society of Nephrology, 11*, 222–231.

Levi, M., Shayman, J. A., Abe, A., Gross, S. K., McCluer, R. H., Biber, J., et al. (1995). Dexamethasone modulates rat renal brush border membrane phosphate transporter mRNA and protein abundance and glycosphingolipid composition. *Journal of Clinical Investigation, 96*, 207–216.

Levy-Litan, V., Hershkovitz, E., Avizov, L., Leventhal, N., Bercovich, D., Chalifa-Caspi, V., et al. (2010). Autosomal-recessive hypophosphatemic rickets is associated with an inactivation mutation in the ENPP1 gene. *American Journal of Human Genetics, 86*, 273–278.

Li, H., Martin, A., David, V., & Quarles, L. D. (2011). Compound deletion of Fgfr3 and Fgfr4 partially rescues the Hyp mouse phenotype. *American Journal of Physiology. Endocrinology and Metabolism, 300*, E508–E517.

Liu, L., Alonso, V., Guo, L., Tourkova, I., Henderson, S. E., Almarza, A. J., et al. (2012). Na^+/H^+-exchanger regulatory factor-1 (NHERF1) directly regulates osteogenesis. *Journal of Biological Chemistry, 287*, 43312–43321.

Liu, S., Guo, R., Simpson, L. G., Xiao, Z. S., Burnham, C. E., & Quarles, L. D. (2003). Regulation of fibroblastic growth factor 23 expression but not degradation by PHEX. *Journal of Biological Chemistry, 278*, 37419–37426.

Liu, S., Tang, W., Fang, J., Ren, J., Li, H., Xiao, Z., et al. (2009). Novel regulators of Fgf23 expression and mineralization in Hyp bone. *Molecular Endocrinology, 23*, 1505–1518.

Liu, S., Tang, W., Zhou, J., Stubbs, J. R., Luo, Q., Pi, M., et al. (2006). Fibroblast growth factor 23 is a counter-regulatory phosphaturic hormone for vitamin D. *Journal of the American Society of Nephrology, 17*, 1305–1315.

Liu, S., Tang, W., Zhou, J., Vierthaler, L., & Quarles, L. D. (2007). Distinct roles for intrinsic osteocyte abnormalities and systemic factors in regulation of FGF23 and bone mineralization in Hyp mice. *American Journal of Physiology. Endocrinology and Metabolism, 293*, E1636–E1644.

Liu, S., Vierthaler, L., Tang, W., Zhou, J., & Quarles, L. D. (2008). FGFR3 and FGFR4 do not mediate renal effects of FGF23. *Journal of the American Society of Nephrology, 19*, 2342–2350.

Loffing, J., Lötscher, M., Kaissling, B., Biber, J., Murer, H., Seikaly, M., et al. (1998). Renal Na/H exchanger NHE-3 and Na-PO_4 cotransporter NaP_i-2 protein expression in glucocorticoid excess and deficient states. *Journal of the American Society of Nephrology, 9*, 1560–1567.

Lorenz-Depiereux, B., Bastepe, M., Benet-Pagès, A., Amyere, M., Wagenstaller, J., Muller-Barth, U., et al. (2006). DMP1 mutations in autosomal recessive hypophosphatemia implicate a bone matrix protein in the regulation of phosphate homeostasis. *Nature Genetics, 38*, 1248–1250.

Lorenz-Depiereux, B., Schnabel, D., Tiosano, D., Hausler, G., & Strom, T. M. (2010). Loss-of-function ENPP1 mutations cause both generalized arterial calcification of infancy and autosomal-recessive hypophosphatemic rickets. *American Journal of Human Genetics, 86*, 267–272.

Lötscher, M., Kaissling, B., Biber, J., Murer, H., Kempson, S. A., & Levi, M. (1996). Regulation of rat renal Na/P_i-cotransporter by parathyroid hormone: Immunohistochemistry. *Kidney International, 49*, 1010–1011.

Lu, Y., Yuan, B., Qin, C., Cao, Z., Xie, Y., Dallas, S. L., et al. (2011). The biological function of DMP-1 in osteocyte maturation is mediated by its 57-kDa C-terminal fragment. *Journal of Bone and Mineral Research: The Official Journal of the American Society for Bone and Mineral Research, 26*(2), 331–340. http://dx.doi.org/10.1002/jbmr.226.

Mahon, M. J. (2008). Ezrin promotes functional expression and parathyroid hormone-mediated regulation of the sodium-phosphate co-transporter 2a in LLC-PK1 cells. *American Journal of Physiology. Renal Physiology, 294*, F667–F675.

Malmstrom, K., Stange, G., & Murer, H. (1988). Intracellular cascades in the parathyroid-hormone-dependent regulation of Na^+/phosphate cotransport in OK cells. *Biochemical Journal, 251*, 207–213.

Mamonova, T., Kurnikova, M., & Friedman, P. A. (2012). Structural basis for NHERF1 PDZ domain binding affinity. *Biochemistry, 51*, 3110–3120.

Mannstadt, M., Bertrand, G., Muresan, M., Weryha, G., Leheup, B., Pulusani, S. R., et al. (2008). Dominant-negative GCMB mutations cause an autosomal dominant form of hypoparathyroidism. *The Journal of Clinical Endocrinology and Metabolism, 93*(9), 3568–3576.

Martin, A., David, V., Laurence, J. S., Schwarz, P. M., Lafer, E. M., Hedge, A. M., et al. (2008). Degradation of MEPE, DMP1, and release of SIBLING ASARM-peptides (minhibins): ASARM-peptide(s) are directly responsible for defective mineralization in HYP. *Endocrinology, 149*, 1757–1772.

Martin, A., Liu, S., David, V., Li, H., Karydis, A., Feng, J. Q., et al. (2011). Bone proteins PHEX and DMP1 regulate fibroblastic growth factor Fgf23 expression in osteocytes through a common pathway involving FGF receptor (FGFR) signaling. *FASEB Journal, 25*, 2551–2562.

Martin, K. J., McConkey, C. L., Jacob, A. K., Gonzalez, E. A., Khan, M., & Baldassare, J. J. (1994). Effect of U-73,122, an inhibitor of phospholipase C, on actions of parathyroid hormone in opossum kidney cells. *American Journal of Physiology, 266*, F254–F258.

Matsumura, Y., Aizawa, H., Shiraki-Iida, T., Nagai, R., Kuro-o, M., & Nabeshima, Y. (1998). Identification of the human klotho gene and its two transcripts encoding membrane and secreted klotho protein. *Biochemical and Biophysical Research Communications, 242*, 626–630.

Miyamoto, K., Segawa, H., Morita, K., Nii, T., Tatsumi, S., Taketani, Y., et al. (1997). Relative contributions of Na^+-dependent phosphate co-transporters to phosphate transport in mouse kidney: RNase H-mediated hybrid depletion analysis. *Biochemical Journal, 327*, 735–739.

Monier-Faugere, M. C., Geng, Z., Mawad, H., Friedler, R. M., Gao, P., Cantor, T. L., et al. (2001). Improved assessment of bone turnover by the PTH-(1-84)/large C-PTH fragments ratio in ESRD patients. *Kidney International, 60*, 1460–1468.

Morales, F. C., Takahashi, Y., Kreimann, E. L., & Georgescu, M. M. (2004). Ezrin-radixin-moesin (ERM)-binding phosphoprotein 50 organizes ERM proteins at the apical membrane of polarized epithelia. *Proceedings of the National Academy of Sciences of the United States of America, 101*, 17705–17710.

Morey, M., Castro-Feijoo, L., Barreiro, J., Cabanas, P., Pombo, M., Gil, M., et al. (2011). Genetic diagnosis of X-linked dominant hypophosphatemic rickets in a cohort study: Tubular reabsorption of phosphate and 1,25(OH)2D serum levels are associated with PHEX mutation type. *BMC Medical Genomics, 12*, 116.

Muff, R., Caulfield, M. P., & Fischer, J. A. (1990). Dissociation of cAMP accumulation and phosphate uptake in opossum kidney (OK) cells with parathyroid hormone (PTH) and parathyroid hormone related protein (PTHrP). *Peptides, 11*, 945–949.

Murer, H., & Biber, J. (1992). Renal tubular phosphate transport. In D. W. Seldin, & G. Giebisch (Eds.), *The kidney: Physiology and pathophysiology: Vol. 2* (2nd ed., pp. 2481–2509). New York: Raven Press, Ltd.

Murer, H., & Biber, J. (2010). Phosphate transport in the kidney. *Journal of Nephrology, 23*(Suppl. 16), S145–S151.

Murer, H., Forster, I., Hernando, N., Lambert, G., Traebert, M., & Biber, J. (1999). Post-transcriptional regulation of the proximal tubule NaPi-II transporter in response to PTH and dietary P_i. *American Journal of Physiology, 277*, F676–F684.

Murer, H., Hernando, N., Forster, I., & Biber, J. (2003). Regulation of Na/Pi transporter in the proximal tubule. *Annual Review of Physiology, 65*, 531–542.

Murphy, T. J., & Bylund, D. B. (1989). Characterization of serotonin-1B receptors negatively coupled to adenylate cyclase in OK cells, a renal epithelial cell line from the opossum. *Journal of Pharmacology and Experimental Therapeutics, 249*, 535–543.

Murray, R., Khundmiri, S., Clark, B., & Lederer, E. (2014). Signaling mechanisms for post-transcriptional regulation of the type IIa sodium-phosphate cotransporter mRNA by parathyroid hormone (892.35). *The FASEB Journal, 28*.

Myakala, K., Motta, S., Murer, H., Wagner, C. A., Koesters, R., Biber, J., et al. (2014). Renal-specific and inducible depletion of NaPi-IIc/Slc34A3, the cotransporter mutated in HHRH, does not affect phosphate or calcium homeostasis in mice. *American Journal of Physiology. Renal Physiology, 306*, F833–F843.

Nagai, S., Okazaki, M., Segawa, H., Bergwitz, C., Dean, T., Potts, J. T., Jr., et al. (2011). Acute down-regulation of sodium-dependent phosphate transporter NPT2a involves predominantly the cAMP/PKA pathway as revealed by signaling-selective parathyroid hormone analogs. *Journal of Biological Chemistry, 286*, 1618–1626.

Nielsen, P. K., Feldt-Rasmussen, U., & Olgaard, K. (1996). A direct effect in vitro of phosphate on PTH release from bovine parathyroid tissue slices but not from dispersed parathyroid cells. *Nephrology, Dialysis, Transplantation, 11*, 1762–1768.

Nowik, M., Picard, N., Stange, G., Capuano, P., Tenenhouse, H. S., Biber, J., et al. (2008). Renal phosphaturia during metabolic acidosis revisited: Molecular mechanisms for decreased renal phosphate reabsorption. *Pflügers Archiv, 457*, 539–549.

Nykjaer, A., Dragun, D., Walther, D., Vorum, H., Jacobsen, C., Herz, J., et al. (1999). An endocytic pathway essential for renal uptake and activation of the steroid 25-(OH) vitamin D_3. *Cell, 96*, 507–515.

Nykjaer, A., Fyfe, J. C., Kozyraki, R., Leheste, J. R., Jacobsen, C., Nielsen, M. S., et al. (2001). Cubilin dysfunction causes abnormal metabolism of the steroid hormone 25(OH) vitamin D(3). *Proceedings of the National Academy of Sciences of the United States of America, 98*, 13895–13900.

Ohyama, Y., Ozono, K., Uchida, M., Shinki, T., Kato, S., Suda, T., et al. (1994). Identification of a vitamin D-responsive element in the 5'-flanking region of the rat 25-hydroxyvitamin D3 24-hydroxylase gene. *Journal of Biological Chemistry, 269*, 10545–10550.

Omasits, U., Ahrens, C. H., Muller, S., & Wollscheid, B. (2014). Protter: Interactive protein feature visualization and integration with experimental proteomic data. *Bioinformatics*, *30*, 884–886.

Omata, K., Abraham, N. G., & Schwartzman, M. L. (1992). Renal cytochrome P-450-arachidonic acid metabolism: Localization and hormonal regulation in SHR. *American Journal of Physiology*, *262*, F591–F599.

Omdahl, J. L., Gray, R. W., Boyle, I. T., Knutson, J., & DeLuca, H. F. (1972). Regulation of metabolism of 25-hydroxycholecalciferol by kidney tissue in vitro by dietary calcium. *Nature: New Biology*, *237*, 63–64.

Palmer, G., Bonjour, J. P., & Caverzasio, J. (1996). Stimulation of inorganic phosphate transport by insulin-like growth factor I and vanadate in opossum kidney cells is mediated by distinct protein tyrosine phosphorylation processes. *Endocrinology*, *137*, 4699–4705.

Parkinson, D. B., & Thakker, R. V. (1992). A donor splice site mutation in the parathyroid hormone gene is associated with autosomal recessive hypoparathyroidism. *Nature Genetics*, *1*(2), 149–152.

Pearce, S. H., Williamson, C., Kifor, O., Bai, M., Coulthard, M. G., Davies, M., et al. (1996). A familial syndrome of hypocalcemia with hypercalciuria due to mutations in the calcium-sensing receptor. *The New England Journal of Medicine*, *335*(15), 1115–1122.

Perry, W., & Stamp, T. C. (1978). Hereditary hypophosphataemic rickets with autosomal recessive inheritance and severe osteosclerosis. A report of two cases. *The Journal of Bone and Joint Surgery. British Volume*, *60-B*(3), 430–434.

Perwad, F., Azam, N., Zhang, M. Y., Yamashita, T., Tenenhouse, H. S., & Portale, A. A. (2005). Dietary and serum phosphorus regulate fibroblast growth factor 23 expression and 1,25-dihydroxyvitamin D metabolism in mice. *Endocrinology*, *146*, 5358–5364.

Pfister, M. F., Forgo, J., Ziegler, U., Biber, J., & Murer, H. (1999). cAMP-dependent and -independent downregulation of type II Na-P_i cotransporters by PTH. *American Journal of Physiology. Renal Physiology*, *276*, F720–F725.

Pfister, M. F., Ruf, I., Stange, G., Ziegler, U., Lederer, E., Biber, J., et al. (1998). Parathyroid hormone leads to the lysosomal degradation of the renal type II Na/P_i cotransporter. *Proceedings of the National Academy of Sciences of the United States of America*, *95*, 1909–1914.

Picard, N., Capuano, P., Stange, G., Mihailova, M., Kaissling, B., Murer, H., et al. (2010). Acute parathyroid hormone differentially regulates renal brush border membrane phosphate cotransporters. *Pflügers Archiv*, *460*, 677–687.

Pollak, M. R., Brown, E. M., Estep, H. L., McLaine, P. N., Kifor, O., Park, J., et al. (1994). Autosomal dominant hypocalcaemia caused by a $Ca(^{2+})$-sensing receptor gene mutation. *Nature Genetics*, *8*(3), 303–307.

Quamme, G., Pfeilschifter, J., & Murer, H. (1989a). Parathyroid hormone inhibition of Na^+/phosphate cotransport in OK cells: Generation of second messengers in the regulatory cascade. *Biochemical and Biophysical Research Communications*, *158*, 951–957.

Quamme, G., Pfeilschifter, J., & Murer, H. (1989b). Parathyroid hormone inhibition of Na^+/phosphate cotransport in OK cells: Requirement of PKC-dependent pathway. *Biochimica et Biophysica Acta*, *1013*, 159–165.

Quarles, L. D. (2012). Skeletal secretion of FGF-23 regulates phosphate and vitamin D metabolism. *Nature Reviews. Endocrinology*, *8*, 276–286.

Rafaelsen, S. H., Raeder, H., Fagerheim, A. K., Knappskog, P., Carpenter T. O., Johansson, S., et al. (2013). Exome sequencing reveals FAM20c mutations associated with fibroblast growth factor 23-related hypophosphatemia, dental anomalies, and ectopic calcification. *Journal of Bone and Mineral Research: The Official Journal of the American Society for Bone and Mineral Research*, *28*(6), 1378–1385. http://dx.doi.org/10.1002/jbmr.1850.

Reshkin, S. J., Forgo, J., & Murer, H. (1990). Functional asymmetry in phosphate transport and its regulation in opossum kidney cells: Parathyroid hormone inhibition. *Pflugers Archiv: European Journal of Physiology*, *416*, 624–631.

Riminucci, M., Collins, M. T., Fedarko, N. S., Cherman, N., Corsi, A., White, K. E., et al. (2003). FGF-23 in fibrous dysplasia of bone and its relationship to renal phosphate wasting. *Journal of Clinical Investigation*, *112*, 683–692.

Rodriguez-Ortiz, M. E., Lopez, I., Munoz-Castaneda, J. R., Martinez-Moreno, J. M., Ramirez, A. P., Pineda, C., et al. (2012). Calcium deficiency reduces circulating levels of FGF23. *Journal of the American Society of Nephrology*, *23*, 1190–1197.

Romero, G., Von Zastrow, M., & Friedman, P. A. (2011). Role of PDZ proteins in regulating trafficking, signaling, and function of GPCRs. Means, motif, and opportunity. *Advances in Pharmacology*, *62*, 279–314.

Rosenblatt, M. (1986). Peptide hormone antagonists that are effective in vivo. Lessons from parathyroid hormone. *New England Journal of Medicine*, *315*, 1004–1013.

Rutsch, F., Ruf, N., Vaingankar, S., Toliat, M. R., Suk, A., Hohne, W., et al. (2003). Mutations in ENPP1 are associated with 'idiopathic' infantile arterial calcification. *Nature Genetics*, *34*, 379–381.

Saito, K., Hirano, K., Kubota, M., Tomino, Y., & Koide, H. (1993). Effect of parathyroid-hormone-related protein on sodium-dependent phosphate transport in renal brush border membrane vesicles in rats. Comparison with parathyroid hormone. *Nephron*, *64*, 600–604.

Saito, A., Sato, H., Iino, N., & Takeda, T. (2010). Molecular mechanisms of receptor-mediated endocytosis in the renal proximal tubular epithelium. *Journal of Biomedicine and Biotechnology*, *2010*, 403272.

Salaun, C., Rodrigues, P., & Heard, J. M. (2001). Transmembrane topology of PiT-2, a phosphate transporter-retrovirus receptor. *Journal of Virology*, *75*, 5584–5592.

Samadfam, R., Richard, C., Nguyen-Yamamoto, L., Bolivar, I., & Goltzman, D. (2009). Bone formation regulates circulating concentrations of fibroblast growth factor 23. *Endocrinology*, *150*, 4835–4845.

Sato, T., Tominaga, Y., Ueki, T., Goto, N., Matsuoka, S., Katayama, A., et al. (2004). Total parathyroidectomy reduces elevated circulating fibroblast growth factor 23 in advanced secondary hyperparathyroidism. *American Journal of Kidney Diseases*, *44*, 481–487.

Scanni, R., vonRotz, M., Jehle, S., Hulter, H. N., & Krapf, R. (2014). The human response to acute enteral and parenteral phosphate loads. *Journal of the American Society of Nephrology*, *25*, 2730–2739.

Schouten, B. J., Hunt, P. J., Livesey, J. H., Frampton, C. M., & Soule, S. G. (2009). FGF23 elevation and hypophosphatemia after intravenous iron polymaltose: A prospective study. *Journal of Clinical Endocrinology and Metabolism*, *94*, 2332–2337.

Segawa, H., Onitsuka, A., Furutani, J., Kaneko, I., Aranami, F., Matsumoto, N., et al. (2009). Npt2a and Npt2c in mice play distinct and synergistic roles in inorganic phosphate metabolism and skeletal development. *American Journal of Physiology. Renal Physiology*, *297*, F671–F678.

Segawa, H., Yamanaka, S., Ohno, Y., Onitsuka, A., Shiozawa, K., Aranami, F., et al. (2007). Correlation between hyperphosphatemia and type II Na-Pi cotransporter activity in klotho mice. *American Journal of Physiology. Renal Physiology*, *292*, F769–F779.

Seikaly, M. G., & Baum, M. (1995). Stimulation of growth hormone secretion in children with X-linked hypophosphatemia. *Pediatric Nephrology*, *9*, 751–752.

Shenolikar, S., Voltz, J. W., Minkoff, C. M., Wade, J. B., & Weinman, E. J. (2002). Targeted disruption of the mouse NHERF-1 gene promotes internalization of proximal tubule sodium-phosphate cotransporter type IIa and renal phosphate wasting. *Proceedings of the National Academy of Sciences of the United States of America*, *99*, 11470–11475.

Shimada, T., Hasegawa, H., Yamazaki, Y., Muto, T., Hino, R., Takeuchi, Y., et al. (2004). FGF-23 is a potent regulator of vitamin D metabolism and phosphate homeostasis. *Journal of Bone and Mineral Research, 19*, 429–435.

Shimada, T., Kakitani, M., Yamazaki, Y., Hasegawa, H., Takeuchi, Y., Fujita, T., et al. (2004). Targeted ablation of Fgf23 demonstrates an essential physiological role of FGF23 in phosphate and vitamin D metabolism. *Journal of Clinical Investigation, 113*, 561–568.

Shimada, T., Mizutani, S., Muto, T., Yoneya, T., Hino, R., Takeda, S., et al. (2001). Cloning and characterization of FGF23 as a causative factor of tumor-induced osteomalacia. *Proceedings of the National Academy of Sciences of the United States of America, 98*, 6500–6505.

Shimada, T., Muto, T., Urakawa, I., Yoneya, T., Yamazaki, Y., Okawa, K., et al. (2002). Mutant FGF-23 responsible for autosomal dominant hypophosphatemic rickets is resistant to proteolytic cleavage and causes hypophosphatemia in vivo. *Endocrinology, 143*, 3179–3182.

Shimada, T., Yamazaki, Y., Takahashi, M., Hasegawa, H., Urakawa, I., Oshima, T., et al. (2005). Vitamin D receptor-independent FGF23 actions in regulating phosphate and vitamin D metabolism. *American Journal of Physiology. Renal Physiology, 289*, F1088–F1095.

Shinki, T., Shimada, H., Wakino, S., Anazawa, H., Hayashi, M., Saruta, T., et al. (1997). Cloning and expression of rat 25-hydroxyvitamin D_3-1α-hydroxylase cDNA. *Proceedings of the National Academy of Sciences of the United States of America, 94*, 12920–12925.

Silverstein, D. M., Spitzer, A., & Barac-Nieto, M. (2000). Hormonal regulation of sodium-dependent phosphate transport in opossum kidney cells. *Hormone Research, 54*, 38–43.

Silverstein, D. M., Spitzer, A., & Barac-Nieto, M. (2005). Parathormone sensitivity and responses to protein kinases in subclones of opossum kidney cells. *Pediatric Nephrology, 20*, 721–724.

Singh, R. J., & Kumar, R. (2003). Fibroblast growth factor 23 concentrations in humoral hypercalcemia of malignancy and hyperparathyroidism. *Mayo Clinic Proceedings, 78*, 826–829.

Slatopolsky, E., Finch, J., Denda, M., Ritter, C., Zhong, M., Dusso, A., et al. (1996). Phosphorus restriction prevents parathyroid gland growth. High phosphorus directly stimulates PTH secretion in vitro. *Journal of Clinical Investigation, 97*, 2534–2540.

Slattery, C., Jenkin, K. A., Lee, A., Simcocks, A. C., McAinch, A. J., Poronnik, P., et al. (2011). Na^+-H^+ exchanger regulatory factor 1 (NHERF1) PDZ scaffold binds an internal binding site in the scavenger receptor megalin. *Cellular Physiology and Biochemistry, 27*, 171–178.

Smith, R. C., O'Bryan, L. M., Farrow, E. G., Summers, L. J., Clinkenbeard, E. L., Roberts, J. L., et al. (2012). Circulating alphaKlotho influences phosphate handling by controlling FGF23 production. *Journal of Clinical Investigation, 122*, 4710–4715.

Sneddon, W. B., Bisello, A., Magyar, C. E., Willick, G. E., Syme, C. A., Galbiati, F., et al. (2004). Ligand-selective dissociation of activation and internalization of the parathyroid hormone receptor. Conditional efficacy of PTH peptide fragments. *Endocrinology, 145*, 2815–2823.

Sneddon, W. B., Gesek, F. A., & Friedman, P. A. (2000). Obligate MAP kinase activation in parathyroid hormone stimulation of calcium transport but not calcium signaling. *Endocrinology, 141*, 4185–4193.

Stock, J. L., Coderre, J. A., Burke, E. M., Danner, D. B., Chipman, S. D., & Shapiro, J. R. (1992). Identification of estrogen receptor mRNA and the estrogen modulation of parathyroid hormone-stimulated cyclic AMP accumulation in opossum kidney cells. *Journal of Cellular Physiology, 150*, 517–525.

Tagliabracci, V. S., Engel, J. L., Wen, J., Wiley, S. E., Worby, C. A., Kinch, L. N., et al. (2012). Secreted kinase phosphorylates extracellular proteins that regulate biomineralization. *Science, 336*, 1150–1153.

Tagliabracci, V. S., Engel, J. L., Wiley, S. E., Xiao, J., Gonzalez, D. J., Nidumanda, A. H., et al. (2014). Dynamic regulation of FGF23 by Fam20C phosphorylation, GalNAc-T3 glycosylation, and furin proteolysis. *Proceedings of the National Academy of Sciences of the United States of America, 111*, 5520–5525.

Takei, Y., Yamamoto, H., Masuda, M., Sato, T., Taketani, Y., & Takeda, E. (2009). Stanniocalcin 2 is positively and negatively controlled by $1,25(OH)_2D_3$ and PTH in renal proximal tubular cells. *Journal of Molecular Endocrinology, 42*, 261–268.

Taketani, Y., Segawa, H., Chikamori, M., Morita, K., Tanaka, K., Kido, S., et al. (1998). Regulation of type II renal Na^+-dependent inorganic phosphate transporters by 1,25-dihydroxyvitamin D_3. Identification of a vitamin D-responsive element in the human NAPi-3 gene. *Journal of Biological Chemistry, 273*, 14575–14581.

Tanaka, Y., & DeLuca, H. F. (1973). The control of 25-hydroxyvitamin D metabolism by inorganic phosphorus. *Archives of Biochemistry and Biophysics, 154*, 566–574.

Tanimura, A., Yamada, F., Saito, A., Ito, M., Kimura, T., Anzai, N., et al. (2011). Analysis of different complexes of type IIa sodium-dependent phosphate transporter in rat renal cortex using blue-native polyacrylamide gel electrophoresis. *Journal of Medical Investigation, 58*, 140–147.

Tebben, P. J., Singh, R. J., Clarke, B. L., & Kumar, R. (2004). Fibroblast growth factor 23, parathyroid hormone, and 1α,25-dihydroxyvitamin D in surgically treated primary hyperparathyroidism. *Mayo Clinic Proceedings, 79*, 1508–1513.

Tenenhouse, H. S., Gauthier, C., Martel, J., Gesek, F. A., Coutermarsh, B. A., & Friedman, P. A. (1998). Na^+-phosphate cotransport in mouse distal convoluted tubule cells: Evidence for *Glvr-1* and *Ram-1* Gene Expression. *Journal of Bone and Mineral Research, 13*, 590–597.

Tenenhouse, H. S., Martel, J., Gauthier, C., Segawa, H., & Miyamoto, K. I. (2003). Differential effects of Npt2a gene ablation and the X-linked Hyp mutation on renal expression of type IIc Na/Pi cotransporter. *American Journal of Physiology. Renal Physiology, 285*, F1271–F1278.

Tenenhouse, H. S., Roy, S., Martel, J., & Gauthier, C. (1998). Differential expression, abundance, and regulation of Na^+-phosphate cotransporter genes in murine kidney. *American Journal of Physiology, 275*, F527–F534.

Topaz, O., Shurman, D. L., Bergman, R., Indelman, M., Ratajczak, P., Mizrachi, M., et al. (2004). Mutations in GALNT3, encoding a protein involved in O-linked glycosylation, cause familial tumoral calcinosis. *Nature Genetics, 36*, 579–581.

Traebert, M., Roth, J., Biber, J., Murer, H., & Kaissling, B. (2000). Internalization of proximal tubular type II Na-P_i cotransporter by PTH: Immunogold electron microscopy. *American Journal of Physiology. Renal Physiology, 278*, F148–F154.

Traebert, M., Völkl, H., Biber, J., Murer, H., & Kaissling, B. (2000). Luminal and contraluminal action of 1-34 and 3-34 PTH peptides on renal type IIa Na-P_i cotransporter. *American Journal of Physiology. Renal Physiology, 278*, F792–F798.

Tsujikawa, H., Kurotaki, Y., Fujimori, T., Fukuda, K., & Nabeshima, Y. (2003). Klotho, a gene related to a syndrome resembling human premature aging, functions in a negative regulatory circuit of vitamin D endocrine system. *Molecular Endocrinology, 17*, 2393–2403.

Uebelhart, B., Herrmann, F., & Rizzoli, R. (2009). Effects of the SERM raloxifene on calcium and phosphate metabolism in healthy middle-aged men. *Clinical Cases in Mineral and Bone Metabolism: The Official Journal of the Italian Society of Osteoporosis, Mineral Metabolism, and Skeletal Diseases, 6*, 163–168.

Uemura, H., Irahara, M., Yoneda, N., Yasui, T., Genjida, K., Miyamoto, K. I., et al. (2000). Close correlation between estrogen treatment and renal phosphate reabsorption capacity. *Journal of Clinical Endocrinology and Metabolism, 85*, 1215–1219.

Uzum, A. K., Salman, S., Telci, A., Boztepe, H., Tanakol, R., Alagol, F., et al. (2010). Effects of vitamin D replacement therapy on serum FGF23 concentrations in vitamin D-deficient women in short term. *European Journal of Endocrinology, 163*, 825–831.

Vallon, V., Muhlbauer, B., & Osswald, H. (2006). Adenosine and kidney function. *Physiological Reviews*, *86*, 901–940.
Varghese, R., Gagliardi, A. D., Bialek, P. E., Yee, S. P., Wagner, G. F., & Dimattia, G. E. (2002). Overexpression of human stanniocalcin affects growth and reproduction in transgenic mice. *Endocrinology*, *143*, 868–876.
Vervloet, M. G., van Ittersum, F. J., Buttler, R. M., Heijboer, A. C., Blankenstein, M. A., & ter Wee, P. M. (2011). Effects of dietary phosphate and calcium intake on fibroblast growth factor-23. *Clinical Journal of the American Society of Nephrology*, *6*, 383–389.
Villa-Bellosta, R., Ravera, S., Sorribas, V., Stange, G., Levi, M., Murer, H., et al. (2009). The Na+-Pi cotransporter PiT-2 (SLC20A2) is expressed in the apical membrane of rat renal proximal tubules and regulated by dietary Pi. *American Journal of Physiology. Renal Physiology*, *296*, F691–699.
Virkki, L. V., Biber, J., Murer, H., & Forster, I. C. (2007). Phosphate transporters: A tale of two solute carrier families. *American Journal of Physiology. Renal Physiology*, *293*, F643–F654.
Voltz, J. W., Brush, M., Sikes, S., Steplock, D., Weinman, E. J., & Shenolikar, S. (2007). Phosphorylation of PDZ1 domain attenuates NHERF-1 binding to cellular targets. *Journal of Biological Chemistry*, *282*, 33879–33887.
Wade, J. B., Liu, J., Coleman, R. A., Cunningham, R., Steplock, D. A., Lee-Kwon, W., et al. (2003). Localization and interaction of NHERF isoforms in the renal proximal tubule of the mouse. *American Journal of Physiology. Cell Physiology*, *285*, C1494–C1503.
Wagner, G. F., Hampong, M., Park, C. M., & Copp, D. H. (1986). Purification, characterization, and bioassay of teleocalcin, a glycoprotein from salmon corpuscles of Stannius. *General and Comparative Endocrinology*, *63*, 481–491.
Wang, C., Li, Y., Shi, L., Ren, J., Patti, M., Wang, T., et al. (2012). Mutations in SLC20A2 link familial idiopathic basal ganglia calcification with phosphate homeostasis. *Nature Genetics*, *44*, 254–256.
Wang, B., Means, C. K., Yang, Y., Mamonova, T., Bisello, A., Altschuler, D. L., et al. (2012). Ezrin-anchored PKA coordinates phosphorylation-dependent disassembly of a NHERF1 ternary complex to regulate hormone-sensitive phosphate transport. *Journal of Biological Chemistry*, *287*, 24148–24163.
Weinman, E. J., Biswas, R. S., Peng, G., Shen, L., Turner, C. L., Xiaofei, E., et al. (2007). Parathyroid hormone inhibits renal phosphate transport by phosphorylation of serine 77 of sodium-hydrogen exchanger regulatory factor-1. *Journal of Clinical Investigation*, *117*, 3412–3420.
Weinman, E. J., Cunningham, R., Wade, J. B., & Shenolikar, S. (2005). The role of NHERF-1 in the regulation of renal proximal tubule sodium-hydrogen exchanger 3 and sodium-dependent phosphate cotransporter 2a. *Journal of Physiology*, *567*, 27–32.
Weinman, E. J., Steplock, D., Shenolikar, S., & Biswas, R. (2011). Fibroblast growth factor-23-mediated inhibition of renal phosphate transport in mice requires sodium-hydrogen exchanger regulatory factor-1 (NHERF-1) and synergizes with parathyroid hormone. *Journal of Biological Chemistry*, *286*, 37216–37221.
Weinman, E. J., Steplock, D., Zhang, Y., Biswas, R., Bloch, R. J., & Shenolikar, S. (2010). Cooperativity between the phosphorylation of Thr95 and Ser77 of NHERF-1 in the hormonal regulation of renal phosphate transport. *Journal of Biological Chemistry*, *285*, 25134–25138.
Weinstein, L. S., Shenker, A., Gejman, P. V., Merino, M. J., Friedman, E., & Spiegel, A. M. (1991). Activating mutations of the stimulatory G protein in the McCune-Albright syndrome. *The New England Journal of Medicine*, *325*(24), 1688–1695.
Weinstein, L. S., Yu, S., Warner, D. R., & Liu, J. (2001). Endocrine manifestations of stimulatory G protein alpha-subunit mutations and the role of genomic imprinting. *Endocrine Reviews*, *22*(5), 675–705.

Winer, K. K., Ko, C. W., Reynolds, J. C., Dowdy, K., Keil, M., Peterson, D., et al. (2003). Long-term treatment of hypoparathyroidism: A randomized controlled study comparing parathyroid hormone-(1–34) versus calcitriol and calcium. *The Journal of Clinical Endocrinology and Metabolism*, *88*(9), 4214–4220.

Wesseling-Perry, K., Harkins, G. C., Wang, H. J., Elashoff, R., Gales, B., Horwitz, M. J., et al. (2010). The calcemic response to continuous parathyroid hormone (PTH)(1-34) infusion in end-stage kidney disease varies according to bone turnover: A potential role for PTH(7-84). *Journal of Clinical Endocrinology and Metabolism*, *95*, 2772–2780.

White, K. E., Cabral, J. M., Davis, S. I., Fishburn, T., Evans, W. E., Ichikawa, S., et al. (2005). Mutations that cause osteoglophonic dysplasia define novel roles for FGFR1 in bone elongation. *American Journal of Human Genetics*, *76*, 361–367.

White, K. E., Carn, G., Lorenz-Depiereux, B., Benet-Pagès, A., Strom, T. M., & Econs, M. J. (2001). Autosomal-dominant hypophosphatemic rickets (ADHR) mutations stabilize FGF-23. *Kidney International*, *60*, 2079–2086.

Winters, R. W., Graham, J. B., Williams, T. F., McFalls, V. W., & Burnett, C. H. (1991). A genetic study of familial hypophosphatemia and vitamin D resistant rickets with a review of the literature. *Medicine (Baltimore) 1958*, *70*, 215–217.

Wu, S., Finch, J., Zhong, M., Slatopolsky, E., Grieff, M., & Brown, A. J. (1996). Expression of the renal 25-hydroxyvitamin D-24-hydroxylase gene: Regulation by dietary phosphate. *American Journal of Physiology*, *271*, F203–F208.

Xu, H., Bai, L., Collins, J. F., & Ghishan, F. K. (2002). Age-dependent regulation of rat intestinal type IIb sodium-phosphate cotransporter by 1,25-(OH)(2) vitamin D(3). *American Journal of Physiology. Cell Physiology*, *282*, C487–C493.

Yamamoto, T., Imanishi, Y., Kinoshita, E., Nakagomi, Y., Shimizu, N., Miyauchi, A., et al. (2005). The role of fibroblast growth factor 23 for hypophosphatemia and abnormal regulation of vitamin D metabolism in patients with McCune-Albright syndrome. *Journal of Bone and Mineral Metabolism*, *23*, 231–237.

Yamamoto, S., Okada, Y., Mori, H., Fukumoto, S., & Tanaka, Y. (2012). Fibroblast growth factor 23-related osteomalacia caused by the prolonged administration of saccharated ferric oxide. *Internal Medicine*, *51*, 2375–2378.

Yamazaki, Y., Okazaki, R., Shibata, M., Hasegawa, Y., Satoh, K., Tajima, T., et al. (2002). Increased circulatory level of biologically active full-length FGF-23 in patients with hypophosphatemic rickets/osteomalacia. *Journal of Clinical Endocrinology and Metabolism*, *87*, 4957–4960.

Yamazaki, Y., Tamada, T., Kasai, N., Urakawa, I., Aono, Y., Hasegawa, H., et al. (2008). Anti-FGF23 neutralizing antibodies show the physiological role and structural features of FGF23. *Journal of Bone and Mineral Research*, *23*, 1509–1518.

Yang, J., Singh, V., Chen, T. E., Sarker, R., Xiong, L., Cha, B., et al. (2014). NHERF2/NHERF3 hetero-dimerization and macro-complex formation are required for the inhibition of NHE3 activity by carbachol. *Journal of Biological Chemistry*, *289*, 20039–20053.

Yoshida, T., Fujimori, T., & Nabeshima, Y. (2002). Mediation of unusually high concentrations of 1,25-dihydroxyvitamin D in homozygous klotho mutant mice by increased expression of renal 1alpha-hydroxylase gene. *Endocrinology*, *143*, 683–689.

Yoshiko, Y., Wang, H., Minamizaki, T., Ijuin, C., Yamamoto, R., Suemune, S., et al. (2007). Mineralized tissue cells are a principal source of FGF23. *Bone*, *40*, 1565–1573.

Yu, X., Sabbagh, Y., Davis, S. I., Demay, M. B., & White, K. E. (2005). Genetic dissection of phosphate- and vitamin D-mediated regulation of circulating Fgf23 concentrations. *Bone*, *36*, 971–977.

CHAPTER TEN

Regulation of Aquaporins by Vasopressin in the Kidney

Masahiro Ikeda*[,1], Toshiyuki Matsuzaki[†]

*Department of Veterinary Pharmacology, University of Miyazaki, Miyazaki, Japan
[†]Department of Anatomy and Cell Biology, Gunma University Graduate School of Medicine, Maebashi, Japan
[1]Corresponding author: e-mail address: a0d302u@cc.miyazaki-u.ac.jp

Contents

Abstract

Vasopressin is the main hormone that regulates water conservation in mammals and one of its major targets is the principal cells in the renal collecting duct. Vasopressin increases the apical water permeability of principal cells, mediated by apical accumulation of aquaporin-2 (AQP2), a water channel protein, thus facilitating water reabsorption by the kidney. The mechanisms underlying the accumulation of AQP2 in response to vasopressin include vesicular trafficking from intracellular storage vesicles

Vitamins and Hormones, Volume 98
ISSN 0083-6729
http://dx.doi.org/10.1016/bs.vh.2014.12.008

expressing AQP2 within several tens of minutes (short-term regulation) and protein expression of AQP2 over a period of hours to days (long-term regulation). This chapter reviews vasopressin signaling in the kidney, focusing on the molecular mechanisms of short- and long-term regulations of AQP2 expression.

1. INTRODUCTION

Cells are enclosed by the plasma membrane lipid bilayer. Although water molecules can permeate the lipid bilayer directly via simple diffusion to some extent, specific membrane water channels are required for tissues and cells handling large amounts of water, such as kidney cells (Takata, Matsuzaki, & Tajika, 2004). In fact, about 10% of the glomerular filtrate (about 18 L/day) in humans is reabsorbed transcellularly via water channel proteins in the renal collecting duct under the regulation of vasopressin, also known as antidiuretic hormone (ADH).

Water channel protein was discovered by Agre's group in 1992, when they were trying to isolate Rh polypeptides from human erythrocytes (Preston & Agre, 1991; Preston, Carroll, Guggino, & Agre, 1992). This water channel was initially named CHIP28 protein, but was later renamed aquaporin-1 (AQP1). Isolation of the cDNA for this newly characterized water channel protein allowed subsequent isolation of other isoforms by PCR-based homologous cloning. Among these isoforms, aquaporin-2 (AQP2) was found in kidney in 1993 and shown to be regulated by vasopressin (Fushimi et al., 1993). Thereafter, AQP3 and AQP4 (originally isolated from brain tissue) were identified in kidney tissue (Ishibashi et al., 1994; Jung et al., 1994). These discoveries have led to a clearer understanding of the urine concentration mechanism at the molecular level (Kortenoeven & Fenton, 2014; Noda, Sohara, Ohta, & Sasaki, 2010). This chapter focuses on the interaction between the vasopressin receptors and the AQP2 water channel through intracellular signaling pathways to regulate water reabsorption and urine concentration in the kidney, also touching on disorders resulting from abnormality of the vasopressin–AQP2 axis.

2. AQUAPORINS IN THE KIDNEY

2.1 General aspects

So far 13 AQP isoforms (AQP0–AQP12) have been characterized in mammals (Ishibashi, Kondo, Hara, & Morishita, 2011), and each has been assigned a sequential number in order of discovery, AQP1 being the first to have been isolated by Agre's group. AQPs are basically composed of ~270 amino acids, spanning the membrane six times with intracellular N- and C-termini (Fig. 1). Each has a pair of highly conserved asparagine-proline-alanine (NPA) motifs in the intracellular loop B and extracellular loop E. These two NPA motifs face close together within the lipid bilayer and form a narrow pore for water permeation. The pore diameter is only slightly larger than that of a water molecule (2.8 Å); for example, the pore diameter of AQP1 (Murata et al., 2000) and AQP4 (Tani & Fujiyoshi, 2014) is 3.0 Å. This narrow pore is critical for specific transfer of water and allows the permeation of 3 billion or more water molecules per second (Tani & Fujiyoshi, 2014). The direction of water flow depends on the osmotic gradient and/or hydrostatic pressure, because AQPs are facilitators of diffusion. The possible regulation of permeability, including some form of gating mechanism, has been investigated by both functional and structural analyses. These have revealed that the permeability of a plant AQP, SoPIP2;1, is increased by phosphorylation (Johansson et al., 1998), and the structural basis of channel gating has been clearly demonstrated by X-ray diffraction analysis (Tornroth-Horsefield et al., 2006). Possible regulation of water permeability has been suggested for not only plant AQPs but also mammalian AQPs such as AQP0, AQP4, and AQP2, but the mechanism still remains elusive (Sachdeva & Singh, 2014).

Mammalian AQPs are subdivided into three subgroups: AQPs, aquaglyceroporins, and unorthodox AQPs (Ishibashi et al., 2011). The

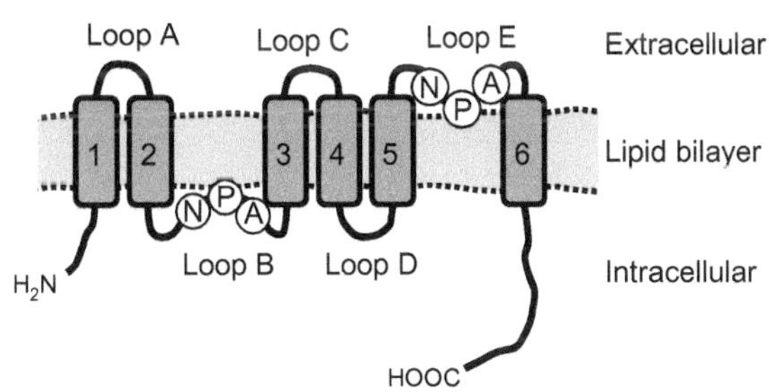

Figure 1 Schematic drawing of the membrane topology of aquaporin. Aquaporins are transmembrane proteins spanning the lipid bilayer six times, both the N- and C-termini being located intracellularly. There are two conserved NPA (asparagine-proline-alanine) motifs in the intracellular loop B between the second and the third transmembrane domains and in the extracellular loop E between the fifth and the sixth transmembrane domains, respectively. Both NPA motifs are embedded in the lipid bilayer from opposite sides to form the pore of the channel.

AQPs, comprising AQP0, AQP1, AQP2, AQP4, AQP5, AQP6, and AQP8, allow selective permeation of water molecules (Ishibashi et al., 2011). The aquaglyceroporins, which include AQP3, AQP7, and AQP9, allow permeation of small molecules such as glycerol and urea, in addition to water molecules. The unorthodox AQPs, AQP11 and AQP12, show a low degree of amino acids' sequence conservation and unusual N-terminal NPA motifs. Although functional aspects of both AQP11 and AQP12 remain obscure, recent evidence suggests that AQP11 is a water-permeable protein (Ikeda et al., 2011; Takahashi et al., 2014; Yakata, Tani, & Fujiyoshi, 2011).

The tissue distributions of the various AQP isoforms in mammals are briefly summarized in Table 1. Details of the tissue distributions have been described in a number of review articles (Matsuzaki et al., 2012, 2004, 2002; Takata et al., 2004).

Table 1 Summary of the tissue distributions of the various aquaporin isoforms

Aquaporin Isoform	Distribution
AQP0	Eye (lens)
AQP1	Capillary, erythrocyte, and kidney (proximal tubule, descending thin limb)
AQP2	Kidney (collecting duct)
AQP3	Kidney (collecting duct), eye, colon, urinary bladder, and skin
AQP4	Brain, kidney (collecting duct), and stomach
AQP5	Salivary gland, lung, lacrimal gland, and sweat gland
AQP6	Kidney (collecting duct)
AQP7	Kidney (proximal tubule), testis, skeletal muscle, and adipose tissue
AQP8	Liver, pancreas, salivary gland, and colon
AQP9	Hepatocyte
AQP10	Intestine
AQP11	Kidney (proximal tubule), liver, and testis
AQP12	Pancreas

2.2 Aquaporins in the kidney

In the kidney, at least seven AQP isoforms, namely, AQP1, AQP2, AQP3, AQP4, AQP6, AQP7, and AQP11, are expressed (Fig. 2; Takata et al., 2005). Among these isoforms, AQP1, AQP2, AQP3, and AQP4 are thought to be important for water reabsorption and are distributed in the proximal tubules, descending limbs of Henle's loop, connecting tubules, and collecting duct system, in accordance with the results from physiologically examined water permeability analyses (reviewed in fig. 1 from Nielsen, Smith, Christensen, Knepper, & Agre, 1993). AQP1 is localized at both the apical and basolateral membranes in the proximal tubules (Nielsen, Smith, et al., 1993) and plays an important role in large amount of water reabsorption in these segments. Also, AQP1 has been shown to be expressed in both the thin descending loop of Henle and the vasa recta, where it is required for countercurrent multiplication and countercurrent exchange to maintain the corticomedullary osmotic gradient (Verkman, 2006; Dantzler, Layton, Layton, & Pannabecker, 2014). AQP2, AQP3, and AQP4 are expressed in the collecting duct system, including the connecting tubules where they are involved in the regulation of urinary concentration. AQP2 provides a water entry pathway through the apical membranes of epithelial cells. On the other hand, basolateral membranes bear AQP3 and AQP4, and AQP2 is also expressed in a restricted region. These AQPs are thought to provide exit pathways for water molecules from epithelial cells toward the renal interstitium. Although the functional significance of the presence of three isoforms at the basolateral membrane remains obscure, it is true that their distributions do not completely overlap. Expression of AQP3 is abundant in the connecting tubules and cortical collecting ducts and then gradually decreases toward the papillary tip (Poulsen, Kim, Frøkiær, Nielsen, & Christensen, 2013; Takata et al., 2005), whereas AQP4 expression is lower in the cortical collecting ducts and increases gradually toward the papillary tip (Poulsen et al., 2013; Takata et al., 2005). Studies of AQP3-null mice and AQP4-null mice have indicated that AQP3 is more important than AQP4 for concentration of urine, suggesting that most of the water is reabsorbed in the collecting duct from the cortex to the outer region of the medulla (Verkman, 2006). Initially, AQP2 was thought to be an apical membrane protein, but subsequent studies have shown that AQP2 is also localized to the basolateral membrane in a restricted area, including connecting tubules and inner medullary collecting ducts (IMCDs; Christensen, Wang, Frøkiær, & Nielsen, 2003).

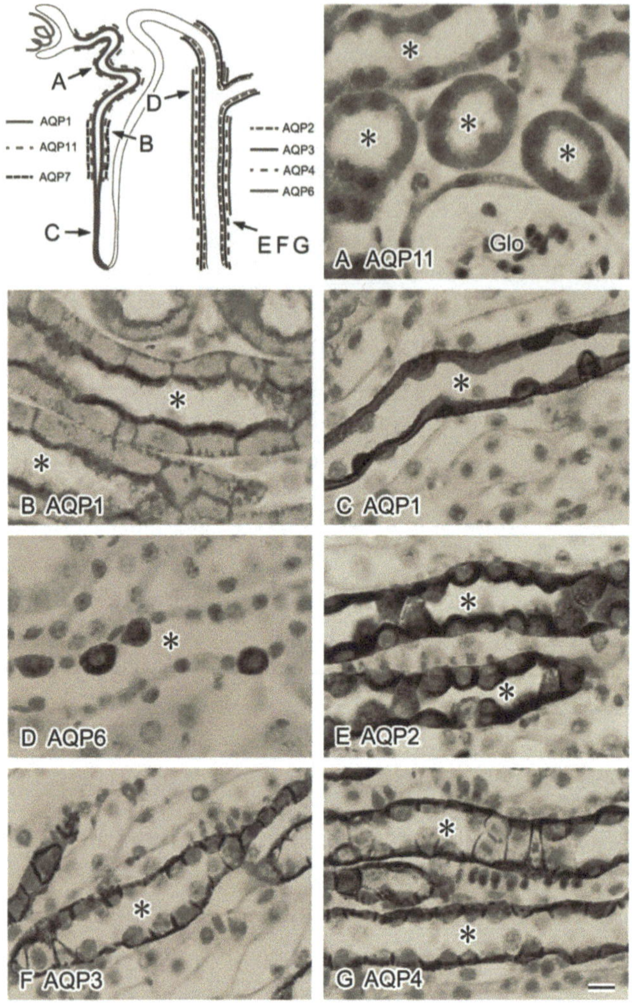

Figure 2 Aquaporins in the kidney. A schematic drawing of the distribution of AQP1, AQP2, AQP3, AQP4, AQP6, AQP7, and AQP11 in the urinary tubule and collecting duct system is shown at the upper left. Micrographs of each region indicated by arrows A–G in mouse kidney immunolabeled with isoform-specific antibodies by the immunoperoxidase method are shown in panels (A)–(G), respectively. Cell nuclei were counterstained with hematoxylin. (A) AQP11 is localized intracellularly in proximal tubular cells. Glo, glomerulus. (B) AQP1 is localized in both the apical and basolateral membranes of proximal tubular cells. (C) AQP1 in the descending limb of the loop of Henle. (D) AQP6 in the cortical collecting duct. AQP6 is localized intracellularly in intercalated cells. (E) AQP2 in the inner medullary collecting duct. Strong labeling is seen in the apical region of principal cells. AQP3 (F) and AQP4 (G) in the inner medullary collecting duct. Both are localized to the basolateral membrane of principal cells. Asterisks indicate the lumen of the immunopositive tubules and ducts. Bar: 10 μm in (A)–(G). (See the color plate.)

Among these three isoforms, AQP2 plays a critical role in vasopressin-regulated urine concentration. It has long been known that the water permeability of the collecting duct varies according to the plasma concentration of vasopressin (Grantham & Burg, 1966). The discovery of AQP2 was a breakthrough in explaining the molecular mechanism by which vasopressin regulates the permeation of water through shuttling of AQP2 between the plasma membrane and intracellular membrane vesicles (Fushimi et al., 1993; Nielsen, DiGiovanni, Christensen, Knepper, & Harris, 1993), i.e., the vasopressin-dependent flow of water molecules into the collecting duct cells through the gate-keeper water channel AQP2, and outflow to the interstitium via basolateral water channels AQP3 and/or AQP4. Because an osmotic gradient provides the driving force for water permeation through water channels, water reabsorption becomes limited when urine equilibrates osmotically with the hypertonic interstitium, even if a large amount of AQP2 remains on the apical surface membrane.

Other AQP isoforms expressed in the kidney, AQP6, AQP7, and AQP11, do not appear to be directly involved in water reabsorption. AQP6 is localized intracellularly in intercalated cells of the collecting ducts and has unique characteristics in which it permeates anion in response to Hg^{2+} or a low pH (Ikeda et al., 2002; Yasui et al., 1999). AQP7 is localized at the apical membrane of the proximal straight tubules and probably plays a role in glycerol absorption from the luminal fluid (Nejsum et al., 2000; Sohara, Rai, Sasaki, & Uchida, 2006). AQP11 has been found to be localized intracellularly, probably at the endoplasmic reticulum, in the renal proximal tubule (Morishita et al., 2005). Surprisingly, AQP11-null mice have been found to develop polycystic kidneys (Morishita et al., 2005), thus shedding light on the physiological role of AQP11 and pathology of polycystic kidney disease.

3. VASOPRESSIN SIGNALING

3.1 Physiology of vasopressin

Vasopressin, the ADH of mammals, is the main hormone that maintains body fluid balance. It is produced mainly by the perikarya of magnocellular neurons in the supraoptic nucleus (SON) and paraventricular nucleus (PVN), and parvicellular neurons in the PVN. In humans, the mature mRNA for vasopressin encodes a preprohormone composed of 168 amino acids (Fig. 3). This preprohormone includes four domains: the N-terminal signal peptide, vasopressin, vasopressin–neurophysin, and

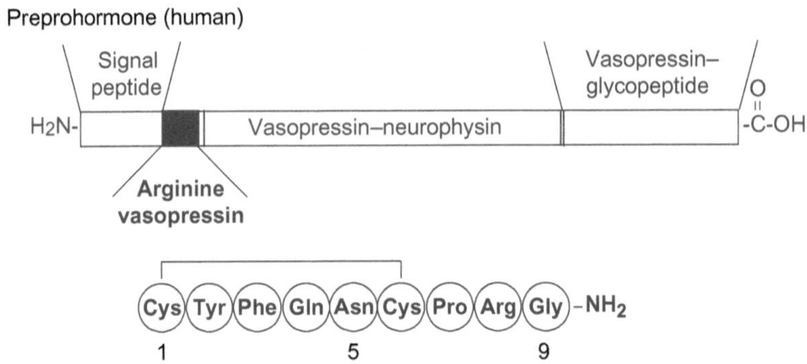

Figure 3 Schematic drawings of human arginine vasopressin and its preprohormone. The preprohormone includes four domains: an N-terminal signal peptide, vasopressin, vasopressin–neurophysin, and vasopressin–glycopeptide. During synthesis, the signal peptide is cleaved from the preprohormone to produce prohormone. The prohormone is then processed by sequential enzymatic cleavages to yield vasopressin. (See the color plate.)

vasopressin–glycopeptide (also known as copeptin). During synthesis, the signal peptide is cleaved from the preprohormone to produce the prohormone. In neurosecretory granules, the prohormone is processed by an endopeptidase, exopeptidase, monooxygenase, and lyase, and axonally transported, leading to the storage of vasopressin in the posterior lobe of the pituitary (Reilly & Jackson, 2011; Verbalis, 2012).

A number of vasopressin isoforms and related peptides (vasopressin-like peptides) have been found in vertebrates and invertebrates. All the peptides identified in vertebrates are nonapeptides and two cysteine residues at positions 1 (Cys1) and 6 (Cys6) are completely conserved, forming a disulfide bridge between the two residues. Most mammals have Arg8-vasopressin (arginine vasopressin) and pigs have Lys8-vasopressin (lypressin; Reilly & Jackson, 2011).

Two physiological stimuli, including an increase in plasma osmolality and hypovolemia/hypotension, are well known to trigger secretion of vasopressin from the posterior pituitary (Robertson, Athar, & Shelton, 1977; Robertson, Mahr, Athar, & Sinha, 1973; Robertson, Shelton, & Athar, 1976). The threshold plasma osmolality for vasopressin secretion is around 280 mOsm/kg. When plasma osmolality is less than this, vasopressin is hardly detectable in serum. However, once the plasma osmolality exceeds the threshold, there is a dramatic vasopressin secretion response. In fact, a 2% elevation of plasma osmolality increases the plasma vasopressin level by 200–300%. Several CNS structures are known to be involved in the perception of plasma osmolality for the secretion of vasopressin. Magnocellular neurons in the SON and PVN are able to perceive increased plasma osmolality, and cells in the subfornical organ, the organum vasculosum of the lamina terminalis, and the median preoptic nucleus are osmosensitive or osmoresponsive (Reilly & Jackson, 2011; Verbalis, 2012).

Vasopressin secretion in response to hypovolemia/hypotension is a defense mechanism against blood loss and/or hypotension. This secretion mechanism is thought to depend on the osmotic regulation mentioned above (Robertson et al., 1977, 1973, 1976). The plasma level of vasopressin is proportional to plasma osmolality above the threshold, the slope of this relationship becoming steeper in proportion to the degree of hypovolemia/hypotension. The neural pathway by which changes in hemodynamic profile lead to alterations in the secretion of vasopressin from the posterior pituitary differs from that operating in osmoregulation (Reilly & Jackson, 2011). When baroreceptors in pulmonary veins, the left atrium, and the left ventricle detect a change in blood volume, and/or when baroreceptors in the aortic arch and carotid sinus sense blood pressure, the signal is initiated and transmitted to brainstem nuclei, mediated by the vagal trunk and glossopharyngeal nerve, and then to the SON and PVN.

3.2 Vasopressin receptors

The effects of vasopressin are mediated through stimulation of vasopressin receptors. So far, three vasopressin receptor subtypes, including V1a, V1b, and V2 receptors, have been identified (Brown & Fenton, 2012; Reilly & Jackson, 2011). V1a receptors are expressed in vascular smooth muscle, cardiac myocytes, brain, testis, the superior cervical ganglion, liver, platelets, and the renal medulla. V1b receptors (also known as V3 receptors) are distributed in the anterior pituitary, pancreas, and adrenal medulla. V2 receptors are highly expressed in the basolateral membrane of principal cells in the renal collecting duct. To a lesser extent, V2 receptors are also known to be located in the thick ascending limb of the loop of Henle and the vascular endothelium.

All of the cloned subtypes belong to the family of seven membrane-spanning domain receptors that couple to heterotrimeric G proteins (GPCRs). Upon stimulation of the V1a or V1b receptor by vasopressin, a Gq–phospholipase C–inositol (1,4,5) trisphosphate pathway is activated, leading to elevation of the intracellular Ca^{2+} concentration and activation of protein kinase C. These intracellular signaling molecules elicit biological outcomes including vasoconstriction, glycogenolysis, platelet activation, and release of adrenocorticotropic hormone (Reilly & Jackson, 2011).

Activation of the V2 receptor upon binding of vasopressin stimulates adenylyl cyclase (AC) activity, mediated by Gs protein (Brown & Fenton, 2012; Reilly & Jackson, 2011). The activation of ACs (AC3 and AC6) results in the production of cAMP. These cAMP molecules bind to the protein kinase A holoenzyme, consisting of two catalytic (C) subunits reversibly bound to a regulatory (R) subunit dimer, to form a heterotetramer complex, and activate the C subunits via a conformational change in the R subunits. The active C subunits have a diverse array of physiological targets, including AQP2. When this signaling cascade is activated in the principal cells of renal collecting ducts, AQP2 is rapidly translocated from intracellular vesicles to the apical plasma membrane, where it accumulates, resulting in an acute increase of water reabsorption (short-term regulation of AQP2). This cascade is also known to increase the expression of AQP2 protein in renal cells mainly through enhancement of transcription, contributing to the late phase of water reabsorption induced by vasopressin (long-term regulation of AQP2). Figure 4 summarizes the short- and long-term regulations of AQP2.

4. AQUAPORINS REGULATED BY VASOPRESSIN (GENERAL ASPECTS)

Among the AQPs expressed in the kidney, AQP2, AQP3, and AQP4 are known to be regulated by vasopressin. The short-term regulation of AQP2 has been well characterized; vasopressin induces the accumulation of AQP2 on the apical surface membrane through vesicular trafficking from intracellular storage vesicles within several tens of minutes (Figs. 4 and 5; Katsura et al., 1995; Tajika et al., 2002; Takata, Matsuzaki, Tajika, Ablimit, & Hasegawa, 2008; Yamamoto et al., 1995). In addition, vasopressin can act directly on AQP2 to increase its water permeability (Eto, Noda, Horikawa, Uchida, & Sasaki, 2010; Kuwahara et al., 1995). Thus, the water permeability of the collecting duct apical membrane is regulated by the apical protein abundance and possibly by the permeability of each channel protein in the short-term regulation. On the other hand, it appears that neither AQP3 nor AQP4 is regulated by vasopressin in the short term, since most of these proteins are expressed continuously on the basolateral membrane (Ecelbarger et al., 1995; Terris, Ecelbarger, Marples, Knepper, & Nielsen, 1995). With regard to long-term regulation, abundance of AQP2 and AQP3 proteins in rats is known to be increased by chronic administration

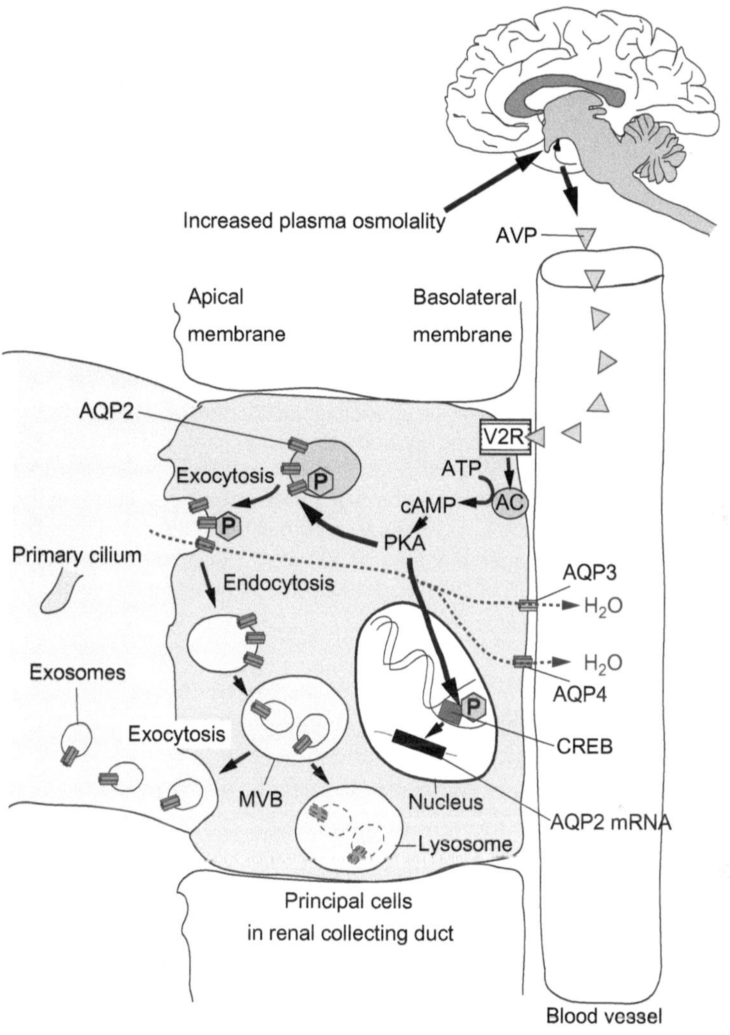

Figure 4 An illustration of the mechanism underlying the short- and long-term regulations of AQP2. Vasopressin released from the posterior pituitary binds to the V2 receptor (V2R) present on the basolateral membrane of principal cells in the renal collecting duct. Activation of the receptor stimulates adenylate cyclase (AC) activity, resulting in the production of cAMP, which in turn activates protein kinase A (PKA), and this protein kinase phosphorylates AQP2. Subsequently, AQP2 accumulates on the apical surface membrane through vesicular trafficking from intracellular storage vesicles expressing AQP2. Thereafter, water molecules flowing into the collecting duct cells through AQP2 are able to flow out to the interstitium via the basolateral water channels AQP3 and/or AQP4. The PKA also phosphorylates CRE-binding protein (CREB), and the phosphorylated CREB then stimulates transcription of the AQP2 gene. Vasopressin also regulates the abundance of renal AQP2 via excretion of exosomal AQP2. When the multivesicular bodies (MVBs) fuse with the plasma membrane in response to vasopressin, intraluminal vesicles, termed exosomes, are released into the urinary space. (See the color plate.)

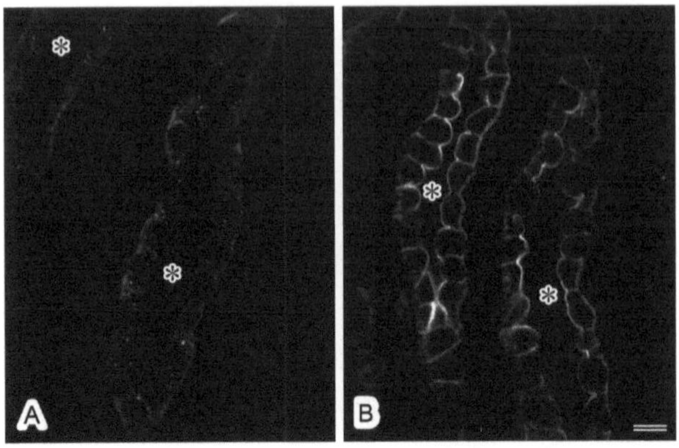

Figure 5 Short-term regulation of the subcellular localization of AQP2. Immunofluorescence microscopy of AQP2-specific antibody staining in kidneys of a water-loaded rat (A) and a vasopressin-administered rat (B). Inner medullary collecting ducts were observed with a laser confocal microscope. AQP2 is localized intracellularly in the kidney of a water-loaded rat (A). AQP2 has markedly accumulated in the apical membrane and slightly in the basolateral membrane at 30 min after vasopressin administration (B). Asterisks indicate the lumen of collecting ducts. Bar: 10 μm. (See the color plate.)

of vasopressin (Terris, Ecelbarger, Nielsen, & Knepper, 1996). AQP4 protein is also increased along the connecting tubule to the initial IMCD (Poulsen et al., 2013). Another water channel, AQP1, for water reabsorption in the proximal tubule is thought not to be regulated by vasopressin in either the short term (Nielsen, Smith, et al., 1993) or the long term (Terris et al., 1996).

5. SHORT-TERM REGULATION OF AQUAPORIN-2 BY VASOPRESSIN

5.1 Phosphorylation of aquaporin-2

Phosphorylation has been shown to play some critical roles in the subcellular localization and water permeability of AQP2 as a form of short-term regulation. The COOH-terminal narrow region (within 14 amino acid residues) of the AQP2 molecule contains four serines: Ser256, Ser261, Ser264, and Ser269. All four of these serines are phosphorylated and dephosphorylated in response to vasopressin (Hoffert et al., 2008; Hoffert, Pisitkun, Wang, Shen, & Knepper, 2006). In isolated IMCD tubules, it has been shown that Ser256, Ser264, and Ser269 are phosphorylated, whereas Ser261 is dephosphorylated, upon stimulation with vasopressin (Hoffert et al., 2008). It has been widely accepted that the phosphorylation of AQP2 at Ser256 is mediated by PKA. Recent phosphoproteomic studies, however, have revealed some other possible kinases for Ser256, including calcium/calmodulin-dependent protein kinase II and protein kinase B (AKT; Bradford et al., 2014). With regard to other serines, kinases that phosphorylate them still remain elusive, although an *in vitro* phosphorylation assay using a synthetic C-terminal peptide has shown that at least PKA does not phosphorylate them (Hoffert et al., 2008).

5.2 Aquaporin-2 phosphorylation and subcellular distribution

Changes in the subcellular localization of AQP2 between intracellular vesicles and the surface plasma membrane occur through vesicular trafficking via exocytosis and endocytosis. Therefore, both exocytotic and endocytotic steps are considered to regulate the amount of AQP2 in the surface plasma membrane (Brown, Bouley, Paunescu, Breton, & Lu, 2012). In collecting duct cells, AQP2 seems to be recycled constitutively in a vasopressin-independent manner between the cell surface and intracellular vesicles and appears to accumulate in the surface plasma membrane upon stimulation with vasopressin by accelerating exocytosis and/or decelerating endocytosis (Brown, 2003; Brown et al., 2012). In constitutive recycling, AQP2 trafficking is thought to take place independently of its phosphorylation; on the other hand, AQP2 phosphorylation should be one of the critical events in AQP2 trafficking by vasopressin for short-term regulation (Brown, Hasler, Nunes, Bouley, & Lu, 2008). However, the underlying mechanisms still remain unclear.

5.2.1 Phosphorylation at Ser256 and exocytosis

The importance of phosphorylation at Ser256 in AQP2 trafficking from intracellular vesicles to the surface membrane has been suggested. Vasopressin acts on the V2 receptor, causing phosphorylation at Ser256 and trafficking of AQP2 to the surface membrane (Fushimi, Sasaki, & Marumo, 1997). It has been shown that incubation of rat kidney papilla slices with vasopressin for 5 min significantly increased phosphorylation at Ser256 (Nishimoto et al., 1999). It has also been revealed that AQP2 is phosphorylated at Ser256 (Christensen, Zelenina, Aperia, & Nielsen, 2000; Promeneur, Kwon, Frøkiær, Knepper, & Nielsen, 2000) and still remains in intracellular vesicles (Christensen et al., 2000) even in Brattleboro rats, which genetically lack vasopressin, thus suggesting that (1) Ser256 may be phosphorylated via some signaling cascade other than that involving vasopressin and (2) phosphorylation at Ser256 is necessary but is not sufficient for trafficking to the surface membrane. There is also an unresolved controversy as to whether the amount of phosphorylation at Ser256 is really increased in response to vasopressin. It has been shown that phosphorylation at Ser256 is not increased in Brattleboro rats 2 h after subcutaneous injection of desmopressin (1-desamino-8-D-arginine vasopressin), an agonist for the V2 receptor (Christensen et al., 2000), and in SD rats 60 min after intravenous administration of desmopressin (Xie et al., 2010). Considering these results, AQP2 trafficking to the surface membrane could require some other critical events in addition to the phosphorylation of AQP2 at Ser256; for example, "turning on" the vesicular trafficking system in response to vasopressin. One possibly important event would be depolymerization of the submembranous actin network. Before AQP2 was identified, vasopressin was reported to depolymerize apical F-actin in rat IMCD cells (Simon, Gao, Franki, & Hays, 1993), suggesting a critical role of actin depolymerization in fusion of subapical vesicles carrying water channels to the surface membrane, as seen in the toad urinary bladder. It was later shown that only depolymerization of the F-actin cytoskeleton causes AQP2 insertion into the surface membrane in cultured IMCD cells (Klussmann et al., 2001). Phosphorylation of AQP2 at Ser256 has also been demonstrated to elevate the affinity of AQP2 for tropomyosin-5b, causing destabilization of the F-actin network and AQP2 trafficking to the surface membrane (Noda et al., 2008). Interestingly, F-actin depolymerization seems to be related to the level of AQP2 expression in AQP2-transfected cultured cells (Yui, Lu, Bouley, & Brown, 2011).

5.2.2 Phosphorylation at Ser256 and endocytosis

The level of AQP2 expression in the apical membrane is also regulated by an endocytotic process, but the effects of phosphorylation at Ser256 on that process also remain elusive. It has been found that dephosphorylation at Ser256 was not required for AQP2 internalization (Nejsum, Zelenina, Aperia, Frøkiær, & Nielsen, 2005). However, the state of phosphorylation at Ser256 appears to affect the endocytotic process, since phosphorylation at Ser256 has been observed to decelerate the endocytosis and to decrease the affinity of the receptor for endocytotic proteins (Moeller, Praetorius, Rützler, & Fenton, 2010).

5.2.3 Phosphorylation at Ser261, Ser264, and Ser269

The effects of phosphorylation at other serines must also be considered. Rapid increases of phosphorylation at Ser264 and Ser269 upon stimulation with vasopressin (Hoffert et al., 2008) suggest that these residues are as important as Ser256. Ser269 has been shown to be unphosphorylated in Brattleboro rats, and phosphorylated Ser269 has only been detected in the cell surface membrane after administration of vasopressin (Moeller, Knepper, & Fenton, 2009). Inhibitory effects of Ser269 phosphorylation on AQP2 internalization and binding to endocytotic proteins have also been demonstrated in cultured cells, similarly to Ser256 (Moeller et al., 2010). On the other hand, Ser261 is phosphorylated without vasopressin stimulation and slowly dephosphorylated upon exposure to vasopressin (Hoffert et al., 2008). In cultured cells, neither inhibitory nor stimulatory effects on exocytotic or endocytotic processes have been observed even if Ser261 is replaced by Ala or Asp, which acts as a molecular mimic for a dephosphorylated or phosphorylated state, respectively (Lu et al., 2008). Little is known about the roles of Ser261 or Ser264 phosphorylation. The subcellular distribution of AQP2 is thought to be regulated by the complex effects of phosphorylation of these four serines—Ser256, Ser261, Ser264, and Ser269—in both exocytosis and endocytosis. Further studies are required in order to clarify these issues.

5.3 Aquaporin-2 phosphorylation and water permeability

According to several studies, the water permeability of AQP2 seems to be directly regulated by phosphorylation. It has been shown that cAMP-dependent phosphorylation increases the water permeability of AQP2 in *Xenopus* oocytes expressing AQP2 or its mutants (Kuwahara et al., 1995). Recently, using purified human AQP2 reconstructed into proteoliposomes, Sasaki's group revealed that phosphorylation at Ser256 increased water permeability (Eto et al., 2010). Phosphorylation at Ser264 or Ser269 does not appear to change water permeability (Moeller, MacAulay, Knepper, & Fenton, 2009). However, in contrast to these observations, using a different strategy, it has been reported that phosphorylation does not alter the water permeability (Lande, Jo, Zeidel, Somers, & Harris, 1996). In terms of the direct regulation of water permeability through AQP2 by phosphorylation, this issue is still being debated (Moeller, Olesen, & Fenton, 2011). Further studies including structural analyses will be necessary in order to clarify the mechanism responsible for direct regulation of AQP2 water permeability.

6. LONG-TERM REGULATION OF AQUAPORIN-2 BY VASOPRESSIN

Human studies in 1950s showed that prolonged dehydration increased maximal urinary-concentrating capacity and chronic water loading reduced it (De Wardener & Herxheimer, 1957; Epstein, Kleeman, & Hendrikx, 1956; Jones & De Wardener, 1956). From these studies, it became clear that a prolonged change in the body's water balance leads to adaptational alterations of urinary concentration. After the discovery of AQP2, since it was well known that the plasma concentration of vasopressin is high in a dehydrated state and low in a hydrated state, many researchers attempted to explain the mechanism underlying this adaptational change in terms of the vasopressin–AQP2 axis, and it is now clear that the regulation of renal AQP2 abundance in response to vasopressin (long-term regulation of AQP2) plays a major role (Nielsen et al., 2002; Radin et al., 2012).

Extensive studies using Brattleboro rats have been performed to understand the long-term regulation of AQP2. It has been reported that the level of renal AQP2 in Brattleboro rats is about 30–50% of that in their parent strain (DiGiovanni, Nielsen, Christensen, & Knepper, 1994; Kishore, Terris, & Knepper, 1996; Promeneur et al., 2000). On the other hand, polyuric diabetes insipidus (DI) +/+ severe male mice, which have genetically high levels of type IV cAMP phosphodiesterase, resulting in a low level of cytosolic cAMP, show greater downregulation of renal AQP2 (up to around 85%) in comparison with control animals (Frøkiaer et al., 1999). This difference in the renal abundance of AQP2 between Brattleboro rats and DI +/+ mice suggests that 50–70% of renal AQP2 in a steady state depends on the vasopressin-V2 receptor–cAMP pathway, whereas the remainder is maintained in a vasopressin-independent manner that includes a cAMP-related and -unrelated processes (Nielsen et al., 2002). So far, it has been suggested that vasopressin-independent mechanism involves oxytocin (Li et al., 2008), secretin (Chu et al., 2007), and prostaglandins (Li et al., 2009). Also since an osmotic gradient is built up in the kidney medulla and the AQP2 mRNA is known to be upregulated and downregulated by hypertonic and hypotonic stimulations, respectively (Kortenoeven, van den Brand, Wetzels, & Deen, 2011; Lam et al., 2004; López-Rodríguez et al., 2004), tonicity might be involved in the vasopressin-independent regulation of AQP2.

Theoretically, it can be considered that the long-term increase in the renal expression of AQP2 protein elicited by vasopressin is mediated by a balance of its synthesis, intracellular degradation, and exit from cells (Fig. 4). In the following sections, these components are discussed separately.

6.1 Aquaporin-2 synthesis

The amount of AQP2 synthesized is thought to depend mainly on the level of intracellular AQP2 mRNA available for translation and its translational rate (Radin et al., 2012). The level of intracellular mRNA is regulated by two mechanisms, i.e., the rates of transcription and degradation.

After it had become clear that AQP2 is regulated by vasopressin, many researchers analyzed the 5′-flanking region that is essential for transcription of the AQP2 gene (Furuno, Uchida, Marumo, & Sasaki, 1996; Hozawa, Holtzman, & Ausiello, 1996; Matsumura, Uchida, Rai, Sasaki, & Marumo, 1997; Uchida, Sasaki, Fushimi, & Marumo, 1994; Yasui, Zelenin, Celsi, & Aperia, 1997). This led to the identification of several *cis*-element motifs and their corresponding transcriptional factors. Furthermore, a recent computational comparison of 1000 bp of the 5′-flanking region of the AQP2 gene among five mammalian species (dog, cow, mouse, rat, and human) has revealed two highly conserved regions centered at −224 and −513 bp upstream from the transcription start site (Yu et al., 2009). The upstream conserved region (centered at −513 bp) includes an SF-1 site, a nuclear factor of activated T cells site, and a forkhead box, whereas the downstream region (centered at −224 bp) contains an AP-2 site, a cAMP-response element (CRE) site, a GATA site, an SRF site, and a HOX site. Among them several *cis*-elements have been reported to have an important role in vasopressin-regulated AQP2 transcription (Schenk et al., 2012).

Many studies of the CRE have confirmed that this *cis*-element is involved in vasopressin-stimulated transcription of the AQP2 gene (Hasler et al., 2002; Hozawa et al., 1996; Rai, Uchida, Marumo, & Sasaki, 1997; Uchida et al., 1994). Binding of vasopressin to the V2 receptor causes an increase of intracellular cAMP in the principal cells of renal collecting duct, thus activating PKA. This protein kinase phosphorylates the CRE-binding protein (CREB), and the phosphorylated CREB then stimulates transcription of the AQP2 gene via the CRE (Fig. 4). On the other hand, although it has been reported that the intracellular cAMP level peaks within 1 h after stimulation of the V2 receptor and then subsequently decreases, the amount of AQP2 remains high for a couple of days thereafter

(Kortenoeven et al., 2012). These data suggest that transcription of the AQP2 gene depends on the PKA–CREB pathway during the early phase of long-term regulation, whereas other factors may be involved in the later phase. Recently, the activation of the exchange factor directly activated by cAMP (Epac) has been reported to be possibly involved in the later phase of long-term regulation in mpkCCD cells (Kortenoeven et al., 2012). Interestingly, it has also been suggested that Epac activation is involved in the translocation of AQP2 to the apical membrane in response to vasopressin (Yip, 2006). The role of Epac in long-term regulation *in vivo* has yet to be elucidated.

A GATA site has been found in the 5′-flanking region of the AQP2 gene (Rai, Uchida, et al., 1997). It has been reported that the nuclear abundance of Gatad2b is increased by the treatment of mpkCCD cells with vasopressin (Schenk et al., 2012). Recently, Yu et al. (2014) have elegantly demonstrated that GATA2 is markedly expressed in principal cells of the renal collecting duct in adult mouse kidney, and that renal tubule cell-specific *Gata2*- knockout causes an increase in urinary volume and a decrease in urinary osmolality, accompanied by a reduction of renal AQP2 protein expression. These data suggest that a GATA box appears to play some roles in the long-term regulation of AQP2.

It has been shown that the nuclear abundance of transcriptional factors such as JunB, Elf3, and Hmbox1 is also altered in response to vasopressin in mpkCCD cells, and that JunB and Elf3 are increased, whereas Hmbox1 is decreased (Schenk et al., 2012). Elf3 mRNA levels have been reported to correlate with the expression of AQP2 in mpkCCD cells. In a promoter assay, vasopressin has also been shown to directly increase the transcription of AQP2 via Elf3 (Yu et al., 2009).

Since miRNAs were first described in 1993 (Lee, Feinbaum, & Ambros, 1993), they have led to major advances in biology, including kidney physiology (Trionfini, Benigni, & Remuzzi, 2015). miRNA is a small RNA species transcribed from noncoding RNA and binds to a target motif in the 3′-untranslated region (3-UTR) of the complementary mRNA, causing translational repression and cleavage-mediated mRNA degradation. Recently, it has been reported that the identification of the vasopressin-responsive miRNAs from kidney IMCD cells, including AQP2-targeting miRNAs (Kim, Jung, & Kwon, 2014). Changes in the expression of 19 miRNAs in response to desmopressin stimulation were observed, and among them they identified four miRNAs targeting 3-UTR of rat AQP2 mRNA using *in silico* analysis with TargetScan 6.2. Subsequently it was demonstrated that two of these four miRNAs were able to attenuate the upregulation of AQP2 mRNA and protein in response to desmopressin in mpkCCDc14 cells, a mouse kidney cortical collecting duct cell line.

6.2 Aquaporin-2 protein degradation

As well as by transcriptional mechanisms, renal AQP2 abundance can be theoretically regulated by the rates of protein degradation and translation. Recently, Knepper's group employed protein mass spectrometry with dynamic stable isotope labeling to measure the half-lives of renal proteins in mpkCCD cells treated with vasopressin (Sandoval et al., 2013). They clearly showed that vasopressin increased the half-life of AQP2 by 145% in comparison with its steady-state half-life in the absence of vasopressin. This suggested that the regulation of renal AQP2 abundance by vasopressin involves a protein degradation mechanism.

Ubiquitination is a multistep enzymatic posttranslational modification process that mediates the covalent conjugation of ubiquitin, a small 8.5-kDa protein composed of 76 amino acids, to target proteins (Komander & Rape, 2012; Pickart & Eddins, 2004). This process involves a hierarchical set of three enzymatic reactions. First, ubiquitin-activating enzyme (also known as E1) activates ubiquitin in an ATP-dependent manner. Next, ubiquitin-conjugating enzyme or ubiquitin-carrier enzyme (UBCs or E2) accepts ubiquitin from E1 through a transthiolation reaction. Finally, ubiquitin ligase (E3) catalyzes the transfer of ubiquitin from the E2 enzyme to the ε-amino group of a lysine residue on the target protein. Covalent ubiquitination of target proteins is reversible, and mediated by deubiquitinating enzymes (DUBs). Endocytosis and various degradation pathways are known to be regulated by ubiquitination of the target proteins. Kamsteeg et al. (2006) have reported that AQP2 is ubiquitinated at Lys270 with one K63-linked chain, and that this ubiquitination is related to either internalization of AQP2 from the apical plasma membrane or its lysosomal degradation. Lee et al. (2011) have observed that upregulation of ubiquitin ligases (E3s) including NEDD4 and cullin-5 (CUL5) in response to desmopressin and the rate of AQP2 degradation is retarded by the treatment of mpkCCDc14 cells with either small interfering RNA of NEDD4 or CUL5. Also, Boulkroun et al. (2008) have shown that mRNA and protein of ubiquitin-specific protease 10, a DUB, are increased in mCCDc11 cells upon treatment with vasopressin. Furthermore, in IMCD cells, a pathway to prevent proteasomal degradation of AQP2 via PKA- and p38-MAPK-dependent mechanisms has been proposed (Nedvetsky et al.,

2010). Together, these findings suggest that ubiquitin-dependent degradation pathway is involved in the long-term regulation of AQP2 expression.

6.3 Secretion of aquaporin-2 via exosomes

Since the discovery of AQP2 in human urine by Kanno et al. (1995), urinary AQP2 excretion has been extensively studied by many independent research groups (Deen, van Aubel, van Lieburg, & van Os, 1996; Elliot, Goldsmith, Knepper, Haughey, & Olson, 1996; Rai, Sekine, et al., 1997; Saito et al., 1997; Wen, Frokiaer, Kwon, & Nielsen, 1999). In 2004, Knepper's group found exosomes, nanosized vesicles derived from various cell types, such as dendritic cells, B-lymphocytes, endothelial cells, and epithelial cells, in urine, and definitively identified AQP2 protein in these urinary exosomes (Pisitkun, Shen, & Knepper, 2004).

Exosomes are released into the extracellular space by fusion of the outer membrane of multivesicular bodies (MVBs), late endosomes, with the cell surface. The endosomal-sorting complex required for transport (ESCRT) machinery, composed of protein complexes, ESCRT-0, -I, -II, and -III, mediates the biogenesis of MVBs (Henne, Buchkovich, & Emr, 2011; Rusten, Vaccari, & Stenmark, 2012). The process involved in biogenesis of MVBs is currently understood as shown in Fig. 6 (Babst, 2011; Henne et al., 2011; Rusten, et al., 2012). ESCRT-0 recognizes and captures ubiquitinated cargo proteins at the endosome membrane and this interaction then initiates the inward movement of a portion of the endosome membrane. Thereafter, the cargo proteins are sequentially influenced by ESCRT-I and ESCRT-II, stabilizing the membrane neck of the forming vesicle. The membrane neck is further narrowed by the action of ESCRT-III and then excised out from the endosome membrane by recruitment of the vacuolar protein sorting 4 complex to ESCRT-III, thereby forming intraluminal vesicles. This endosome containing nanosized intraluminal vesicles is known as the MVB. Once MVBs are fused with lysosomes, degradation of the cargo proteins occurs. When the MVBs are fused with the plasma membrane, the intraluminal vesicles, termed exosomes, are released into the extracellular space.

As it has been suggested that renal protein abundance is regulated by the rate of urinary exosomal excretion (Sonoda et al., 2009), vasopressin is considered to regulate renal AQP2 abundance through the excretion of exosomal AQP2. In this context, Dear's group has shown that treatment of mCCDc11 cells with desmopressin increases the exosomal secretion of

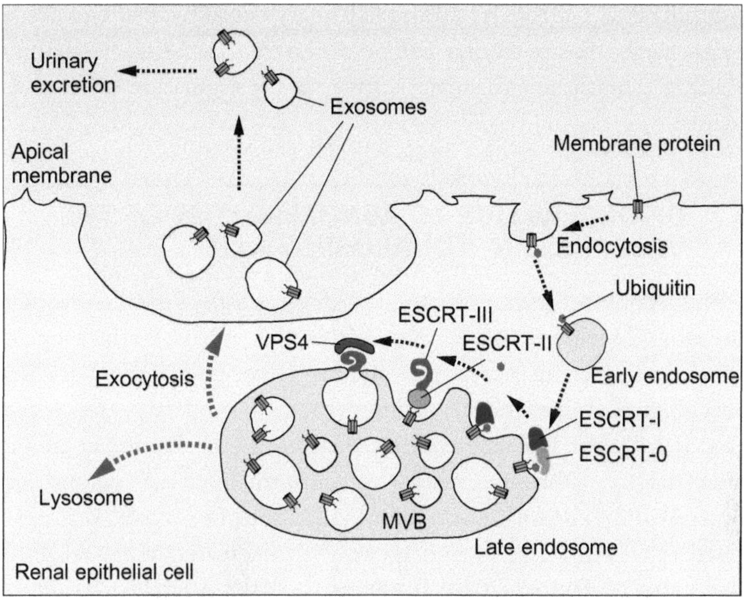

Figure 6 Biogenesis of exosomes in renal epithelial cells. The endosome containing nanosized intraluminal vesicles is known as the multivesicular body (MVB). The formation of this MVB is mediated by the endosomal-sorting complex required for transport (ESCRT) machinery, including the ESCRT-0, -I, -II, and -III protein complexes. Ubiquitination of cargo proteins such as membrane proteins causes endocytosis. Then, ESCRT-0 captures the ubiquitinated proteins at the endosome membrane, and this interaction then initiates the invagination of a portion of the endosome membrane. Thereafter, the cargo proteins are sequentially influenced by ESCRT-I, ESCRT-II, and ESCRT-III and then finally excised from the endosome membrane by the action of vacuolar protein sorting 4 (VPS4), resulting in the formation of intraluminal vesicles (exosomes). Once MVBs have fused with lysosomes, the cargo proteins are degraded, and when the MVBs have fused with the plasma membrane, exosomes are released into the extracellular space. (See the color plate.)

AQP2 from cells into culture medium, and that when rats are treated with desmopressin, the ratio of exosomal AQP2 relative to flotilin-1, commonly used as a marker of exosomes in urine, is significantly increased. Furthermore, the same group has also reported that exosomes are able to transfer AQP2 in vasopressin-stimulated mCCDc11 cells to unstimulated cells, suggesting an exosomal intercellular signaling pathway (Street et al., 2011). Recently, our group has reported that treatment of rats with diuretics elevates the levels of plasma vasopressin and urinary exosomal AQP2, accompanied by a decrease in renal cortical AQP2 protein, suggesting that abundance of renal AQP2 is regulated by excretion of exosomal AQP2

(Higashijima et al., 2013). As studies of urinary exosomal proteins are still in their early stages, future studies will be necessary to assess more definitively the involvement of the exosomal pathway in the regulation of renal AQP2 abundance.

7. DISORDERS DUE TO ABNORMALITIES OF THE VASOPRESSIN–AQUAPORIN-2 AXIS

Numerous studies have focused on diseases caused by alterations of the vasopressin-V2 receptor–AQP2 axis. As this topic is beyond the scope of the present chapter, we will merely present a brief overview of these diseases and pharmacological interventions against them (for details, see excellent reviews as follows: Kortenoeven & Fenton, 2014; Radin et al., 2012; Verbalis, 2012). Diseases related to abnormalities of vasopressin-V2 receptor–AQP2 axis can be roughly divided into two types, diseases with increased renal AQP2 level and those with decreased renal AQP2 level (Kortenoeven & Fenton, 2014; Radin et al., 2012; Verbalis, 2012).

Diseases with increased renal AQP2 level include syndrome of inappropriate secretion of ADH (SIADH) and congestive heart failure. SIADH is caused by disturbance of osmoregulation due to vascular, infectious, or neoplastic abnormalities in the lung or central nervous system. Congestive heart failure is caused by malfunction of heart and/or blood vessels. These diseases are characterized by the increased level of plasma vasopressin, hyponatremia, and hypervolemia. Antagonists for the V2 receptor including mozavaptan and tolvaptan have been reported to correct hyponatoremia and hypervolemia in these diseases. Hepatic cirrhosis resembles SIADH and congestive heart failure in terms of increased vasopressin, hyponatremia, and hypervolemia. However, the level of renal or urinary AQP2 varies considerably between reported cases. Nephrogenic syndrome of inappropriate antidiuresis is a rare inherited disease due to mutations of V2 receptor. These mutations are gain-of-function mutations causing constitutive activation of the receptor, and in turn very low levels of plasma vasopressin.

Diseases with decreased renal or urinary AQP2 level include central DI, gestational DI, and nephrogenic DI. These diseases are manifested by hypotonic polyuria. Central DI is characterized by impairment of normal AVP production, caused by abnormal function of the neurohypophysis because of inherited mutations of the AVP–neurophysin gene (congenital type) or acquired conditions such as drug exposure, neoplastic disorders, inflammation, and trauma. Desmopressin is known to be effective for patients with central DI. Gestational DI is caused by an increased rate of vasopressin metabolism by oxytocinase/vasopressinase produced by the placenta. Since desmopressin is known to be resistant to degradation by the enzyme, it is useful for the control of gestational DI. Nephrogenic DI is classified into two major types: congenital and acquired. More than 90% of cases of congenital nephrogenic DI are due to mutations of the V2 receptor and the remainder is caused by mutations of AQP2. Mutations of both V2 receptor and AQP2 are loss-of-function mutations. The causes of acquired nephrogenic DI include intoxication (due to lithium, demeclocycline, cisplatin, etc.), hypercalcemia, hypokalemia, and obstruction of the urinary tract. Most of the acquired causes are known to affect the long-term regulation of AQP2 by vasopressin. Since desmopressin has little effect on nephrogenic DI, some patients are treated with thiazide diuretics, which are known to have a paradoxical antidiuretic effect in such patients. However, an insufficient effect of thiazide is sometimes observed, and there are also nonresponders to thiazide. Prostaglandin synthase inhibitors have been reported to reduce urine volume and free water clearance in patients with nephrogenic DI. This effect of prostaglandin synthase inhibitors is thought to be mediated via enhanced intracellular cAMP production.

As current pharmacological interventions are insufficient for the control of nephrogenic DI, several promising new drug therapies have been reported (Kortenoeven & Fenton, 2014). One of them is pharmacological chaperones, which are considered to enhance the apical membrane expression of ER-retained V2 receptor mutants. In fact, it has been shown that SR49059 rescues the plasma membrane expression of V2 receptor mutants in *in vitro* experiments, and also it reduces urine volume in patients with mutation of the V2 receptor (Bernier et al., 2006). As elevation of intracellular cGMP concentration has been reported to increase the apical membrane expression of AQP2, drugs that are able to exert such an effect have been highlighted (Kortenoeven & Fenton, 2014). Sildenafil, an inhibitor of phosphodiesterase-5, prevents the degradation of cGMP and has been reported to correct polyuria in a rat model of acquired nephrogenic DI (Bouley et al., 2005; Sanches et al., 2012). Another treatment strategy is the use of statins, which were originally formulated as antihypercholesterolemia drugs, acting via inhibition of HMG-CoA reductase. Recent studies have shown that statins are able to inhibit an endocytosis pathway of AQP2 in *in vitro* studies and also to increase urine osmolality in a rat model of nephrogenic DI (Danilovic et al., 2012; Li et al., 2011).

8. CONCLUSION AND FUTURE DIRECTIONS

The discovery of AQPs has facilitated a deeper understanding of water homeostasis in the mammalian body, and it is now clear that the vasopressin-V2 receptor–AQP2 axis plays a critical role in the regulation of urinary concentration. Studies conducted over the last decade have helped to clarify how the vasopressin signaling pathway regulates AQP2 at the molecular level, including short- and long-term regulations. This accumulated knowledge has led to the development of possible therapeutic strategies. However, the mechanisms regulating the pathways of protein degradation and excretion are still not well understood. Furthermore, although it has been pointed out that chromatin immunoprecipitation experiments followed by sequencing is a promising strategy for revealing variability in the functions of DNA-binding proteins across genomes, and leading to a better understanding of the mechanisms of AQP2 transcription elicited by vasopressin, it is still a relatively new technique (Furey, 2012; Mimura, Kanki, Kodama, & Nangaku, 2013). Future studies employing recently developed and rapidly evolving technologies will contribute to a deeper understanding of the regulation of the vasopressin-V2 receptor–AQP2 axis, leading to novel therapeutic strategies for disorders attributable to axis impairment.

ACKNOWLEDGMENTS

This work was supported by JSPS KAKENHI, 24380160 (M. I.), 25660241 (M. I.), 25221205 (M. I.), 22790191 (T. M.), 26460267 (T. M.), and 25463068 (T. M.).

REFERENCES

Babst, M. (2011). MVB vesicle formation: ESCRT-dependent, ESCRT-independent and everything in between. *Current Opinion in Cell Biology*, *23*, 452–457.

Bernier, V., Morello, J. P., Zarruk, A., Debrand, N., Salahpour, A., Lonergan, M., et al. (2006). Pharmacologic chaperones as a potential treatment for X-linked nephrogenic diabetes insipidus. *Journal of the American Society of Nephrology*, *17*, 232–243.

Bouley, R., Pastor-Soler, N., Cohen, O., McLaughlin, M., Breton, S., & Brown, D. (2005). Stimulation of AQP2 membrane insertion in renal epithelial cells in vitro and in vivo by the cGMP phosphodiesterase inhibitor sildenafil citrate (Viagra). *American Journal of Physiology—Renal Physiology*, *288*, F1103–F1112.

Boulkroun, S., Ruffieux-Daidié, D., Vitagliano, J. J., Poirot, O., Charles, R. P., Lagnaz, D., et al. (2008). Vasopressin-inducible ubiquitin-specific protease 10 increases ENaC cell surface expression by deubiquitylating and stabilizing sorting nexin 3. *American Journal of Physiology—Renal Physiology*, *295*, F889–F900.

Bradford, D., Raghuram, V., Wilson, J. L. L., Chou, C. L., Hoffert, J. D., Knepper, M. A., et al. (2014). Use of LC–MS/MS and Bayes' theorem to identify protein kinases that phosphorylate aquaporin-2 at Ser256. *American Journal of Physiology—Cell Physiology, 307*, C123–C139.

Brown, D. (2003). The ins and outs of aquaporin-2 trafficking. *American Journal of Physiology—Renal Physiology, 284*, F893–F901.

Brown, D., Bouley, R., Paunescu, T. G., Breton, S., & Lu, H. A. J. (2012). New insights into the dynamic regulation of water and acid–base balance by renal epithelial cells. *American Journal of Physiology—Cell Physiology, 302*, C1421–C1433.

Brown, D., & Fenton, R. A. (2012). The cell biology of vasopressin action. In M. W. Taal, G. M. Chertow, P. A. Marsden, K. Skorecki, A. S. L. Yu, & B. M. Barry (Eds.), *Brenner & Rector's the kidney* (pp. 353–383). Philadelphia: Saunders, an imprint of Elsevier Inc.

Brown, D., Hasler, U., Nunes, P., Bouley, R., & Lu, H. A. J. (2008). Phosphorylation events and the modulation of aquaporin 2 cell surface expression. *Current Opinion in Nephrology and Hypertension, 17*, 491–498.

Christensen, B. M., Wang, W., Frøkiær, J., & Nielsen, S. (2003). Axial heterogeneity in basolateral AQP2 localization in rat kidney: Effect of vasopressin. *American Journal of Physiology—Renal Physiology, 284*, F701–F717.

Christensen, B. M., Zelenina, M., Aperia, A., & Nielsen, S. (2000). Localization and regulation of PKA-phosphorylated AQP2 in response to V2-receptor agonist/antagonist treatment. *American Journal of Physiology—Renal Physiology, 278*, F29–F42.

Chu, J. Y., Chung, S. C., Lam, A. K., Tam, S., Chung, S. K., & Chow, B. K. (2007). Phenotypes developed in secretin receptor-null mice indicated a role for secretin in regulating renal water reabsorption. *Molecular and Cellular Biology, 27*, 2499–2511.

Danilovic, A., Lopes, R. I., Sanches, T. R., Shimizu, M. H., Oshiro, F. M., Andrade, L., et al. (2012). Atorvastatin prevents the downregulation of aquaporin-2 receptor after bilateral ureteral obstruction and protects renal function in a rat model. *Urology, 80*, 485.e15–485.e20.

Dantzler, W. H., Layton, A. T., Layton, H. E., & Pannabecker, T. L. (2014). Urine-concentrating mechanism in the inner medulla: Function of the thin limbs of the loops of Henle. *Clinical Journal of the American Society of Nephrology, 9*, 1781–1789.

De Wardener, H. E., & Herxheimer, A. (1957). The effect of a high water intake on the kidney's ability to concentrate the urine in man. *The Journal of Physiology, 139*, 42–52.

Deen, P. M., van Aubel, R. A., van Lieburg, A. F., & van Os, C. H. (1996). Urinary content of aquaporin 1 and 2 in nephrogenic diabetes insipidus. *Journal of the American Society of Nephrology, 7*, 836–841.

DiGiovanni, S. R., Nielsen, S., Christensen, E. I., & Knepper, M. A. (1994). Regulation of collecting duct water channel expression by vasopressin in Brattleboro rat. *Proceedings of the National Academy of Sciences of the United States of America, 91*, 8984–8988.

Ecelbarger, C. A., Terris, J., Frindt, G., Echevarria, M., Marples, D., Nielsen, S., et al. (1995). Aquaporin-3 water channel localization and regulation in rat kidney. *American Journal of Physiology—Renal Physiology, 269*, F663–F672.

Elliot, S., Goldsmith, P., Knepper, M., Haughey, M., & Olson, B. (1996). Urinary excretion of aquaporin-2 in humans: A potential marker of collecting duct responsiveness to vasopressin. *Journal of the American Society of Nephrology, 7*, 403–409.

Epstein, F. H., Kleeman, C. R., & Hendrikx, A. (1956). The influence of bodily hydration on the renal concentrating process. *The Journal of Clinical Investigation, 36*, 629–634.

Eto, K., Noda, Y., Horikawa, S., Uchida, S., & Sasaki, S. (2010). Phosphorylation of aquaporin-2 regulates its water permeability. *The Journal of Biological Chemistry, 285*, 40777–40784.

Frøkiaer, J., Marples, D., Valtin, H., Morris, J. F., Knepper, M. A., & Nielsen, S. (1999). Low aquaporin-2 levels in polyuric DI +/+ severe mice with constitutively high cAMP-phosphodiesterase activity. *American Journal of Physiology—Renal Physiology, 276*, F179–F190.

Furey, T. S. (2012). ChIP-seq and beyond: New and improved methodologies to detect and characterize protein–DNA interactions. *Nature Reviews Genetics*, *13*, 840–852.

Furuno, M., Uchida, S., Marumo, F., & Sasaki, S. (1996). Repressive regulation of the aquaporin-2 gene. *American Journal of Physiology*, *271*, F854–F860.

Fushimi, K., Sasaki, S., & Marumo, F. (1997). Phosphorylation of serine 256 is required for cAMP-dependent regulatory exocytosis of the aquaporin-2 water channel. *The Journal of Biological Chemistry*, *272*, 14800–14804.

Fushimi, K., Uchida, S., Hara, Y., Hirata, Y., Marumo, F., & Sasaki, S. (1993). Cloning and expression of apical membrane water channel of rat kidney collecting tubule. *Nature*, *361*, 549–552.

Grantham, J. J., & Burg, M. B. (1966). Effect of vasopressin and cyclic AMP on permeability of isolated collecting tubules. *American Journal of Physiology*, *211*, 255–259.

Hasler, U., Mordasini, D., Bens, M., Bianchi, M., Cluzeaud, F., Rousselot, M., et al. (2002). Long term regulation of aquaporin-2 expression in vasopressin-responsive renal collecting duct principal cells. *The Journal of Biological Chemistry*, *277*, 10379–10386.

Henne, W. M., Buchkovich, N. J., & Emr, S. D. (2011). The ESCRT pathway. *Developmental Cell*, *21*, 77–91.

Higashijima, Y., Sonoda, H., Takahashi, S., Kondo, H., Shigemura, K., & Ikeda, M. (2013). Excretion of urinary exosomal AQP2 in rats is regulated by vasopressin and urinary pH. *American Journal of Physiology—Renal Physiology*, *305*, F1412–F1421.

Hoffert, J. D., Fenton, R. A., Moeller, H. B., Simons, B., Tchapyjnikov, D., McDill, B. W., et al. (2008). Vasopressin-stimulated increase in phosphorylation at Ser269 potentiates plasma membrane retention of aquaporin-2. *The Journal of Biological Chemistry*, *283*, 24617–24627.

Hoffert, J. D., Pisitkun, T., Wang, G., Shen, R. F., & Knepper, M. A. (2006). Quantitative phosphoproteomics of vasopressin-sensitive renal cells: Regulation of aquaporin-2 phosphorylation at two sites. *Proceedings of the National Academy of Sciences of the United States of America*, *103*, 7159–7164.

Hozawa, S., Holtzman, E. J., & Ausiello, D. A. (1996). cAMP motifs regulating transcription in the aquaporin 2 gene. *American Journal of Physiology*, *270*, C1695–C1702.

Ikeda, M., Andoo, A., Shimono, M., Takamatsu, N., Taki, A., Muta, K., et al. (2011). The NPC motif of aquaporin-11, unlike the NPA motif of known aquaporins, is essential for full expression of molecular function. *The Journal of Biological Chemistry*, *286*, 3342–3350.

Ikeda, M., Beitz, E., Kozono, D., Guggino, W. B., Agre, P., & Yasui, M. (2002). Characterization of aquaporin-6 as a nitrate channel in mammalian cells. Requirement of pore-lining residue threonine 63. *The Journal of Biological Chemistry*, *277*, 39873–39879.

Ishibashi, K., Kondo, S., Hara, S., & Morishita, Y. (2011). The evolutionary aspects of aquaporin family. *American Journal of Physiology—Regulatory, Integrative and Comparative Physiology*, *300*, R566–R576.

Ishibashi, K., Sasaki, S., Fushimi, K., Uchida, S., Kuwahara, M., Saito, H., et al. (1994). Molecular cloning and expression of a member of the aquaporin family with permeability to glycerol and urea in addition to water expressed at the basolateral membrane of kidney collecting duct cells. *Proceedings of the National Academy of Sciences of the United States of America*, *91*, 6269–6273.

Johansson, I., Karlsson, M., Shukla, V. K., Chrispeels, M. J., Larsson, C., & Kjellbom, P. (1998). Water transport activity of the plasma membrane aquaporin PM28A is regulated by phosphorylation. *The Plant Cell*, *10*, 451–459.

Jones, R. V. H., & De Wardener, H. E. (1956). Urine concentration after fluid deprivation or pitressin tannate in oil. *British Medical Journal*, *1*, 271–274.

Jung, J. S., Bhat, R. V., Preston, G. M., Guggino, W. B., Baraban, J. M., & Agre, P. (1994). Molecular characterization of an aquaporin cDNA from brain: Candidate osmoreceptor and regulator of water balance. *Proceedings of the National Academy of Sciences of the United States of America, 91*, 13052–13056.

Kamsteeg, E. J., Hendriks, G., Boone, M., Konings, I. B., Oorschot, V., van der Sluijs, P., et al. (2006). Short-chain ubiquitination mediates the regulated endocytosis of the aquaporin-2 water channel. *Proceedings of the National Academy of Sciences of the United States of America, 103*, 18344–18349.

Kanno, K., Sasaki, S., Hirata, Y., Ishikawa, S., Fushimi, K., Nakanishi, S., et al. (1995). Urinary excretion of aquaporin-2 in patients with diabetes insipidus. *The New England Journal of Medicine, 332*, 1540–1545.

Katsura, T., Verbavatz, J. M., Farinas, J., Ma, T., Ausiello, D., Verkman, A. S., et al. (1995). Constitutive and regulated membrane expression of aquaporin 1 and aquaporin 2 water channels in stably transfected LLC-PK1 epithelial cells. *Proceedings of the National Academy of Sciences of the United States of America, 92*, 7212–7216.

Kim, J. E., Jung, H. J., & Kwon, T. H. (2014). Vasopressin-responsive miRNAs and AQP2-targeting miRNAs in kidney collecting duct cells. *Journal of the American Society of Nephrology, 25*, (Abstract), 103A.

Kishore, B. K., Terris, J. M., & Knepper, M. A. (1996). Quantitation of aquaporin-2 abundance in microdissected collecting ducts: Axial distribution and control by AVP. *American Journal of Physiology, 271*, F62–F70.

Klussmann, E., Tamma, G., Lorenz, D., Wiesner, B., Maric, K., Hofmann, F., et al. (2001). An inhibitory role of Rho in the vasopressin-mediated translocation of aquaporin-2 into cell membranes of renal principal cells. *The Journal of Biological Chemistry, 276*, 20451–20457.

Komander, D., & Rape, M. (2012). The ubiquitin code. *Annual Review of Biochemistry, 81*, 203–229.

Kortenoeven, M. L. A., & Fenton, R. A. (2014). Renal aquaporins and water balance disorders. *Biochimica et Biophysica Acta, 1840*, 1533–1549.

Kortenoeven, M. L., Trimpert, C., van den Brand, M., Li, Y., Wetzels, J. F., & Deen, P. M. (2012). In mpkCCD cells, long-term regulation of aquaporin-2 by vasopressin occurs independent of protein kinase A and CREB but may involve Epac. *American Journal of Physiology—Renal Physiology, 302*, F1395–F1401.

Kortenoeven, M. L., van den Brand, M., Wetzels, J. F., & Deen, P. M. (2011). Hypotonicity-induced reduction of aquaporin-2 transcription in mpkCCD cells is independent of the tonicity responsive element, vasopressin, and cAMP. *The Journal of Biological Chemistry, 286*, 13002–13010.

Kuwahara, M., Fushimi, K., Terada, Y., Bai, L., Marumo, F., & Sasaki, S. (1995). cAMP-dependent phosphorylation stimulates water permeability of aquaporin-collecting duct water channel protein expressed in *Xenopus* oocytes. *The Journal of Biological Chemistry, 270*, 10384–10387.

Lam, A. K., Ko, B. C., Tam, S., Morris, R., Yang, J. Y., Chung, S. K., et al. (2004). Osmotic response element-binding protein (OREBP) is an essential regulator of the urine concentrating mechanism. *The Journal of Biological Chemistry, 279*, 48048–48054.

Lande, M. B., Jo, I., Zeidel, M. L., Somers, M., & Harris, H. W., Jr. (1996). Phosphorylation of aquaporin-2 does not alter the membrane water permeability of rat papillary water channel-containing vesicles. *The Journal of Biological Chemistry, 271*, 5552–5557.

Lee, R. C., Feinbaum, R. L., & Ambros, V. (1993). The C. elegans heterochronic gene lin-4 encodes small RNAs with antisense complementarity to lin-14. *Cell, 75*, 843–854.

Lee, Y. J., Lee, J. E., Choi, H. J., Lim, J. S., Jung, H. J., Baek, M. C., et al. (2011). E3 ubiquitin-protein ligases in rat kidney collecting duct: Response to vasopressin stimulation and withdrawal. *American Journal of Physiology—Renal Physiology, 301*, F883–F896.

Li, J. H., Chou, C. L., Li, B., Gavrilova, O., Eisner, C., Schnermann, J., et al. (2009). A selective EP4 PGE2 receptor agonist alleviates disease in a new mouse model of

X-linked nephrogenic diabetes insipidus. *The Journal of Clinical Investigation, 119*, 3115–3126.

Li, C., Wang, W., Summer, S. N., Westfall, T. D., Brooks, D. P., Falk, S., et al. (2008). Molecular mechanisms of antidiuretic effect of oxytocin. *Journal of the American Society of Nephrology, 19*, 225–232.

Li, W., Zhang, Y., Bouley, R., Chen, Y., Matsuzaki, T., Nunes, P., et al. (2011). Simvastatin enhances aquaporin-2 surface expression and urinary concentration in vasopressin-deficient Brattleboro rats through modulation of Rho GTPase. *American Journal of Physiology—Renal Physiology, 301*, F309–F318.

López-Rodríguez, C., Antos, C. L., Shelton, J. M., Richardson, J. A., Lin, F., Novobrantseva, T. I., et al. (2004). Loss of NFAT5 results in renal atrophy and lack of tonicity-responsive gene expression. *Proceedings of the National Academy of Sciences of the United States of America, 101*, 2392–2397.

Lu, H. A. J., Matsuzaki, T., Bouley, R., Hasler, U., Qin, Q. H., & Brown, D. (2008). The phosphorylation state of serine 256 is dominant over that of serine 261 in the regulation of AQP2 trafficking in renal epithelial cells. *American Journal of Physiology—Renal Physiology, 295*, F290–F294.

Matsumura, Y., Uchida, S., Rai, T., Sasaki, S., & Marumo, F. (1997). Transcriptional regulation of aquaporin-2 water channel gene by cAMP. *Journal of the American Society of Nephrology, 8*, 861–867.

Matsuzaki, T., Susa, T., Shimizu, K., Sawai, N., Suzuki, T., Aoki, T., et al. (2012). Function of the membrane water channel aquaporin-5 in the salivary gland. *Acta Histochemica et Cytochemica, 45*, 251–259.

Matsuzaki, T., Tajika, Y., Ablimit, A., Aoki, T., Hagiwara, H., & Takata, K. (2004). Aquaporins in the digestive system. *Medical Electron Microscopy, 37*, 71–80.

Matsuzaki, T., Tajika, Y., Tserentsoodol, N., Suzuki, T., Aoki, T., Hagiwara, H., et al. (2002). Aquaporins: A water channel family. *Anatomical Science International, 77*, 85–93.

Mimura, I., Kanki, Y., Kodama, T., & Nangaku, M. (2013). Revolution of nephrology research by deep sequencing: ChIP-seq and RNA-seq. *Kidney International, 85*, 31–38.

Moeller, H. B., Knepper, M. A., & Fenton, R. A. (2009). Serine 269 phosphorylated aquaporin-2 is targeted to the apical membrane of collecting duct principal cells. *Kidney International, 75*, 295–303.

Moeller, H. B., MacAulay, N., Knepper, M. A., & Fenton, R. A. (2009). Role of multiple phosphorylation sites in the COOH-terminal tail of aquaporin-2 for water transport: Evidence against channel gating. *American Journal of Physiology—Renal Physiology, 296*, F649–F657.

Moeller, H. B., Olesen, E. T., & Fenton, R. A. (2011). Regulation of the water channel aquaporin-2 by posttranslational modification. *American Journal of Physiology—Renal Physiology, 300*, F1062–F1073.

Moeller, H. B., Praetorius, J., Rützler, M. R., & Fenton, R. A. (2010). Phosphorylation of aquaporin-2 regulates its endocytosis and protein–protein interactions. *Proceedings of the National Academy of Sciences of the United States of America, 107*, 424–429.

Morishita, Y., Matsuzaki, T., Hara-Chikuma, M., Andoo, A., Shimono, M., Matsuki, A., et al. (2005). Disruption of aquaporin-11 produces polycystic kidneys following vacuolization of the proximal tubule. *Molecular and Cellular Biology, 25*, 7770–7779.

Murata, K., Mitsuoka, K., Hirai, T., Walz, T., Agre, P., Heymann, J. B., et al. (2000). Structural determinants of water permeation through aquaporin-1. *Nature, 407*, 599–605.

Nedvetsky, P. I., Tabor, V., Tamma, G., Beulshausen, S., Skroblin, P., Kirschner, A., et al. (2010). Reciprocal regulation of aquaporin-2 abundance and degradation by protein kinase A and p38-MAP kinase. *Journal of the American Society of Nephrology, 21*, 1645–1656.

Nejsum, L. N., Elkjær, M. L., Hager, J., Frøkiær, J., Kwon, T. H., & Nielsen, S. (2000). Localization of aquaporin-7 in rat and mouse kidney using RT-PCR, immunoblotting, and immunocytochemistry. *Biochemical and Biophysical Research Communications, 277*, 164–170.

Nejsum, L. N., Zelenina, M., Aperia, A., Frøkiær, J., & Nielsen, S. (2005). Bidirectional regulation of AQP2 trafficking and recycling: Involvement of AQP2–S256 phosphorylation. *American Journal of Physiology—Renal Physiology, 288*, F930–F938.

Nielsen, S., DiGiovanni, S. R., Christensen, E. I., Knepper, M. A., & Harris, H. W. (1993). Cellular and subcellular immunolocalization of vasopressin-regulated water channel in rat kidney. *Proceedings of the National Academy of Sciences of the United States of America, 90*, 1663–1667.

Nielsen, S., Frøkiaer, J., Marples, D., Kwon, T. H., Agre, P., & Knepper, M. A. (2002). Aquaporins in the kidney: From molecules to medicine. *Physiological Reviews, 82*, 205–244.

Nielsen, S., Smith, B. L., Christensen, E. I., Knepper, M. A., & Agre, P. (1993). CHIP28 water channels are localized in constitutively water-permeable segments of the nephron. *The Journal of Cell Biology, 120*, 371–383.

Nishimoto, G., Zelenina, M., Li, D., Yasui, M., Aperia, A., Nielsen, S., et al. (1999). Arginine vasopressin stimulates phosphorylation of aquaporin-2 in rat renal tissue. *American Journal of Physiology, 276*, F254–F259.

Noda, Y., Horikawa, S., Kanda, E., Yamashita, M., Meng, H., Eto, K., et al. (2008). Reciprocal interaction with G-actin and tropomyosin is essential for aquaporin-2 trafficking. *The Journal of Cell Biology, 182*, 587–601.

Noda, Y., Sohara, E., Ohta, E., & Sasaki, S. (2010). Aquaporins in kidney pathophysiology. *Nature Reviews Nephrology, 6*, 168–178.

Pickart, C. M., & Eddins, M. J. (2004). Ubiquitin: Structures, functions, mechanisms. *Biochimica et Biophysica Acta, Molecular Cell Research, 1695*, 55–72.

Pisitkun, T., Shen, R. F., & Knepper, M. A. (2004). Identification and proteomic profiling of exosomes in human urine. *Proceedings of the National Academy of Sciences of the United States of America, 101*, 13368–13373.

Poulsen, S. B., Kim, Y. H., Frøkiær, J., Nielsen, S., & Christensen, B. M. (2013). Long-term vasopressin-V2-receptor stimulation induces regulation of aquaporin 4 protein in renal inner medulla and cortex of Brattleboro rats. *Nephrology, Dialysis, Transplantation, 28*, 2058–2065.

Preston, G. M., & Agre, P. (1991). Isolation of the cDNA for erythrocyte integral membrane protein of 28 kilodaltons: Member of an ancient channel family. *Proceedings of the National Academy of Sciences of the United States of America, 88*, 11110–11114.

Preston, G. M., Carroll, T. P., Guggino, W. B., & Agre, P. (1992). Appearance of water channels in Xenopus oocytes expressing red cell CHIP28 protein. *Science, 256*, 385–387.

Promeneur, D., Kwon, T. H., Frøkiær, J., Knepper, M. A., & Nielsen, S. (2000). Vasopressin V2-receptor-dependent regulation of AQP2 expression in Brattleboro rats. *American Journal of Physiology—Renal Physiology, 279*, F370–F382.

Radin, M. J., Yu, M. J., Stoedkilde, L., Miller, R. L., Hoffert, J. D., Frokiaer, J., et al. (2012). Aquaporin-2 regulation in health and disease. *Veterinary Medicine and Science, 41*, 455–470.

Rai, T., Sekine, K., Kanno, K., Hata, K., Miura, M., Mizushima, A., et al. (1997). Urinary excretion of aquaporin-2 water channel protein in human and rat. *Journal of the American Society of Nephrology, 8*, 1357–1362.

Rai, T., Uchida, S., Marumo, F., & Sasaki, S. (1997). Cloning of rat and mouse aquaporin-2 gene promoters and identification of a negative cis-regulatory element. *American Journal of Physiology—Renal Physiology, 273*, F264–F273.

Reilly, R. F., & Jackson, E. K. (2011). Regulation of renal function and vascular volume. In L. L. Brunton (Ed.), *Goodman & Gilman's the pharmacological basis of therapeutics* (pp. 671–719). New York: The McGraw-Hill Companies, Inc.

Robertson, G. L., Athar, S., & Shelton, R. L. (1977). Osmotic control of vasopressin function. In T. E. Andreoli, J. J. Grantham, & F. C. Rector (Eds.), *Disturbance in body fluid osmolality* (pp. 125–148). Bethesda: The American Physiological Society.

Robertson, G. L., Mahr, E. A., Athar, S., & Sinha, T. (1973). Development and clinical application of new method for the radioimmunoassay of arginine vasopressin in human plasma. *The Journal of Clinical Investigation, 52*, 2340–2352.

Robertson, G. L., Shelton, R. L., & Athar, S. (1976). The osmoregulation of vasopressin. *Kidney International, 10*, 25–37.

Rusten, T. E., Vaccari, T., & Stenmark, H. (2012). Shaping development with ESCRTs. *Nature Cell Biology, 14*, 38–45.

Sachdeva, R., & Singh, B. (2014). Insights into structural mechanisms of gating induced regulation of aquaporins. *Progress in Biophysics and Molecular Biology, 114*(69–79), 2014.

Saito, T., Ishikawa, S. E., Sasaki, S., Nakamura, T., Rokkaku, K., Kawakami, A., et al. (1997). Urinary excretion of aquaporin-2 in the diagnosis of central diabetes insipidus. *The Journal of Clinical Endocrinology & Metabolism, 82*, 1823–1827.

Sanches, T. R., Volpini, R. A., Massola Shimizu, M. H., Bragança, A. C., Oshiro-Monreal, F., Seguro, A. C., et al. (2012). Sildenafil reduces polyuria in rats with lithium-induced NDI. *American Journal of Physiology—Renal Physiology, 302*, F216–F225.

Sandoval, P. C., Slentz, D. H., Pisitkun, T., Saeed, F., Hoffert, J. D., & Knepper, M. A. (2013). Proteome-wide measurement of protein half-lives and translation rates in vasopressin-sensitive collecting duct cells. *Journal of the American Society of Nephrology, 24*, 1793–1805.

Schenk, L. K., Bolger, S. J., Luginbuhl, K., Gonzales, P. A., Rinschen, M. M., Yu, M. J., et al. (2012). Quantitative proteomics identifies vasopressin-responsive nuclear proteins in collecting duct cells. *Journal of the American Society of Nephrology, 23*, 1008–1018.

Simon, H., Gao, Y., Franki, N., & Hays, R. M. (1993). Vasopressin depolymerizes apical F-actin in rat inner medullary collecting duct. *American Journal of Physiology—Cell Physiology, 265*, C757–C762.

Sohara, E., Rai, T., Sasaki, S., & Uchida, S. (2006). Physiological roles of AQP7 in the kidney: Lessons from AQP7 knockout mice. *Biochimica et Biophysica Acta, 1758*, 1106–1110.

Sonoda, H., Yokota-Ikeda, N., Oshikawa, S., Kanno, Y., Yoshinaga, K., Uchida, K., et al. (2009). Decreased abundance of urinary exosomal aquaporin-1 in renal ischemia–reperfusion injury. *American Journal of Physiology—Renal Physiology, 297*, F1006–F1016.

Street, J. M., Birkhoff, W., Menzies, R. I., Webb, D. J., Bailey, M. A., & Dear, J. W. (2011). Exosomal transmission of functional aquaporin 2 in kidney cortical collecting duct cells. *The Journal of Physiology, 589*, 6119–6127.

Tajika, Y., Matsuzaki, T., Suzuki, T., Aoki, T., Hagiwara, H., Tanaka, S., et al. (2002). Immunohistochemical characterization of the intracellular pool of water channel aquaporin-2 in the rat kidney. *Anatomical Science International, 77*, 189–195.

Takahashi, S., Muta, K., Sonoda, H., Kato, A., Abdeen, A., & Ikeda, M. (2014). The role of cysteine 227 in subcellular localization, water permeability, and multimerization of aquaporin-11. *FEBS Open Bio, 4*, 315–320.

Takata, K., Matsuzaki, T., & Tajika, Y. (2004). Aquaporins: Water channel proteins of the cell membrane. *Progress in Histochemistry and Cytochemistry, 39*, 1–84.

Takata, K., Matsuzaki, T., Tajika, Y., Ablimit, A., & Hasegawa, T. (2008). Localization and trafficking of aquaporin 2 in the kidney. *Histochemistry and Cell Biology, 130*, 197–209.

Takata, K., Matsuzaki, T., Tajika, Y., Ablimit, A., Suzuki, T., Aoki, T., et al. (2005). Aquaporin water channels in the kidney. *Acta Histochemica et Cytochemica, 38*, 199–207.

Tani, K., & Fujiyoshi, Y. (2014). Water channel structures analysed by electron crystallography. *Biochimica et Biophysica Acta, 1840*, 1605–1613.
Terris, J., Ecelbarger, C. A., Marples, D., Knepper, M. A., & Nielsen, S. (1995). Distribution of aquaporin-4 water channel expression within rat kidney. *American Journal of Physiology—Renal Physiology, 269*, F775–F785.
Terris, J., Ecelbarger, C. A., Nielsen, S., & Knepper, M. A. (1996). Long-term regulation of four renal aquaporins in rats. *American Journal of Physiology, 271*, F414–F422.
Tornroth-Horsefield, S., Wang, Y., Hedfalk, K., Johanson, U., Karlsson, M., Tajkhorshid, E., et al. (2006). Structural mechanism of plant aquaporin gating. *Nature, 439*, 688–694.
Trionfini, P., Benigni, A., & Remuzzi, G. (2015). MicroRNAs in kidney physiology and disease. *Nature Reviews Nephrology, 11*, 23–33.
Uchida, S., Sasaki, S., Fushimi, K., & Marumo, F. (1994). Isolation of human aquaporin-CD gene. *The Journal of Biological Chemistry, 269*, 23451–23455.
Verbalis, J. G. (2012). Disorders of water balance. In M. W. Taal, G. M. Chertow, P. A. Marsden, K. Skorecki, A. S. L. Yu, & B. M. Barry (Eds.), *Brenner & Rector's the kidney* (pp. 540–594). Philadelphia: Saunders, an imprint of Elsevier Inc.
Verkman, A. S. (2006). Roles of aquaporins in kidney revealed by transgenic mice. *Seminars in Nephrology, 26*, 200–208.
Wen, H., Frokiaer, J., Kwon, T. H., & Nielsen, S. (1999). Urinary excretion of aquaporin-2 in rat is mediated by a vasopressin-dependent apical pathway. *Journal of the American Society of Nephrology, 10*, 1416–1429.
Xie, L., Hoffert, J. D., Chou, C. L., Yu, M. J., Pisitkun, T., Knepper, M. A., et al. (2010). Quantitative analysis of aquaporin-2 phosphorylation. *American Journal of Physiology—Renal Physiology, 298*, F1018–F1023.
Yakata, K., Tani, K., & Fujiyoshi, Y. (2011). Water permeability and characterization of aquaporin-11. *Journal of Structural Biology, 174*, 315–320.
Yamamoto, T., Sasaki, S., Fushimi, K., Ishibashi, K., Yaoita, E., Kawasaki, K., et al. (1995). Vasopressin increases AQP-CD water channel in apical membrane of collecting duct cells in Brattleboro rats. *American Journal of Physiology—Cell Physiology, 268*, C1546–C1551.
Yasui, M., Hazama, A., Kwon, T. H., Nielsen, S., Guggino, W. B., & Agre, P. (1999). Rapid gating and anion permeability of an intracellular aquaporin. *Nature, 402*, 184–187.
Yasui, M., Zelenin, S. M., Celsi, G., & Aperia, A. (1997). Adenylate cyclase-coupled vasopressin receptor activates AQP2 promoter via a dual effect on CRE and AP1 elements. *American Journal of Physiology—Renal Physiology, 272*, F443–F450.
Yip, K. P. (2006). Epac-mediated Ca(2+) mobilization and exocytosis in inner medullary collecting duct. *American Journal of Physiology—Renal Physiology, 291*, F882–F890.
Yu, M. J., Miller, R. L., Uawithya, P., Rinschen, M. M., Khositseth, S., Braucht, D. W., et al. (2009). Systems-level analysis of cell-specific AQP2 gene expression in renal collecting duct. *Proceedings of the National Academy of Sciences of the United States of America, 106*, 2441–2446.
Yu, L., Moriguchi, T., Souma, T., Takai, J., Satoh, H., Morito, N., et al. (2014). GATA2 regulates body water homeostasis through maintaining aquaporin 2 expression in renal collecting ducts. *Molecular and Cellular Biology, 34*, 1929–1941.
Yui, N., Lu, H. A. J., Bouley, R., & Brown, D. (2011). AQP2 is necessary for vasopressin- and forskolin-mediated filamentous actin depolymerization in renal epithelial cells. *Biology Open, 1*, 101–108.

CHAPTER ELEVEN

The Structure and Function of the Dopamine Transporter and its Role in CNS Diseases

Patrick C. McHugh[1], David A. Buckley
Department of Pharmacy, School of Applied Sciences, University of Huddersfield, Huddersfield, United Kingdom
[1]Corresponding author: e-mail address: p.c.mchugh@hud.ac.uk

Contents

Abstract

In this chapter, we explore the basic science of the dopamine transporter (DAT), an integral component of a system that regulates dopamine homeostasis. Dopamine is a key neurotransmitter for several brain functions including locomotor control and reward systems. The transporter structure, function, mechanism of action, localization, and distribution, in addition to gene regulation, are discussed. Over many years, a wealth of information concerning the DAT has been accrued and has led to increased interest in the role of the DAT in a plethora of central nervous system diseases. These DAT

Vitamins and Hormones, Volume 98
ISSN 0083-6729
http://dx.doi.org/10.1016/bs.vh.2014.12.009

characteristics are explored in relation to a range of neurological and neuropsychiatric diseases, with a particular focus on the genetics of the DAT. In addition, we discuss the pharmacology of the DAT and how this relates to disease and addiction.

1. INTRODUCTION

Dopamine is an evolutionarily ancient catecholamine present in most eukaryotes (Callier et al., 2003). In many animals, dopamine plays an essential role as a modulatory neurotransmitter in many behavioral and decision-making processes, such as aggression, sexual behavior, reward, learning, and memory (Dalley & Everitt, 2009). The dopamine transporter (DAT) has a critical role in dopamine homeostasis; loss of proper function and regulation of the transporter has been implicated in several dopamine-related diseases. For example, decreased striatal DAT binding has been reported in clinical depression (Laasonen-Balk et al., 1999) and schizophrenic patients (Mateos et al., 2005). Moreover, genetic differences in the dopamine transporter gene (DAT1), including a variable number tandem repeat (VNTR), have been identified as risk factors for several neuropsychiatric disorders (Joyce, Stephenson, Kennedy, Mulder, & McHugh, 2014; Mazei-Robison, Couch, Shelton, Stein, & Blakely, 2005). The DAT is also well known for its role in addiction, including alcoholism (Agrawal et al., 2013) and the abuse of psychostimulants such as amphetamine (AMPH; Giros, Jaber, Jones, Wightman, & Caron, 1996), which acts on the transporter to elicit their behavioral effects. The relationship between the DAT and the central nervous system (CNS) diseases will be discussed.

2. THE DOPAMINERGIC SYSTEM

Dopamine, a catecholamine neurotransmitter, is a precursor in the synthesis of the neurotransmitter noradrenaline. Dopamine is synthesized through a series of enzymatic reactions, beginning with the hydroxylation of the amino acid tyrosine to L-DOPA via tyrosine hydroxylase (TH; Fig. 1). L-DOPA is subsequently decarboxylated by aromatic amino acid decarboxylase to produce dopamine, which is then packaged into synaptic vesicles by the vesicular monoamine transporter 2 (VMAT2) and released at nerve terminals into the synapse upon stimulation (Fig. 1). Several factors influence dopaminergic neurotransmission, such as the amount of dopamine synthesized and released, the number of dopaminergic receptors (DRs) at

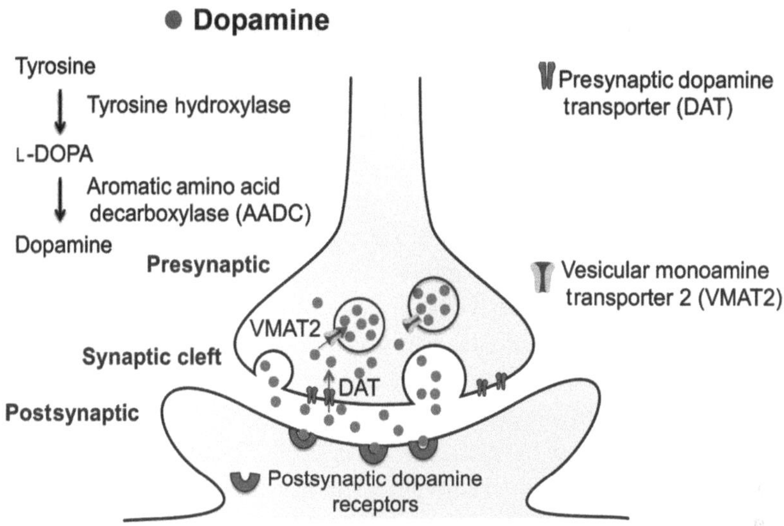

Figure 1 A schematic drawing of the post- and presynaptic regions of the dopamine neuron. Dopamine is synthesized in the presynaptic neuron starting from tyrosine to L-DOPA, and then the aromatic amino acid decarboxylase (AADC) catalyzes the formation of dopamine. *Reproduced by permission of* The Royal Society of Chemistry *(http://pubs.rsc.org/en/content/articlehtml/2014/cs/c3cs60430f).*

the synapse, and the amount of time dopamine spends in the synaptic space. Released dopamine then binds to DRs to evoke a response in the postsynaptic cell. Dopamine is then cleared from the synapse primarily by the DAT, where it reenters the presynaptic neuron to be recycled and repackaged into vesicles.

As a precursor of noradrenaline, dopamine was originally thought to be an intermediate of noradrenaline production and not to possess its own signaling properties. It was later shown by Carlsson and Hillarp (1958) that dopamine had signaling activities of its own. The authors depleted the catecholamines in rabbits by administering reserpine, which irreversibly blocks VMAT2, an isoform of VMAT expressed in CNS monoaminergic cells and transports intracellular noradrenaline and dopamine into presynaptic vesicles for subsequent release into the synaptic cleft (Fig. 1). As a consequence of blocking this process, free neurotransmitters are subsequently metabolized. It was demonstrated through the administration of L-DOPA to rabbits that this process could be reversed for dopamine but not for noradrenaline, resulting in increases in dopamine, but not noradrenaline. Subsequent work revealed regions of the brain enriched with dopamine and distinct dopamine signaling pathways were identified (Fuxe, 1965).

There are four main dopaminergic pathways: the mesolimbic pathway, the mesocortical pathway, the nigrostriatal pathway, and the tuberoinfundibular pathway (Fig. 2). The mesolimbic pathway originates in the ventral tegmental area (VTA) and innervates the ventral striatum, also known as the nucleus accumbens. This pathway is implicated in reward and pleasure. The mesocortical pathway also originates from the VTA but instead projects to the frontal lobes of the cerebrum, particularly the prefrontal cortex, and is involved in cognition and emotion. Both the mesolimbic and mesocortical pathways have been linked to addiction, depression, and schizophrenia. The nigrostriatal pathway consists of neurons whose cell bodies originate in the substantia nigra and terminate in the dorsal striatum. This area is implicated in movement, since degeneration of these projections has been shown to cause Parkinson's disease (PD), characterized by tremors, rigidity, and overall improper movement (Barbeau, 1962). It has also been demonstrated that this region is important in feeding behavior (Robinson, Rainwater, Hnasko, & Palmiter, 2007). Lastly, neurons of the tuberoinfundibular pathway, which refers to a group of dopaminergic

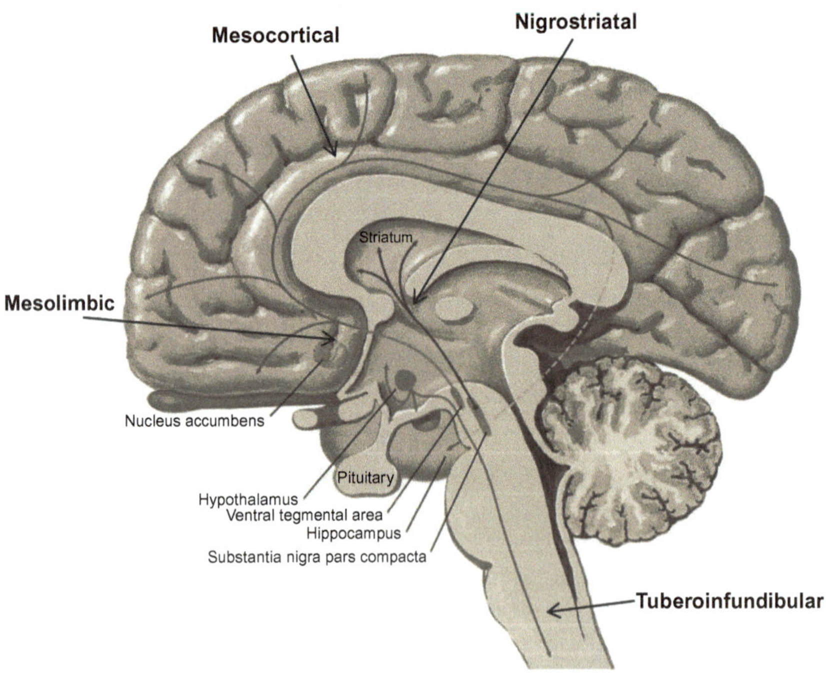

Figure 2 Dopaminergic pathways in the human brain. *Adapted from Scarr, Gibbons, Neo, Udawela, and Dean (2013).*

neurons in the arcuate nucleus of the hypothalamus that project to the median eminence, control prolactin secretion from the anterior pituitary gland (Weiner & Ganong, 1978). Dopamine is involved in a number of physiological and behavioral processes including cognition, locomotion, mood, motivation, and reward. Abnormalities in the central dopaminergic systems contribute to many CNS diseases (Kienast & Heinz, 2006).

3. THE DOPAMINE TRANSPORTER

3.1 Structure

Initial attempts to characterize the function and structure of the rat dopamine transporter (rDAT) elucidated a topological conformation consisting of 12 transmembrane domains (Giros, el Mestikawy, Bertrand, & Caron, 1991) and a molecular weight of approximately 80 kDa (Vaughan, Brown, McCoy, & Kuhar, 1996; Fig. 3). The rDAT exhibits intracellular amino and carboxyl-termini and notable sequence homology to the noradrenaline transporter (NET; 64%) and, to a lesser degree, the gamma-aminobutyric acid (GABA) transporter (30%; Giros et al., 1991). In 1992, a human DAT1 3.5-kb cDNA clone was successfully isolated from a human substantia nigra cDNA library with the corresponding gene designated to the distal end of chromosome 5. The subsequent protein, 620 amino acids in length, similarly consists of 12 putative transmembrane domains as determined by hydrophobicity analysis, with 92% homogeneity to the rDAT and 84% homogeneity to the bovine DAT (Giros et al., 1992). Aside from the five intracellular and six extracellular loops, the DAT also exhibits a cytoplasmic amino and carboxyl-termini that modulate uptake and substrate

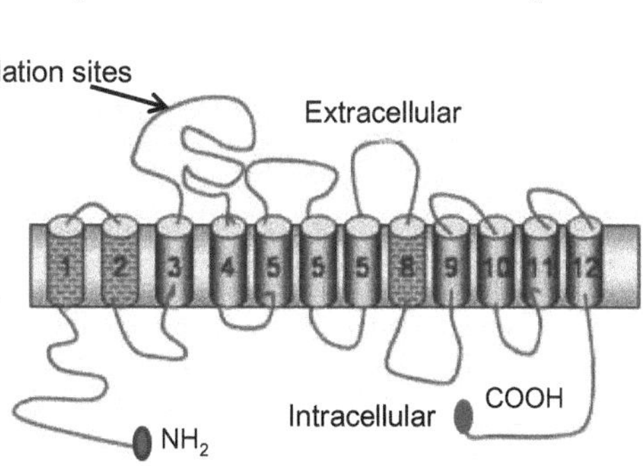

Figure 3 A schematic representation of the DAT showing the 12 transmembrane domains. The intracellular amino- and carboxyl-termini and glycosylation sites are shown. The textured transmembrane domains 1, 2, and 8 are conserved across DAT, NET, and SERT. *Adapted from Madras, Miller, and Fischman (2005).*

recognition with multiple protein kinase A, C, and Ca^{2+}–calmodulin phosphorylation sites (Fog et al., 2006). Structurally, three N-linked glycosylation sites are situated within the large, second extracellular loop. These sites have been shown to be crucial for stable membrane expression and substrate transport efficiency of the DAT (Li et al., 2004), as phosphorylation can facilitate downregulation of transporter function. A disulfide bond is also localized to the second extracellular loop. Research has highlighted that this is not related to transporter function but is instead implicated in DAT biosynthesis (Chen et al., 2007). Analysis of leucine zipper-like motifs, which are present within the DAT structure, demonstrated that the motifs located on transmembrane domain 2, but not 9, are crucial for plasma membrane localization of the DAT (Torres et al., 2003; Fig. 3). Latterly, the crystal structure of the *Drosophila melanogaster* DAT has been determined. The structure, which is bound with the tricyclic antidepressant nortriptyline, provides insights into the mechanism of interaction between the antidepressants and the solute carrier 6 (SLC6) family of neurotransmitter transporters (Penmatsa, Wang, & Gouaux, 2013).

3.2 Function

Termination of dopaminergic neurotransmission occurs through one of two distinct mechanisms, either by enzymatic degradation, often mediated by monoamine oxidase, or by dopamine translocation across the plasma membrane via the DAT. Upon dopamine release into the synaptic cleft as a consequence of calcium-mediated fusion of vesicles with the presynaptic membrane (Egana et al., 2009), the plasma membrane DAT utilizes ionic gradients between the synaptic cleft and the presynaptic neuron to drive the transport of dopamine. Upon reuptake, dopamine is subsequently processed into vesicles by VMAT2. The uptake and subsequent localization of dopamine within the nerve terminal terminates neurotransmission and permits the recycling of neurotransmitter for subsequent release (Fig. 1). Classified as a secondary active transporter, the DAT is a member of a group of SLCs that predominantly permit the thermodynamically unfavorable translocation of a substrate with the electrochemically favorable cotransport of sodium; in contrast, some transporters use ATP hydrolysis as a facilitator for substrate transport (Kristensen et al., 2011). This reliance on sodium for neurotransmitter reuptake, coupled with membrane topology, places the DAT within the sodium- and chloride-dependent SLC6 family of monoamine transporters, which includes the serotonin transporter (SERT),

NET, and multiple GABA and glycine transporters. Variations in the density, population, and function of membrane-localized DAT therefore impact directly upon the synaptic concentration and ultimately the availability of dopamine (Shumay, Fowler, & Volkow, 2010). The DAT is therefore a key modulator of intrasynaptic dopamine levels (Laasonen-Balk et al., 1999), and through dopamine homeostasis, it preserves normal neurological function within the dopaminergic pathways of the CNS.

3.3 Mechanism

As a member of the SLC family, the DAT uses a sodium gradient, which is maintained by the sodium/potassium ATPase pump, to facilitate the unidirectional rapid reuptake of dopamine alongside sodium and chloride ions, which can occur against very large concentration gradients (Kristensen et al., 2011). The necessity of sodium for neurotransmitter transport was first deduced by *in vitro* analysis of transient rDAT cDNA expression in the COS7 cell line which highlighted sodium-dependent dopamine uptake (Giros et al., 1991). Although the function of chloride ion transportation by the DAT is not fully understood, it is suggested that chloride may function to compensate for the sodium ion charge, a hypothesis derived from mutagenic studies involving the introduction of a negatively charged amino acid in close proximity to the transporters sodium binding sites and thereby resulting in chloride-independent transport (Zomot et al., 2007). The kinetics of neurotransmitter transport within the SLC neurotransmitter family of transporters observes the Michaelis–Menton model. In addition, substrate-ion stoichiometry follows a 1:2:1 ratio for substrate, sodium, and chloride, respectively (Gu, Wall, & Rudnick, 1994; Kristensen et al., 2011). However, substrate specificity across the monoamine transporters does not exhibit exclusivity, as the DAT is also able to transport noradrenaline (Gether, Andersen, Larsson, & Schousboe, 2006), and vice versa (Moron, Brockington, Wise, Rocha, & Hope, 2002). The lack of specificity is of particular relevance in the prefrontal cortex, wherein dopamine reuptake is largely undertaken by the NET.

Mechanistic analysis of monoamine transport has been largely undertaken using a bacterial homolog of sodium-/chloride-dependent neurotransmitter transporters, LeuT, a leucine transporter isolated from *Aquifex aeolicus* with a well-defined crystal structure (Yamashita, Singh, Kawate, Jin, & Gouaux, 2005). The use of the LeuT, which possesses 55–67% sequence homology to the SLC6 neurotransmitter transporters in the regions that are considered to be core components of substrate transport, is a valid facilitator of neurotransmitter transporter studies, including substrate and inhibitor binding analysis (Kristensen et al., 2011). Three-dimensional models of the DAT have also been generated using LeuT as a basis for modeling and substrate binding prediction (Indarte, Madura, & Surratt, 2008). Strong similarities in the binding of dopamine, serotonin, and noradrenaline to their respective transporters have been proposed, with ionic interactions between dopamine and the DAT Asp79 residue, while the *meta*-hydroxyl group of dopamine localized within a hydrophilic pocket containing Gly153, Ser422, and Ala423 residues, and latterly, an additional transporter hydrophilic group is present for accommodation of the dopamine *para*-hydroxyl group (Koldso, Christiansen, Sinning, & Schiott, 2013). The Ser-528 residue, however, is thought to play a crucial role in maintaining DAT efficiency by stabilizing the transporter and thereby promoting a conformation that reduces the likelihood of outward dopamine transport (Chen & Justice, 2000).

3.4 Location and distribution

The DAT is not ubiquitously expressed, but is localized to dopaminergic neurons within established neurological systems implicating dopamine neurotransmission (Kristensen et al., 2011), including the four main dopaminergic pathways (Fig. 2). DAT1 mRNA exhibited diverse patterns of expression, located solely in cell bodies of dopaminergic neurons and predominantly within the substantia nigra/VTAs (Richtand, Kelsoe, Segal, & Kuczenski, 1995). A significant degree of correspondence between dopaminergic neurons expressing both DAT and TH mRNA is observed. Using immunohistochemistry facilitated by highly specific monoclonal antibodies, areas highly populated by the rDAT were located within the striatum and nucleus accumbens (Ciliax et al., 1999), while the DAT has also been located in the midbrain and prefrontal cortex (Fig. 4). The determination of DAT expression therefore indicates the presence of dopaminergic neurons in the CNS (Kristensen et al., 2011).

Within the dopaminergic system, however, the rDAT is predominantly localized to dendrites, perikaryon, axons, and terminals, but was not found within areas of synaptic activity in the striatum and appeared localized to dendrites and axons within the substantia nigra. This indicates that the reuptake of dopamine proceeds neurotransmitter diffusion from the synapse (Nirenberg, Vaughan, Uhl, Kuhar, & Pickel, 1996). Nirenberg and

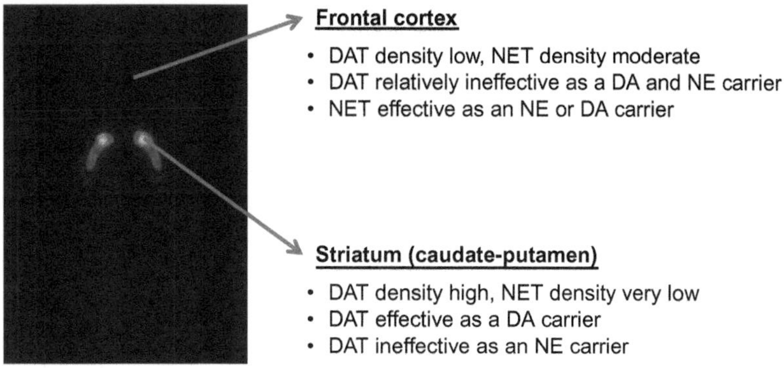

Figure 4 Areas of DAT density and distribution within the human brain. Areas of DAT density and distribution illustrated using positron emission tomography with [^{11}C] altropane. *Adapted from Madras et al. (2005).*

colleagues went further to illustrate the diversity of DAT distribution. In the VTAs of the rat brain, the majority of the rDAT was found localized within intracellular membranes of the perikarya and large proximal dendrites, while in small to medium diameter dendrites and unmyleinated axons, the DAT was primarily found in the plasma membrane (Nirenberg et al., 1997). These findings, coupled with a significant variation in DAT expression between subpopulations of dopaminergic neurons within the human CNS (Ciliax et al., 1999) and variations in rDAT immunoreactivity within the rat brain, lead to deductions that the variation is likely to be of functional significance (Freed et al., 1995).

However, multiple factors have been shown to correlate to changes in DAT expression. Therefore, significant interindividual variation in DAT expression exists. DAT levels adjust according to the intensity of dopamine signaling; when synaptic dopamine is abundant, outer membrane bound DAT presence increases to promote reuptake and vice versa (Shumay et al., 2010). Furthermore, DAT1 mRNA expression was found to be notably reduced, up to 75% in localized regions in persons of advancing age. However, compensatory mechanisms, inclusive of transporter redistribution, have been observed in rodent studies (Cruz-Muros et al., 2009). Age-related changes in dopaminergic neurotransmission may, therefore, be a consequence of reduced DAT expression (Bannon & Whitty, 1997). In addition, alterations in DAT expression have been associated with the pathogenesis of HIV-associated dementia (Wang et al., 2004), while drug-induced changes, encompassing both cocaine and AMPH, have been shown to decrease and increase membrane DAT expression, respectively (Zahniser & Sorkin, 2004).

3.5 Gene structure and regulation

Situated on chromosome 5 (5p.15.32), the gene encoding for the DAT spans over 60 kb and consists of 15 exons and 14 introns, a structure comparable to that of the NET. However, the protein-coding component of the DAT1 excludes exon 1 and terminates within exon 15 with a single transcription start site, as determined by 5-RACE and RNase protection assays (Kawarai, Kawakami, Yamamura, & Nakamura, 1997). In addition to the absence of variable polyadenylation sites, there are no observed splice variants of the DAT1 with distinct RNA splicing sequences present at each intron–exon junction. The DAT1-coding region, and subsequently the protein sequence, displays high interspecies conservation, which reinforces the DAT as an integral component of dopaminergic neurotransmission. However, the high protein sequence homology does not correlate to conservation within the DAT1 locus. Aside from the interspecies similarities within the conserved regions of the gene promoter, splice sites, and coding regions, significant differences are observed between species in other loci (Shumay et al., 2010).

Analysis of the GC-rich 5′-DAT1-flanking region highlighted a locus void of the TATA and CAT *cis*-regulatory elements, although E box and multiple SP1 transcription factor (SP1) binding sites were identified, and thereby displaying similar structural characteristics to the human D1A dopamine receptor gene (Kawarai et al., 1997). The functional importance of the GC-rich transcription factor SP1 motif has been demonstrated in a SP-deficient cell line. Through electrophoretic mobility shift assays, SP1 was found to associate with the GC-rich region and, subsequently, induce transcriptional activity (Wang & Bannon, 2005).

However, a highly conserved sequence of 180 nucleotides, located immediately 5′ of the transcription initiation site, exhibits significant sequence homology between the human and rat regions. This GC-rich sequence exhibits a relative abundance of transcription factor binding sites, including AP1 and CREB, which provides further evidence for transcriptional regulation within this locus (Donovan et al., 1995). Promoter analysis has identified CCAAT sequences, suggesting that DAT1 transcription may be under the influence of the CCAAT element and consequently regulated by NF-Y (Dolfini, Zambelli, Pavesi, & Mantovani, 2009), a heterotrimeric transcription factor shown to be a prerequisite for RNA polymerase II binding. NF-Y is also known to interact with the transcription factors SP1, GATA-1, CREB, and c-Myc, all of which are predicted, through bioinformatic analysis, to interact with the DAT1 5′ region (Shumay et al., 2010).

In addition, nuclear receptor-related I protein (Nurr1), a transcription factor essential for the regulation of multiple genes within the midbrain dopaminergic system (Simon, Bhatt, Gherbassi, Sgado, & Alberi, 2003), has been shown, through the luciferase reporter assay system, to promote transcription by interacting with DAT1 promoter constructs incorporating NGFI-B response elements (Sacchetti, Brownschidle, Granneman, & Bannon, 1999). The crucial role of Nurr1 has been demonstrated in animal models wherein the loss of Nurr1 during mouse development resulted in a distinct absence of midbrain dopaminergic neurons, while reductions in Nurr1 function after maturation have been associated with the progressive loss of striatal dopamine (Luo, 2012). Crucially however, differential gene regulation is a prominent feature of the DAT1, which is diversely expressed within specific dopaminergic systems, and thereby reinforces the transporter's distinct role in dopamine homeostasis. The neuron-restrictive silencer factor (NRSF) is typically localized within nonneuronal cells and functions as a transcriptional repressor of neuron-specific genes. Using DAT1 promoter-luciferase constructs, the presence of a neuron-restrictive silencer element within the construct results in significantly reduced reporter gene expression (Sacchetti et al., 1999). This suggests that NRSF may possess an integral role in the silencing of DAT expression in nonneuronal cells.

3.6 Genetic variation

The fundamental role of the DAT within the dopaminergic system has led to a plethora of association studies underpinned by the prominent genotypic variations exhibited within the DAT1 locus. Mutations and single nucleotide polymorphisms (SNPs) have been described throughout the DAT1 and its flanking regions. Located within the DAT1 core promoter region, the polymorphism -67A/T (rs2975226) has been analyzed in association studies relating to attention deficit hyperactivity disorder (ADHD) (Xu et al., 2009), in addition to another promoter SNP, -839C/T (rs2652511; de Azeredo et al., 2014). The DAT1 also exhibits a VNTR within intron 8, consisting of either five or six copies of a 30-nucleotide; the 6-repeat allele has a frequency of 38% in Caucasians. In addition, several polymorphic variants within DAT1 exons were identified, but only two mutations resulted in a change in the protein composition. This reinforces the high level of conservation within the DAT1 protein coding regions, which suggests that variations in gene expression may result from polymorphisms within the DAT1 promoter, enhancer, or other noncoding regions (Vandenbergh et al., 2000).

A DAT1 3′-untranslated region (UTR) VNTR has been the focus of multiple research groups in an attempt to determine an association between the VNTR and several CNS disorders (Mercier, Turpin, & Lucotte, 1999). The polymorphism, which is absent in the rDAT and thereby suggests a relatively recent genotypic modification (VanNess, Owens, & Kilts, 2005), consists of a 40-basepair tandem repeat which is typically observed as 3–11 copies (Vandenbergh et al., 1992) with alleles consisting of 9 and 10 repeats representing the most common genotypes. Although the VNTR is not located within the exonic regions or splice sites of the DAT1, it is established that polymorphisms in noncoding loci may result in differential gene expression through a variety of mechanisms, including changes in RNA stability, transport, and subsequent translation (VanNess et al., 2005). A multitude of *in vitro* functional assays have been undertaken in order to elucidate the potential influence of the VNTR on gene expression. Functional analysis, using the luciferase assay reporter system, demonstrated that allelic variants of the DAT1 VNTR altered luciferase expression (Fuke et al., 2001), yet subsequent functional determination illustrated that regions within introns 9, 12, and 14 promote transcription through reputed enhancer elements, but luciferase expression was not significantly different between the VNTR 9- and 10-repeat alleles (Greenwood & Kelsoe, 2003). The acquisition of inconsistent observations was continued by VanNess and coworkers who utilized a combination of binding assays, immunoblots, and stable cell line transfection. It was deduced that HEK293 cells transfected with the 10-repeat allele exhibited notably greater DAT density compared to the 9-repeat allele (VanNess et al., 2005). Mechanistically however, the VNTR has been associated with DAT1 repression mediated by interactions with the basic helix–loop–helix (bHLH) transcriptional factor Hesr1. Analysis using Hesr1-knockout mice highlighted, through quantitative PCR, that the DAT1 and other dopamine-related genes were significantly upregulated, a functional consequence of reduced Hesr1 binding to the DAT1 3′-noncoding region (Fuke et al., 2006). A *cis*- and *trans*-acting inhibition of DAT1 expression was also observed by HESR1 and HESR2 through interactions with the DAT1 core promoter and VNTR, respectively, with differential expression depending on the VNTR allele

(Kanno & Ishiura, 2011). The variability and inconsistencies in association studies have therefore led to conclusions that the DAT1 VNTR may not have a significant functional implication on DAT expression, but through linkage disequilibrium, the VNTR may form a haplotype with other polymorphic variants that results in a phenotype suggestive of a neuropsychiatric disorder (Mill, Asherson, Craig, & D'Souza, 2005).

4. THE DOPAMINE TRANSPORTER AND DISEASE

4.1 Parkinson's disease

PD, also known as idiopathic PD, primary Parkinsonism, hypokinetic rigid syndrome, or paralysis agitans, is a degenerative disorder of the CNS implicating the dopaminergic neurons of the basal ganglia, in particular the *pars compacta* component of the substantia nigra, and is the dominant form of Parkinsonism. Others include dementia with Lewy bodies, vascular PD, and drug-induced PD. It is typified by motor symptoms such as tremor, bradykinesia, postural instability, and rigidity. The neuropsychiatric effects also evident in affected individuals with PD may include cognitive defects encompassing deficits in memory and language function, psychosis-related symptoms such as hallucinations that are primarily visual in presentation, and disturbances in sleep and wakefulness relating to REM behavior disorder (Weintraub & Burn, 2011).

The development of PD is multifactorial; both environmental and genetic predispositions facilitate disease development. Early interpretation attributed isolated cases of chronic PD to 1-methyl-4-phenylpyridinium (MPP^+) exposure, a neurotoxic metabolite of a contaminant found in the synthetic opiate 1-methyl-4-phenyl-4-propionoxypiperidine. This association prompted extensive environmental analysis of PD causation, which linked the disease to several environmental and agricultural factors often related to direct or indirect exposure to pesticides, herbicides, and solvents (Pezzoli & Cereda, 2013). However, mixed results have been observed (Lock, Zhang, & Checkoway, 2013), potentially due to the variability in the nature and duration of exposure. Similarly, multiple genetic factors are also associated with PD development, including haplotypes within the α-synuclein gene (Simon-Sanchez et al., 2009). *In vitro* mechanistic analysis has demonstrated that α-synuclein complexes with the intracellular carboxyl terminal of the DAT to induce a presynaptic clustering of the transporter and thereby facilitating increased dopamine uptake and, ultimately, dopamine-induced apoptosis (Lee, Liu, Pristupa, & Niznik, 2001). The complex pathophysiological processes of PD are further extended as both activated microglia and reactive astrocytes are localized at sites of neuronal loss, both of which are involved in inflammatory processes and demonstrate the potential for dopamine uptake (Chotibut, Apple, Jefferis, & Salvatore, 2012).

Phenotypic variations in both physical and cognitive PD symptoms, coupled with differences in disease pathology and L-DOPA response, suggest that there is a significant underlying genetic component to the disease. Familial cases are thought to comprise of approximately 15% of all PD cases (Vahedi et al., 2010). The DAT1 VNTR was analyzed in Caucasians, and of the five alleles detected, no significance was observed between genotype and PD, although a small sample size left the study underpowered to detect rare but significant associations (Mercier et al., 1999). Proceeding meta-analysis of case–control studies to determine an association between the DAT1 and PD demonstrated a somewhat conflicting association. The DAT1 VNTR 10-repeat allele was shown to be neuroprotective in East Asians, but the reputed interaction was not replicated in a Caucasian population. However, the G allele and GG genotype within the DAT1 promoter region (rs2652510) were shown to be a significant risk factor for PD development (Zhai, Li, Zhao, & Lin, 2014). On the other hand, assessment of the 5′-region of the DAT1 determined six common haplotypes with notable differences in *in vitro* reporter gene assays, although none were found to associate with a PD phenotype (Kelada et al., 2005).

Considering that dopaminergic nerve terminals are typically reduced by up to 50% in early PD (Nutt, Carter, & Sexton, 2004), the implications on the DAT are striking, as neuronal regression directly corresponds to the loss of the DAT (Seeman & Niznik, 1990). This was further highlighted with Western blot analysis which elucidated a considerable and progressive reduction in DAT immunoreactivity in the putamen, caudate, and nucleus accumbens in patients with PD (Miller et al., 1997). Positron emission tomography (PET) highlighted that, in PD, greater DAT expression correlates to a stabilization in synaptic dopamine levels. This observation, coupled with a progressive reduction in DAT levels throughout disease progression, may ultimately contribute to fluctuations in motor function (Sossi et al., 2007). Moreover, with the use of PET methodology, research has highlighted a notable age-related decline in the DAT, typically around 5.6–6.1% per decade, which has been attributed to dopaminergic neuron loss within the substantia nigra (Ishibashi et al., 2009). In PD, however, similar rates of DAT decline have been observed annually, but variations are observed depending on the location of the dopaminergic neuronal circulatory (Ishibashi, Oda, Ishiwata, & Ishii, 2014). The use of PET and single-photon computerized emission tomography (SPECT) has therefore been utilized to determine the striatal dopaminergic composition. Using the DAT as a marker of dopaminergic nerve terminal presence, PET analysis has shown that radiolabeled tropane derivatives, such as [^{11}C]cocaine, provide an avenue for diagnosing and monitoring the progressive loss of the dopaminergic system associated with PD (Shih, Hoexter, Andrade, & Bressan, 2006).

4.2 Borderline personality disorder

Borderline personality disorder (BPD) is a severe disorder of personality, described as a psychiatric diagnosis in the *Diagnostic and Statistical Manual of Mental Disorders, Fourth Edition* (*DSM-IV 4th Ed.*; American Psychiatric Association, 2000). BPD is considered to be a prolonged disturbance of personality function characterized by depth and variability of moods. It is often comorbid in patients who have a diagnosis of depression. Individuals with BPD often present during adolescence or young adulthood and exhibit general impulsivity, impulsive-aggression traits, paranoia, and dissociation, with a tendency toward attempted suicide (Soloff, Lynch, Kelly, Malone, & Mann, 2000). Although the etiology is not clear, twin studies have highlighted the potential for genetic predisposition to BPD (Distel et al., 2008), with variants in serotonergic and dopaminergic pathways considered potential sources for a BPD phenotype (Joyce et al., 2014). Dysfunction within the dopaminergic system, which is involved in cognitive and emotional processing and the regulation of impulsive behaviors, has been associated with BPD symptoms (Friedel, 2004). This has been demonstrated in animal research which suggests that repeated exposure to social stress has long-term effects on DAT density (Lucas et al., 2004), while DAT1-knockout mice demonstrate abnormalities of social interaction (Rodriguiz, Chu, Caron, & Wetsel, 2004). These studies highlight the role of the dopaminergic system, in particular the DAT, in the neurobiology of BPD. It has been shown that DAT1 VNTR genotypes may confer a significant genetic risk factor, with BPD patients having higher frequencies of the DAT1 9,9 and 9,10 genotypes (Joyce et al., 2009, 2006, 2014). However, others have deduced no association between the DAT1 polymorphisms and BPD, but did find significance after independent replication with the dopamine D4 receptor polymorphism -616 CC and borderline traits in young adults (Nemoda et al., 2010).

4.3 Schizophrenia

In a similar fashion to borderline personality disorder, symptoms suggestive of schizophrenia often become apparent during early adulthood with an overall prevalence of 0.7–0.8% (Sullivan, 2005). Patients may present with a variety of positive and negative symptoms, often experiencing periods of exacerbation and remission with functional impairment. On average, persons diagnosed with schizophrenia live 12 years fewer than their counterparts, a deduction which may be attributed to the increased incidence of substance abuse, cardiovascular disease, and type 2 diabetes mellitus (Lambert, Velakoulis, & Pantelis, 2003). The cause of schizophrenia is, however, multifactorial. Environmental aspects, such as infection, depression, and complications within the prenatal and perinatal period, have been postulated as risk factors (Maki et al., 2005). Family history of schizophrenia is the predominant risk factor, exceeding both prenatal bereavement and rubella infection (Sullivan, 2005). It is also considered that, although individual genes have a relatively low effect on the schizophrenia phenotype, the genetic component may be highly significant when considering the amalgamated contribution of a multitude of genes (Gilmore, 2010). Schizophrenia may therefore typically be considered a polygenic mental disorder comprising of thousands of common genetic variants.

As schizophrenia is considered to be pathophysiologically underpinned by hyperdopaminergic function (Gainetdinov, 2008), the contribution of the DAT to the schizophrenia phenotype has been examined. DAT density did not significantly vary between young adults with the disorder and controls with no apparent density changes in patients who had been subjected to antipsychotic treatment (Lavalaye et al., 2001). Chen and coworkers reinforced this deduction in a similar study finding no changes in striatal DAT levels in antipsychotic-naive young adults with schizophrenia (Chen et al., 2013). This is further compounded since DAT alterations in the striatum have generally not been associated with schizophrenia (Laruelle et al., 2000). The lack of association is further reinforced by meta-analysis, which found no evidence between striatal dopaminergic nerve terminal density and schizophrenia with striatal DAT density independent of antipsychotic choice and the duration of illness (Fusar-Poli & Meyer-Lindenberg, 2013). On the other hand, the intronic polymorphisms consisting of the haplotype rs2975223 (G) and rs2455391(C) have been shown to increase the risk of schizophrenia (Zheng, Shen, & Xu, 2012), while the AT and TT alleles of a DAT1 core promoter polymorphism (-67A/T) have been shown to be more common in male Iranian schizophrenia patients (Khodayari et al., 2004). Similar results were obtained in an association study within Han Chinese, which determined significance between a haplotype within the DAT1 promoter region and schizophrenia (Huang et al., 2010). The role of the DAT in schizophrenia is therefore somewhat undetermined; larger scale independently replicated association studies would be required to validate the aforementioned analysis of DAT1 polymorphisms.

It has been recently shown that schizophrenia and bipolar disorder, two disorders with shared symptoms, have a common genetic relationship (Cross-Disorder Group of the Psychiatric Genomics et al., 2013). A DAT1 haplotype spanning from exon 9 to 15 has been found in linkage disequilibrium with bipolar disorder (Greenwood et al., 2001). This was expanded utilizing independent samples which deduced that multiple DAT1 variants corresponded to increase susceptibility to bipolar disorder (Greenwood, Schork, Eskin, & Kelsoe, 2006), with others deducing significance between the 3′-UTR SNP rs27072 and bipolar disorder (Pinsonneault et al., 2011). Further exploration would be required to fully elucidate these genetic associations and whether they are relevant to endophenotypes common within these two disorders.

4.4 Obsessive compulsive disorder

Obsessive compulsive disorder (OCD) is defined as a chronic psychiatric anxiety disorder, characterized by the presence of obsessive thoughts and/or ritualistic behaviors. Common obsessions exhibited by those with OCD include those relating to hygiene and anxiety pertaining to contamination, while compulsions can involve excessive hand-washing and repetitive checking (Parmet, Lynm, & Golub, 2011). Negative social and occupational effects of OCD are commonly observed due to the time-consuming effects of these obsessions and compulsions (*DSM-IV 4th Ed.*, American Psychiatric Association, 2000).

The development of OCD is thought to arise from a combination of familial and environmental factors with an early age of onset potentially indicating OCD with a genetic predominance in origin (Cath, van Grootheest, Willemsen, van Oppen, & Boomsma, 2008). The neurobiology of OCD was originally thought to be explained by the serotonin hypothesis (Barr, Goodman, & Price, 1993), yet this fails to account for around 40% of individuals unresponsive to selective-serotonin reuptake inhibitors (SSRIs; Miguita, Cordeiro, Shavitt, Miguel, & Vallada, 2011). The role of the dopaminergic system in OCD is not a new concept (Goodman et al., 1990). Sufficient evidence exists to implicate the dopaminergic system in the pathophysiology of OCD. Pharmacological studies have shown that therapeutic regimens incorporating the combination of an SSRI with a dopaminergic antagonist, or atypical antipsychotic, are effective in the treatment of resistant OCD, therefore implicating both serotonergic and dopaminergic systems (Hesse et al., 2005). Analysis using SPECT methodology reaffirms the potential role of dopaminergic involvement in OCD. Significant increases in DAT binding ratios were observed in the left caudate and left putamen in comparison to healthy controls (van der Wee et al., 2004), while others have determined that drug-naive OCD patients exhibited notably reduced striatal DAT (Hesse et al., 2005).

4.5 Attention deficit hyperactivity disorder

ADHD, a neuropsychiatric disorder frequently diagnosed in childhood, is typically observed as a symptomatic amalgamation of impulsivity, inattentiveness, and hyperactivity. Although ADHD can be exacerbated by environmental triggers and psychosocial events, the disorder is considered to have a genetic component, a deduction originating from adoption and twin studies. One such study reported that the rates of ADHD for monozygotic and dizygotic twins were 81% and 29%, respectively (Gillis, Gilger, Pennington, & DeFries, 1992). The genes commonly thought to contribute to the development of ADHD predominantly incorporate those involved in the catecholamine pathways. These include genes encoding for the DAT, NET, and among others, the D4 and D5 dopamine receptors (Madras et al., 2005). The potential role of the DAT in forming the ADHD phenotype is further reinforced when considering that the commonly used psychostimulants, AMPH and methylphenidate, both target the transporter to increase extracellular dopamine levels (Mazei-Robinson & Blakely, 2006).

As a consequence of the hypodopaminergic theory relating to the pathophysiology of ADHD (Gill, Daly, Heron, Hawi, & Fitzgerald, 1997), multiple studies have scrutinized the role of the DAT and the influence of genetic variants. Meta-analysis researching the density of the DAT within the striatum of ADHD patients determined that exposure to the psychostimulants correlated to increased DAT density, highlighting adaptations to pharmacological intervention (Fusar-Poli, Rubia, Rossi, Sartori, & Balottin, 2012). In terms of DAT1 variants, however, Roman and coworkers found that, although no association between the 10-repeat allele of the DAT1 VNTR and ADHD could be determined alone, the disorder was shown to be related to both the DAT1 VNTR and the 7-repeat allele of the DRD4 48-nucleotide VNTR (Roman et al., 2001). The 480-bp (10-repeat) VNTR allele in the DAT1 3′-region trended toward an association with ADHD, but when analyzed concomitantly with intron 9 and exon 9 variants, associations were significant. However, Ala559Val and Glu602Gly, both polymorphisms resulting in amino acid changes, were not observed in the study (Barr et al., 2001). Familial analysis, however, has shown that the 10-repeat allele of the DAT1 VNTR significantly correlated to ADHD (Gill et al., 1997).

4.6 Alcoholism

Alcoholism is a term generally used to mean compulsive and uncontrolled consumption of alcohol. The American Medical Association considers alcoholism as a disease. The CNS depressant effects of alcohol are well detailed, ranging from disinhibition and euphoria to motor impairment, vomiting, and at excessive levels, respiratory failure. The effects of alcohol have been in part attributed to its action on $GABA_A$, glycine, and NMDA receptors, wherein alcohol-induced alternations in neuronal excitation lead to changes in the expression of these receptors (Davies, 2003). Alcoholism and the associated implications, particularly those apparent upon alcohol withdrawal, have therefore been the subject of a multitude of association studies focusing on serotonergic, GABAergic, and among others, dopaminergic pathways (Gorwood et al., 2003). The role of genetics in alcoholism has therefore been considered, and it has been observed that alcohol-related seizures may be the subject of genetic predisposition (Schaumann et al., 1994). Alcohol consumption has also been shown to interfere with dopaminergic pathways. PET analysis in alcoholic subjects highlighted a significant reduction in striatal D2 receptors, but not the DAT (Volkow et al., 1996). It has also been deduced that within the early stages of alcohol abstinence, the binding of a cocaine analogue with the DAT significantly increased, which infers the potential for alcohol-induced disruption of dopaminergic activity and the subsequent potential for relapse upon cessation of alcohol use (Laine et al., 1999).

In terms of genetic susceptibility to alcoholism, however, there are mixed results which may correlate to a lack of power within the genetic association studies (van der Zwaluw et al., 2009). A reputed association was found between the DAT1 VNTR 9-repeat allele and alcoholism, but this did not achieve replication in a separate cohort (Bhaskar, Thangaraj, Wasnik, Singh, & Raghavendra Rao, 2012). Others have determined that the 9-repeat allele is significantly more abundant among alcoholics than healthy control subjects, yet no association was found between genotype, alcohol intake levels, and withdrawal symptoms (Kohnke et al., 2005). In addition, mixed results have also been accrued from studies relating to the DAT1 genotype and alcohol withdrawal symptoms with the DAT1 VNTR and two SNPs, rs27072 and rs27048, exhibiting a significant association with withdrawal seizures (Le Strat et al., 2008). Meta-analysis has summarized that a link between the DAT1 VNTR 9-repeat allele, alcohol withdrawal symptoms, and delirium tremens may exist (Du, Nie, Li, & Wan, 2011).

5. PHARMACOLOGICAL TARGETING OF THE DAT

Pharmacological targeting of the dopaminergic pathways within the CNS forms the basis of the treatment of numerous diseases implicating dopamine neurotransmission. It is well established that both cocaine and AMPH interfere with dopaminergic reward pathways to elevate extracellular dopamine levels, a process which facilitates drug addiction. AMPH has been shown to interact both directly and indirectly with the DAT to induce this increase in synaptic dopamine. The psychostimulant facilitates a nonvesicular efflux of cytosolic dopamine via the DAT through two distinct mechanisms, by a slow exchange system and by a rapid channel type efflux. The latter, which causes a stark increase in synaptic dopamine similar to the release of vesicular dopamine through exocytosis, is considered to be responsible for the psychostimulant properties of AMPH (Kahlig et al., 2005). In addition to competitive AMPH uptake by the DAT, the stimulant also causes the internalization of the DAT (Zahniser & Sorkin, 2004), a mechanism which may assist in the prevention of dopamine-induced cytotoxicity (Saunders et al., 2000). Moreover, further *in vitro* mechanistic analysis deduced that a 22-amino acid truncation of the cytoplasmic DAT amino terminus reduced AMPH-induced dopamine efflux by 80% in HEK293 cells, with similar results obtained through mutation of five amino-terminal serine residues to alanine, as deduced by amperometry and patch-clamp methodology. This highlighted that protein kinase C-mediated phosphorylation may be necessary to permit dopamine efflux, but has no implications for dopamine reuptake. This illustrates independent mechanisms for reuptake and efflux (Khoshbouei et al., 2004). The determination of independent mechanisms, coupled with the ability of AMPH to induce phosphorylation (Giambalvo, 2003), therefore, demonstrates a potential target for treating drug addiction. Response to d-AMPH, however, has been shown to be subject to variations. Individuals homozygous for the 9-repeat allele of the 3′-UTR VNTR reported a response comparable to that of placebo, which lends support to pharmacogenetic research into methylphenidate response, whereby patients showed an uncharacteristic subdued response to dose escalation (Stein et al., 2005).

Cocaine, a widely used psychomotor stimulant sourced from the *Erythroxylon coca* plant, produces a variety of physical and neuropsychiatric effects including euphoria, locomotion stimulation, and a sensation of reward (Tilley et al., 2007), which underpins the alkaloids ability to invoke addiction. The signs of acute cocaine administration, coupled with a multitude of adverse effects, are often related to high-affinity inhibition of the DAT and subsequent increases in extracellular dopamine (Ritz, Lamb, Goldberg, & Kuhar, 1987), although cocaine also inhibits the reuptake of the monoamines serotonin and noradrenaline (Izenwasser, 2004). It has also been demonstrated, using LeuT, that the DAT binding sites for cocaine, dopamine, AMPH, and benzatropine-like DAT inhibitors display crossover (Beuming et al., 2008). PET analysis of DAT occupancy after administration has demonstrated that doses of cocaine corresponding to those commonly used by drug abusers resulted in DAT occupancy of 60–77% with a minimum of 47% required to induce the typical signs and sensations of cocaine use, which infers that high DAT occupancy would be required to block the effects of cocaine (Volkow et al., 1997). The role of DAT inhibition by cocaine has been illustrated in mouse models carrying a cocaine-insensitive mutant DAT wherein the stimulant did not produce reward-related behavioral changes or biochemical changes in extracellular dopamine (Chen et al., 2006). It has been suggested that both genetic and pharmacological alterations in DAT expression may influence addiction susceptibility (Cagniard, Sotnikova, Gainetdinov, & Zhuang, 2014). In addition to cocaine-mediated inhibition of the DAT, the tropane alkaloid also induces changes to the distribution of the DAT. After cocaine treatment, increases in

DAT immunoreactivity were observed on the plasma membrane, alongside decreases in intracellular DAT, which suggest cocaine-induced alterations in DAT trafficking (Little, Elmer, Zhong, Scheys, & Zhang, 2002). This is in contrast to AMPH, which causes internalization of the DAT (Zahniser & Sorkin, 2004).

A multitude of other compounds have also been shown to interact with the DAT. These include the antidepressants, nomifensine and nortriptyline, and the tetracyclic compound, mazindol, which inhibits both the DAT and the SERT (Severinsen et al., 2014). In addition to WIN35428, a commonly utilized dopamine reuptake inhibitor in research (Valchar & Hanbauer, 1993), vanoxerine (GBR-12909), which displays 500-fold greater affinity for the DAT than cocaine, has been shown to reduce and, at higher doses, prevent the self-administration of cocaine in primates (Preti, 2000). However, the development of vanoxerine was ultimately suspended in clinical trials due to its effects upon QTc (Rothman, Baumann, Prisinzano, & Newman, 2008).

6. CONCLUSION

This chapter has explored the role of the DAT in the dopaminergic system. This system has a wide range of functions within the CNS and is inevitably implicated in a multitude of diseases pertaining to the CNS. In particular, the DAT, an integral component of the dopaminergic system, is discussed in relation to these CNS diseases, including PD, schizophrenia, and ADHD. We have elucidated various features of the DAT, including its structure, function, and regulation, as well as DAT1 genetics, and how these characteristics relate to neurological and neuropsychiatric diseases. We also explored pharmacological targeting of the transporter. In summary, there are relationships between the DAT and the aforementioned CNS diseases; however, much more work is required to fully elucidate these relationships. This will allow us to understand not only how the DAT and the dopaminergic system relate to disease, but also how we can improve diagnosis and treatment.

ACKNOWLEDGMENTS

Financial support is through the University of Huddersfield Research Fund. Dr. Patrick McHugh is a Senior Lecturer at the University of Huddersfield, and Mr. David Buckley a Pharmacist and predoctoral student at the University of Huddersfield. There are no conflicts of interest.

REFERENCES

Agrawal, A., Wetherill, L., Bucholz, K. K., Kramer, J., Kuperman, S., Lynskey, M. T., et al. (2013). Genetic influences on craving for alcohol. *Addictive Behaviors*, *38*(2), 1501–1508. http://dx.doi.org/10.1016/j.addbeh.2012.03.021.

Bannon, M. J., & Whitty, C. J. (1997). Age-related and regional differences in dopamine transporter mRNA expression in human midbrain. *Neurology*, *48*(4), 969–977.

Barbeau, A. (1962). The pathogenesis of Parkinson's disease: A new hypothesis. *Canadian Medical Association Journal*, *87*, 802–807.

Barr, L. C., Goodman, W. K., & Price, L. H. (1993). The serotonin hypothesis of obsessive compulsive disorder. *International Clinical Psychopharmacology*, *8*(Suppl. 2), 79–82.

Barr, C. L., Xu, C., Kroft, J., Feng, Y., Wigg, K., Zai, G., et al. (2001). Haplotype study of three polymorphisms at the dopamine transporter locus confirm linkage to attention-deficit/hyperactivity disorder. *Biological Psychiatry*, *49*(4), 333–339.

Beuming, T., Kniazeff, J., Bergmann, M. L., Shi, L., Gracia, L., Raniszewska, K., et al. (2008). The binding sites for cocaine and dopamine in the dopamine transporter overlap. *Nature Neuroscience*, *11*(7), 780–789. http://dx.doi.org/10.1038/nn.2146.

Bhaskar, L. V., Thangaraj, K., Wasnik, S., Singh, L., & Raghavendra Rao, V. (2012). Dopamine transporter (DAT1) VNTR polymorphism and alcoholism in two culturally different populations of south India. *The American journal on addictions/American Academy of Psychiatrists in Alcoholism and Addictions*, *21*(4), 343–347. http://dx.doi.org/10.1111/j.1521-0391.2012.00244.x.

Cagniard, B., Sotnikova, T. D., Gainetdinov, R. R., & Zhuang, X. (2014). The dopamine transporter expression level differentially affects responses to cocaine and amphetamine. *Journal of Neurogenetics*, *28*(1–2), 112–121. http://dx.doi.org/10.3109/01677063.2014.908191.

Callier, S., Snapyan, M., Le Crom, S., Prou, D., Vincent, J. D., & Vernier, P. (2003). Evolution and cell biology of dopamine receptors in vertebrates. *Biology of the Cell*, *95*(7), 489–502.

Carlsson, A., & Hillarp, N. A. (1958). On the state of the catechol amines of the adrenal medullary granules. *Acta Physiologica Scandinavica*, *44*(2), 163–169. http://dx.doi.org/10.1111/j.1748-1716.1958.tb01618.x.

Cath, D. C., van Grootheest, D. S., Willemsen, G., van Oppen, P., & Boomsma, D. I. (2008). Environmental factors in obsessive-compulsive behavior: Evidence from discordant and concordant monozygotic twins. *Behavior Genetics*, *38*(2), 108–120. http://dx.doi.org/10.1007/s10519-007-9185-9.

Chen, N., & Justice, J. B. (2000). Differential effect of structural modification of human dopamine transporter on the inward and outward transport of dopamine. *Brain Research. Molecular Brain Research*, *75*(2), 208–215.

Chen, R., Tilley, M. R., Wei, H., Zhou, F., Zhou, F. M., Ching, S., et al. (2006). Abolished cocaine reward in mice with a cocaine-insensitive dopamine transporter. *Proceedings of the National Academy of Sciences of the United States of America*, *103*(24), 9333–9338. http://dx.doi.org/10.1073/pnas.0600905103.

Chen, R., Wei, H., Hill, E. R., Chen, L., Jiang, L., Han, D. D., et al. (2007). Direct evidence that two cysteines in the dopamine transporter form a disulfide bond. *Molecular and Cellular Biochemistry*, *298*(1–2), 41–48. http://dx.doi.org/10.1007/s11010-006-9348-7.

Chen, K. C., Yang, Y. K., Howes, O., Lee, I. H., Landau, S., Yeh, T. L., et al. (2013). Striatal dopamine transporter availability in drug-naive patients with schizophrenia: A case-control SPECT study with [(99 m)Tc]-TRODAT-1 and a meta-analysis. *Schizophrenia Bulletin*, *39*(2), 378–386. http://dx.doi.org/10.1093/schbul/sbr163.

Chotibut, T., Apple, D. M., Jefferis, R., & Salvatore, M. F. (2012). Dopamine transporter loss in 6-OHDA Parkinson's model is unmet by parallel reduction in dopamine uptake. *PLoS One*, *7*(12), e52322. http://dx.doi.org/10.1371/journal.pone.0052322.

Ciliax, B. J., Drash, G. W., Staley, J. K., Haber, S., Mobley, C. J., Miller, G. W., et al. (1999). Immunocytochemical localization of the dopamine transporter in human brain. *Journal of Comparative Neurology*, *409*(1), 38–56.

Cross-Disorder Group of the Psychiatric Genomics, C., Lee, S. H., Ripke, S., Neale, B. M., Faraone, S. V., Purcell, S. M., et al (2013). Genetic relationship between five psychiatric disorders estimated from genome-wide SNPs. *Nature Genetics*, *45*(9), 984–994. http://dx.doi.org/10.1038/ng.2711.

Cruz-Muros, I., Afonso-Oramas, D., Abreu, P., Perez-Delgado, M. M., Rodriguez, M., & Gonzalez-Hernandez, T. (2009). Aging effects on the dopamine transporter expression and compensatory mechanisms. *Neurobiology of Aging*, *30*(6), 973–986. http://dx.doi.org/10.1016/j.neurobiolaging.2007.09.009.

Dalley, J. W., & Everitt, B. J. (2009). Dopamine receptors in the learning, memory and drug reward circuitry. *Seminars in Cell and Developmental Biology*, *20*(4), 403–410. http://dx.doi.org/10.1016/j.semcdb.2009.01.002.

Davies, M. (2003). The role of GABAA receptors in mediating the effects of alcohol in the central nervous system. *Journal of Psychiatry and Neuroscience*, *28*(4), 263–274.

de Azeredo, L. A., Rovaris, D. L., Mota, N. R., Polina, E. R., Marques, F. Z., Contini, V., et al. (2014). Further evidence for the association between a polymorphism in the promoter region of SLC6A3/DAT1 and ADHD: Findings from a sample of adults. *European Archives of Psychiatry and Clinical Neuroscience*, *264*(5), 401–408.

Distel, M. A., Trull, T. J., Derom, C. A., Thiery, E. W., Grimmer, M. A., Martin, N. G., et al. (2008). Heritability of borderline personality disorder features is similar across three countries. *Psychological Medicine*, *38*(9), 1219–1229. http://dx.doi.org/10.1017/S0033291707002024.

Dolfini, D., Zambelli, F., Pavesi, G., & Mantovani, R. (2009). A perspective of promoter architecture from the CCAAT box. *Cell Cycle*, *8*(24), 4127–4137.

Donovan, D. M., Vandenbergh, D. J., Perry, M. P., Bird, G. S., Ingersoll, R., Nanthakumar, E., et al. (1995). Human and mouse dopamine transporter genes: Conservation of 5'-flanking sequence elements and gene structures. *Brain Research. Molecular Brain Research*, *30*(2), 327–335.

Du, Y., Nie, Y., Li, Y., & Wan, Y. J. (2011). The association between the SLC6A3 VNTR 9-repeat allele and alcoholism-a meta-analysis. *Alcoholism: Clinical and Experimental Research*, *35*(9), 1625–1634. http://dx.doi.org/10.1111/j.1530-0277.2011.01509.x.

Egana, L. A., Cuevas, R. A., Baust, T. B., Parra, L. A., Leak, R. K., Hochendoner, S., et al. (2009). Physical and functional interaction between the dopamine transporter and the synaptic vesicle protein synaptogyrin-3. *Journal of Neuroscience*, *29*(14), 4592–4604. http://dx.doi.org/10.1523/JNEUROSCI.4559-08.2009.

Fog, J. U., Khoshbouei, H., Holy, M., Owens, W. A., Vaegter, C. B., Sen, N., et al. (2006). Calmodulin kinase II interacts with the dopamine transporter C terminus to regulate amphetamine-induced reverse transport. *Neuron*, *51*(4), 417–429. http://dx.doi.org/10.1016/j.neuron.2006.06.028.

Freed, C., Revay, R., Vaughan, R. A., Kriek, E., Grant, S., Uhl, G. R., et al. (1995). Dopamine transporter immunoreactivity in rat brain. *Journal of Comparative Neurology*, *359*(2), 340–349. http://dx.doi.org/10.1002/cne.903590211.

Friedel, R. O. (2004). Dopamine dysfunction in borderline personality disorder: A hypothesis. *Neuropsychopharmacology*, *29*(6), 1029–1039. http://dx.doi.org/10.1038/sj.npp.1300424.

Fuke, S., Minami, N., Kokubo, H., Yoshikawa, A., Yasumatsu, H., Sasagawa, N., et al. (2006). Hesr1 knockout mice exhibit behavioral alterations through the dopaminergic nervous system. *Journal of Neuroscience Research*, *84*(7), 1555–1563. http://dx.doi.org/10.1002/jnr.21062.

Fuke, S., Suo, S., Takahashi, N., Koike, H., Sasagawa, N., & Ishiura, S. (2001). The VNTR polymorphism of the human dopamine transporter (DAT1) gene affects gene expression. *Pharmacogenomics Journal*, *1*(2), 152–156.

Fusar-Poli, P., & Meyer-Lindenberg, A. (2013). Striatal presynaptic dopamine in schizophrenia, part I: Meta-analysis of dopamine active transporter (DAT) density. *Schizophrenia Bulletin*, *39*(1), 22–32. http://dx.doi.org/10.1093/schbul/sbr111.

Fusar-Poli, P., Rubia, K., Rossi, G., Sartori, G., & Balottin, U. (2012). Striatal dopamine transporter alterations in ADHD: Pathophysiology or adaptation to psychostimulants? A meta-analysis. *The American Journal of Psychiatry*, *169*(3), 264–272. http://dx.doi.org/10.1176/appi.ajp.2011.11060940.

Fuxe, K. (1965). Evidence for the existence of monoamine neurons in the central nervous system. Iv. Distribution of monoamine nerve terminals in the central nervous system. *Acta Physiologica Scandinavica. Supplementum*, (Suppl. 247), 237.

Gainetdinov, R. R. (2008). Dopamine transporter mutant mice in experimental neuropharmacology. *Naunyn-Schmiedeberg's Archives of Pharmacology*, *377*(4–6), 301–313. http://dx.doi.org/10.1007/s00210-007-0216-0.

Gether, U., Andersen, P. H., Larsson, O. M., & Schousboe, A. (2006). Neurotransmitter transporters: Molecular function of important drug targets. *Trends in Pharmacological Sciences*, *27*(7), 375–383. http://dx.doi.org/10.1016/j.tips.2006.05.003.

Giambalvo, C. T. (2003). Differential effects of amphetamine transport vs. dopamine reverse transport on particulate PKC activity in striatal synaptoneurosomes. *Synapse*, *49*(2), 125–133. http://dx.doi.org/10.1002/syn.10223.

Gill, M., Daly, G., Heron, S., Hawi, Z., & Fitzgerald, M. (1997). Confirmation of association between attention deficit hyperactivity disorder and a dopamine transporter polymorphism. *Molecular Psychiatry*, *2*(4), 311–313.

Gillis, J. J., Gilger, J. W., Pennington, B. F., & DeFries, J. C. (1992). Attention deficit disorder in reading-disabled twins: Evidence for a genetic etiology. *Journal of Abnormal Child Psychology*, *20*(3), 303–315.

Gilmore, J. H. (2010). Understanding what causes schizophrenia: A developmental perspective. *The American Journal of Psychiatry*, *167*(1), 8–10. http://dx.doi.org/10.1176/appi.ajp.2009.09111588.

Giros, B., el Mestikawy, S., Bertrand, L., & Caron, M. G. (1991). Cloning and functional characterization of a cocaine-sensitive dopamine transporter. *FEBS Letters*, *295*(1–3), 149–154.

Giros, B., el Mestikawy, S., Godinot, N., Zheng, K., Han, H., Yang-Feng, T., et al. (1992). Cloning, pharmacological characterization, and chromosome assignment of the human dopamine transporter. *Molecular Pharmacology*, *42*(3), 383–390.

Giros, B., Jaber, M., Jones, S. R., Wightman, R. M., & Caron, M. G. (1996). Hyperlocomotion and indifference to cocaine and amphetamine in mice lacking the dopamine transporter. *Nature*, *379*(6566), 606–612. http://dx.doi.org/10.1038/379606a0.

Goodman, W. K., McDougle, C. J., Price, L. H., Riddle, M. A., Pauls, D. L., & Leckman, J. F. (1990). Beyond the serotonin hypothesis: A role for dopamine in some forms of obsessive compulsive disorder? *Journal of Clinical Psychiatry*, (Suppl. 51), 36–43, discussion 55–38.

Gorwood, P., Limosin, F., Batel, P., Hamon, M., Ades, J., & Boni, C. (2003). The A9 allele of the dopamine transporter gene is associated with delirium tremens and alcohol-withdrawal seizure. *Biological Psychiatry*, *53*(1), 85–92.

Greenwood, T. A., Alexander, M., Keck, P. E., McElroy, S., Sadovnick, A. D., Remick, R. A., et al. (2001). Evidence for linkage disequilibrium between the dopamine transporter and bipolar disorder. *American Journal of Medical Genetics*, *105*(2), 145–151.

Greenwood, T. A., & Kelsoe, J. R. (2003). Promoter and intronic variants affect the transcriptional regulation of the human dopamine transporter gene. *Genomics*, *82*(5), 511–520.

Greenwood, T. A., Schork, N. J., Eskin, E., & Kelsoe, J. R. (2006). Identification of additional variants within the human dopamine transporter gene provides further evidence for an association with bipolar disorder in two independent samples. *Molecular Psychiatry*, *11*(2), 125–133. http://dx.doi.org/10.1038/sj.mp.4001764, 115.

Gu, H., Wall, S. C., & Rudnick, G. (1994). Stable expression of biogenic amine transporters reveals differences in inhibitor sensitivity, kinetics, and ion dependence. *Journal of Biological Chemistry*, *269*(10), 7124–7130.

Hesse, S., Muller, U., Lincke, T., Barthel, H., Villmann, T., Angermeyer, M. C., et al. (2005). Serotonin and dopamine transporter imaging in patients with obsessive-compulsive disorder. *Psychiatry Research*, *140*(1), 63–72. http://dx.doi.org/10.1016/j.pscychresns.2005.07.002.

Huang, S. Y., Chen, H. K., Ma, K. H., Shy, M. J., Chen, J. H., Lin, W. C., et al. (2010). Association of promoter variants of human dopamine transporter gene with schizophrenia in Han Chinese. *Schizophrenia Research*, *116*(1), 68–74. http://dx.doi.org/10.1016/j.schres.2009.10.004.

Indarte, M., Madura, J. D., & Surratt, C. K. (2008). Dopamine transporter comparative molecular modeling and binding site prediction using the LeuT(Aa) leucine transporter as a template. *Proteins*, *70*(3), 1033–1046. http://dx.doi.org/10.1002/prot.21598.

Ishibashi, K., Ishii, K., Oda, K., Kawasaki, K., Mizusawa, H., & Ishiwata, K. (2009). Regional analysis of age-related decline in dopamine transporters and dopamine D2-like receptors in human striatum. *Synapse*, *63*(4), 282–290. http://dx.doi.org/10.1002/syn.20603.

Ishibashi, K., Oda, K., Ishiwata, K., & Ishii, K. (2014). Comparison of dopamine transporter decline in a patient with Parkinson's disease and normal aging effect. *Journal of the Neurological Sciences*, *339*(1–2), 207–209. http://dx.doi.org/10.1016/j.jns.2014.01.015.

Izenwasser, S. (2004). The role of the dopamine transporter in cocaine abuse. *Neurotoxicity Research*, *6*(5), 379–383.

Joyce, P. R., McHugh, P. C., Light, K. J., Rowe, S., Miller, A. L., & Kennedy, M. A. (2009). Relationships between angry-impulsive personality traits and genetic polymorphisms of the dopamine transporter. *Biological Psychiatry*, *66*(8), 717–721. http://dx.doi.org/10.1016/j.biopsych.2009.03.005.

Joyce, P. R., McHugh, P. C., McKenzie, J. M., Sullivan, P. F., Mulder, R. T., Luty, S. E., et al. (2006). A dopamine transporter polymorphism is a risk factor for borderline personality disorder in depressed patients. *Psychological Medicine*, *36*(6), 807–813.

Joyce, P. R., Stephenson, J., Kennedy, M., Mulder, R. T., & McHugh, P. C. (2014). The presence of both serotonin 1A receptor (HTR1A) and dopamine transporter (DAT1) gene variants increase the risk of borderline personality disorder. *Frontiers in Genetics*, *4*, 313. http://dx.doi.org/10.3389/fgene.2013.00313.

Kahlig, K. M., Binda, F., Khoshbouei, H., Blakely, R. D., McMahon, D. G., Javitch, J. A., et al. (2005). Amphetamine induces dopamine efflux through a dopamine transporter channel. *Proceedings of the National Academy of Sciences of the United States of America*, *102*(9), 3495–3500. http://dx.doi.org/10.1073/pnas.0407737102.

Kanno, K., & Ishiura, S. (2011). Differential effects of the HESR/HEY transcription factor family on dopamine transporter reporter gene expression via variable number of tandem repeats. *Journal of Neuroscience Research*, *89*(4), 562–575. http://dx.doi.org/10.1002/jnr.22593.

Kawarai, T., Kawakami, H., Yamamura, Y., & Nakamura, S. (1997). Structure and organization of the gene encoding human dopamine transporter. *Gene*, *195*(1), 11–18, doi: S0378-1119(97)00131-5 [pii].

Kelada, S. N., Costa-Mallen, P., Checkoway, H., Carlson, C. S., Weller, T. S., Swanson, P. D., et al. (2005). Dopamine transporter (SLC6A3) 5' region haplotypes significantly affect transcriptional activity in vitro but are not associated with Parkinson's disease. *Pharmacogenetics and Genomics*, *15*(9), 659–668.

Khodayari, N., Garshasbi, M., Fadai, F., Rahimi, A., Hafizi, L., Ebrahimi, A., et al. (2004). Association of the dopamine transporter gene (DAT1) core promoter polymorphism—67 T variant with schizophrenia. *American Journal of Medical Genetics Part B: Neuropsychiatric Genetics*, *129B*(1), 10–12. http://dx.doi.org/10.1002/ajmg.b.30067.

Khoshbouei, H., Sen, N., Guptaroy, B., Johnson, L., Lund, D., Gnegy, M. E., et al. (2004). N-terminal phosphorylation of the dopamine transporter is required for amphetamine-induced efflux. *PLoS Biology*, *2*(3), E78. http://dx.doi.org/10.1371/journal.pbio.0020078.

Kienast, T., & Heinz, A. (2006). Dopamine and the diseased brain. *CNS & Neurological Disorders: Drug Targets*, *5*(1), 109–131.

Kohnke, M. D., Batra, A., Kolb, W., Kohnke, A. M., Lutz, U., Schick, S., et al. (2005). Association of the dopamine transporter gene with alcoholism. *Alcohol and Alcoholism*, *40*(5), 339–342. http://dx.doi.org/10.1093/alcalc/agh179.

Koldso, H., Christiansen, A. B., Sinning, S., & Schiott, B. (2013). Comparative modeling of the human monoamine transporters: Similarities in substrate binding. *ACS Chemical Neuroscience*, *4*(2), 295–309. http://dx.doi.org/10.1021/cn300148r.

Kristensen, A. S., Andersen, J., Jorgensen, T. N., Sorensen, L., Eriksen, J., Loland, C. J., et al. (2011). SLC6 neurotransmitter transporters: Structure, function, and regulation. *Pharmacological Reviews*, *63*(3), 585–640. http://dx.doi.org/10.1124/pr.108.000869.

Laasonen-Balk, T., Kuikka, J., Viinamaki, H., Husso-Saastamoinen, M., Lehtonen, J., & Tiihonen, J. (1999). Striatal dopamine transporter density in major depression. *Psychopharmacology*, *144*(3), 282–285.

Laine, T. P., Ahonen, A., Torniainen, P., Heikkila, J., Pyhtinen, J., Rasanen, P., et al. (1999). Dopamine transporters increase in human brain after alcohol withdrawal. *Molecular Psychiatry*, *4*(2), 189–191, 104–185.

Lambert, T. J., Velakoulis, D., & Pantelis, C. (2003). Medical comorbidity in schizophrenia. *The Medical Journal of Australia*, (Suppl.178), S67–S70.

Laruelle, M., Abi-Dargham, A., van Dyck, C., Gil, R., D'Souza, D. C., Krystal, J., et al. (2000). Dopamine and serotonin transporters in patients with schizophrenia: An imaging study with [(123)I]beta-CIT. *Biological Psychiatry*, *47*(5), 371–379.

Lavalaye, J., Linszen, D. H., Booij, J., Dingemans, P. M., Reneman, L., Habraken, J. B., et al. (2001). Dopamine transporter density in young patients with schizophrenia assessed with [123]FP-CIT SPECT. *Schizophrenia Research*, *47*(1), 59–67.

Lee, F. J., Liu, F., Pristupa, Z. B., & Niznik, H. B. (2001). Direct binding and functional coupling of alpha-synuclein to the dopamine transporters accelerate dopamine-induced apoptosis. *The FASEB Journal*, *15*(6), 916–926.

Le Strat, Y., Ramoz, N., Pickering, P., Burger, V., Boni, C., Aubin, H. J., et al. (2008). The 3' part of the dopamine transporter gene DAT1/SLC6A3 is associated with withdrawal seizures in patients with alcohol dependence. *Alcoholism: Clinical and Experimental Research*, *32*(1), 27–35. http://dx.doi.org/10.1111/j.1530-0277.2007.00552.x.

Li, L. B., Chen, N., Ramamoorthy, S., Chi, L., Cui, X. N., Wang, L. C., et al. (2004). The role of N-glycosylation in function and surface trafficking of the human dopamine transporter. *Journal of Biological Chemistry*, *279*(20), 21012–21020. http://dx.doi.org/10.1074/jbc.M311972200.

Little, K. Y., Elmer, L. W., Zhong, H., Scheys, J. O., & Zhang, L. (2002). Cocaine induction of dopamine transporter trafficking to the plasma membrane. *Molecular Pharmacology*, *61*(2), 436–445.

Lock, E. A., Zhang, J., & Checkoway, H. (2013). Solvents and Parkinson disease: A systematic review of toxicological and epidemiological evidence. *Toxicology and Applied Pharmacology*, *266*(3), 345–355. http://dx.doi.org/10.1016/j.taap.2012.11.016.

Lucas, L. R., Celen, Z., Tamashiro, K. L., Blanchard, R. J., Blanchard, D. C., Markham, C., et al. (2004). Repeated exposure to social stress has long-term effects on indirect markers of dopaminergic activity in brain regions associated with motivated behavior.

Neuroscience, *124*(2), 449–457. http://dx.doi.org/10.1016/j.neuroscience.2003.12.009, S0306452203009199 [pii].

Luo, Y. (2012). The function and mechanisms of Nurr1 action in midbrain dopaminergic neurons, from development and maintenance to survival. *International Review of Neurobiology*, *102*, 1–22. http://dx.doi.org/10.1016/B978-0-12-386986-9.00001-6.

Madras, B. K., Miller, G. M., & Fischman, A. J. (2005). The dopamine transporter and attention-deficit/hyperactivity disorder. *Biological Psychiatry*, *57*(11), 1397–1409. http://dx.doi.org/10.1016/j.biopsych.2004.10.011.

Maki, P., Veijola, J., Jones, P. B., Murray, G. K., Koponen, H., Tienari, P., et al. (2005). Predictors of schizophrenia—A review. *British Medical Bulletin*, *73–74*, 1–15. http://dx.doi.org/10.1093/bmb/ldh046.

Mateos, J. J., Lomena, F., Parellada, E., Font, M., Fernandez, E., Pavia, J., et al. (2005). Decreased striatal dopamine transporter binding assessed with [123I] FP-CIT in first-episode schizophrenic patients with and without short-term antipsychotic-induced parkinsonism. *Psychopharmacology*, *181*(2), 401–406. http://dx.doi.org/10.1007/s00213-005-2250-2.

Mazei-Robinson, M. S., & Blakely, R. D. (2006). ADHD and the dopamine transporter: Are there reasons to pay attention? In H. Sitte & M. Freissmuth (Eds.), *Neurotransmitter transporters*: *Vol. 175* (pp. 373–415). Berlin Heidelberg: Springer.

Mazei-Robison, M. S., Couch, R. S., Shelton, R. C., Stein, M. A., & Blakely, R. D. (2005). Sequence variation in the human dopamine transporter gene in children with attention deficit hyperactivity disorder. *Neuropharmacology*, *49*(6), 724–736. http://dx.doi.org/10.1016/j.neuropharm.2005.08.003.

Mercier, G., Turpin, J. C., & Lucotte, G. (1999). Variable number tandem repeat dopamine transporter gene polymorphism and Parkinson's disease: No association found. *Journal of Neurology*, *246*(1), 45–47.

Miguita, K., Cordeiro, Q., Shavitt, R. G., Miguel, E. C., & Vallada, H. (2011). Association study between genetic monoaminergic polymorphisms and OCD response to clomipramine treatment. *Arquivos de Neuro-Psiquiatria*, *69*(2B), 283–287.

Mill, J., Asherson, P., Craig, I., & D'Souza, U. M. (2005). Transient expression analysis of allelic variants of a VNTR in the dopamine transporter gene (DAT1). *BMC Genetics*, *6*, 3. http://dx.doi.org/10.1186/1471-2156-6-3.

Miller, G. W., Staley, J. K., Heilman, C. J., Perez, J. T., Mash, D. C., Rye, D. B., et al. (1997). Immunochemical analysis of dopamine transporter protein in Parkinson's disease. *Annals of Neurology*, *41*(4), 530–539. http://dx.doi.org/10.1002/ana.410410417.

Moron, J. A., Brockington, A., Wise, R. A., Rocha, B. A., & Hope, B. T. (2002). Dopamine uptake through the norepinephrine transporter in brain regions with low levels of the dopamine transporter: Evidence from knock-out mouse lines. *Journal of Neuroscience*, *22*(2), 389–395.

Nemoda, Z., Lyons-Ruth, K., Szekely, A., Bertha, E., Faludi, G., & Sasvari-Szekely, M. (2010). Association between dopaminergic polymorphisms and borderline personality traits among at-risk young adults and psychiatric inpatients. *Behavioral and Brain Functions*, *6*, 4. http://dx.doi.org/10.1186/1744-9081-6-4.

Nirenberg, M. J., Chan, J., Vaughan, R. A., Uhl, G. R., Kuhar, M. J., & Pickel, V. M. (1997). Immunogold localization of the dopamine transporter: An ultrastructural study of the rat ventral tegmental area. *Journal of Neuroscience*, *17*(14), 5255–5262.

Nirenberg, M. J., Vaughan, R. A., Uhl, G. R., Kuhar, M. J., & Pickel, V. M. (1996). The dopamine transporter is localized to dendritic and axonal plasma membranes of nigrostriatal dopaminergic neurons. *Journal of Neuroscience*, *16*(2), 436–447.

Nutt, J. G., Carter, J. H., & Sexton, G. J. (2004). The dopamine transporter: Importance in Parkinson's disease. *Annals of Neurology*, *55*(6), 766–773. http://dx.doi.org/10.1002/ana.20089.

Parmet, S., Lynm, C., & Golub, R. M. (2011). AMA patient page. Obsessive-compulsive disorder. *JAMA: The Journal of the American Medical Association*, *305*(18), 1926. http://dx.doi.org/10.1001/jama.305.18.1926.

Penmatsa, A., Wang, K. H., & Gouaux, E. (2013). X-ray structure of dopamine transporter elucidates antidepressant mechanism. *Nature*, *503*(7474), 85–90. http://dx.doi.org/10.1038/nature12533.

Pezzoli, G., & Cereda, E. (2013). Exposure to pesticides or solvents and risk of Parkinson disease. *Neurology*, *80*(22), 2035–2041. http://dx.doi.org/10.1212/WNL.0b013e318294b3c8.

Pinsonneault, J. K., Han, D. D., Burdick, K. E., Kataki, M., Bertolino, A., Malhotra, A. K., et al. (2011). Dopamine transporter gene variant affecting expression in human brain is associated with bipolar disorder. *Neuropsychopharmacology*, *36*(8), 1644–1655. http://dx.doi.org/10.1038/npp.2011.45.

Preti, A. (2000). Vanoxerine National Institute on drug abuse. *Current Opinion in Investigational Drugs*, *1*(2), 241–251.

Richtand, N. M., Kelsoe, J. R., Segal, D. S., & Kuczenski, R. (1995). Regional quantification of dopamine transporter mRNA in rat brain using a ribonuclease protection assay. *Neuroscience Letters*, *200*(2), 73–76.

Ritz, M. C., Lamb, R. J., Goldberg, S. R., & Kuhar, M. J. (1987). Cocaine receptors on dopamine transporters are related to self-administration of cocaine. *Science*, *237*(4819), 1219–1223.

Robinson, S., Rainwater, A. J., Hnasko, T. S., & Palmiter, R. D. (2007). Viral restoration of dopamine signaling to the dorsal striatum restores instrumental conditioning to dopamine-deficient mice. *Psychopharmacology*, *191*(3), 567–578. http://dx.doi.org/10.1007/s00213-006-0579-9.

Rodriguiz, R. M., Chu, R., Caron, M. G., & Wetsel, W. C. (2004). Aberrant responses in social interaction of dopamine transporter knockout mice. *Behavioural Brain Research*, *148*(1–2), 185–198.

Roman, T., Schmitz, M., Polanczyk, G., Eizirik, M., Rohde, L. A., & Hutz, M. H. (2001). Attention-deficit hyperactivity disorder: A study of association with both the dopamine transporter gene and the dopamine D4 receptor gene. *American Journal of Medical Genetics*, *105*(5), 471–478.

Rothman, R. B., Baumann, M. H., Prisinzano, T. E., & Newman, A. H. (2008). Dopamine transport inhibitors based on GBR12909 and benztropine as potential medications to treat cocaine addiction. *Biochemical Pharmacology*, *75*(1), 2–16. http://dx.doi.org/10.1016/j.bcp.2007.08.007.

Sacchetti, P., Brownschidle, L. A., Granneman, J. G., & Bannon, M. J. (1999). Characterization of the 5'-flanking region of the human dopamine transporter gene. *Brain Research. Molecular Brain Research*, *74*(1–2), 167–174.

Saunders, C., Ferrer, J. V., Shi, L., Chen, J., Merrill, G., Lamb, M. E., et al. (2000). Amphetamine-induced loss of human dopamine transporter activity: An internalization-dependent and cocaine-sensitive mechanism. *Proceedings of the National Academy of Sciences of the United States of America*, *97*(12), 6850–6855. http://dx.doi.org/10.1073/pnas.110035297.

Scarr, E., Gibbons, A. S., Neo, J., Udawela, M., & Dean, B. (2013). Cholinergic connectivity: It's implications for psychiatric disorders. *Frontiers in Cellular Neuroscience*, 7, 55. http://dx.doi.org/10.3389/fncel.2013.00055.

Schaumann, B. A., Annegers, J. F., Johnson, S. B., Moore, K. J., Lubozynski, M. F., & Salinsky, M. C. (1994). Family history of seizures in posttraumatic and alcohol-associated seizure disorders. *Epilepsia*, *35*(1), 48–52.

Seeman, P., & Niznik, H. B. (1990). Dopamine receptors and transporters in Parkinson's disease and schizophrenia. *The FASEB Journal*, *4*(10), 2737–2741.

Severinsen, K., Koldso, H., Thorup, K. A., Schjoth-Eskesen, C., Moller, P. T., Wiborg, O., et al. (2014). Binding of mazindol and analogs to the human serotonin and dopamine transporters. *Molecular Pharmacology*, *85*(2), 208–217. http://dx.doi.org/10.1124/mol.113.088922.

Shih, M. C., Hoexter, M. Q., Andrade, L. A., & Bressan, R. A. (2006). Parkinson's disease and dopamine transporter neuroimaging—A critical review. *Sau Paulo Medical Journal*, *124*(3), 168–175.

Shumay, E., Fowler, J. S., & Volkow, N. D. (2010). Genomic features of the human dopamine transporter gene and its potential epigenetic states: Implications for phenotypic diversity. *PLoS One*, *5*(6), e11067. http://dx.doi.org/10.1371/journal.pone.0011067.

Simon, H. H., Bhatt, L., Gherbassi, D., Sgado, P., & Alberi, L. (2003). Midbrain dopaminergic neurons: Determination of their developmental fate by transcription factors. *Annals of the New York Academy of Sciences*, *991*, 36–47.

Simon-Sanchez, J., Schulte, C., Bras, J. M., Sharma, M., Gibbs, J. R., Berg, D., et al. (2009). Genome-wide association study reveals genetic risk underlying Parkinson's disease. *Nature Genetics*, *41*(12), 1308–1312. http://dx.doi.org/10.1038/ng.487.

Soloff, P. H., Lynch, K. G., Kelly, T. M., Malone, K. M., & Mann, J. J. (2000). Characteristics of suicide attempts of patients with major depressive episode and borderline personality disorder: A comparative study. *The American Journal of Psychiatry*, *157*(4), 601–608.

Sossi, V., de la Fuente-Fernandez, R., Schulzer, M., Troiano, A. R., Ruth, T. J., & Stoessl, A. J. (2007). Dopamine transporter relation to dopamine turnover in Parkinson's disease: A positron emission tomography study. *Annals of Neurology*, *62*(5), 468–474. http://dx.doi.org/10.1002/ana.21204.

Stein, M. A., Waldman, I. D., Sarampote, C. S., Seymour, K. E., Robb, A. S., Conlon, C., et al. (2005). Dopamine transporter genotype and methylphenidate dose response in children with ADHD. *Neuropsychopharmacology*, *30*(7), 1374–1382. http://dx.doi.org/10.1038/sj.npp.1300718.

Sullivan, P. F. (2005). The genetics of schizophrenia. *PLoS Medicine*, *2*(7), e212. http://dx.doi.org/10.1371/journal.pmed.0020212.

Tilley, M. R., Cagniard, B., Zhuang, X., Han, D. D., Tiao, N., & Gu, H. H. (2007). Cocaine reward and locomotion stimulation in mice with reduced dopamine transporter expression. *BMC Neuroscience*, *8*, 42. http://dx.doi.org/10.1186/1471-2202-8-42.

Torres, G. E., Carneiro, A., Seamans, K., Fiorentini, C., Sweeney, A., Yao, W. D., et al. (2003). Oligomerization and trafficking of the human dopamine transporter. Mutational analysis identifies critical domains important for the functional expression of the transporter. *Journal of Biological Chemistry*, *278*(4), 2731–2739. http://dx.doi.org/10.1074/jbc.M201926200.

Vahedi, S., Rajabian, M., Misaghian, A., Grbec, D., Simon, H. H., & Alavian, K. N. (2010). Parkinson's disease candidate gene prioritization based on expression profile of midbrain dopaminergic neurons. *Journal of Biomedical Science*, *17*, 66. http://dx.doi.org/10.1186/1423-0127-17-66.

Valchar, M., & Hanbauer, I. (1993). Comparison of [3H]WIN 35,428 binding, a marker for dopamine transporter, in embryonic mesencephalic neuronal cultures with striatal membranes of adult rats. *Journal of Neurochemistry*, *60*(2), 469–476.

Vandenbergh, D. J., Persico, A. M., Hawkins, A. L., Griffin, C. A., Li, X., Jabs, E. W., et al. (1992). Human dopamine transporter gene (DAT1) maps to chromosome 5p15.3 and displays a VNTR. *Genomics*, *14*(4), 1104–1106.

Vandenbergh, D. J., Thompson, M. D., Cook, E. H., Bendahhou, E., Nguyen, T., Krasowski, M. D., et al. (2000). Human dopamine transporter gene: Coding region conservation among normal, Tourette's disorder, alcohol dependence and attention-deficit hyperactivity disorder populations. *Molecular Psychiatry*, *5*(3), 283–292.

van der Wee, N. J., Stevens, H., Hardeman, J. A., Mandl, R. C., Denys, D. A., van Megen, H. J., et al. (2004). Enhanced dopamine transporter density in psychotropic-naive patients with obsessive-compulsive disorder shown by [123I]{beta}-CIT SPECT. *The American Journal of Psychiatry*, *161*(12), 2201–2206. http://dx.doi.org/10.1176/appi.ajp.161.12.2201.

van der Zwaluw, C. S., Engels, R. C., Buitelaar, J., Verkes, R. J., Franke, B., & Scholte, R. H. (2009). Polymorphisms in the dopamine transporter gene (SLC6A3/DAT1) and alcohol dependence in humans: A systematic review. *Pharmacogenomics*, *10*(5), 853–866. http://dx.doi.org/10.2217/pgs.09.24.

VanNess, S. H., Owens, M. J., & Kilts, C. D. (2005). The variable number of tandem repeats element in DAT1 regulates in vitro dopamine transporter density. *BMC Genetics*, *6*, 55. http://dx.doi.org/10.1186/1471-2156-6-55.

Vaughan, R. A., Brown, V. L., McCoy, M. T., & Kuhar, M. J. (1996). Species- and brain region-specific dopamine transporters: Immunological and glycosylation characteristics. *Journal of Neurochemistry*, *66*(5), 2146–2152.

Volkow, N. D., Wang, G. J., Fischman, M. W., Foltin, R. W., Fowler, J. S., Abumrad, N. N., et al. (1997). Relationship between subjective effects of cocaine and dopamine transporter occupancy. *Nature*, *386*(6627), 827–830. http://dx.doi.org/10.1038/386827a0.

Volkow, N. D., Wang, G. J., Fowler, J. S., Logan, J., Hitzemann, R., Ding, Y. S., et al. (1996). Decreases in dopamine receptors but not in dopamine transporters in alcoholics. *Alcoholism: Clinical and Experimental Research*, *20*(9), 1594–1598.

Wang, J., & Bannon, M. J. (2005). Sp1 and Sp3 activate transcription of the human dopamine transporter gene. *Journal of Neurochemistry*, *93*(2), 474–482. http://dx.doi.org/10.1111/j.1471-4159.2005.03051.x.

Wang, G. J., Chang, L., Volkow, N. D., Telang, F., Logan, J., Ernst, T., et al. (2004). Decreased brain dopaminergic transporters in HIV-associated dementia patients. *Brain*, *127*(Pt. 11), 2452–2458. http://dx.doi.org/10.1093/brain/awh269.

Weiner, R. I., & Ganong, W. F. (1978). Role of brain monoamines and histamine in regulation of anterior pituitary secretion. *Physiological Reviews*, *58*(4), 905–976.

Weintraub, D., & Burn, D. J. (2011). Parkinson's disease: The quintessential neuropsychiatric disorder. *Movement Disorders*, *26*(6), 1022–1031. http://dx.doi.org/10.1002/mds.23664.

Xu, X., Mill, J., Sun, B., Chen, C. K., Huang, Y. S., Wu, Y. Y., et al. (2009). Association study of promoter polymorphisms at the dopamine transporter gene in attention deficit hyperactivity disorder. *BMC Psychiatry*, *9*, 3. http://dx.doi.org/10.1186/1471-244X-9-3.

Yamashita, A., Singh, S. K., Kawate, T., Jin, Y., & Gouaux, E. (2005). Crystal structure of a bacterial homologue of Na^+/Cl^- dependent neurotransmitter transporters. *Nature*, *437*(7056), 215–223. http://dx.doi.org/10.1038/nature03978.

Zahniser, N. R., & Sorkin, A. (2004). Rapid regulation of the dopamine transporter: Role in stimulant addiction? *Neuropharmacology*, *47*(Suppl. 1), 80–91. http://dx.doi.org/10.1016/j.neuropharm.2004.07.010.

Zhai, D., Li, S., Zhao, Y., & Lin, Z. (2014). SLC6A3 is a risk factor for Parkinson's disease: A meta-analysis of sixteen years' studies. *Neuroscience Letters*, *564*, 99–104. http://dx.doi.org/10.1016/j.neulet.2013.10.060.

Zheng, C., Shen, Y., & Xu, Q. (2012). Association of intron 1 variants of the dopamine transporter gene with schizophrenia. *Neuroscience Letters*, *513*(2), 137–140. http://dx.doi.org/10.1016/j.neulet.2012.02.015.

Zomot, E., Bendahan, A., Quick, M., Zhao, Y., Javitch, J. A., & Kanner, B. I. (2007). Mechanism of chloride interaction with neurotransmitter: Sodium symporters. *Nature*, *449*(7163), 726–730. http://dx.doi.org/10.1038/nature06133.

CHAPTER TWELVE

Regulation of the Norepinephrine Transporter by Endothelins: A Potential Therapeutic Target

Marcelo S. Vatta[*,1], **Liliana G. Bianciotti**[†], **María J. Guil**[*], **Sandra I. Hope**[*]

[*]Cátedra de Fisiología e Instituto de la Química y Metabolismo del Fármaco (IQUIMEFA-CONICET), Facultad de Farmacia y Bioquímica, Universidad de Buenos Aires, Buenos Aires, Argentina

[†]Cátedra de Fisiopatología, Facultad de Farmacia y Bioquímica, Universidad de Buenos Aires, Instituto de Inmunología, Genética y Metabolismo (INIGEM-CONICET), Buenos Aires, Argentina

[1]Corresponding author: e-mail address: mvatta@ffyb.uba.ar

Contents

Abstract

Neuronal norepinephrine (NE) uptake is a crucial step in noradrenergic neurotransmission that regulates NE concentration in the synaptic cleft. It is a key mechanism mediated by the NE transporter (NET) which takes the neurotransmitter into the presynaptic neuron terminal or the adrenal medulla chromaffin cell. The activity of NET is short and long terms modulated by phosphorylation mediated by protein kinases A, C, and G and calcium–calmodulin-dependent protein kinase, whereas the transporter availability at the cell surface is regulated by glycosylation. Several neuropeptides like angiotensins II, III, and 1–7, bradykinin, natriuretic peptides, as well as endothelins (ETs) regulate a

Vitamins and Hormones, Volume 98
ISSN 0083-6729
http://dx.doi.org/10.1016/bs.vh.2014.12.013

wide variety of biological effects, including noradrenergic transmission and in particular neuronal NE uptake. Diverse reports, including studies from our laboratory, show that ETs differentially modulate the activity and expression of NET not only in normal conditions but also in diverse cardiovascular diseases such as congestive heart failure and hypertension. Current literature supports a key role for the interaction between ETs and NE in maintaining neurotransmission homeostasis and further suggests that this interaction may represent a potential therapeutic target for various diseases, particularly hypertension.

1. INTRODUCTION

The present review focuses on the participation of the endothelinergic system in the regulation of neuronal norepinephrine (NE) uptake which is the crucial step that ends NE activity at the synaptic cleft. The modulation of the NE transporter (NET) activity and expression by endothelin (ETs) is reviewed in normal animals and in different cardiovascular diseases focusing on the differential regulation exerted by ETs in diverse areas of the central nervous system (CNS) and peripheral nervous system. In addition, the contribution of other neuropeptides to the regulation of NET is briefly described. This review is divided into three sections: the first section describes the main features of NET regulation; the second section involves the description of ETs biosynthesis, receptors, and coupled intracellular signaling pathways and main biological actions; and the third section focuses on the regulation of NET by ETs and other neuropeptides.

2. NEURONAL NE UPTAKE

2.1 General aspects

Catecholamines (CAs) are classic neurotransmitters of low-molecular-weight that share a catechol group in their structure. The biosynthetic pathway begins with the stereospecific uptake of the CA precursor L-tyrosine into the cytoplasm where it is transformed to L-DOPA by tyrosine hydroxylase (TH). This is the limiting step in CA biosynthetic pathway. Then, by the action of L-aromatic amino acid decarboxylase, L-DOPA is converted to dopamine that enters into the synaptic vesicle through a monoamine vesicular transporter, where it is transformed to NE by dopamine-β-hydroxylase. In response to diverse stimuli, NE is released from vesicles through a complex exocytotic calcium-dependent mechanism to the synaptic cleft where it binds to specific adrenergic receptors localized at the pre- and/or postsynaptic levels to exert diverse biological effects. The amine is then removed mainly by uptake mechanisms which will be discussed in more detail in the following section (Fig. 1; Flatmark, 2000; Kvetnansky, Sabban, & Palkovits, 2009).

2.2 NE inactivation

NE exerts its actions on target cells by binding to and activating adrenergic receptors. The target cell expression of different receptor types determines the ultimate cellular effect, and thus NE has different actions on different cell types. The signal is terminated by the removal of NE through distinct uptake mechanisms that may eventually be followed by enzymatic degradation. Although part of the neurotransmitter may be lost through its diffusion across the synaptic cleft to the blood, it is a route of minor importance for NE inactivation (Brandao, 1976; Kvetnansky et al., 2009). The extracellular uptake of NE into the cytosol, either by presynaptically (uptake 1) or by nonneuronal cells in the vicinity (uptake 2), represents the major mechanisms for NE inactivation (Hahn & Blakely, 2007; Kvetnansky et al., 2009). Part of the NE is transported by the monoamine vesicular transporter into synaptic vesicles to be further released upon stimuli, but uptake may be followed by rapid degradation to various metabolites through monoamine oxidase (MAO) or catechol-*O*-methyl transferase (COMT) enzymes (Eisenhofer, Kopin, & Goldstein, 2004; Kopin, 1994; Youdim & Riederer, 1988). In neurons, MAO converts NE to the corresponding aldehyde which is then nonenzymatically further oxidized (Eisenhofer et al., 2004; Kopin, 1994). Similarly, but in nonneuronal cells, COMT methylates the m-hydroxyl group of the phenyl ring of NE, rendering it less active (Eisenhofer et al., 2004; Youdim & Riederer, 1988).

2.2.1 Nonneuronal uptake or uptake 2

Nonneuronal uptake occurs in nonneuronal cells like myocytes, hepatocytes, chromaffin cells, as well as glial, renal, and smooth muscle cells (Hughes, 1972; Iversen & Salt, 1970; Kvetnansky et al., 2009; Trendelenburg, 1988; Vatta et al., 1997). It is mediated by the classic corticosterone-sensitive extraneuronal MAO transporter and two organic cation transporters (OCT1 and OCT2) localized on extraneuronal structures (Eisenhofer, 2001; Friedgen, Wolfel, Russ, Schomig, & Graefe, 1996; Haag et al., 2004). It is a nonstereo-specific, low-affinity, and

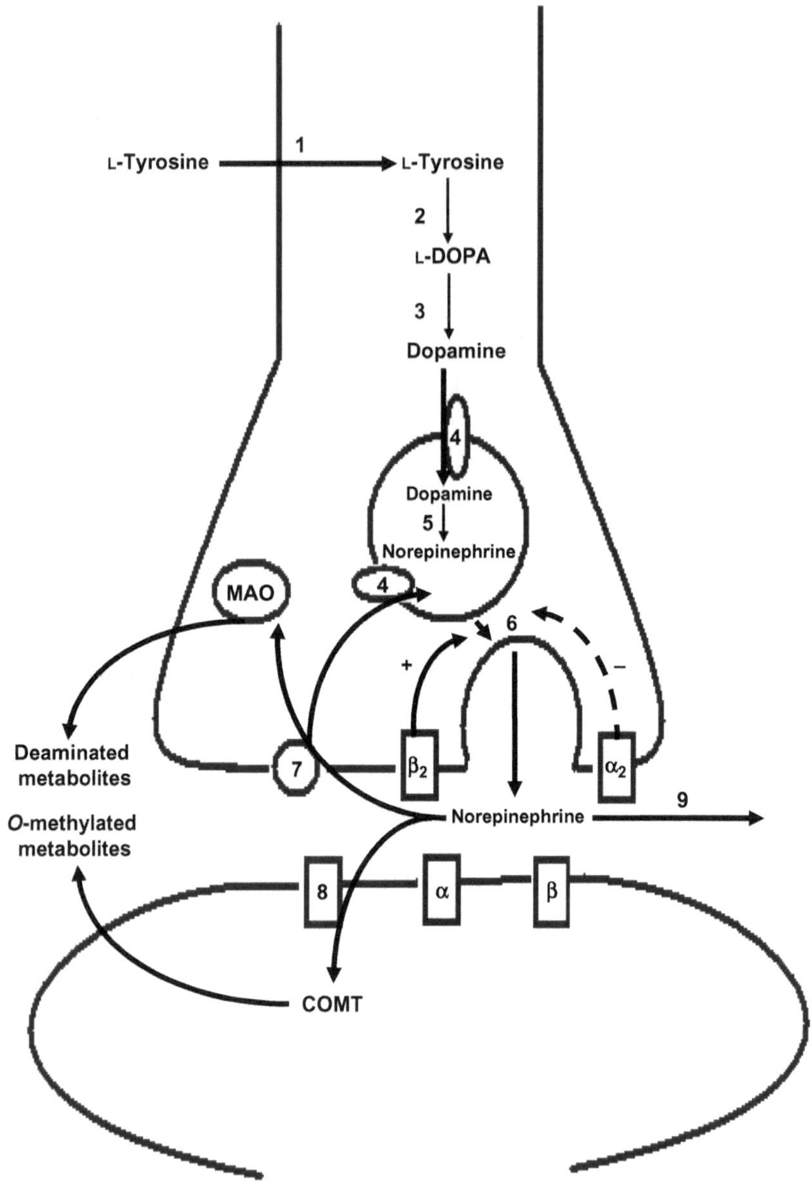

Figure 1 Schematic representation of the events occuring at the noradrenergic terminal and an effector cell. A Na^+-dependent cotransport of aromatic amino acid uptakes L-tyrosine (1) to be converted to L-DOPA by the tyrosine hydroxylase enzyme (2). Subsequently, the L-aromatic amino acid decarboxylase enzyme converts L-DOPA to dopamine (3) which by a vesicular transport of amines (4) is taken back into vesicles, where it is finally converted to norepinephrine (NE) by the dopamine-β-hydroxylase enzyme (5). Upon appropriate stimuli, NE is released by a calcium-dependent mechanism into the

high-capacity mechanism that it is independent of Na^+ and Cl^- (Eisenhofer, 2001). Nonneuronal uptake is not inhibited by usual inhibitors of neuronal uptake but by metanephrine and corticosteroids like hydrocortisone (Olivier, Soudijn, & van Wijngaarden, 2000). This mechanism primarily clears circulating CA, rather than neuronal NE, although it may contribute to the clearance of synaptic NE under conditions of diminished NET function. Nonneuronal uptake is followed by enzymatic degradation of CA.

2.2.2 Neuronal uptake or uptake 1

Neuronal uptake of CA is a high-affinity and Na^+- and Cl^--dependent system mediated by the NET, dopamine transporter (DAT), and serotonin transporter (SERT) proteins (Iversen, 2006). Uptake of NE into the presynaptic neuron by NET represents the primary mechanism for NE inactivation (Bönish & Brüss, 2006; Wersinger, Jeannotte, & Sidhu, 2006; Zahniser & Doolen, 2001). Approximately, 80–90% of the released NE to the synaptic cleft is transported into neurons by NET (Esler et al., 1990). In the late 1950s and early 1960s, Julius Axelrod revealed the existence of neuronal uptake as a key mechanism for CA inactivation (Axelrod, 1957, 1962). Before this finding, it was accepted that inactivation of CA was mediated by rapid enzymatic degradation at the synaptic cleft like it occurs with acetylcholine which is hydrolyzed by acetylcholinesterase. At present, it is well established that CA inactivation is primarily mediated by transporter proteins belonging to the solute carrier 6 (SLC6) transport family (Kristensen et al., 2011; Pramod, Foster, Carvelli, & Henry, 2013).

2.3 NE transporter

The NET has a pivotal role in the regulation of synaptic NE in the brain and in peripheral tissues. This transporter is highly relevant given that it is a molecular target for diverse antidepressants, like the tricyclics, and abused

synaptic cleft (6), where NE binds to adrenoceptors localized at the pre- and postsynaptic membrane. The α2 and β2 receptors placed at the presynaptic membrane down- or upregulate neuronal NE release, respectively. The biological inactivation of NE occurs not only by neuronal uptake or uptake 1 mediated by the norepinephrine uptake transporter (NET) (7), but also by extraneuronal uptake or uptake 2 (8) or diffusion into the extracellular space (9). The neurotransmitter transported by NET is either internalized into vesicles by the amine vesicular transporter (4) or degraded by the monoamine oxidase (MAO) enzyme to deaminated metabolites. NE undergoing nonneuronal uptake is metabolized to *O*-methylated derivatives by catechol-*O*-methyltransferase (COMT).

substances as cocaine and amphetamines (Iversen, 2000; Kristensen et al., 2011; Schlessingera et al., 2011).

The NET is localized on the membrane of noradrenergic neurons, and it not only regulates NE availability at the synaptic cleft but also plays an important role in the neurotransmitter homeostasis between presynaptic and postsynaptic nerve endings (Liang, 2007; Matthies et al., 2009; Xu et al., 2000). The NET can also transport dopamine and epinephrine, and although it has higher affinity for dopamine than NE, it is not expressed in dopaminergic nerve terminals (Apparsundaram, Moore, Malone, Hartzell, & Blakely, 1997; Buck & Amara, 1994; Giros et al., 1994). Nevertheless, it is believed that it may contribute to clear dopamine in certain brain areas (Carboni, Tanda, Frau, & Di Chiara, 1990; Moron, Brockington, Wise, Rocha, & Hope, 2002). The affinity rank order for NET is dopamine > NE > epinephrine.

NE uptake by NET is a secondary active transport dependent on Na^+ and Cl^- ions. Uptake is driven by an inwardly directed Na^+ gradient maintained by the action of the Na^+–K^+-ATPase (Bönish & Brüss, 2006; Kristensen et al., 2011; Kvetnansky et al., 2009). The NET belongs to the SLC6 transport family, which consists of four groups based on the substrate they transport: the neurotransmitter transporters which besides NET (SLC6A2) include DAT, γ-aminobutyric acid (GABA) transporter, SERT, and glycine transporter; the amino acid transporters; the osmolyte transporters; and creatine transporters (Kristensen et al., 2011; Pramod et al., 2013).

2.3.1 NET: Structure and function

The human and rat NET contains 617 amino acids, and its distribution is consistent with the distribution of noradrenergic pathways. The human gene encoding NET is localized to chromosome 16q12.2 (Brüss, Kunz, Lingen, & Bönisch, 1993; Gelernter et al., 1993). Alternative mRNA splicing variants of the transporter which results in the expression of NET that differ in carboxyl-terminal regions have been reported (Pörzgen, Bönisch, Hammermann, & Brüss, 1998; Sogawa et al., 2007). These variants from alternative splicing may influence maturation as well as trafficking of the transporter to the cell surface. In addition, several polymorphisms in the coding and noncoding regions of human NET gene have also been shown (Hahn & Blakely, 2002; Iwasa, Kurabayashi, Nagai, Nakamura, & Tanaka, 2001; Shannon et al., 2000; Stöber et al., 1996; Tellioglu & Robertson, 2001).

The NET is mostly expressed in the *locus coeruleus* that sends projections to cerebellum, cortex, and thalamus and in the A1 and A2 cell groups that project to the hypothalamus, preoptic area, and forebrain (Kvetnansky et al., 2009; Matthies et al., 2009; Schroeter et al., 2000). In addition, NET is present in the sympathetic neurons of the autonomic nervous system and the chromaffin cells of the adrenal medulla (Kippenberger et al., 1999; Schroeter et al., 2000; Wehrwein et al., 2008).

Sequence analysis combined with biochemical approaches revealed that these transporters have 12 hydrophobic transmembrane spanning domains with intracellular amino and carboxy termini and a large extracellular loop between transmembrane helices TM3 and TM4 that contains three N-glycosylation sites (Kristensen et al., 2011; Pramod et al., 2013; Ramamoorthy, Shippenberg, & Jayanthi, 2011; Yamashita, Singh, Kaeate, Jin, & Gouaux, 2005). Another structural feature is a pair of Cys residues in extracellular loop 2 that form an intraloop disulfide bridge. The N-glycosylation sites modulate the activity and the structural stability of the transporter. The structure has an inverted symmetry given by the TM1–TM5 and TM6–TM10 domains (Kristensen et al., 2011; Pramod et al., 2013; Yamashita et al., 2005). Within the amino and carboxy termini and within its internal domain, the NET contains consensus sites for serine/threonine phosphorylation (Ramamoorthy et al., 2011). The NET shares high homology with the other monoamine transporters (DAT and SERT), particularly at TM1, TM2, and TM4–8 domains (Pramod et al., 2013).

The ligand-binding site is in close vicinity with Na^+ which allows a coupled movement between NE and the ion (Hahn & Blakely, 2007; Torres & Anara, 2007; Yamashita et al., 2005). Kinetic studies show that the entry of Na^+ and Cl^- is a requisite for NE binding (Apparsundaram, 2011; Bönish & Brüss, 2006; Gouaux, 2009). Removal of Na^+ or Cl^- abolishes NET-mediated NE uptake. NET is believed to translocate one substrate molecule with two Na^+ and one Cl^- (Kristensen et al., 2011). The transporter is inhibited by cocaine, desipramine, and nisoxetine among other drugs (Apparsundaram, 2011; Hahn & Blakely, 2007; Robinson, 2003).

The activity and expression of NET is regulated by short- and long-term mechanisms. Short-term regulation involves phosphorylation, trafficking, and protein–protein interaction, whereas long-term modulation involves transcriptional modifications.

The abundance of NET on the plasma membrane depends on NE release rate thus avoiding asynchrony between the release and uptake of the neurotransmitter (Oaks & Sidhu, 2011; Quick, 2006; Ramamoorthy et al.,

2011). NET can be internalized into endosomal compartments or recycled to the membrane upon demand (Oaks & Sidhu, 2011). The rate of recycling/internalization can be as high as 3–5% transporters/min.

SNARE proteins (soluble *N*-ethylmaleimide-sensitive factor attachment protein receptor) present in the vesicles (v-SNARES) and in the plasma membrane (t-SNARE) regulate neuronal release and inactivation of NE (Oaks & Sidhu, 2011; Quick, 2006; Ramamoorthy et al., 2011). Synaptobrevin or vesicle associated membrane proteins (VAMP) (v-SNARE) as well as syntaxin and SNAP-25 (t-SNARES) are involved in NE release as well as the NET recycling/internalization process.

A stable ternary complex (1:1:1) is formed by SNARE proteins which interact with N- and P/Q-type calcium channels to allow the fusion of vesicles with the plasma membrane and the eventual release of NE (Chen & Scheller, 2001; Jena, 2011; Rizo & Südhof, 2002; Sheng, Rettig, Cook, & Catterall, 1996). Syntaxin A1 interacts with NET to regulate the availability of the transporter in the membrane and to inhibit NE release (Sung & Blakely, 2007). If the syntaxin/NET complex is dissociated, NE inactivation is retarded (Sung et al., 2003).

High calcium concentrations favor not only NET insertion in the membrane but also NET/syntaxin complex formation, whereas the absence of calcium reduces NET recycling (Sung & Blakely, 2007). Protein kinase C (PKC) activation leads to decreased transport (reduced V_{max}) as a result of redistribution of the transporter from the surface to intracellular compartments (Apparsundaram, Galli, DeFelice, Hartzell, & Blakely, 1998; Jayanthi, Samuvel, & Ramamoorthy, 2004). Mutation of Thr258 and Ser259 to Ala significantly reduces transporter phosphorylation and prevents phorbol 12-myristate 13-acetate-induced decrease in NE uptake and NET internalization, an effect that was not observed upon substitution of the other potential phosphorylation sites in the transporter (Jayanthi, Annamalai, Samuvel, Gether, & Ramamoorthy, 2006). Stimulation of calcium–calmodulin-dependent protein kinase (CaMKII) activity in PC12 cells, which endogenously express NET, correlates with enhanced NET transport activity (Ramamoorthy et al., 2011; Sung & Blakely, 2007; Uchida, Kiuchi, Ohno, Yura, & Oguchi, 1998). Protein phosphatases like protein phosphatases 1 and 2 play a relevant role in maintaining NET in a relatively dephosphorylated state (Bauman et al., 2000). The phosphorylation sites appear to be potential sites for the presynaptic modulation of NET activity. Nevertheless, despite numerous studies, the molecular basis of the processes by which phosphorylation controls the transporter function still remains to be fully elucidated.

2.3.2 Involvement of the NET in disease

Impairment of NET has been associated with diverse neurological diseases like Alzheimer and Parkinson; psychiatric disorders like depression and attention-deficit hyperactivity disorder; as well as cardiovascular diseases like congestive heart failure, blood pressure elevation, and postural tachycardia syndrome (Bönish & Brüss, 2006; Esler et al., 2006; Kristensen et al., 2011; Liang, 2007; Schroeder & Jordan, 2012; Shannon et al., 2000; Tellioglu & Robertson, 2001; Whiskey & Taylor, 2013). Whether NET impairment is involved in the genesis and/or maintenance of these pathophysiological conditions remains to be further elucidated.

3. ENDOTHELINS

3.1 General aspects

ETs are a family of vasoactive isopeptides comprised by ET-1, ET-2, and ET-3 that were first evidenced by Hickey, Rubanyi, Paul, and Highsmith (1985). They provided the first evidence for a potent vasoconstrictor substance derived from bovine aortic cells (Hickey et al., 1985). In 1988, these substances were then isolated, purified, and further characterized from porcine and rat aortic endothelial cells (Itoh et al., 1988; Yanagisawa, Kurihara, et al., 1988). ETs are 21 amino acid peptides with a hydrophobic C-terminus and two cysteine bridges at the N-terminus, which are essential for the biological activity (Kimura et al., 1988; Yanagisawa, Inoue, et al., 1988). ET-1, the first characterized member of the family, induces a strong long-lasting pressor response when intravenously injected (Yanagisawa & Masaki, 1989). ET-1, ET-2, and ET-3 are encoded in rodents, porcine, and humans by different independent genes localized to the 6, 1, and 20 chromosomes, respectively (Arinami et al., 1991; Inoue et al., 1989; Yanagisawa, Inoue, et al., 1988). Phylogenetic studies show that ETs are highly conserved in different species like humans, fishes, insects, mollusks, and annelids (Kasuya, Kobayashi, & Uemura, 1991; Kuwaki et al., 1997). ET-1 gene expression is enhanced by stress, and it plays a relevant role in diverse systemic disorders associated with endothelium injury like hypertension, uremic hemolytic syndrome, thrombocytopenic purpura, and heart failure (Hynynen & Khalil, 2006; Itoh et al., 1988). Immunoreactive ET-1 and mRNA ET-1 were reported in the blood vessels, heart, liver, kidney, peripheral nervous system, and CNS (Giaid et al., 1989; Kuddus, Nalesnik, Subbotin, Rao, & Gandhi, 2000; Naicker & Bhoola, 2001; Naidoo, Mahabeer, & Raidoo, 2001; Naidoo, Naidoo, Mahabeer, &

Raidoo, 2004; Rubanyi & Botelho, 1991; Zhan & Rockey, 2011). ET-1 is the peptide with most potent vasoconstrictor property known (Yanagisawa, Inoue, et al., 1988; Yanagisawa, Kurihara, et al., 1988).

ET-2 is principally expressed in vascular endothelium, and minor in kidney, heart, placenta, uterus, CNS, and in the gastrointestinal tract (Takizawa et al., 2005; Uchide, Adur, Fukamachi, & Saida, 2000), whereas ET-3 is primarily localized in the CNS, and also in pituitary, intestine, pancreas, and liver (Matsumoto, Suzuki, Onda, & Fujino, 1989; Sluck, Lin, Katolik, Jeng, & Lehmann, 1999). ET-2 and ET-3 differ from ET-1 in the amino acid sequence but share similar biological properties (Fig. 2).

The venom of the burrowing asp *Atractaspis engaddensis* contains several 21 amino acid residue peptides known as sarafotoxins. The sarafotoxins are homologous to the mammalian ET family, and they have similar biological activities (Kloog et al., 1988; Kochva, Bdolah, & Wollberg, 1993).

The precursors of ETs are processed by peptidases to create mature active forms. The 203-residue pre–pro-ETs (212 in human) are cleaved by furin-like endopeptidases to form inactive intermediates of 39 amino acids termed big-ETs or pro-ETs (Fig. 3; Sluck et al., 1999). Processing is mediated by zinc metalloproteases from the neprilysin superfamily, termed endothelin-converting enzymes (ECEs) that cleave big-ETs into the biological active

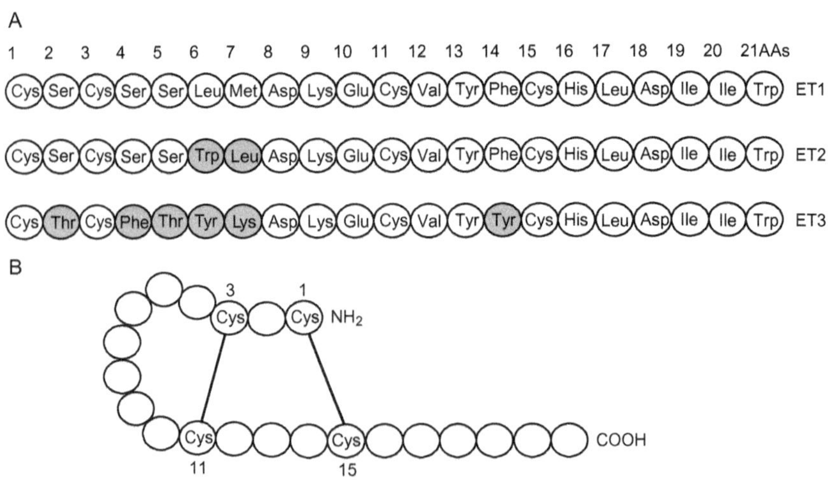

Figure 2 Endothelins structure. (A) The amino acid sequence of the endothelin family isoforms. Filled circles represent the amino acid residues different from ET-1 sequence. (B) Cyclic endothelin structure that includes two disulfide bridges joining the cysteine residues in positions 1–15 and 3–11. AAs: amino acids and ET-1, ET-2, and ET-3: endothelins 1, 2, and 3.

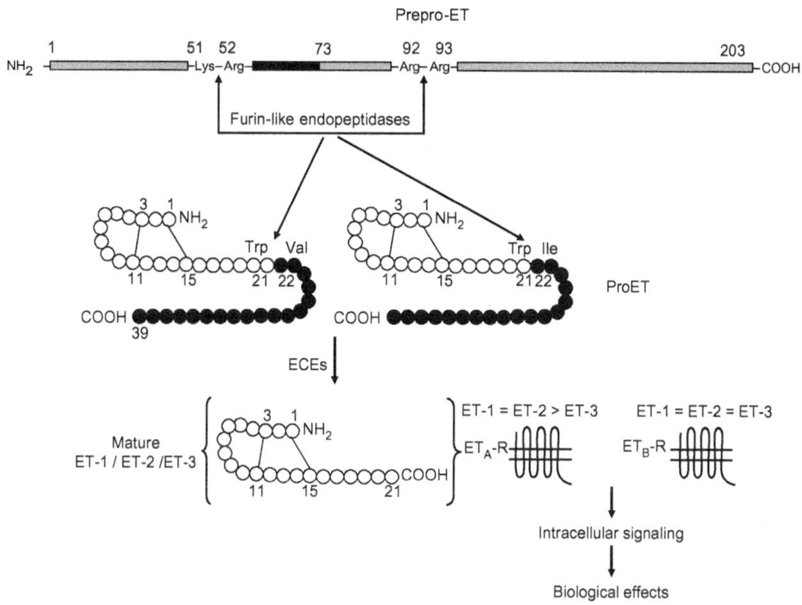

Figure 3 Endothelin biosynthetic pathway and affinity for endothelin receptors. ET-1, ET-2, and ET-3: endothelins 1, 2, and 3; ECEs: endothelin-converting enzymes; and ET_A-R and ET_B-R: endothelin receptor type A and endothelin receptor type B.

peptides (Opgenorth, Wu-Wong, & Shiosaki, 1992; Sluck et al., 1999). Although several isoforms of ECEs have been described, isoforms ECE-1, ECE-2, and ECE-3 are the most important in ETs' cleavage (Hasegawa et al., 1998; Kawanabe & Nauli, 2011; Opgenorth et al., 1992). ECE-1 is a membrane-bound metalloprotease that acts at neutral pH, whereas ECE-2 acts at acidic intracellular pH (Emoto & Yanagisawa, 1995). While ECE-1 and ECE-2 cleave pro-ET-1, these enzymes display low affinity for big-ET-2 and big-ET-3. ECE-3, originally purified from iris microsomes, would be specific for the cleavage of big-ET-3 to ET-3 (Hasegawa et al., 1998).

Once synthetized, ETs are secreted through two distinct secretory pathways. One is the classical exocytotic pathway activated in response to stimuli like hypothermia, stretch, and diverse agonists (Khimji & Rockey, 2010; Macarthur, Warner, Wood, Corder, & Vane, 1994; Russell & Davenport, 1999; Yoshitomi, Kojima, Ogi, & Kuramochi, 1998). The other secretory via is constitutive secretion that results in elevated ETs release as a consequence of increased biosynthesis without peptide package into vesicles (Khimji & Rockey, 2010; Ohkita, Tawa, Kitada, & Matsumura, 2012; Russell & Davenport, 1999).

3.2 Receptors and intracellular signaling pathways

ETs mediate their biological actions through two well-characterized G protein-coupled receptors (GPCRs) termed ET_A and ET_B (Fig. 3; Sokolovsky, 1995a, 1995b). The ET_A receptor displays higher affinity for ET-1 and ET-2 than for ET-3, whereas ET_B binds the three isopeptides with similar affinity (Fig. 3; D'Orleans-Juste et al., 2002; Meyers & Sethna, 2013).

In rodents, ET_A is highly expressed in the heart, blood vessels, and lungs whereas to a lesser extent in the liver, kidney, CNS, and the reproductive system (King, Gude, Di Iulio, & Brennecke, 1995; Kohan, Rossi, Inscho, & Pollock, 2011; Koyama, 2013; Kurokawa, Yamada, & Ochi, 1997; Kuwaki et al., 1997; Opgenorth, 1995). Conversely, ET_B is expressed mainly in the CNS, although it is also found in lungs, pancreas, heart, kidneys, and the endocrine and reproductive systems (D'Orleans-Juste et al., 2002; King et al., 1995; Koyama, 2013; Kuwaki et al., 1997; Yamamoto & Uemura, 1998).

Both ET receptors can be pharmacologically identified by selective antagonists and agonists (Fig. 4). However, several studies support the existence of receptors other than ET_A and ET_B given that diverse biological effects mediated by ETs fail to be mimicked by ET agonists or inhibited by ET antagonists. These receptors were termed *atypical receptors* or ET_{AX} and ET_{BX} (di Nunzio, Legaz, Rodano, Bianciotti, & Vatta, 2004; Nabhen et al., 2009, 2011; Nambi et al., 1997; Pate et al., 1999; Perfume et al., 2007, 2008; Sokolovsky, 1995a, 1995b). Whether they represent ET_A and ET_B receptor subtypes or additional ET receptors remains to be investigated.

A third receptor subtype termed ET_C with higher affinity for ET-3 was cloned in *Xenopus laevis* (Karne, Jayawickreme, & Lerner, 1993; Kumar et al., 1994). Although it has not been cloned in mammals so far, it is believed that this receptor coupled to the phosphoinositide pathway would mediate various ET-3 biological effects like the neuronal release of NE in the posterior hypothalamus (di Nunzio et al., 2004). Diverse studies show that ET receptors may associate between themselves and with other receptors. It was shown that ET receptors interact to form homodimers and heterodimers (Evans & Walker, 2008; Gregan, Jürgensen, et al., 2004; Gregan, Schaefer, Rosenthal, & Oksche, 2004; Watts, 2010). Associations of ET receptors and other non-ET receptor were also reported, like dimerization between ET receptors and dopamine D3 receptor (D_3) or angiontensin type 1 receptors (AT1) (Watts, 2010; Yu et al., 2009; Zeng et al., 2005).

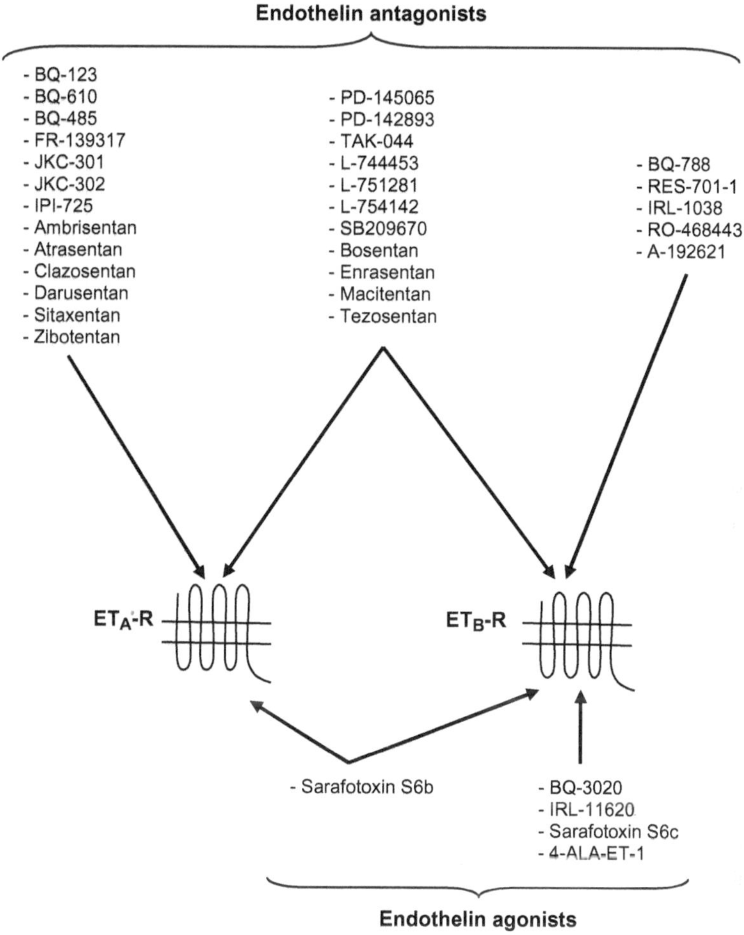

Figure 4 Endothelin receptor antagonists and agonists. ET_A-R and ET_B-R: endothelin receptor type A and endothelin receptor type B.

ET_A and ET_B can activate different G proteins (Gq, Gs, Go, or Gi) and thus trigger multiple signaling pathways depending on the ligand concentration and cell type involved (Hersch, Huang, Grider, & Murthy, 2004; Kitamura et al., 1999; Kuwaki et al., 1999; Shraga-Levine & Sokolovsky, 2000; Sokolovsky, 1995a, 1995b).

ETs activate phospholipase C (PLC) resulting in the cleavage of the phospholipid phosphatidylinositol 4,5-biphosphate into diacylglycerol which stimulates PKC and inositol 1,4,5-triphosphate (IP_3) which binds to IP_3 receptors in the smooth endoplasmic reticulum causing cytosolic calcium concentration to increase (Deacon & Knox, 2010; Jain, Olovsson, Burton, & Yung, 2012; Kato et al., 2013; Kuwaki et al., 1997; Sokolovsky, 1995a, 1995b).

ETs can also stimulate phospholipase A_2, which releases arachidonic acid, a precursor of eicosanoids including prostaglandins and leukotrienes and phospholipase D that catalyzes the hydrolysis of phosphatidylcholine to generate phosphatidic acid and choline (Deacon & Knox, 2010; Kuwaki et al., 1997; Robin, Chouayekh, Bole-Feysot, Leiber, & Tanfin, 2005; Sokolovsky, 1995a, 1995b).

The activation of ET receptors also increases cytosolic calcium through the Na^+/H^+ cotransporter and the calcium voltage and/or ligand channels that induce a sustained calcium influx. The increase in cytosolic calcium activates CaMK and also nitric oxide synthases (NOSs) which through cGMP activate protein kinase G (PKG). ETs can also inhibit or stimulated adenylyl cyclase by activating either Gi or Gs (Kuwaki et al., 1997, 1999; Pollock, Keith, & Highsmith, 1995).

3.3 Biological actions of ETs

ETs and ET receptors are widely distributed in different tissues and cell types supporting a broad spectrum of biological effects. The main effects of ETs with focus on the cardiovascular system and the CNS are discussed in the following sections.

3.3.1 Effects of ETs on the cardiovascular function

Intravenous or intraarterial infusion of ETs causes a sustained vasoconstriction preceded by a transient vasodilation (Cocks, Broughton, Dib, Sudnir, & Angus, 1989; Haynes, Ferro, & Webb, 1995; Haynes, Strachan, & Webb, 1995). It is now known that ET_A receptor activation is responsible for the ETs vasoconstrictor property (Clozel, Gray, Breu, Löfler, & Osterwalder, 1992; Harrison, Randiantsoa, & Schoffter, 1992; Moreland, McMullen, Delaney, Lee, & Hunt, 1992). ET_B agonists can also elicit vasoconstriction, but it depends on the vessel type and size as well as the specie involved (King et al., 1995).

ETs infusion to anesthetized rats or chemically denervated rats, either normotensive or spontaneously hypertensive, results in a rapid and transient decrease in blood pressure followed by a sustained increase (Winquist, Bunting, Garsky, Lumma, & Schofield, 1989). The initial vasodepressor response is mediated by ET_B receptor activation and endothelium-derived vasodilator substances like nitric oxide (NO) and prostaglandins, whereas the sustained pressor response would be direct or indirect through sympathetic activation but mediated by ET_A receptors (Knuepfer, O'Brien, Hoang, Gan, & Song, 1994).

ETs are involved in maintaining the vascular tone not only in physiological but also in pathophysiological conditions like vascular spasm, diabetes mellitus, insulin resistance, and hypertension (Hall et al., 2012; Haynes & Webb, 1998; Ivey, Osman, & Little, 2008; Meyers & Sethna, 2013; Nasser & El-Mas, 2014; Schiffrin, 1998; Schneider, Boesen, & Pollock, 2007).

ETs also induce hemodynamic changes in different vascular beds when applied to the brain. Recent reports in agreement with studies from our laboratory suggest that the central cardiovascular effects elicited by ETs are mediated by the activation of noradrenergic activity in the brain (Chen et al., 2012; di Nunzio, Jaureguiberry, Rodano, Bianciotti, & Vatta, 2002; di Nunzio et al., 2004; Hope, Schmipp, Rossi, Bianciotti, & Vatta, 2008; Kuwaki et al., 1999; Morgazo et al., 2005; Nabhen et al., 2009, 2011; Perfume et al., 2007, 2008).

3.3.2 Effects of ETs on the CNS

ET receptors are expressed in different brain regions and nuclei supporting that ETs mediate numerous biological actions (Kohan et al., 2011; Kurokawa et al., 1997; Kuwaki et al., 1997, 1999; Stojilkovic & Catt, 1996).

ETs in the CNS regulate cardiovascular and renal functions as well as the synthesis and release of diverse hormones and neurohormones like vasopressin, renin, aldosterone, follicle-stimulating hormone, growth hormone, and prolactin (Chan, Tang, & O, 2008; Kanyicska, Sellix, & Freeman, 2003; Kuhlmann, Amann, Schlotzer-Schrehardt, Kruse, & Cyrsiefen, 2005; Kuwaki et al., 1997; ThanThan et al., 2010; Yamamoto et al., 1991).

The intracerebroventricular injection of ET-1 elicits a sustained rise in blood pressure (Gulati, Rebello, Roy, & Saxena, 1995; Lu et al., 2007; Nishimura et al., 1990, 1991; Ouchi et al., 1989; Sirén & Feuerstein, 1989). Similar effects elicit ET-1 microinjections in the area postrema of anesthetized rats (Yamamoto et al., 1991). Adrenergic receptor antagonists inhibit ET-1 pressor response, supporting that the peptide applied to the brain activates the catecholaminergic system in diverse areas of the CNS (Ferguson & Smith, 1990). Furthermore, ET-1 applied to the area postrema increases plasma CA levels (Matsumura et al., 1994), showing a close relationship between the endothelinergic and catecholaminergic systems. The administration of ET-1 low doses to areas like the rostral ventrolateral medulla increases blood pressure, and renal sympathetic activity, but in some animals induces cardiorespiratory collapse and death (Kumada, Cao, & Kuwaki, 2003; Mosqueda-García, Inagami, Appalsamy, Sugiura, & Robertson, 1993). The pressor response following ET-1 administration is associated with sympathetic activation (Chen et al., 2012). Low doses of ET-3 applied to the solitarii tract nucleus also increase blood pressure and heart rate (Mosqueda-García et al., 1993).

Several studies suggest that ETs act as putative neurotransmitters or regulatory neuropeptides in the CNS (di Nunzio et al., 2002, 2004; Hope et al., 2008; Jaureguiberry, di Nunzio, Dattilo, Bianciotti, & Vatta, 2004; Kurokawa, Yamada, Liu, & Kudo, 2000; Kurokawa et al., 1997; Morgazo et al., 2005; Nabhen et al., 2009, 2011; Perfume et al., 2007, 2008; Yamada & Kurokawa, 1998). ETs enhance the release of monoamines from nerve endings that activate the renin–angiotensin system (Oparil et al., 1995; Yamada & Kurokawa, 1998). Webber, Pennefather, Head, and van den Buuse (1998) reported that ET-1 through ET_B receptors induces dopamine release in the striatum. It was also reported that ETs also stimulate substance P secretion from the hypothalamus and adenohypophysis (Calvo et al., 1990). In addition, both ET-1 and ET-3 increase NOS activity in the hypothalamus (di Nunzio et al., 2002; Jaureguiberry et al., 2004).

The hypothalamic areas and nuclei present high density of mRNA ET as well as ET receptors (Kurokawa et al., 1997; Kuwaki et al., 1997; Stojilkovic & Catt, 1996). Studies carried out in our laboratory show that ETs differentially regulate NE release in the anterior and the posterior hypothalamic regions of normotensive rats. The anterior and posterior hypothalamic areas are considered sympathoinhibitory and sympathoexcitatory, respectively. In the anterior hypothalamus, ET-1 and ET-3 diminish neuronal NE release through the activation of ET_B receptors coupled to the NO/soluble guanylyl cyclase/cGMP/PKG/$GABA_A$ pathway (di Nunzio et al., 2002; Jaureguiberry et al., 2004). However, in the posterior hypothalamus, ET-1 and ET-3 increase neuronal NE release through different intracellular pathways. ET-1 activates nonconventional (*atypical*) ET receptors and triggers different intracellular signaling involving a cross talk among PLC, PKC, and the adenylyl cyclase pathway, whereas ET-3 activates ET_C receptors and stimulates phosphoinositide signaling (di Nunzio et al., 2004). We also reported that ET-1 and ET-3 are involved in the short- and long-term modulations of activity and expression of TH in the anterior and posterior hypothalamic regions. The regulation of TH by ETs is very complex and involves different ET receptor subtypes, including nonconventional or *atypical* receptors and multiple signaling pathways (Karne et al., 1993; Perfume et al., 2007, 2008). In the rat anterior hypothalamic region, ET-1 and

ET-3 decrease TH activity through ET_B receptors coupled to the activation of the phosphoinositide, CaMKII, and NO/GMPc/PKG pathways (Morgazo et al., 2005; Perfume et al., 2008). ETs do not modify TH expression levels in the short term, but at long term, both peptides decrease it. In the posterior hypothalamus, ET-1 and ET-3 at short term decrease TH activity through a nonconventional receptor coupled to the phosphoinositide, CaMKII, and O/cGMP/PKG pathways, whereas TH protein level remains unaltered (Perfume et al., 2007). On the other hand, in the posterior hypothalamus, ET-1 and ET-3 increase TH activity and expression at long term through an *atypical* receptor coupled to protein kinase A (PKA), phosphoinositide, and CaMKII activation (Perfume et al., 2008). Studies from our laboratory also show that ET receptors are expressed in the rat olfactory bulb, and they regulate the activity and expression of TH. Short-term modulation of TH by ET-1 and ET-3 in the olfactory bulb results in a significant increase in the enzyme activity without changes in the enzyme expression (Nabhen et al., 2009). Both ETs also participate in the long-term modulation of TH by increasing its activity through different ET receptors and intracellular signaling pathways, its mRNA, and the phosphorylation of the enzyme at serine 19, 31, and 40 sites (Nabhen et al., 2011). Short-term modulation of TH results in the enzyme activation, whereas long-term regulation results in increased synthesis (Nabhen et al., 2009, 2011). Enhanced TH activity results from its phosphorylation at diverse serine sites. Ser-40 is the most promiscuous site given that it can be phosphorylated by PKA, PKC, CaMKII, and PKG. We also showed that ET-1 and ET-3 regulate neuronal NE release in the anterior and posterior hypothalamus and the olfactory bulbs of normotensive rats (di Nunzio et al., 2002, 2004; Nabhen et al., 2009, 2011).

ET-1 and ET-3 also regulate noradrenergic neurotransmission in pathophysiological conditions like hypertension. Recent studies from our laboratory show that both ETs differentially regulate NE neurotransmission in brain regions (anterior and posterior hypothalamus, and olfactory bulb) of desoxycortosterone acetate (DOCA)-salt hypertensive rats (Abramoff, T. et al., unpublished data).

Both ETs are also involved in the modulation of NET as will be further discussed in the present review.

3.3.3 Other biological effects

ETs behave as paracrine and/or autocrine factors. Autocrine effects include endothelium proliferation and NO synthesis and release (Goto et al., 2013; Hirata et al., 1993; King et al., 1995; Tsukahara, Ende, Magazine, Bahou, & Goligorsky, 1994). DNA synthesis as well as the proliferation, migration, and invasion of endothelial cells are mediated by ET_B receptor activation (Herman & Simonson, 1995; MacCumber, Ross, & Snyder, 1990; Morbidelli, Orlando, Maggi, Ledda, & Ziche, 1995).

Paracrine actions include the contraction and proliferation of smooth muscle cells (Eddahibi et al., 2006; Horinouchi, Terada, Higashi, & Miwa; 2013; Hynynen & Khalil, 2006; King et al., 1995; Opgenorth, 1995). ET-1 increases the expression of several proto-oncogenes (c-myc, c-fos, and c-jun) in the vascular smooth muscle (Chen, Qiong, & Gardner, 2006; Herman & Simonson, 1995; Hynynen & Khalil, 2006). Radioautographic studies show the presence of binding sites for ETs in endocrine and neuroendocrine organs like the hypothalamus, pituitary, and the adrenal gland supporting ETs modulation of the endocrine function (Filosa et al., 2012; Hatae et al., 2007; Hynynen & Khalil, 2006; Kurokawa et al., 2000; Kuwaki et al., 1997; Sluck et al., 1999; Yamamoto & Uemura, 1998).

Other studies from our laboratory show that centrally applied ET-1 and ET-3 induce either cholestasis or choleresis depending on the dose (Rodríguez et al., 2005, 2006). However, when ETs are peripherally infused in doses that not induce changes in portal venous pressure and blood flow, they induce choleresis mediated by ET_B receptors coupled to NOS activation and vago-vagal reflexes (Rodríguez et al., 2013).

4. ET AND NE NEURONAL UPTAKE INTERACTION

4.1 General aspects

Neuropeptides are defined as small protein-like molecules produced and released by neurons through the regulated secretory route and acting on neural substrates (Belzung, Yalcin, Griebel, Surget, & Leman, 2006; Burbach, 2011). They are the most diverse class of signaling molecules in the brain and the peripheral nervous system engaged in many physiological functions. The neuropeptide families involved in the regulation of cardiovascular function include angiotensins (Ang) II and III, kallikreins, natriuretic peptides, and ETs (AbdAlla, Abdel-Baset, Lother, el Massiery, & Quitterer, 2005; Duchene & Ahluwalia, 2009; Ganten, Paul, & Lang, 1991; Kohan et al., 2011; Kuwaki et al., 1997; Szczepańska-Sadowska, 2006; Vatta, Bianciotti, Perfume, Nabhen, & Hope, 2009).

Neuropeptides, also called high-molecular-weight neurotransmitters, colocalize with classic neurotransmitters in neurons and synapsis regulating the efficiency of neuronal communication (de Lartigue, 2014; Merighi, 2002; Tasker, Oliet, Bains, Brown, & Stern, 2012).

Classic neurotransmitters are synthetized in the axon terminal, whereas neuropeptides are produced in the neuronal soma and packed in vesicles which travel by fast axonal transport down the axon (Hökfelt et al., 2000; Shakiryanova, Tully, Hewes, Deitcher, & Levitan, 2005). Neuropeptide synthesis is a dynamic process that easily adjusts to the activity of the neurons where they operate (Hökfelt, Bartfai, & Bloom, 2003; Hökfelt et al., 2000). However, it may undergo sustained adaptive changes in response to sustained environmental modifications (Barakat et al., 2006; Nostramo, Tillinger, Serova, Kvetnansky, & Sabban, 2013).

No membrane uptake system for neuropeptides was identified in neurons, so it is believed that they are inactivated by active peptidases. Neuropeptides remain in the synaptic cleft longer than classic neurotransmitters because their enzymatic degradation is slower thus allowing a more prolonged interaction with their receptors (Szczepańska-Sadowska, 2006).

Neuropeptides regulate not only neurons but also glial cells where their receptors are also expressed in those cell types (Filosa et al., 2012; Garrido-Gil, Rodríguez-Pallares, Dominguez-Meijide, Guerra, & Labandeira-Garcia, 2013; Hökfelt et al., 2003, 2000; Stern & Filosa, 2013). They regulate neuronal circuits involved in relevant biological functions like learning, memory, cognition and emotion, body temperature, thirst, appetite, metabolism, hypothalamic hormone release, as well as respiratory and cardiovascular activities (Bourque, Ciura, Trudel, Stachniak, & Sharif-Naeini, 2007; de Wied, Diamant, & Fodor, 1993; Dimicco & Zaretsky, 2007; Garcia-Segura, Lorenz, & DonCarlos, 2008; Guyenet, 2006; Pfaff, Kieffer, & Swanson, 2008; Plant, 2008; Wyss & Carlson, 2001).

Intense and sustained regulation by neuropeptides occur in diverse pathophysiological situations like addiction, cardiovascular diseases, energy disturbances, and sodium and water imbalances (Boutrel, 2008; Cottrell et al., 2009; de Lecea et al., 2006; Penna et al., 2006; Stepniakowski, Budzikowski, Loń, & Szczepańska-Sadowska, 1994; Szczepańska-Sadowska, Paczwa, Loń, & Ganten, 1998). These findings reveal the importance of studies aiming to understand the molecular mechanisms underlying neuropeptide regulation and the interaction with classical neurotransmitters.

4.2 Interaction with the endothelinergic system

Neuropeptides regulate diverse aspects of neurotransmission like the inactivation of neurotransmitters in the synaptic cleft. In the case of NE, its neuronal uptake is a complex mechanism highly regulated by neuropeptides.

The interaction between ETs and NE was first described by Backs, Bresch, Lutz, Kristen, and Haass (2005). These authors, with the aim to determine the underlying causes of sympathetic cardiac activation in congestive heart failure, showed that ET-1 through the activation of ET_A receptors inhibits in a concentration- and time-dependent fashion [^{3}H]-NE uptake in isolated perfused hearts. In the same study they showed that the ET_A antagonist, darusetan, reduces [^{3}H]-NE uptake and the number of NET-binding sites in rats with transverse aortic constriction (an animal model of congestive heart failure; Backs et al., 2005).

In 2008, the first reports showing an interaction between NE uptake and ETs in the brain were published. It was shown that ET-1 and ET-3 reduce NE uptake without affecting nonneuronal uptake in the posterior hypothalamus of normotensive rats (Hope et al., 2008). The internalization of NET is the underlying mechanism responsible for ETs' response. Conversely, in the anterior hypothalamus, ET-3 stimulates neuronal NE uptake by enhancing NET recycling to the plasma membrane without increasing the *de novo* synthesis of the transporter (Hope et al., 2008). In contrast, ET-1 elicits the same response as that observed in the posterior hypothalamus (Hope et al., 2008). Further studies in the anterior hypothalamus show that ET-1 reduces NE uptake through the activation of ET_B receptors coupled to the cAMP/PKA pathway, whereas ET-3 increases the uptake of the amine through an *atypical* ET GPCR coupled to the PLC/PKC/IP_3 and cAMP/PKA pathways without involving NO generation (Abramoff, T. et al., unpublished data).

On the other hand, in the posterior hypothalamus both ETs diminish neuronal NE uptake through different signaling pathways. ET-1 activates ET_B receptors, whereas ET-3 an *atypical* ET GPCR, although both receptors are coupled to neuronal NOS activation. In addition, the *atypical* ET GPCR is coupled to the phosphoinositide pathway (Hope, Nabhen, Soria, Bianciotti, & Vatta, 2010).

We recently reported that ETs also modulate NE uptake in the olfactory bulbs. Both ET-1 and ET-3 dose dependently diminish the amine uptake through the activation of *atypical* ET GPCRs coupled to PKC, PKG, and CaMKII activation (Abramoff, T. et al., unpublished data).

Increasing reports in the literature support the involvement of NET in different pathophysiological conditions. In the last years, our laboratory focused the studies on the underlying mechanisms of ETs–NE interaction in experimental hypertension. In the DOCA-salt animal model of hypertension, we observed that ET-1 and ET-3 diminish neuronal NE uptake in the anterior hypothalamus. These findings reveal a distinct ET-3 response in normotensive and hypertensive animals in this hypothalamic area. Another interesting finding was that ETs decrease both glycosylated (expressed in plasma membrane) and nonglycosylated NET (internalized transporter). Taken together, these findings support that ETs induce NET down-regulation in the anterior hypothalamus of DOCA-salt hypertensive rats (Abramoff, T. et al., unpublished data).

On the other hand, ET-1 and ET-3 increase neuronal NE uptake in the posterior hypothalamus and olfactory bulbs of hypertensive animals due to an increase in the glycosylated form of NET (Abramoff, T. et al., unpublished data).

The evidence presented support that ETs play a relevant role in the regulation of NET activity, kinetic, and internalization.

4.3 Interaction with other neuropeptides

The renin–angiotensin system, and particularly Ang II, has been extensively studied regarding NE uptake. Palaic and Khairallah (1967a, 1967b, 1968) were the first to show that Ang II modulates NE uptake in the brain and the heart. These studies were later confirmed and extended by other authors. Ang II stimulates NE uptake in the hypothalamus and brain stem neuronal cultures (Sumners & Raizada, 1986; Sumners, Shalit, Kalberg, & Raizada, 1987). Conversely, studies from our laboratory show that Ang II and Ang III inhibit neuronal NE uptake in diverse brain areas and in the adrenal medulla (Fernández et al., 1990; Papouchado, Vatta, Escalada, Bianciotti, & Fernández, 1995; Vatta, Bianciotti, Locatelli, Papouchado, & Fernández, 1992; Vatta, Bianciotti, Papouchado, Locatelli, & Fernández, 1991). In addition, it was recently reported that Ang 1–7 increases the uptake of NE in the brain (Lopez-Verrilli et al., 2012).

Bradykinin also regulates NE uptake, although few reports exist in the literature. It was shown that bradykinin increases NE release without inhibiting NE uptake (Seyedi, Win, Lander, & Levi, 1997). However, studies from our laboratory report that bradykinin enhances the uptake of the amine in the hypothalamus and medulla oblongata and in the adrenal medulla of normotensive rats (Fernández, Vatta, & Bianciotti, 1993; Vatta, Bianciotti, & Fernández, 1993a).

Other neuropeptides involved in the regulation of NE uptake are natriuretic peptides. This family is comprised by atrial natriuretic peptide (ANP), B-type natriuretic peptide (BNP), and C-type natriuretic peptide (CNP). Different studies show that ANP, BNP, and CNP increase NE uptake in the adrenal medulla and different areas and regions of the brain (Fermepín, Vatta, Bianciotti, Wolovich, & Fernández, 2000;

Fernández et al., 1990; Vatta, Bianciotti, & Fernández, 1993b; Vatta et al., 1991, 1997; Vatta, Presas, Bianciotti, Zarrabeitia, & Fernández, 1996; Vatta et al., 1995, 1994).

5. CONCLUSION

Since the discovery of ETs in the late 1980s, most studies centered on their involvements in the regulation of cardiovascular and renal activities. From all the studies on ETs, only 10% focused on the role of these peptides in the brain, and approximately 1% focused on the interaction between ETs and noradrenergic neurotransmission. Future studies will surely unveil further aspects on the molecular mechanisms underlying the interaction between the endothelinergic system and noradrenergic transmission, and in particular NE uptake by NET in the brain. The role of this interaction is relevant not only in physiological but also in pathophysiological conditions like cardiovascular and neurodegenerative diseases, and mood disorders as being evidenced by emergic reports in the literature. Current evidence suggests that the interaction between ETs and NE uptake may represent a potential therapeutic target for various diseases, particularly hypertension.

ACKNOWLEDGMENTS

The studies from our laboratory cited in this review were supported by grants from the Agencia Nacional de Promoción Científica y Tecnológica (ANPCyT), Consejo Nacional de Investigaciones Científicas y Técnicas (CONICET), and Universidad de Buenos Aires (UBACyT).

REFERENCES

AbdAlla, S., Abdel-Baset, A., Lother, H., el Massiery, A., & Quitterer, U. (2005). Mesangial AT1/B2 receptor heterodimers contribute to angiotensin II hyperresponsiveness in experimental hypertension. *Journal of Molecular Neuroscience*, *26*, 185–192.

Apparsundaram, S. (2011). Function and regulation of monoamine transporters: Focus on the norepinephrine transporter. *CNS Spectrums*, *6*(671–4), 677–678.

Apparsundaram, S., Galli, A., DeFelice, L. J., Hartzell, H. C., & Blakely, R. D. (1998). Acute regulation of norepinephrine transport: I. Protein kinase C-linked muscarinic receptors influence transport capacity and transporter density in SK-N-SH cells. *The Journal of Pharmacology and Experimental Therapeutics*, *287*, 733–743.

Apparsundaram, S., Moore, K. R., Malone, M. D., Hartzell, H. C., & Blakely, R. D. (1997). Molecular cloning and characterization of an L-epinephrine transporter from sympathetic ganglia of the bullfrog, Rana catesbiana. *The Journal of Neuroscience*, *17*, 2691–2702.

Arinami, T., Ishikawa, M., Inoue, A., Yanagisawa, M., Masaki, T., Yoshida, M. C., et al. (1991). Chromosomal assignments of the human endothelin family genes: The endothelin-1 gene (EDN1) to 6p23-p24, the endothelin-2 gene (EDN2) to 1p34, and the endothelin-3 gene (EDN3) to 20q13.2-q13.3. *American Journal of Human Genetics, 48*, 990–996.

Axelrod, J. (1957). *O*-methylation of (epinephrine and other catecholamines) catecholamines in vitro and in vivo. *Science, 126*, 400–401.

Axelrod, J. (1962). Purification and properties of phenylethanolamine-*N*-methyl transferase. *The Journal of Biological Chemistry, 237*, 1657–1660.

Backs, J., Bresch, E., Lutz, M., Kristen, A. V., & Haass, M. (2005). Endothelin-1 inhibits the neuronal norepinephrine transporter in hearts of male rats. *Cardiovascular Research, 67*, 283–290.

Barakat, Y., Pape, J. R., Boutahricht, M., El Ouezzani, S., Alaoui, A., Chaigniau, M., et al. (2006). Vasopressin-containing neurons of the hypothalamic parvocellular paraventricular nucleus of the jerboa: Plasticity related to immobilization stress. *Neuroendocrinology, 84*, 396–404.

Bauman, A. L., Apparsundaram, S., Ramamoorthy, S., Wadzinski, B. E., Vaughan, R. A., & Blakely, R. D. (2000). Cocaine and antidepressant-sensitive biogenic amine transporters exist in regulated complexes with protein phosphatase. *The Journal of Neuroscience, 20*, 7571–7578.

Belzung, C., Yalcin, I., Griebel, G., Surget, A., & Leman, S. (2006). Neuropeptides in psychiatric diseases: An overview with a particular focus on depression and anxiety disorders. *CNS & Neurological Disorders: Drug Targets, 5*, 135–145.

Bönish, H., & Brüss, M. (2006). The norepinephrine transporter in physiology and disease. *Handbook of Experimental Pharmacology, 175*, 485–524.

Bourque, C. W., Ciura, S., Trudel, E., Stachniak, T. J., & Sharif-Naeini, R. (2007). Neurophysiological characterization of mammalian osmosensitive neurones. *Experimental Physiology, 92*, 499–505.

Boutrel, B. (2008). A neuropeptide-centric view of psychostimulant addiction. *British Journal of Pharmacology, 154*, 343–357.

Brandao, F. (1976). A comparative study of the role played by some inactivation pathways in the disposition of the transmitter in the rabbit aorta and the saphenous vein of the dog. *Blood Vessels, 13*, 309–318.

Brüss, M., Kunz, J., Lingen, B., & Bönisch, H. (1993). Chromosomal mapping of the human gene for the tricyclic antidepressant sensitive noradrenaline transporter. *Human Genetics, 91*, 278–280.

Buck, K. J., & Amara, S. G. (1994). Chimeric dopamine–norepinephrine transporters delineate structural domains influencing selectivity for catecholamines and 1-methyl-4-phenyl-pyridinium. *Proceedings of the National Academy of Sciences of the United States of America, 91*, 12584–12588.

Burbach, J. P. (2011). What are neuropeptides? *Methods in Molecular Biology, 789*, 1–36.

Calvo, J. J., González, R., De Carvalho, L. F., Takahashi, K., Kanse, S. M., Hart, G. R., et al. (1990). Release of substance P from rat hypothalamus and pituitary by endothelin. *Endocrinology, 126*, 2288–2295.

Carboni, E., Tanda, G. L., Frau, R., & Di Chiara, G. (1990). Blockade of the noradrenaline carrier increases extracellular dopamine concentrations in the prefrontal cortex: Evidence that dopamine is taken up in vivo by noradrenergic terminals. *Journal of Neurochemistry, 55*, 1067–1070.

Chan, Y. F., Tang, F., & O, W. S. (2008). Adrenomedullin in the rat testis. II: Its production, actions on inhibin secretion, regulation by follicle-stimulating hormone, and its interaction with endothelin 1 in the Sertoli cell. *Biology of Reproduction, 78*, 780–785.

Chen, S., Qiong, Y., & Gardner, D. G. (2006). A role for p38 mitogen-activated protein kinase and c-myc in endothelin-dependent rat aortic smooth muscle cell proliferation. *Hypertension, 47*, 252–258.

Chen, Y. A., & Scheller, R. H. (2001). SNARE-mediated membrane fusion. *Nature Reviews. Molecular Cell Biology*, *2*, 98–106.

Chen, A. D., Xiong, X. Q., Gan, X. B., Zhang, F., Zhou, Y. B., Gao, X. Y., et al. (2012). Endothelin-1 in paraventricular nucleus modulates cardiac sympathetic afferent reflex and sympathetic activity in rats. *PLoS One*, *7*, e40748.

Clozel, M., Gray, G. A., Breu, V., Löfler, B. M., & Osterwalder, R. (1992). The endothelin ETB receptor mediates both vasodilation and vasoconstriction in vivo. *Biochemical and Biophysical Research Communications*, *186*, 867–873.

Cocks, T. M., Broughton, A., Dib, M., Sudnir, K., & Angus, J. A. (1989). Endothelin is blood vessel selective: Studies on a variety of human and dog vessels in vitro and on regional blood flow in the conscious rabbit. *Clinical and Experimental Pharmacology & Physiology*, *16*, 243–246.

Cottrell, E. C., Cripps, R. L., Duncan, J. S., Barrett, P., Mercer, J. G., Herwig, A., et al. (2009). Developmental changes in hypothalamic leptin receptor: Relationship with the postnatal leptin surge and energy balance neuropeptides in the postnatal rat. *American Journal of Physiology. Regulatory, Integrative and Comparative Physiology*, *296*, R631–R639.

D'Orleans-Juste, P., Labonté, J., Bkaily, G., Choufani, S., Plante, M., & Honoré, J. C. (2002). Function of the endothelin (B) receptor in cardiovascular physiology and pathophysiology. *Pharmacology & Therapeutics*, *95*, 221–238.

de Lartigue, G. (2014). Putative roles of neuropeptides in vagal afferent signaling. *Physiology & Behavior*, *136C*, 155–169, pii: S0031-9384(14)00145-0.

de Lecea, L., Jones, B. E., Boutrel, B., Borgland, S. L., Nishino, S., Bubser, M., et al. (2006). Addiction and arousal: Alternative roles of hypothalamic peptides. *The Journal of Neuroscience*, *26*, 10372–10375.

de Wied, D., Diamant, M., & Fodor, M. (1993). Central nervous system effects of the neurohypophyseal hormones and related peptides. *Frontiers in Neuroendocrinology*, *14*, 251–302.

Deacon, K., & Knox, A. J. (2010). Endothelin-1 (ET-1) increases the expression of remodeling genes in vascular smooth muscle through linked calcium and cAMP pathways: Role of a phospholipase A(2)(cPLA(2)/cyclooxygenase-2 (COX-2)/prostacyclin receptor-dependent autocrine loop. *The Journal of Biological Chemistry*, *285*, 25913–25927.

di Nunzio, A. S., Jaureguiberry, M. S., Rodano, V., Bianciotti, L. G., & Vatta, M. S. (2002). Endothelin-1 and -3 diminish neuronal NE release through an NO mechanism in rat anterior hypothalamus. *American Journal of Physiology, Regulatory, Integrative and Comparative Physiology*, *283*, R615–R622.

di Nunzio, A. S., Legaz, G., Rodano, V., Bianciotti, L. G., & Vatta, M. S. (2004). Modulatory effect of endothelin-1 and -3 on neuronal norepinephrine release in the rat posterior hypothalamus. *Regulatory Peptides*, *118*, 51–59.

Dimicco, J. A., & Zaretsky, D. V. (2007). The dorsomedial hypothalamus: A new player in thermoregulation. *American Journal of Physiology. Regulatory, Integrative and Comparative Physiology*, *292*, R47–R63.

Duchene, J., & Ahluwalia, A. (2009). The kinin B(1) receptor and inflammation: New therapeutic target for cardiovascular disease. *Current Opinion in Pharmacology*, *9*, 125–131.

Eddahibi, S., Guignabert, C., Barlier-Mur, A. M., Dewachter, L., Fadel, E., Dartevelle, P., et al. (2006). Cross talk between endothelial and smooth muscle cells in pulmonary hypertension: Critical role for serotonin-induced smooth muscle hyperplasia. *Circulation*, *113*, 1857–1864.

Eisenhofer, G. (2001). The role of neuronal and extraneuronal plasma membrane transporters in the inactivation of peripheral catecholamines. *Pharmacology & Therapeutics*, *91*, 35–62.

Eisenhofer, G., Kopin, I. J., & Goldstein, D. S. (2004). Catecholamine metabolism: A contemporary view with implications for physiology and medicine. *Pharmacological Reviews*, *56*, 331–349.

Emoto, N., & Yanagisawa, M. (1995). Endothelin-converting enzyme-2 is a membrane-bound, phosphoramidon-sensitive metalloprotease with acidic pH optimum. *The Journal of Biological Chemistry*, *270*, 15262–16268.

Esler, M., Alvarenga, M., Pier, C., Richards, J., El-Osta, A., Barton, D., et al. (2006). The neuronal noradrenaline transporter, anxiety and cardiovascular disease. *Journal of Psychopharmacology*, *20*(4 Suppl.), 60–66.

Esler, M., Jennings, G., Lambert, G., Meredith, I., Horne, M., & Eisenhofer, G. (1990). Overflow of catecholamine neurotransmitters to the circulation: Source, fate, and functions. *Physiological Reviews*, *70*, 963–985.

Evans, N. J., & Walker, J. W. (2008). Endothelin receptor dimers evaluated by FRET, ligand binding, and calcium mobilization. *Biophysical Journal*, *95*, 483–492.

Ferguson, A. V., & Smith, P. (1990). Cardiovascular responses induced by endothelin microinjection into area postrema. *Regulatory Peptides*, *27*, 75–85.

Fermepín, M. R., Vatta, M. S., Bianciotti, L. G., Wolovich, T. J., & Fernández, B. E. (2000). B-type and C-type natriuretic peptides modify norepinephrine uptake in discrete encephalic nuclei of the rat. *Cellular and Molecular Neurobiology*, *20*, 763–771.

Fernández, B. E., Dominguez, A. E., Vatta, M. S., Mendez, M. A., Bianciotti, L. G., & Martinez-Seeber, A. (1990). Atrial natriuretic peptide and angiotensin II interaction on noradrenaline uptake in the central nervous system. *Archives Internationales de Pharmacodynamie et de Thérapie*, *307*, 11–17.

Fernández, B. E., Vatta, M. S., & Bianciotti, L. G. (1993). Comparative effects of bradykinin and atrial natriuretic factor on neuronal and non-neuronal noradrenaline uptake in the central nervous system of the rat. *Archives Internationales de Physiologie et de Biochimie*, *101*, 337–340.

Filosa, J. A., Naskar, K., Perfume, G., Iddings, J. A., Biancardi, V. C., Vatta, M. S., et al. (2012). Endothelin-mediated calcium responses in supraoptic nucleus astrocytes influence magnocellular neurosecretory firing activity. *Journal of Neuroendocrinology*, *24*, 378–392.

Flatmark, T. (2000). Catecholamine biosynthesis and physiological regulation in neuroendocrine cells. *Acta Physiologica Scandinavica*, *168*, 1–17.

Friedgen, B., Wolfel, R., Russ, H., Schomig, E., & Graefe, K. H. (1996). The role of extraneuronal amine transport systems for the removal of extracellular catecholamines in the rabbit. *Naunyn-Schmiedeberg's Archives of Pharmacology*, *354*, 275–286.

Ganten, D., Paul, M., & Lang, R. E. (1991). The role of neuropeptides in cardiovascular regulation. *Cardiovascular Drugs and Therapy*, *5*, 119–130.

Garcia-Segura, L. M., Lorenz, B., & DonCarlos, L. L. (2008). The role of glia in the hypothalamus: Implications for gonadal steroid feedback and reproductive neuroendocrine output. *Reproduction*, *135*, 419–429.

Garrido-Gil, P., Rodríguez-Pallares, J., Dominguez-Meijide, A., Guerra, M. J., & Labandeira-Garcia, J. L. (2013). Brain angiotensin regulates iron homeostasis in dopaminergic neurons and microglial cells. *Experimental Neurology*, *250*, 384–396.

Gelernter, J., Kruger, S., Pakstis, A. J., Pacholczyk, T., Sparkes, R. S., Kidd, K. K., et al. (1993). Assignment of the norepinephrine transporter protein (NET1) locus to chromosome 16. *Genomics*, *18*, 690–692.

Giaid, A., Gibson, S. J., Ibraim, B. N., Legon, S., Bloom, S. R., Yanagisawa, M., et al. (1989). Endothelin 1, an endothelium-derived peptide, is expressed in neurons of the human spinal cord and dorsal root ganglia. *Proceedings of the National Academy of Sciences of the United States of America*, *86*, 7634–7638.

Giros, B., Wang, Y. M., Suter, S., McLeskey, S. B., Pifl, C., & Caron, M. G. (1994). Delineation of discrete domains for substrate, cocaine, and tricyclic antidepressant interactions using chimeric dopamine–norepinephrine transporters. *The Journal of Biological Chemistry*, *269*, 15985–15988.

Goto, A., Sumiyama, K., Kamioka, Y., Nakasyo, E., Ito, K., Iwasaki, M., et al. (2013). GDNF and endothelin 3 regulate migration of enteric neural crest-derived cells via protein kinase A and Rac1. *The Journal of Neuroscience*, *33*, 4901–4912.

Gouaux, E. (2009). The molecular logic of sodium-coupled neurotransmitter transporters. *Philosophical Transaction of the Royal Society of London Series B: Biological Sciences*, *364*, 149–154.

Gregan, B., Jürgensen, J., Papsdorf, G., Furkert, J., Schaefer, M., Beyermann, M., et al. (2004). Ligand-dependent differences in the internalization of endothelin A and endothelin B receptor heterodimers. *The Journal of Biological Chemistry*, *279*, 27679–27687.

Gregan, B., Schaefer, M., Rosenthal, W., & Oksche, A. (2004). Fluorescence resonance energy transfer analysis reveals the existence of endothelin-A and endothelin-B receptor homodimers. *Journal of Cardiovascular Pharmacology*, *44*, S30–S33.

Gulati, A., Rebello, S., Roy, S., & Saxena, P. R. (1995). *Journal of Cardiovascular Pharmacology*, *26*, S244–S246.

Guyenet, P. G. (2006). The sympathetic control of blood pressure. *Nature Reviews. Neuroscience*, *7*, 335–346.

Haag, C., Berkels, R., Gründemann, D., Lazar, A., Taubert, D., & Schömig, E. (2004). The localisation of the extraneuronal monoamine transporter (EMT) in rat brain. *Journal of Neurochemistry*, *88*, 291–297.

Hahn, M. K., & Blakely, R. D. (2002). Monoamine transporter gene structure and polymorphisms in relation to psychiatric disorders and other complex disorders. *Pharmacogenomics Journal*, *2*, 217–235.

Hahn, M. K., & Blakely, R. D. (2007). The functional impact of SLC6 transporter genetic variation. *Annual Review of Pharmacology and Toxicology*, *47*, 401–441.

Hall, J. E., Granger, J. P., do Carmo, J. M., da Silva, A. A., Dubinion, J., George, E., et al. (2012). Hypertension: Physiology and pathophysiology. *Comprehensive Physiology*, *2*, 2393–2442.

Harrison, V. J., Randiantsoa, A., & Schoffter, P. (1992). Heterogeneity of endothelin–sarafotoxin receptors mediating contraction of pig coronary artery. *British Journal of Pharmacology*, *105*, 511–513.

Hasegawa, H., Hiki, K., Sawamura, T., Aoyama, T., Okamoto, Y., Miwa, S., et al. (1998). Purification of a novel endothelin-converting enzyme specific for big endothelin-3. *FEBS Letters*, *428*, 304–308.

Hatae, N., Aksentijevich, N., Zemkova, H. W., Kretschmannova, K., Tomic, M., & Stojilkovic, S. S. (2007). Cloning and functional identification of novel endothelin receptor type A isoforms in pituitary. *Molecular Endocrinology*, *21*, 1192–1204.

Haynes, W. G., Ferro, C. E., & Webb, D. J. (1995). Physiologic role of endothelin in maintenance of vascular tone in humans. *Journal of Cardiovascular Pharmacology*, *3*, S183–S185.

Haynes, W. G., Strachan, F. E., & Webb, D. J. (1995). Endothelin ETA and ETB receptor cause vasoconstriction of human resistance and capacitance vessels in vivo. *Circulation*, *92*, 357–363.

Haynes, W. G., & Webb, D. J. (1998). Endothelin as a regulator of cardiovascular function in health and disease. *Journal of Hypertension*, *16*, 1081–1098.

Herman, W. H., & Simonson, M. S. (1995). Nuclear signaling by endothelin—1. A Ras pathway for activation of the c-fos serum response element. *The Journal of Biological Chemistry*, *270*, 11654–11661.

Hersch, E., Huang, J., Grider, J. R., & Murthy, K. S. (2004). Gq/G13 signaling by ET-1 in smooth muscle: MYPT1 phosphorylation via ETA and CPI-17 dephosphorylation via ETB. *American Journal of Physiology. Cell Physiology*, *287*, C1209–C1218.

Hickey, K. A., Rubanyi, G., Paul, R. J., & Highsmith, R. F. (1985). Characterization of a coronary vasoconstrictor produced by culture endothelial cells. *American Journal of Physiology. Cell Physiology*, *248*, C550–C556.

Hirata, Y., Emori, T., Eguchi, S., Kanno, K., Imai, T., Ohta, K., et al. (1993). Endothelin receptor subtype B mediates synthesis of nitric oxide by cultured bovine endothelial cells. *The Journal of Clinical Investigation*, *91*, 1367–1373.

Hökfelt, T., Bartfai, T., & Bloom, F. (2003). Neuropeptides: Opportunities for drug discovery. *The Lancet Neurology*, *2*, 463–472.

Hökfelt, T., Broberger, C., Xu, Z. Q., Sergeyev, V., Ubink, R., & Diez, M. (2000). Neuropeptides—An overview. *Neuropharmacology*, *39*, 1337–1356.

Hope, S. I., Nabhen, S. L., Soria, C., Bianciotti, L. G., & Vatta, M. S. (2010). Endothelin-1 and -3 modulate the neuronal norepinephrine transporter through multiple signalling pathways in the rat posterior hypothalamus. *Neurochemistry International*, *57*, 306–313.

Hope, S. I., Schmipp, J., Rossi, A. H., Bianciotti, L. G., & Vatta, M. S. (2008). Regulation of the neuronal norepinephrine transporter by endothelin-1 and -3 in the rat anterior and posterior hypothalamus. *Neurochemistry International*, *53*, 207–213.

Horinouchi, T., Terada, K., Higashi, T., & Miwa, S. (2013). Endothelin receptor signalling: New insight into its regulatory mechanisms. *Journal of Pharmacological Sciences*, *123*, 85–101.

Hughes, J. (1972). Evaluation of mechanisms controlling the release and inactivation of the adrenergic transmitter in the rabbit portal vein and vas deferens. *British Journal of Pharmacology*, *44*, 472–491.

Hynynen, M. M., & Khalil, R. A. (2006). The vascular endothelin system in hypertension—Recent patents and discoveries. *Recent Patents on Cardiovascular Drug Discovery*, *1*, 95–108.

Inoue, A., Yanagisawa, M., Kimura, S., Kasuya, Y., Miyauchi, T., Goto, K., et al. (1989). The human endothelin family: Three structurally and pharmacologically distinct isopeptides predicted by three separate genes. *Proceedings of the National Academy of Sciences of the United States of America*, *86*, 2863–2867.

Itoh, Y., Yanagisawa, M., Ohkubo, S., Kimura, C., Kosaka, T., Inoue, A., et al. (1988). Cloning and sequence analysis of cDNA encoding the precursor of a human endothelium-derived vasoconstrictor peptide, endothelin: Identity of human and porcine endothelin. *FEBS Letters*, *231*, 440–444.

Iversen, L. (2000). Neurotransmitter transporters: Fruitful targets for CNS drug discovery. *Molecular Psychiatry*, *5*, 357–362.

Iversen, L. (2006). Neurotransmitter transporters and their impact on the development of psychopharmacology. *British Journal of Pharmacology*, *147*, S82–S88.

Iversen, L. L., & Salt, P. J. (1970). Inhibition of catecholamine Uptake2 by steroids in the isolated rat heart. *British Journal of Pharmacology*, *40*, 528–530.

Ivey, M. E., Osman, N., & Little, P. J. (2008). Endothelin-1 signalling in vascular smooth muscle: Pathways controlling cellular functions associated with atherosclerosis. *Atherosclerosis*, *199*, 237–247.

Iwasa, H., Kurabayashi, M., Nagai, R., Nakamura, Y., & Tanaka, T. (2001). Genetic variations in five genes involved in the excitement of cardiomyocytes. *Journal of Human Genetics*, *46*, 549–552.

Jain, A., Olovsson, M., Burton, G. J., & Yung, H. W. (2012). Endothelin-1 induces endoplasmic reticulum stress by activating the PLC-IP(3) pathway: Implications for placental pathophysiology in preeclampsia. *American Journal of Pathology*, *180*, 2309–2320.

Jaureguiberry, M. S., di Nunzio, A. S., Dattilo, M. A., Bianciotti, L. G., & Vatta, M. S. (2004). Endothelin 1 and 3 enhance neuronal nitric oxide synthetase activity through ETB receptors involving multiple signaling pathways in the rat anterior hypothalamus. *Peptides*, *25*, 1133–1138.

Jayanthi, L. D., Annamalai, B., Samuvel, D. J., Gether, U., & Ramamoorthy, S. (2006). Phosphorylation of the norepinephrine transporter at threonine 258 and serine 259 is linked to protein kinase C-mediated transporter internalization. *The Journal of Biological Chemistry*, *281*, 23326–23340.

Jayanthi, L. D., Samuvel, D. L., & Ramamoorthy, S. (2004). Regulated internalization and phosphorylation of the native norepinephrine transporter in response to phorbol esters. Evidence for localization in lipids rafts and lipid raft-mediated internalization. *The Journal of Biological Chemistry*, *279*, 19315–19326.

Jena, B. P. (2011). Role of SNAREs in membrane fusion. *Advances in Experimental Medicine and Biology*, *713*, 13–32.

Kanyicska, B., Sellix, M. T., & Freeman, M. E. (2003). Autocrine regulation of prolactin secretion by endothelins throughout the estrous cycle. *Endocrine*, *20*, 53–58.

Karne, S., Jayawickreme, C. K., & Lerner, M. R. (1993). Cloning and characterization of an endothelin-3 specific receptor (ETC receptor) from *Xenopus laevis* dermal melanophores. *The Journal of Biological Chemistry*, *268*, 19126–19133.

Kasuya, Y., Kobayashi, H., & Uemura, H. (1991). Endothelin-like immunoreactivity in the nervous system of invertebrates and fish. *Journal of Cardiovascular Pharmacology*, *7*, S463–S466.

Kato, K., Okamura, K., Hatta, M., Morita, H., Kajioka, S., Naito, S., et al. (2013). Involvement of IP3-receptor activation in endothelin-1-induced Ca(2+) influx in rat pulmonary small artery. *European Journal of Pharmacology*, *720*, 255–263.

Kawanabe, Y., & Nauli, S. M. (2011). Endothelin. *Cellular and Molecular Life Sciences*, *68*, 195–203.

Khimji, A., & Rockey, D. C. (2010). Endothelin—Biology and disease. *Cellular Signalling*, *22*, 1615–1625.

Kimura, S., Kasuya, Y., Sawamura, T., Shinmi, O., Sugita, Y., Yanagisawa, M., et al. (1988). Structure–activity relationships of endothelin: Importance of the C-terminal moiety. *Biochemical and Biophysical Research Communications*, *156*, 1182–1186.

King, R. G., Gude, N. M., Di Iulio, J. L., & Brennecke, S. P. (1995). Regulation of human placental fetal vessel tone: Role of nitric oxide. *Reproduction, Fertility, and Development*, *7*, 1407–1411.

Kippenberger, A. G., Palmer, D. J., Comer, A. M., Lipski, J., Burton, L. D., & Christie, D. L. (1999). Localization of then noradrenaline transporter in rat adrenal medulla and PC12 cells: Evidence for its association with secretory granules in PC12 cells. *Journal of Neurochemistry*, *73*, 1024–1032.

Kitamura, K., Shirashi, N., Singer, W. D., Handlogten, M. E., Tomita, K., & Miller, R. T. (1999). Endothelin B receptors activate Galpha13. *American Journal of Physiology. Cell Physiology*, *276*, C930–C937.

Kloog, Y., Ambar, I., Sokolovsky, M., Kochva, E., Wollberg, Z., & Bdolah, A. (1988). Sarafotoxin, a novel vasoconstrictor peptide: Phosphoinositide hydrolysis in rat heart and brain. *Science*, *242*, 268–270.

Knuepfer, M. M., O'Brien, D., Hoang, D., Gan, Q., & Song, C. (1994). Central sympathetic control of spinal endothelin release in the rat. *European Journal of Pharmacology*, *259*, 305–308.

Kochva, E., Bdolah, A., & Wollberg, Z. (1993). Sarafotoxins and endothelins: Evolution, structure and function. *Toxicon*, *31*, 541–568.

Kohan, D. E., Rossi, N. F., Inscho, E. W., & Pollock, D. M. (2011). Regulation of blood pressure and salt homeostasis by endothelin. *Physiological Reviews*, *91*, 1–77.

Kopin, I. J. (1994). Monoamine oxidase and catecholamine metabolism. *Journal of Neural Transmission*, *41*, 57–67.

Koyama, Y. (2013). Endothelin systems in the brain: Involvement in pathophysiological responses of damaged nerve tissues. *Biomolecular Concepts*, *4*, 335–347.

Kristensen, A. S., Andersen, J., Jørgensen, T. N., Sørensen, L., Eriksen, J., Loland, C. J., et al. (2011). SLC6 neurotransmitter transporters: Structure, function, and regulation. *Pharmacological Reviews*, *63*, 585–640.

Kuddus, R. H., Nalesnik, M. A., Subbotin, V. M., Rao, A. S., & Gandhi, C. R. (2000). Enhanced synthesis and reduced metabolism of endothelin-1 (ET-1) by hepatocytes—An important mechanism of increased endogenous levels of ET-1 in liver cirrhosis. *Journal of Hepatology*, *33*, 725–732.

Kuhlmann, A., Amann, K., Schlotzer-Schrehardt, U., Kruse, F. E., & Cyrsiefen, C. (2005). Endothelin-1 and ETA/ETB receptor protein and mRNA: Expression in normal and vascularized human corneas. *Cornea*, *24*, 837–844.

Kumada, M., Cao, W., & Kuwaki, T. (2003). Effect of endothelin on vasomotor and respiratory neurons in the rostral ventrolateral medulla in rats. *Cellular and Molecular Neurobiology*, *23*, 691–707.

Kumar, C., Mwangi, V., Nuthulaganti, P., Wu, H. L., Pullen, M., Brun, K., et al. (1994). Cloning and characterization of a novel endothelin receptor from Xenopus heart. *The Journal of Biological Chemistry*, *269*, 13414–13420.

Kurokawa, K., Yamada, H., Liu, Y., & Kudo, M. (2000). Immunohistochemical distribution of the endothelin converting enzyme-1 in the rat hypothalamo–pituitary axis. *Neuroscience Letters*, *284*, 81–84.

Kurokawa, K., Yamada, H., & Ochi, J. (1997). Topographical distribution of neurons containing endothelin type A receptor in the rat brain. *The Journal of Comparative Neurology*, *389*, 348–360.

Kuwaki, T., Kurihara, H., Cao, W. H., Kurihara, Y., Unekawa, M., Yazaki, Y., et al. (1997). Physiological role of brain endothelin in the central autonomic control: From neuron to knockout mouse. *Progress in Neurobiology*, *51*, 545–579.

Kuwaki, T., Ling, G. Y., Onodera, M., Ishii, T., Nakamura, A., Ju, K. H., et al. (1999). *Clinical and Experimental Pharmacology & Physiology*, *26*, 989–994.

Kvetnansky, R., Sabban, E. L., & Palkovits, M. (2009). Catecholaminergic system in stress: Structural and molecular genetic approaches. *Physiological Reviews*, *89*, 535–606.

Liang, C. S. (2007). Cardiac sympathetic nerve terminal function in congestive heart failure. *Acta Pharmacologica Sinica*, *28*, 921–927.

Lopez-Verrilli, M. A., Rodríguez-Fermepín, M., Longo Carbajosa, N., Landa, S., Cerrato, B. D., García, S., et al. (2012). Angiotensin-(1–7) through Mas receptor up-regulates neuronal norepinephrine transporter via Akt and Erk1/2-dependent pathways. *Journal of Neurochemistry*, *120*, 46–55.

Lu, Y., Wang, L. G., Liao, Z., Tang, C. S., Wang, W. Z., & Yuan, W. J. (2007). Cardiovascular effects of centrally applied endothelin-1 1–31 and its relationship to endothelin-1 1–21 in rats. *Autonomic Neuroscience*, *133*, 146–152.

Macarthur, H., Warner, T. D., Wood, E. G., Corder, R., & Vane, J. R. (1994). Endothelin-1 release from endothelial cells in culture is elevated both acutely and chronically by short periods of mechanical stretch. *Biochemical and Biophysical Research Communications*, *200*(1), 395–400.

MacCumber, M. W., Ross, C. A., & Snyder, S. H. (1990). Endothelin in the brain: Receptors, mitogenesis, and biosynthesis in glial cells. *Proceedings of the National Academy of Sciences of the United States of America*, *87*, 2359–2363.

Matsumoto, H., Suzuki, N., Onda, H., & Fujino, M. (1989). Abundance of endothelin-3 in rat intestine, pituitary gland, and brain. *Biochemical and Biophysical Research Communications*, *164*, 74–80.

Matsumura, K., Abe, I., Fukuhara, M., Tominaga, M., Tsuchihashi, T., Kobayashi, K., et al. (1994). Naloxone augments sympathetic outflow induced by centrally administered endothelin in conscious rabbits. *American Journal of Physiology. Renal Physiology*, *266*, R1403–R1410.

Matthies, H. J. G., Han, Q., Shields, A., Wright, J., Moore, J. L., Danny, G., et al. (2009). Subcellular localization of the antidepressant sensitive norepinephrine transporter. *BMC Neuroscience*, *10*, 65.

Merighi, A. (2002). Costorage and coexistence of neuropeptides in the mammalian CNS. *Progress in Neurobiology*, *66*, 161–190.

Meyers, K. E., & Sethna, C. (2013). Endothelin antagonists in hypertension and kidney disease. *Pediatric Nephrology, 28*, 711–720.

Morbidelli, L., Orlando, C., Maggi, C. A., Ledda, F., & Ziche, M. (1995). Proliferation and migration of endothelial cells is promoted by endothelins via activation of ETB receptors. *American Journal of Physiology Heart and Circulatory Physiology, 269*, H686–H695.

Moreland, S., McMullen, D. M., Delaney, C. L., Lee, V. G., & Hunt, J. T. (1992). Venous smooth muscle contains vasoconstrictor ETB-like receptors. *Biochemical and Biophysical Research Communications, 184*, 100–106.

Morgazo, C., Perfume, G., Legaz, G., di Nunzio, A. S., Hope, S. I., Bianciotti, L. G., et al. (2005). Involvement of nitric oxide pathways in short term modulation of tyrosine hydroxylase activity by endothelins 1 and 3 in the rat anterior hypothalamus. *Biochemical and Biophysical Research Communications, 334*, 796–802.

Moron, J. A., Brockington, A., Wise, R. A., Rocha, B.-A., & Hope, B. T. (2002). Dopamine uptake through the norepinephrine transporter in brain regions with low levels of the dopamine transporter: Evidence from knock-out mouse lines. *The Journal of Neuroscience, 22*, 389–395.

Mosqueda-García, R., Inagami, T., Appalsamy, M., Sugiura, M., & Robertson, R. M. (1993). Endothelin as a neuropeptide. Cardiovascular effects in the brainstem of normotensive rats. *Circulation Research, 72*, 20–35.

Nabhen, S. L., Morales, V. P., Guil, M. J., Hocht, C., Bianciotti, L. G., & Vatta, M. S. (2011). Mechanisms involved in the long-term modulation of tyrosine hydroxylase by endothelins in the olfactory bulb of normotensive rats. *Neurochemistry International, 58*, 196–205.

Nabhen, S. L., Perfume, G., Battistone, M. A., Rossi, A., Abramoff, T., Bianciotti, L. G., et al. (2009). Short-term effects of endothelins on tyrosine hydroxylase activity and expression in the olfactory bulb of normotensive rats. *Neurochemical Research, 34*, 953–963.

Naicker, S., & Bhoola, K. D. (2001). Endothelins: Vasoactive modulators of renal function in health and disease. *Pharmacology & Therapeutics, 90*, 61–88.

Naidoo, V., Mahabeer, R., & Raidoo, D. M. (2001). Cellular distribution of endothelin-1 mRNA in human brain by in situ RT-PCR. *Metabolic Brain Disease, 16*, 207–218.

Naidoo, V., Naidoo, S., Mahabeer, R., & Raidoo, D. M. (2004). Cellular distribution of the endothelin system in the human brain. *Journal of Chemical Neuroanatomy, 27*, 87–98.

Nambi, P., Pullen, M., Kincad, J., Nuthulaganti, P., Aiyar, N., Brooks, D. P., et al. (1997). Identification and characterization of a novel endothelin receptor that binds both ETA- and ETB-selective ligands. *Molecular Pharmacology, 52*, 582–589.

Nasser, S. A., & El-Mas, M. M. (2014). Endothelin ETA receptor antagonism in cardiovascular disease. *European Journal of Pharmacology, 737*, 210–213.

Nishimura, M., Takahashi, H., Matsusawa, M., Ikegaki, I., Nakanishi, T., Hirabayashi, M., et al. (1990). Intracerebroventricular injections of endothelins increase arterial pressure in conscious rats. *Japanese Circulation Journal, 54*, 662–670.

Nishimura, M., Takahashi, H., Matsusawa, M., Ikegaki, I., Sakamoto, M., Nakanishi, T., et al. (1991). Chronic intracerebroventricular infusions of endothelin elevate arterial pressure in rats. *Journal of Hypertension, 9*, 71–76.

Nostramo, R., Tillinger, A., Serova, L., Kvetnansky, R., & Sabban, E. L. (2013). Bradykinin B2 receptor in the adrenal medulla of male rats and mice: Glucocorticoid-dependent increase with immobilization stress. *Endocrinology, 154*, 3729–3738.

Oaks, A. W., & Sidhu, A. (2011). Sinuclein modulation of monoamine transporters. *FEBS Letters, 585*, 1001–1006.

Ohkita, M., Tawa, M., Kitada, K., & Matsumura, Y. (2012). Pathophysiological roles of endothelin receptors in cardiovascular diseases. *Journal of Pharmacological Sciences, 119*, 302–313.

Olivier, B., Soudijn, W., & van Wijngaarden, I. (2000). Serotonin, dopamine and norepinephrine transporters in the central nervous system and their inhibitors. *Progress in Drug Research, 54*, 59–119.

Oparil, S., Chen, S. J., Meng, Q. C., Elton, T. S., Yano, M., & Chen, Y. F. (1995). Endothelin A receptor antagonist prevents acute hypoxia-induced pulmonary hypertension in the rat. *American Journal of Physiology. Lung Cellular and Molecular Physiology, 268*, L95–L100.

Opgenorth, T. J. (1995). Endothelin receptor antagonism. *Advances in Pharmacology, 33*, 1–65.

Opgenorth, T. J., Wu-Wong, J. R., & Shiosaki, K. (1992). Endothelin-converting enzymes. *FASEB Journal, 6*, 2653–2659.

Ouchi, Y., Kim, S., Souza, A. C., Iijima, S., Hattori, A., Orimo, H., et al. (1989). Central effect of endothelin on blood pressure in conscious rats. *American Journal of Physiology. Heart and Circulatory Physiology, 256*, H1747–H1751.

Palaic, D., & Khairallah, P. A. (1967a). Inhibition on noradrenaline uptake by angiotensin. *The Journal of Pharmacy and Pharmacology, 19*, 396–397.

Palaic, D., & Khairallah, P. A. (1967b). Effect of angiotensin on uptake and release of norepinephrine by brain. *Biochemical Pharmacology, 16*, 2291–2298.

Palaic, D., & Khairallah, P. A. (1968). Inhibition of norepinephrine re-uptake by angiotensin in brain. *Journal of Neurochemistry, 15*, 1195–1202.

Papouchado, M. L., Vatta, M. S., Escalada, A., Bianciotti, L. G., & Fernández, B. E. (1995). Angiotensin III modulates noradrenaline uptake and release in the rat hypothalamus. *Journal of Autonomic Pharmacology, 15*, 1–8.

Pate, M. A., Chester, A. H., Crabbe, D. S., Amrani, M., Brown, T. J., Roach, A. G., et al. (1999). Characterization of constrictor endothelin receptors in the human internal thoracic artery and saphenous vein. *Journal of Cardiovascular Pharmacology, 33*, 567–572.

Penna, C., Rastaldo, R., Mancardi, D., Cappello, S., Pagliaro, P., Westerhof, N., et al. (2006). Effect of endothelins on the cardiovascular system. *Journal of Cardiovascular Medicine, 7*, 645–652.

Perfume, G., Morgazo, C., Nabhen, S., Battistone, A., Hope, S. I., Bianciotti, L. G., et al. (2007). Short-term regulation of tyrosine hydroxylase activity and expression by endothelin-1 and endothelin-3 in the rat posterior hypothalamus. *Regulatory Peptides, 142*, 69–77.

Perfume, G., Nabhen, S. L., Barrera, K. R., Otero, M. G., Bianciotti, L. G., & Vatta, M. S. (2008). Long-term modulation of tyrosine hydroxylase activity and expression by endothelin-1 and -3 in the rat anterior and posterior hypothalamus. *American Journal of Physiology. Regulatory, Integrative and Comparative Physiology, 294*, R905–R914.

Pfaff, D. W., Kieffer, B. L., & Swanson, L. W. (2008). Mechanisms for the regulation of state changes in the central nervous system: An introduction. *Annals of the New York Academy of Sciences, 1129*, 1–7.

Plant, T. M. (2008). Hypothalamic control of the pituitary gonadal axis in higher primates: Key advances over the last two decades. *Journal of Neuroendocrinology, 20*, 719–726.

Pollock, D. M., Keith, T. L., & Highsmith, R. F. (1995). Endothelin receptors and calcium signaling. *FASEB Journal, 9*, 1196–1204.

Pörzgen, P., Bönisch, H., Hammermann, R., & Brüss, M. (1998). The human noradrenaline transporter gene contains multiple polyadenylation sites and two alternatively spliced C-terminal exons. *Biochimica et Biophysica Acta, 1398*, 365–370.

Pramod, A. B., Foster, J., Carvelli, L., & Henry, L. K. (2013). SLC6 transporters: Structure, function, regulation, disease association and therapeutics. *Molecular Aspects of Medicine, 34*, 197–219.

Quick, M. W. (2006). The role of SNARE proteins in trafficking and function of neurotransmitter transporters. *Handbook of Experimental Pharmacology, 175*, 181–196.

Ramamoorthy, S., Shippenberg, T. S., & Jayanthi, L. D. (2011). Regulation of monoamine transporter: Role of transporter phosphorylation. *Pharmacology & Therapeutics, 129*, 220–238.

Rizo, J., & Südhof, T. C. (2002). SNAREs and Munc18 in synaptic vesicle fusion. *Nature Reviews. Neuroscience, 3*, 641–653.
Robin, P., Chouayekh, S., Bole-Feysot, C., Leiber, D., & Tanfin, Z. (2005). Contribution of phospholipase D in endothelin-1-mediated extracellular signal-regulated kinase activation and proliferation in rat uterine leiomyoma cells. *Biology of Reproduction, 72*, 69–77.
Robinson, M. B. (2003). Signaling pathways take aim at neurotransmitter transporter. *Science's STKE, 2003*(207), pe50.
Rodríguez, M. R., Sabbatini, M. E., Santella, G., Dabas, P., Villagra, A., Vatta, M. S., et al. (2005). Endothelin-3 applied to the brain evokes opposite effects on bile secretion mediated by a central nitric oxide pathway. *Peptides, 26*, 1219–1227.
Rodríguez, M. R., Sabbatini, M. E., Santella, G., Vescina, C., Vatta, M. S., & Bianciotti, L. G. (2006). Vagally mediated cholestatic and choleretic effects of centrally applied Endothelin-1 through ETA receptors. *Regulatory Peptides, 135*, 54–62.
Rodríguez, M. R., Soria, L. R., Ventimiglia, M. S., Najenson, A. C., Di María, A., Dabas, P., et al. (2013). Endothelin-1 and -3 induce choleresis in the rat through ETB receptors coupled to nitric oxide and vago-vagal reflexes. *Clinical Science, 125*, 521–532.
Rubanyi, G. M., & Botelho, L. H. (1991). Endothelins. *FASEB Journal, 5*, 2713–2720.
Russell, F. D., & Davenport, A. P. (1999). Secretory pathways in endothelin synthesis. *British Journal of Pharmacology, 126*, 391–398.
Schiffrin, E. L. (1998). Endothelin and endothelin antagonists in hypertension. *Journal of Hypertension, 16*, 1891–1895.
Schlessingera, A., Geiera, E., Fana, H., Irwinb, J. J., Shoichet, B. K., Giacominia, K. M., et al. (2011). Structure-based discovery of prescription drugs that interact with the norepinephrine transporter, NET. *Proceedings of the National Academy of Sciences of the United States of America, 108*, 15810–15815.
Schneider, M. P., Boesen, E. I., & Pollock, D. M. (2007). Contrasting actions of endothelin ET(A) and ET(B) receptors in cardiovascular disease. *Annual Review of Pharmacology and Toxicology, 47*, 731–759.
Schroeder, C., & Jordan, J. (2012). Norepinephrine function and human cardiovascular disease. *American Journal of Physiology. Heart and Circulatory Physiology, 303*, H1273–H1282.
Schroeter, S., Apparsundaram, S., Wiley, R. G., Miner, L. H., Sesack, S. R., & Blakely, R. D. (2000). Immunolocalization of the cocaine- and antidepressant-sensitive l-norepinephrine transporter. *The Journal of Comparative Neurology, 420*, 211–232.
Seyedi, N., Win, T., Lander, H. M., & Levi, R. (1997). Bradykinin B2-receptor activation augments norepinephrine exocytosis from cardiac sympathetic nerve endings. Mediation by autocrine/paracrine mechanisms. *Circulation Research, 81*, 774–784.
Shakiryanova, D., Tully, A., Hewes, R. S., Deitcher, D. L., & Levitan, E. S. (2005). Activity-dependent liberation of synaptic neuropeptide vesicles. *Nature in Neuroscience, 8*, 173–178.
Shannon, J. R., Flattem, N. L., Jordan, J., Jacob, G., Black, B. K., Biaggioni, I., et al. (2000). Orthostatic intolerance and tachycardia associated with norepinephrine-transporter deficiency. *The New England Journal of Medicine, 342*, 541–549.
Sheng, Z. H., Rettig, J., Cook, T., & Catterall, W. A. (1996). Calcium-dependent interaction of N-type calcium channels with the synaptic core complex. *Nature, 376*, 451–454.
Shraga-Levine, Z., & Sokolovsky, M. (2000). Functional coupling of G proteins to endothelin receptors is ligand and receptor subtype specific. *Cellular and Molecular Neurobiology, 20*, 305–317.
Sirén, A. L., & Feuerstein, G. (1989). Hemodynamic effects of endothelin after systemic and central nervous system administration in the conscious rat. *Neuropeptides, 14*, 231–236.

Sluck, J. M., Lin, R. C., Katolik, L. I., Jeng, A. Y., & Lehmann, J. C. (1999). Endothelin converting enzyme-1-, endothelin-1-, and endothelin-3-like immunoreactivity in the rat brain. *Neuroscience*, *91*, 1483–1497.

Sogawa, C., Kumagai, K., Sogawa, N., Morita, K., Dohi, T., & Kitayama, S. (2007). C-terminal region regulates the functional expression of human noradrenaline transporter splice variants. *The Biochemical Journal*, *401*, 185–195.

Sokolovsky, M. (1995a). Endothelin receptor subtypes and their role in transmembrane signaling mechanisms. *Pharmacology & Therapeutics*, *68*, 435–471.

Sokolovsky, M. (1995b). Endothelin receptor heterogeneity, G-proteins, and signalling via cAMP and cGMP cascades. *Cellular and Molecular Neurobiology*, *15*, 561–571.

Stepniakowski, K., Budzikowski, A., Loń, S., & Szczepańska-Sadowska, E. (1994). Central cardiovascular effects of AVP and ANP in normotensive and spontaneously hypertensive rats. *Journal of the Autonomic Nervous System*, *47*, 33–43.

Stern, J. E., & Filosa, J. A. (2013). Bidirectional neuro-glial signaling modalities in the hypothalamus: Role in neurohumoral regulation. *Autonomic Neuroscience*, *175*, 51–60.

Stöber, G., Nöthen, M. M., Pörzgen, P., Brüsss, M., Knapp, M., Beckmann, H., et al. (1996). Systematic search for variation in the human norepinephrine transporter gene: Identification of five naturally occurring missense mutations and study of association with major psychiatric disorders. *American Journal of Medical Genetics*, *67*(6), 523–532.

Stojilkovic, S. S., & Catt, K. J. (1996). Expression and signal transduction pathways of endothelin receptors in neuroendocrine cells. *Frontiers in Neuroendocrinology*, *17*, 327–369.

Sumners, C., & Raizada, M. K. (1986). Angiotensin II stimulates norepinephrine uptake in hypothalamus–brain stem neuronal cultures. *American Journal of Physiology. Cell Physiology*, *250*, C236–C244.

Sumners, C., Shalit, S. L., Kalberg, C. J., & Raizada, M. K. (1987). Norepinephrine metabolism in neuronal cultures is increased by angiotensin II. *American Journal of Physiology. Cell Physiology*, *252*, C650–C656.

Sung, U., Apparsundaram, S., Galli, A., Kahling, K. M., Savchenko, V., Quick, M. W., et al. (2003). A regulated interaction syntaxin 1A with antidepressant-sensitive norepinephrine transporter establishes catecholamine clearance capacity. *The Journal of Neuroscience*, *23*, 1697–1709.

Sung, U., & Blakely, R. D. (2007). Calcium-dependent interactions of the human norepinephrine transporter with syntaxin 1A. *Molecular and Cellular Neurosciences*, *34*, 251–260.

Szczepańska-Sadowska, E. (2006). Neuropeptides in neurogenic disorders of the cardiovascular control. *Journal of Physiology and Pharmacology*, *57*(Suppl. 11), 31–53.

Szczepańska-Sadowska, E., Paczwa, P., Loń, S., & Ganten, D. (1998). Increased pressor function of central vasopressinergic system in hypertensive renin transgenic rats. *Journal of Hypertension*, *16*, 1505–1514.

Takizawa, S., Uchide, T., Adur, J., Kozakai, T., Kotake-Nara, E., Quan, J., et al. (2005). Differential expression of endothelin-2 along the mouse intestinal tract. *Journal of Molecular Endocrinology*, *35*, 201–209.

Tasker, J. G., Oliet, S. H., Bains, J. S., Brown, C. H., & Stern, J. E. (2012). Glial regulation of neuronal function: From synapse to systems physiology. *Journal of Neuroendocrinology*, *24*, 566–576.

Tellioglu, T., & Robertson, D. (2001). Genetic or acquired deficits in the norepinephrine transporter: Current understanding of clinical implications. *Expert Reviews in Molecular Medicine*, *2001*, 1–10.

ThanThan, S., Mekaru, C., Seki, N., Hidaka, K., Ueno, A., ThidarMyint, H., et al. (2010). Endogenous ghrelin released in response to endothelin stimulates growth hormone secretion in cattle. *Domestic Animal Endocrinology*, *38*, 1–12.

Torres, G., & Anara, S. G. (2007). Glutamate and monoamine transporters: New visions of form and function. *Current Opinion in Neurobiology*, *17*, 304–312.
Trendelenburg, U. (1988). The extraneuronal uptake and metabolism of catecholamines. In U. Trendelenburg (Ed.), *Handbook of experimental pharmacology* (pp. 279–319). Berlin: Springer-Verlag.
Tsukahara, H., Ende, H., Magazine, H. I., Bahou, W. F., & Goligorsky, M. S. (1994). Molecular and functional characterization of the non-isopeptide-selective ETB receptor in endothelial cells. Receptor coupling to nitric oxide synthase. *The Journal of Biological Chemistry*, *269*, 21778–21785.
Uchide, T., Adur, J., Fukamachi, T., & Saida, K. (2000). Quantitative analysis of endothelin-1 and vasoactive intestinal contractor/endothelin-2 gene expression in rats by real-time reverse transcriptase polymerase chain reaction. *Journal of Cardiovascular Pharmacology*, *36*, S5–S8.
Uchida, J., Kiuchi, Y., Ohno, M., Yura, A., & Oguchi, K. (1998). Ca(2+)-dependent enhancement of [3H]noradrenaline uptake in PC12 cells through calmodulin-dependent kinases. *Brain Research*, *809*, 155–164.
Vatta, M. S., Bianciotti, L. G., & Fernández, B. E. (1993a). Atrial natriuretic peptide and bradykinin interaction on norepinephrine uptake in rat adrenal medulla. *Archives Internationales de Physiologie et de Biochimie*, *101*, 129–132.
Vatta, M. S., Bianciotti, L. G., & Fernández, B. E. (1993b). Influence of atrial natriuretic factor on uptake, intracellular distribution, and release of norepinephrine in rat adrenal medulla. *Canadian Journal of Physiology and Pharmacology*, *71*, 195–200.
Vatta, M. S., Bianciotti, L. G., Locatelli, A. S., Papouchado, M. L., & Fernández, B. E. (1992). Monophasic and biphasic effects of angiotensin II and III on norepinephrine uptake and release in rat adrenal medulla. *Canadian Journal of Physiology and Pharmacology*, *70*, 821–825.
Vatta, M. S., Bianciotti, L. G., Papouchado, M. L., Locatelli, A. S., & Fernández, B. E. (1991). Effects of atrial natriuretic peptide and angiotensin III on the uptake and intracellular distribution of norepinephrine in medulla oblongata of the rat. *Comparative Biochemistry and Physiology Part C: Comparative Pharmacology and Toxicology*, *99*, 293–297.
Vatta, M. S., Bianciotti, L. G., Perfume, G., Nabhen, S. L., & Hope, S. I. (2009). Endothelins: A family of peptides with multiple biological functions. In G. Rodriguez de Lores Arnaiz (Ed.), *Function of Neuropepties at Central Nervus System* (pp. 149–169). Kerala: Research Signpost.
Vatta, M. S., Presas, M. F., Bianciotti, L. G., Rodríguez-Fermepín, M., Ambros, R., & Fernandez, B. E. (1997). B and C types natriuretic peptides modify norepinephrine uptake and release in the rat adrenal medulla. *Peptides*, *18*, 1483–1489.
Vatta, M. S., Presas, M., Bianciotti, L. G., Zarrabeitia, V., & Fernández, B. E. (1996). B and C natriuretic peptides modulate norepinephrine uptake and release in the rat hypothalamus. *Regulatory Peptides*, *65*, 175–184.
Vatta, M., Rodríguez-Fermepín, M., Bianciotti, L., Perazzo, J., Monserrat, A., & Fernández, B. (1995). Atrial natriuretic factor enhances norepinephrine uptake in circumventricular organs, locus coeruleus and nucleus tractus solitarii of the rat. *Neuroscience Letters*, *197*, 29–32.
Vatta, M., Travaglianti, M., Bianciotti, L., Coll, C., Perazzo, J., & Fernández, B. (1994). Atrial natriuretic factor effects on norepinephrine uptake in discrete telencephalic and diencephalic nuclei of the rat. *Brain Research*, *646*, 324–326.
Watts, S. W. (2010). Endothelin receptors: What's new and what do we need to know? *American Journal of Physiology. Regulatory, Integrative and Comparative Physiology*, *298*, R254–R260.
Webber, K. M., Pennefather, J. M., Head, G. A., & van den Buuse, M. (1998). Endothelin induces dopamine release from rat striatum via endothelin-B receptors. *Neuroscience*, *86*, 1173–1180.
Wehrwein, E. A., Parker, L. M., Wright, A. A., Spitsbergen, J. M., Novotny, M., Babankova, D., et al. (2008). Cardiac norepinephrine transporter protein expression is

inversely correlated to chamber norepinephrine content. *American Journal of Physiology. Regulatory, Integrative and Comparative Physiology*, *295*, R857–R863.
Wersinger, C., Jeannotte, A., & Sidhu, A. (2006). Attenuation of the norepinephrine transporter activity and trafficking via interactions with alpha-synuclein. *The European Journal of Neuroscience*, *24*, 3141–3152.
Whiskey, E., & Taylor, D. (2013). A review of the adverse effects and safety of noradrenergic antidepressants. *Journal of Psychopharmacology*, *27*, 732–739.
Winquist, R. J., Bunting, P. B., Garsky, V. M., Lumma, P. K., & Schofield, T. L. (1989). Prominent depressor response to endothelin in spontaneously hypertensive rats. *European Journal of Pharmacology*, *163*, 199–203.
Wyss, J. M., & Carlson, S. H. (2001). The role of the nervous system in hypertension. *Current Hypertension Reports*, *3*, 255–262.
Xu, F., Gainetdinov, R. R., Wetsel, W. C., Jones, S. R., Bohn, L. M., Miller, G. W., et al. (2000). Mice lacking the norepinephrine transporter are supersensitive to psychostimulants. *Nature in Neuroscience*, *3*, 465–471.
Yamada, H., & Kurokawa, K. (1998). Histochemical studies on endothelin and the endothelin-A receptor in the hypothalamus. *Journal of Cardiovascular Pharmacology*, *31*, S215–S218.
Yamamoto, T., Kimura, T., Ota, K., Shoji, M., Inoue, M., Sato, K., et al. (1991). Central effects of endothelin-1 on vasopressin and atrial natriuretic peptide release and cardiovascular and renal function in conscious rats. *Journal of Cardiovascular Pharmacology*, *7*, S316–S318.
Yamamoto, T., & Uemura, H. (1998). Distribution of endothelin-B receptor like immunoreactivity in rat brain, kidney, and pancreas. *Journal of Cardiovascular Pharmacology*, *31*, S207–S2011.
Yamashita, A., Singh, S. K., Kaeate, T., Jin, Y., & Gouaux, E. (2005). Crystal structure of a bacterial homologue of Na^+/Cl^--dependent neurotransmitter transporters. *Nature*, *437*, 215–223.
Yanagisawa, M., Inoue, A., Ishikawa, T., Kasuya, Y., Kimura, S., Kumaagaye, S., et al. (1988). Primary structure, synthesis, and biological activity of rat endothelin, an endothelium-derived vasoconstrictor peptide. *Proceedings of the National Academy of Sciences of the United States of America*, *85*, 6964–6967.
Yanagisawa, M., Kurihara, H., Kimura, S., Tomobe, Y., Kobayashi, M., Mitsui, Y., et al. (1988). A novel potent vasoconstrictor peptide produced by vascular endothelial cells. *Nature*, *332*, 411–415.
Yanagisawa, M., & Masaki, T. (1989). Endothelin, a novel endothelium-derived peptide. Pharmacological activities, regulation and possible role in cardiovascular control. *Biochemical Pharmacology*, *38*, 1877–1883.
Yoshitomi, Y., Kojima, S., Ogi, M., & Kuramochi, M. (1998). Acute renal failure in accidental hypothermia of cold water immersion. *American Journal of Kidney Diseases*, *31*, 856–859.
Youdim, M. B., & Riederer, P. (1988). Monoamine oxidase B; A misunderstood and misjudged enzyme. *Pharmacological Research Communications*, *20*(Suppl. 4), 9–14.
Yu, C., Yang, Z., Ren, H., Zhang, Y., Han, Y., He, D., et al. (2009). D3 dopamine receptor regulation of ETB receptors in renal proximal tubule cells from WKY and SHRs. *American Journal of Hypertension*, *22*, 877–883.
Zahniser, N. R., & Doolen, S. (2001). Chronic and acute regulation of Na^+/Cl^--dependent neurotransmitter transporters: Drugs, substrates, presynaptic receptors, and signaling systems. *Pharmacology & Therapeutics*, *92*, 21–55.
Zeng, C., Hopfer, U., Asico, L. D., Eisner, G. M., Felder, R. A., & Jose, P. A. (2005). Altered AT1 receptor regulation of ETB receptors in renal proximal tubule cells of spontaneously hypertensive rats. *Hypertension*, *46*, 926–931.
Zhan, S., & Rockey, D. C. (2011). Tumor necrosis factor α stimulates endothelin-1 synthesis in rat hepatic stellate cells in hepatic wound healing through a novel IKK/JNK pathway. *Experimental Cell Research*, *317*, 1040–1048.

CHAPTER THIRTEEN

Vitamin D-Enhanced Duodenal Calcium Transport

Kannikar Wongdee[*,†], **Narattaphol Charoenphandhu**[†,‡,1]

[*]Office of Academic Management, Faculty of Allied Health Sciences, Burapha University, Chonburi, Thailand
[†]Center of Calcium and Bone Research (COCAB), Faculty of Science, Mahidol University, Bangkok, Thailand
[‡]Department of Physiology, Faculty of Science, Mahidol University, Bangkok, Thailand
[1]Corresponding author: e-mail address: naratt@narattsys.com

Contents

Abstract

For humans and rodents, duodenum is a very important site of calcium absorption since it is exposed to ionized calcium released from dietary complexes by gastric acid. Calcium traverses the duodenal epithelium via both transcellular and paracellular pathways in a vitamin D-dependent manner. After binding to the nuclear vitamin D receptor, 1,25-dihydroxyvitamin D_3 [$1{,}25(OH)_2D_3$] upregulates the expression of several calcium transporter genes, e.g., TRPV5/6, calbindin-D_{9k}, plasma membrane Ca^{2+}-$ATPase_{1b}$, and NCX1, thereby enhancing the transcellular calcium transport. This action has been reported to be under the regulation of parathyroid–kidney–intestinal and bone–kidney–intestinal

Vitamins and Hormones, Volume 98
ISSN 0083-6729
http://dx.doi.org/10.1016/bs.vh.2014.12.010

axes, in which the plasma calcium and fibroblast growth factor-23 act as negative feedback regulators, respectively. $1{,}25(OH)_2D_3$ also modulates the expression of tight junction-related genes and convective water flow, presumably to increase the paracellular calcium permeability and solvent drag-induced calcium transport. However, vitamin D-independent calcium absorption does exist and plays an important role in calcium homeostasis under certain conditions, particularly in neonatal period, pregnancy, and lactation as well as in naturally vitamin D-impoverished subterranean mammals.

1. INTRODUCTION

Calcium is fundamental to a number of cellular and organ functions, such as intracellular signal transduction, neurotransmitter release, muscle contraction, cardiac contractility, and bone metabolism. In mammals, ~99% of total body calcium is stored in bone mainly as hydroxyapatite nanocrystals and to a smaller extent as amorphous calcium phosphate and free-ionized calcium in bone extracellular fluid. Extraosseous calcium (~1% of total body calcium) exists in three major forms, i.e., albumin-bound form, free-ionized calcium, and calcium complexes with phosphate or sulfate. A drastic change in the plasma calcium concentration, either a decrease (hypocalcemia) or increase (hypercalcemia), is lethal; therefore, calcium level is tightly regulated by several hormones, which are three classical calcium-regulating hormones, namely parathyroid hormone (PTH), 1,25-dihydroxyvitamin D_3 [$1{,}25(OH)_2D_3$], and calcitonin, as well as some other endocrine or paracrine factors [e.g., estrogen, prolactin, insulin-like growth factor (IGF)-1, and fibroblast growth factor (FGF)-23]. Besides plasma calcium level, body calcium metabolism as a whole is also controlled by these calcium-regulating hormones to assure that dietary calcium is adequately absorbed into the body, properly stored in bone, while the excess is excreted via the kidney. Under certain conditions, for instance pregnancy and lactation, a large amount of calcium is utilized for fetal growth and milk production, respectively; therefore, the intestinal calcium absorption is markedly enhanced to match a high-calcium demand (Charoenphandhu et al., 2009).

Both humans and rodents respond to $1{,}25(OH)_2D_3$ treatments by raising calcium absorption by 200–300% (Fleet, Eksir, Hance, & Wood, 2002; Kutuzova et al., 2008). However, mild-to-moderate vitamin D insufficiency rarely diminishes intestinal calcium absorption since PTH effectively induces conversion of 25-hydroxyvitamin D_3 [$25(OH)D_3$] into

$1,25(OH)_2D_3$ for stimulation of calcium absorption (Need & Nordin, 2008). It is noteworthy that simple vitamin D_3 supplement [not $1,25(OH)_2D_3$ supplement] cannot increase calcium absorption in either vitamin D-deficient or rachitic volunteers (Gallagher, Jindal, & Smith, 2014; Thacher & Abrams, 2010). Dietary calcium traverses the intestinal epithelium by two major pathways, i.e., transcellular and paracellular pathways in a $1,25(OH)_2D_3$-dependent manner. Both humans and rodents respond to $1,25(OH)_2D_3$ treatments by raising calcium absorption by 200–300% (Fleet et al., 2002; Kutuzova et al., 2008). In the transcellular pathway, free-ionized calcium diffuses across the apical plasma membrane, cytoplasm, and then the basolateral membrane before entering the interstitial fluid and blood circulation. Luminal calcium also moves paracellularly across the tight junction into the lateral intercellular space—also known as paracellular space—before entering the interstitium.

In humans and rodents, although both calcium transport mechanisms take place in all segments of the small intestine, the transcellular calcium transport is predominant in the proximal part, particularly the duodenum, and is of importance during low-calcium intake (Armbrecht, Zenser, Gross, & Davis, 1980). Regular diet without dairy products is generally considered low normal calcium diet, which necessitates the presence of an active uphill transport mechanism. Paracellular calcium transport, on the other hand, is present along the entire length of the small intestine. However, it is believed that a considerable amount of lumen-to-plasma paracellular calcium flux also occurs in the duodenum since it is exposed to high luminal concentration of free-ionized calcium, which is liberated from insoluble complexes by gastric acidic environment (Kopic & Geibel, 2013), especially after high-calcium meal or oral calcium supplementation. Our investigation has provided evidence that the paracellular calcium flux in the rat duodenum is much greater than the transcellular calcium flux (up to 10:1 when luminal calcium >5 mmol/L) (Charoenphandhu, Tudpor, Pulsook, & Krishnamra, 2006).

2. SOURCES OF $1,25(OH)_2D_3$ FOR STIMULATION OF DUODENAL CALCIUM TRANSPORT

There are two possible sources of $1,25(OH)_2D_3$, i.e., from plasma and *de novo* synthesis in the intestinal cells, the latter of which is also known as intracrine $1,25(OH)_2D_3$ (Balesaria, Sangha, & Walters, 2009; Kopic & Geibel, 2013). Plasma $1,25(OH)_2D_3$ is predominantly synthesized in the renal proximal tubular cells by 25(OH)D_3 1α-hydroxylase (CYP27B1) (Kopic & Geibel, 2013). Regarding the intracrine 1,25$(OH)_2D_3$ synthesis, the duodenal absorptive cells have been shown to express 1α-hydroxylase that is capable of converting 25(OH)D_3 to 1,25$(OH)_2D_3$, which later diffuses from the cytoplasm to activate nuclear vitamin D receptor (VDR) (Balesaria et al., 2009). The *de novo* synthesis of 1α-hydroxylase in human duodenal tissue is upregulated after 25(OH)D_3 exposure. It is noted that an increase in 1α-hydroxylase transcripts shows a positive correlation with the expression of transient receptor potential vanilloid calcium channel (TRPV)-6 (Balesaria et al., 2009). It is, therefore, possible that human duodenal absorptive cells can convert 25(OH)D_3 into 1,25$(OH)_2D_3$, which may subsequently upregulate calcium absorption through TRPV6 (Balesaria et al., 2009; Cui et al., 2009; Koszewski, Horst, & Goff, 2012). Finally, the action of 1,25$(OH)_2D_3$ is terminated by 24-hydroxylase (CYP24), which is also under the regulation of 1,25$(OH)_2D_3$. Kutuzova and DeLuca (2004) reported that, in vitamin D-deficient rats, a single-dose intrajugular injection of 1,25$(OH)_2D_3$ (730 ng/kg body weight) markedly upregulated the duodenal expression of CYP24 transcript within 3 h.

Alternatively, circulating 1,25$(OH)_2D_3$ can exert its biological actions by binding to either plasma membrane receptor 1,25D_3-MARRS (membrane-associated, rapid response steroid-binding) protein or VDR. Activation of 1,25D_3-MARRS is responsible for the 1,25$(OH)_2D_3$-enhanced rapid calcium absorption across the duodenal epithelium (nongenomic action, minute-to-minute regulation) (Nemere, Garbi, Hämmerling, & Khanal, 2010; Nemere & Norman, 1990). However, the underlying cellular mechanism and intracellular signaling of 1,25D_3-MARRS are not completely understood. In contrast, VDR is a nuclear receptor that forms a heterodimer with retinoid X receptor (RXR). Thereafter, the 1,25$(OH)_2D_3$–VDR–RXR complex—now being a transcription factor—binds to the vitamin D-responsive elements (VDREs) to regulate transcription of vitamin D-sensitive genes, including *Trpv6* and *Cyp24a1* (genomic action, long-term regulation) (Meyer, Zella, Nerenz, & Pike, 2007).

3. VITAMIN D-ENHANCED TRANSCELLULAR CALCIUM TRANSPORT

Since ionized calcium cannot freely move across the lipid bilayer of the plasma membrane, the polarized duodenal absorptive cells abundantly express a number of calcium transporters in both apical and basolateral plasma membranes as well as in the cytoplasm, most of which are under a tight regulation of $1{,}25(OH)_2D_3$ and VDR signaling (Fig. 1). Thus, the intestinal calcium absorption is dramatically decreased in intestine-specific VDR knockout mice as compared to normal mice (Lieben et al., 2012). The vitamin D-enhanced transcellular calcium transport is an active transport process, which requires cellular energy in the form of ATP. This transport mechanism is also sex and age dependent. For instance, female mice have higher calcium absorption and expression of calcium transporters than male mice, and a single $1{,}25(OH)_2D_3$ injection in female mice induces a

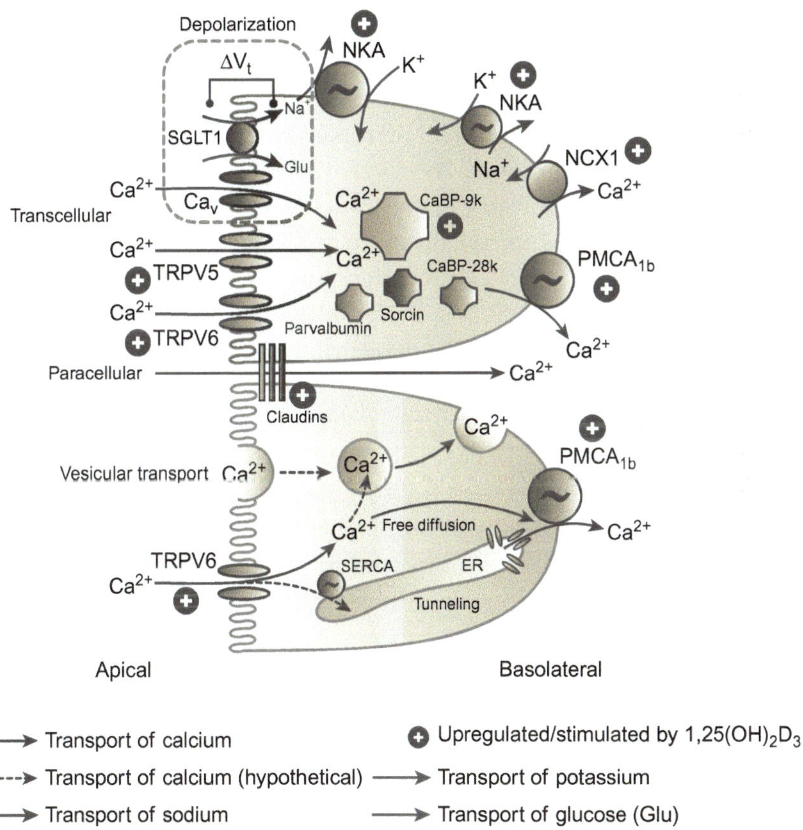

Figure 1 A schematic diagram shows possible mechanisms of transcellular and paracellular calcium transport in the rat duodenum. $1{,}25(OH)_2D_3$ is capable of stimulating both paracellular and transcellular calcium transport [TRPV5/6-mediated calcium entry, calcium-binding protein (CaBP)-facilitated calcium diffusion, and basolateral active calcium extrusion], as indicated by red plus signs. ΔV_t, change in membrane potential. Please see text for details. Glu, glucose. (See the color plate.)

greater upregulation of calcium transporter expression than in male mice (Song & Fleet, 2004). As for age dependency, the $1,25(OH)_2D_3$-induced duodenal calcium absorption is relatively high in young adult rodents before gradually decreased with age (Wood, Fleet, Cashman, Bruns, & DeLuca, 1998).

3.1 Apical calcium entry

Free-ionized calcium traverses the duodenal apical membrane via calcium channels by simple diffusion toward electrochemical potential (i.e., luminal calcium concentrations of ~2–6 mmol/L vs. intracellular concentrations of ~0.1–0.3 μmol/L, and more negative potential inside) (Wasserman, 2004). Two families of calcium channels, i.e., TRPV5 and 6 ($P_{Ca}/P_{Na} > 100$) and L-type voltage-dependent calcium channel (Ca_v; conductance 11–25 pS), are expressed in the apical (brush border) membrane of duodenal absorptive cells (Morgan, Mace, Helliwell, Affleck, & Kellett, 2003; van de Graaf et al., 2003), suggesting a redundancy of apical calcium entry mechanisms. In addition, the absence of one apical calcium transporter can lead to a compensatory upregulation of other calcium transporter expression (Gkika et al., 2006). Under normal conditions, TRPV6, and to a lesser extent TRPV5, plays an important role in the duodenal calcium entry in a vitamin D-dependent manner (Kellett, 2011). Opening of apical calcium channels is generally dependent on membrane potential (apical resting potential of −47 mV). Ca_v opens at a depolarizing potential with activation threshold of around −25 mV, whereas TRPV6 fully functions under a relatively hyperpolarized potential of <−50 mV (Kellett, 2011). In renal epithelial cells, the T-type Ca_v has been reported to mediate apical calcium entry. For example, T-type Ca_v opening in response to testosterone was found to enhance calcium reabsorption in the rabbit distal renal tubule, but not proximal tubule, via mitogen-activated protein kinase/extracellular signal-regulated kinase (MEK)- and tyrosine kinase-dependent mechanisms (Couchourel, Leclerc, Filep, & Brunette, 2004).

During postprandial period, L-type Ca_v, particularly the 1.3 subtype ($Ca_v1.3$), helps to enhance apical calcium entry because the $Ca_v1.3$-mediated calcium transport is dependent on luminal glucose (Kellett, 2011; Morgan, Mace, Affleck, & Kellett, 2007). Specifically, glucose together with other substrates of sodium-dependent glucose transporter (SGLT)-1, such as galactose, initially induces sodium entry, thereby depolarizing the apical plasma membrane from −47 to −20 mV (Kellett,

2011). Depolarization then triggers the opening of $Ca_v1.3$ for calcium to diffuse down its concentration gradient into the cytoplasm. This $Ca_v1.3$-mediated calcium transport is markedly diminished by L-type calcium channel blockers, including dihydropyridine (e.g., nifedipine) and phenylalkylamine (e.g., verapamil) (Morgan et al., 2003; Thongon, Nakkrasae, Thongbunchoo, Krishnamra, & Charoenphandhu, 2009).

It has been widely known that $1,25(OH)_2D_3$ potently enhances the expression of TRPV5 and TRPV6, thus increasing their activities (Khuituan et al., 2012; Okano, Tsugawa, Morishita, & Kato, 2004). A single-dose subcutaneous injection of 2 μg/kg $1,25(OH)_2D_3$ in mice was found to upregulate the duodenal mRNA expression of both TRPV5 and TRPV6 within 3 h and peaked at 6 h before returning to the baseline levels at 24 h (Khuituan et al., 2012). On the other hand, the $Ca_v1.3$-mediated calcium transport is independent of $1,25(OH)_2D_3$ (Kellett, 2011). At the molecular level, the regulation of TRPV6 by $1,25(OH)_2D_3$ has been identified in human intestinal cells. Meyer, Watanuki, Kim, Shevde, and Pike (2006) demonstrated in human intestine-like Caco-2 cells using chromatin immunoprecipitation and found five putative VDREs located upstream to *Trpv6* coding region, and mutagenesis within the VDRE abrogated all responses to the $1,25(OH)_2D_3$. It was also reported that heterodimer of VDR and RXR could bind to *Trpv6* gene, resulting in both recruitment of chromatin remodeling coactivator steroid receptor coactivator 1 as well as modification of histone acetylation. Hence, *Trpv6* gene was broadly acetylated in response to $1,25(OH)_2D_3$ specifically at the VDR/RXR-binding sites.

3.2 Cytoplasmic translocation

Theoretically, free-ionized calcium can freely diffuse across cytoplasmic aqueous medium at higher rate than proteins. Specifically, the diffusion coefficient of free-ionized calcium in cytoplasm is 223 $\mu m^2/s$ (Allbritton, Meyer, & Stryer, 1992), whereas the diffusion coefficient of calbindin-D_{9k} is ~1/14 that of calcium. However, cells usually prevent bulk flow of free calcium across the cytoplasm, while utilizing facilitated calcium diffusion in protein-bound form. Since elevated intracellular calcium may trigger undesirable intracellular signal transductions and apoptosis, the duodenal absorptive cells use several mechanisms to keep the cytoplasmic free-ionized calcium level as low as possible. First, excessive free-ionized calcium and certain calcium-laden calcium-binding proteins (e.g., calmodulin) close to the apical membrane deactivate TRPV6 and perhaps Ca_v, which can, in turn, restrict calcium entry (Derler et al., 2006; Kellett, 2011; Lee et al., 2006; Nilius et al., 2001). Second, most calcium is buffered by cytoplasmic calcium-binding proteins, particularly calbindin-D_{9k} (S100g), and probably also by calbindin-D_{28k}, parvalbumin, and calmodulin (Schröder, Schlumbohm, Kaune, & Breves, 1996; Timmermans, Bindels, & Van Os, 1995; Walters, 1989). Meanwhile, calbindin-D_{9k} with two EF-hand calcium-binding domains indeed elevates the total intracellular concentration of calcium (bound form plus ionized form) by ~50-fold, thereby facilitating translational diffusion (ferry) of calcium across the cytoplasm (Feher, 1983). The unoccupied calcium-binding proteins in close vicinity to apical ion channels prevent channel inactivation, leading to persistent calcium entry (Fig. 2).

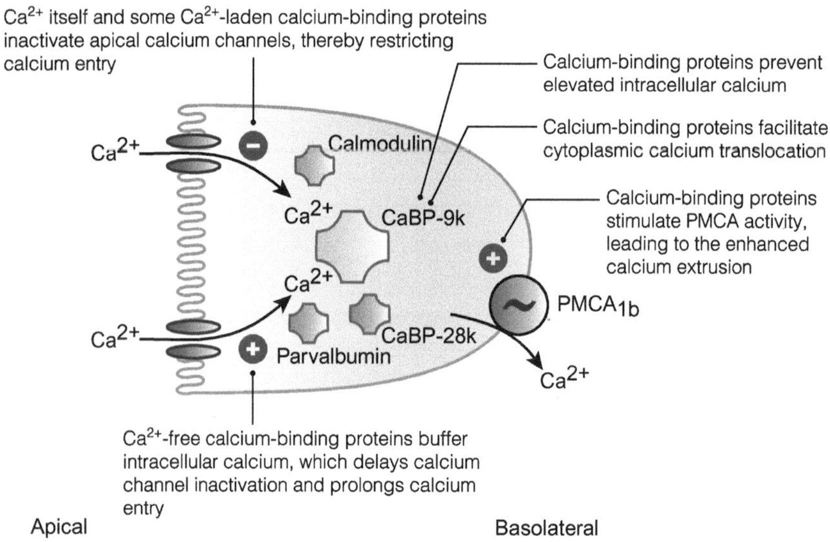

Figure 2 Cytoplasmic calcium-binding proteins (CaBP) in the duodenal enterocytes show a high degree of redundancy and have a number of functions. Calcium-free calcium-binding proteins (e.g., calbindin-D_{9k} and -D_{28k}) buffer the intracellular calcium, which can, in turn, delay calcium channel inactivation and prolong calcium entry. On the other hand, some calcium-laden calcium-binding proteins (e.g., calmodulin and calbindin-D_{28k}) as well as intracellular calcium itself lead to calcium-dependent inactivation of apical calcium channels, thereby restricting calcium entry. Several calcium-binding proteins, especially calbindin-D_{9k}, ferry the intracellular calcium from apical-to-basolateral compartment by a process known as translational diffusion rather than bucket brigade. Finally, they potentiate PMCA activity to accelerate the basolateral calcium extrusion. Ca^{2+}, free-ionized calcium. (See the color plate.)

The levels of calbindin-D_{9k} mRNA and protein in the duodenal absorptive cells of human and rodents are strongly dependent on $1,25(OH)_2D_3$ (Fleet et al., 2002; Khuituan et al., 2012). Indeed, the expressions of other calcium-binding proteins and/or their transcripts, e.g., calbindin-D_{28k}, parvalbumin, calmodulin, and sorcin, are also $1,25(OH)_2D_3$-dependent (Bindels, Timmermans, Hartog, Coers, & Van Os, 1991; Khanal & Nemere, 2008a; Wood, Tchack, Angelo, Pratt, & Sonna, 2004) although their exact contribution to the cytoplasmic calcium translocation remains enigmatic. The presence of transcellular active calcium transport in the small intestine of calbindin-D_{9k} knockout mice (Benn et al., 2008) indicates functional redundancy of cytoplasmic calcium-binding proteins and thus strengthens the physiological role of other proteins. Upregulation of calbindin-D_{28k} and parvalbumin mRNA expression is evident under certain conditions with high-calcium demand, such as pregnancy and lactation (Charoenphandhu, Wongdee, Teerapornpuntakit, Thongchote, & Krishnamra, 2008; Teerapornpuntakit, Klanchui, Karoonuthaisiri, Wongdee, & Charoenphandhu, 2014), suggesting that they play a role in cytoplasmic facilitated calcium transport. In avian small intestine and mammalian renal tubular cells, calbindin-D_{28k} is the principal protein for intracellular apical-to-basolateral transport (Hall & Norman, 1990).

Cytoplasmic calcium translocation in both avian and mammals may also be mediated by a mechanism called vesicular calcium transport (Fig. 1), in which calcium is captured inside an endocytic vesicle or later pumped into an intracellular membrane-bound vesicle before being extruded at the basolateral membrane by exocytosis (Khanal & Nemere, 2008b; Nemere, Leathers, & Norman, 1986). In the duodenal enterocytes of white Leghorn cockerels, vitamin D-dependent calbindin-D_{28k} is localized in the lysosome-like vesicles, presumably to help buffer intravesicular calcium (Nemere et al., 1986). Vesicular calcium transport underlies the rapid nongenomic $1,25(OH)_2D_3$-enhanced vectorial calcium absorption—previously known as transcaltachia (Nemere & Norman, 1990).

In addition, calcium is hypothesized to translocate across the cytoplasm by tunneling through endoplasmic reticulum (ER) (Fig. 1). This mechanism was identified in pancreatic cells but has never been studied in the duodenal enterocytes. The responsible transporters in the duodenal ER membrane are unknown. In pancreatic acinar cells, Mogami, Nakano, Tepikin, and Petersen (1997) showed recharging of intracellular calcium stores in response to calcium depletion mediated by movement of calcium through a tunnel from the basal to secretory poles. Since VDR has been shown to interact with sarcoendoplasmic reticulum Ca^{2+}-ATPase (SERCA) in cardiomyocytes (Zhao & Simpson, 2010), it is tempting to speculate that $1,25(OH)_2D_3$ might directly modulate SERCA activity, leading to the enhanced calcium tunneling across the duodenal cytoplasm.

3.3 Basolateral extrusion

Since electrochemical gradient across the duodenal basolateral membrane is thermodynamically unfavorable for simple diffusion [i.e., ~0.1–1 μmol/L cytoplasmic calcium vs. 1.2–1.3 mmol/L plasma calcium, and positive potential difference (PD) of ~58 mV (extracellular positive voltage)], metabolically energized active transport processes are required for basolateral uphill calcium extrusion. The cellular energy of ~9.3 kcal (39 kJ) is required to transport one mole of calcium against this electrochemical gradient (Wasserman & Fullmer, 1995). Although both plasma membrane Ca^{2+}-ATPase (PMCA) subtype 1b and Na^+/Ca^{2+}-exchanger (NCX)-1 play an important role in the basolateral calcium transport, the detailed cellular and molecular mechanisms remain controversial. $PMCA_{1b}$ (*Atp2b1*) is a primary active transporter (P-type ATPase) that can directly hydrolyze ATP to drive its transport activity (K_m ~0.2 μmol/L in the presence of calmodulin), whereas NCX1 (*Slc8a1*; Na^+/Ca^{2+} ratio 3:1) is coupled with Na^+/K^+-ATPase (NKA) that generates sodium gradient for NCX1-mediated calcium efflux (Fig. 3A). In rats, the $PMCA_{1b}$ activity is highest in the duodenum as compared to other small intestinal segments (Ghijsen & Van Os, 1982). $PMCA_{1b}$-mediated calcium transport is approximately fivefold greater than the NCX1-mediated transport (Ghijsen, de Jong, & Van Os, 1983). It was, therefore, previously believed that $PMCA_{1b}$ was the principal active transporter for basolateral calcium extrusion. However, our group recently demonstrated a complete abolishment of the $1,25(OH)_2D_3$-enhanced active calcium transport in the mouse duodenum when inhibited by NCX inhibitor (100 μmol/L KB-R7943) alone and calmodulin-dependent PMCA inhibitor (100 μmol/L trifluoperazine) alone as well as KB-R7943 plus trifluoperazine (Khuituan et al., 2013), indicating that in response to $1,25(OH)_2D_3$, both transporters might function in an interdependent manner. In other words, once $PMCA_{1b}$ is inhibited, calcium transport via NCX1 eventually decreases, and vice versa. Similar finding is also observed when the duodenal active calcium transport is stimulated by prolactin (Dorkkam, Wongdee, Suntornsaratoon, Krishnamra, & Charoenphandhu, 2013).

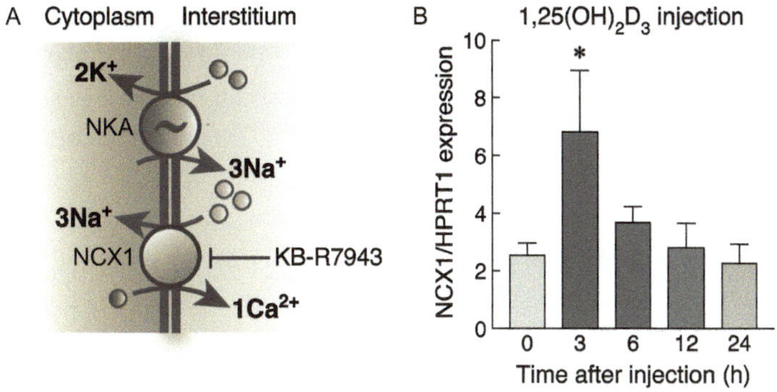

Figure 3 (A) Basolateral calcium extrusion through Na^{+}/Ca^{2+}-exchanger (NCX)-1. The NCX1 is coupled with Na^{+}/K^{+}-ATPase (NKA). NKA first moves three sodium ions out of the cell and moves two potassium ions into the cell, thereby generating sodium gradient for NCX1 function. NCX1 then allows influx of three sodium ions and efflux of one calcium ion. NCX1 activity can be inhibited by KB-R7943. (B) Time-dependent expression of NCX1 in duodenal epithelial cells of mice subcutaneously injected with a single-dose 2 μg/kg $1,25(OH)_2D_3$, as determined by quantitative real-time PCR. The duodenal epithelial cells were collected at 0, 3, 6, 12, and 24 h after $1,25(OH)_2D_3$ injection. Values are means ± SE of NCX1/HPRT1 expression ratios. The results show that, at 3 h postinjection, $1,25(OH)_2D_3$ significantly increases the mRNA level of NCX1 by 2.69-fold (*$P<0.05$ vs. 0 h; one-way analysis of variance with Dunnett's posttest) before returning to the baseline thereafter. HPRT1, hypoxanthine phosphoribosyltransferase-1 (a housekeeping gene). (See the color plate.)

Expression of $PMCA_{1b}$ transcripts is significantly upregulated by $1,25(OH)_2D_3$ in subclones of human Caco-2 cells (Fleet et al., 2002), whereas vitamin D deficiency decreases the mRNA level of PMCA (Zelinski, Sykes, & Weiser, 1991). Since $1,25(OH)_2D_3$ supplementation markedly increases the activities of PMCA, but not NCX, in vitamin D-deficient rats (Ghijsen et al., 1983), the NCX1-mediated calcium extrusion is thought to be vitamin D independent. However, in the duodenum of chicks (*Gallus domesticus*), vitamin D deficiency leads to a decrease in NCX activity, which is fully restored by $1,25(OH)_2D_3$ administration (Centeno, Picotto, Pérez, Alisio, & Tolosa de Talamoni, 2011), whereas low-calcium diet enhances both PMCA and NCX activities (Centeno et al., 2004). $1,25(OH)_2D_3$ administration also upregulates the mRNA and protein expression of NCX1 in the chick duodenum (Centeno et al., 2011). Similarly, in the duodenum of vitamin D-replete mouse (Khuituan et al., 2012, 2013), $1,25(OH)_2D_3$ enhances the NCX1-mediated calcium transport presumably by upregulating NCX1 transcription (Fig. 3B). Recently, another family of the basolateral calcium transporters, namely potassium-dependent sodium calcium exchangers (e.g., NCKX3 and 4), has been suggested to contribute to calcium extrusion (Kopic & Geibel, 2013), but whether $1,25(OH)_2D_3$ is capable of stimulating these transporters in the small intestine is not known. A study in mouse dendritic cells indicated that $1,25(OH)_2D_3$ could increase the membrane abundance of NCKX proteins as well as their activities (Shumilina et al., 2010).

The basolateral calcium extrusion is indeed coupled with apical calcium uptake and cytoplasmic calcium translocation. Claassen, Coetzer, de Winter, Haag, and Kruger (1996) reported a linear relationship between the apical calcium uptake and PMCA activity in the duodenum of male rats. In addition, vitamin D-dependent calbindin-D_{9k} as well as some other calcium-binding proteins, such as calmodulin, calbindin-D_{28k}, and parvalbumin, appears to be salient activators of PMCA activity (Timmermans et al., 1995; Walters, 1989). This coupling mechanism thus ensures that calcium absorption occurs at the highest efficiency without intracellular calcium accumulation, which not only blocks apical calcium uptake but may also induce undesirable responses, e.g., inhibition of sodium absorption and apoptosis.

4. VITAMIN D-ENHANCED PARACELLULAR CALCIUM TRANSPORT

In the leaky epithelia with transepithelial resistance of $<100\ \Omega\ cm^2$ like the small and large intestine, luminal calcium is able to traverse the epithelial sheet through the paracellular pathway—an electrically charged, watery space between the two epithelial cells (also known as the lateral intercellular space). In the rat duodenum, calcium movement is driven by free energy of electrochemical gradient (passive diffusion) or solvent drag (Charoenphandhu & Krishnamra, 2007; Charoenphandhu, Wongdee, & Krishnamra, 2010). Charge- and size-selective properties of the junctional complex, particularly tight junction (effective pore diameter ~0.5–3.5 nm), also affect the paracellular calcium transport (Charoenphandhu et al., 2010).

4.1 Paracellular calcium transport driven by electrochemical gradient

Dostal and Toverud (1984) reported a reduction in the passive nonsaturable component of duodenal calcium absorption in vitamin D-deficient rats, suggesting that the passive paracellular calcium transport is indeed dependent on $1,25(OH)_2D_3$ action. Generally, this mode of calcium transport is considered unsaturable and increases linearly with luminal free-ionized calcium concentration. It was previously believed that the paracellular space was merely a simple water-filled channel, where transepithelial fluxes in the apical-to-basolateral and basolateral-to-apical directions were identical when the driving transepithelial calcium gradients had the same magnitude. However, our group has demonstrated in human intestinal epithelium-like Caco-2 monolayer that the paracellular space—or more precisely the tight junction—is a rectifying channel (Thongon et al., 2009). Therefore, under certain conditions (e.g., short-circuit condition; transepithelial PD = 0 mV), the apical-to-basolateral paracellular calcium flux is slightly greater than that in the opposite direction (Thongon et al., 2009). In humans and rats fed calcium-replete diet, free-ionized calcium concentration in the duodenal lumen is as high as 2–6 mmol/L, which can diffuse down the concentration gradient via the paracellular pathway into the interstitial fluid or plasma (plasma free-ionized calcium concentration ~1.25 mmol/L) (Duflos, Bellaton, Pansu, & Bronner, 1995; Wasserman, 2004). Our investigation in Ussing chamber showed that a significant amount of the duodenal paracellular passive flux occurred when the apical calcium concentration was 5 mmol/L, but not 2.5 mmol/L, versus the basolateral concentration of 1.25 mmol/L (Charoenphandhu et al., 2006).

In addition to calcium concentration gradient, the voltage gradient or PD can also drive the paracellular calcium flux. Under normal conditions, the duodenal lumen is negative compared to plasma; therefore, plasma calcium diffuses down the voltage gradient into the lumen (i.e., calcium secretion). However, since PD across the duodenal epithelium is relatively low (~2–5 mV, negative luminal potential) and since elimination of PD by short-circuit current does not much affect transepithelial calcium flux, electrodiffusion of calcium may be considered negligible (Charoenphandhu, Limlomwongse, & Krishnamra, 2001). It has been reported that $1,25(OH)_2D_3$ is capable of enhancing the activities of electrogenic transporters, such as NKA and SGLT-1, which, in turn, alters the PD (Elstein & Silver, 1986; Rexhepaj et al., 2011). Hence, $1,25(OH)_2D_3$ might also indirectly modulate the voltage-dependent calcium transport, but the effect would still be relatively small.

4.2 Solvent drag-induced paracellular calcium transport

A correlation between transepithelial paracellular water flow and tracer solutes substantiates physiological significance of solvent drag, in which ions and small hydrophilic molecules traverse the tight junction and the remaining part of the paracellular space along with the stream of water (Mullen, Muller, & Van Bruggen, 1985). Our group has provided evidence that rat duodenal epithelium is able to absorb calcium via the solvent drag-induced mechanism (Tanrattana, Charoenphandhu, Limlomwongse, & Krishnamra, 2004). Nevertheless, direct visualization of water flow across the tight junction (e.g., using fluorescent marker and confocal microscope) has been unsuccessful (Kovbasnjuk, Leader, Weinstein, & Spring, 1998), and most solvent drag studies rely upon the use of paracellular markers (e.g., mannitol, polyethylene glycol, and inulin)—substances not absorbed transcellularly—and/or measurement of transepithelial $[^3H]_2O$ flux. To obtain a substantial magnitude of solvent drag-induced calcium transport, the duodenal epithelial cells have to activate SGLT-1 and NKA. Specifically, sodium entering through SGLT-1 (and perhaps through some other sodium-coupled apical transports, e.g., amino acid transporters) is pumped into the paracellular space by NKA, thereby increasing the sodium concentration in the space for osmotic water flow. The presence of SGLT-1 substrates, such as glucose and galactose, in the apical compartment thus augments the duodenal calcium absorption (Suntornsaratoon et al., 2014).

In the duodenum of male mice, $1,25(OH)_2D_3$ was found to significantly increase the transepithelial $[^3H]_2O$ flow, but the paracellular transport of mannitol (molecular radius ~0.4 nm) was not observed until there was a widening of tight junction by tumor necrosis factor-α (Khuituan et al., 2013). Such an increase in water flow can be explained by the stimulatory effects of $1,25(OH)_2D_3$ on SGLT-1 and NKA (Elstein & Silver, 1986; Rexhepaj et al., 2011). Indeed, the solvent drag-induced duodenal calcium transport occurs very rapidly within 60 min after a direct exposure to 10–1000 nmol/L $1,25(OH)_2D_3$, which is probably mediated by nongenomic pathways involving $1,25D_3$-MARRS, phosphatidylinositol 3-kinase (PI3K), protein kinase C (PKC), and MEK (Tudpor, Teerapornpuntakit, Jantarajit, Krishnamra, & Charoenphandhu, 2008). In contrast, Nemere et al. (1986) reported no difference in the serum tritium levels after adding $[^3H]_2O$ into the duodenal lumen of vitamin D-deficient and $1,25(OH)_2D_3$-treated chicks, suggesting that $1,25(OH)_2D_3$ might not regulate the lumen-to-plasma water flow in avian. Alternatively, $1,25(OH)_2D_3$ might have used other mechanisms to increase solvent drag-induced calcium transport, for example, by inducing contraction of perijunctional actomyosin ring to widen the tight junction pores (i.e., modulating size selectivity; for review, please see Turner, 2000), or by increasing cation permselectivity of the tight junction (Tudpor et al., 2008).

4.3 Charge- and size-selective properties of tight junction

The duodenum epithelium is basically cation selective with permeability ratio of sodium and chloride (P_{Na}/P_{Cl}) between 1.2 and 2 (Charoenphandhu et al., 2006; Tudpor et al., 2008). It allows paracellular movement of small monovalent and divalent cations, e.g., Na^+ and Ca^{2+}, both of which have ionic radii of ~115 pm. In general, alterations of charge/size-selective properties and/or cation permselectivity of tight junction markedly affect paracellular ion transport via passive diffusion and solvent drag-induced mechanisms. Our group has reported that although a rapid exposure (60 min) to 100 nmol/L $1,25(OH)_2D_3$ did not change the charge-selective property of the rat duodenal epithelium as indicated by no change in P_{Na}/P_{Cl}, the actual values of P_{Na} and P_{Cl} were significantly increased (Tudpor et al., 2008). Therefore, the $1,25(OH)_2D_3$-exposed duodenal epithelium is likely to have higher paracellular permeability to both cations (e.g., calcium) and anions.

At the molecular level, $1,25(OH)_2D_3$ alters tight junction permselectivity of intestinal epithelium by regulating the expression of tight junction-related genes, particularly claudins. Fujita et al. (2008) have demonstrated by using RNA interference and overexpression techniques that claudin-2 and -12 were upregulated by $1,25(OH)_2D_3$ through the VDR, resulting in an increase in paracellular calcium transport in human enterocyte-like Caco-2 cells. *In vivo* experiment also revealed downregulation of claudin-2 and -12 in the duodenum, ileum, and colon of the VDR knockout mice at both transcriptional and translational levels (Fujita et al., 2008). Similarly, Hwang et al. (2013) showed downregulation of claudin-2 as well as claudin-15 mRNAs and proteins in the duodenum of calcium/vitamin D-deficient calbindin-D_{9k} knockout mice. On the other hand, high concentration of $1,25(OH)_2D_3$ conversely suppressed claudin-3 mRNA expression in the rat duodenum as determined by Affymetrix microarrays (Kutuzova & DeLuca, 2004). The aforementioned evidence thus suggested that $1,25(OH)_2D_3$ regulates duodenal paracellular calcium transport through a combination of different claudins, i.e., claudin-2, -3, -12, and/or -15. In addition, other charge-selective claudins, such as claudin-10 and -16, the latter of which has been reported to be calcium-permeable and is expressed in the rat small intestine, may have roles in calcium absorption (Inai et al., 2005; Wongdee, Teerapornpuntakit, Siangpro, Chaipai, & Charoenphandhu, 2013).

$1,25(OH)_2D_3$ also modulates the expression of duodenal nonclaudin tight junction-related proteins, such as *zonula occludens* protein (ZO) and occludin (Hwang et al., 2013). Hwang et al. (2013) found the expressions of occludin and ZO-1 transcripts to be downregulated in calbindin-D_{9k} knockout mice given calcium- and vitamin D-deficient diet, but not normal diet. Western blot and immunohistochemical analyses of ZO-1 in duodenum also showed similar trend as mRNA results. E-cadherin—the critical component of the adherens junction for maintaining integrity of the epithelium—was also upregulated in response to $1,25(OH)_2D_3$ in human intestinal epithelium-like Caco-2 cells via RhoA–Rho-associated coiled-coil kinase (RhoA–ROCK) activation (Ordóñez-Morán et al., 2008).

5. REGULATION OF CALCIUM TRANSPORT BY THE PARATHYROID–KIDNEY–INTESTINAL AXIS

Since dietary calcium intake and daily calcium requirement vary largely from day to day and in different phases of life, the homeostatic system has to adjust calcium storage in bone as well as renal and intestinal calcium transport to meet the body requirements with a minimal change in plasma calcium level (Peacock, 2010). However, plasma calcium fluctuation may occur in several circumstances, e.g., changes in intestinal calcium absorption and renal calcium excretion. Such transient change of plasma calcium is recovered by an integrative response of three calcium-regulating organs organized as the parathyroid–kidney–intestinal axis (Fig. 4) (Kopic & Geibel, 2013; Peacock, 2010). PTH secretion from the parathyroid gland is inhibited after calcium binds to calcium sensing receptor (CaSR) in the plasma membrane of parathyroid chief cells. In a classical loop of regulation, low plasma calcium level curtails this CaSR-induced suppression, thereby increasing PTH secretion. PTH, in turn, raises plasma calcium level by (i) enhancing calcium reabsorption in the thick ascending limb of the Henle's loop and distal renal tubule, which occurs within minutes, and (ii) stimulating osteoblasts to secrete RANKL, which activates osteoclast-mediated bone resorption within minutes to hours (Favus, Bushinsky, & Lemann, 2006; Kopic & Geibel, 2013). PTH can further induce bone calcium release by osteocytes through a poorly defined mechanism, known as osteocytic osteolysis (Tazawa et al., 2004).

PTH and low plasma calcium also independently stimulate 1α-hydroxylase in the proximal renal tubule, thus increasing the production of $1,25(OH)_2D_3$, which potently stimulates intestinal calcium absorption

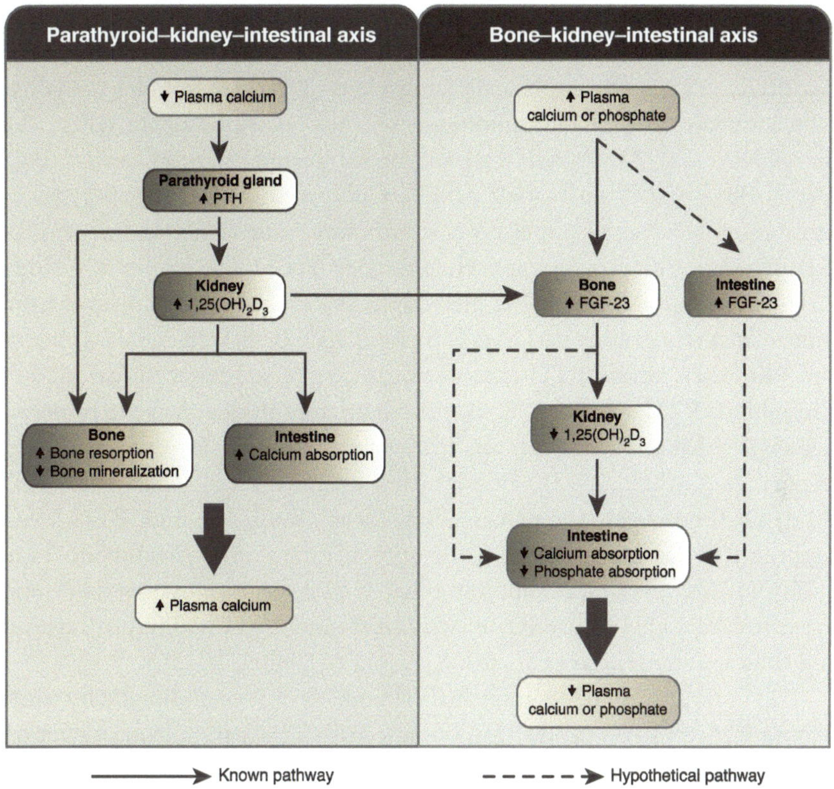

Figure 4 Interrelationship between the parathyroid–kidney–intestinal (left panel) and bone–kidney–intestinal axes (right panel) in regulating calcium homeostasis. In the parathyroid–kidney–intestinal axis, low plasma calcium level is detected by parathyroid glands, thereby stimulating PTH release. PTH stimulates 1α-hydroxylase in the kidney to convert $25(OH)D_3$ into $1,25(OH)_2D_3$. An increase in $1,25(OH)_2D_3$ level further enhances bone resorption, inhibits bone mineralization, and promotes intestinal calcium absorption, thus increasing plasma calcium level. In the bone–kidney–intestinal axis, high plasma calcium leads to an increase in circulating level of fibroblast growth factor (FGF)-23 produced by bone and perhaps by the intestine. FGF-23 suppresses renal $1,25(OH)_2D_3$ production by inhibiting 1α-hydroxylase, thereby decreasing intestinal calcium and phosphate absorption, which, in turn, lower plasma calcium and phosphate levels. An increase in circulating $1,25(OH)_2D_3$ level in the parathyroid–kidney–intestinal axis can also increase bone FGF-23 production, which links the two axes together. Furthermore, FGF-23 is hypothesized to exert a direct inhibitory effect on the intestine. Circulating FGF-23 might directly suppress intestinal calcium and phosphate absorption, whereas local FGF-23 production from the enterocytes might serve to prevent excessive intestinal calcium/phosphate absorption. (See the color plate.)

within 24 h after PTH secretion (Favus et al., 2006). An increase in plasma calcium induces a negative feedback that inhibits PTH secretion from the parathyroid chief cells. At the cellular level, ionized calcium binds to CaSR, which signals through phospholipase C, inositol triphosphate (IP_3), and diacylglycerol (DAG). CaSR can also be stimulated by other cations, e.g., Mg^{2+}, Pb^{2+}, Cd^{2+}, Fe^{2+}, Ba^{2+}, Ni^{2+}, Co^{2+}, and Gd^{3+}, as well as polycationic molecules, e.g., spermine, spermidine, putrescine, and neomycin (Handlogten, Shiraishi, Awata, Huang, & Miller, 2000; Kopic & Geibel, 2013; Quinn et al., 1997). An increase in IP_3 causes calcium release from intracellular store, whereas DAG activates PKC. Both intracellular calcium and PKC then inhibit PTH granule release. Increased intracellular calcium also inhibits PTH synthesis. Thus, plasma calcium itself acts as the principal negative feedback regulator in the parathyroid–kidney–intestinal axis (Kopic & Geibel, 2013). Besides the calcium level, PTH and 1,25 $(OH)_2D_3$ themselves take part in the feedback loop; i.e., high PTH levels suppress PTH gene expression while upregulating CaSR. Meanwhile, high $1,25(OH)_2D_3$ levels also suppress renal CYP27B1 (1α-hydroxylase) and increase CYP24 (24-hydroxylase) activities to lower its own circulating level and thus activity (Favus et al., 2006; Kopic & Geibel, 2013).

PTH not only increases intestinal calcium absorption indirectly by stimulating $1,25(OH)_2D_3$ production but also has a direct stimulatory effect on the intestinal calcium absorption. Nemere and Szego (1981) demonstrated that 0.01 nM PTH enhanced ^{45}Ca uptake in the rat intestine. Picotto, Massheimer, and Boland (1997) similarly revealed a stimulatory effect of PTH on calcium influx into the rat duodenal cells. This direct effect of PTH on enterocytic calcium influx is potently blocked by L-type calcium channel blockers, verapamil and nitrendipine (Picotto et al., 1997), suggesting that PTH induces calcium entry via L-type Ca_V. The direct stimulatory effect of PTH on intestinal calcium transport was also demonstrated in other species, e.g., chick and sea bream (Fuentes, Figueiredo, Power, & Canário, 2006; Nemere & Norman, 1986).

At the cellular and molecular levels, several lines of evidence have revealed the expression of PTH/PTH-related peptide (PTHrP) receptor transcripts in nonclassical PTH target tissues, e.g., blood vessels, adrenal gland, lung, liver, and intestine (Li et al., 1995; Ureña et al., 1993). Li et al. (1995) by using RT-PCR and Northern blot analysis reported that PTH/PTHrP receptor transcripts were widely expressed in the rat small intestine as well as in rat and human intestinal cell lines, such as IEC-6 and LoVo, respectively. *In situ* hybridization and immunohistochemistry showed nuclear localization of PTHrP and its receptor—which can bind to both PTH and PTHrP—in the epithelial cells lining villus and crypts, and in smooth muscles surrounding the gut in rat (Watson et al., 2000). The presence of PTH, PTHrP, and their receptors in the same intestinal regions/cell lines, therefore, implicated the possibility of autocrine and paracrine actions of PTH and/or PTHrP (Li et al., 1995; Nemere & Larsson, 2002). Besides calciotropic action, PTH also plays an important role in regulating intestinal transport of other ions, such as phosphate and bicarbonate (Laohapitakworn, Thongbunchoo, Nakkrasae, Krishnamra, & Charoenphandhu, 2011; Nemere, 1996). By using perfused duodenal loop, PTH was found to stimulate phosphate absorption in the chick duodenum (Nemere, 1996), while an *in vitro* experiment in Caco-2 monolayer revealed that PTH rapidly and directly stimulated apical bicarbonate secretion through cystic fibrosis transmembrane conductance regulator in protein kinase A- and PI3K-dependent manner (Charoenphandhu et al., 2011; Laohapitakworn et al., 2011).

As for calcitonin, although it is one of the major calcium-regulating hormones, little is known regarding its effects on intestinal calcium absorption and the available data are controversial. Several investigators have suggested the inhibitory effect of calcitonin on intestinal calcium absorption. For instance, Olson, DeLuca, and Potts (1972) determined calcium absorption in vascularly perfused isolated rat small intestine and found that acute infusion of calcitonin immediately decreased intestinal calcium absorption. Cramer (1973), on the other hand, showed that physiological and pharmacological doses of calcitonin did not alter calcium absorption in rat. A study of the calcitonin effect on intestinal calcium absorption in adult Merino sheep found that intravenous infusion of high-dose calcitonin initially increased calcium absorption associated with hypocalcemia, followed by a marked reduction in calcium absorption after a delay of few days (Swaminathan, Ker, & Care, 1974). Chronic infusion of calcitonin (0.2 U/h for 12 days in rats) paradoxically increased plasma calcium by 50%, which likely resulted from the increased intestinal calcium absorption since this effect disappeared after removal of calcium from the diet (Jaeger, Jones, Clemens, & Hayslett, 1986). However, the exact role(s) and underlying mechanism of calcitonin on intestinal calcium transport are waiting for an extensive systematic investigation.

6. NOVEL CONCEPT OF THE BONE–KIDNEY–INTESTINAL AXIS OF CALCIUM REGULATION

A prolonged increase in the duodenal calcium influx may lead to calcium-dependent inactivation of the transcellular transport process, which appears to occur from single-channel level (e.g., closure of calcium channel) to systemic level (e.g., release of hypocalcemic hormone) in order to help restrict excessive calcium uptake into the body. The molecular mechanism of this phenomenon in the duodenal enterocytes is not well understood, but can be explained, in part, by inhibition of apical calcium entry through TRPV6 and Ca_v by ionized calcium as well as calcium-laden calcium-binding proteins, particularly calmodulin—known as calcium-dependent inactivation of calcium channels (Derler et al., 2006; Lee et al., 2006). Inhibition of $1,25(OH)_2D_3$ production by elevated plasma ionized calcium (Kopic & Geibel, 2013) after enhanced intestinal calcium absorption certainly contributes to the inactivation of transcellular calcium transport process. Moreover, certain local and systemic humoral factor(s) may negatively regulate the duodenal calcium transport in a calcium and/or $1,25(OH)_2D_3$-dependent manner.

FGF-23 has been known as osteocyte/osteoblast-derived phosphate-regulating hormone although it is also abundantly expressed in other cell types, such as kidney, brain, lung, liver, spleen, and duodenal enterocytes (Khuituan et al., 2012; Kolek et al., 2005). $1,25(OH)_2D_3$ increases FGF-23 expression in both bone and intestine (Khuituan et al., 2012; Kolek et al., 2005). In healthy individuals, an increase in serum phosphate level induces PTH and FGF-23 release. PTH and FGF-23 prevents hyperphosphatemic spikes, which could have undesirable consequences, such as ectopic calcification. The hypophosphatemic action of both hormones is by inducing phosphaturia through inhibition of NaPi-2a and -2c activities in the proximal renal tubules and by suppressing CYP27B1, an enzyme for $1,25(OH)_2D_3$ synthesis (Jüppner, 2011; Kolek et al., 2005). FGF-23 not only controls phosphate homeostasis but also acts as a calcium-regulating hormone. Shimada et al. (2005) have demonstrated using VDR knockout mice that calcium is a potent stimulator of FGF-23 production via a VDR-independent pathway. In normal diet-fed VDR knockout mice, serum FGF-23 levels are very low, whereas calcium supplement significantly increases both serum FGF-23 and mRNA expression in bone. Furthermore, in a study using PTH knockout and PTH/CaSR double knockout mice, FGF-23 level was associated with interactions between serum calcium × phosphorus products and threshold levels for calcium and phosphorus (Quinn et al., 2013). Specifically, the calcium-mediated increase in serum FGF-23 level required a threshold phosphate level of 5 mg/dL (i.e., serum phosphate >5 mg/dL), whereas the phosphate-induced increase in serum FGF-23 required serum calcium levels of >8 mg/dL. Serum FGF-23 begins to increase exponentially when the calcium × phosphorus products were >50 mg^2/dL^2 (Quinn et al., 2013). The aforementioned findings have indicated that the regulation of FGF-23 by calcium and phosphate is fundamentally important in coordinating the serum levels of both mineral ions, and vice versa, to ensure that the calcium × phosphate product is maintained within a normal range (Quinn et al., 2013).

Regarding the calcium-regulating action of FGF-23, our laboratory has demonstrated, for the first time, a novel role of FGF-23 as a negative feedback regulator for the $1{,}25(OH)_2D_3$-enhanced duodenal calcium absorption in mice (Khuituan et al., 2012, 2013). This finding provides an alternative explanation of how the duodenal enterocytes restrict excessive calcium transport and thus prevent lethal hypercalcemia. Therein, mice injected with 1 μg/kg $1{,}25(OH)_2D_3$ once a day for 3 days showed an approximately 1.8-fold increase in duodenal calcium transport, which was completely abolished by concurrent intravenous injection of 140 μg/kg FGF-23 (Khuituan et al., 2012). FGF-23 directly inhibited the $1{,}25(OH)_2D_3$-induced calcium transport via MAPK/ERK, p38 MAPK, and PKC signaling pathways (Khuituan et al., 2012), and perhaps indirectly via renal inhibition of $1{,}25(OH)_2D_3$ synthesis (Shimada et al., 2005). The observed action of FGF-23 markedly affected both paracellular and transcellular transport, the latter of which apparently resulted from the downregulation of calcium transporter expression, i.e., TRPV5, TRPV6, and calbindin-D_{9k} (Khuituan et al., 2012). As for the paracellular pathway, FGF-23 modestly but significantly decreased the $1{,}25(OH)_2D_3$-induced paracellular calcium flux and calcium permeability (Khuituan et al., 2013). FGF-23 is, therefore, recognized as a part of the bone–kidney–intestinal axis, as depicted in Fig. 4.

In addition, the calcium-regulating role of FGF-23 may shed a new light on the search for a biomarker of body calcium status and efficiency of oral calcium supplementation. Specifically, positive calcium balance or efficient calcium supplement should be accompanied by a relatively high circulating level of FGF-23. Up until now, clinicians often use plasma levels of calcium, inorganic phosphate, PTH, $25(OH)D_3$, $1{,}25(OH)_2D_3$, and bone turnover markers (e.g., C-telopeptide of type I collagen) to represent body calcium status. Other promising candidates as indicators of calcium deficiency and calcium intake in humans include urinary concentrations of pseudouridine and citrate as revealed by a metabonomics analysis (Wang et al., 2013).

7. VITAMIN D-INDEPENDENT INTESTINAL CALCIUM TRANSPORT

Although $1,25(OH)_2D_3$ is of the utmost importance for avian and mammalian intestinal calcium absorption, the vitamin D-independent calcium transport does have considerable impact under certain conditions (e.g., young growing period) as well as in some mammalian species (e.g., underground mole-rats). In other words, the duodenal calcium absorption sometimes requires a different set of calciotropic hormones independent of $1,25(OH)_2D_3$. In some circumstances, such as pregnancy and lactation, $1,25(OH)_2D_3$ is not the main calciotropic hormone, but it exerts a permissive action or works cooperatively with other hormones to regulate calcium metabolism.

7.1 Calcium absorption in neonatal period

In 14-day-old suckling rats, the duodenal calcium absorption that mostly occurs through the paracellular (nonsaturable) pathway is similar under vitamin D-replete and vitamin D-deficient conditions, indicating the vitamin D-independent nature of neonatal calcium transport (Dostal & Toverud, 1984). The vitamin D-dependent calcium absorption is usually seen after weaning (~18- to 21-day-old rats) (Dostal & Toverud, 1984; Halloran & DeLuca, 1980). Interestingly, Halloran and DeLuca (1980) demonstrated that intrajugular injection of $1,25(OH)_2D_3$ enhanced neither passive nor active calcium absorption in vitamin D-deficient suckling rats (14 days postpartum), but markedly stimulated active calcium absorption in weaning rats (28 days postpartum). An investigation on vitamin D-deficient newborn piglets (<6 days postpartum) showed that duodenal active calcium transport was not different from that in vitamin D-replete piglets, and 2-day $1,25(OH)_2D_3$ supplementation could not enhance calcium absorption (Schroeder, Dahl, & Breves, 1998). Since a number of nutrients (e.g., casein phosphopeptides, lactose, glucose, and galactose) and bioactive peptides (e.g., prolactin, IGF-1, calcitonin, and PTHrP)—all of which are also present in milk—have been reported to augment calcium transport (Kocián, Skála, & Bakos, 1973; Pahuja & DeLuca, 1981; Schuette, Knowles, & Ford, 1989; Zhou, Nemere, & Norman, 1992), these milk ingredients may be alternative factors that stimulate duodenal calcium absorption in the neonates. Certain milk peptides, especially prolactin, may be able to exert its calciotropic action from the blood side after being absorbed as a whole peptide in newborn animals (Gonnella, Harmatz, & Walker, 1989; Whitworth & Grosvenor, 1978).

7.2 Calcium absorption in pregnant and lactating periods

During pregnancy, the duodenal calcium absorption is markedly stimulated to supply calcium for the development of fetus and mineral accretion in maternal bone, the latter of which serves to expand the calcium storage pool in bone that is later resorbed during lactation, a period of high-calcium demand for milk production (Charoenphandhu et al., 2010). Calcium hyperabsorption persists until the end of lactation (Charoenphandhu et al., 2010). Bone is also an important source of calcium for lactogenesis, particularly when dietary calcium intake is inadequate (Bowman & Miller, 2001; Kovacs, 2005). It has been known for several decades that the intestinal calcium absorption during pregnancy and lactations are independent of vitamin D (Pahuja & DeLuca, 1981). The duodenal TRPV6 mRNA expression and calcium transport are significantly increased in pregnant VDR-null mice, similar to those found in wild-type mice (Fudge & Kovacs, 2010). Although the transcellular active calcium transport does occur in vitamin D-deficient dams, lactating rodents sometimes show an elevated $1,25(OH)_2D_3$ level (Douard et al., 2012), which is capable of potentiating the action of other calciotropic hormone (e.g., prolactin), and may also be essential for calcium mobilization into milk (Ajibade et al., 2010). Brommage, Baxter, and Gierke (1990) estimated ~23% of milk calcium transfer being vitamin D-dependent.

Our group has elaborated that the pituitary hormone prolactin is the salient calciotropic hormone for pregnancy and lactation (for reviews, please see Charoenphandhu et al., 2010; Wongdee & Charoenphandhu, 2013). Long-term hyperprolactinemia (~75–200 ng/mL; baseline level in nonpregnant rat ~7–10 ng/mL), which occurs physiologically during these reproductive periods, leads to upregulation of transcellular calcium transporter expression (e.g., TRPV5, TRPV6, calbindin-D_{9k}, and $PMCA_{1b}$) as well as the expression of tight junction genes, thereby increasing both transcellular and paracellular duodenal calcium transport (Charoenphandhu et al., 2009; Teerapornpuntakit et al., 2014). During lactation, the suckling-induced prolactin surge (~400–800 ng/mL) further increases the duodenal calcium flux to compensate for calcium loss in milk, possibly by sensitizing the $Ca_v1.3$-mediated transcellular calcium transport (Charoenphandhu et al., 2009; Nakkrasae, Thongon, Thongbunchoo,

Krishnamra, & Charoenphandhu, 2010; Suntornsaratoon et al., 2014). Besides $1,25(OH)_2D_3$ and prolactin, other endocrine factors with elevated levels during pregnancy and/or lactation, especially placental lactogen, estrogen, calcitonin, and PTHrP, may also contribute to the enhanced calcium absorption (Colin et al., 1999; Fuentes et al., 2006; Jaeger et al., 1986; Takeuchi, Morikawa, Ueda, & Mochizuki, 1988).

7.3 Calcium absorption in naturally vitamin D-impoverished mammals

Not all adult mammals require vitamin D for intestinal calcium absorption. For example, the subterranean mole-rats (such as *Cryptomys damarensis* and *Heterocephalus glaber*) live in an underground maze without exposure to sunlight and usually consume vitamin D-depleted herbivorous diet. Therefore, they are naturally vitamin D-deficient with undetectable $25(OH)D_3$ and remarkably low $1,25(OH)_2D_3$ levels (Pitcher & Buffenstein, 1995). However, the mole-rats have efficient calcium absorption despite unclear cellular and molecular mechanisms. The duodenal calcium absorption is paracellular, whereas the transcellular active calcium absorption is found only in the cecum and proximal colon (Pitcher & Buffenstein, 1994). The 3-day oral vitamin D supplementation is not able to enhance calcium absorption in either foregut or hindgut of mole-rats (Pitcher & Buffenstein, 1995). The subterranean mole-rats appear to compensate for the vitamin D-impoverished state by depositing extra calcium in the chisel tooth and consuming calcium-rich plants (Buffenstein, Laundy, Pitcher, & Pettifor, 1995).

8. CONCLUSION AND PERSPECTIVES

Duodenal calcium transport exhibits redundancy in term of calcium transporter expression, possibly to ensure that dietary calcium uptake is always adequate to meet calcium demand. For example, a number of duodenal transporters, e.g., TRPV5, TRPV6, and $Ca_v1.3$, work cooperatively and complementarily for apical calcium uptake. The redundancy thus explains as to why the transcellular active calcium transport is not completely abolished by TRPV6 knockout, calbindin-D_{9k} knockout, or TRPV6/calbindin-D_{9k} double knockout (Benn et al., 2008). Our group has also reported that TRPV5/TRPV6 double knockdown is unable to nullify apical-to-basolateral calcium transport across human intestinal epithelium-like Caco-2 monolayer (Nakkrasae et al., 2010). Calbindin-D_{28k}-null mice manifest normal bone and teeth development, suggesting no disturbance to calcium transport in the absence of calbindin-D_{28k} (Turnbull et al., 2004). Therefore, further investigation is required to demonstrate contribution of each transporter to the total calcium absorption.

The expressions and/or activities of most proteins related to intestinal calcium absorption in humans and rodents (i.e., TRPV5, TRPV6, calbindin-D_{9k}, $PMCA_{1b}$, NCX1, and several claudins) are dependent on $1,25(OH)_2D_3$. Both transcellular (saturable) and paracellular (nonsaturable) calcium transport are thus markedly stimulated by $1,25(OH)_2D_3$. Unlike the well-established $1,25(OH)_2D_3$-induced transcellular mechanism, little is known regarding the cellular and molecular mechanisms of the $1,25(OH)_2D_3$-induced paracellular calcium transport, especially in humans and nonhuman primates. Furthermore, the vitamin D-independent calcium absorption is evident under certain conditions, such as neonatal period, but responsible endocrine and paracrine regulators remain to be investigated. Species bias—i.e., information is mainly from human, rodent, and chick—precludes drawing a solid conclusion that $1,25(OH)_2D_3$ is the utmost important regulator of intestinal calcium transport in avian and mammalian species. Data from foregut fermenters (e.g., dairy cow), horses, and rabbits are scant although some species (e.g., rabbits) have been postulated to possess similar calcium metabolism and $1,25(OH)_2D_3$-dependent intestinal calcium absorption as in humans and rodents (Norris, Pettifor, Gray, & Buffenstein, 2001).

Indeed, duodenum is not the only site of calcium absorption, but it has been intensively studied due to its vigorous active calcium transport, calcium transport efficiency, and vitamin D responsiveness as well as abundance of calcium transporter expression (Lee et al., 1981; Teerapornpuntakit, Dorkkam, Wongdee, Krishnamra, & Charoenphandhu, 2009; Wasserman, 2004). Most absorbable dietary calcium (>80% of absorbed calcium) is transported by the jejunum and ileum (lower transport rate but greater sojourn time of 40–50 min compared to the duodenal sojourn time of 2–3 min) using the paracellular passive transport rather than the transcellular active transport (Duflos et al., 1995; Wasserman, 2004). In normal rats, the cecum—the most proximal part of the large intestine—absorbs calcium at a higher rate than the duodenum (>2-fold), and surgical removal of cecum (cecectomy) leads to fecal calcium wasting, compensatory increase in colonic calcium transport, and osteopenia (Charoenphandhu, Suntornsaratoon, Jongwattanapisan, Wongdee, & Krishnamra, 2012; Jongwattanapisan et al., 2012). Calcium transport in the distal small intestine and large intestine (cecum and colon) is also $1,25(OH)_2D_3$-dependent (Grinstead, Pak, &

Krejs, 1984; Karbach, 1991; Karbach & Feldmeier, 1993; Lee et al., 1981). However, exogenous 1,25$(OH)_2D_3$ cannot further increase cecal calcium transport (Karbach & Feldmeier, 1993). Intraperitoneal 1,25$(OH)_2D_3$ injection was found to similarly increase dietary calcium uptake in intact and cecectomized vitamin D-replete male rats (Brommage, Binacua, & Carrié, 1995), suggesting that 1,25$(OH)_2D_3$ is not the main regulator of cecal calcium absorption. Hence, the underlying molecular mechanisms and endocrine regulatory axes as well as paracrine/autocrine regulation of calcium absorption in the intestinal segments distal to the duodenum are important issues worth exploring.

ACKNOWLEDGMENTS

The authors thank Prof. Nateetip Krishnamra for critical reading of the manuscript, and Dr. Jarinthorn Teerapornpuntakit for providing data regarding the effect of 1,25$(OH)_2D_3$ on duodenal NCX1 expression. This work was supported by Mahidol University (to N. C.), the Cluster and Program Management Office (CPMO), National Science and Technology Development Agency (NSTDA, P-13-00100 to N. C.), the Thailand Research Fund (TRF)–Mahidol University through the TRF Senior Research Scholar Grant (RTA5780001 to N. C.), the TRF–Faculty of Allied Health Sciences, Burapha University (RSA5780041 to K. W.), the Faculty of Allied Health Sciences, Burapha University Research Grant of Fiscal Year 2015 (to K. W.), and Burapha University Research Grant of Fiscal Year 2013 (to K. W.).

Conflict of interest statement: The authors declare no conflict of interest.

REFERENCES

Ajibade, D. V., Dhawan, P., Fechner, A. J., Meyer, M. B., Pike, J. W., & Christakos, S. (2010). Evidence for a role of prolactin in calcium homeostasis: Regulation of intestinal transient receptor potential vanilloid type 6, intestinal calcium absorption, and the 25-hydroxyvitamin D_3 1α hydroxylase gene by prolactin. *Endocrinology*, *151*, 2974–2984.

Allbritton, N. L., Meyer, T., & Stryer, L. (1992). Range of messenger action of calcium ion and inositol 1,4,5-trisphosphate. *Science*, *258*, 1812–1815.

Armbrecht, H. J., Zenser, T. V., Gross, C. J., & Davis, B. B. (1980). Adaptation to dietary calcium and phosphorus restriction changes with age in the rat. *American Journal of Physiology*, *239*, E322–E327.

Balesaria, S., Sangha, S., & Walters, J. R. (2009). Human duodenum responses to vitamin D metabolites of TRPV6 and other genes involved in calcium absorption. *American Journal of Physiology. Gastrointestinal and Liver Physiology*, *297*, G1193–G1197.

Benn, B. S., Ajibade, D., Porta, A., Dhawan, P., Hediger, M., Peng, J. B., et al. (2008). Active intestinal calcium transport in the absence of transient receptor potential vanilloid type 6 and calbindin-D_{9k}. *Endocrinology*, *149*, 3196–3205.

Bindels, R. J., Timmermans, J. A., Hartog, A., Coers, W., & Van Os, C. H. (1991). Calbindin-D_{9k} and parvalbumin are exclusively located along basolateral membranes in rat distal nephron. *Journal of the American Society of Nephrology*, *2*, 1122–1129.

Bowman, B. M., & Miller, S. C. (2001). Skeletal adaptations during mammalian reproduction. *Journal of Musculoskeletal and Neuronal Interactions*, *1*, 347–355.

Brommage, R., Baxter, D. C., & Gierke, L. W. (1990). Vitamin D-independent intestinal calcium and phosphorus absorption during reproduction. *American Journal of Physiology, 259*, G631–G638.

Brommage, R., Binacua, C., & Carrié, A. L. (1995). The cecum does not participate in the stimulation of intestinal calcium absorption by calcitriol. *Journal of Steroid Biochemistry and Molecular Biology, 54*, 71–73.

Buffenstein, R., Laundy, M. T., Pitcher, T., & Pettifor, J. M. (1995). Vitamin D_3 intoxication in naked mole-rats (*Heterocephalus glaber*) leads to hypercalcaemia and increased calcium deposition in teeth with evidence of abnormal skin calcification. *General and Comparative Endocrinology, 99*, 35–40.

Centeno, V. A., Díaz de Barboza, G. E., Marchionatti, A. M., Alisio, A. E., Dallorso, M. E., Nasif, R., et al. (2004). Dietary calcium deficiency increases Ca^{2+} uptake and Ca^{2+} extrusion mechanisms in chick enterocytes. *Comparative Biochemistry and Physiology. Part A: Molecular & Integrative Physiology, 139*, 133–141.

Centeno, V., Picotto, G., Pérez, A., Alisio, A., & Tolosa de Talamoni, N. (2011). Intestinal Na^+/Ca^{2+} exchanger protein and gene expression are regulated by $1,25(OH)_2D_3$ in vitamin D-deficient chicks. *Archives of Biochemistry and Biophysics, 509*, 191–196.

Charoenphandhu, N., & Krishnamra, N. (2007). Prolactin is an important regulator of intestinal calcium transport. *Canadian Journal of Physiology and Pharmacology, 85*, 569–581.

Charoenphandhu, N., Laohapitakworn, S., Kraidith, K., Nakkrasae, L. I., Jongwattanapisan, P., Tharabenjasin, P., et al. (2011). Electrogenic Na^+/HCO_3^- co-transporter-1 is essential for the parathyroid hormone-stimulated intestinal HCO_3^- secretion. *Biochemical and Biophysical Research Communications, 409*, 775–779.

Charoenphandhu, N., Limlomwongse, L., & Krishnamra, N. (2001). Prolactin directly stimulates transcellular active calcium transport in the duodenum of female rats. *Canadian Journal of Physiology and Pharmacology, 79*, 430–438.

Charoenphandhu, N., Nakkrasae, L. I., Kraidith, K., Teerapornpuntakit, J., Thongchote, K., Thongon, N., et al. (2009). Two-step stimulation of intestinal Ca^{2+} absorption during lactation by long-term prolactin exposure and suckling-induced prolactin surge. *American Journal of Physiology. Endocrinology and Metabolism, 297*, E609–E619.

Charoenphandhu, N., Suntornsaratoon, P., Jongwattanapisan, P., Wongdee, K., & Krishnamra, N. (2012). Enhanced trabecular bone resorption and microstructural bone changes in rats after removal of the cecum. *American Journal of Physiology. Endocrinology and Metabolism, 303*, E1069–E1075.

Charoenphandhu, N., Tudpor, K., Pulsook, N., & Krishnamra, N. (2006). Chronic metabolic acidosis stimulated transcellular and solvent drag-induced calcium transport in the duodenum of female rats. *American Journal of Physiology. Gastrointestinal and Liver Physiology, 291*, G446–G455.

Charoenphandhu, N., Wongdee, K., & Krishnamra, N. (2010). Is prolactin the cardinal calciotropic maternal hormone? *Trends in Endocrinology and Metabolism, 21*, 395–401.

Charoenphandhu, N., Wongdee, K., Teerapornpuntakit, J., Thongchote, K., & Krishnamra, N. (2008). Transcriptome responses of duodenal epithelial cells to prolactin in pituitary-grafted rats. *Molecular and Cellular Endocrinology, 296*, 41–52.

Claassen, N., Coetzer, H., de Winter, R., Haag, M., & Kruger, M. C. (1996). Relationship between duodenal calcium uptake and Ca^{2+}-Mg^{2+}-ATPase activity. *Medical Science Research, 24*, 809–811.

Colin, E. M., Van Den Bemd, G. J., Van Aken, M., Christakos, S., De Jonge, H. R., DeLuca, H. F., et al. (1999). Evidence for involvement of 17β-estradiol in intestinal calcium absorption independent of 1,25-dihydroxyvitamin D_3 level in the rat. *Journal of Bone and Mineral Research, 14*, 57–64.

Couchourel, D., Leclerc, M., Filep, J., & Brunette, M. G. (2004). Testosterone enhances calcium reabsorption by the kidney. *Molecular and Cellular Endocrinology, 222*, 71–81.
Cramer, C. F. (1973). Effect of salmon calcitonin on in vivo calcium absorption in rats. *Calcified Tissue Research, 13*, 169–172.
Cui, M., Zhao, Y., Hance, K. W., Shao, A., Wood, R. J., & Fleet, J. C. (2009). Effects of MAPK signaling on 1,25-dihydroxyvitamin D-mediated CYP24 gene expression in the enterocyte-like cell line, Caco-2. *Journal of Cellular Physiology, 219*, 132–142.
Derler, I., Hofbauer, M., Kahr, H., Fritsch, R., Muik, M., Kepplinger, K., et al. (2006). Dynamic but not constitutive association of calmodulin with rat TRPV6 channels enables fine tuning of Ca^{2+}-dependent inactivation. *Journal of Physiology, 577*, 31–44.
Dorkkam, N., Wongdee, K., Suntornsaratoon, P., Krishnamra, N., & Charoenphandhu, N. (2013). Prolactin stimulates the L-type calcium channel-mediated transepithelial calcium transport in the duodenum of male rats. *Biochemical and Biophysical Research Communications, 430*, 711–716.
Dostal, L. A., & Toverud, S. U. (1984). Effect of vitamin D_3 on duodenal calcium absorption in vivo during early development. *American Journal of Physiology, 246*, G528–G534.
Douard, V., Suzuki, T., Sabbagh, Y., Lee, J., Shapses, S., Lin, S., et al. (2012). Dietary fructose inhibits lactation-induced adaptations in rat 1,25-$(OH)_2D_3$ synthesis and calcium transport. *Federation of American Societies for Experimental Biology, 26*, 707–721.
Duflos, C., Bellaton, C., Pansu, D., & Bronner, F. (1995). Calcium solubility, intestinal sojourn time and paracellular permeability codetermine passive calcium absorption in rats. *Journal of Nutrition, 125*, 2348–2355.
Elstein, D., & Silver, J. (1986). Na,K-ATPase activity and 25(OH)vitamin D_3 hydroxylation in rat proximal tubules. *Pflügers Archiv: European Journal of Physiology, 407*, 451–455.
Favus, M. J., Bushinsky, D., & Lemann, J., Jr. (2006). Regulation of calcium, magnesium, and phosphate metabolism. In M. J. Favus (Ed.), *Primer on the metabolic bone diseases and disorders of mineral metabolism* (6th ed., pp. 76–83). Washington, DC: American Society for Bone and Mineral Research.
Feher, J. J. (1983). Facilitated calcium diffusion by intestinal calcium-binding protein. *American Journal of Physiology, 244*, C303–C307.
Fleet, J. C., Eksir, F., Hance, K. W., & Wood, R. J. (2002). Vitamin D-inducible calcium transport and gene expression in three Caco-2 cell lines. *American Journal of Physiology. Gastrointestinal and Liver Physiology, 283*, G618–G625.
Fudge, N. J., & Kovacs, C. S. (2010). Pregnancy up-regulates intestinal calcium absorption and skeletal mineralization independently of the vitamin D receptor. *Endocrinology, 151*, 886–895.
Fuentes, J., Figueiredo, J., Power, D. M., & Canário, A. V. (2006). Parathyroid hormone-related protein regulates intestinal calcium transport in sea bream (*Sparus auratus*). *American Journal of Physiology. Regulatory, Integrative and Comparative Physiology, 291*, R1499–R1506.
Fujita, H., Sugimoto, K., Inatomi, S., Maeda, T., Osanai, M., Uchiyama, Y., et al. (2008). Tight junction proteins claudin-2 and -12 are critical for vitamin D-dependent Ca^{2+} absorption between enterocytes. *Molecular Biology of the Cell, 19*, 1912–1921.
Gallagher, J. C., Jindal, P. S., & Smith, L. M. (2014). Vitamin D does not increase calcium absorption in young women: A randomized clinical trial. *Journal of Bone and Mineral Research, 29*, 1081–1087.
Ghijsen, W. E., de Jong, M. D., & Van Os, C. H. (1983). Kinetic properties of Na^+/Ca^{2+} exchange in basolateral plasma membranes of rat small intestine. *Biochimica et Biophysica Acta, 730*, 85–94.
Ghijsen, W. E., & Van Os, C. H. (1982). 1α,25-Dihydroxy-vitamin D_3 regulates ATP-dependent calcium transport in basolateral plasma membranes of rat enterocytes. *Biochimica et Biophysica Acta, 689*, 170–172.

Gkika, D., Hsu, Y. J., van der Kemp, A. W., Christakos, S., Bindels, R. J., & Hoenderop, J. G. (2006). Critical role of the epithelial Ca^{2+} channel TRPV5 in active Ca^{2+} reabsorption as revealed by TRPV5/calbindin-D_{28k} knockout mice. *Journal of the American Society of Nephrology, 17*, 3020–3027.

Gonnella, P. A., Harmatz, P., & Walker, W. A. (1989). Prolactin is transported across the epithelium of the jejunum and ileum of the suckling rat. *Journal of Cellular Physiology, 140*, 138–149.

Grinstead, W. C., Pak, C. Y., & Krejs, G. J. (1984). Effect of 1,25-dihydroxyvitamin D_3 on calcium absorption in the colon of healthy humans. *American Journal of Physiology, 247*, G189–G192.

Hall, A. K., & Norman, A. W. (1990). Regulation of calbindin-D_{28K} gene expression in the chick intestine: Effects of serum calcium status and 1,25-dihydroxyvitamin D_3. *Journal of Bone and Mineral Research, 5*, 331–336.

Halloran, B. P., & DeLuca, H. F. (1980). Calcium transport in small intestine during early development: Role of vitamin D. *American Journal of Physiology, 239*, G473–G479.

Handlogten, M. E., Shiraishi, N., Awata, H., Huang, C., & Miller, R. T. (2000). Extracellular Ca^{2+}-sensing receptor is a promiscuous divalent cation sensor that responds to lead. *American Journal of Physiology. Renal Physiology, 279*, F1083–F1091.

Hwang, I., Yang, H., Kang, H. S., Ahn, C., Hong, E. J., An, B. S., et al. (2013). Alteration of tight junction gene expression by calcium- and vitamin D-deficient diet in the duodenum of calbindin-null mice. *International Journal of Molecular Sciences, 14*, 22997–23010.

Inai, T., Sengoku, A., Guan, X., Hirose, E., Iida, H., & Shibata, Y. (2005). Heterogeneity in expression and subcellular localization of tight junction proteins, claudin-10 and -15, examined by RT-PCR and immunofluorescence microscopy. *Archives of Histology and Cytology, 68*, 349–360.

Jaeger, P., Jones, W., Clemens, T. L., & Hayslett, J. P. (1986). Evidence that calcitonin stimulates 1,25-dihydroxyvitamin D production and intestinal absorption of calcium in vivo. *Journal of Clinical Investigation, 78*, 456–461.

Jongwattanapisan, P., Suntornsaratoon, P., Wongdee, K., Dorkkam, N., Krishnamra, N., & Charoenphandhu, N. (2012). Impaired body calcium metabolism with low bone density and compensatory colonic calcium absorption in cecectomized rats. *American Journal of Physiology. Endocrinology and Metabolism, 302*, E852–E863.

Jüppner, H. (2011). Phosphate and FGF-23. *Kidney International, 79*(Suppl. 121), S24–S27.

Karbach, U. (1991). Segmental heterogeneity of cellular and paracellular calcium transport across the rat duodenum and jejunum. *Gastroenterology, 100*, 47–58.

Karbach, U., & Feldmeier, H. (1993). The cecum is the site with the highest calcium absorption in rat intestine. *Digestive Diseases and Sciences, 38*, 1815–1824.

Kellett, G. L. (2011). Alternative perspective on intestinal calcium absorption: Proposed complementary actions of $Ca_v1.3$ and TRPV6. *Nutrition Reviews, 69*, 347–370.

Khanal, R. C., & Nemere, I. (2008a). Regulation of intestinal calcium transport. *Annual Review of Nutrition, 28*, 179–196.

Khanal, R. C., & Nemere, I. (2008b). Endocrine regulation of calcium transport in epithelia. *Clinical and Experimental Pharmacology and Physiology, 35*, 1277–1287.

Khuituan, P., Teerapornpuntakit, J., Wongdee, K., Suntornsaratoon, P., Konthapakdee, N., Sangsaksri, J., et al. (2012). Fibroblast growth factor-23 abolishes 1,25-dihydroxyvitamin D_3-enhanced duodenal calcium transport in male mice. *American Journal of Physiology. Endocrinology and Metabolism, 302*, E903–E913.

Khuituan, P., Wongdee, K., Jantarajit, W., Suntornsaratoon, P., Krishnamra, N., & Charoenphandhu, N. (2013). Fibroblast growth factor-23 negates $1,25(OH)_2D_3$-induced intestinal calcium transport by reducing the transcellular and paracellular calcium fluxes. *Archives of Biochemistry and Biophysics, 536*, 46–52.

Kocián, J., Skála, I., & Bakos, K. (1973). Calcium absorption from milk and lactose-free milk in healthy subjects and patients with lactose intolerance. *Digestion, 9*, 317–324.

Kolek, O. I., Hines, E. R., Jones, M. D., LeSueur, L. K., Lipko, M. A., Kiela, P. R., et al. (2005). 1α,25-Dihydroxyvitamin D_3 upregulates FGF23 gene expression in bone: The final link in a renal-gastrointestinal-skeletal axis that controls phosphate transport. *American Journal of Physiology. Gastrointestinal and Liver Physiology, 289*, G1036–G1042.

Kopic, S., & Geibel, J. P. (2013). Gastric acid, calcium absorption, and their impact on bone health. *Physiological Reviews, 93*, 189–268.

Koszewski, N. J., Horst, R. L., & Goff, J. P. (2012). Importance of apical membrane delivery of 1,25-dihydroxyvitamin D_3 to vitamin D-responsive gene expression in the colon. *American Journal of Physiology. Gastrointestinal and Liver Physiology, 303*, G870–G878.

Kovacs, C. S. (2005). Calcium and bone metabolism during pregnancy and lactation. *Journal of Mammary Gland Biology and Neoplasia, 10*, 105–118.

Kovbasnjuk, O., Leader, J. P., Weinstein, A. M., & Spring, K. R. (1998). Water does not flow across the tight junctions of MDCK cell epithelium. *Proceedings of the National Academy of Sciences of the United States of America, 95*, 6526–6530.

Kutuzova, G. D., & DeLuca, H. F. (2004). Gene expression profiles in rat intestine identify pathways for 1,25-dihydroxyvitamin D_3 stimulated calcium absorption and clarify its immunomodulatory properties. *Archives of Biochemistry and Biophysics, 432*, 152–166.

Kutuzova, G. D., Sundersingh, F., Vaughan, J., Tadi, B. P., Ansay, S. E., Christakos, S., et al. (2008). TRPV6 is not required for 1α,25-dihydroxyvitamin D_3-induced intestinal calcium absorption in vivo. *Proceedings of the National Academy of Sciences of the United States of America, 105*, 19655–19659.

Laohapitakworn, S., Thongbunchoo, J., Nakkrasae, L. I., Krishnamra, N., & Charoenphandhu, N. (2011). Parathyroid hormone (PTH) rapidly enhances CFTR-mediated HCO_3^- secretion in intestinal epithelium-like Caco-2 monolayer: A novel ion regulatory action of PTH. *American Journal of Physiology. Cell Physiology, 301*, C137–C149.

Lee, D., Obukhov, A. G., Shen, Q., Liu, Y., Dhawan, P., Nowycky, M. C., et al. (2006). Calbindin-D_{28k} decreases L-type calcium channel activity and modulates intracellular calcium homeostasis in response to K^+ depolarization in a rat beta cell line RINr1046-38. *Cell Calcium, 39*, 475–485.

Lee, D. B., Walling, M. M., Levine, B. S., Gafter, U., Silis, V., Hodsman, A., et al. (1981). Intestinal and metabolic effect of 1,25-dihydroxyvitamin D_3 in normal adult rat. *American Journal of Physiology, 240*, G90–G96.

Li, H., Seitz, P. K., Thomas, M. L., Selvanayagam, P., Rajaraman, S., & Cooper, C. W. (1995). Widespread expression of the parathyroid hormone-related peptide and PTH/PTHrP receptor genes in intestinal epithelial cells. *Laboratory Investigation, 73*, 864–870.

Lieben, L., Masuyama, R., Torrekens, S., Van Looveren, R., Schrooten, J., Baatsen, P., et al. (2012). Normocalcemia is maintained in mice under conditions of calcium malabsorption by vitamin D-induced inhibition of bone mineralization. *Journal of Clinical Investigation, 122*, 1803–1815.

Meyer, M. B., Watanuki, M., Kim, S., Shevde, N. K., & Pike, J. W. (2006). The human transient receptor potential vanilloid type 6 distal promoter contains multiple vitamin D receptor binding sites that mediate activation by 1,25-dihydroxyvitamin D_3 in intestinal cells. *Molecular Endocrinology, 20*, 1447–1461.

Meyer, M. B., Zella, L. A., Nerenz, R. D., & Pike, J. W. (2007). Characterizing early events associated with the activation of target genes by 1,25-dihydroxyvitamin D_3 in mouse kidney and intestine in vivo. *Journal of Biological Chemistry, 282*, 22344–22352.

Mogami, H., Nakano, K., Tepikin, A. V., & Petersen, O. H. (1997). Ca^{2+} flow via tunnels in polarized cells: Recharging of apical Ca^{2+} stores by focal Ca^{2+} entry through basal membrane patch. *Cell, 88*, 49–55.

Morgan, E. L., Mace, O. J., Affleck, J., & Kellett, G. L. (2007). Apical GLUT2 and $Ca_v1.3$: Regulation of rat intestinal glucose and calcium absorption. *Journal of Physiology, 580*, 593–604.

Morgan, E. L., Mace, O. J., Helliwell, P. A., Affleck, J., & Kellett, G. L. (2003). A role for $Ca_v1.3$ in rat intestinal calcium absorption. *Biochemical and Biophysical Research Communications, 312*, 487–493.

Mullen, T. L., Muller, M., & Van Bruggen, J. T. (1985). Role of solute drag in intestinal transport. *Journal of General Physiology, 85*, 347–363.

Nakkrasae, L. I., Thongon, N., Thongbunchoo, J., Krishnamra, N., & Charoenphandhu, N. (2010). Transepithelial calcium transport in prolactin-exposed intestine-like Caco-2 monolayer after combinatorial knockdown of TRPV5, TRPV6 and $Ca_v1.3$. *Journal of Physiological Sciences, 60*, 9–17.

Need, A. G., & Nordin, B. E. (2008). Misconceptions—Vitamin D insufficiency causes malabsorption of calcium. *Bone, 42*, 1021–1024.

Nemere, I. (1996). Parathyroid hormone rapidly stimulates phosphate transport in perfused duodenal loops of chicks: Lack of modulation by vitamin D metabolites. *Endocrinology, 137*, 3750–3755.

Nemere, I., Garbi, N., Hämmerling, G. J., & Khanal, R. C. (2010). Intestinal cell calcium uptake and the targeted knockout of the $1,25D_3$-MARRS (membrane-associated, rapid response steroid-binding) receptor/PDIA3/Erp57. *Journal of Biological Chemistry, 285*, 31859–31866.

Nemere, I., & Larsson, D. (2002). Does PTH have a direct effect on intestine? *Journal of Cellular Biochemistry, 86*, 29–34.

Nemere, I., Leathers, V., & Norman, A. W. (1986). 1,25-Dihydroxyvitamin D_3-mediated intestinal calcium transport. Biochemical identification of lysosomes containing calcium and calcium-binding protein (calbindin-D_{28K}). *Journal of Biological Chemistry, 261*, 16106–16114.

Nemere, I., & Norman, A. W. (1986). Parathyroid hormone stimulates calcium transport in perfused duodena from normal chicks: Comparison with the rapid (transcaltachic) effect of 1,25-dihydroxyvitamin D_3. *Endocrinology, 119*, 1406–1408.

Nemere, I., & Norman, A. W. (1990). Transcaltachia, vesicular calcium transport, and microtubule-associated calbindin-D_{28K}: Emerging views of 1,25-dihydroxyvitamin D_3 mediated intestinal calcium absorption. *Mineral and Electrolyte Metabolism, 16*, 109–114.

Nemere, I., & Szego, C. M. (1981). Early actions of parathyroid hormone and 1,25-dihydroxycholecalciferol on isolated epithelial cells from rat intestine: I. Limited lysosomal enzyme release and calcium uptake. *Endocrinology, 108*, 1450–1462.

Nilius, B., Prenen, J., Vennekens, R., Hoenderop, J. G., Bindels, R. J., & Droogmans, G. (2001). Modulation of the epithelial calcium channel, ECaC, by intracellular Ca^{2+}. *Cell Calcium, 29*, 417–428.

Norris, S. A., Pettifor, J. M., Gray, D. A., & Buffenstein, R. (2001). Calcium metabolism and bone mass in female rabbits during skeletal maturation: Effects of dietary calcium intake. *Bone, 29*, 62–69.

Okano, T., Tsugawa, N., Morishita, A., & Kato, S. (2004). Regulation of gene expression of epithelial calcium channels in intestine and kidney of mice by 1α,25-dihydroxyvitamin D_3. *Journal of Steroid Biochemistry and Molecular Biology, 89–90*, 335–338.

Olson, E. B., Jr., DeLuca, H. F., & Potts, J. T., Jr. (1972). Calcitonin inhibition of vitamin D-induced intestinal calcium absorption. *Endocrinology, 90*, 151–157.

Ordóñez-Morán, P., Larriba, M. J., Pálmer, H. G., Valero, R. A., Barbáchano, A., Duñach, M., et al. (2008). RhoA–ROCK and p38MAPK–MSK1 mediate vitamin D effects on gene expression, phenotype, and Wnt pathway in colon cancer cells. *Journal of Cell Biology, 183*, 697–710.

Pahuja, D. N., & DeLuca, H. F. (1981). Stimulation of intestinal calcium transport and bone calcium mobilization by prolactin in vitamin D-deficient rats. *Science*, *214*, 1038–1039.

Peacock, M. (2010). Calcium metabolism in health and disease. *Clinical Journal of the American Society of Nephrology*, *5*(Suppl. 1), S23–S30.

Picotto, G., Massheimer, V., & Boland, R. (1997). Parathyroid hormone stimulates calcium influx and the cAMP messenger system in rat enterocytes. *American Journal of Physiology*, *273*, C1349–C1353.

Pitcher, T., & Buffenstein, R. (1994). Passive uptake in the small intestine and active uptake in the hindgut contribute to the highly efficient mineral metabolism of the common mole-rat, *Cryptomys hottentotus*. *British Journal of Nutrition*, *71*, 573–582.

Pitcher, T., & Buffenstein, R. (1995). Intestinal calcium transport in mole-rats (*Cryptomys damarensis* and *Heterocephalus glaber*) is independent of both genomic and non-genomic vitamin D mediation. *Experimental Physiology*, *80*, 597–608.

Quinn, S. J., Thomsen, A. R., Pang, J. L., Kantham, L., Bräuner-Osborne, H., Pollak, M., et al. (2013). Interactions between calcium and phosphorus in the regulation of the production of fibroblast growth factor 23 in vivo. *American Journal of Physiology. Endocrinology and Metabolism*, *304*, E310–E320.

Quinn, S. J., Ye, C. P., Diaz, R., Kifor, O., Bai, M., Vassilev, P., et al. (1997). The Ca^{2+}-sensing receptor: A target for polyamines. *American Journal of Physiology*, *273*, C1315–C1323.

Rexhepaj, R., Alesutan, I., Gu, S., Pelzl, L., Eichenmüller, M., Pathare, G., et al. (2011). SGK1-dependent stimulation of intestinal SGLT1 activity by vitamin D. *Pflügers Archiv: European Journal of Physiology*, *462*, 489–494.

Schröder, B., Schlumbohm, C., Kaune, R., & Breves, G. (1996). Role of calbindin-D_{9k} in buffering cytosolic free Ca^{2+} ions in pig duodenal enterocytes. *Journal of Physiology*, *492*, 715–722.

Schroeder, B., Dahl, M. R., & Breves, G. (1998). Duodenal Ca^{2+} absorption is not stimulated by calcitriol during early postnatal development of pigs. *American Journal of Physiology*, *275*, G305–G313.

Schuette, S. A., Knowles, J. B., & Ford, H. E. (1989). Effect of lactose or its component sugars on jejunal calcium absorption in adult man. *American Journal of Clinical Nutrition*, *50*, 1084–1087.

Shimada, T., Yamazaki, Y., Takahashi, M., Hasegawa, H., Urakawa, I., Oshima, T., et al. (2005). Vitamin D receptor-independent FGF23 actions in regulating phosphate and vitamin D metabolism. *American Journal of Physiology. Renal Physiology*, *289*, F1088–F1095.

Shumilina, E., Xuan, N. T., Matzner, N., Bhandaru, M., Zemtsova, I. M., & Lang, F. (2010). Regulation of calcium signaling in dendritic cells by 1,25-dihydroxyvitamin D_3. *Federation of American Societies for Experimental Biology*, *24*, 1989–1996.

Song, Y., & Fleet, J. C. (2004). 1,25 Dihydroxycholecalciferol-mediated calcium absorption and gene expression are higher in female than in male mice. *Journal of Nutrition*, *134*, 1857–1861.

Suntornsaratoon, P., Kraidith, K., Teerapornpuntakit, J., Dorkkam, N., Wongdee, K., Krishnamra, N., et al. (2014). Pre-suckling calcium supplementation effectively prevents lactation-induced osteopenia in rats. *American Journal of Physiology. Endocrinology and Metabolism*, *306*, E177–E188.

Swaminathan, R., Ker, J., & Care, D. (1974). Calcitonin and intestinal calcium absorption. *Journal of Endocrinology*, *61*, 83–94.

Takeuchi, K., Morikawa, H., Ueda, Y., & Mochizuki, M. (1988). Studies on the effects of placental lactogen on calcium metabolism during pregnancy. *Nihon Naibunpi Gakkai Zasshi*, *64*, 1175–1186.

Tanrattana, C., Charoenphandhu, N., Limlomwongse, L., & Krishnamra, N. (2004). Prolactin directly stimulated the solvent drag-induced calcium transport in the duodenum of female rats. *Biochimica et Biophysica Acta, 1665*, 81–91.

Tazawa, K., Hoshi, K., Kawamoto, S., Tanaka, M., Ejiri, S., & Ozawa, H. (2004). Osteocytic osteolysis observed in rats to which parathyroid hormone was continuously administered. *Journal of Bone and Mineral Metabolism, 22*, 524–529.

Teerapornpuntakit, J., Dorkkam, N., Wongdee, K., Krishnamra, N., & Charoenphandhu, N. (2009). Endurance swimming stimulates transepithelial calcium transport and alters the expression of genes related to calcium absorption in the intestine of rats. *American Journal of Physiology. Endocrinology and Metabolism, 296*, E775–E786.

Teerapornpuntakit, J., Klanchui, A., Karoonuthaisiri, N., Wongdee, K., & Charoenphandhu, N. (2014). Expression of transcripts related to intestinal ion and nutrient absorption in pregnant and lactating rats as determined by custom-designed cDNA microarray. *Molecular and Cellular Biochemistry, 391*, 103–116.

Thacher, T. D., & Abrams, S. A. (2010). Relationship of calcium absorption with 25(OH)D and calcium intake in children with rickets. *Nutrition Reviews, 68*, 682–688.

Thongon, N., Nakkrasae, L. I., Thongbunchoo, J., Krishnamra, N., & Charoenphandhu, N. (2009). Enhancement of calcium transport in Caco-2 monolayer through PKC_ζ-dependent $Ca_v1.3$-mediated transcellular and rectifying paracellular pathways by prolactin. *American Journal of Physiology. Cell Physiology, 296*, C1373–C1382.

Timmermans, J. A., Bindels, R. J., & Van Os, C. H. (1995). Stimulation of plasma membrane Ca^{2+} pump by calbindin-D_{28k} and calmodulin is additive in EGTA-free solutions. *Journal of Nutrition, 125*, 1981S–1986S.

Tudpor, K., Teerapornpuntakit, J., Jantarajit, W., Krishnamra, N., & Charoenphandhu, N. (2008). 1,25-Dihydroxyvitamin D_3 rapidly stimulates the solvent drag-induced paracellular calcium transport in the duodenum of female rats. *Journal of Physiological Sciences, 58*, 297–307.

Turnbull, C. I., Looi, K., Mangum, J. E., Meyer, M., Sayer, R. J., & Hubbard, M. J. (2004). Calbindin independence of calcium transport in developing teeth contradicts the calcium ferry dogma. *Journal of Biological Chemistry, 279*, 55850–55854.

Turner, J. R. (2000). 'Putting the squeeze' on the tight junction: Understanding cytoskeletal regulation. *Seminars in Cell and Developmental Biology, 11*, 301–308.

Ureña, P., Kong, X. F., Abou-Samra, A. B., Jüppner, H., Kronenberg, H. M., Potts, J. T., Jr., et al. (1993). Parathyroid hormone (PTH)/PTH-related peptide receptor messenger ribonucleic acids are widely distributed in rat tissues. *Endocrinology, 133*, 617–623.

van de Graaf, S. F., Hoenderop, J. G., Gkika, D., Lamers, D., Prenen, J., Rescher, U., et al. (2003). Functional expression of the epithelial Ca^{2+} channels (TRPV5 and TRPV6) requires association of the S100A10-annexin 2 complex. *EMBO Journal, 22*, 1478–1487.

Walters, J. R. (1989). Calbindin-D_{9k} stimulates the calcium pump in rat enterocyte basolateral membranes. *American Journal of Physiology, 256*, G124–G128.

Wang, M., Yang, X., Wang, F., Li, R., Ning, H., Na, L., et al. (2013). Calcium-deficiency assessment and biomarker identification by an integrated urinary metabonomics analysis. *BMC Medicine, 11*, 86. http://dx.doi.org/10.1186/1741-7015-11-86.

Wasserman, R. H. (2004). Vitamin D and the dual processes of intestinal calcium absorption. *Journal of Nutrition, 134*, 3137–3139.

Wasserman, R. H., & Fullmer, C. S. (1995). Vitamin D and intestinal calcium transport: Facts, speculations and hypotheses. *Journal of Nutrition, 125*, 1971S–1979S.

Watson, P. H., Fraher, L. J., Hendy, G. N., Chung, U. I., Kisiel, M., Natale, B. V., et al. (2000). Nuclear localization of the type 1 PTH/PTHrP receptor in rat tissues. *Journal of Bone and Mineral Research, 15*, 1033–1044.

Whitworth, N. S., & Grosvenor, C. E. (1978). Transfer of milk prolactin to the plasma of neonatal rats by intestinal absorption. *Journal of Endocrinology, 79*, 191–199.

Wongdee, K., & Charoenphandhu, N. (2013). Regulation of epithelial calcium transport by prolactin: From fish to mammals. *General and Comparative Endocrinology, 181*, 235–240.
Wongdee, K., Teerapornpuntakit, J., Siangpro, C., Chaipai, S., & Charoenphandhu, N. (2013). Duodenal villous hypertrophy and upregulation of claudin-15 protein expression in lactating rats. *Journal of Molecular Histology, 44*, 103–109.
Wood, R. J., Fleet, J. C., Cashman, K., Bruns, M. E., & DeLuca, H. F. (1998). Intestinal calcium absorption in the aged rat: Evidence of intestinal resistance to 1,25$(OH)_2$ vitamin D. *Endocrinology, 139*, 3843–3848.
Wood, R. J., Tchack, L., Angelo, G., Pratt, R. E., & Sonna, L. A. (2004). DNA microarray analysis of vitamin D-induced gene expression in a human colon carcinoma cell line. *Physiological Genomics, 17*, 122–129.
Zelinski, J. M., Sykes, D. E., & Weiser, M. M. (1991). The effect of vitamin D on rat intestinal plasma membrane Ca-pump mRNA. *Biochemical and Biophysical Research Communications, 179*, 749–755.
Zhao, G., & Simpson, R. U. (2010). Interaction between vitamin D receptor with caveolin-3 and regulation by 1,25-dihydroxyvitamin D_3 in adult rat cardiomyocytes. *Journal of Steroid Biochemistry and Molecular Biology, 121*, 159–163.
Zhou, L. X., Nemere, I., & Norman, A. W. (1992). A parathyroid-related peptide induces transcaltachia (the rapid, hormonal stimulation of intestinal Ca^{2+} transport). *Biochemical and Biophysical Research Communications, 186*, 69–73.

CHAPTER FOURTEEN

Endocannabinoid Transport Revisited

Simon Nicolussi, Jürg Gertsch[1]
Institute of Biochemistry and Molecular Medicine, NCCR TransCure, University of Bern, Bern, Switzerland
[1]Corresponding author: e-mail address: gertsch@ibmm.unibe.ch

Contents

Abstract

Endocannabinoids are arachidonic acid-derived endogenous lipids that activate the endocannabinoid system which plays a major role in health and disease. The primary endocannabinoids are anandamide (AEA, *N*-arachidonoylethanolamine) and 2-arachidonoyl glycerol. While their biosynthesis and metabolism have been studied in detail, it remains unclear how endocannabinoids are transported across the cell membrane. In this review, we critically discuss the different models of endocannabinoid trafficking, focusing on AEA cellular uptake which is best studied. The evolution of the current knowledge obtained with different AEA transport inhibitors is reviewed and the confusions caused by the lack of their specificity discussed. A comparative summary of the most important AEA uptake inhibitors and the studies involving their use is provided. Based on a comprehensive literature analysis, we propose a model of facilitated AEA membrane transport followed by intracellular shuttling and sequestration. We conclude that novel and more specific probes will be essential to identify the missing targets involved in endocannabinoid membrane transport.

Vitamins and Hormones, Volume 98
ISSN 0083-6729
http://dx.doi.org/10.1016/bs.vh.2014.12.011

1. THE ENDOCANNABINOID SYSTEM

With the identification of the molecular target of Δ^9-tetrahydrocannabinol (Δ^9-THC), the main psychoactive principle in *Cannabis sativa* L. in 1990 (Gaoni & Mechoulam, 1964; Matsuda, Lolait, Brownstein, Young, & Bonner, 1990), the search for the different molecular components of the endocannabinoid system (ECS) was initiated. The ECS is an evolutionary ancient lipid signaling system which consists of the two G protein-coupled cannabinoid receptors CB_1 and CB_2, their arachidonic acid-derived endogenous ligands called endocannabinoids (Fig. 1), and multiple enzymes and proteins involved in their biosynthesis, transport, and degradation (Howlett et al., 2002; Ligresti, Petrosino, & Di Marzo, 2009; Mechoulam & Parker, 2013; Pertwee, 2014; Pertwee et al., 2010). The ECS appears to accomplish a broad spectrum of modulatory functions in the central nervous system (CNS) primarily via CB_1 receptors and in the immune system primarily via CB_2 receptors, while being involved in keeping up general cellular homeostasis under pathological conditions (Di Marzo, 2008; Ligresti et al., 2009; Mechoulam & Parker, 2013; Pacher & Kunos, 2013). Under physiological conditions, the ECS may act as a protective system that interlinks with other lipid signaling pathways of the arachidonic acid network (Nomura et al., 2011). In the CNS, CB_1 receptor signaling modulates synaptic plasticity, neuronal excitability, reward and motivation, emotions, learning and memory, locomotion, appetite, and thermoregulation (Lutz, 2009; Mechoulam & Parker, 2013; Skaper & Di Marzo, 2012). Peripheral effects of ECS activation include not only differential modulation of the immune response but also regulation of blood

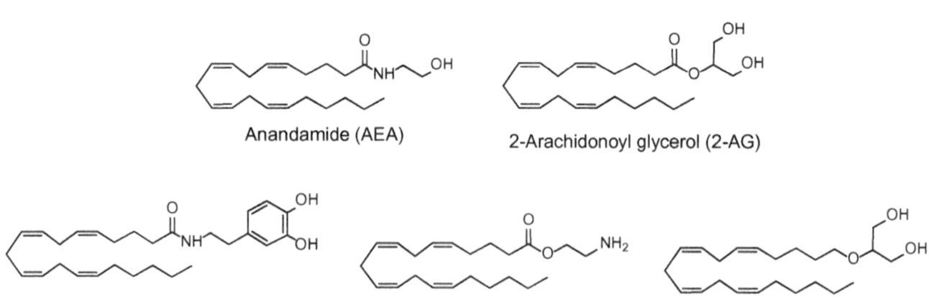

Figure 1 Chemical structures of the major endocannabinoids anandamide (AEA) and 2-arachidonoyl glycerol (2-AG) as well as the proposed endocannabinoids *N*-arachidonoyl dopamine (NADA), virodhamine, and noladin ether (2-AGE). All known endocannabinoids to date are derived from arachidonic acid, though not all arachidonic acid derivatives are endocannabinoids.

pressure, gastric and intestinal motility, lipogenesis, bone development, fertility, and acute or inflammatory nociception (Bab, Ofek, Tam, Rehnelt, & Zimmer, 2008; Di Marzo, Bifulco, & De Petrocellis, 2004; Di Marzo, Ligresti, Morera, Nalli, & Ortar, 2004; Guindon & Hohmann, 2009; Ligresti et al., 2009; Pacher & Kunos, 2013). CB_2 receptors are primarily expressed in immune cells but may be found in many peripheral cell types (Kleyer et al., 2012). While CB_2 receptor activation in general mediates immunosuppressive effects, which limit inflammation and associated tissue injury in large number of pathological conditions, in only few known disease states, its activation may potentially enhance tissue damage (Pacher & Mechoulam, 2011). Overall, there is emerging evidence that lipid endocannabinoid signaling through cannabinoid CB_2 receptors represents a fundamental tissue-protective mechanism. Endocannabinoids may also interact with other GPCRs like GPR55 (Brown, 2007; Ryberg et al., 2007). In addition to GPCRs, endocannabinoids also interact with numerous other intra- and extracellular hydrophobic receptor sites, including peroxisome proliferator-activated receptors (PPAR) and ion channels, with differential binding affinities and kinetics, thus leading to a highly complex and dynamic lipid network (Bernabò, Barboni, & Maccarrone, 2013). Current efforts in academic and applied research are dedicated to the understanding of the apparently complex tissue-specific regulation of the ECS and its underlying molecular machinery in health and disease. As outlined in this review, endocannabinoid transport across cellular membranes is of key importance for the regulation of extra- versus intracellular receptor targets which determine the cell-specific functional repertoire of endocannabinoids.

The major endocannabinoids 2-arachidonoyl glycerol (2-AG) and anandamide (AEA) are produced upon demand triggered by cellular activation (e.g., via calcium influx) at the inner leaflet of the cellular membrane from different phospholipid precursor molecules. Unlike neurotransmitters, these lipids are not stored in vesicles but segregate and interact with hydrophobic surfaces including phospholipid domains and lipid acceptors like serum albumin. Endocannabinoids can access the CB receptor-binding sites from the extracellular space. They are primarily degraded in intracellular domains by the widespread hydrolases, fatty acid amide hydrolase (FAAH) and monoacylglycerol lipase (MAGL), among others (Blankman, Simon, & Cravatt, 2007). Quantitative GC/MS- and LC/MS-based studies show that 2-AG levels are significantly higher than AEA levels in different tissues like brain, spleen, and liver, but AEA is more easily detected in serum and is often increased on stress (Buczynski & Parsons, 2010; Gachet, Rhyn, Bosch, Quednow, & Gertsch, 2014; Marazzi, Kleyer, Paredes, & Gertsch, 2011;

Schreiber et al., 2007). The distinct endocannabinoid levels reflect differential production, trafficking, and degradation in different tissues. Although additional endocannabinoids have been proposed (Fig. 1), namely *N*-arachidonoyl dopamine (NADA), virodhamine, and noladin ether (Bisogno et al., 2000; Hanuš et al., 2001; Porter et al., 2002), their detection is less robust and their physiological role remains unclear. To activate the ECS in many instances seems to be beneficial (Di Marzo, 2008). However, the direct pharmacological targeting of CB receptors may lead to extra-physiological effects, as exemplified, e.g., by the pronounced side effects of CB1 activation by small organic molecules such as Δ^9-THC. Hence, the modulation of the endocannabinoid "tone" may be an alternative more physiological strategy to therapeutically exploit the ECS. While much effort has been dedicated to the development of endocannabinoid degradation inhibitors (i.e., FAAH and MAGL inhibitors; Blankman & Cravatt, 2013), the concept of specifically inhibiting endocannabinoid cellular uptake has been convoluted by the fact that robust and validated lipid transport assays are not trivial to establish (Oddi et al., 2010), and most of the apparently selective AEA uptake inhibitors also inhibit FAAH to different degrees and/or target other proteins within the ECS. In the following, a chronological overview and a discussion on the current state of knowledge on endocannabinoid cellular uptake and transport (trafficking) are provided. The focus of this review is on AEA uptake and its pharmacological inhibition, though also the cellular uptake of 2-AG and the other suggested endocannabinoids is comprehensively reviewed.

2. AEA CELLULAR UPTAKE AND INTRACELLULAR TRANSPORT—A PRIMER

More than 20 years after the discovery of AEA in porcine brain (Devane et al., 1992), the underlying molecular mechanisms of endocannabinoid membrane transport (i.e., release into the extracellular space and cellular uptake) are still not completely understood. While numerous studies on AEA uptake have been performed, few data are available on the uptake of 2-AG and the other proposed endocannabinoids. One reason for this is that transport assay conditions for 2-AG uptake are greatly hampered by the fact that 2-AG is degraded rapidly by both intracellular and extracellular hydrolases, which are differentially expressed in different cell types (*vide infra*). To date, several hypothetical models have been proposed and reviewed for AEA uptake (Felder, Dickason-Chesterfield, & Moore, 2006; Yates & Barker, 2009a, 2009b). Due to the difficulty of transport assay handling, differential transport kinetics as a function of concentration, the lack of an identified endocannabinoid membrane transporter (EMT), and the biases of researchers, there is an ongoing thought-provoking debate on the transport mechanism of endocannabinoids across the plasma membrane (Fowler, 2012, 2013; Yates & Barker, 2009a, 2009b).

In the 1990s, the first reports on the cellular uptake of AEA described a temperature and time-dependent transport, which was linked to the enzymatic hydrolysis by FAAH in C6 glioma cells, N18TG2 neuroblastoma cells, and primary neuronal cells (Deutsch & Chin, 1993; Di Marzo et al., 1994). This uptake process of AEA was rapid ($t_{1/2}=2.5$ min), saturable, and did not compete with closely related *N*-acylethanolamines (NAEs) such as *N*-stearoylethanolamine, *N*-linolenylethanolamine, or *N*-palmitoylethanolamine (PEA; Di Marzo et al., 1994). That NAEs do not compete with AEA uptake was also confirmed by Chicca, Marazzi, Nicolussi, and Gertsch (2012). Unlike [^{3}H]AEA, [^{3}H]PEA did not show a time-dependent clearance from extracellular medium during the uptake process into primary neuronal cells. These data implicated a specific mechanism for cellular AEA uptake following immediate hydrolysis by its major degrading enzyme FAAH (Cravatt, Giang, & Mayfield, 1996; Deutsch & Chin, 1993; Di Marzo et al., 1994). The liberated arachidonic acid and ethanolamine were shown to be incorporated into phospholipids such as phosphatidylethanolamine (Di Marzo et al., 1994; Chicca et al., 2012). Interestingly, the fate of ethanolamine generated by AEA hydrolysis can be distinct in different cell types (e.g., monocytes vs. macrophages; Chicca et al., 2012). The early investigations in the 1990s suggested a carrier-mediated uptake process for AEA. Several research groups then focussed on the elucidation of the AEA uptake mechanisms and provided evidence that this process is not dependent on either ATP or coupled to an ion gradient (Na^+, Cl^-, H^+; Beltramo et al., 1997; Hillard, Edgemond, Jarrahian, & Campbell, 1997; Hillard & Jarrahian, 2000). This pointed to a fundamentally different uptake process (i.e., facilitated or energy-independent membrane transport) than seen for catecholamine or amino acid transport across cell membranes (Hillard & Jarrahian, 2000). Still, the transport process of AEA displayed high-affinity Michaelis–Menten constants in astrocytes ($K_m=0.31$ μM), cortical neurons ($K_m=1.2$ μM), and in the human neuroblastoma cell line CHP100 ($K_m=0.2$ μM; Beltramo et al., 1997; Maccarrone et al., 1998). These values were comparable to those obtained with the transporters of serotonin ($K_m=0.3$–0.5 μM), dopamine ($K_m=0.9$–1.2 μM), and noradrenaline ($K_m=0.45$ μM; Masson, Sagné, Hamon, & El Mestikawy, 1999; Piomelli, 2003). However, among more than 25 cell lines, taking into account different assay protocols and confounding factors such as sticking of lipids to plastic and vials, the range of the apparent K_m values for AEA uptake spreads from 0.13 to 45 μM (Felder et al., 2006; Oddi et al., 2010). Moreover, the given apparent K_m values represent an average affinity constant of AEA to the cellular uptake and transport machinery, which is also affected by other binding proteins. Nevertheless, using an identical assay protocol, different cells show different K_m values for AEA uptake that do not seem to be affected to the same degree as the V_{max} values by FAAH activity. Clearly, the phenomenon of cellular AEA uptake (i.e., either via passive diffusion or via a facilitated membrane transport) and FAAH activity is intrinsically coupled processes in most cell lines (Deutsch et al., 2001; Ligresti et al., 2004; Nicolussi, Chicca, et al., 2014). Any inhibitor of FAAH will thus also inhibit AEA uptake. The existence of cell lines that do not express FAAH but still show AEA uptake that can be inhibited by small organic molecules is fortunate and may help to discover more selective probes (Dickason-Chesterfield et al., 2006; Nicolussi, Chicca, et al., 2014; Nicolussi, Viveros-Paredes, et al., 2014). More recently, different cytoplasmic AEA-binding proteins have been proposed to regulate or to be involved in AEA transport and to be the target of some of the reported AEA uptake inhibitors (Kaczocha, Vivieca, Sun, Glaser, & Deutsch, 2012; Maccarrone, Dainese, & Oddi, 2010). These findings led to spirited conclusions like: "AEA uptake can be adequately described as a diffusion process across the plasma membrane followed by intracellular carrier-mediated transport to effector molecules, catabolic enzymes and sequestration sites" (Fowler, 2013), thus excluding the possibility of a putative EMT. Only recently, new probes that clearly do not inhibit FAAH or bind to known cytoplasmic-binding proteins, but still show potent AEA uptake inhibition have been reported (Nicolussi, Chicca, et al., 2014; Nicolussi, Viveros-Paredes, et al., 2014). Taking into account the different assay conditions used for AEA uptake and FAAH in independent reports, below we discuss the evidence in favor of or against an EMT, starting with the potential inhibitors of this putative protein.

2.1 Pharmacological inhibition of AEA cellular uptake—A history of confusions

A small rock holds back a great wave.

HOMER, the Odyssey

In nature, many biochemical processes are spatially separated. This is particularly true for membrane lipid-derived signaling molecules like endocannabinoids that can initiate cellular signaling at both extracellularly accessible (CB receptors) and intracellularly accessible sites (e.g., TRPVs, PPARs, $GABA_A$; Baur, Gertsch, & Sigel, 2013; Goodfellow & Glass, 2009; Ross, 2003; Sigel et al., 2011; Watanabe et al., 2003). The situation is picturesque: the endocannabinoid is generated in one compartment and then moves by diffusion or shuttling to another domain (e.g., the receptor site), where it triggers a signaling cascade. Since it is a highly lipophilic molecule ($c \log P > 5$), it will likely be associated with hydrophobic protein surfaces/cavities or phospholipid membranes. To terminate signaling, the ligand is shuttled to yet another compartment where the enzyme machinery recognizes it as major substrate, leading to either hydrolysis or further metabolism (e.g., oxygenation). Thus, signaling is orchestrated like a molecular symphony starting with the biosynthesis and release of the ligand, which then moves and finds its receptor (the climax), followed by the separation and parting, culminating in the tragedy of attrition and disappearance. There may well be numerous climaxes as endocannabinoids are promiscuous molecules. The odyssey (from the molecular perspective) of an endocannabinoid is affected by many circumstances, such as segregation into and crossing of membrane domains and barriers. Upon interaction of the lipid with its receptor at a binding site accessible from the extracellular space (as found in many GPCRs), the subsequent coordinated movement involves the transport or diffusion across cell membranes. In this context, the evolution of a membrane protein that facilitates this passage (hence facilitated transport) would make sense. The recent discovery of apparently specific inhibitors of AEA uptake (*vide infra*) may suggest such a gating mechanism. To use an analogy from Greek mythology, the putative EMT acts like Charon, the ferryman of Hades, who brings the ligands into the underworld (Hades could represent the intracellular binding proteins). Charon, fact or fiction, plays an important role as he brings the ligands to the Olympian gods. The underworld is connected to different rivers that carry the lipids to different enzymes like FAAH, MAGL, NAAA, COX-2, CYP450, etc. where they are degraded or transformed, acting like the fields of Asphodel and punishment (FAAH and MAGL) or Elysium (where endocannabinoids are metabolized to other bioactive lipids). As nicely exemplified by neurotransmitter reuptake inhibitors (e.g., selective serotonin reuptake inhibitors), the inhibition of cellular transport can lead to increased receptor activation and prolonged endogenous stimulation and therapeutic effects (Haenisch & Boenisch, 2011). To keep with our analogy, disabling or distracting Charon can prolong the life of a molecule on death row. The interference with the movement of endocannabinoids therefore has great potential to influence pathophysiological processes. Specific inhibitors for particular transport mechanisms (of Charon or Hades type) have the potential to serve as therapeutic agents (Pertwee, 2014; Batista, Gobira, Viana, Aguiar, & Moreira, 2014). Since GPCRs greatly amplify cellular signaling, a small rock (uptake inhibitor) can hold back a great wave of signaling molecules and lead to dramatic changes.

A primary characteristic of a carrier-mediated transport process is its blockage by chemical probes. In the 1990s, the first AEA uptake inhibitors were synthesized. Based on the observed substrate specificity, initially mainly structural analogs of AEA were synthesized and tested for [^{3}H] AEA uptake inhibition in rat brain neurons and astrocytes (Beltramo et al., 1997; Khanolkar et al., 1996). The very first inhibitor of AEA cellular uptake was the *N*-(4-hydroxyphenyl)-arachidonamide AM404, which resulted in an $IC_{50} = 1\ \mu M$ in neurons and an $IC_{50} = 5\ \mu M$ in astrocytes (Table 1; Beltramo et al., 1997). This probe, which was later discovered to be a bioactive arachidonic acid-conjugated metabolite of the drug paracetamol (Högestätt et al., 2005), was reported to be selective over FAAH inhibition ($IC_{50} > 30\ \mu M$). It thus provided the first pharmacological evidence for a specific AEA membrane transporter (AMT; Beltramo et al., 1997). AM404 was later suggested to be a competitive inhibitor of AEA uptake and it was found to be transported as a pseudo-substrate of the putative AMT (Beltramo & Piomelli, 1999). Interestingly, other structural analogs of AEA were not transported, indicating the existence of at least two distinct sets of structural requirements in the function of a putative EMT, one for substrate recognition and another for substrate translocation (Piomelli et al., 1999). Over the years, AM404 was shown to inhibit FAAH by different groups with IC_{50} values close to the IC_{50} values obtained for AEA uptake inhibition (see Table 1), thus questioning the selectivity of this probe. AM404 may also interact with other targets of the ECS such as TRPV1 (Zygmunt, Chuang, Movahed, Julius, & Högestätt, 2000). However, the reversed amide analog AM1172 apparently solved the problem of selectivity over FAAH being an equally potent inhibitor of AEA cellular uptake as AM404, but resistant to hydrolysis by the enzyme (Fegley et al., 2004).

The quest for better AEA cellular uptake inhibitors continued with the aim to increase their potency and evaluate the structure–activity relationship for the inhibition of the putative EMT target (Di Marzo, Bifulco, et al., 2004; Di Marzo, Ligresti, et al., 2004; Jarrahian et al., 2000). However,

Table 1 Comparative summary of selected inhibitors of anandamide cellular uptake (EMT), intracellular transport (FABP5), or AEA degradation (FAAH)

Compound	AEA cellular uptake (EMT)		FAAH	FABP5
	IC_{50} value	Cell type, AEA cellular uptake kinetics	IC_{50} value	K_i value
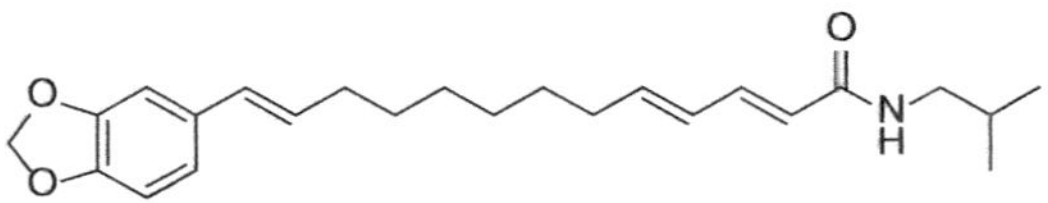 **Guineensine**	0.29 μM[12] 0.37 μM[12] 0.62 μM[12]	U937 [8]$K_m = 0.31 \pm 0.03$ μM, $V_{max} = 427.9 \pm 11.8$ pmol/min/mg U937 HMC-1[a] [11]$K_m = 0.2 \pm 0.2$ μM, $V_{max} = 25 \pm 3$ pmol/min/mg	46.8 μM[12]	>100 μM[12]
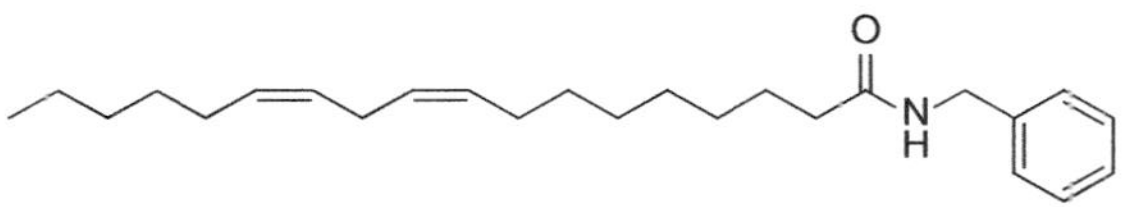 **Macamide 7**	0.67 ± 0.13 μM[32]	U937	4.1 μM[32]	Not reported
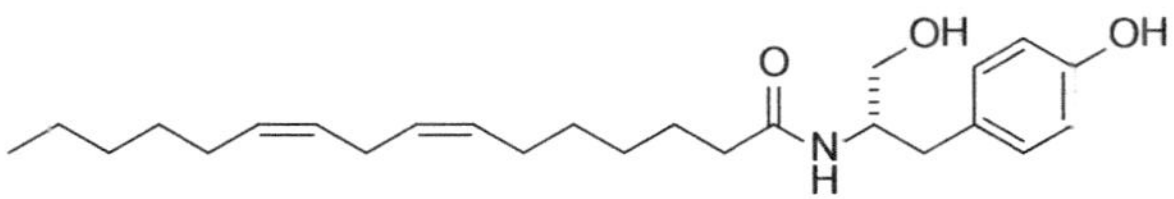 **Farinosone-C derivative (BSL-34)**	0.23 ± 0.14 μM[36]	U937	>10 μM[36]	Not reported
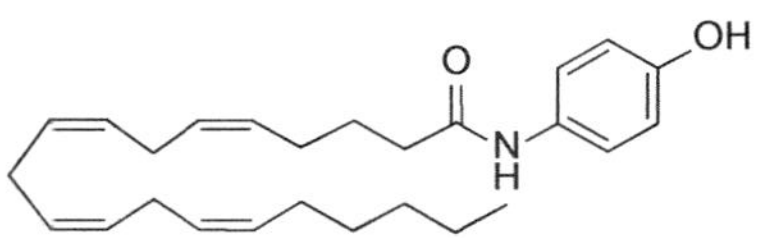 **AM404**	1 μM[1]	Rat cortical neurons [1]$K_m = 1.2$ μM, $V_{max} = 90.9$ pmol/min/mg	5.9 μM[10]	0.39 ± 0.08 μM[17]
	5 μM[1]	Rat cortical astrocytes [1]$K_m = 0.32 \pm 0.1$ μM, $V_{max} = 171$ pmol/min/mg	3.6 μM[10]	
	2.2 ± 0.2 μM[2]	Astrocytoma [2]$K_m = 0.6 \pm 0.1$ μM, $V_{max} = 14.7 \pm 1.5$ pmol/min/mg	0.5 μM[9]	
	10.2 ± 0.7 μM[3]	C6 glioma [4]$K_m = 0.7 \pm 0.1$ μM, $V_{max} = 0.39$ fmol/min/cell	0.8 μM[8]	
	10.9 μM[10]	RBL-2H3 [21]$K_m = 11.4 \pm 2.3$ μM, $V_{max} = 17.5 \pm 2.1 \times 10^{-17}$ mol/min/cell	>30 μM[1]	
	8.1 ± 2.6 μM[3]	RBL-2H3		

Continued

Table 1 Comparative summary of selected inhibitors of anandamide cellular uptake (EMT), intracellular transport (FABP5), or AEA degradation (FAAH)—cont'd

	AEA cellular uptake (EMT)		FAAH	FABP5
Compound	IC_{50} value	Cell type, AEA cellular uptake kinetics	IC_{50} value	K_i value
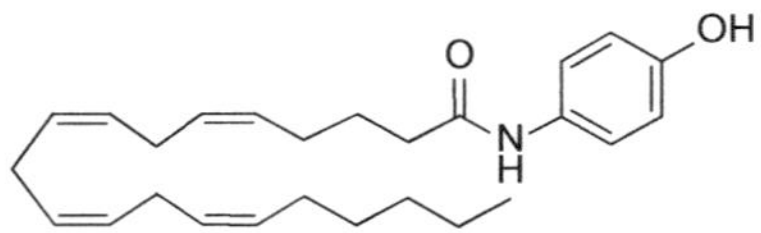	10.2 μM[10]	RBL-2H3	K_i = 0.60 μM[19]	
	4 ± 2 μM[6]	RBL-2H3	6 μM[31]	
	14.9 ± 1.2 μM[27]	RBL-2H3		
	20 μM[35]	RBL-2H3		
	1.8 ± 0.1 μM[8]	U937 [7] $K_m = 0.13 \pm 0.01$ μM, $V_{max} = 140 \pm 15$ pmol/min/mg		
AM404—cont'd	4.9 ± 1.7 μM[19]	Cerebellar granular neurons [22] $K_m = 41$ μM, $V_{max} = 61$ fmol/min/cell		
	>100 μM[8]	HMC-1[a]		
	44.4 ± 4.9 μM[27]	HeLa[a]		
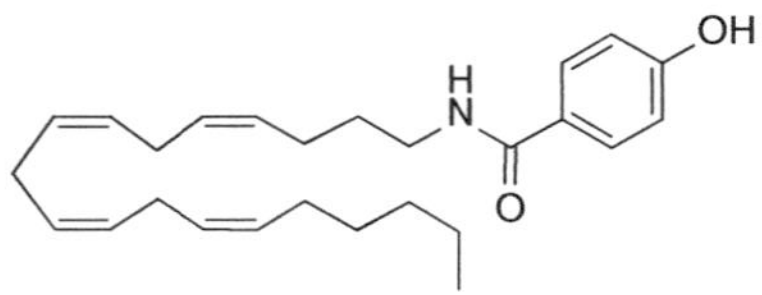	2.5 ± 0.9 μM[33]	Astrocytoma CCF-STTG1	>5 μM[33]	Not reported
	2.1 ± 0.4 μM[33]	Rat cortical neurons	K_i = 3.18 μM[19]	
	24.0 ± 1.9 μM[19]	Cerebellar granular neurons	>100 μM[27]	
	86.6 ± 3.6 μM[27]	RBL-2H3	>50 μM[34]	
	68.9 ± 26.1 μM[27]	HeLa[a]		
AM1172				
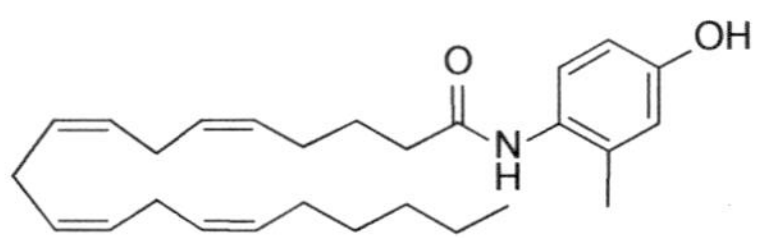	10.2 ± 1.3 μM[3]	C6 glioma	2.0 μM[10]	1.75 ± 0.21 μM[17]
	6.1 μM[10]	C6 glioma	1.2 μM[10]	
	11.2 ± 1.0 μM[3]	RBL-2H3	0.4 μM[8]	
	9.9 ± 1.5 μM[27]	RBL-2H3	K_i = 0.44 μM[19]	
	1.1 ± 0.1 μM[8]	U937	11.1 μM[27]	
	5.5 ± 1.2 μM[19]	Cerebellar granular neurons	5 μM[34]	
VDM11	>100 μM[8]	HMC-1[a]	1.6–2.9 μM[37]	
	23.8 ± 5.5 μM[27]	HeLa[a]		

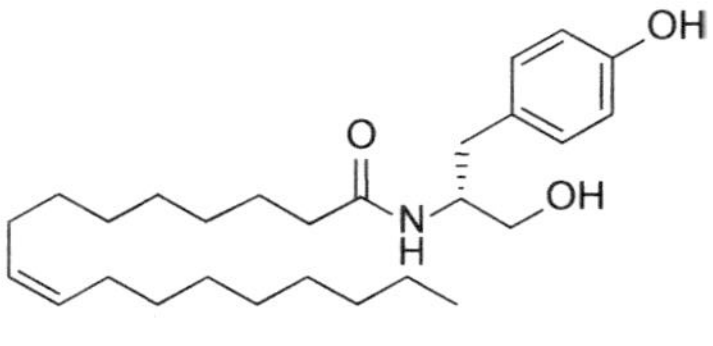 OMDM-2	K_i = 3.0 ± 0.6 μM[18]	RBL-2H3	>50 μM[18]	3.85 ± 0.25 μM[17]
	16.6 μM[10]	C6 glioma	54 μM[10]	>100 μM[8]
	3.2 μM[10]	RBL-2H3	23.4 μM[12]	
	9.1 ± 3.4 μM[27]	RBL-2H3	>100 μM[10]	
	3.93 μM[12]	U937	K_i = 9.7 μM[19]	
	5.2 ± 0.6 μM[14]	U937	>100 μM[27]	
	3.2 μM[8]	U937	>50 μM[34]	
	3.1 μM[8]	HMC-1[a]		
	>100 μM[27]	HeLa[a]		
	4.9 ± 0.5 μM[19]	Cerebellar granular neurons		
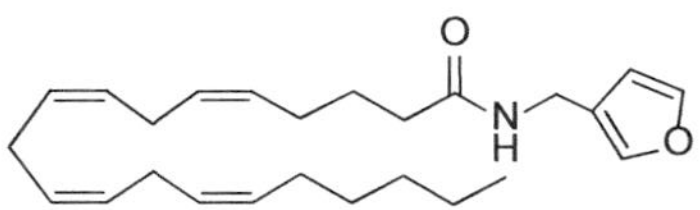 UCM707	0.8 ± 0.4 μM[13]	U937	30 μM[13]	25.8 μM[8]
	1.34 μM[12]	U937	8.32 μM[12]	
	1.8 ± 0.8 μM[14]	U937	>100 μM[12]	
	41 μM[10]	C6 glioma	K_i = 0.37 μM[19]	
	25 μM[10]	RBL-2H3	20.5 μM[27]	
	20.1 ± 5.4 μM[27]	RBL-2H3	50 μM[34]	
	3.6 μM[8]	HMC-1[a]		
	56.4 ± 8.7 μM[27]	HeLa[a]		
	30.3 ± 9.3 μM[19]	Cerebellar granular neurons		
	4 ± 1 μM[24]	Cortical neurons wt [23]K_m = 1.1 μM, V_{max} = 151 pmol/min/mg		
	3 ± 1 μM[24]	Cortical neurons FAAH (−/−) [23]K_m = 1.3 μM, V_{max} = 157 pmol/min/mg		
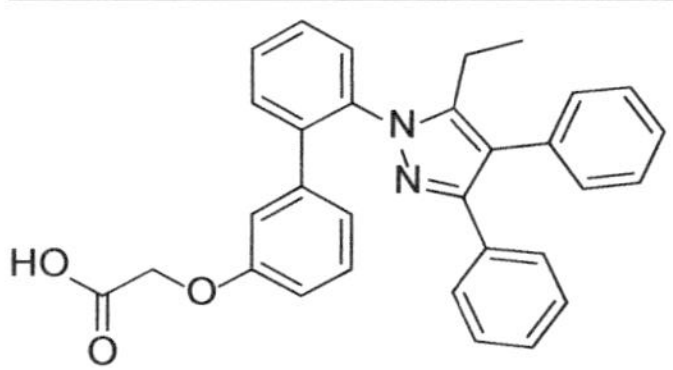 BMS309403	~100 μM (57% inh.)[25]	N18TG2 [26]K_m = 1.8 μM, V_{max} = 1.74 fmol/min/cell	>100 μM[8]	0.35 ± 0.003 μM[16]
	>100 μM (48% inh.)[25]	COS-7-FAAH-eGFP	>100 μM[25]	0.75 ± 0.16 μM[15]
	24.7 ± 6.8 μM[8]	U937		0.89 ± 0.31 μM[17]
	25 ± 11 μM[30]	U937		
	>100 μM (30% inh.)[8]	HMC-1[a]		

Continued

Table 1 Comparative summary of selected inhibitors of anandamide cellular uptake (EMT), intracellular transport (FABP5), or AEA degradation (FAAH)—cont'd

	AEA cellular uptake (EMT)		FAAH	FABP5
Compound	**IC_{50} value**	**Cell type, AEA cellular uptake kinetics**	**IC_{50} value**	**K_i value**
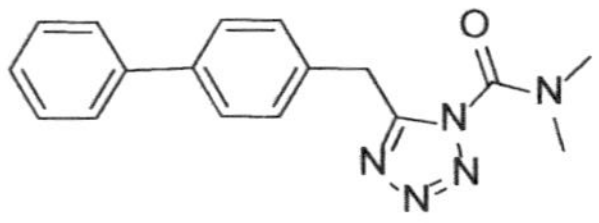 **LY2183240**	270 ± 29 pM[5] 15 nM[28] 1 ± 0.5 nM[30] 1.93 ± 0.47 μM[8] 29.7 ± 6.4 μM[27]	RBL-2H3 [5]$K_m = 4.69 \pm 0.46$ μM, $V_{max} = 0.02$ fmol/min/cell RBL-2H3 U937 HMC-1[a] HeLa[a]	14 ± 0.7 nM[27] 2.1 nM[28] 12.4 nM[29] 1 ± 0.2 nM[30]	Not reported
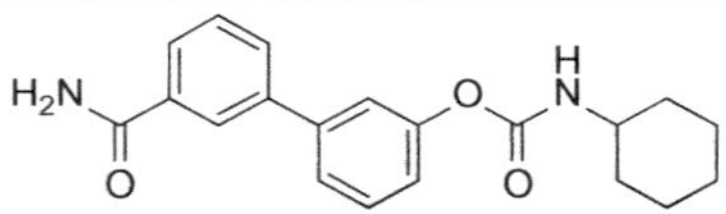 **URB597**	1.5 ± 0.2 nM[27] 948 ± 246 pM[30] 65.1 ± 13.0 μM[27] 9.02 ± 2.91 μM[30]	RBL-2H3 U937 HeLa[a] HMC-1[a]	4.6 ± 1.6 nM[20] 157 ± 5 nM[27] 2 ± 1 nM[12] 2 ± 0.2 nM[30] 2 ± 0.2 nM[34]	Not reported
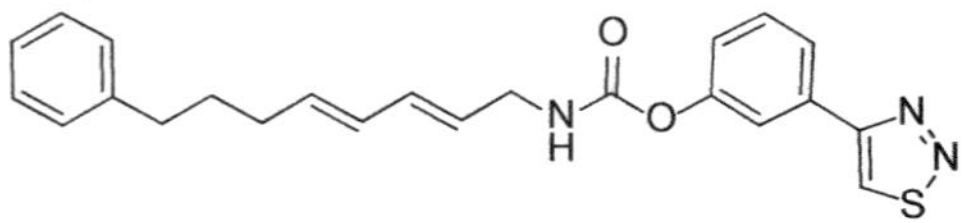 **WOBE492**	479 ± 112 fM[30] 0.12 ± 0.01 μM[30]	U937 HMC-1[a]	14 ± 1 pM[30]	Not reported
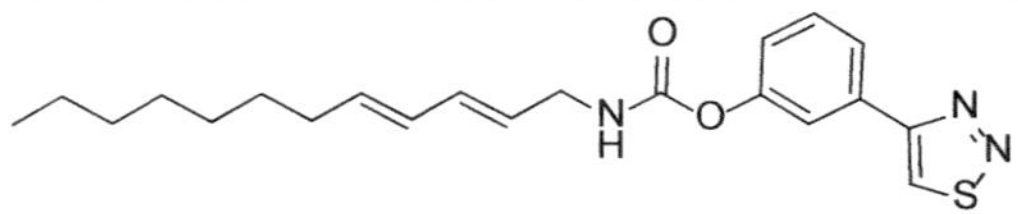 **WOBE498**	5 ± 1 pM[30] 0.25 ± 0.03 μM[30]	U937 HMC-1[a]	15 ± 1 pM[30]	Not reported

[a]Cell type lacking FAAH activity (Day, Rakhshan, Deutsch, & Barker, 2001; Maccarrone, Bari, et al., 2000; Maccarrone, Fiorucci, et al., 2000; Nicolussi, Viveros-Paredes, et al., 2014).
[1]Beltramo et al. (1997), [2]Piomelli et al. (1999), [3]De Petrocellis, Bisogno, Davis, Pertwee, and Di Marzo (2000), [4]Deutsch et al. (2001), [5]Moore et al. (2005), [6]López-Rodríguez et al. (2003), [7]Maccarrone et al. (1998), [8]Nicolussi (2014), [9]Jarrahian, Manna, Edgemond, Campbell, and Hillard (2000), [10]Fowler, Tiger, Ligresti, López-Rodríguez, and Di Marzo (2004), [11]Maccarrone, Bari, et al. (2000), Maccarrone, Fiorucci, et al. (2000), [12]Nicolussi, Viveros-Paredes, et al. (2014), [13]López-Rodríguez et al. (2003) [14]Chicca et al. (2012), [15]Berger et al. (2012), [16]Sulsky et al. (2007), [17]Kaczocha et al. (2012), [18]Ortar, Ligresti, De Petrocellis, Morera, and Di Marzo (2003), [19]Hillard, Shi, Tuniki, Falck, and Campbell (2007), [20]Mor et al. (2004), [21]Rakhshan, Day, Blakely, and Barker (2000), [22]Hillard et al. (1997), [23]Fegley et al. (2004), [24]Ortega-Gutiérrez, Hawkins, Viso, López-Rodríguez, and Cravatt (2004), [25]Kaczocha, Glaser, and Deutsch (2009), [26]Deutsch et al. (2001), [27]Dickason-Chesterfield et al. (2006), [28]Ortar et al. (2008), [29]Alexander and Cravatt (2006), [30]Nicolussi, Chicca, et al. (2014), [31]Glaser et al. (2003), [32]Hajdu, Nicolussi, and Rau (2014), [33]Fegley et al. (2004), [34]Kaczocha, Hermann, Glaser, Bojesen, and Deutsch (2006), [35] Porter et al. (2002), [36]Burch, Chicca, Gertsch, & Gademann, et al. (2014), and [37]Vandevoorde and Fowler (2005).

at the same time, evidence increased that the inhibition of FAAH also leads to an inhibition in [^{3}H]AEA uptake and the aspect of selectivity over FAAH became crucial (Day et al., 2001; Deutsch et al., 2001). Besides the well-studied AEA uptake inhibitor VDM11 (De Petrocellis et al., 2000), more selective inhibitors such as the oleic acid derivatives OMDM-1 and OMDM-2 (Ortar et al., 2003) or UCM707 (López-Rodríguez et al., 2003, 2001) were synthesized (Table 1).

The search for good chemical probes was difficult and the potency of the developed inhibitors did not succeed the low micromolar range. Even worse, despite their initially published selectivity over FAAH, some of these inhibitors were later revealed to inhibit FAAH with similar or even overlapping IC_{50} values with those seen for AEA uptake inhibition (Fowler et al., 2004; Hillard et al., 2007; Vandevoorde & Fowler, 2005). Accordingly, the highly potent FAAH inhibitors LY2183240 and URB597 (see Table 1) resulted in pronounced AEA cellular uptake inhibition in different cell types (Dickason-Chesterfield et al., 2006; Moore et al., 2005; Mor et al., 2004; Nicolussi, Chicca, et al., 2014), essentially representative of all FAAH inhibitors. Not surprisingly, the effect of FAAH inhibitors on AEA uptake was found to be dramatically reduced in cell types lacking FAAH activity, such as HeLa or HMC-1 cells (Dickason-Chesterfield et al., 2006; Nicolussi, Chicca, et al., 2014; Nicolussi, Viveros-Paredes, et al., 2014). It was correctly concluded that selective inhibition of AEA cellular uptake and its enzymatic hydrolysis is difficult to separate and that a relative comparability of different compounds is only given by using the same assay protocols (Fowler et al., 2004). Obviously, care needs to be taken when comparing IC_{50} values between different assays related to AEA transport. For example, FAAH inhibitors might show time dependency and their potency is highly dependent on the assay conditions like preincubation time, enzyme concentration, albumin content in the assay buffer, and substrate concentration (Hillard et al., 2007; Nicolussi, Chicca, et al., 2014; Vandevoorde & Fowler, 2005). One very pronounced example brings this to the point. While in many independent studies, UCM707 was shown to be a very weak inhibitor of FAAH with IC_{50} values in the micromolar range (see Table 1), in one particular case a striking $K_i = 0.37\ \mu M$ was reported (Hillard et al., 2007). The authors of the study state that their FAAH assay conditions were designed to be very sensitive to inhibitors (substrate concentration [^{3}H]AEA = 2 nM), but unfortunately, no positive control was presented (Hillard et al., 2007). Methodological differences of FAAH assays have for sure further confused the development and handling of specific inhibitors of AEA cellular uptake.

Nevertheless, additional indirect evidence for the existence of a carrier-mediated (facilitated) AEA uptake mechanism was provided by the demonstration of AEA uptake in neuronal preparations of FAAH knockout mice (Fegley et al., 2004; Ligresti et al., 2004; Ortega-Gutiérrez et al., 2004). Known AEA uptake inhibitors like UCM707 still reduced the accumulation of AEA, but it could also be shown that the uptake efficacy was much lower in cells lacking FAAH compared to neuronal cells derived from wild-type mice (Fegley et al., 2004; Ligresti et al., 2004; Ortega-Gutiérrez et al., 2004). These studies led to the correct conclusion that FAAH acts as the major driving force for AEA cellular uptake by creating an inward concentration gradient upon the hydrolytic cleavage of AEA from the cytosolic pool. Yet, FAAH cannot be fully responsible for AEA uptake and additional proteins involved in AEA transport across the plasma membrane were proposed (Ligresti et al., 2004). Cell type-dependent differences in which AEA uptake and FAAH-mediated hydrolysis are more or less tightly coupled were described (Yates & Barker, 2009a, 2009b). In agreement with the view that FAAH is not the only player in AEA transport, cells lacking FAAH such as like HMC-1 cells (Maccarrone, Bari, et al., 2000; Maccarrone, Fiorucci, et al., 2000; Nicolussi, Chicca, et al., 2014; Nicolussi, Viveros-Paredes, et al., 2014) show robust AEA uptake kinetics, although with a lower V_{max} than in FAAH-expressing cells.

In 2009, the identification of intracellular carrier proteins and lipid droplets as potential sequestration domains for AEA added a new chapter to the story of AEA transport research (Kaczocha et al., 2009; Oddi et al., 2009, 2008). As will be discussed below, intracellular endocannabinoid-binding proteins show K_d values significantly higher than the K_m values for AEA observed in transport kinetics, suggesting that the endocannabinoid-binding kinetics and stoichiometry of these proteins may be in disagreement with AEA transport. It is also noted that overexpression of any cytosolic AEA-binding protein will have an impact on its uptake kinetics, even though the physiological intracellular concentration may not play the same role. Indeed, some cell types that show AEA uptake do not express known AEA-binding proteins, though the presence of other proteins can of course not be ruled out. Nevertheless, the presence of intracellular endocannabinoid-binding proteins at least suggests that these lipids are spatially organized into domains and that they do not randomly diffuse. In the following, the different more or less feasible models of AEA cellular uptake and transport are discussed one by one, culminating in, according to our opinion, the most conclusive model of facilitated diffusion across the plasma membrane and carrier-mediated intracellular transport of AEA (see Section 2.2.5).

2.2 The different models of AEA cellular uptake and transport

2.2.1 Model of simple diffusion driven by FAAH activity

Early studies have proposed a model in which AEA uptake is mediated exclusively via simple diffusion across the plasma membrane (Fasia, Karava, & Siafaka-Kapadai, 2003; Glaser et al., 2003). Unfortunately, little attention has been paid to AEA assay concentrations, and too often, experiments were carried out using AEA concentrations 10–100 × higher than the K_m value obtained in this cell line. In this model, the enzymatic activity of FAAH creates an inward concentration gradient and is the major driving force for AEA diffusion (Day et al., 2001; Deutsch et al., 2001; Glaser et al., 2003; Kaczocha et al., 2006). The paradoxical inhibition of diffusion by small organic molecule inhibitors as the primary evidence of the carrier-mediated model was readily refuted in this study because inhibitors like AM404 did not inhibit AEA cellular uptake at short incubation times (<40 s) and resulted to be inhibitors of FAAH (Glaser et al., 2003). It was also criticized that previous studies showing a saturable AEA accumulation process were carried out at too long time points at which downstream effects like metabolism and intracellular sequestration become operative (Glaser et al., 2003; Glaser, Kaczocha, & Deutsch, 2005). In another study by the same group, a similar result was reported, measuring AEA uptake after 25 s in RBL-2H3 cells where no inhibition could be observed by the AEA transporter inhibitors UCM707, OMDM-2, and VDM11 (Kaczocha et al., 2006). In this assay setup, the AEA uptake and FAAH hydrolysis profiles of the inhibitors mirrored one of the FAAH inhibitor URB597 at low concentration. These data indicated that weak inhibition of FAAH can have a pronounced effect on AEA cellular uptake (Kaczocha et al., 2006). It was suggested that the target of these inhibitors is not a membrane transporter but FAAH itself or a yet uncharacterized intracellular component that delivers AEA to FAAH (Kaczocha et al., 2006). In contrast, Ligresti et al. (2004) did convincingly show saturable AEA uptake within 90 s not only in RBL-2H3 and C6 glioma cell lines but also in mouse brain synaptosomes from FAAH (−/−) mice. What could be a possible reason for these discrepancies? In the study by Glaser et al. (2003) measuring simple diffusion of AEA, very high nonphysiological AEA concentrations (1–100 μM) were used which may easily mask the transport kinetics seen with concentrations of 50–500 nM. As a note, at $\geq$1 μM AEA simple diffusion kinetics can be readily measured, yet such high concentrations are not found in tissues. Thus, the explanation that the dissimilar results were due to cell type-specific mechanisms or experimental variability within the assays (Yates & Barker, 2009a, 2009b) could be complemented with the different AEA concentrations used. The model of simple AEA diffusion across the cell membrane has in fact been repetitively challenged by numerous biochemical studies probing AEA cellular uptake with the existing older and newer inhibitors (Table 1).

2.2.2 Model of endocytosis-mediated AEA uptake and intracellular sequestration

Another possible explanation how AEA might enter a cell was described as carrier-mediated caveolae-related endocytotic process (Felder et al., 2006; McFarland & Barker, 2004; McFarland et al., 2004). By pharmacological inhibition of caveolae-related endocytosis or by preventing caveolae formation, a significant decrease in AEA accumulation was shown in RBL-2H3 cells (McFarland et al., 2004). The authors proposed that extracellular AEA binds to an ancillary binding protein in caveolin-rich regions of the plasma membrane and is taken up via a rapid endocytotic event (McFarland et al., 2004; Yates & Barker, 2009a, 2009b). Lipid rafts are specialized microdomains within the plasma membrane, which are enriched with cholesterol (Brown & London, 2000). AEA was shown to be able to form complexes with this steroid and treatment of C6 glioma cells with cholesterol doubled the catalytic efficiency of AEA uptake (Bari, Paradisi, Pasquariello, & Maccarrone, 2005; Di Pasquale, Chahinian, Sanchez, & Fantini, 2009). By disruption of the caveolae-rich membrane domain formation with cholesterol-depleting agents, a 50% reduction in AEA uptake was measured (McFarland et al., 2004). Moreover, siRNA-mediated knockdown of the dynamin-2 (Dyn2) protein resulted in decreased uptake of the fluorescent AEA analog SKM4-45-1 in mouse neuronal CAD cells. However, the knockdown of Dyn2 did not affect the uptake of radiolabeled AEA, which seriously questioned the relevance of the finding (McFarland, Bardell, & Yates, 2008).

The model of endocytosis-mediated AEA uptake proposes that extracellular AEA binds to a membrane carrier protein located within caveolae lipid rafts and that caveolae-derived vesicle formation and subsequent endocytosis of the membrane-packaged endocannabinoids are induced (Yates & Barker, 2009a, 2009b). The authors state that this model supports the existence of a plasma membrane-localized AEA-binding protein and is in line with intracellular sequestration by a protein or lipid compartment for AEA (Hillard &

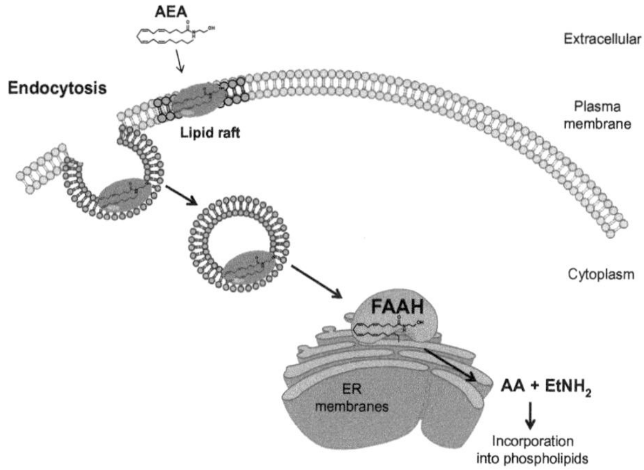

Figure 2 Model of endocytosis-mediated AEA uptake and intracellular sequestration. (See the color plate.)

Jarrahian, 2003; McFarland et al., 2008). Additional support for this model was provided by the demonstration that exogenously applied AEA and also 2-AG localized in lipid raft regions in dorsal root ganglia and microglia cell lines (Rimmerman et al., 2012, 2008).

This model (Fig. 2) may have some relevance as endocytosis is a fundamental process of membrane recovery in neurons upon calcium-related exocytosis of secretory vesicles (Houy et al., 2013). Intracellular sequestration of ECs, which would provide an intracellular compartment that is not free to equilibrate with the extracellular AEA pool, was suggested (Hillard & Jarrahian, 2000, 2005). Related experiments applying biotinylated AEA showed lipid droplets (adiposomes) as subcellular compartments capable of trapping AEA (Oddi et al., 2008). The endocytosis model can be considered as a feasible route of endocannabinoid uptake, which might take place under certain circumstances (e.g., also in caveolin independent massive endocytosis). However, as will be shown below, this model cannot explain why AEA uptake inhibitors can also inhibit the efflux of AEA. Besides, the stoichiometry between added AEA and the presence of ancillary binding proteins, which simply bind the lipid, would have to be calculated.

2.2.3 Model of passive diffusion across the plasma membrane following carrier-mediated intracellular transport and sequestration of AEA

Based on the "simple diffusion model" described in Section 2.2.1, an updated model for AEA cellular uptake was proposed in 2009, right after

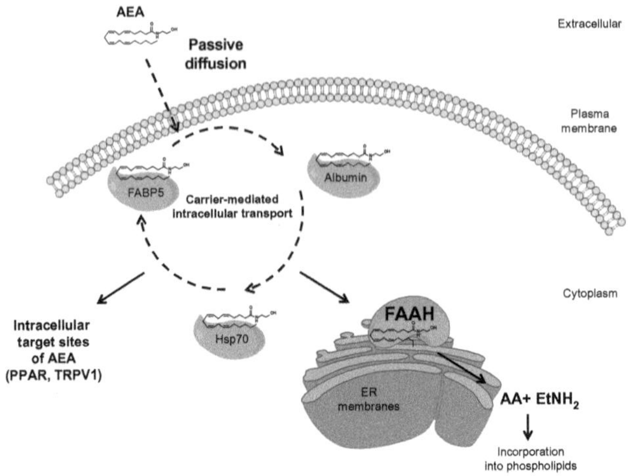

Figure 3 Model of passive diffusion of AEA across the plasma membrane following intracellular sequestration and/or carrier-mediated transport. (See the color plate.)

the simultaneous identification of different cytosolic AEA carrier proteins (FAPB5, Hsp70 (heat-shock protein 70), and albumin; Kaczocha et al., 2009; Oddi et al., 2009). This model (Fig. 3) describes a passive diffusion process of AEA across the plasma membrane followed by intracellular carrier-mediated transport to effector proteins, catabolic enzymes, and sequestration sites (Fowler, 2012, 2013). Noteworthy, the involvement of fatty acid binding proteins (FABPs) in the transport of endocannabinoids was presented already one year earlier in the context of intracellular PPAR activation (Sun, Alexander, Kendall, & Bennett, 2008). Consequently, their role in AEA cellular uptake was investigated and FABP5 and FABP7 were shown to mediate AEA intracellular transport from the plasma membrane to FAAH in artificial COS-7-FAAH-eGFP and N18TG2 neuroblastoma cells (Kaczocha et al., 2009). Moreover, overexpression of COS-7-FAAH-eGFP cells with FABP5 and FABP7 increased AEA uptake and hydrolysis by 32% and 35%, respectively (Kaczocha et al., 2009). N18TG2 cells showed an increase of 36% upon FABP5 and 42% upon FABP7 overexpression. Using these artificial cell constructs, it was shown that a reduction of AEA uptake and hydrolysis could be monitored after preincubation of the cells with the FABP5 inhibitor BMS309403 (see also Table 1). Interestingly, the efficacy of inhibition resulted to be moderate. Very high concentrations of 20–100 μM were needed to reach significant effects (Kaczocha et al., 2009). Intriguingly, while BMS309403 exhibits a $K_i = 350 \pm 3$ nM for FABP5 binding, a concentration of 100 μM of this probe was needed to reach ~50% inhibition of cellular AEA uptake (see also Table 1; Furuhashi & Hotamisligil, 2008; Kaczocha et al., 2009; Sulsky et al., 2007). In another study, FABP5 was suggested to be the main target of the previously reported AEA uptake inhibitors OMDM-2, VDM11, and AM404 (Kaczocha et al., 2012). In Table 1, an overview of these compounds is presented based on the published literature data. These inhibitors showed binding affinities to FABP5 comparable to the published K_i values for AEA-FABP5 binding ($K_i = 1.26 \pm 0.18$ μM (Kaczocha et al., 2012) and $K_i = 8.92$ μM (Nicolussi, 2014; Nicolussi, Viveros-Paredes, et al., 2014)). Surprisingly, the endogenous fatty acid ligand arachidonic acid showed a strong affinity to FABP5, one order of magnitude higher ($K_i = 0.12 \pm 0.01$ μM (Kaczocha et al., 2012)). It is generally accepted that arachidonic acid does not affect AEA cellular uptake up to a concentration of 100 μM (Beltramo et al., 1997; Chicca et al., 2012; Hillard et al., 1997; Piomelli et al., 1999), a finding that would challenge the role of FABP5 in AEA transport. The development of FABP5 inhibitors to modulate AEA cellular transport is ongoing (Berger et al., 2012) and a first *in vivo* evaluation of one of these compounds was recently reported (Kaczocha et al., 2014). However, it is noted that the high concentrations needed in *in vitro* AEA cellular uptake inhibition experiments do not match the affinity of the used compounds to FABP5. The further unaddressed points are the expression levels of FABPs in nontransfected cell lines or in mouse brain neurons and whether they correlate with AEA cellular uptake. As already pointed out above, the K_i values obtained for AEA binding to FABP5 do not match the K_m values in many cells that show AEA transport. The physiological role, especially of FABP5 in AEA uptake and intracellular trafficking, has to be further investigated. Recently, it was shown that FABP5 both promotes the hydrolysis of AEA into arachidonic acid and thus reduces brain AEA levels and directly shuttles arachidonic acid to the nucleus where it delivers it to PPARβ/δ, enabling its activation (Yu, Levi, Casadesus, Kunos, & Noy, 2014). In FABP5 (−/−) versus wild-type mice, a 50% higher level of AEA could be detected in whole brains. However, another study mentioned that in adult neurons, neither FABP5 nor FABP7 is expressed in significant amounts (Liedhegner, Vogt, Sem, Cunningham, & Hillard, 2014).

In the group of Maccarrone, albumin and Hsp70 have been identified as cytosolic AEA-binding proteins in mouse skin keratinocytes using proteomics (Oddi et al., 2009). These two proteins were proposed to act as "AEA chaperons" at the inner plasma membrane, where they promote the desorption of the endocannabinoid and facilitate its transport through the cytoplasm (Oddi et al., 2009). These findings are interesting and further support the model of carrier-mediated intracellular transport of AEA. Additional biophysical data will be needed to assess the importance of these binding proteins in cellular systems. Another candidate in the list of intracellular carrier proteins for AEA is the recently identified FAAH-like AEA transporter (FLAT; Fu et al., 2011). It was proposed to be a partially cytosolic, catalytically silent variant of the AEA-degrading enzyme FAAH. However, the role and existence of FLAT as an AEA transporter was strongly questioned as no expression in mouse brain, spinal cord, or dorsal root ganglia could be detected by an independent group (Fowler, 2014; Leung, Elmes, Glaser, Deutsch, & Kaczocha, 2013). Furthermore, a certain enzymatic activity could still be detected in artificial FLAT-transfected HeLa cells (Leung et al., 2013). An inhibitor of FLAT (ARN272) showed promising indirect cannabimimetic effects in a mouse model of nausea and vomiting (O'Brien et al., 2013). Further studies are awaited confirming the tissue-specific expression of FLAT and the pharmacological profiling of ARN272.

In a recent docking study, the sterol carrier protein-2 (SCP-2) was proposed to be yet another potential endocannabinoid carrier protein (Liedhegner et al., 2014). Although an increase of AEA accumulation could be detected in SCP-2-transfected HEK-293 cells, competition experiments with AM404 and 2-AG did not show a significant difference in their IC_{50} values. It was concluded that SCP-2 is a low-affinity binding protein for AEA (micromolar range) and that it might facilitate AEA cellular uptake to a minor degree. Given the nanomolar concentrations of AEA in the brain, this protein (and possibly others) does not contribute significantly to the intracellular sequestration of AEA, which is a fair statement (Liedhegner et al., 2014). In theory, any protein with hydrophobic surfaces/cavities may serve as an acceptor for lipids like AEA and other endocannabinoids. Because in most studies little attention is paid to stoichiometry, the question whether the K_d value and stoichiometry match usually remains unanswered. To our opinion, the different organizational forms of phospholipid membranes may exhibit stronger binding affinity to AEA (though no K_d was determined) than do some of the proposed binding proteins with micromolar K_i values.

The identification of intracellular carrier proteins for AEA has clearly provided a missing link to explain how endocannabinoids are able to cross the cytosol, which constitutes a hydrophilic barrier for these lipophilic compounds (Fegley et al., 2004; Fowler, 2012, 2013; Glaser et al., 2005;

Hillard & Jarrahian, 2003; Hillard et al., 2007; Kaczocha et al., 2009; Oddi et al., 2009). The physiological role of intracellular carriers in AEA transport needs to be further elucidated and it has to be demonstrated how exactly these proteins interact with the cell plasma membrane. The final question left open is whether a specific membrane transporter protein is involved in the AEA uptake process across the plasma membrane.

2.2.4 The putative EMT—What is the evidence?

Prior to 2009, the vast majority of data suggested that AEA crosses the plasma membrane via a facilitated diffusion process mediated by a putative EMT (Beltramo et al., 1997; Di Marzo et al., 1994; Fegley et al., 2004; Hillard et al., 1997; Ligresti et al., 2004; Maccarrone, Bari, et al., 2000; Maccarrone, Fiorucci, et al., 2000; McFarland & Barker, 2004; Moore et al., 2005; Ortega-Gutiérrez et al., 2004; Piomelli, 2003; Rakhshan et al., 2000; Yates & Barker, 2009a, 2009b). However, a specific membrane transporter protein for endocannabinoids has to date not been identified. As mentioned in Section 2.1 and summarized in Table 1, different inhibitors of AEA cellular uptake including AM404, VDM11, OMDM-2, or UCM707 provided strong indirect evidence for such a target by pharmacological inhibition. For example, the inhibitor UCM707 was investigated in neuronal preparations of FAAH (−/−) mice and still showed an $IC_{50} = 3 \pm 1$ μM for AEA accumulation (Ortega-Gutiérrez et al., 2004). This was in agreement with its selectivity for AEA uptake inhibition over FAAH (López-Rodríguez et al., 2003). A direct comparison of the data obtained from neuronal cells of FAAH (+/+) mice nicely demonstrated that AEA cellular uptake is a facilitated process including multiple components, whereof a specific "UCM707-binding protein" was proposed to participate with a relative contribution of at least 30% (Ortega-Gutiérrez et al., 2004). FABP5 as an intracellular endocannabinoid carrier protein (Kaczocha et al., 2012; Sanson et al., 2014) was therefore an intuitive suggestion for such a UCM707 protein target. In a recent study, the affinity of UCM707 to FAPB5 was evaluated (see Table 1) and resulted in a $K_i = 25.8$ μM (19.5–44.7 μM; Nicolussi, 2014). This low-affinity interaction with FAB5 does not match the determined IC_{50} value for AEA cellular uptake. Moreover, given that UCM707 still works in FAAH-lacking cells (*vide supra*) and synergizes with FAAH inhibitors for AEA uptake inhibition and also inhibits AEA efflux (Chicca et al., 2012), the possibility that UCM707 targets the putative EMT is left open.

We recently identified the plant-natural *N*-isobutylamide guineensine as the first published nanomolar and strongly selective inhibitor of AEA cellular uptake over FAAH inhibition and other ECS targets (Nicolussi, Viveros-Paredes, et al., 2014). In a high-content screening of medicinal plant extracts, a *Piper nigrum* extract (5 μg/mL) was found to block [^{3}H]AEA uptake in U937 cells comparable to the positive controls UCM707 and OMDM-2 which were used at a concentration of 10 μM. Bioactivity-guided isolation finally revealed guineensine as the main active constituent with an IC_{50}=290 nM (190–440 nM) for AEA cellular uptake. Guineensine was characterized in some detail and did not show a relevant inhibition of FAAH or FABP5 binding (K_i>100 μM). Intriguingly, the nanomolar inhibition of AEA cellular uptake could be reproduced also in HMC-1 cells lacking FAAH. In Table 1, guineensine is shown in comparison to the most important AEA cellular uptake inhibitors known to date. However, the primary target of guineensine has not yet been identified though the pharmacological data (indirect cannabimimetic effects) strongly suggest that it targets the putative EMT. Another recently identified compound in the list of plant-derived natural AEA uptake inhibitors is the *N*-benzyl-(9*Z*,12*Z*)-octadecadieneamide (macamide 7 in Table 1), which not only exhibits a nanomolar IC_{50} value for AEA uptake inhibition but also inhibits FAAH at low micromolar concentrations (Hajdu et al., 2014). Furthermore, an analog of the natural product farinosone-C (BSL-34 in Table 1) was found to be a more selective inhibitor of AEA uptake (IC_{50} value=232 nM (131–394 nM)) over FAAH inhibition (IC_{50}>10 μM), with close structural similarity to OMDM-2 (Burch et al., 2014). More potent and specific probes are currently being developed in our laboratory. Recently, starting from the natural product-derived 2*E*,4*E*-dodecadiene scaffold, a series of diverse *N*-alkylcarbamates were prepared which showed both ultrapotent FAAH inhibition, leading to hyperpotent (fM) AEA uptake inhibitors into U937 cells (Nicolussi, Chicca, et al., 2014). Interestingly, several *N*-alkylcarbamates (e.g., WOBE492 and WOBE498) showed a FAAH-independent AEA uptake inhibition in HMC-1 cells with IC_{50} values below 300 nM.

Additional evidence for the existence of the putative EMT involved in the translocation of AEA across the plasma membrane was provided in 2005 by the pharmaceutical company Eli Lilly. They have developed the highly potent tetrazole inhibitor called LY2183240 with an astonishing potency of IC_{50}=270±29 pM in RBL-2H3 cells for AEA cellular uptake (Moore et al., 2005; see Table 1). Using the modified radiolabeled probe

[^{125}I]-LY2318912, a high-affinity membrane-binding site involved in the transport of AEA could be identified, curiously also in FAAH-lacking HeLa cells ($K_d = 7.06 \pm 1.69$ nM, $B_{max} = 32.2 \pm 2.98$ fmol/mg). In human FAAH-transfected HeLa cells, neither the binding affinity (K_d) nor the B_{max} value changed significantly, indicating that the binding site is independent of FAAH (Moore et al., 2005). Having raised high hopes for the expected molecular identification of the putative EMT (Mechoulam & Deutsch, 2005), a follow-up study was never published. On the contrary, in the same year, LY2183240 was shown to be an ultrapotent, irreversible, and non-specific inhibitor of FAAH, MAGL, and other serine hydrolases (Alexander & Cravatt, 2006). The irreversible inhibition of FAAH clearly explained the high potency of the compound for AEA cellular uptake inhibition, but the question of the identity of the reported membrane target site was left open.

In the following, experiments that are strongly in favor of the putative EMT are presented. In several cell types including cerebellar granule cells, HUVEC, striatal neurons, and U937 cells, AEA efflux could be monitored after preloading of the cells with [^{3}H]AEA or stimulation of AEA biosynthesis (Chicca et al., 2012; Gerdeman, Ronesi, & Lovinger, 2002; Hillard et al., 1997; Ligresti et al., 2004; Maccarrone, Bari, Battista, & Finazzi-Agrò, 2002; Ronesi, Gerdeman, & Lovinger, 2004). Ionomycin (300 nM) was used to stimulate the production of AEA in cerebellar granule cells and led to a 56% reduction in the uptake of extracellularly added [^{3}H]AEA (Hillard & Jarrahian, 2003). In this context, it was also demonstrated that the transport inhibitor VDM11 inhibited the release of *de novo* biosynthesized AEA in HEK-293 cells (Ligresti et al., 2004). These studies pointed toward a bidirectional membrane transport mechanism for AEA shown by independent groups (Chicca et al., 2012; Hillard et al., 1997; Ligresti et al., 2004; Maccarrone et al., 2002). AEA as well as 2-AG release was assessed in an electrophysiological study measuring striatal long-term depression in acute brain slice preparation (Ronesi et al., 2004). By postsynaptically blocking the endocannabinoid membrane transport using VDM11, a disruption of endocannabinoid release could be achieved (Ronesi et al., 2004). In another study, OMDM-2 and AM404 increased activity-dependent AEA and 2-AG levels in the hypothalamus and inhibited the synaptically driven spiking activity in postsynaptic neurons upon enhanced retrograde signaling (Di et al., 2005). Most importantly, the blocked postsynaptic-spiking activity after extracellular treatment with OMDM-2 recovered after 30 min of washout with artificial cerebrospinal fluid (Di et al., 2005). Such an effect is hard to explain without the participation of a membrane transporter protein (EMT), because the washout should not affect intracellular carriers. Taken together, AEA transport across the plasma membrane has the characteristics of facilitated diffusion in a bidirectional way, which can be inhibited by known AEA cellular uptake inhibitors (Table 1; Chicca et al., 2012; Gerdeman et al., 2002; Hillard & Jarrahian, 2000; Ligresti et al., 2004). Additionally, it shows the phenomenon of the "*trans* effect of flux coupling" which occurs within the first 20 s of [^{3}H]AEA incubation in cerebellar granule cells (Hillard & Jarrahian, 2000). In principle, a transport protein accumulates and orients itself toward the extracellular space in response to accumulated intracellular AEA (Hillard & Jarrahian, 2000; Yates & Barker, 2009a, 2009b). Consequently, AEA can be transported against the concentration gradient, which could be monitored by [^{3}H] AEA that was added extracellularly (Hillard & Jarrahian, 2000).

Based on an uptake assay in endothelial cells using the fluorescent AEA derivative SKM4-45-1, a role for TRPV1 in AEA uptake was suggested (Hofmann et al., 2014). However, in this study, several confounding factors were introduced. For example, it was not clear whether the observed uptake process was more an event of SKM4-45-1 binding to TRPV1 following internalization. The misleading overlap between the ligand recognition properties of TRPV1 and the putative EMT has been studied already years ago (Beltramo & Piomelli, 1999; De Petrocellis et al., 2000; Melck et al., 1999).

In summary, selective pharmacological inhibition, bidirectional transport across the plasma membrane, inhibition of endocannabinoid release by AEA uptake inhibitors, prolonged retrograde signaling in electrophysiological studies, the "*trans* effect of flux coupling, and the identification of a high-affinity membrane-binding site distinct from FAAH are findings which cannot be explained by intracellular carrier proteins alone. Although the indirect evidence in favor of a plasma membrane transporter has recently been challenged, there are still many unanswered questions and too many gaps to explain the mechanism of how AEA exactly crosses the plasma membrane of a cell. The molecular identification of the proposed EMT would have massive implications in the field and would finally terminate the ongoing debate on endocannabinoid membrane transport.

2.2.5 Model of facilitated diffusion across the plasma membrane and carrier-mediated intracellular transport of AEA

The following model integrates the indirect evidence provided in favor of a transporter-mediated uptake of AEA at the plasma membrane

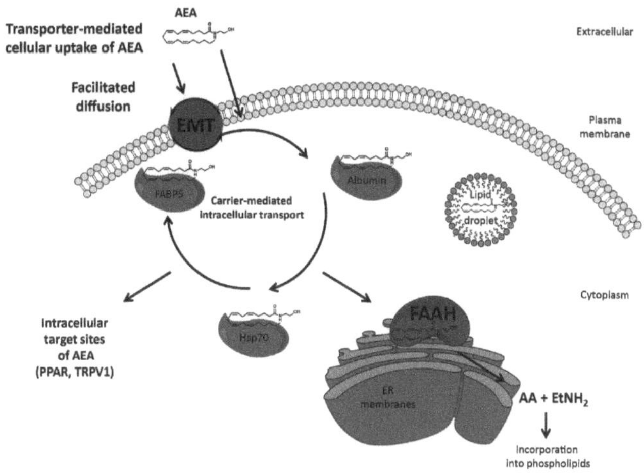

Figure 4 Model of facilitated diffusion (passive transport) across the plasma membrane, mediated by a putative endocannabinoid membrane transporter (EMT) followed by intracellular carrier-mediated transport.

(EMT; summarized in Section 2.2.4) and the cytosolic shuttling of the endocannabinoid (Fig. 4). As discussed above, intracellular carrier proteins exemplified by FABP5, Hsp70 and albumin may explain how AEA is transported through the hydrophilic environment of the cytosol to reach intracellular targets (PPAR, TRPV1) and degradation sites (FAAH; reviewed by Maccarrone et al., 2010).

This comprehensive model suggests that AEA associates with the plasma membrane and accumulates in certain domains where it binds to a specific high-affinity membrane transporter protein that facilitates translocation of its substrate rapidly toward the inner leaflet of the lipid bilayer. Noteworthy, as expected for a passive EMT, this process is likely bidirectional. Entering the cytosol, AEA is delivered to intracellular carrier proteins, which differentially shuttle the endocannabinoid to its target or degradation sites in inner membrane compartments (Kaczocha et al., 2009; Oddi et al., 2009). In endoplasmic membranes, hydrolytic degradation by the FAAH takes places and arachidonic acid and ethanolamine are released. Depending on cell types, ethanolamine is rapidly incorporated into phospholipids (Chicca et al., 2012; Di Marzo et al., 1994).

No matter which model is currently favored, it is clear that different cell types are likely to utilize different mechanisms of endocannabinoid transport (Fowler, 2013). We believe that primarily neurons and immune cells make use of a putative EMT. A more general model without the identification of the putative EMT and knowledge about its expression and regulation is therefore unrealistic.

3. AEA AND 2-AG TRANSPORT AT THE SYNAPSE

2-AG is the major endocannabinoid in the brain. 2-AG levels reach in average 12 nmol/g, while the concentration of AEA is 19 pmol/g in rat whole brain (Buczynski & Parsons, 2010). 2-AG is 100- to 1000-fold more abundant than AEA, depending on extraction protocols, tissues, and species. In contrast to AEA, which is present at low nanomolar concentrations, 2-AG was estimated to reach low micromolar brain concentrations (Di Marzo, 2006). Curiously, relatively high levels of 2-AG can be detected in batches of fetal bovine sera, which are routinely used in cellular experiments (Marazzi et al., 2011). The main route of 2-AG biosynthesis is executed by the diacylglycerol lipase (DAGL; Basavarajappa, 2007; Bisogno et al., 2003; Kondo et al., 1998; Stella, Schweitzer, & Piomelli, 1997). There is a general consensus about the role of the serine hydrolase MAGL as the major degrading enzyme of 2-AG (Blankman et al., 2007; Dinh, Kathuria, & Piomelli, 2004; Saario et al., 2005). In two studies, the overall contribution of MAGL in 2-AG hydrolysis was at least 50% in rat brain and 85% in mouse brain (Blankman et al., 2007; Dinh et al., 2004). At the synapse, MAGL shows a complementary localization to DAGL, in agreement with the role of 2-AG in CB_1 receptor-mediated retrograde signaling (Bisogno et al., 2003; Castillo, Younts, Chávez, & Hashimotodani, 2012; Gulyas et al., 2004). In a comprehensive profile of mouse brain enzymes able to hydrolyze 2-AG, two other important serine hydrolases, the α/β hydrolase 6 (ABHD6) and α/β hydrolase 12 (ABHD12), were reported to mediate the remaining MAGL-independent 2-AG hydrolase activity (Blankman et al., 2007). While ABHD12 participates by ~9%, ABHD12 cleaves 2-AG by only ~4%. The remaining 2% might be mediated by FAAH and other enzymes (Blankman et al., 2007). The different routes of 2-AG and AEA biosynthesis and degradation have been reviewed elsewhere (Muccioli, 2010) and are only briefly mentioned in this context because of the link between EC cellular uptake and degradation. In Fig. 5, the enzymatic inactivation and transformation of 2-AG and AEA are illustrated.

In case of AEA, FAAH is the major degrading enzyme in the brain (Cravatt et al., 1996). Although different routes of AEA transformation exist, which in theory might influence its cellular uptake, their contributions seem

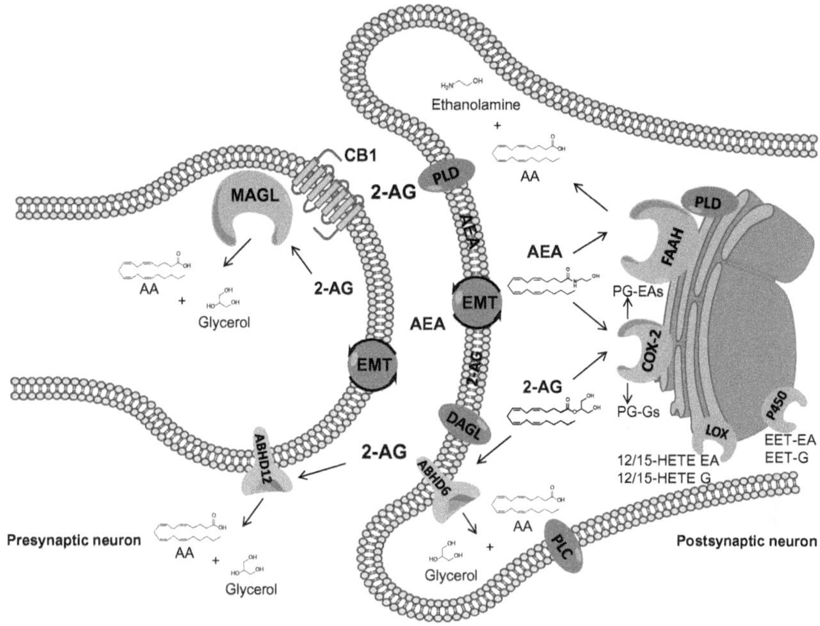

Figure 5 Transport, inactivation, and transformation of the endocannabinoids, 2-AG and AEA, at the synapse. AA = arachidonic acid, AEA = anandamide, 2-AG = 2-arachidonoyl glycerol, ABHD6/12 = α/β hydrolase 6/12, CB1 = cannabinoid receptor type 1, COX-2 = cyclooxygenase 2, DAGL = diacylglycerol lipase, EET-EA/-G = epoxyeicosatrienoic acid ethanolamide/glycerol, EMT = putative endocannabinoid membrane transporter, FAAH = fatty acid amide hydrolase, 12/15 HETE EA/G = 12/15-hydroxyeicosatetraenoic acid ethanolamide/glycerol, LOX = lipoxygenase, P450 = cytochrome enzymes, PLC = phospholipase C, PLD = phospholipase D, and PG-EAs/Gs = prostaglandin ethanolamides/glycerol esters. (See the color plate.)

insignificant. In a recent study, the role of P450 cytochrome enzymes was investigated in the context of [^{3}H]AEA uptake. The CYP3A4 inhibitors nefazodone and ketoconazole inhibited AEA cellular uptake in C6 glioma cells, but only at very high concentrations and in an FAAH-dependent manner (Björklund, Larsson, Jacobsson, & Fowler, 2014). In another study, it was concluded that COX-2 inhibition does not lead to an inhibition of AEA cellular uptake (Nicolussi, Viveros-Paredes, et al., 2014).

However, in the case of 2-AG, the situation is much more complex. Although MAGL dominates 2-AG degradation, at least eight different enzymes are involved in its inactivation (Marrs et al., 2010; Muccioli, 2010; Murataeva, Straiker, & Mackie, 2013). The membrane-associated serine hydrolases MAGL, but also ABHD6 and particularly ABHD12, with an extracellular hydrolytic domain, might hydrolyze 2-AG before it even enters the cytoplasmic space. Consequently, investigations of 2-AG cellular uptake are hampered and a full analytical control over transport assays are necessary.

4. TRANSPORT OF 2-AG AND OTHER SUGGESTED ENDOCANNABINOIDS

4.1 2-AG cellular uptake and intracellular transport

While extensive research was performed to elucidate the underlying mechanisms of AEA cellular uptake, few studies address the transport processes of 2-AG (Fig. 1). During the development of the first inhibitors of endocannabinoid transport, comparable $[^3H]$2-AG uptake kinetics to AEA were found in the human astrocytoma cell line CCF-STTG1 ($K_m = 0.7 \pm 0.1$ μM, $V_{max} = 28 \pm 6$ pmol/min/mg; Beltramo & Piomelli, 2000; see also Table 1). In a competitive uptake experiment to $[^3H]$AEA (30 nM), 2-AG resulted in an $IC_{50} = 18.5 \pm 0.7$ μM (Piomelli et al., 1999). Vice versa, $[^3H]$2-AG cellular uptake was inhibited also by AEA with an $IC_{50} = 4.2 \pm 0.3$ μM, while AM404 resulted in an $IC_{50} = 1.8 \pm 0.1$ μM. The cellular uptake process of $[^3H]$2-AG was rapid, saturable, and Na^+- and energy independent (Beltramo & Piomelli, 2000; Beltramo et al., 1997). More recently, it was shown that in human monocytic leukemia cells (U937), UCM707 and OMDM-2 inhibit the uptake of 2-AG and AEA with identical potency (Chicca et al., 2012). Furthermore, both endocannabinoids competed for cellular uptake in experiments using both radioactive and analytical methods of detection (Chicca et al., 2012). Competition experiments suggested that the putative EMT might have a higher affinity for AEA than 2-AG as in equimolar experiments the ratio of $[^3H]$AEA/2-AG is approximately 1:10, while the ratio of $[^3H]$2-AG/AEA is approximately 1:5 (Chicca et al., 2012), though potential extracellular 2-AG hydrolysis could confound these values. Based on these data, a common carrier-mediated mechanism was suggested for the cellular uptake of 2-AG and AEA. Further, TLC analyses revealed that in astrocytoma cells, $[^3H]$2-AG was rapidly metabolized to glycerol and $[^3H]$arachidonic acid, which was found in phospholipids, diacylglycerols, and triacyglycerols (Beltramo & Piomelli, 2000). An excess of arachidonic acid (100 μM) or inhibiting the acyl-coenzyme A synthetase (triacsin C) reduced the uptake of $[^3H]$2-AG in human astrocytoma cells. This suggested that interfering with the membrane incorporation of 2-AG hydrolysis-derived $[^3H]$

arachidonic acid might activate a negative feedback loop which eventually slows down the cellular uptake of [^{3}H]2-AG (Beltramo & Piomelli, 2000). At that time, the role of MAGL in the inactivation of 2-AG had just been discovered (Di Marzo et al., 1999; Goparaju, Ueda, Taniguchi, & Yamamoto, 1999), while the contribution of the serine hydrolases ABHD6 and ABHD12 to 2-AG hydrolysis (see Fig. 5) was not yet known. These serine hydrolases might act as guardians of 2-AG signaling at the plasma membrane (Savinainen, Saario, & Laitinen, 2011) and mediate the inactivation of 2-AG by enzymatic cleavage. However, also in C6 glioma cells the cellular uptake of [^{3}H]2-AG could be monitored with similar affinities to [^{14}C]AEA and the two endocannabinoids were again shown to compete with each other for a common carrier-mediated uptake mechanism. Already in these initial studies, it was noted that the uptake efficacy of [^{14}C]AEA was higher than the one of [^{3}H]2-AG, when coincubated in C6 glioma cells. The authors demonstrated sensitivity of the 2-AG cellular uptake process to AM404 ($K_i = 10.2 \pm 1.7$ μM) with the same potency as for AEA uptake inhibition. They showed saturability, time and temperature dependence, and enhancement of the cellular uptake process of [^{3}H]2-AG by nitric oxide donors (Bisogno et al., 2001). Although in an earlier, preliminary study no evidence for a facilitated diffusion process for 2-AG could be found in rat basophilic leukemia RBL-2H3 and mouse neuroblastoma N18TG2 cells (Di Marzo, Bisogno, Suguira, Melck, & De Petrocellis, 1998), the same conclusions were drawn as previously suggested by Beltramo and Piomelli (2000). The existence of a "2-AG/AMT" here called putative EMT was suggested. The characteristics of 2-AG transport (temperature dependence, pharmacological inhibition by AM404, competitiveness with AEA) and further evidence for the existence of the putative EMT were provided in a study with primary rat cortical neurons and in electrophysiological rat brain slice experiments. In one study, AEA inhibited [^{3}H]2-AG cellular accumulation and AM404 was shown to enhance the 2-AG-mediated reduction of hippocampal inhibitory postsynaptic currents (Hajos, Kathuria, Dinh, Piomelli, & Freud, 2004). Most of the electrophysiological studies investigated the role of AEA and 2-AG in parallel, with respect to their action as CB_1 receptor agonists in the retrograde signaling (see Fig. 6). As indicated in Section 2.2.4 with AEA, AM404 (1 μM) resulted in a fivefold increase of extracellular 2-AG levels after stimulation of neuroendocrine cells of the hypothalamic supraoptic nucleus in rat brain slices (Di et al., 2005). OMDM-2 (5 μM) and AM404 (1 μM) enhanced the inhibition of synaptically driven spiking activity in postsynaptic neurons upon enhanced

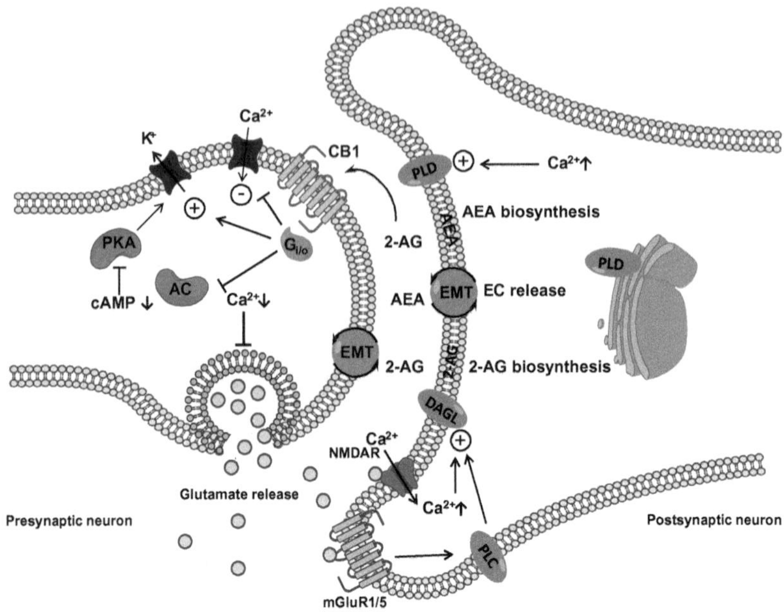

Figure 6 Schematic representation of the ECS in pre- and postsynaptic neurons mediating retrograde signaling at the glutamatergic synapse. (See the color plate.)

retrograde signaling by 2-AG. In a washout experiment (30 min with artificial cerebrospinal fluid after bath application (extracellular) of OMDM-2, 5 μM), the spiking activity in SON magnocellular neurons recovered (Di et al., 2005). In a different study, similar results were found after the intracellular application of AM404 (2 μM) and VDM11 (10 μM) in postsynaptic striatal medium spiny neurons (Ronesi et al., 2004). These studies provided additional evidence for the suggested transporter also under more physiological conditions. Importantly, they showed that this putative target of EC uptake inhibitors is also involved in the postsynaptic release of 2-AG and AEA during retrograde signaling at the synapse.

A so far undressed point was whether 2-AG cellular uptake was also regulated by its enzymatic hydrolysis, a correlation which could be found for AEA uptake and FAAH (Chicca et al., 2014; Deutsch et al., 2001; Fowler et al., 2004; Kaczocha et al., 2006; Nicolussi, Chicca, et al., 2014; Nicolussi, Viveros-Paredes, et al., 2014). While the unspecific MAGL inhibitor MAFP (methylarachidonoyl fluorophosphonate) (Goparaju et al., 1999) potently inhibited [^{3}H]2-AG hydrolysis in RBL-2H3, PC3, AT-1, and Neuro2a cells, it did not lead to an inhibition of

[^{3}H]2-AG cellular uptake in case of the first three cell lines (Fowler & Ghafouri, 2008). In Neuro2a cells, a moderate inhibition of [^{3}H]2-AG uptake could be reached using 1 μM MAFP. FAAH blockage did not affect [^{3}H]2-AG uptake. It was concluded that the inhibition of [^{3}H]2-AG hydrolysis does not regulate its cellular uptake (Fowler & Ghafouri, 2008). In a recent study using the human leukemic monocyte lymphoma cell line U937, the effect of JZL184, a specific inhibitor of MAGL (Long et al., 2009), was investigated in comparison to UCM707 and OMDM-2 in [^{3}H]2-AG cellular uptake experiments (Chicca et al., 2012). Therefore, a sophisticated multiphase assay was used which simultaneously monitored not only the intracellular and extracellular levels of [^{3}H]2-AG but also the production of [^{3}H]glycerol upon enzymatic cleavage. JZL184 resulted in a concentration-dependent increase of the intracellular [^{3}H]2-AG levels with an IC_{50} comparable to the inhibition of MAGL (IC_{50} = 6 nM; Long et al., 2009). The extracellular [^{3}H]2-AG levels increased, while the [^{3}H] glycerol production decreased accordingly. On the contrary, the endocannabinoid uptake inhibitors UCM707 and OMDM-2 resulted in a decrease of [^{3}H]2-AG cellular uptake with comparable potency seen for [^{3}H]AEA uptake inhibition. Additionally, [^{3}H]2-AG efflux experiments were performed, where not only UCM707 and OMDM-2, but also LY2183240 blocked the release of [^{3}H]2-AG from preloaded U937 cells. These set of data showed that the blockage of [^{3}H]2-AG hydrolysis and the inhibition of the putative EMT lead to opposite readouts, clearly suggesting that two separate mechanisms are involved in [^{3}H]2-AG uptake (Chicca et al., 2012). The same was also true for AEA uptake when a mass spectrometric method of detection was used. On the contrary, when [^{3}H] AEA uptake was measured, the differential effects of inhibiting FAAH and the putative EMT could not be appreciated due to the high incorporation of the hydrolysis-derived [^{3}H]ethanolamine into phospholipids, which rendered the two mechanisms methodologically indistinguishable in U937 cells (Chicca et al., 2012). On the other hand, [^{3}H]2-AG and 2-AG uptake did not show any methodologically discrepancy because the hydrolysis-derived [^{3}H]glycerol is negligibly incorporated into phospholipids in U937 cells (Chicca et al., 2012). The dataset was supplemented with a complete [^{3}H]2-AG uptake competition panel, in which all the other endocannabinoids (AEA, NADA, virodhamine, and 2-AGE, shown in Fig. 1) competed with [^{3}H]2-AG uptake. This study showed for the first time comprehensively that all ECs, but not other NAEs, share a common cellular membrane uptake mechanism (Chicca et al., 2012).

Recently, the first evidence for intracellular carriers of 2-AG was provided by two independent groups. The known cytosolic AEA shuttle FABP5 was shown also to bind 2-AG. Using fluorescence polarization and an NBD-labeled fatty acid probe which was displaced from FABP5, a $K_i = 8.72\ \mu M$ (7.77–$9.78\ \mu M$) was determined (Nicolussi, Viveros-Paredes, et al., 2014). Simultaneously, a crystallographic study of FABP5 as an intracellular carrier protein of endocannabinoids confirmed the binding data (Sanson et al., 2014). Again, concentration matters and the fact that a lipid can be cocrystallized *in vitro* with a protein is not evidence of its physiological role, unless the biophysical parameters (K_d and concentration) are in agreement with the cellular environment. Clearly, the K_d for 2-AG binding to FABP5 more closely matches the K_m for 2-AG transport than in the case of AEA.

In summary, 2-AG shares common mechanisms of cellular uptake with AEA, which are protein-facilitated, saturable, temperature-dependent, and can be inhibited by known EC transporter inhibitors (Table 1). Not surprisingly, for 2-AG this is independent of the enzymatic activity of FAAH. The efficient uptake of not only 2-AG but also AEA may involve different mechanisms, including facilitated diffusion, interaction with intracellular binding proteins (Hermann, Kaczocha, & Deutsch, 2006; Nicolussi, Chicca, et al., 2014; Nicolussi, Viveros-Paredes, et al., 2014; Sanson et al., 2014). Because 2-AG is rapidly degraded at the plasma membrane, 2-AG uptake might be influenced by the serine hydrolases MAGL, ABHD6, or ABHD12. Further investigations are necessary to clarify the reported mechanisms of cellular accumulation, which seems to be in common for all ECs.

4.2 Transport of virodhamine, noladin ether, and NADA

Least studied are the transport processes of the remaining suggested endocannabinoids virodhamine, noladin ether, and NADA (see Fig. 1; Bisogno et al., 2000; Fezza et al., 2002; Hanuš et al., 2001; Porter et al., 2002). To our knowledge, there are only two studies that mention virodhamine in the context of AEA cellular uptake. Virodhamine consists of an arachidonoyl moiety joined to ethanolamide by an ester linkage. Given the similarity to AEA, a competitive effect on the transport of AEA was suggested early. In both studies, competitive EC uptake experiments against radiolabeled AEA were performed. In RBL-2H3 cells, virodhamine showed an $IC_{50} = 123\ \mu M$ using $[^{14}C]$AEA ($5\ \mu M$) as substrate for cellular uptake (Porter et al., 2002). Because of the very high concentrations of AEA used, virodhamine might have resulted in this extraordinary high IC_{50} value, and the actual transport kinetics might have been masked (*vide supra*). Comparing the potency of the positive control AM404 in the same setup, an $IC_{50}=20$ μM was found. This is the highest ever reported IC_{50} value for AM404 in RBL-2H3 cells (see Table 1), suggesting that these initial uptake experiments may have been carried out under suboptimal conditions, likely measuring also simple diffusion with a reduced signal to background. Even though the evidence was preliminary, it was concluded for the first time that virodhamine and AEA share the same mechanism of cellular uptake. This finding was confirmed more recently by our group in U937 cells in a comparative study with all of the five endocannabinoids (Fig. 1). In short-time cellular uptake experiments using 100 nM [^{3}H]AEA and 1 μM [^{3}H]2-AG, virodhamine resulted in nanomolar IC_{50} values of 206 and 850 nM, respectively (Chicca et al., 2012). The efficacy of inhibition was lower compared to those of the investigated EC uptake inhibitors UCM707 and OMDM-2. These data are in line with a common mechanism of cellular uptake (membrane passage) for all endocannabinoids, which is independent of their enzymatic cleavage (Chicca et al., 2012).

Noladin ether (2-AGE, 2-arachidonyl glycerol ether) belongs to the group of glycerol-type endocannabinoids like 2-AG and acts as CB_1 receptor-selective full agonist (Fezza et al., 2002; Hanuš et al., 2001). The inactivation of [^{3}H]2-AGE was studied in rat C6 glioma and RBL-2H3 cells. A rapid ($t_{1/2} \cong 5$ min) cellular uptake process was described, which was time, concentration, and temperature-dependent. The apparent K_m and V_{max} values resulted in 12.8 ± 2.1 μM and 0.21 ± 0.08 nmol/min/mg protein in C6 glioma cells, comparable to the uptake kinetics found for 2-AG in this cell type (Bisogno et al., 2001; Fezza et al., 2002). A minor part of the transported [^{3}H]2-AGE was degraded and partially incorporated into phospholipids. Competitive uptake experiments of 2-AGE against [^{14}C]AEA or [^{3}H]2-AG resulted K_i values of 15.6 and 22.6 μM respectively, which was in a comparable range like previously observed for AEA and 2-AG against each other (Bisogno et al., 2001). Again, the arisen opinion of a common cellular transport mechanism for AEA, 2-AG, and 2-AGE was substantiated (Fezza et al., 2002). Affirmative data were generated in U937 cells, where 2-AGE competed with [^{3}H]AEA and [^{3}H]2-AG for cellular uptake with IC_{50} values of 1.1 and 1.4 μM, respectively (Chicca et al., 2012).

NADA is a CB_1 receptor agonist (CB_1 $K_i=250$ nM) with a 40-fold selectivity over CB_2 (Bisogno et al., 2000). At the same time, it acts as a potent TRPV1 activator (Huang et al., 2002; Marinelli et al., 2007). In rat C6 glioma cells, a temperature-dependent and rapid cellular uptake was demonstrated for [^{3}H]NADA ($K_m = 55 \pm 8$ μM, $B_{max} = 1.9 \pm 0.3$ nmol/min/mg protein). The substrate affinity in this cell line was lower than the one of AEA ($K_m = 11.0 \pm 2.0$ μM, $B_{max} = 1.7 \pm 0.3$ nmol/min/mg protein), but comparable transport efficiencies were found (Bisogno et al., 2001; Huang et al., 2002). Another study demonstrated the effect of NADA and a series of *N*-acyl-dopamines on the AMT and its degrading enzyme FAAH (Bisogno et al., 2000). Rat C6 glioma and RBL-2H3 cells were used to study the uptake of [^{14}C]AEA in the presence of synthesized NADAs. NADA inhibited the uptake of [^{14}C]AEA with an $IC_{50} = 21.5 \pm 9.1$ μM in RBL-2H3 and $IC_{50} = 17.5 \pm 1.7$ μM in C6 glioma cells. Comparable results were found for α-linolenoyl-DA, pinolenoyl-DA, eicosapentaenoyl-DA, and docosapentaenoyl-DA ($IC_{50} = 17.5$–33 μM). In addition, the inhibitory effect of these NADAs on FAAH was tested in N18TG2 membranes and slightly right shifted inhibitory potencies compared to the inhibition of [^{14}C]AEA uptake were found. The authors concluded that FAAH inhibition might contribute to a limited extent to the inhibition of AEA-facilitated transport in agreement with previous findings that *N*-acyl-vanillyl-amides are selective inhibitors of the AEA transporter (Bisogno et al., 2000; Melck et al., 1999). NADA was investigated in an electrophysiological study where the focus was on TRPV1 and CB_1 receptor activation in dopaminergic neurons. A key function for the putative EMT in promoting the stimulation of TRPV1 of CB_1 was described, because it could be shown that cellular NADA uptake could be inhibited with OMDM-1 (enantiomer of OMDM-2). In substantia nigra, brain slices were shown to accumulate [^{3}H]NADA an effect which also could be blocked with the selective inhibitor of the "NADA transporter," which may be identical to the EMT (Marinelli et al., 2007). It was suggested that sensory neurons possess an EMT-like mechanism which is involved in the pharmacological action of AEA and NADA at native TRPV1 channels (Price, Patwardhan, Flores, & Hargreaves, 2005). In another study, it was proposed that transport of both NADA and capsaicin into dorsal root ganglion neurons and the subsequent activation of TRPV1 are partly governed by FAAH-dependent mechanisms, as well as via the putative AMT (Millns et al., 2006).

Recently, binding interactions of virodhamine, 2-AGE, and NADA with FABP5 were measured, showing that 2-AGE ($K_i = 12.9$ μM (10.3–18.2 μM)) and virodhamine ($K_i = 55.8$ μM (23.3–79.0 μM)) only moderately bind to FABP5, while NADA did not show any significant binding (K_i > 100 μM; Nicolussi, 2014). Since endocannabinoids are mainly found at low nanomolar tissue concentrations, FABP5-binding interactions with endocannabinoids cannot explain most of the phenomena of common endocannabinoid transport. Taken together, the endocannabinoids virodhamine, 2-AGE, and NADA (Fig. 1) clearly share common mechanisms of cellular uptake and exhibit comparable transport kinetics to AEA and 2-AG. They are suggested to be substrates of the putative EMT and are sensitive to pharmacological inhibition by known AEA uptake inhibitors. Further studies are necessary to clarify their interplay with intracellular binding proteins and to elucidate their exact membrane transport mechanism.

5. CONCLUSIONS

Due to the experimental difficulty of studying lipid transport and the unfortunate confusions caused by different assay setups, the history of endocannabinoid transport research is not straightforward. Also, the literature is dispersed and seldom is the whole picture portrayed. We acknowledge that without adequate analytical tools (suitable chemical probes, mass spectrometry, proteomics) and educated biophysical insights (e.g., concentration and time matters), the puzzle of endocannabinoid transport will not be solved. The subject matter is too complex to be tackled by small projects and requires joint research efforts. With the recent discovery of cytoplasmic endocannabinoid-binding proteins (FABPs, albumin, Hsp70), the mechanisms of intracellular shuttling can now be addressed experimentally, but the molecular mechanism of membrane transport remains unexplained. To quote T.S. Eliot, "The journey not the arrival matters," endocannabinoids are likely part of highly ordered trafficking mechanisms, both extracellularly and intracellularly, that determine their journey. A careful review of the literature suggests additional yet unknown membrane proteins that facilitate and control the membrane passage of both AEA and 2-AG (and possibly the other endocannabinoids). Thus, the putative EMT should not be refuted because biophysical data (e.g., significantly different K_m values in different cells independent of FAAH and intracellular protein targets) clearly implicate its existence. Moreover, certain apparently selective inhibitors for AEA uptake (e.g., guineensine) also inhibit 2-AG uptake but do not interact with any of the known metabolic enzymes or AEA-binding proteins, suggesting an additional common target that is competitive with endocannabinoid membrane transport. While excellent tool compounds have been synthesized for most targets within the ECS (Blankman & Cravatt, 2013), novel and more potent (ideally low nanomolar) chemical probes are needed to study endocannabinoid membrane transport. The use of cell lines lacking FAAH (e.g., HMC-1 cells) to develop AEA transport inhibitors might be helpful. A chemical biology approach could thus advance the field and bring novel solutions to an old problem. If the EMT stands for Charon, the ferryman of Hades (*vide supra*), the chemical probe that could lead to the identification of the elusive EMT is a fake obol that upsets Charon and irreversibly sticks to his hands. The idea, though not entirely new, is launched and the quest may begin.

ACKNOWLEDGMENTS

We thank Dr. Andrea Chicca for reading the chapter and for helpful suggestions. Our research on endocannabinoid transport is financially supported by grants from the NCCR TransCure and Swiss EIN Roche programme.

REFERENCES

Alexander, J. P., & Cravatt, B. F. (2006). The putative endocannabinoid transport blocker LY2183240 is a potent inhibitor of FAAH and several other brain serine hydrolases. *Journal of the American Chemical Society*, *128*, 9699–9704.

Bab, I., Ofek, O., Tam, J., Rehnelt, J., & Zimmer, A. (2008). Endocannabinoids and the regulation of bone metabolism. *Journal of Neuroendocrinology*, *20*(Suppl. 1), 69–74.

Bari, M., Paradisi, A., Pasquariello, N., & Maccarrone, M. (2005). Cholesterol-dependent modulation of type 1 cannabinoid receptors in nerve cells. *Journal of Neuroscience Research*, *81*, 275–283.

Basavarajappa, B. (2007). Critical enzymes involved in endocannabinoid metabolism. *Protein and Peptide Letters*, *14*, 237–246.

Batista, L. A., Gobira, P. H., Viana, T. G., Aguiar, D. C., & Moreira, F. A. (2014). Inhibition of endocannabinoid neuronal uptake and hydrolysis as strategies for developing anxiolytic drugs. *Behavioural Pharmacology*, *25*, 425–433.

Baur, R., Gertsch, J., & Sigel, E. (2013). Do N-arachidonyl-glycine (NA-glycine) and 2-arachidonoyl glycerol (2-AG) share mode of action and the binding site on the β2 subunit of GABAA receptors? *PeerJ*, *1*, e149.

Beltramo, M., & Piomelli, D. (1999). Anandamide transport inhibition by the vanilloid agonist olvanil. *European Journal of Pharmacology*, *364*, 75–78.

Beltramo, M., & Piomelli, D. (2000). Carrier-mediated transport and enzymatic hydrolysis of the endogenous cannabinoid 2-arachidonylglycerol. *Neuroreport*, *11*, 1231–1235.

Beltramo, M., Stella, N., Calignano, A., Lin, S. Y., Makriyannis, A., & Piomelli, D. (1997). Functional role of high-affinity anandamide transport, as revealed by selective inhibition. *Science*, *277*, 1094–1097.

Berger, W. T., Ralph, B. P., Kaczocha, M., Sun, J., Balius, T. E., Rizzo, R. C., et al. (2012). Targeting fatty acid binding protein (FABP) anandamide transporters—A novel strategy for development of anti-inflammatory and anti-nociceptive drugs. *PLoS One*, 7, e50968.

Bernabò, N., Barboni, B., & Maccarrone, M. (2013). Systems biology analysis of the endocannabinoid system reveals a scale-free network with distinct roles for anandamide and 2-arachidonoylglycerol. *Omics: A Journal of Integrative Biology*, *17*, 646–654.

Bisogno, T., Howell, F., Williams, G., Minassi, A., Cascio, M. G., Ligresti, A., et al. (2003). Cloning of the first sn1-DAG lipases points to the spatial and temporal regulation of endocannabinoid signaling in the brain. *The Journal of Cell Biology*, *163*, 463–468.

Bisogno, T., Maccarrone, M., De Petrocellis, L., Jarrahian, A., Finazzi-Agrò, A., Hillard, C., et al. (2001). The uptake by cells of 2-arachidonoylglycerol, an endogenous agonist of cannabinoid receptors. *European Journal of Biochemistry*, *268*, 1982–1999.

Bisogno, T., Melck, D., Bobrov, M., Gretskaya, N. M., Bezuglov, V., De Petrocellis, L., et al. (2000). N-acyl-dopamines: Novel synthetic CB(1) cannabinoid-receptor ligands and inhibitors of anandamide inactivation with cannabimimetic activity in vitro and in vivo. *The Biochemical Journal*, *351*, 817–824.

Björklund, E., Larsson, T. N. L., Jacobsson, S. O. P., & Fowler, C. J. (2014). Ketoconazole inhibits the cellular uptake of anandamide via inhibition of FAAH at pharmacologically relevant concentrations. *PLoS One*, *9*, e87542.

Blankman, J. L., & Cravatt, B. F. (2013). Chemical probes of endocannabinoid metabolism. *Pharmacological Reviews*, *65*, 849–871.

Blankman, J. L., Simon, G. M., & Cravatt, B. F. (2007). A comprehensive profile of brain enzymes that hydrolyze the endocannabinoid 2-arachidonoylglycerol. *Chemistry & Biology*, *14*, 1347–1356.

Brown, A. J. (2007). Novel cannabinoid receptors. *British Journal of Pharmacology*, *152*, 567–575.

Brown, D. A., & London, E. (2000). Structure and function of sphingolipid- and cholesterol-rich membrane rafts. *The Journal of Biological Chemistry*, *275*, 17221–17224.

Buczynski, M. W., & Parsons, L. H. (2010). Quantification of brain endocannabinoid levels: Methods, interpretations and pitfalls. *British Journal of Pharmacology*, *160*, 423–442.

Burch, P., Chicca, A., Gertsch, J., & Gademann, K. (2014). Functionally optimized neuritogenic farinosone C analogs: SAR-study and investigations on their mode of action. *ACS Medicinal Chemistry Letters*, *5*, 172–177.

Castillo, P. E., Younts, T. J., Chávez, A. E., & Hashimotodani, Y. (2012). Endocannabinoid signaling and synaptic function. *Neuron*, *76*, 70–81.

Chicca, A., Caprioglio, D., Minassi, A., Petrucci, V., Appendino, G., Taglialatela-Scafati, O., et al. (2014). Functionalization of β-caryophyllene generates novel polypharmacology in the endocannabinoid system. *ACS Chemical Biology*, *9*, 1499–1507.

Chicca, A., Marazzi, J., Nicolussi, S., & Gertsch, J. (2012). Evidence for bidirectional endocannabinoid transport across cell membranes. *The Journal of Biological Chemistry*, *287*, 34660–34682.

Cravatt, B., Giang, D., & Mayfield, S. (1996). Molecular characterization of an enzyme that degrades neuromodulatory fatty-acid amides. *Nature*, *384*, 83–87.

Day, T., Rakhshan, F., Deutsch, D. G., & Barker, E. L. (2001). Role of fatty acid amide hydrolase in the transport of the endogenous cannabinoid anandamide. *Molecular Pharmacology*, *59*, 1369–1375.

De Petrocellis, L., Bisogno, T., Davis, J. B., Pertwee, R. G., & Di Marzo, V. (2000). Overlap between the ligand recognition properties of the anandamide transporter and the VR1 vanilloid receptor: Inhibitors of anandamide uptake with negligible capsaicin-like activity. *FEBS Letters*, *483*, 52–56.

Deutsch, D. G., & Chin, S. (1993). Enzymatic synthesis and degradation of anandamide, a cannabinoid receptor agonist. *Biochemical Pharmacology*, *46*, 791–796.

Deutsch, D. G., Glaser, S. T., Howell, J. M., Kunz, J. S., Puffenbarger, R., Hillard, C. J., et al. (2001). The cellular uptake of anandamide is coupled to its breakdown by fatty-acid amide hydrolase. *The Journal of Biological Chemistry*, *276*, 6967–6973.

Devane, W. A., Hanuš, L., Breuer, A., Pertwee, R. G., Stevenson, L. A., Griffin, G., et al. (1992). Isolation and structure of a brain constituent that binds to the cannabinoid receptor. *Science*, *258*, 1946–1949.

Di, S., Boudaba, C., Popescu, I. R., Weng, F. J., Harris, C., Marcheselli, V. L., et al. (2005). Activity-dependent release and actions of endocannabinoids in the rat hypothalamic supraoptic nucleus. *The Journal of Physiology*, *569*, 751–760.

Dickason-Chesterfield, A. K., Kidd, S. R., Moore, S. A., Schaus, J. M., Liu, B., Nomikos, G. G., et al. (2006). Pharmacological characterization of endocannabinoid transport and fatty acid amide hydrolase inhibitors. *Cellular and Molecular Neurobiology*, *26*, 407–423.

Di Marzo, V. (2006). Endocannabinoids: Synthesis and degradation. *Reviews of Physiology, Biochemistry and Pharmacology*, *160*, 1–24.

Di Marzo, V. (2008). Targeting the endocannabinoid system: To enhance or reduce? *Nature Reviews Drug Discovery*, *7*, 438–455.

Di Marzo, V., Bifulco, M., & De Petrocellis, L. (2004). The endocannabinoid system and its therapeutic exploitation. *Nature Reviews Drug Discovery*, *3*, 771–784.

Di Marzo, V., Bisogno, T., De Petrocellis, L., Melck, D., Orlando, P., Wagner, J. A., et al. (1999). Biosynthesis and inactivation of the endocannabinoid 2-arachidonoylglycerol in circulating and tumoral macrophages. *European Journal of Biochemistry*, *264*, 258–267.

Di Marzo, V., Bisogno, T., Suguira, T., Melck, D., & De Petrocellis, L. (1998). The novel endogenous cannabinoid 2-arachidonoylglycerol is inactivated by neuronal- and basophil-like cells: Connections with anandamide. *Biochemical Journal*, *331*, 15–19.

Di Marzo, V., Fontana, A., Cadas, H., Schinelli, S., Cimino, G., Schwartz, J., et al. (1994). Formation and inactivation of endogenous cannabinoid anandamide in central neurons. *Nature*, *372*, 686–691.

Di Marzo, V., Ligresti, A., Morera, E., Nalli, M., & Ortar, G. (2004). The anandamide membrane transporter. Structure–activity relationships of anandamide and oleoylethanolamine analogs with phenyl rings in the polar head group region. *Bioorganic & Medicinal Chemistry*, *12*, 5161–5169.

Dinh, T. P., Kathuria, S., & Piomelli, D. (2004). RNA interference suggests a primary role for monoacylglycerol lipase in the degradation of the endocannabinoid 2-arachidonoylglycerol. *Molecular Pharmacology*, *66*, 1260–1264.

Di Pasquale, E., Chahinian, H., Sanchez, P., & Fantini, J. (2009). The insertion and transport of anandamide in synthetic lipid membranes are both cholesterol-dependent. *PLoS One*, *4*, e4989.

Fasia, L., Karava, V., & Siafaka-Kapadai, A. (2003). Uptake and metabolism of [3H]anandamide by rabbit platelets. Lack of transporter? *European Journal of Biochemistry*, *270*, 3498–3506.

Fegley, D., Kathuria, S., Mercier, R., Li, C., Goutopoulos, A., Makriyannis, A., et al. (2004). Anandamide transport is independent of fatty-acid amide hydrolase activity and is blocked by the hydrolysis-resistant inhibitor AM1172. *Proceedings of the National Academy of Sciences of the United States of America*, *101*, 8756–8761.

Felder, C. C., Dickason-Chesterfield, A. K., & Moore, S. A. (2006). Cannabinoid biology. *Molecular Interventions*, *6*, 149–161.

Fezza, F., Bisogno, T., Minassi, A., Appendino, G., Mechoulam, R., & Di, V. (2002). Noladin ether, a putative novel endocannabinoid: Inactivation mechanisms and a sensitive method for its quantification in rat tissues. *FEBS Letters*, *513*, 294–298.

Fowler, C. J. (2012). Anandamide uptake explained? *Trends in Pharmacological Sciences*, *33*, 181–185.

Fowler, C. J. (2013). Transport of endocannabinoids across the plasma membrane and within the cell. *The FEBS Journal*, *280*, 1895–1904.

Fowler, C. J. (2014). Has FLAT fallen flat? *Trends in Pharmacological Sciences*, *35*, 51–52.

Fowler, C. J., & Ghafouri, N. (2008). Does the hydrolysis of 2-arachidonoylglycerol regulate its cellular uptake? *Pharmacological Research: The Official Journal of the Italian Pharmacological Society, 58*, 72–76.

Fowler, C. J., Tiger, G., Ligresti, A., López-Rodríguez, M. L., & Di Marzo, V. (2004). Selective inhibition of anandamide cellular uptake versus enzymatic hydrolysis—A difficult issue to handle. *European Journal of Pharmacology, 492*, 1–11.

Fu, J., Bottegoni, G., Sasso, O., Bertorelli, R., Rocchia, W., Masetti, M., et al. (2011). A catalytically silent FAAH-1 variant drives anandamide transport in neurons. *Nature Neuroscience, 15*, 64–69.

Furuhashi, M., & Hotamisligil, G. S. (2008). Fatty acid-binding proteins: Role in metabolic diseases and potential as drug targets. *Nature Reviews Drug Discovery, 7*, 489–503.

Gachet, S. M., Rhyn, P., Bosch, O., Quednow, B., & Gertsch, J. (2014). A quantitiative LC–MS/MS method for the measurement of arachidonic acid, prostanoids, endocannabinoids, N-acylethanolamines and steroids in human plasma. *Journal of Chromatography B, 976–977*, 6–18.

Gaoni, Y., & Mechoulam, R. (1964). Isolation, structure, and partial synthesis of an active constituent of hashish. *Journal of the American Chemical Society, 86*, 1646–1647.

Gerdeman, G. L., Ronesi, J., & Lovinger, D. M. (2002). Postsynaptic endocannabinoid release is critical to long-term depression in the striatum. *Nature Neuroscience, 5*, 446–451.

Glaser, S. T., Abumrad, N. A., Fatade, F., Kaczocha, M., Studholme, K. M., & Deutsch, D. G. (2003). Evidence against the presence of an anandamide transporter. *Proceedings of the National Academy of Sciences of the United States of America, 100*, 4269–4274.

Glaser, S. T., Kaczocha, M., & Deutsch, D. G. (2005). Anandamide transport: A critical review. *Life Sciences, 77*, 1584–1604.

Goodfellow, C. E., & Glass, M. (2009). Anandamide receptor signal transduction. *Vitamins and Hormones, 81*, 79–110.

Goparaju, S. K., Ueda, N., Taniguchi, K., & Yamamoto, S. (1999). Enzymes of porcine brain hydrolyzing 2-arachidonoylglycerol, an endogenous ligand of cannabinoid receptors. *Biochemical Pharmacology, 57*, 417–423.

Guindon, J., & Hohmann, A. (2009). The endocannabinoid system and pain. *CNS & Neurological Disorders Drug Targets, 8*, 403–421.

Gulyas, A., Cravatt, B. F., Bracey, M. H., Dinh, T. P., Piomelli, D., Boscia, F., et al. (2004). Segregation of two endocannabinoid-hydrolyzing enzymes into pre- and postsynaptic compartments in the rat hippocampus, cerebellum and amygdala. *The European Journal of Neuroscience, 20*, 441–458.

Haenisch, B., & Boenisch, H. (2011). Depression and antidepressants: Insights from knockout of dopamine, serotonin or noradrenaline re-uptake transporters. *Pharmacology and Therapeutics, 129*, 352–368.

Hajdu, Z., Nicolussi, S., & Rau, M. (2014). Identification of endocannabinoid system-modulating N-alkylamides from Heliopsis helianthoides var. scabra and Lepidium meyenii. *Journal of Natural Products*, 77(7), 1663–1669.

Hajos, N., Kathuria, S., Dinh, T., Piomelli, D., & Freud, T. (2004). Endocannabinoid transport tightly controls 2-arachidonoyl glycerol actions in the hippocampus: Effects of low temperature and the transport inhibitor AM404. *European Journal of Neuroscience, 19*, 2991–2996.

Hanuš, L., Abu-Lafi, S., Fride, E., Breuer, A., Vogel, Z., Shalev, D. E., et al. (2001). 2-Arachidonyl glyceryl ether, an endogenous agonist of the cannabinoid CB1 receptor. *Proceedings of the National Academy of Sciences of the United States of America, 98*, 3662–3665.

Hermann, A., Kaczocha, M., & Deutsch, D. G. (2006). 2-Arachidonoylglycerol (2-AG) membrane transport: History and outlook. *The AAPS Journal, 8*, E409–E412.

Hillard, C. J., Edgemond, W. S., Jarrahian, A., & Campbell, W. B. (1997). Accumulation of N-arachidonoylethanolamine (anandamide) into cerebellar granule cells occurs via facilitated diffusion. *Journal of Neurochemistry*, *69*, 631–638.

Hillard, C. J., & Jarrahian, A. (2000). The movement of N-arachidonoylethanolamine (anandamide) across cellular membranes. *Chemistry and Physics of Lipids*, *108*, 123–134.

Hillard, C. J., & Jarrahian, A. (2003). Cellular accumulation of anandamide: Consensus and controversy. *British Journal of Pharmacology*, *140*, 802–808.

Hillard, C. J., & Jarrahian, A. (2005). Accumulation of anandamide: Evidence for cellular diversity. *Neuropharmacology*, *48*, 1072–1078.

Hillard, C. J., Shi, L., Tuniki, V. R., Falck, J. R., & Campbell, W. B. (2007). Studies of anandamide accumulation inhibitors in cerebellar granule neurons: Comparison to inhibition of fatty acid amide hydrolase. *Journal of Molecular Neuroscience*, *33*, 18–24.

Hofmann, N. A., Barth, S., Waldeck-Weiermair, M., Klec, C., Strunk, D., Malli, R., et al. (2014). TRPV1 mediates cellular uptake of anandamide and thus promotes endothelial cell proliferation and network-formation. *Biology Open*, *3*, 1164–1172.

Högestätt, E. D., Jönsson, B. A. G., Ermund, A., Andersson, D. A., Björk, H., Alexander, J. P., et al. (2005). Conversion of acetaminophen to the bioactive N-acylphenolamine AM404 via fatty acid amide hydrolase-dependent arachidonic acid conjugation in the nervous system. *The Journal of Biological Chemistry*, *280*, 31405–31412.

Houy, S., Croisé, P., Gubar, O., Chasserot-Golaz, S., Tryoen-Tóth, P., Bailly, Y., et al. (2013). Exocytosis and endocytosis in neuroendocrine cells: Inseparable membranes!. *Frontiers in Endocrinology*, *4*, 135.

Howlett, A. C., Barth, F., Bonner, T. I., Cabral, G., Casellas, P., Devane, W. A., et al. (2002). International Union of Pharmacology. XXVII. Classification of cannabinoid receptors. *Pharmacological Reviews*, *54*, 161–202.

Huang, S. M., Bisogno, T., Trevisani, M., Al-Hayani, A., De Petrocellis, L., Fezza, F., et al. (2002). An endogenous capsaicin-like substance with high potency at recombinant and native vanilloid VR1 receptors. *Proceedings of the National Academy of Sciences of the United States of America*, *99*, 8400–8405.

Jarrahian, A., Manna, S., Edgemond, W., Campbell, W., & Hillard, C. (2000). Structure–activity relationships among N-arachidonylethanolamine (anandamide) head group analogues for the anandamide transporter. *Journal of Neurochemistry*, *74*, 2597–2606.

Kaczocha, M., Glaser, S. T., & Deutsch, D. G. (2009). Identification of intracellular carriers for the endocannabinoid anandamide. *Proceedings of the National Academy of Sciences of the United States of America*, *106*, 6375–6380.

Kaczocha, M., Hermann, A., Glaser, S. T., Bojesen, I. N., & Deutsch, D. G. (2006). Anandamide uptake is consistent with rate-limited diffusion and is regulated by the degree of its hydrolysis by fatty acid amide hydrolase. *The Journal of Biological Chemistry*, *281*, 9066–9075.

Kaczocha, M., Rebecchi, M. J., Ralph, B. P., Teng, Y.-H. G., Berger, W. T., Galbavy, W., et al. (2014). Inhibition of fatty acid binding proteins elevates brain anandamide levels and produces analgesia. *PLoS One*, *9*, e94200.

Kaczocha, M., Vivieca, S., Sun, J., Glaser, S. T., & Deutsch, D. G. (2012). Fatty acid-binding proteins transport N-acylethanolamines to nuclear receptors and are targets of endocannabinoid transport inhibitors. *The Journal of Biological Chemistry*, *287*, 3415–3424.

Khanolkar, A. D., Abadji, V., Lin, S., Hill, W. A., Taha, G., Abouzid, K., et al. (1996). Head group analogs of arachidonylethanolamide, the endogenous cannabinoid ligand. *Journal of Medicinal Chemistry*, *39*, 4515–4519.

Kleyer, J., Nicolussi, S., Taylor, P., Simonelli, D., Furger, E., Anderle, P., et al. (2012). Cannabinoid receptor trafficking in peripheral cells is dynamically regulated by a binary biochemical switch. *Biochemical Pharmacology*, *83*, 1393–1412.

Kondo, S., Kondo, H., Nakane, S., Kodaka, T., Tokumura, A., Waku, K., et al. (1998). 2-Arachidonoylglycerol, an endogenous cannabinoid receptor agonist: Identification as one of the major species of monoacylglycerols in various rat tissues, and evidence for its generation through Ca^{2+}-dependent and -independent mechanisms. *FEBS Letters*, *429*, 152–156.

Leung, K., Elmes, M. W., Glaser, S. T., Deutsch, D. G., & Kaczocha, M. (2013). Role of FAAH-like anandamide transporter in anandamide inactivation. *PLoS One*, *8*, e79355.

Liedhegner, E. S., Vogt, C. D., Sem, D. S., Cunningham, C. W., & Hillard, C. J. (2014). Sterol carrier protein-2: Binding protein for endocannabinoids. *Molecular Neurobiology*, *50*, 149–158.

Ligresti, A., Morera, E., Van Der Stelt, M., Monory, K., Lutz, B., Ortar, G., et al. (2004). Further evidence for the existence of a specific process for the membrane transport of anandamide. *The Biochemical Journal*, *380*, 265–272.

Ligresti, A., Petrosino, S., & Di Marzo, V. (2009). From endocannabinoid profiling to "endocannabinoid therapeutics" *Current Opinion in Chemical Biology*, *13*, 321–331.

Long, J. Z., Li, W., Booker, L., Burston, J. J., Kinsey, S. G., Schlosburg, J. E., et al. (2009). Selective blockade of 2-arachidonoylglycerol hydrolysis produces cannabinoid behavioral effects. *Nature Chemical Biology*, *5*, 37–44.

López-Rodríguez, M. L., Viso, A., Ortega-Gutiérrez, S., Fowler, C. J., Tiger, G., De Lago, E., et al. (2003a). Design, synthesis, and biological evaluation of new inhibitors of the endocannabinoid uptake: Comparison with effects on fatty acid amidohydrolase. *Journal of Medicinal Chemistry*, *46*, 1512–1522.

López-Rodríguez, M. L., Viso, A., Ortega-Gutiérrez, S., Fowler, C. J., Tiger, G., De Lago, E., et al. (2003b). Design, synthesis and biological evaluation of new endocannabinoid transporter inhibitors. *European Journal of Medicinal Chemistry*, *38*, 403–412.

López-Rodríguez, M. L., Viso, A., Ortega-Gutiérrez, S., Lastres-Becker, I., González, S., Fernández-Ruiz, J., et al. (2001). Design, synthesis and biological evaluation of novel arachidonic acid derivatives as highly potent and selective endocannabinoid transporter inhibitors. *Journal of Medicinal Chemistry*, *44*, 4505–4508.

Lutz, B. (2009). Endocannabinoid signals in the control of emotion. *Current Opinion in Pharmacology*, *9*, 46–52.

Maccarrone, M., Bari, M., Battista, N., & Finazzi-Agrò, A. (2002). Estrogen stimulates arachidonoylethanolamide release from human endothelial cells and platelet activation. *Blood*, *100*, 4040–4048.

Maccarrone, M., Bari, M., Lorenzon, T., Bisogno, T., Di Marzo, V., & Finazzi-Agrò, A. (2000). Anandamide uptake by human endothelial cells and its regulation by nitric oxide. *The Journal of Biological Chemistry*, *275*, 13484–13492.

Maccarrone, M., Dainese, E., & Oddi, S. (2010). Intracellular trafficking of anandamide: New concepts for signaling. *Trends in Biochemical Sciences*, *35*, 601–608.

Maccarrone, M., Fiorucci, L., Erba, F., Bari, M., Finazzi-Agrò, A., & Ascoli, F. (2000). Human mast cells take up and hydrolyze anandamide under the control of 5-lipoxygenase and do not express cannabinoid receptors. *FEBS Letters*, *468*, 176–180.

Maccarrone, M., van der Stelt, M., Rossi, A., Veldink, G., Vliegenthart, J. F., & Agrò, A. (1998). Anandamide hydrolysis by human cells in culture and brain. *The Journal of Biological Chemistry*, *273*, 32332–32339.

Marazzi, J., Kleyer, J., Paredes, J. M., & Gertsch, J. (2011). Endocannabinoid content in fetal bovine sera—Unexpected effects on mononuclear cells and osteoclastogenesis. *Journal of Immunological Methods*, *373*, 219–228.

Marinelli, S., Di Marzo, V., Florenzano, F., Fezza, F., Viscomi, M. T., van der Stelt, M., et al. (2007). N-arachidonoyl-dopamine tunes synaptic transmission onto dopaminergic neurons by activating both cannabinoid and vanilloid receptors. *Neuropsychopharmacology*, *32*, 298–308.

Marrs, W. R., Blankman, J. L., Horne, E. A., Thomazeau, A., Lin, Y. H., Coy, J., et al. (2010). The serine hydrolase ABHD6 controls the accumulation and efficacy of 2-AG at cannabinoid receptors. *Nature Neuroscience*, *13*, 951–957.

Masson, J., Sagné, C., Hamon, M., & El Mestikawy, S. (1999). Neurotransmitter transporters in the central nervous system. *Pharmacological Reviews*, *51*, 439–464.

Matsuda, L., Lolait, S., Brownstein, M., Young, A. C., & Bonner, T. (1990). Structure of a cannabinoid receptor and functional expression of the cloned cDNA. *Nature*, *346*, 561–564.

McFarland, M., Bardell, T., & Yates, M. (2008). RNA interference-mediated knockdown of dynamin 2 reduces endocannabinoid uptake into neuronal dCAD cells. *Molecular Pharmacology*, *74*, 101–108.

McFarland, M. J., & Barker, E. L. (2004). Anandamide transport. *Pharmacology & Therapeutics*, *104*, 117–135.

McFarland, M. J., Porter, A. C., Rakhshan, F. R., Rawat, D. S., Gibbs, R. A., & Barker, E. L. (2004). A role for caveolae/lipid rafts in the uptake and recycling of the endogenous cannabinoid anandamide. *The Journal of Biological Chemistry*, *279*, 41991–41997.

Mechoulam, R., & Deutsch, D. G. (2005). Toward an anandamide transporter. *Proceedings of the National Academy of Sciences of the United States of America*, *102*, 17541–17542.

Mechoulam, R., & Parker, L. A. (2013). The endocannabinoid system and the brain. *Annual Review of Psychology*, *64*, 21–47.

Melck, D., Bisogno, T., De Petrocellis, L., Chuang, H., Julius, D., Bifulco, M., et al. (1999). Unsaturated long-chain N-acyl-vanillyl-amides (N-AVAMs): Vanilloid receptor ligands that inhibit anandamide-facilitated transport and bind to CB1 cannabinoid receptors. *Biochemical and Biophysical Research Communications*, *262*, 275–284.

Millns, P. J., Chimenti, M., Ali, N., Ryland, E., de Lago, E., Fernandez-Ruiz, J., et al. (2006). Effects of inhibition of fatty acid amide hydrolase vs. the anandamide membrane transporter on TRPV1-mediated calcium responses in adult DRG neurons; the role of CB receptors. *European Journal of Neuroscience*, *24*, 3489–3495.

Moore, S. A., Nomikos, G. G., Dickason-Chesterfield, A. K., Schober, D. A., Schaus, J. M., Ying, B., et al. (2005). Identification of a high-affinity binding site involved in the transport of endocannabinoids. *Proceedings of the National Academy of Sciences of the United States of America*, *102*, 17852–17857.

Mor, M., Rivara, S., Lodola, A., Plazzi, P. V., Tarzia, G., Duranti, A., et al. (2004). Cyclohexylcarbamic acid 3′- or 4′-substituted biphenyl-3-yl esters as fatty acid amide hydrolase inhibitors: Synthesis, quantitative structure–activity relationships, and molecular modeling studies. *Journal of Medicinal Chemistry*, *47*, 4998–5008.

Muccioli, G. G. (2010). Endocannabinoid biosynthesis and inactivation, from simple to complex. *Drug Discovery Today*, *15*, 474–483.

Murataeva, N., Straiker, A., & Mackie, K. (2013). Parsing the players: 2-AG synthesis and degradation in the CNS. *British Journal of Pharmacology*, *171*, 1379–1391.

Nicolussi, S. (2014). *Identification and characterization of potent and selective inhibitors of endocannabinoid uptake* (pp. 1–173). Doctoral dissertation. Available from GCB Graduate School, University of Bern, Switzerland.

Nicolussi, S., Chicca, A., Rau, M., Rihs, S., Soeberdt, M., Abels, C., et al. (2014). Correlating FAAH and anandamide cellular uptake inhibition using N-alkylcarbamate inhibitors: From ultrapotent to hyperpotent. *Biochemical Pharmacology*, *92*, 669–689.

Nicolussi, S., Viveros-Paredes, J. M., Gachet, M. S., Rau, M., Flores-Soto, M. E., Blunder, M., et al. (2014). Guineensine is a novel inhibitor of endocannabinoid uptake showing cannabimimetic behavioral effects in BALB/c mice. *Pharmacological Research: The Official Journal of the Italian Pharmacological Society*, *80C*, 52–65.

Nomura, D. K., Morrison, B. E., Blankman, J. L., Long, J. Z., Kinsey, S. G., Marcondes, M. C. G., et al. (2011). Endocannabinoid hydrolysis generates brain prostaglandins that promote neuroinflammation. *Science*, *334*, 809–813.

O'Brien, L. D., Limebeer, C. L., Rock, E. M., Bottegoni, G., Piomelli, D., & Parker, L. A. (2013). Anandamide transport inhibition by ARN272 attenuates nausea-induced behaviour in rats, and vomiting in shrews (Suncus murinus). *British Journal of Pharmacology*, *170*, 1130–1136.

Oddi, S., Fezza, F., Catanzaro, G., De Simone, C., Pucci, M., Piomelli, D., et al. (2010). Pitfalls and solutions in assaying anandamide transport in cells. *Journal of Lipid Research*, *51*, 2435–2444.

Oddi, S., Fezza, F., Pasquariello, N., D'Agostino, A., Catanzaro, G., De Simone, C., et al. (2009). Molecular identification of albumin and Hsp70 as cytosolic anandamide-binding proteins. *Chemistry & Biology*, *16*, 624–632.

Oddi, S., Fezza, F., Pasquariello, N., De Simone, C., Rapino, C., Dainese, E., et al. (2008). Evidence for the intracellular accumulation of anandamide in adiposomes. *Cellular and Molecular Life Sciences*, *65*, 840–850.

Ortar, G., Cascio, M. G., Moriello, A. S., Camalli, M., Morera, E., Nalli, M., et al. (2008). Carbamoyl tetrazoles as inhibitors of endocannabinoid inactivation: A critical revisitation. *European Journal of Medicinal Chemistry*, *43*, 62–72.

Ortar, G., Ligresti, A., De Petrocellis, L., Morera, E., & Di Marzo, V. (2003). Novel selective and metabolically stable inhibitors of anandamide cellular uptake. *Biochemical Pharmacology*, *65*, 1473–1481.

Ortega-Gutiérrez, S., Hawkins, E. G., Viso, A., López-Rodríguez, M. L., & Cravatt, B. F. (2004). Comparison of anandamide transport in FAAH wild-type and knockout neurons: Evidence for contributions by both FAAH and the CB1 receptor to anandamide uptake. *Biochemistry*, *43*, 8184–8190.

Pacher, P., & Kunos, G. (2013). Modulating the endocannabinoid system in human health and disease—Successes and failures. *The FEBS Journal*, *280*, 1918–1943.

Pacher, P., & Mechoulam, R. (2011). Is lipid signaling through cannabinoid 2 receptors part of a protective system? *Progress in Lipid Research*, *50*, 193–211.

Pertwee, R. G. (2014). Elevating endocannabinoid levels: Pharmacological strategies and potential therapeutic applications. *The Proceedings of the Nutrition Society*, *73*, 96–105.

Pertwee, R. G., Howlett, A. C., Abood, M. E., Alexander, S. P., Di Marzo, V., Elphick, M. R., et al. (2010). International Union of Basic and Clinical Pharmacology. LXXIX. Cannabinoid receptors and their ligands: Beyond CB1 and CB2. *Pharmacological Reviews*, *62*, 588–631.

Piomelli, D. (2003). The molecular logic of endocannabinoid signalling. *Nature Reviews Neuroscience*, *4*, 873–884.

Piomelli, D., Beltramo, M., Glasnapp, S., Lin, S. Y., Goutopoulos, A., Xie, X. Q., et al. (1999). Structural determinants for recognition and translocation by the anandamide transporter. *Proceedings of the National Academy of Sciences of the United States of America*, *96*, 5802–5807.

Porter, A. C., Sauer, J., Knierman, M. D., Becker, G. W., Berna, M. J., Bao, J., et al. (2002). Characterization of a novel endocannabinoid, virodhamine, with antagonist activity at the CB1 receptor. *The Journal of Pharmacology and Experimental Therapeutics*, *301*, 1020–1024.

Price, T. J., Patwardhan, A. M., Flores, C. M., & Hargreaves, K. M. (2005). A role for the anandamide membrane transporter in TRPV1-mediated neurosecretion from trigeminal sensory neurons. *Neuropharmacology*, *49*, 25–39.

Rakhshan, F., Day, T., Blakely, R., & Barker, E. (2000). Carrier-mediated uptake of the endogenous cannabinoid anandamide in RBL-2H3 cells. *The Journal of Pharmacology and Experimental Therapeutics*, *292*, 960–967.

Rimmerman, N., Bradshaw, H. B., Kozela, E., Levy, R., Juknat, A., & Vogel, Z. (2012). Compartmentalization of endocannabinoids into lipid rafts in a microglial cell line devoid of caveolin-1. *British Journal of Pharmacology, 165*, 2436–2449.

Rimmerman, N., Hughes, H. V., Bradshaw, H. B., Pazos, M. X., Mackie, K., Prieto, A. L., et al. (2008). Compartmentalization of endocannabinoids into lipid rafts in a dorsal root ganglion cell line. *British Journal of Pharmacology, 153*, 380–389.

Ronesi, J., Gerdeman, G. L., & Lovinger, D. M. (2004). Disruption of endocannabinoid release and striatal long-term depression by postsynaptic blockade of endocannabinoid membrane transport. *The Journal of Neuroscience: The Official Journal of the Society for Neuroscience, 24*, 1673–1679.

Ross, R. A. (2003). Anandamide and vanilloid TRPV1 receptors. *British Journal of Pharmacology, 140*, 790–801.

Ryberg, E., Larsson, N., Sjögren, S., Hjorth, S., Hermansson, N. O., Leonova, J., et al. (2007). The orphan receptor GPR55 is a novel cannabinoid receptor. *British Journal of Pharmacology, 152*, 1092–1101.

Saario, S. M., Salo, O. M. H., Nevalainen, T., Poso, A., Laitinen, J. T., Järvinen, T., et al. (2005). Characterization of the sulfhydryl-sensitive site in the enzyme responsible for hydrolysis of 2-arachidonoyl-glycerol in rat cerebellar membranes. *Chemistry & Biology, 12*, 649–656.

Sanson, B., Wang, T., Sun, J., Wang, L., Kaczocha, M., Ojima, I., et al. (2014). Crystallographic study of FABP5 as an intracellular endocannabinoid transporter. *Acta Crystallographica Section D: Biological Crystallography, 70*, 290–298.

Savinainen, J. R., Saario, S. M., & Laitinen, J. T. (2011). The serine hydrolases MAGL, ABHD6 and ABHD12 as guardians of 2-arachidonoylglycerol signalling through cannabinoid receptors. *Acta Physiologica (Oxford, England), 204*, 267–276.

Schreiber, D., Harlfinger, S., Nolden, B. M., Gerth, C. W., Jaehde, U., Schömig, E., et al. (2007). Determination of anandamide and other fatty acyl ethanolamides in human serum by electrospray tandem mass spectrometry. *Analytical Biochemistry, 361*, 162–168.

Sigel, E., Baur, R., Racz, I., Marazzi, J., Smart, T. G., Zimmer, A., et al. (2011). The major central endocannabinoid directly acts at GABAA receptors. *Proceedings of the National Academy of Sciences of the United States of America, 108*, 18150–18155.

Skaper, S. D., & Di Marzo, V. (2012). Endocannabinoids in nervous system health and disease: The big picture in a nutshell. *Philosophical Transactions of the Royal Society B, 367*, 3193–3200.

Stella, N., Schweitzer, P., & Piomelli, D. (1997). A second endogenous cannabinoid that modulates long-term potentiation. *Nature, 388*, 773–778.

Sulsky, R., Magnin, D. R., Huang, Y., Simpkins, L., Taunk, P., Patel, M., et al. (2007). Potent and selective biphenyl azole inhibitors of adipocyte fatty acid binding protein (aFABP). *Bioorganic & Medicinal Chemistry Letters, 17*, 3511–3515.

Sun, Y., Alexander, S. P., Kendall, D. A., & Bennett, A. J. (2008). Involvement of fatty acid binding proteins in the transport of endocannabinoids to peroxisome proliferator activated receptors. In *18th annual symposium of the international cannabinoid research society (ICRS), 1, 17*, (Program and Abstracts).

Vandevoorde, S., & Fowler, C. J. (2005). Inhibition of fatty acid amide hydrolase and monoacylglycerol lipase by the anandamide uptake inhibitor VDM11: Evidence that VDM11 acts as an FAAH substrate. *British Journal of Pharmacology, 145*, 885–893.

Watanabe, H., Vriens, J., Prenen, J., Droogmans, G., Voets, T., & Nilius, B. (2003). Anandamide and arachidonic acid use epoxyeicosatrienoic acids to activate TRPV4 channels. *Nature, 424*, 434–438.

Yates, M., & Barker, E. L. (2009a). Inactivation and biotransformation of the endogenous cannabinoids anandamide and 2-arachidonoylglycerol. *Molecular Pharmacology, 76*, 11–17.

Yates, M. L., & Barker, E. L. (2009b). Organized trafficking of anandamide and related lipids. *Vitamins and Hormones*, *81*, 25–53.

Yu, S., Levi, L., Casadesus, G., Kunos, G., & Noy, N. (2014). Fatty acid-binding protein 5 (FABP5) regulates cognitive function both by decreasing anandamide levels and by activating the nuclear receptor peroxisome proliferator-activated receptor β/δ (PPARβ/δ) in the brain. *Journal of Biological Chemistry*, *289*, 12748–12758.

Zygmunt, P. M., Chuang, H., Movahed, P., Julius, D., & Högestätt, E. D. (2000). The anandamide transport inhibitor AM404 activates vanilloid receptors. *European Journal of Pharmacology*, *396*, 39–42.

CHAPTER FIFTEEN

Adenosine Transporters and Receptors: Key Elements for Retinal Function and Neuroprotection

Alexandre dos Santos-Rodrigues, Mariana R. Pereira, Rafael Brito, Nádia A. de Oliveira, Roberto Paes-de-Carvalho[1]
Program of Neurosciences, Fluminense Federal University, Niterói, Rio de Janeiro, Brazil
[1]Corresponding author: e-mail address: robpaes@vm.uff.br

Contents

Vitamins and Hormones, Volume 98
ISSN 0083-6729
http://dx.doi.org/10.1016/bs.vh.2014.12.014

Abstract

Adenosine is an important neuroactive substance in the central nervous system, including in the retina where subclasses of adenosine receptors and transporters are expressed since early stages of development. Here, we review some evidence showing that adenosine plays important functions in the mature as well as in the developing tissue. Adenosine transporters are divided into equilibrative and concentrative, and the major transporter subtype present in the retina is the ENT1. This transporter is responsible for a bidirectional transport of adenosine and the uptake or release of this nucleoside appears to be regulated by different signaling pathways that are also controlled by activation of adenosine receptors. Adenosine receptors are also key players in retina physiology regulating a variety of functions in the mature and developing tissue. Regulation of excitatory neurotransmitter release and neuroprotection are the main functions played be adenosine in the mature tissue, while regulation of cell survival and neurogenesis are some of the functions played by adenosine in developing retina. Since adenosine is neuroprotective against excitotoxic and metabolic dysfunctions observed in neurological and ocular diseases, the search for adenosine-related drugs regulating adenosine transporters and receptors can be important for advancement of therapeutic strategies against these diseases.

1. INTRODUCTION

1.1 The retina and its neurotransmitters

The retina is a specialized tissue of the central nervous system (CNS), which is responsible for the reception and transduction of light stimuli derived from the outside environment. Visual information processed in the retina is transmitted and processed in higher brain structures including the visual cortex. The retina is highly organized but contains relatively few cell types: the primarily photosensitive cells named photoreceptors (rods and cones), neuronal cells named horizontal, bipolar, amacrine, ganglion, and in some species, interplexiform cells, as well as glial cells, composed of Müller and microglial cells. The retina is organized in layers separating cell bodies from plexus as in other parts of the CNS (Fig. 1). The outer nuclear layer is composed of photoreceptor cell bodies and the outer plexiform layer contains the processes from photoreceptors, horizontal, and bipolar cells. The cell bodies of the two latter neuronal cells as well as from amacrine and Müller cells constitute

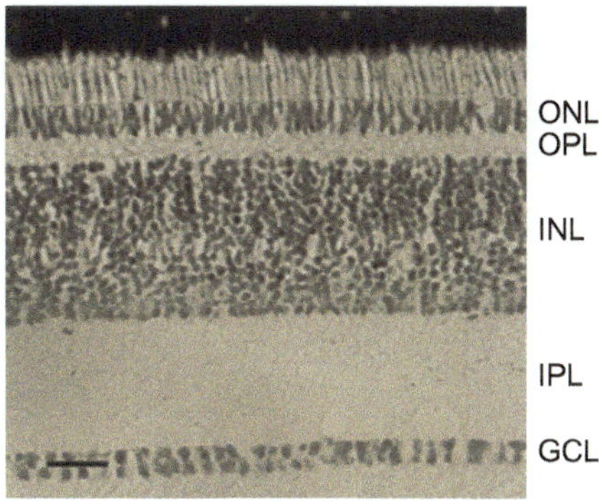

Figure 1 Schematic organization of chicken retina. Representative image of a post-hatching chicken retina stained with cresyl violet. The retina is highly organized in layers (nuclear and plexiform layers) with different cell types: the primarily photosensitive cells named photoreceptors (rods and cones); neuronal cells named horizontal, bipolar, amacrine, and ganglion; and glial cells composed of Müller and microglial cells. ONL, outer nuclear layer; OPL, outer plexiform layer; INL, inner nuclear layer; IPL, inner plexiform layer; and GCL, ganglion cell layer. Scale bar = 20 μm.

the inner nuclear layer. The inner plexiform layer is large and composed of processes from bipolar, amacrine, and ganglion cells. The ganglion cell layer contains cell bodies from ganglion as well as from displaced amacrine cells. The axons from ganglion cells constitute the optic nerve that carry the information processed in the retina to higher CNS structures.

1.2 The chicken retina as a model for neurochemical studies

The avian retina, especially from *Gallus gallus*, is a very convenient model for neurochemical studies of the CNS for many reasons. First, it is very easy to isolate the tissue free from contamination with other tissues during most of the embryonic period of development. Second, the neurogenesis in the chicken retina is very well known and the cells from the early developing tissue can be dissociated to prepare cultures, where many of the neurochemical properties are maintained as in the intact tissue. Three types of cultures are especially useful for the study of retinal neurochemistry: the mixed cultures containing neurons and glial cells, and purified cultures of neurons or glial cells (Fig. 2).

Most, if not all, neurotransmitters and neuromodulators present in other areas of the CNS are also present in the retina, such as glutamate, dopamine, GABA, acetylcholine, and adenosine. Acetylcholine was the first neurotransmitter identified in the chick retina (Lindeman, 1947), but a detailed study of several aspects of chick retinal development and neurogenesis came

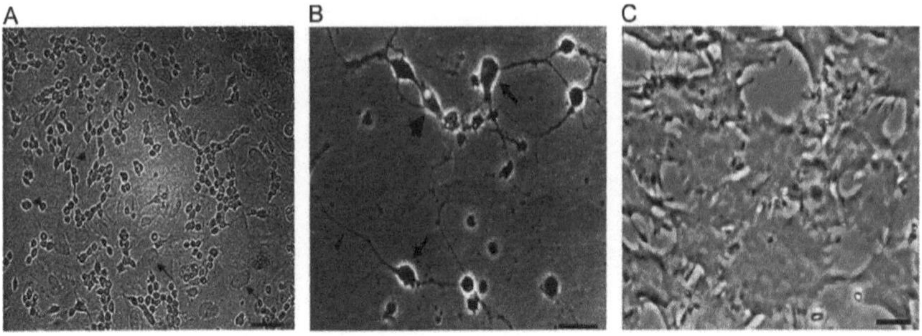

Figure 2 Types of cultures from chick retina. (A) Phase-contrast micrograph of a mixed culture with neurons (arrowheads) and glial cells (long arrows). (B) Phase-contrast micrograph of a purified neuronal culture showing the presence of neurons (long arrows) and photoreceptors (large arrowhead) and the absence of glial cells. (C) Phase-contrast photomicrograph of a purified culture of glial cells. Scale bars = 30 μm.

up in the 1950s and 1960s (Coulombre, 1955; Fujita & Horii, 1963; Witkovsky, 1963). The pioneer studies by Moscona and collaborators used the chicken retina model to study the regulation of glutamine synthetase activity induced by corticoids (Kirk & Moscona, 1963; Moscona & Kirk, 1965; Moscona & Moscona, 1963; Moscona & Piddington, 1966). As stated above, cell cultures of the chicken retina were developed and used in a variety of neurochemistry studies. Cultures of retinal cell aggregates were first developed (Sheffield & Moscona, 1969, 1970) and used to study glutamine synthetase induction and its dependence on cell interactions (Morris & Moscona, 1970, 1971). Thereafter, different studies using distinct types of retinal cultures, including monolayer cultures, were also performed, showing the properties of GABA uptake and synthesis (Tunnicliff, Cho, & Martin, 1974; Tunnicliff, Firneisz, Ngo, & Martin, 1975). Important studies at this period showed the sequential appearance of neurons and the formation of synapses during chick retinal development (Hughes & LaVelle, 1974; Kahn, 1974). Acetylcholine, GABA, and glutamate receptors were studied during retinal development (Lopez-Colome, 1981; Vogel, Daniels, & Nirenberg, 1976; Yazulla & Brecha, 1980). Dopamine receptors, coupled to cAMP production, were also studied during chick retinal development and in monolayer cultures (de Mello, 1978; de Mello, Ventura, Paes-de-Carvalho, Klein, & de Mello, 1982). As specified above, many neurotransmitter and neuromodulator systems are expressed in the chicken retina. Studies of GABA release as well as other amino acids induced by glutamate highlighted the importance of these amino acids in retinal physiology

(Campochiaro, Ferkany, & Coyle, 1984, 1985). One important compound present in the retina is adenosine and the presence of adenosine transporters and receptors in this tissue has consequences for retinal physiology and development that will be considered in detail along the next sections of this chapter.

2. THE NUCLEOSIDE ADENOSINE IN THE CNS

Adenosine is an important nucleoside component of the purinergic system. It is present in all vertebrates' tissues, including the CNS, modulating several physiological processes (Cunha, 2001). Initial studies demonstrating that adenosine could act as a signaling molecule were performed by Drury and Szent-Gyorgyi (1929) who showed that extracts of mammalian heart muscle injected in animals were able to induce a decrease in heart rate. The compound responsible for these effects was isolated and its chemical properties corresponded to AMP. However, animals injected with adenosine showed the same effects.

Although several studies have indicated that adenosine and ATP act like transmitter molecules, only in 1972 the purinergic term was accepted (Burnstock, 1972). In 1978, Burnstock proposed the existence of purinergic receptors and divided them in P1, selective for adenosine and P2, selective for ATP and ADP (Burnstock, 2009).

The adenosine receptors are metabotropic receptors named A1, A2a, A2b, and A3 receptors. A1 and A3 receptors are coupled to G_i/G_o protein and inhibit adenylyl cyclase activity, while A2a and A2b receptors are coupled to G_s protein and increase the enzyme activity (Ribeiro, Sebastião, & de Mendonca, 2002; Fig. 3A). However, adenosine receptors also can activate the PLC pathway (Abbracchio et al., 1995; Biber, Klotz, Berger, Gebicke-Harter, & van Calker, 1997; Gao, Chen, Weber, & Linden, 1999; Offermanns & Simon, 1995).

2.1 Adenosine in the retina

Several studies in the literature point to the existence of an adenosine system during development of the vertebrate retina, probably indicating an important role for this nucleoside in normal retinal development (Paes-de-Carvalho, 1990; Paes-de-Carvalho & de Mello, 1982, 1985). As we just mentioned above, adenosine receptors classically regulate cAMP levels. In the chick retina, this phenomenon was first shown by Paes-de-Carvalho and de Mello (1982) who showed that retinas from 17-day-old embryos

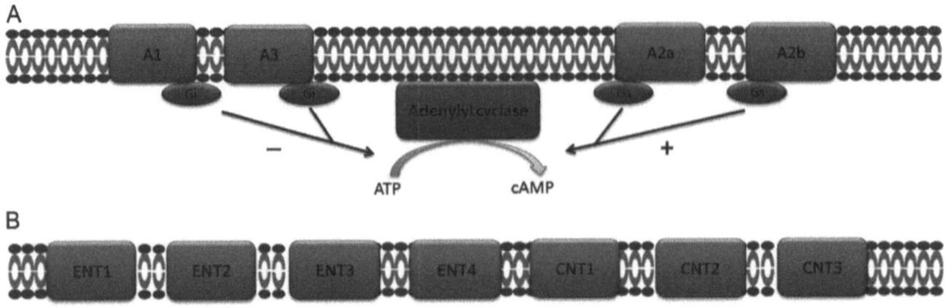

Figure 3 Adenosine receptors and nucleoside transporters. (A) There are four subtypes of adenosine receptors: A1 and A3 that inhibit adenylyl cyclase decreasing intracellular cAMP levels, and A2a and A2b that activate adenylyl cyclase increasing intracellular cAMP levels. (B) Equilibrative nucleoside transporters (ENTs 1–4) and concentrative nucleoside transporters (CNTs 1–3). In chicken retina, only ENT1 and ENT2 were described to be present based on pharmacological assays.

(E17) were able to accumulate cAMP when stimulated with adenosine or its nonhydrolysable and nonselective adenosine receptor agonist 2-chloroadenosine (Paes-de-Carvalho & de Mello, 1982). Interestingly, this effect showed a variation according to the developmental stage of the tissue. In retinas from E8 to E13, adenosine was not able to induce an increase of cAMP levels. This effect was observed only after E14 reaching a maximum in E17 retinas. In posthatching animals, adenosine could also induce cAMP formation, although with a reduced increase when compared to E15 and E18 embryos (Paes-de-Carvalho & de Mello, 1982). In E12 retinas, dopamine also had the ability to enhance the accumulation of cAMP via activation of D1 receptors and this effect was blocked by increasing doses of adenosine A1 receptor agonists. The same inhibition pattern was found when retinas were incubated with 2-chloroadenosine (Paes-de-Carvalho & de Mello, 1985). In addition to the chick retina, retinas from other vertebrate species also accumulate cAMP when stimulated with adenosine. Stimulation with adenosine, dopamine, or norepinephrine promoted cAMP accumulation in the rabbit retina. However, only the effect of adenosine was blocked by IBMX, an adenosine receptor antagonist (Blazynski, Kinscherf, Geary, & Ferrendelli, 1986).

Autoradiography for [^{3}H]-adenosine and [^{3}H]-cyclohexyladenosine (CHA), a selective adenosine A1 receptor agonist, showed that the distribution of subpopulations of retina cells that accumulate adenosine is similar in rabbits, mice, and squirrels (Blazynski, Mosinger, & Cohen, 1989). Moreover, for all three species, cells localized to the ganglion cell layer accumulated adenosine and exhibited adenosine-like immunoreactivity. A smaller proportion of cells localized in the inner nuclear layer presented adenosine-like immunoreactivity, while a larger proportion of cells in this layer accumulated adenosine (Blazynski et al., 1989). In rabbit retinas, uptake of [^{3}H]-adenosine as well as the selective adenosine A1 receptor agonist [^{3}H]-R-phenylisopropyladenosine (R-PIA) into retinal cells was assessed autoradiographically, in the presence and absence of the purine nucleoside transport inhibitor nitrobenzylthioinosine (NBMPR). Under control conditions, both purine nucleosides accumulated in cell bodies localized to the ganglion cell and the inner nuclear layers. In presence of NBMPR, significantly less accumulation of nucleosides within cell bodies was observed, particularly within the inner nuclear layer (Blazynski, 1991).

The presence of nucleoside transporters and cAMP accumulation in the chick embryo retina suggests that adenosine can be endogenously produced in this tissue. Accordingly, immunohistochemical markers showed the presence of adenosine at distinct layers at different stages. At early stages of development, as E8, when few cells migrated to their final positions, there was no labeling for adenosine. However, the presence of endogenous adenosine was observed at E12 in the inner and outer nuclear as well as in the ganglion cell and inner plexiform layers (Paes-de-Carvalho, Braas, Adler, & Snyder, 1992). In subsequent ages, as E15 and 5-day-old posthatching animals, the labeling was also found in these layers but with a higher intensity in the ganglion cell layer in posthatching animals (Paes-de-Carvalho et al., 1992). In the inner nuclear layer, adenosine immunoreactivity was restricted to cell bodies of amacrine cells, whereas in the outer nuclear layer both types of photoreceptors appeared to have endogenous adenosine. On the other hand, not all cells showed adenosine labeling in the ganglion cell layer, indicating heterogeneity in this cell population (Paes-de-Carvalho et al., 1992). Adenosine is also present in other vertebrate retinas. In the rat, cat, and guinea pig retina, adenosine is present in the ganglion cell and inner nuclear layers (Blazynski et al., 1989; Braas, Zarbin, & Snyder, 1987). In rabbit retinas, adenosine was observed in the ganglion cell layer and a low intensity staining was found in cell bodies of amacrine cells in the inner nuclear layer (Blazynski et al., 1989).

2.2 Actions of adenosine in the retina

Adenosine regulates several events in CNS such as neurotransmitter release and neuroprotection (Cunha, 2001). These events can also be modulated by adenosine in the retina. Previous work has shown that adenosine inhibits acetylcholine release stimulated by light in rabbit retinas and that a pretreatment of retinas with adenosine deaminase, which is an enzyme that converts adenosine into inosine, produced a 30% increase of acetylcholine release (Blazynski, Woods, & Mathews, 1992). Moreover, adenosine displays a neuroprotective effect against ganglion cell death induced by axotomy in rat retinas and neuronal death induced by glutamate in chick embryo retinas (Ferreira & Paes-de-Carvalho, 2001; Paes-de-Carvalho, Maia, & Ferreira, 2003; Perigolo-Vicente et al., 2014, 2013).

Adenosine is also implicated in anomalies and diseases related to the visual system, such as myopia (Cui et al., 2010), glaucoma (Zhong, Yang, Huang, & Luo, 2013), diabetic retinopathy (DR) (Wurm et al., 2008), and ischemic events (Dreixler et al., 2009; Li & Roth, 1999). Some features of these mechanisms between adenosine and these disorders are discussed later in this chapter.

2.3 Adenosine A1 receptors in the retina

The expression of adenosine receptors in the chick embryo retina has been characterized throughout development. Through autoradiography studies using [^{3}H]-CHA, it was shown the expression of these receptors at early stages of development (E10), but at very low levels (Paes-de-Carvalho, 1990). However, with the progress of development, receptor levels grow dramatically reaching a peak at E16. Interestingly, A1 receptor levels decrease in animals after hatching, but remain high when compared to age E12 (Paes-de-Carvalho, 1990). Later, autoradiography studies elucidated the tissue localization of this receptor type. Autoradiography of adenosine A1 receptors using L-[^{3}H]-PIA showed the presence of grains over the inner and outer plexiform layers at E12 with a gradual increase in intensity in these layers up to E18. In posthatching animals, it was found a small reduction in labeling intensity compared to earlier stages of development (Paes-de-Carvalho et al., 1992). According to the results found in 1990, autoradiography data at E8 revealed no grains over the tissue. In addition, very little specific binding was found in the inner and outer nuclear layers and in the ganglion cell layer at the ages studied (Paes-de-Carvalho et al., 1992). The presence of A1 receptors have been described in retinas from other species. In rabbit and mouse retinas, A1 receptors are present at all layers with a high expression in the inner plexiform and inner nuclear layers (Blazynski, 1990). Rabbit retinal homogenates showed increased adenylyl cyclase activity when challenged with forskolin, a direct activator of the enzyme (Blazynski, 1987). Interestingly, preincubation of homogenates with low concentrations of CHA or phenylisopropyladenosine (R-PIA), both A1 receptor agonists, promoted an inhibition of the increase of adenylyl cyclase activity induced by forskolin (Blazynski, 1987). Autoradiography in rat retina using L-[^{3}H]PIA showed an apparent presence of grains over the nerve fiber, ganglion cell, and inner plexiform layers as well as in the inner portion of the inner nuclear layer. In the monkey retina, binding was more diffusely distributed from the nerve fiber to the outer plexiform layers. The binding was observed throughout the whole human retina, but it was enriched in regions of the inner retina, especially in the ganglion cell layer (Braas et al., 1987). Although the autoradiographic grain densities associated with specific retinal lamina varied between rat, monkey, and human tissues, the ganglion cell and nerve fiber layers contain high densities of binding sites in all species (Braas et al., 1987). Recently, Li and coworkers showed the heterologous expression of A1 receptors in retinas of transgenic mice using different molecular tools (Li et al., 2007).

2.4 Adenosine A2a receptors in the retina

Adenosine A2a receptors have been demonstrated to be present in retinas of rats (Dreixler et al., 2009; Huang et al., 2014; Kvanta, Seregard, Sejersen, Kull, & Fredholm, 1997), mice (Blazynski, 1990; Blazynski & Perez, 1991; Li et al., 2013), bovines (Blazynski, 1993; Blazynski & McIntosh, 1993; McIntosh & Blazynski, 1994), dogs (Taomoto, McLeod, Merges, & Lutty, 2000), guinea pigs (Cui et al., 2010), and salamanders (Alfinito, Alli, & Townes-Anderson, 2002). In rat retina, the A2a receptor mRNA was found in the ganglion cell, inner nuclear, and outer nuclear layers, as well as in the choriocapillaris (Kvanta et al., 1997). Immunohistochemical studies in the developing mouse retina revealed the presence of A2a receptors in inner plexiform and ganglion cell layers, in accordance with the mRNA expression (Huang et al., 2014). A2a receptor immunoreactivity was also found in nonneuronal cell types, such as the retinal pigment epithelium and choroid. Moreover, high magnification images showed that A2aR immunoreactivity is localized to wave-generating starburst amacrine cells (SACs) in retinal cross-sections. Further immunostaining in dissociated SACs confirmed the localization of A2aR, indicating that this receptor is expressed in SACs (Huang et al., 2014). Binding assays for [^{3}H]-NECA revealed the existence of A2a receptors in the mouse retina, which were observed primarily over the retinal pigmented epithelium and the outer and inner segments of photoreceptors (Blazynski, 1990). While virtually all of the [^{3}H]-NECA binding was displaced by an excess of unlabeled NECA, displacement with antagonist or a large excess of CHA revealed that approximately 30% of the [^{3}H]-NECA binding was not to A1, but to A2 receptor sites. Therefore, the majority of the binding in the outer retina represents A2 receptor sites (Blazynski, 1990). Binding assays for [^{3}H]-NECA and [^{3}H]-CGS21680 also revealed the presence of A2a receptors in membranes from bovine retinas (Blazynski, 1993; Blazynski & McIntosh, 1993) as well as in bovine rod outer segments (McIntosh & Blazynski, 1994). In guinea pigs, the expression of A2a receptors is mainly distributed in the cytoplasm and extracellular matrix of the sclera and retina (Cui et al., 2010).

In the chicken retina, recent evidence from our group demonstrates the expression of A2a receptors at specific ages (Vardiero, E., Pereira, M.R., & Paes-de-Carvalho, R. unpublished data). However, more studies are necessary in order to identify the cell types expressing this class of receptors during retinal development.

2.5 Adenosine A3 and A2b receptors in the retina

Until the present moment, there is no evidence for the expression of A3 receptors in the developing chick retina. However, there is evidence for the expression of mRNA for this receptor in rat ganglion cells (Zhang et al., 2006) and in guinea pig retina (Cui et al., 2010), although initial studies have not detected A3 receptor mRNA in rat retinas (Kvanta et al., 1997). Interestingly, Zhang et al. (2010) showed that activation of A3 receptors modulates NMDA-dependent calcium influx in ganglion cells. Thus, there is some evidence for a functional role for A3 receptors in the retina but additional experiments are necessary to confirm these findings.

Regarding the A2b receptor, little is known about its presence in the retina of different animals, although it seems to be expressed in the sclera and retina of guinea pigs (Cui et al., 2010).

3. NEUROMODULATORY ACTIONS OF ADENOSINE IN THE RETINA

3.1 Modulation of ionic channels by adenosine receptors

The presence of adenosine receptors in retinas from different species indicates a possible role for these receptors in retinal physiology. Indeed, several works were published trying to elucidate this role. In tiger salamander retinas, for example, adenosine or the selective A1 receptor agonist CHA is able to reduce the amplitude of voltage-dependent calcium channel currents in ganglion cells *in vitro* (Sun, Barnes, & Baldridge, 2002). This effect was blocked and reduced, respectively, by treatment with the selective A1 antagonist DPCPX and administration of ω-conotoxin GVIA, an N-type calcium channel blocker. In addition, adenosine reduced the calcium influx induced by glutamate and voltage-gated calcium currents in rat retinal ganglion cells cultures, an effect which was blocked by A1 but not A2a receptor antagonists. Similar results were observed in ganglion cells from intact retinas stimulated with NMDA (Hartwick, Lalonde, Barnes, & Baldridge, 2004). On the other hand, calcium influx through L-type calcium channels induced by depolarization of rod photoreceptors is inhibited by treatment with adenosine or the selective A2a agonist DPMA (Stella, Bryson, & Thoreson, 2002). As stated above, it was demonstrated that activation of A3 receptors inhibits the calcium rise induced by glutamate or NMDA in rat retinal ganglion cells cultures (Zhang et al., 2010). Taken together, these data suggest that adenosine may have a direct effect on the release of neurotransmitters in the retina.

Pericytes located in the mouse retinal microvasculature, when challenged with adenosine, exhibit a hyperpolarization-dependent opening of ATP-sensitive potassium (K_{ATP}) channels (Li & Puro, 2001). Experiments with selective agonists and antagonists indicated that adenosine A1 and A2a receptors provided effective pathways for activating K_{ATP} currents in pericytes recorded under normal metabolic conditions. However, during chemical ischemia, the A1 receptor pathway rapidly became ineffective, while activation of adenosine A2a receptors continued to open K_{ATP} channels in ischemic pericytes (Li & Puro, 2001). These results suggest that adenosine serves as a vasoactive signal in the retinal microvasculature. Indeed, injection of adenosine or nonselective agonists in rabbit, cat, and marmoset retinas induces vasodilatation and hemorrhage in the eye (Campochiaro & Sen, 1989). Furthermore, it was shown that adenosine has a relaxing effect on porcine retinal arterioles, an effect blocked by A2a and A2b receptor antagonists, but potentiated by A1 and A3 receptor antagonists (Riis-Vestergaard, Misfeldt, & Bek, 2014).

3.2 Modulation of neurotransmitter release by adenosine receptors

An important function of adenosine receptors is the regulation of neurotransmitter release. A1 receptor activation inhibits acetylcholine release evoked by high concentration of KCl through the inhibition of voltage-sensitive Ca^{2+} channels in cultures of chick retinal neurons (Santos,

Santos, Carvalho, & Duarte, 1998), which are enriched in cholinergic amacrine-like neurons. A subpopulation of these cholinergic neurons also store GABA. However, the regulation of neurotransmitter release by A1 receptors in these cultures seems to be selective to acetylcholine since A1 receptor activation does not affect GABA release evoked by high KCl (Santos, Caramelo, Carvalho, & Duarte, 2000). The release of excitatory amino acids can also be modulated by adenosine receptors (Rego, Agostinho, Melo, Cunha, & Oliveira, 2000). Glutamate and aspartate release can be stimulated with glycolysis inhibitors in cultures of chick embryo retinas. The release of both compounds significantly increases with A2a but not A1 receptor antagonists. These observations indicate that tonic activation of A2a receptors by adenosine decreases excitatory amino acid release. This effect probably occurs through reversion of adenosine transporters because metabolic inhibition induces an increase of extracellular adenosine levels (Rego, Santos, & Oliveira, 1997). Regarding the uptake of aspartate, this was not changed by A2a receptor inhibition, indicating that A2a receptor-dependent modulation of excitatory amino acid release does not involve uptake regulation. GABA release was also analyzed and blocking A2a receptors does not modify the release stimulated by glycolysis inhibitors (Rego et al., 2000). Moreover, oxidative stress also increases aspartate release induced by high KCl, an effect that decreases with the incubation with adenosine deaminase or A1 receptor agonists and increases with A2a receptor agonists. In addition, oxidative stress increases adenosine extracellular levels induced by high KCl. These data indicate that oxidative stress induces adenosine release that, through A1 and A2a receptor activation, modulates aspartate release (Agostinho et al., 2000). Taken together, these results indicate that adenosine receptors differentially regulate neurotransmitter release.

3.3 A1 receptors regulate axonal growth

Another important function of adenosine receptors appears to be the modulation of axonal growth as A1 receptors participate in the guidance of retinal ganglion cell axons to the optic tectum. This process involves *Engrailed* transcription factors which stimulate ATP synthesis and release by growth cones. Secreted ATP is converted to adenosine and activates A1 receptors that are more expressed in temporal growth cones. A1 receptor activation then enhances ephrin A5 signaling and produces a collapse of temporal growth cones. These results demonstrate the importance of A1 receptors in increasing the precision of retinal projection map (Stettler et al., 2012).

3.4 Adenosine receptors in Müller cells and regulation of cell volume homeostasis

Adenosine receptors are also expressed in Müller glial cells where they can regulate several cellular functions through different signaling pathways. In purified cultures of Müller cells from chick embryo retinas, stimulation of A1 receptors increases ERK phosphorylation in a PKC and Src kinase-dependent manner (dos Santos-Rodrigues, A., Pereira, M.R., da Silva, I.L.A., Rodrigues, S.A., Leão-Ferreira, L.R., & Paes-de-Carvalho, R. unpublished results).

In the literature, you also can find reports about adenosine receptors expressed in Müller cells regulating cell volume homeostasis. In hyposmotic conditions, Müller cells show a cellular swelling and adenosine inhibits this effect via A1 receptor activation. Swelling was observed in retinal slices from adenosine A1 receptor-deficient mice (A1AR$^{-/-}$) perfused with hyposmolar solution, indicating that A1 receptors are important to regulate Müller cell volume (Wurm et al., 2009). Müller cell swelling is also observed in pathologic situations such as diabetic retinopathy and retinal ischemia (Uckermann et al., 2006; Wurm et al., 2008). Perfusion of retinal slices from diabetic rats with hypotonic solution induces a time-dependent swelling of Müller cells that is not observed in retinal slices from control rats. The osmotic swelling was inhibited by triamcinolone, an anti-inflammatory glucocorticoid clinically used for diabetic macular edema, and by adenosine treatment. The effects of triamcinolone and adenosine were blocked by A1, but not A2a receptor antagonists. Indeed, the effect of triamcinolone on Müller cell swelling involves an increase of extracellular adenosine levels via adenosine transporters and ATP extracellular metabolism. Then, activation of adenosine A1 receptors induces K^+ and Cl^- efflux, preventing osmotic swelling of Müller cells (Wurm et al., 2008). Regarding retinal ischemia, slices of postischemic rat retina treated with a hypotonic solution showed Müller cell swelling which was not observed in retinal slices from control rats. Neuropeptide Y inhibits this effect through stimulation of glutamate release and consequent activation of metabotropic glutamate receptors that stimulate adenosine release. Then, A1 receptors are stimulated and inhibit osmotic swelling via regulation of K^+ and Cl^- channels (Uckermann et al., 2006). Similar results were observed with atrial natriuretic factor (Kalisch et al., 2006). These data indicate that glial cells in retinal slices from diabetic or ischemic animals are more sensitive to osmotic stress than cells from control animals.

In support of the above-mentioned data, another report reported that cultures of rat Müller cells exposed to high glucose (HG) medium showed an increase of caspase-1 activation. This effect was reduced by apyrase, an enzyme that metabolizes ATP, and by adenosine deaminase, suggesting the involvement of ATP and adenosine. The activation of adenosine receptors with a nonselective agonist, or inhibition of adenosine transporters, also increased caspase-1 activation. Furthermore, a selective A2b receptor antagonist reduced caspase-1 activation induced by HG, demonstrating that this effect involves A2b receptor activation (Trueblood, Mohr, & Dubyak, 2011).

3.5 A2a and A2b receptors modulate TNF-α production by microglia and phagocytosis of photoreceptor outer segments

Adenosine has been shown to regulate inflammation in different experimental models. In rat retinal microglia cultures, in which A2a receptors are highly expressed, the activation of these receptors decreases TNF-α production induced by LPS. Moreover, microglial cells treatment with cannabidiol, a nonpsychotropic cannabinoid, enhanced the adenosine effect on the decrease of TNF-α production through inhibition of adenosine uptake. Cannabidiol also decreases TNF-α production induced by LPS and this effect is blocked by an A2a antagonist. These results demonstrated the anti-inflammatory effect of A2a receptor in rat retina (Liou et al., 2008).

Adenosine receptors are also expressed in human retinal pigment epithelium cells in culture. Both the mRNA and adenosine receptor protein were detected and the A2b receptor was found to be more expressed in these cells (Wan et al., 2011). Adenosine can regulate an important function of retinal pigment epithelium, which is the fagocytosis of photoreceptor outer segments. This process was inhibited by adenosine and this inhibition was reduced by 8-phenyltheophylline, an adenosine receptor antagonist, and potentiated in the presence of dipyridamole, an adenosine transporter blocker. Furthermore, a large accumulation of cAMP was observed with the nonselective agonist NECA, suggesting an involvement of A2a or b receptors in this event (Gregory, Abrams, & Hall, 1994).

3.6 Regulation of adenosine receptor expression

The regulation of adenosine receptor expression was widely studied in CNS (Castillo, Leon, Ballesteros-Yanez, Albasanz, & Martin, 2010; Lopes, Cunha, & Ribeiro, 1999; Lopez-Zapata, Leon, Castillo, Albasanz, & Martin, 2011; Lorenzo et al., 2010). However, the importance and the factors involved in this regulation are poorly understood. In mixed cultures of chick embryo retina, A2a receptors are expressed in neuronal and Müller cells and A1 receptors are mostly expressed in neuronal cells. In addition, the chronic activation of A2a receptors in these cultures promotes an increase of A1 receptor expression in a cAMP/PKA pathway-dependent manner (Pereira, Hang, Vardiero, de Mello, & Paes-de-Carvalho, 2010). Preliminary results showed that this regulation is also dependent on NFkB activation, a finding which is in agreement with data demonstrating the regulation of A1 receptor expression by this transcription factor (Hammond, Bonnet, Kemp, Yates, & Bowmer, 2004; Hammond, Stolk, Archer, & McConnell, 2004; Jhaveri, Reichensperger, Toth, Sekino, & Ramkumar, 2007; Nie et al., 1998). Brito, Pereira, Paes-de-Carvalho, & Calaza (2012) showed that chick embryo retinas at E16 exposed since E14 to CGS21680 (A2a agonist) and ZM241385 (A2a antagonist) have, respectively, decreased or increased levels of A2a receptors when compared to control. Peculiarly, treatment with CGS21680 also induced an increase of A1 receptor expression. On the other hand, treatment with the selective A2a antagonists ZM241385 or SCH58261 led to a reduction of A1 receptor expression, indicating that endogenous adenosine released in the retinal environment is able to activate A2a receptors and induce the increase of A1 receptor expression (Brito et al., 2012). In addition, treatment with the nucleoside transporter blocker NBMPR was able to mimic the effects of SCH58261 and ZM241385, demonstrating the critical role of nucleoside transporters in the regulation of A1 receptor expression (Brito et al., 2012). In summary, the data indicate that normal A1 receptor expression depends on the activation of A2a receptors and that endogenous adenosine and nucleoside transporters are key regulators of this phenomenon.

Overall, these results indicate the need to look more closely at the consequences of long-term pharmacological treatments with adenosine receptor agonists and antagonists in retina and other CNS areas.

4. NUCLEOSIDE TRANSPORTERS

Adenosine levels found in extracellular and intracellular media are regulated by two different families of nucleoside transporters, which are known as equilibrative nucleoside transporters (ENTs) family and concentrative nucleoside transporters (CNTs) family (Fig. 3B). These transporters are integral membrane proteins and can transport adenosine itself as well as to carry other nucleosides, nucleobases, and some drugs which are nucleoside analogs.

Nucleosides are essential for the synthesis of nucleotides and this means that they are extremely important to cells, which are in the process of cell division, when nucleic acids are being produced, and also in cells that have a high metabolic rate (Abdulla & Coe, 2007). The physiological importance of nucleosides as substrates for nucleic acid synthesis has promoted a development of nucleoside analog drugs, which are useful for the treatment of some types of cancers and viral diseases. Nucleoside drugs are antimetabolites which, once phosphorylated, are able to interfere with the synthesis of new nucleoside molecules and also in the biosynthesis of nucleotides. These analogs are typically hydrophilic and request the presence of transporters to get in cells. Little information is known about the structure, function, and regulation of these transporters and therefore more studies are necessary to a better optimization of chemotherapeutic treatments based on the analogs mentioned above (Abdulla & Coe, 2007).

4.1 Equilibrative nucleoside transporters (ENTs)

ENTs transport nucleosides bidirectionally according to their concentrations in extracellular and intracellular milieu, while CNTs (concentrative nucleoside transporters) promote influx of nucleosides against the concentration gradient using the energy from sodium concentration gradient across cellular membranes (Podgorska, Kocbuch, & Pawelczyk, 2005). At present, there are four isoforms of ENTs described, which are referred as ENT1, ENT2, ENT3, and ENT4. They are also divided into sensitive and insensitive to an inhibitor called NBMPR. This compound binds ENT1 with high affinity (K_d 1–10 nM) through a noncovalent interaction at a high-affinity binding pocket. On the other hand, ENT2 is not affected by NBMPR in the nanomolar concentration range, being inhibited only with higher NBMPR concentrations (>10 μM) (Kong, Engel, & Wang, 2004). NBMPR is supposed to interact with the region spanning the third and sixth transmembrane domain (Baldwin et al., 2004). These transporters can be glycosylated and have 11 transmembrane domains with a cytoplasmatic N-terminal and an extracellular C-terminal, with a big extracellular loop linking the transmembrane domains 1 and 2 and a big intracellular loop between transmembrane domains 6 and 7.

Besides the transport of nucleosides, ENT1 is unable to transport nucleobases such as adenine, guanine, and hypoxanthine. However, ENT2 is able to transport these nucleobases. Transmembrane regions 5 and 6 of ENT2 appear to be important for recognition of these nucleobases, because the insertion of this region from an ENT2 rat isoform to an ENT1 rat isoform made this transporter capable of transporting nucleobases. (For a review, see Kong et al., 2004.)

ENT1 is a widely expressed transporter, mainly found at the plasma membrane, being considered the main regulator of homeostatic maintenance of adenosine levels (Bone, Robillard, Stolk, & Hammond, 2007). In the literature, there are several reports that prove that ENT1 inhibition can potentiate neuroprotective and cardioprotective effects induced by adenosine (Bone et al., 2007). These transporters were cloned from at least rats, mice, humans, and canines (Hammond, Bonnet, et al., 2004; Hammond, Stolk, et al., 2004). In mice, two ENT1 isoforms were described, known as mENT1a and mENT1b (Bone et al., 2007; Kiss et al., 2000). The only differences between these isoforms are located in the central intracellular loop, which connects transmembrane domains 6 and 7. The mENT1b has a serine in position 254 followed by a sequence lysine–glycine, while the mENT1a possesses an arginine in position 254 and the sequence lysine–glycine is deleted in this isoform. This Ser^{254} is part of a potential consensus sequence for phosphorylation by casein kinase II (CKII) (Bone et al., 2007). This differential feature between these two isoforms suggests a possibility of differential modulation with respect to CKII.

4.2 Concentrative nucleoside transporters (CNTs)

CNTs promote the influx of nucleosides against concentration gradient using the energy from sodium concentration gradient across cell membranes (Podgorska et al., 2005). These transporters are subdivided into three isoforms, known as CNT1, CNT2, and CNT3. These transporters can be glycosylated and possess 13 transmembrane domains with a cytoplasmatic N-terminal and an extracellular C-terminal.

In terms of mechanisms of regulation of CNTs, there is little information available. A report from 2010 (Fernandez-Calotti & Pastor-Anglada, 2010) showed that all-*trans*-retinoic acid increased the insertion rate of CNT3 in plasmatic membrane through a p38, TGF-β1, and ERK 1/2 mechanism. Another report by Errasti-Murugarren, Molina-Arcas, Casado, and Pastor-Anglada (2010) identified that a wild-type isoform of CNT3 can be found in lipid rafts and in nonraft domains, but a CNT3 variant, known as CNT3C602R, was found in lower levels in lipid rafts when compared with wild-type CNT3.

4.3 Nucleoside transporters in the retina

Several works in the literature report the presence of nucleoside transporters in retinas from different species (Blazynski, 1991; Braas et al., 1987; Schaeffer & Anderson, 1981). It is already known the presence of an adenosine uptake system in retinas from goldfish (Studholme & Yazulla, 1997), rabbits (Perez, Ehinger, Lindstrom, & Fredholm, 1986), rats (Schaeffer & Anderson, 1981), and chicken (Paes-de-Carvalho, Braas, Snyder, & Adler, 1990; Perez & Bruun, 1987), among others.

In the chicken retina, using autoradiography assays with (^{3}H) NBMPR to label NBMPR-sensitive nucleoside transporters, Paes-de-Carvalho and colleagues (1992) found a large labeling throughout the E8 retinal neuroblastic layer, in accordance with the lack of cellular organization in specific layers at this stage. The analysis of retinas at older stages of development as E12, E15, E18, and posthatching animals also showed [^{3}H]-NBMPR labeling but, however, it was lower compared to E8 and restricted to the outer and inner plexiform layers (Paes-de-Carvalho et al., 1992), which are regions rich in synapses. These results feature a similar localization found for A1 adenosine receptors in the chicken retina (Paes-de-Carvalho, 1990). This colocalization has already been observed in other CNS structures (Jennings et al., 2001), thereby correlating with a modulatory activity of this transporter in the activation of A1 receptors by adenosine. Overall, these data demonstrate the expression of adenosine transporters at all stages of chick retina development.

In cultured chick retinal neurons, it was demonstrated the presence of a high-affinity system for adenosine uptake. Under a depolarizing stimulus, there is a large increase of purines release, mostly as inosine (Paes-de-Carvalho et al., 1990). In this same work, the authors report that incubation of cultures with the ENT1 inhibitor NBMPR (10 nM) induced an adenosine uptake inhibition higher than 80%. This result indicates that ENT1 is the main transporter expressed in these cells, since there is no inhibition of other nucleoside transporters using such low concentration of NBMPR. Adenosine uptake in the cultures was sodium independent, indicating that retinal cells do not express CNTs. Moreover, the uptake was strongly blocked when [^{3}H]-adenosine was incubated with adenosine deaminase, an enzyme that converts adenosine to inosine, before being added to the cultures, meaning that nucleoside transporters expressed in these retinal cells have low affinity for inosine.

Also in cultured chick retinal neurons, another report showed that dopamine is able to promote an increase of purine release (Paes-de-Carvalho,

2002). Using mixed cultures of chick retina cells, the presence of ENT1 transporter was detected by binding experiments using the high-affinity ligand [^{3}H]-NBMPR. The authors also reported that a long-term incubation of these cultures with adenosine plus EHNA, an adenosine deaminase inhibitor, did promote a significant decrease in B_{max} transporter levels, suggesting that somehow the activation of adenosine receptors modulates the number of nucleoside transporters.

In mixed cultures of chick retina cells, more than 90% of [^{3}H]-adenosine taken up by cells is converted into adenine nucleotides, while around 80% of purine release stimulated by activation of ionotropic glutamate receptors is found as inosine and hypoxanthine (Paes-de-Carvalho et al., 2005). Interestingly, similar results were previously described in purified cultures of chick retinal neurons (Paes-de-Carvalho et al., 1990). This stimulatory effect on purine release induced by glutamate is blocked by the ENT1 blocker NBMPR (Paes-de-Carvalho et al., 2005). These results are similar in some terms with a study made by Perez and colleagues (1986), who demonstrated that the major amount of [^{3}H]-adenosine taken up by rabbit retina cells is converted into adenine nucleotides and, in the presence of a depolarizing stimulus, an increase of purine release is observed, mainly as hypoxanthine, xanthine, and inosine. This release was partially blocked by dipyridamole, an inhibitor of nucleoside transporters.

4.4 Regulation of ENTs by protein kinases

In this section, we will focus on different cellular mechanisms shown to be able to regulate the activity and/or expression of nucleoside transporters, mainly the equilibrative ones. Back in the 1990s, hENT1 was the first ENT to be described in the literature (Griffiths et al., 1997). Before cloning, the transport activity was ascribed to an es transporter (equilibrative, sensitive), based on sensitivity to inhibition by NBMPR. Nagy, Diamond, Casso, Franklin, and Gordon (1990) demonstrated that an acute exposure of cultured cells (including those from the CNS) to ethanol led to decreased adenosine uptake and that this effect was primarily due to effects on the es transporter. Moreover, this ethanol sensitivity was PKA- and PKC-dependent suggesting that es (subsequently confirmed to be ENT1) is proned to posttranslational regulation by kinases (Coe, Dohrman, Constantinescu, Diamond, & Gordon, 1996; Coe, Yao, Diamond, & Gordon, 1996; Nagy, Diamond, & Gordon, 1991). Intriguingly, another study showed that PKC activation induced a reduction of adenosine uptake in cultured chromaffin cells (Delicado, Sen, & Miras-Portugal, 1991).

PKA and PKC activators also had similar results on adenosine uptake in single chromaffin and neuroblastoma cells (Sen et al., 1998; Sen, Delicado, & Miras-Portugal, 1999). On the other hand, PKC activation is also able to increase hENT1-dependent nucleoside flux (Coe, Zhang, McKenzie, & Naydenova, 2002) in certain cells and this may occur via activation of adenosine receptors and the MAP kinase pathway (Grden et al., 2008). In hippocampal synaptosomes, A2a receptor activation induced an increase of adenosine uptake, in a PKC-dependent manner (Pinto-Duarte, Coelho, Cunha, Ribeiro, & Sebastião, 2005), a nonclassical cellular signaling pathway triggered by A2a receptor activation.

Other studies have demonstrated PKC involvement, besides nitric oxide (NO) and ERKs, in the inhibition of adenosine transport mediated by glucose in human fetal endothelial cells (Montecinos et al., 2000) and in B-lymphocytes (Sakowicz, Szutowicz, & Pawelczyk, 2005). An additional report also described a role of NO in the regulation of ENT1 protein levels (Vega et al., 2009) and more recently, another study showed that peroxynitrite regulates ENT1 activity in microvascular endothelial cells (Bone, Antic, Vilas, & Hammond, 2014). In chick retina, inhibition of ERKs was also able to decrease adenosine uptake (dos Santos-Rodrigues, Ferreira, & Paes-de-Carvalho, 2011) although the underlying mechanism is not clear.

To sum up, three kinases seem to be the major ENT1 regulators found up to this date: CKII, PKA, and PKC. Indeed, there are several reports showing putative CKII phosphorylation sites in both ENT1 and ENT2 sequences (Hammond, Bonnet, et al., 2004; Hammond, Stolk, et al., 2004; Kiss et al., 2000; Robillard, Bone, Park, & Hammond, 2008; Stolk, Cooper, Vilk, Litchfield, & Hammond, 2005) and an mENT1 splice variant (mENT1b) is predicted to have a potential CKII phosphorylation site within the large intracellular loop located between transmembrane domains 6 and 7 (Handa et al., 2001; Kiss et al., 2000). Bone and colleagues (2007) showed that activation of CKII induced an increase in the number and activity of ENTs at the membrane, but the underlying mechanism induced by CKII was not described so far.

While our understanding of the regulatory pathways and physiological relevance of kinase-dependent regulation of ENTs continues to be obscure, we now know that ENT1 can be directly phosphorylated, *in vitro*, within the large intracellular loop between transmembrane domains 6 and 7 by both PKC and PKA, showing that this is a potential mechanism of regulation of ENTs by kinase-dependent pathways (Reyes et al., 2011). Concerning the other ENT subclasses, there are even less available data about posttranslational regulatory mechanisms. Noteworthy, mENT2 possesses potential PKC phosphorylation sites in the first and third intracellular loops (Kiss et al., 2000). Based on analysis by NetPhosK (Blom, Sicheritz-Ponten, Gupta, Gammeltoft, & Brunak, 2004), human ENT2 has putative PKC, PKA, and CKII phosphorylation sites, some of which are highly conserved among species, suggesting that direct phosphorylation may be a mechanism of regulation of this isoform. Accordingly with these *in silico* analyses, Lu and colleagues (2010) showed that chronic morphine treatment induced an increase of extracellular adenosine levels, an effect mediated by a PKC-mediated decrease in ENT2 activity.

5. ADENOSINE AND NEUROPROTECTION IN THE RETINA

5.1 Adenosine neuroprotection in glaucoma disease

Glaucoma is an eye disease that is a common cause for blindness. At present, the visual defects found in glaucoma are avoidable, but not reversible. The elevated intraocular pressure (IOP) is considered as an important risk factor for the outset and prognostics of this disorder (Zhong et al., 2013). The high IOP usually leads to axonal degeneration simultaneously to death of retinal ganglion cells, which interrupts the visual transmission (Quigley, 2011). Despite the importance of IOP in glaucoma development, there are some cases where even with IOP under control the blindness keeps progressing, suggesting that there is some IOP-independent mechanisms in the disease (Daugeliene, Yamamoto, & Kitazawa, 1999; Gliklich, Steinmann, & Spaeth, 1989).

IOP is generated by the aqueous humor circulation system, and the humor is secreted from ciliary epithelium cells (Zhong et al., 2013). Adenosine is present in aqueous humor, and all the four adenosine receptors are present in those epithelium cells. In patients with ocular hypertension, the mean level of adenosine in the aqueous humor is elevated in comparison with normotensive patients (Daines, Kent, McAleer, & Crosson, 2003). It has been demonstrated that adenosine receptors can modulate IOP (Fig. 4). In rabbits, activation of adenosine A1 receptors reduces the IOP response against the hypertensive effect of A2a receptor activation (Crosson & Gray, 1994). The stimulation of A1 receptors reduces, while stimulation of A2a and A3 receptors exacerbates the IOP in mice. The opposite effect was observed when it was used the respective adenosine receptor antagonists (Avila, Stone, & Civan, 2001). When adenosine was

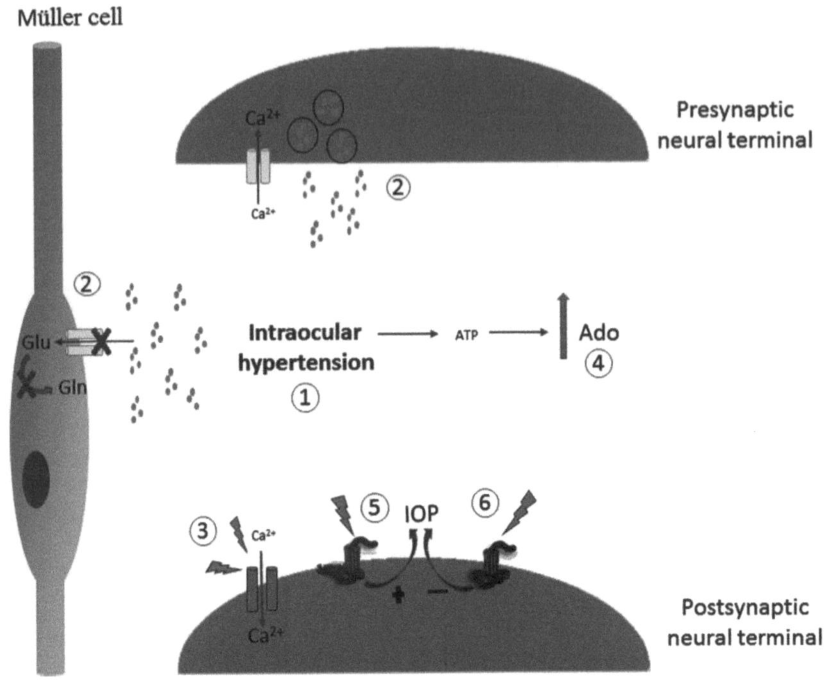

Figure 4 Death pathways induced by intraocular hypertension in glaucoma disease and the protection afforded by adenosine. Intraocular hypertension (1) leads to an increase of extracellular glutamate (2) and NMDA and non-NMDA receptor overstimulation (3) activating an excitotoxicity pathway. Intraocular hypertension (1) also elevates adenosine (Ado) levels (4) that may activate its receptors. A3 receptor activation exacerbates IOP (5), while A1 activation prevents hypertension (6), protecting retinal neurons.

applied to the eyes, IOP was abruptly increased and this effect was nearly 70% inhibited by A3 receptor blockade (Avila et al., 2001). Actually, A3 receptors appear to be important in aqueous humor maintenance and regulation.

A3 receptor knockout (A3AR$^{-/-}$) mice presented a lower baseline for IOP against wild-type (A3AR$^{+/+}$) mice and displayed no effects when challenged with selective A3 receptor agonists and antagonists (Avila, Stone, & Civan, 2002). Moreover, adenosine produced a smaller effect (~7- to 10-fold lower) in A3AR$^{-/-}$ mice in comparison with A3AR$^{+/+}$ mice, which was not prevented by selective adenosine A3 receptor antagonists (Avila et al., 2002). In different species, adenosine can act on aqueous humor dynamic through their receptors, presenting opposite responses (Zhong et al., 2013).

5.2 Adenosine neuroprotection in diabetic retinopathy

Diabetic retinopathy is the most common complication of diabetes and can aggravate to blindness (Yau et al., 2012). Evidence has increased showing that retina neurodegeneration may appear earlier than circulatory issues (Carrasco, Hernandez, de Torres, Farres, & Simo, 2008; Carrasco et al., 2007). Diabetic retinopathy impairs the glutamatergic transmission in retinal Müller cells, leading to accumulation of glutamate, overactivation of NMDA receptors, followed by microglial activation and destruction of blood–retinal barrier, ischemia, and neuronal cell death (Fig. 5; Cervantes-Villagrana, Garcia-Roman, Gonzalez-Espinosa, & Lamas, 2010; Liou, Ahmad, Naime, Fatteh, & Ibrahim, 2011).

Adenosine signaling exerts a critical role on retinal neuroprotection against diabetes. Activation of adenosine A2a receptors was able to block retinal responses to hyperglycemia, reducing apoptosis, mainly in the ganglion

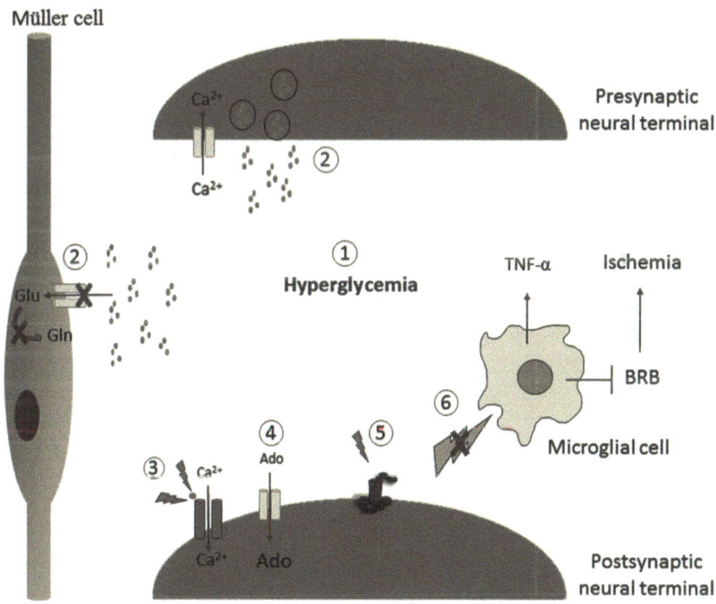

Figure 5 Death pathways induced by high glucose in diabetic retinopathy and the protection afforded by adenosine. Hyperglycemia (1) leads to an increase of extracellular glutamate (2) and NMDA and non-NMDA receptor overstimulation (3) activating an excitotoxicity pathway. High glucose levels induce an increase on ENT1 expression, raising adenosine (Ado) uptake (4) and decreasing its protective activity. Stimulation of A2a receptor (5) inhibits microglial activation (6) induced by glutamatergic excitotoxicity, preventing the release of proinflammatory cytokines, blood–retinal barrier (BRB) breakdown and an ischemic response.

cells layer, decreasing the release of proinflammatory cytokines, like TNF-α, and inhibiting microglial activation (Fig. 5; Ibrahim, El-Shishtawy, Zhang, Caldwell, & Liou, 2011). Moreover, the diabetic phenotype is exacerbated in diabetic A2a knockout mice (Ibrahim et al., 2011). An experimental model of diabetes with exposure of human aortic smooth muscle cells to HG demonstrated that HG may upregulate ENT1 mRNA and protein levels, rising adenosine uptake rates (Fig. 5; Leung, Man, & Tse, 2005). These data suggest that the maintenance of extracellular adenosine levels is important to restrain retina damage during diabetes.

5.3 Adenosine neuroprotection in ischemia

Diseases that cause visual damage and blindness are usually related to retinal ischemia (Ghiardi, Gidday, & Roth, 1999; Roth, 2004). Lower levels of blood supply can lead to ischemia and oxygen deprivation inducing an increase in energy consumption, rate of ATP breakdown to adenosine, and a higher release of excitatory amino acids like glutamate (Fig. 6; Ghiardi et al., 1999). The excess of extracellular glutamate causes excitotoxicity and leads to cell death. The severity of damages to retinal function depends on the time of insult.

It is well known that during ischemic events occurs a large increase of extracellular adenosine and glutamate levels. HPLC analysis in rat retina showed that adenosine and its metabolites (inosine, hypoxanthine, and xanthine) increase in response to ischemia followed by reperfusion and that larger is the increase observed as ischemia progresses (Roth, Park, et al., 1997; Roth, Rosenbaum, et al., 1997). The effect of EHNA (an adenosine deaminase inhibitor) on enhancing the recovery of electroretinogram and on preventing retinal layers postischemia thinning suggests that adenosine is a significant endogenous protective agent (Larsen & Osborne, 1996).

An interesting fact about adenosine metabolism is that in conditions with low oxygen tension, for instance, during ischemia, xanthine dehydrogenase (XDH), an enzyme responsible for metabolizing hypoxanthine to xanthine and uric acid, is modified to xanthine oxidase and this reaction ends up in the generation of oxygen free radical species (Fig. 6; Phillis, 1994; Roth, Rosenbaum, et al., 1997). Therefore, during ischemia–reperfusion, metabolism of adenosine may contribute to retina damage. Accordingly, when Roth, Park, et al. (1997) and Roth, Rosenbaum, et al. (1997) inhibited XDH/oxidase activity, they noticed a higher rate of recovery of retinal electric activity after an ischemia–reperfusion event (Roth, Park, et al., 1997).

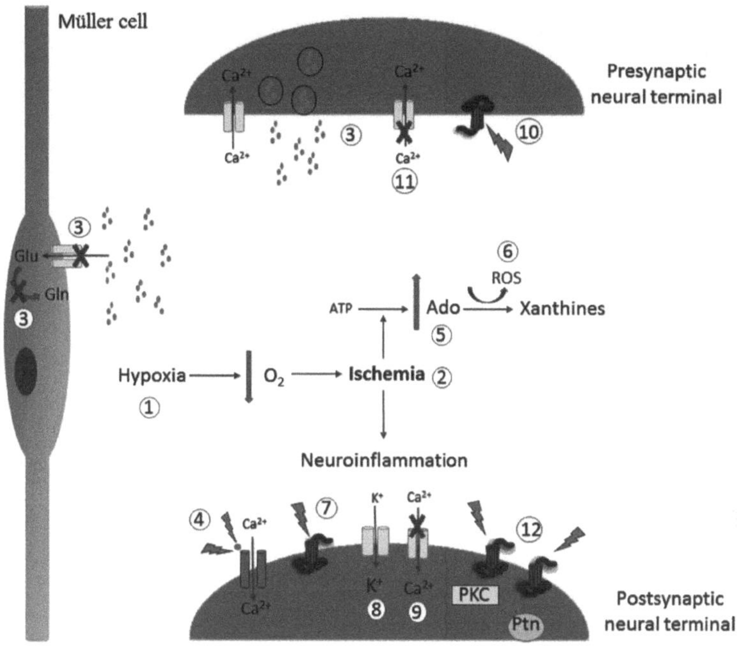

Figure 6 Death pathway induced by hypoxia/ischemia and glutamatergic excitotoxicity and the protection afforded by adenosine. Hypoxia/ischemia insult (1, 2) leads to an increase of extracellular glutamate (3) and NMDA and non-NMDA receptor overstimulation (4), activating an excitotoxicity pathway. Ischemia also induces depletion of ATP, increasing adenosine (Ado) levels (5), which acts as a protective factor. Metabolism of Ado to xanthines in low O_2 conditions release ROS (6), leading to oxidative stress and neuronal death. A1 receptor activation may be protective against glutamate excitotoxicity since postsynaptic activation (7) leads to cell hyperpolarization (8), inhibiting Ca^{2+} influx (9) and the downstream apoptosis pathway, while presynaptic activation (10) inhibits Ca^{2+} influx (11) and glutamate release. Moreover, the stimulation of A1 and A2a receptors (12) mimics the IPC protective effect in retinal neurons, activating the PKC pathway and enhancing protein synthesis.

The retina, as well as other tissues, presents an endogenous protective capacity against ischemic events. For example, ischemic preconditioning (IPC) is a neuroprotective strategy that works in the retina (Roth, 2004). An IPC is a short period of ischemia that does not cause any injury and develops a tolerance against a later stronger ischemia event (Roth, 2004). IPC completely prevented the functional and histological impairment promoted by prolonged and harmful ischemia in a time-related way (Roth et al., 1998). Adenosine is involved in this endogenous neuroprotection, since the activation of A1 and A2a adenosine receptors mimics the preconditioning effect (Fig. 6; Li, Yang, Rosenbaum, & Roth, 2000).

The adenosine response to IPC clearly depends on the synthesis *de novo* of proteins, with concomitant activation of PKC and K^+-ATP channels downstream to A1 and A2a receptor activation (Li et al., 2000; Roth et al., 1998).

5.4 Adenosine neuroprotection in excitotoxicity

Glutamate is the major excitatory neurotransmitter in the retina. High concentrations of this amino acid can induce cell death through overstimulation of its receptors, an event known as excitotoxicity (Fig. 6). However, these toxic effects are not always due to enhanced release of glutamate. Glutamate levels are regulated by uptake by glial cells, followed by enzymatic degradation (Ishikawa, 2013). The impairment of glutamate transporter activity is even more critical than the glutamate exocytosis by synaptic vesicles (Izumi et al., 2002).

Excitotoxicity induced by glutamate seems to be involved in a variety of retinal diseases, such as ischemia/hypoxia–reperfusion, glaucoma, and diabetic retinopathy (Ishikawa, 2013; Osborne et al., 2004). In these conditions, the expression and function of glutamate transporters as well as glutamine synthetase activity in glial cells, factors known to be responsible for glial glutamate intracellular metabolism, are critical to protect the retina (Barnett, Pow, & Bull, 2001; Harada et al., 2007; Ishikawa, 2013; Naskar, Vorwerk, & Dreyer, 2000). These changes in retina caused by exacerbated stimulation of NMDA receptors show many similarities with glaucoma (Shen, Liu, & Yang, 2006). Moreover, diabetes can induce loss of ganglion cells in the retina and alteration in the expression of glutamate receptor subunits (Lau, Kroes, Moskal, & Linsenmeier, 2013).

The mechanism of excitotoxicity by glutamate and other excitatory amino acids involves excessive membrane depolarization followed by an enhancement of intracellular Ca^{2+} levels, which activates cell death pathways (Wardas, 2002). Activation of presynaptic adenosine A1 receptors classically inhibits voltage-dependent Ca^{2+} channels, preventing the release of neurotransmitters such as glutamate (Fig. 3; Sperlagh & Vizi, 2011; Wardas, 2002). Postsynaptic A1 receptors are known to activate the opening of K^+ channels, hyperpolarizing the postsynaptic neurons, and inhibiting Ca^{2+} influx through voltage-dependent Ca^{2+} channels, then avoiding the triggering of cell death pathways related to Ca^{2+} (Fig. 6; Wardas, 2002). The major neuroprotective mechanism mediated by adenosine A2a receptors is through vasodilatation and suppression of oxidative stress (de Mendonca, Sebastião, & Ribeiro, 2000). However, activation of A2a receptors is usually considered as detrimental to cellular viability due to its trend to release neurotransmitters (Sperlagh & Vizi, 2011; Wardas, 2002). Nevertheless, some reports suggest that A2a receptor agonists can reduce neuronal damage by glutamate (Jones, Smith, & Stone, 1998).

5.5 A neuroprotective model in chick retina

Our group has demonstrated that the adenosine players, such as receptors and transporters, clearly regulate different cellular physiological functions during development of the chick retina (Paes-de-Carvalho, 2002). Besides this endogenous function, adenosine also exerts a huge role against damage events in this model. In 2001, Ferreira and Paes-de-Carvalho showed that cultured isolated neurons from E8 chick embryo retina were sensitive to chronic stimulation of glutamate receptors, especially NMDA receptors, presenting a cellular death about 80% higher than untreated cells (Ferreira & Paes-de-Carvalho, 2001). At this same work, they also demonstrated that adenosine is protective against glutamate excitotoxicity and the maximal effect was only observed after 24 h of pretreatment of cultures with adenosine or an ENT1 inhibitor (NBMPR) (Ferreira & Paes-de-Carvalho, 2001). These data suggest that the protective role of adenosine on retinal neurons is through long-term events, as well as synthesis and release of neurotrophic factors or regulation of protein expression. This same work also found out that adenosine protects those retinal neurons from glutamate excitotoxicity by activating A2a receptors and recruiting the cAMP pathway (Ferreira & Paes-de-Carvalho, 2001). A very similar neuroprotection mechanism mediated by adenosine in cultured chick retinal neurons was observed when these cultures were challenged to death with a protocol of medium replacement (Paes-de-Carvalho et al., 2003). An interesting result was that a conditioned medium from sister cultures, when used instead of fresh medium, promoted no increase in cell death (Paes-de-Carvalho et al., 2003). Accordingly, it is possible to speculate that there are some neuroprotective factors that can be released into the culture medium and this protective effect is lost when medium is replaced.

According to Socodato and colleagues, the protection by adenosine is determined by the stage of chick retina development. When mitotic retinal progenitor cells from E6 are kept in culture, adenosine promotes cell death by apoptosis instead of cell survival (Socodato, Brito, Calaza, & Paes-de-Carvalho, 2011). Interestingly, this effect also depends on activations of A2a receptors but the signaling pathway does not involve cAMP/PKA, as in the protection of E8 cultured neurons by adenosine. Indeed, the death effect is mediated through the PLC/PKC pathway and a decrease of CREB activation, which is a transcriptional factor associated with cell survival (Socodato et al., 2011). Then, there is a switch of signaling pathways activated by A2a receptors during development: at E6, activation of A2a receptors promotes an increase of PKC activity and a decrease of CREB phosphorylation, increasing cell death; and at E8, adenosine A2a receptors activate PKA and increase CREB phosphorylation leading to neuroprotection. One curious fact about these findings is that this shift happens in a very short time window of retina development.

6. CONCLUDING REMARKS

In this chapter, we reviewed available information on the neuromodulatory roles of the nucleoside adenosine especially in the retina. Adenosine receptors and transporters are expressed in the retina since early stages of embryonic development in specific cell types and cellular locations. Both receptors and transporters appear to work in a concatenated way in order to assure important actions and functions of the nucleoside in the tissue. While in the mature tissue, adenosine plays important functions such as the regulation of excitatory transmitter release and neuroprotection from ischemic and excitotoxic insults, adenosine also plays important roles during development, such as regulation of cell survival, neurogenesis, and axonal growth. We also briefly reviewed some eye pathologies in which the activation of adenosine receptors or regulation of adenosine transport could have important benefits, highlighting a perspective for the use of adenosine-related drugs in therapeutic strategies for the treatment of different CNS diseases.

REFERENCES

Abbracchio, M. P., Brambilla, R., Ceruti, S., Kim, H. O., von Lubitz, D. K., Jacobson, K. A., et al. (1995). G protein-dependent activation of phospholipase C by adenosine A3 receptors in rat brain. *Molecular Pharmacology*, *48*(6), 1038–1045.

Abdulla, P., & Coe, I. R. (2007). Characterization and functional analysis of the promoter for the human equilibrative nucleoside transporter gene, hENT1. *Nucleosides, Nucleotides & Nucleic Acids*, *26*(1), 99–110.

Agostinho, P., Caseiro, P., Rego, A. C., Duarte, E. P., Cunha, R. A., & Oliveira, C. R. (2000). Adenosine modulation of D-[3H]aspartate release in cultured retina cells exposed to oxidative stress. *Neurochemistry International*, *36*, 255–265.

Alfinito, P. D., Alli, R., & Townes-Anderson, E. (2002). Adenosine A(2a) receptor-mediated inhibition of rod opsin mRNA expression in tiger salamander. *Journal of Neurochemistry*, *83*, 665–672.

Avila, M. Y., Stone, R. A., & Civan, M. M. (2001). A(1)-, A(2A)- and A(3)-subtype adenosine receptors modulate intraocular pressure in the mouse. *British Journal of Pharmacology*, *134*(2), 241–245.
Avila, M. Y., Stone, R. A., & Civan, M. M. (2002). Knockout of A3 adenosine receptors reduces mouse intraocular pressure. *Investigative Ophthalmology & Visual Science*, *43*(9), 3021–3026.
Baldwin, S. A., Beal, P. R., Yao, S. Y., King, A. E., Cass, C. E., & Young, J. D. (2004). The equilibrative nucleoside transporter family, SLC29. *Pflügers Archiv*, *447*(5), 735–743.
Barnett, N. L., Pow, D. V., & Bull, N. D. (2001). Differential perturbation of neuronal and glial glutamate transport systems in retinal ischaemia. *Neurochemistry International*, *39*(4), 291–299.
Biber, K., Klotz, K. N., Berger, M., Gebicke-Harter, P. J., & van Calker, D. (1997). Adenosine A1 receptor-mediated activation of phospholipase C in cultured astrocytes depends on the level of receptor expression. *The Journal of Neuroscience: The Official Journal of the Society for Neuroscience*, *17*, 4956–4964.
Blazynski, C. (1987). Adenosine A1 receptor-mediated inhibition of adenylate cyclase in rabbit retina. *The Journal of Neuroscience*, *7*, 2522–2528.
Blazynski, C. (1990). Discrete distributions of adenosine receptors in mammalian retina. *Journal of Neurochemistry*, *54*, 648–655.
Blazynski, C. (1991). The accumulation of [3H]phenylisopropyl adenosine ([3H]PIA) and [3H]adenosine in to rabbit retinal neurons is inhibited by nitrobenzylthioinosine (NBI). *Neuroscience Letters*, *121*, 1–4.
Blazynski, C. (1993). Characterization of adenosine A2 receptors in bovine retinal pigment epithelial membranes. *Experimental Eye Research*, *56*, 595–599.
Blazynski, C., Kinscherf, D. A., Geary, K. M., & Ferrendelli, J. A. (1986). Adenosine-mediated regulation of cyclic AMP levels in isolated incubated retinas. *Brain Research*, *366*, 224–229.
Blazynski, C., & McIntosh, H. (1993). Characterization of adenosine A2 receptors in bovine retinal membranes. *Experimental Eye Research*, *56*, 585–593.
Blazynski, C., Mosinger, J. L., & Cohen, A. L. (1989). Comparison of adenosine uptake and endogenous adenosine-containing cells in mammalian retina. *Visual Neuroscience*, *2*, 109–111.
Blazynski, C., & Perez, M. T. (1991). Adenosine in vertebrate retina: Localization, receptor characterization, and function. *Cellular and Molecular Neurobiology*, *11*, 463–484.
Blazynski, C., Woods, C., & Mathews, G. C. (1992). Evidence for the action of endogenous adenosine in the rabbit retina: Modulation of the light-evoked release of acetylcholine. *Journal of Neurochemistry*, *58*, 761–767.
Blom, N., Sicheritz-Ponten, T., Gupta, R., Gammeltoft, S., & Brunak, S. (2004). Prediction of post-translational glycosylation and phosphorylation of proteins from the amino acid sequence. *Proteomics*, *4*(6), 1633–1649.
Bone, D. B., Antic, M., Vilas, G., & Hammond, J. R. (2014). Oxidative stress modulates nucleobase transport in microvascular endothelial cells. *Microvascular Research*, *95*, 68–75.
Bone, D. B., Robillard, K. R., Stolk, M., & Hammond, J. R. (2007). Differential regulation of mouse equilibrative nucleoside transporter 1 (mENT1) splice variants by protein kinase CK2. *Molecular Membrane Biology*, *24*(4), 294–303.
Braas, K. M., Zarbin, M. A., & Snyder, S. H. (1987). Endogenous adenosine and adenosine receptors localized to ganglion cells of the retina. *Proceedings of the National Academy of Sciences of the United States of America*, *84*, 3906–3910.
Brito, R., Pereira, M. R., Paes-de-Carvalho, R., & Calaza, C. (2012). Expression of A1 adenosine receptors in the developing avian retina: In vivo modulation by A(2A) receptors and endogenous adenosine. *Journal of Neurochemistry*, *123*, 239–249.
Burnstock, G. (1972). Purinergic nerves. *Pharmacological Reviews*, *24*(3), 509–581.

Burnstock, G. (2009). Purinergic signalling: Past, present and future. *Brazilian Journal of Medical and Biological Research, 42*(1), 3–8.

Campochiaro, P., Ferkany, J. W., & Coyle, J. T. (1984). The dissociation of evoked release of [3H]-GABA and of endogenous GABA from chick retina in vitro. *Experimental Eye Research, 39*(3), 299–305.

Campochiaro, P., Ferkany, J. W., & Coyle, J. T. (1985). Excitatory amino acid analogs evoke release of endogenous amino acids and acetyl choline from chick retina in vitro. *Vision Research, 25*(10), 1375–1386.

Campochiaro, P. A., & Sen, H. A. (1989). Adenosine and its agonists cause retinal vasodilation and hemorrhages. Implications for ischemic retinopathies. *Archives of Ophthalmology, 107*, 412–416.

Carrasco, E., Hernandez, C., de Torres, I., Farres, J., & Simo, R. (2008). Lowered cortistatin expression is an early event in the human diabetic retina and is associated with apoptosis and glial activation. *Molecular Vision, 14*, 1496–1502.

Carrasco, E., Hernandez, C., Miralles, A., Huguet, P., Farres, J., & Simo, R. (2007). Lower somatostatin expression is an early event in diabetic retinopathy and is associated with retinal neurodegeneration. *Diabetes Care, 30*(11), 2902–2908.

Castillo, C. A., Leon, D. A., Ballesteros-Yanez, I., Albasanz, J. L., & Martin, M. (2010). Glutamate differently modulates excitatory and inhibitory adenosine receptors in neuronal and glial cells. *Neurochemistry International, 57*(1), 33–42.

Cervantes-Villagrana, A. R., Garcia-Roman, J., Gonzalez-Espinosa, C., & Lamas, M. (2010). Pharmacological inhibition of N-methyl d-aspartate receptor promotes secretion of vascular endothelial growth factor in Muller cells: Effects of hyperglycemia and hypoxia. *Current Eye Research, 35*(8), 733–741.

Coe, I. R., Dohrman, D. P., Constantinescu, A., Diamond, I., & Gordon, A. S. (1996). Activation of cyclic AMP-dependent protein kinase reverses tolerance of a nucleoside transporter to ethanol. *The Journal of Pharmacology and Experimental Therapeutics, 276*(2), 365–369.

Coe, I. R., Yao, L., Diamond, I., & Gordon, A. S. (1996). The role of protein kinase C in cellular tolerance to ethanol. *The Journal of Biological Chemistry, 271*(46), 29468–29472.

Coe, I., Zhang, Y., McKenzie, T., & Naydenova, Z. (2002). PKC regulation of the human equilibrative nucleoside transporter, hENT1. *FEBS Letters, 517*(1–3), 201–205.

Coulombre, A. J. (1955). Correlations of structural and biochemical changes in the developing retina of the chick. *The American Journal of Anatomy, 96*(1), 153–189.

Crosson, C. E., & Gray, T. (1994). Modulation of intraocular pressure by adenosine agonists. *Journal of Ocular Pharmacology, 10*(1), 379–383.

Cui, D., Trier, K., Zeng, J., Wu, K., Yu, M., & Ge, J. (2010). Adenosine receptor protein changes in guinea pigs with form deprivation myopia. *Acta Ophthalmologica, 88*, 759–765.

Cunha, R. A. (2001). Adenosine as a neuromodulator and as a homeostatic regulator in the nervous system: Different roles, different sources and different receptors. *Neurochemistry International, 38*(2), 107–125.

Daines, B. S., Kent, A. R., McAleer, M. S., & Crosson, C. E. (2003). Intraocular adenosine levels in normal and ocular-hypertensive patients. *Journal of Ocular Pharmacology and Therapeutics, 19*(2), 113–119.

Daugeliene, L., Yamamoto, T., & Kitazawa, Y. (1999). Risk factors for visual field damage progression in normal-tension glaucoma eyes. *Graefe's Archive for Clinical and Experimental Ophthalmology, 237*(2), 105–108.

de Mello, F. G. (1978). The ontogeny of dopamine-dependent increase of adenosine 3′,5′-cyclic monophosphate in the chick retina. *Journal of Neurochemistry, 31*(4), 1049–1053.

de Mello, M. C., Ventura, A. L., Paes-de-Carvalho, R., Klein, W. L., & de Mello, F. G. (1982). Regulation of dopamine- and adenosine-dependent adenylate cyclase systems

of chicken embryo retina cells in culture. *Proceedings of the National Academy of Sciences of the United States of America*, *79*(18), 5708–5712.
de Mendonca, A., Sebastião, A. M., & Ribeiro, J. A. (2000). Adenosine: Does it have a neuroprotective role after all? *Brain Research. Brain Research Reviews*, *33*(2–3), 258–274.
Delicado, E. G., Sen, R. P., & Miras-Portugal, M. T. (1991). Effects of phorbol esters and secretagogues on nitrobenzylthioinosine binding to nucleoside transporters and nucleoside uptake in cultured chromaffin cells. *The Biochemical Journal*, *279*(Pt 3), 651–655.
dos Santos-Rodrigues, A., Ferreira, J. M., & Paes-de-Carvalho, R. (2011). Differential adenosine uptake in mixed neuronal/glial or purified glial cultures of avian retinal cells: Modulation by adenosine metabolism and the ERK cascade. *Biochemical and Biophysical Research Communications*, *414*(1), 175–180.
Dreixler, J. C., Hemmert, J. W., Shenoy, S. K., Shen, Y., Lee, H. T., Shaikh, A. R., et al. (2009). The role of Akt/protein kinase B subtypes in retinal ischemic preconditioning. *Experimental Eye Research*, *88*, 512–521.
Drury, A. N., & Szent-Gyorgyi, A. (1929). The physiological activity of adenine compounds with especial reference to their action upon the mammalian heart. *The Journal of Physiology*, *68*(3), 213–237.
Errasti-Murugarren, E., Molina-Arcas, M., Casado, F. J., & Pastor-Anglada, M. (2010). The human concentrative nucleoside transporter-3 C602R variant shows impaired sorting to lipid rafts and altered specificity for nucleoside-derived drugs. *Molecular Pharmacology*, *78*(2), 157–165.
Fernandez-Calotti, P., & Pastor-Anglada, M. (2010). All-trans-retinoic acid promotes trafficking of human concentrative nucleoside transporter-3 (hCNT3) to the plasma membrane by a TGF-beta1-mediated mechanism. *The Journal of Biological Chemistry*, *285*(18), 13589–13598.
Ferreira, J. M., & Paes-de-Carvalho, R. (2001). Long-term activation of adenosine A(2a) receptors blocks glutamate excitotoxicity in cultures of avian retinal neurons. *Brain Research*, *900*(2), 169–176.
Fujita, S., & Horii, M. (1963). Analysis of cytogenesis in chick retina by tritiated thymidine autoradiography. *Archivum Histologicum Japonicum*, *23*, 359–366.
Gao, Z., Chen, T., Weber, M. J., & Linden, J. (1999). A2B adenosine and P2Y2 receptors stimulate mitogen-activated protein kinase in human embryonic kidney-293 cells. Cross-talk between cyclic AMP and protein kinase c pathways. *The Journal of Biological Chemistry*, *274*(9), 5972–5980.
Ghiardi, G. J., Gidday, J. M., & Roth, S. (1999). The purine nucleoside adenosine in retinal ischemia-reperfusion injury. *Vision Research*, *39*(15), 2519–2535.
Gliklich, R. E., Steinmann, W. C., & Spaeth, G. L. (1989). Visual field change in low-tension glaucoma over a five-year follow-up. *Ophthalmology*, *96*(3), 316–320.
Grden, M., Podgorska, M., Kocbuch, K., Rzepko, R., Szutowicz, A., & Pawelczyk, T. (2008). High glucose suppresses expression of equilibrative nucleoside transporter 1 (ENT1) in rat cardiac fibroblasts through a mechanism dependent on PKC-zeta and MAP kinases. *Journal of Cellular Physiology*, *215*(1), 151–160.
Gregory, C. Y., Abrams, T. A., & Hall, M. O. (1994). Stimulation of A2 adenosine receptors inhibits the ingestion of photoreceptor outer segments by retinal pigment epithelium. *Investigative Ophthalmology & Visual Science*, *35*(3), 819–825.
Griffiths, M., Beaumont, N., Yao, S. Y., Sundaram, M., Boumah, C. E., Davies, A., et al. (1997). Cloning of a human nucleoside transporter implicated in the cellular uptake of adenosine and chemotherapeutic drugs. *Nature Medicine*, *3*(1), 89–93.
Hammond, L. C., Bonnet, C., Kemp, P. J., Yates, M. S., & Bowmer, C. J. (2004). Chronic hypoxia up-regulates expression of adenosine A1 receptors in DDT1-MF2 cells. *Biochemical Pharmacology*, *67*(3), 421–426.

Hammond, J. R., Stolk, M., Archer, R. G., & McConnell, K. (2004). Pharmacological analysis and molecular cloning of the canine equilibrative nucleoside transporter 1. *European Journal of Pharmacology*, *491*(1), 9–19.

Handa, M., Choi, D. S., Caldeiro, R. M., Messing, R. O., Gordon, A. S., & Diamond, I. (2001). Cloning of a novel isoform of the mouse NBMPR-sensitive equilibrative nucleoside transporter (ENT1) lacking a putative phosphorylation site. *Gene*, *262*(1–2), 301–307.

Harada, T., Harada, C., Nakamura, K., Quah, H. M., Okumura, A., Namekata, K., et al. (2007). The potential role of glutamate transporters in the pathogenesis of normal tension glaucoma. *The Journal of Clinical Investigation*, *117*(7), 1763–1770.

Hartwick, A. T., Lalonde, M. R., Barnes, S., & Baldridge, W. H. (2004). Adenosine A1-receptor modulation of glutamate-induced calcium influx in rat retinal ganglion cells. *Investigative Ophthalmology & Visual Science*, *45*(10), 3740–3748.

Huang, P. C., Hsiao, Y. T., Kao, S. Y., Chen, C. F., Chen, Y. C., Chiang, C. W., et al. (2014). Adenosine A(2A) receptor up-regulates retinal wave frequency via starburst amacrine cells in the developing rat retina. *PLoS One*, *9*, e95090.

Hughes, W. F., & LaVelle, A. (1974). On the synaptogenic sequence in the chick retina. *The Anatomical Record*, *179*(3), 297–301.

Ibrahim, A. S., El-Shishtawy, M. M., Zhang, W., Caldwell, R. B., & Liou, G. I. (2011). A ((2)A) adenosine receptor (A((2)A)AR) as a therapeutic target in diabetic retinopathy. *The American Journal of Pathology*, *178*(5), 2136–2145.

Ishikawa, M. (2013). Abnormalities in glutamate metabolism and excitotoxicity in the retinal diseases. *Scientifica*, *2013*, 528940.

Izumi, Y., Shimamoto, K., Benz, A. M., Hammerman, S. B., Olney, J. W., & Zorumski, C. F. (2002). Glutamate transporters and retinal excitotoxicity. *Glia*, *39*(1), 58–68.

Jennings, L. L., Hao, C., Cabrita, M. A., Vickers, M. F., Baldwin, S. A., Young, J. D., et al. (2001). Distinct regional distribution of human equilibrative nucleoside transporter proteins 1 and 2 (hENT1 and hENT2) in the central nervous system. *Neuropharmacology*, *40*(5), 722–731.

Jhaveri, K. A., Reichensperger, J., Toth, L. A., Sekino, Y., & Ramkumar, V. (2007). Reduced basal and lipopolysaccharide-stimulated adenosine A1 receptor expression in the brain of nuclear factor-kappaB p50$^{-/-}$ mice. *Neuroscience*, *146*(1), 415–426.

Jones, P. A., Smith, R. A., & Stone, T. W. (1998). Protection against kainate-induced excitotoxicity by adenosine A2A receptor agonists and antagonists. *Neuroscience*, *85*(1), 229–237.

Kahn, A. J. (1974). An autoradiographic analysis of the time of appearance of neurons in the developing chick neural retina. *Developmental Biology*, *38*(1), 30–40.

Kalisch, F., Wurm, A., Iandiev, I., Uckermann, O., Dilsiz, N., Reichenbach, A., et al. (2006). Atrial natriuretic peptide inhibits osmotical glial cell swelling in the ischemic rat retina: Dependence on glutamatergic-purinergic signaling. *Experimental Eye Research*, *83*(4), 962–971.

Kirk, D. L., & Moscona, A. A. (1963). Synthesis of experimentally induced glutamine synthetase (glutamotransferase activity) in embryonic chick retina in vitro. *Developmental Biology*, *8*, 341–357.

Kiss, A., Farah, K., Kim, J., Garriock, R. J., Drysdale, T. A., & Hammond, J. R. (2000). Molecular cloning and functional characterization of inhibitor-sensitive (mENT1) and inhibitor-resistant (mENT2) equilibrative nucleoside transporters from mouse brain. *The Biochemical Journal*, *352*(Pt 2), 363–372.

Kong, W., Engel, K., & Wang, J. (2004). Mammalian nucleoside transporters. *Current Drug Metabolism*, *5*(1), 63–84.

Kvanta, A., Seregard, S., Sejersen, S., Kull, B., & Fredholm, B. B. (1997). Localization of adenosine receptor messenger RNAs in the rat eye. *Experimental Eye Research, 65*, 595–602.

Larsen, A. K., & Osborne, N. N. (1996). Involvement of adenosine in retinal ischemia. Studies on the rat. *Investigative Ophthalmology & Visual Science, 37*(13), 2603–2611.

Lau, J. C., Kroes, R. A., Moskal, J. R., & Linsenmeier, R. A. (2013). Diabetes changes expression of genes related to glutamate neurotransmission and transport in the Long-Evans rat retina. *Molecular Vision, 19*, 1538–1553.

Leung, G. P., Man, R. Y., & Tse, C. M. (2005). D-Glucose upregulates adenosine transport in cultured human aortic smooth muscle cells. *American Journal of Physiology. Heart and Circulatory Physiology, 288*(6), H2756–H2762.

Li, Q., & Puro, D. G. (2001). Adenosine activates ATP-sensitive K(+) currents in pericytes of rat retinal microvessels: Role of A1 and A2a receptors. *Brain Research, 907*, 93–99.

Li, B., & Roth, S. (1999). Retinal ischemic preconditioning in the rat: Requirement for adenosine and repetitive induction. *Investigative Ophthalmology & Visual Science, 40*, 1200–1216.

Li, N., Salom, D., Zhang, L., Harris, T., Ballesteros, J. A., Golczak, M., et al. (2007). Heterologous expression of the adenosine A1 receptor in transgenic mouse retina. *Biochemistry, 46*, 8350–8359.

Li, B., Yang, C., Rosenbaum, D. M., & Roth, S. (2000). Signal transduction mechanisms involved in ischemic preconditioning in the rat retina in vivo. *Experimental Eye Research, 70*(6), 755–765.

Li, H., Zhang, Z., Blackburn, M. R., Wang, S. W., Ribelayga, C. P., & O'Brien, J. (2013). Adenosine and dopamine receptors coregulate photoreceptor coupling via gap junction phosphorylation in mouse retina. *The Journal of Neuroscience, 33*, 3135–3150.

Lindeman, V. F. (1947). The formation of acetylcholine in the developing retina of the chick embryo. *Federation Proceedings, 6*(1 Pt 2), 153.

Liou, G. I., Ahmad, S., Naime, M., Fatteh, N., & Ibrahim, A. S. (2011). Role of adenosine in diabetic retinopathy. *Journal of Ocular Biology, Diseases and Informatics, 4*(1–2), 19–24.

Liou, G. I., Auchampach, J. A., Hillard, C. J., Zhu, G., Yousufzai, B., Mian, S., et al. (2008). Mediation of cannabidiol anti-inflammation in the retina by equilibrative nucleoside transporter and A2A adenosine receptor. *Investigative Ophthalmology & Visual Science, 49*(12), 5526–5531.

Lopes, L. V., Cunha, R. A., & Ribeiro, J. A. (1999). Cross talk between A(1) and A(2A) adenosine receptors in the hippocampus and cortex of young adult and old rats. *Journal of Neurophysiology, 82*(6), 3196–3203.

Lopez-Colome, A. M. (1981). High affinity binding of L-glutamate to chick retinal membranes. *Neurochemical Research, 6*(9), 1019–1033.

Lopez-Zapata, A., Leon, D., Castillo, C. A., Albasanz, J. L., & Martin, M. (2011). Maternal glutamate intake during gestation and lactation regulates adenosine A(1) and A(2A) receptors in rat brain from mothers and neonates. *Neuroscience, 199*, 133–142.

Lorenzo, A. M., Leon, D., Castillo, C. A., Ruiz, M. A., Albasanz, J. L., & Martin, M. (2010). Maternal caffeine intake during gestation and lactation down-regulates adenosine A1 receptor in rat brain from mothers and neonates. *Journal of Neuroscience Research, 88*(6), 1252–1261.

Lu, G., Zhou, Q. X., Kang, S., Li, Q. L., Zhao, L. C., Chen, J. D., et al. (2010). Chronic morphine treatment impaired hippocampal long-term potentiation and spatial memory via accumulation of extracellular adenosine acting on adenosine A1 receptors. *The Journal of Neuroscience, 30*(14), 5058–5070.

McIntosh, H. H., & Blazynski, C. (1994). Characterization and localization of adenosine A2 receptors in bovine rod outer segments. *Journal of Neurochemistry, 62*, 992–997.

Montecinos, V. P., Aguayo, C., Flores, C., Wyatt, A. W., Pearson, J. D., Mann, G. E., et al. (2000). Regulation of adenosine transport by D-glucose in human fetal endothelial cells: Involvement of nitric oxide, protein kinase C and mitogen-activated protein kinase. *The Journal of Physiology, 529*(Pt 3), 777–790.

Morris, J. E., & Moscona, A. A. (1970). Induction of glutamine synthetase in embryonic retina: Its dependence on cell interactions. *Science, 167*(3926), 1736–1738.

Morris, J. E., & Moscona, A. A. (1971). The induction of glutamine synthetase in cell aggregates of embryonic neural retina: Correlations with differentiation and multicellular organization. *Developmental Biology, 25*(3), 420–444.

Moscona, A. A., & Kirk, D. L. (1965). Control of glutamine synthetase in the embryonic retina in vitro. *Science, 148*(3669), 519–521.

Moscona, M. H., & Moscona, A. A. (1963). Inhibition of adhesiveness and aggregation of dissociated cells by inhibitors of protein and RNA synthesis. *Science, 142*(3595), 1070–1071.

Moscona, A. A., & Piddington, R. (1966). Stimulation by hydrocortisone of premature changes in the developmental pattern of glutamine synthetase in embryonic retina. *Biochimica et Biophysica Acta, 121*(2), 409–411.

Nagy, L. E., Diamond, I., Casso, D. J., Franklin, C., & Gordon, A. S. (1990). Ethanol increases extracellular adenosine by inhibiting adenosine uptake via the nucleoside transporter. *The Journal of Biological Chemistry, 265*(4), 1946–1951.

Nagy, L. E., Diamond, I., & Gordon, A. S. (1991). cAMP-dependent protein kinase regulates inhibition of adenosine transport by ethanol. *Molecular Pharmacology, 40*(5), 812–817.

Naskar, R., Vorwerk, C. K., & Dreyer, E. B. (2000). Concurrent downregulation of a glutamate transporter and receptor in glaucoma. *Investigative Ophthalmology & Visual Science, 41*(7), 1940–1944.

Nie, Z., Mei, Y., Ford, M., Rybak, L., Marcuzzi, A., Ren, H., et al. (1998). Oxidative stress increases A1 adenosine receptor expression by activating nuclear factor kappa B. *Molecular Pharmacology, 53*(4), 663–669.

Offermanns, S., & Simon, M. I. (1995). G alpha 15 and G alpha 16 couple a wide variety of receptors to phospholipase C. *The Journal of Biological Chemistry, 270*(25), 15175–15180.

Osborne, N. N., Casson, R. J., Wood, J. P., Chidlow, G., Graham, M., & Melena, J. (2004). Retinal ischemia: Mechanisms of damage and potential therapeutic strategies. *Progress in Retinal and Eye Research, 23*(1), 91–147. http://dx.doi.org/10.1016/j.preteyeres.2003.12.001.

Paes-de-Carvalho, R. (1990). Development of A1 adenosine receptors in the chick embryo retina. *Journal of Neuroscience Research, 25*(2), 236–242.

Paes-de-Carvalho, R. (2002). Adenosine as a signaling molecule in the retina: Biochemical and developmental aspects. *Anais da Academia Brasileira de Ciências, 74*(3), 437–451.

Paes-de-Carvalho, R., Braas, K. M., Adler, R., & Snyder, S. H. (1992). Developmental regulation of adenosine A1 receptors, uptake sites and endogenous adenosine in the chick retina. *Brain Research. Developmental Brain Research, 70*(1), 87–95.

Paes-de-Carvalho, R., Braas, K. M., Snyder, S. H., & Adler, R. (1990). Analysis of adenosine immunoreactivity, uptake, and release in purified cultures of developing chick embryo retinal neurons and photoreceptors. *Journal of Neurochemistry, 55*(5), 1603–1611.

Paes-de-Carvalho, R., & de Mello, F. G. (1982). Adenosine-elicited accumulation of adenosine 3′, 5′- cyclic monophosphate in the chick embryo retina. *Journal of Neurochemistry, 38*, 493–500.

Paes-de-Carvalho, R., & de Mello, F. G. (1985). Expression of A1 adenosine receptor modulating dopamine-dependent cyclic AMP accumulation in the chick embryo retina. *Journal of Neurochemistry, 44*, 845–851.

Paes-de-Carvalho, R., Dias, B. V., Martins, R. A., Pereira, M. R., Portugal, C. C., & Lanfredi, C. (2005). Activation of glutamate receptors promotes a calcium-dependent and transporter-mediated release of purines in cultured avian retinal cells: Possible involvement of calcium/calmodulin-dependent protein kinase II. *Neurochemistry International, 46*(6), 441–451.

Paes-de-Carvalho, R., Maia, G. A., & Ferreira, J. M. (2003). Adenosine regulates the survival of avian retinal neurons and photoreceptors in culture. *Neurochemical Research, 28*(10), 1583–1590.

Pereira, M. R., Hang, V. R., Vardiero, E., de Mello, F. G., & Paes-de-Carvalho, R. (2010). Modulation of A1 adenosine receptor expression by cell aggregation and long-term activation of A2a receptors in cultures of avian retinal cells: Involvement of the cyclic AMP/PKA pathway. *Journal of Neurochemistry, 113*(3), 661–673.

Perez, M. T., & Bruun, A. (1987). Colocalization of (3H)-adenosine accumulation and GABA immunoreactivity in the chicken and rabbit retinas. *Histochemistry, 87*(5), 413–417.

Perez, M. T., Ehinger, B. E., Lindstrom, K., & Fredholm, B. B. (1986). Release of endogenous and radioactive purines from the rabbit retina. *Brain Research, 398*(1), 106–112.

Perigolo-Vicente, R., Ritt, K., Goncalves-de-Albuquerque, C. F., Castro-Faria-Neto, H. C., Paes-de-Carvalho, R., & Giestal-de-Araujo, E. (2014). IL-6, A1 and A2aR: A crosstalk that modulates BDNF and induces neuroprotection. *Biochemical and Biophysical Research Communications, 449*(4), 477–482.

Perigolo-Vicente, R., Ritt, K., Pereira, M. R., Torres, P. M., Paes-de-Carvalho, R., & Giestal-de-Araujo, E. (2013). IL-6 treatment increases the survival of retinal ganglion cells in vitro: The role of adenosine A1 receptor. *Biochemical and Biophysical Research Communications, 430*(2), 512–518.

Phillis, J. W. (1994). A "radical" view of cerebral ischemic injury. *Progress in Neurobiology, 42*(4), 441–448.

Pinto-Duarte, A., Coelho, J. E., Cunha, R. A., Ribeiro, J. A., & Sebastião, A. M. (2005). Adenosine A2A receptors control the extracellular levels of adenosine through modulation of nucleoside transporters activity in the rat hippocampus. *Journal of Neurochemistry, 93*(3), 595–604.

Podgorska, M., Kocbuch, K., & Pawelczyk, T. (2005). Recent advances in studies on biochemical and structural properties of equilibrative and concentrative nucleoside transporters. *Acta Biochimica Polonica, 52*(4), 749–758.

Quigley, H. A. (2011). Glaucoma. *Lancet, 377*(9774), 1367–1377.

Rego, A. C., Agostinho, P., Melo, J., Cunha, R. A., & Oliveira, C. R. (2000). Adenosine A2A receptors regulate the extracellular accumulation of excitatory amino acids upon metabolic dysfunction in chick cultured retinal cells. *Experimental Eye Research, 70*(5), 577–587.

Rego, A. C., Santos, M. S., & Oliveira, C. R. (1997). Adenosine triphosphate degradation products after oxidative stress and metabolic dysfunction in cultured retinal cells. *Journal of Neurochemistry, 69*(3), 1228–1235.

Reyes, G., Nivillac, N. M., Karim, M. Z., Desouza, L., Siu, K. W., & Coe, I. R. (2011). The Equilibrative Nucleoside Transporter (ENT1) can be phosphorylated at multiple sites by PKC and PKA. *Molecular Membrane Biology, 28*(6), 412–426.

Ribeiro, J. A., Sebastião, A. M., & de Mendonca, A. (2002). Adenosine receptors in the nervous system: Pathophysiological implications. *Progress in Neurobiology, 68*(6), 377–392.

Riis-Vestergaard, M. J., Misfeldt, M. W., & Bek, T. (2014). Dual effects of adenosine on the tone of porcine retinal arterioles in vitro. *Investigative Ophthalmology & Visual Science, 55*, 1630–1636.

Robillard, K. R., Bone, D. B., Park, J. S., & Hammond, J. R. (2008). Characterization of mENT1Delta11, a novel alternative splice variant of the mouse equilibrative nucleoside transporter 1. *Molecular Pharmacology, 74*(1), 264–273.

Roth, S. (2004). Endogenous neuroprotection in the retina. *Brain Research Bulletin, 62*(6), 461–466.

Roth, S., Li, B., Rosenbaum, P. S., Gupta, H., Goldstein, I. M., Maxwell, K. M., et al. (1998). Preconditioning provides complete protection against retinal ischemic injury in rats. *Investigative Ophthalmology & Visual Science*, *39*(5), 777–785.

Roth, S., Park, S. S., Sikorski, C. W., Osinski, J., Chan, R., & Loomis, K. (1997). Concentrations of adenosine and its metabolites in the rat retina/choroid during reperfusion after ischemia. *Current Eye Research*, *16*(9), 875–885.

Roth, S., Rosenbaum, P. S., Osinski, J., Park, S. S., Toledano, A. Y., Li, B., et al. (1997). Ischemia induces significant changes in purine nucleoside concentration in the retina-choroid in rats. *Experimental Eye Research*, *65*(6), 771–779.

Sakowicz, M., Szutowicz, A., & Pawelczyk, T. (2005). Differential effect of insulin and elevated glucose level on adenosine transport in rat B lymphocytes. *International Immunology*, *17*(2), 145–154.

Santos, P. F., Caramelo, O. L., Carvalho, A. P., & Duarte, C. B. (2000). Adenosine A1 receptors inhibit Ca2+ channels coupled to the release of ACh, but not of GABA, in cultured retina cells. *Brain Research*, *852*(1), 10–15.

Santos, P. F., Santos, M. S., Carvalho, A. P., & Duarte, C. B. (1998). Modulation of [3H] acetylcholine release from cultured amacrine-like neurons by adenosine A1 receptors. *Journal of Neurochemistry*, *71*(3), 1086–1094.

Schaeffer, J. M., & Anderson, S. M. (1981). Nucleoside uptake by rat retina cells. *Life Sciences*, *29*(9), 939–946.

Sen, R. P., Delicado, E. G., Alvarez, A., Brocklebank, A. M., Wiley, J. S., & Miras-Portugal, M. T. (1998). Flow cytometric studies of nucleoside transport regulation in single chromaffin cells. *FEBS Letters*, *422*(3), 368–372.

Sen, R. P., Delicado, E. G., & Miras-Portugal, M. T. (1999). Differential modulation of nucleoside transport types in neuroblastoma cells by protein kinase activation. *Neuropharmacology*, *38*(7), 1009–1015.

Sheffield, J. B., & Moscona, A. A. (1969). Early stages in the reaggregation of embryonic chick neural retina cells. *Experimental Cell Research*, *57*(2), 462–466.

Sheffield, J. B., & Moscona, A. A. (1970). Electron microscopic analysis of aggregation of embryonic cells: The structure and differentiation of aggregates of neural retina cells. *Developmental Biology*, *23*(1), 36–61.

Shen, Y., Liu, X. L., & Yang, X. L. (2006). N-methyl-D-aspartate receptors in the retina. *Molecular Neurobiology*, *34*(3), 163–179.

Socodato, R., Brito, R., Calaza, K. C., & Paes-de-Carvalho, R. (2011). Developmental regulation of neuronal survival by adenosine in the in vitro and in vivo avian retina depends on a shift of signaling pathways leading to CREB phosphorylation or dephosphorylation. *Journal of Neurochemistry*, *116*(2), 227–239.

Sperlagh, B., & Vizi, E. S. (2011). The role of extracellular adenosine in chemical neurotransmission in the hippocampus and Basal Ganglia: Pharmacological and clinical aspects. *Current Topics in Medicinal Chemistry*, *11*(8), 1034–1046.

Stella, S. L., Jr., Bryson, E. J., & Thoreson, W. B. (2002). A2 adenosine receptors inhibit calcium influx through L-type calcium channels in rod photoreceptors of the salamander retina. *Journal of Neurophysiology*, *87*, 351–360.

Stettler, O., Joshi, R. L., Wizenmann, A., Reingruber, J., Holcman, D., Bouillot, C., et al. (2012). Engrailed homeoprotein recruits the adenosine A1 receptor to potentiate ephrin A5 function in retinal growth cones. *Development*, *139*(1), 215–224.

Stolk, M., Cooper, E., Vilk, G., Litchfield, D. W., & Hammond, J. R. (2005). Subtype-specific regulation of equilibrative nucleoside transporters by protein kinase CK2. *The Biochemical Journal*, *386*(Pt 2), 281–289.

Studholme, K. M., & Yazulla, S. (1997). 3H-adenosine uptake selectively labels rod horizontal cells in goldfish retina. *Visual Neuroscience*, *14*(2), 207–212.

Sun, X., Barnes, S., & Baldridge, W. H. (2002). Adenosine inhibits calcium channel currents via A1 receptors on salamander retinal ganglion cells in a mini-slice preparation. *Journal of Neurochemistry, 81*, 550–556.

Taomoto, M., McLeod, D. S., Merges, C., & Lutty, G. A. (2000). Localization of adenosine A2a receptor in retinal development and oxygen-induced retinopathy. *Investigative Ophthalmology & Visual Science, 41*, 230–243.

Trueblood, K. E., Mohr, S., & Dubyak, G. R. (2011). Purinergic regulation of high-glucose-induced caspase-1 activation in the rat retinal Muller cell line rMC-1. *American Journal of Physiology. Cell Physiology, 301*(5), C1213–C1223.

Tunnicliff, G., Cho, Y. D., & Martin, R. O. (1974). Kinetic properties of the GABA uptake system in cultures of chick retina. *Neurobiology, 4*(1), 38–42.

Tunnicliff, G., Firneisz, G., Ngo, T. T., & Martin, R. O. (1975). Developmental changes in the kinetics of gamma-aminobutyric acid transport by chick retina. *Journal of Neurochemistry, 25*(5), 649–652.

Uckermann, O., Wolf, A., Kutzera, F., Kalisch, F., Beck-Sickinger, A. G., Wiedemann, P., et al. (2006). Glutamate release by neurons evokes a purinergic inhibitory mechanism of osmotic glial cell swelling in the rat retina: Activation by neuropeptide Y. *Journal of Neuroscience Research, 83*(4), 538–550.

Vega, J. L., Puebla, C., Vasquez, R., Farias, M., Alarcon, J., Pastor-Anglada, M., et al. (2009). TGF-beta1 inhibits expression and activity of hENT1 in a nitric oxide-dependent manner in human umbilical vein endothelium. *Cardiovascular Research, 82*(3), 458–467.

Vogel, Z., Daniels, M. P., & Nirenberg, M. (1976). Synapse and acetylcholine receptor synthesis by neurons dissociated from retina. *Proceedings of the National Academy of Sciences of the United States of America, 73*(7), 2370–2374.

Wan, W. J., Cui, D. M., Yang, X., Hu, J. M., Li, C. X., Hu, S. L., et al. (2011). Expression of adenosine receptors in human retinal pigment epithelium cells in vitro. *Chinese Medical Journal, 124*(8), 1139–1144.

Wardas, J. (2002). Neuroprotective role of adenosine in the CNS. *Polish Journal of Pharmacology, 54*(4), 313–326.

Witkovsky, P. (1963). An ontogenetic study of retinal function in the chick. *Vision Research, 61*, 341–355.

Wurm, A., Iandiev, I., Hollborn, M., Wiedemann, P., Reichenbach, A., Zimmermann, H., et al. (2008). Purinergic receptor activation inhibits osmotic glial cell swelling in the diabetic rat retina. *Experimental Eye Research, 87*, 385–393.

Wurm, A., Lipp, S., Pannicke, T., Linnertz, R., Farber, K., Wiedemann, P., et al. (2009). Involvement of A(1) adenosine receptors in osmotic volume regulation of retinal glial cells in mice. *Molecular Vision, 15*, 1858–1867.

Yau, J. W., Rogers, S. L., Kawasaki, R., Lamoureux, E. L., Kowalski, J. W., Bek, T., et al. (2012). Global prevalence and major risk factors of diabetic retinopathy. *Diabetes Care, 35*(3), 556–564.

Yazulla, S., & Brecha, N. (1980). Binding and uptake of the GABA analogue, 3H-muscimol, in the retinas of goldfish and chicken. *Investigative Ophthalmology & Visual Science, 19*(12), 1415–1426.

Zhang, M., Budak, M. T., Lu, W., Khurana, T. S., Zhang, X., Laties, A. M., et al. (2006). Identification of the A3 adenosine receptor in rat retinal ganglion cells. *Molecular Vision, 16*, 937–948.

Zhang, M., Hu, H., Zhang, X., Lu, W., Lim, J., Eysteinsson, T., et al. (2010). The A3 adenosine receptor attenuates the calcium rise triggered by NMDA receptors in retinal ganglion cells. *Neurochemistry International, 56*(1), 35–41.

Zhong, Y., Yang, Z., Huang, W. C., & Luo, X. (2013). Adenosine, adenosine receptors and glaucoma: An updated overview. *Biochimica et Biophysica Acta, 1830*, 2882–2890.

INDEX

Note: Page numbers followed by "*f*" indicate figures and "*t*" indicate tables.

B

F

G

H

M

N

T

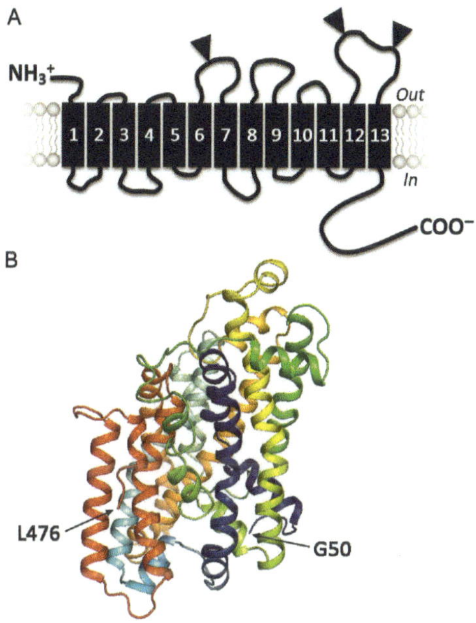

Juan Pablo Nicola *et al.*, Figure 1 NIS secondary and tertiary structure. (A) Secondary structure. NIS secondary structure model showing the 13 transmembrane segments from the extracellular amino terminus to the intracellular carboxy terminus. Black triangles mark N-linked glycosylation sites at N225, N485, and N497. (B) Tertiary structure. Membrane plane of the NIS homology model built using the rat NIS sequence including residues G50 through L476 (Paroder-Belenitsky et al., 2011), based on the X-ray structure of vSGLT. The NIS homology model is shown as a ribbon representation and rainbow colored by sequence, from the amino terminus (blue) to the carboxy terminus (red).

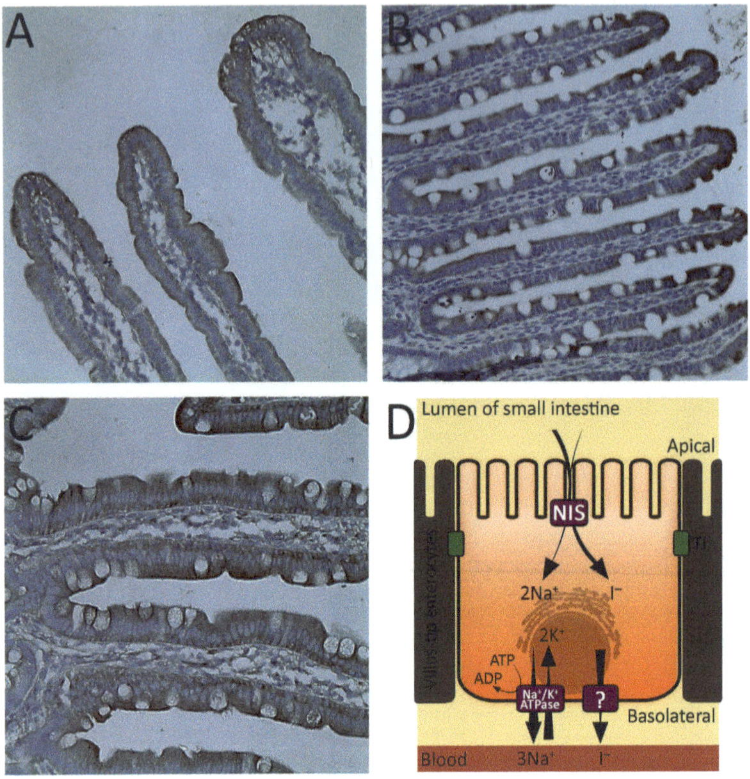

Juan Pablo Nicola *et al.*, Figure 3 NIS expression in rat small intestine. Immunohistochemistry analysis demonstrating NIS expression in the three sections of the rat small intestine. NIS-specific immunostaining is evident in the apical membrane of the absorptive enterocytes (Nicola et al., 2009). (A) Duodenum, (B) jejunum, and (C) ileum. All pictures are presented at 40 × magnification. (D) Schematic representation of NIS-mediated I^- absorption in villus-tip small intestine enterocytes. NIS mediates transcellular apical transport of dietary I^- against its concentration gradient by coupling it to Na^+ transport. The Na^+ gradient is generated by the Na^+/K^+ ATPase in the basolateral membrane, maintaining the Na^+ electrochemical gradient that favors the Na^+-dependent accumulation of substrates. Transport of I^- into the blood may be facilitated by still uncharacterized channels or transporters. It is not known whether I^- is translocated across the intestinal epithelium through the intercellular space between the enterocytes (paracellular transport). TJ, tight junctions.

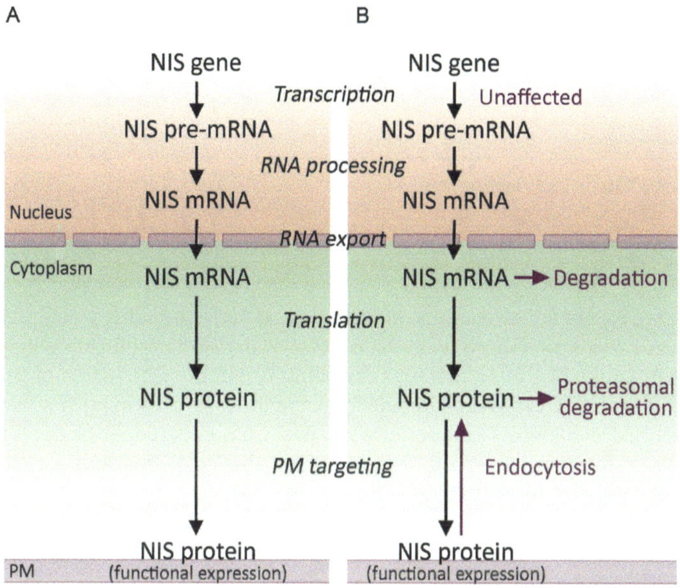

Juan Pablo Nicola *et al.*, Figure 4 Different levels of intestinal NIS regulation induced by high concentrations of I^-. (A) Under normal conditions, the NIS gene is transcribed into NIS pre-mRNA. Then introns are removed and exons reconnected to generate mature NIS mRNA, which is exported to the cytoplasm and translated into protein. NIS protein is glycosylated and targeted to the plasma membrane of small intestine absorptive enterocytes where it mediates I^- transport. (B) Under I^- excess, although the NIS gene is normally transcribed, mature NIS mRNA levels are decreased due to a reduction in its stability. NIS protein expression levels are diminished in response to increased proteasomal degradation, and NIS expression at the plasma membrane is down-regulated due to increased amiloride-sensitive internalization. As a result, I^- excess decreased NIS-mediated accumulation of I^- in enterocytes (Nicola et al., 2012).

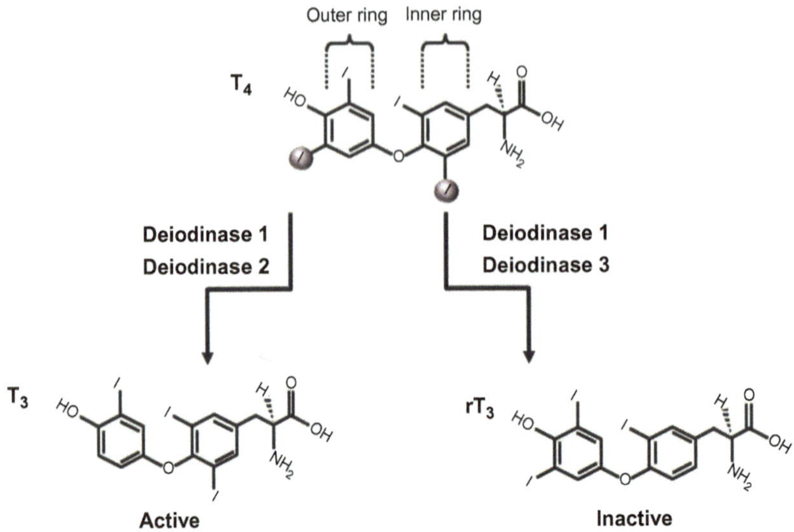

Peying Fong, Figure 2 Deiodinases regulate thyroid hormone activity at target tissues. Deiodinases 2 and 3 primarily govern this process, with deiodinase 2 catalyzing removal of I^- from the outer ring of T_4, thereby generating T_3 (left branch). Action of deiodinase 3, on the other hand, catalyzes removal from the inner ring, forming the inactive rT_3 (right branch). Deiodinase 1 can perform removal from either inner or outer ring, but its function in humans is obscure (see text). The shaded I^- residues indicate those removed by action of respective deiodinases depicted.

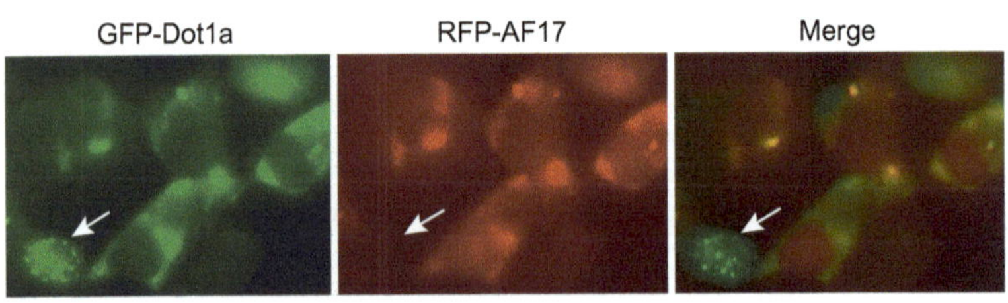

Lihe Chen *et al.*, Figure 3 Representative images showing that AF17 facilitates Dot1a cytoplasmic distribution. 293T cells were transiently transfected with GFP-Dot1a and RFP-AF17, and examined by epifluorescence microscopy (20×). Arrow: a cell showing typical nuclear localization pattern of GFP-Dot1a in the absence of detectable RFP-AF17. *This figure was originally published in Reisenauer et al. (2009). © The American Society for Biochemistry and Molecular Biology.*

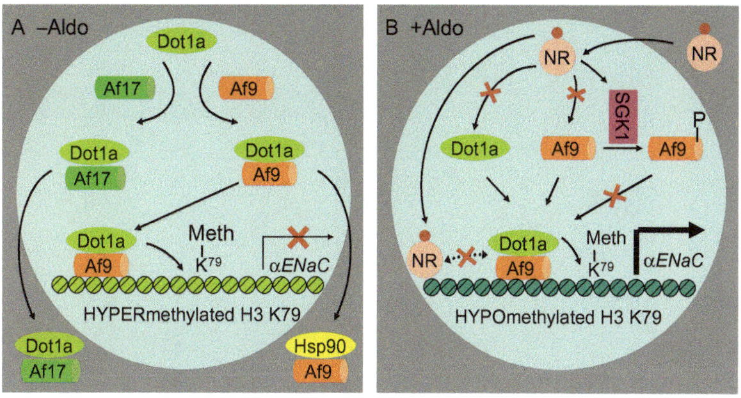

Lihe Chen *et al.*, Figure 4 Model for Dot1a–Af9-mediated repression of *αENaC* transcription. Under basal conditions (−Aldo), Af9 recruits Dot1a to form a nuclear complex, which indirectly or directly through Af9 DNA-binding activity binds specific sites of the *αENaC* promoter, leading to hypermethylation of histone H3 K79 and repression of *αENaC* transcription (Zhang, Xia, Jalal, et al., 2006; Zhang, Xia, Reisenauer, et al., 2006; Zhang, Yu, et al., 2013). Af17 relieves the repression by competing with Af9 for binding Dot1a and promoting Dot1a redistribution from the nucleus to cytoplasm (Reisenauer et al., 2009; Wu et al., 2011). Hsp90 also interacts with Af9 and apparently promotes its cytoplasmic distribution to derepress the *αENaC* promoter (Lin & Hemenway, 2010). Aldosterone (+Aldo) stimulates *αENaC* transcription by multiple mechanisms. Through the classical action, aldosterone binds and activates the nuclear hormone receptors (NR) that are either glucocorticoid receptor or mineralocorticoid receptor homo- or heterodimers to bind the glucocorticoid response element in the *αENaC* promoter and transactivate *αENaC*. In parallel, aldosterone releases Dot1a–Af9-mediated repression by limiting the formation of the complex through (A) downregulating Dot1a (Zhang, Xia, Jalal, et al., 2006) and Af9 expression (Zhang, Xia, Reisenauer, et al., 2006), presumably via nuclear receptor-dependent or -independent (not shown) mechanisms, (B) decreasing the Dot1a–Af9 interaction via Sgk1-mediated phosphorylation of Af9 at S435 (Zhang et al., 2007), and (C) counterbalancing Dot1a–Af9 complex by activating MR to compete for binding Af9 (Zhang, Zhou, et al., 2013). These concerted actions lead to histone H3 K79 hypomethylation at specific subregions of the *αENaC* promoter. Sgk1 might join and downregulate the Dot1a–Af9 complex associated with the *αENaC* promoter (not shown for simplicity). In all cases, Af9-free Dot1a binds DNA nonspecifically and catalyzes histone H3 K79 methylation throughout the genome under basal conditions (not shown). Meth, methylation.

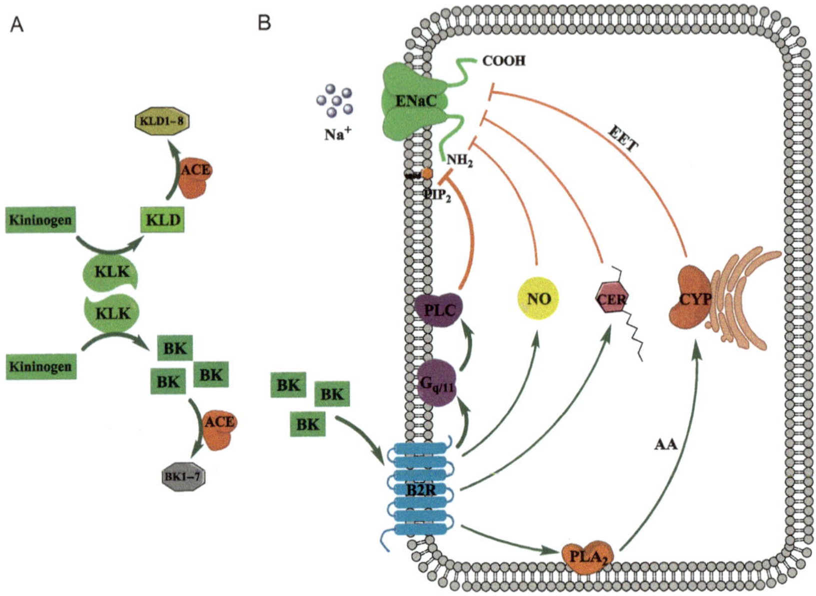

Mykola Mamenko *et al.*, Figure 1 Kallikrein–kinin system in regulation of ENaC activity in the distal nephron. (A) Principal scheme of Bradykinin synthesis and degradation. KLK, kallikrein; KLD, kallidin; KLD1–8, kallidin (1–8); BK, bradykinin; BK1–7, bradykinin (1–7); ACE, angiotensin converting enzyme. (B) Intracellular signaling pathways coupling BK with changes in ENaC activity. Bold arrows represent signaling pathways supported by direct experimental evidence; and thin arrows indicate additional possible mechanisms of regulation. BK, bradykinin; B2R, bradykinin type 2 receptor; ENaC, epithelial Na^+ channel; $G_{q/11}$, heterotrimeric G protein Q/11 subtype; PLC, phospholipase C; PLA_2, phospholipase A2; NO, nitric oxide; CYP, cytochrome P450; AA, arachidonic acid; CER, ceramide; EET, epoxyeicosatrienoic acids; PIP_2, phosphatidylinositol 4,5-bisphosphate. NH_2 and COOH demonstrate amino- and carboxyl-termini of ENaC subunits, respectively. Green and red arrows represent stimulatory and inhibitory actions, respectively.

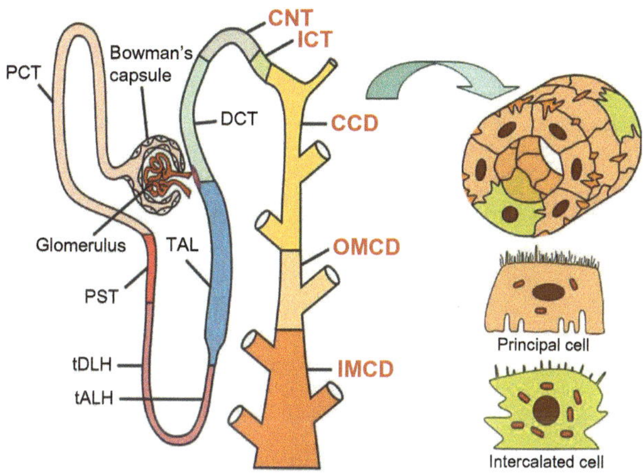

Andrey Sorokin and Alexander Staruschenko, Figure 1 Structure of the nephron and specific segments involved in effects of ET-1 on ENaC. Every nephron contains a renal corpuscle (glomerulus and Bowman's capsule), a proximal tubule (proximal convoluted and straight tubules; PCT and PST, respectively), a loop of Henle (thin descending limb of Henle's loop (tDLH), thin ascending limb of Henle's loop (tALH), and thick ascending limb of Henle's loop (TAL)), a distal convoluted tubule (DCT), connecting tubule (CNT), and the collecting duct system, which includes the initial collecting tubule (ICT), the cortical collecting duct (CCD), the outer medullary collecting duct (OMCD), and the inner medullary collecting duct (IMCD). Structure of the CCD as a cross section and schematic presentation of principal and intercalated cells that comprise these segments are also shown. *Modified from Staruschenko (2012).*

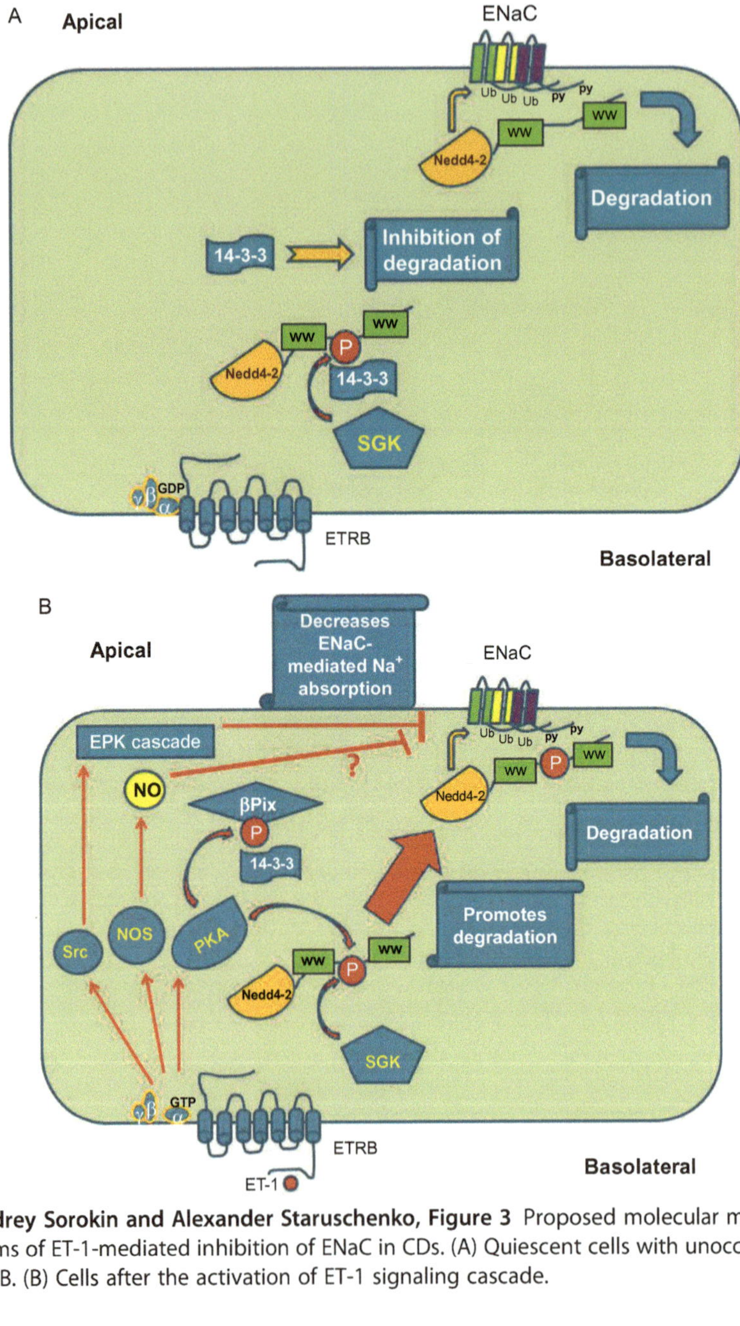

Andrey Sorokin and Alexander Staruschenko, Figure 3 Proposed molecular mechanisms of ET-1-mediated inhibition of ENaC in CDs. (A) Quiescent cells with unoccupied ETRB. (B) Cells after the activation of ET-1 signaling cascade.

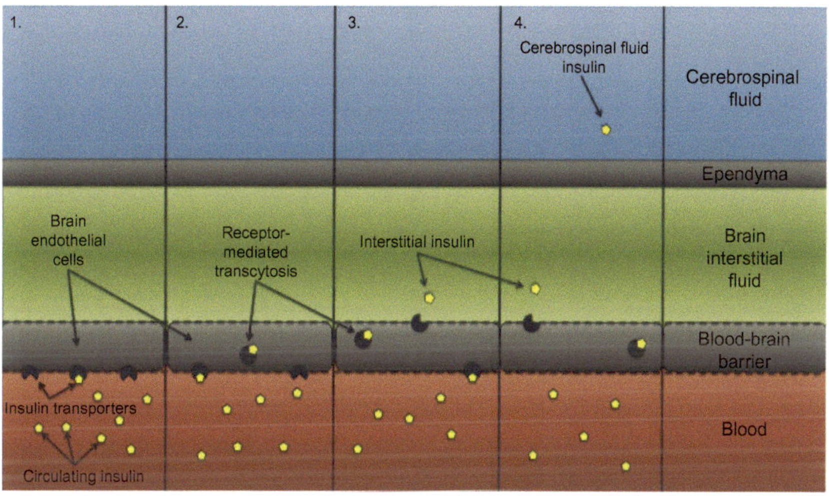

Denovan P. Begg, Figure 1 Transport of circulating insulin from the blood into the central nervous system. (1) Circulating insulin, released into the blood pancreatic β cells, binds to transporters on endothelial cells of the blood–brain barrier. (2) Insulin is transported through endothelial cells by receptor-mediated transcytosis. (3) Insulin is released into brain interstitial fluid, here it can act on neuronal insulin receptors. (4) Insulin then moves through the ependyma into the cerebrospinal fluid.

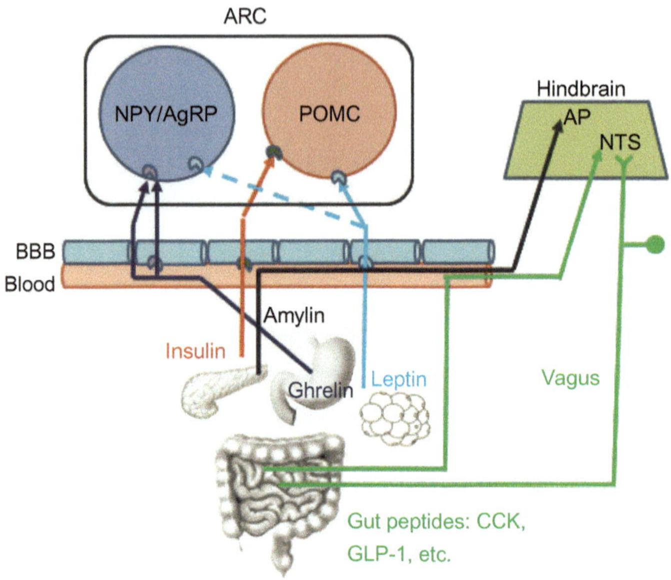

Denovan P. Begg, Figure 2 Numerous hormonal signals from the periphery influence different areas of the brain involved in food intake. The adiposity signal insulin, from the pancreas, interacts with a transporter on capillary endothelial cells comprising the blood–brain barrier (BBB), is transported into the brain and stimulates pro-opiomelanocortin (POMC) cells in the arcuate nucleus of the hypothalamus (ARC). The other adiposity signal leptin, produced by adipocytes, enters the brain via active transport and stimulates POMC cells while inhibiting neuropeptide-Y (NPY)/agouti-related peptide (AgRP) cells in the ARC. Ghrelin from the stomach has been reported to enter the brain not only via active transport, like insulin, but also via fenestrated capillaries during fasting, it stimulates NPY/AgRP cells in the ARC. Other factors act via the hindbrain to alter food intake, pancreatic amylin circulates directly to the area postrema (AP) in the hindbrain. Gut peptides such as glucagon-like peptide-1 (GLP-1) and cholecystokinin (CCK) either stimulate the nucleus of the solitary tract (NTS) directly or stimulate vagal afferent nerves whose axons end in the NTS.

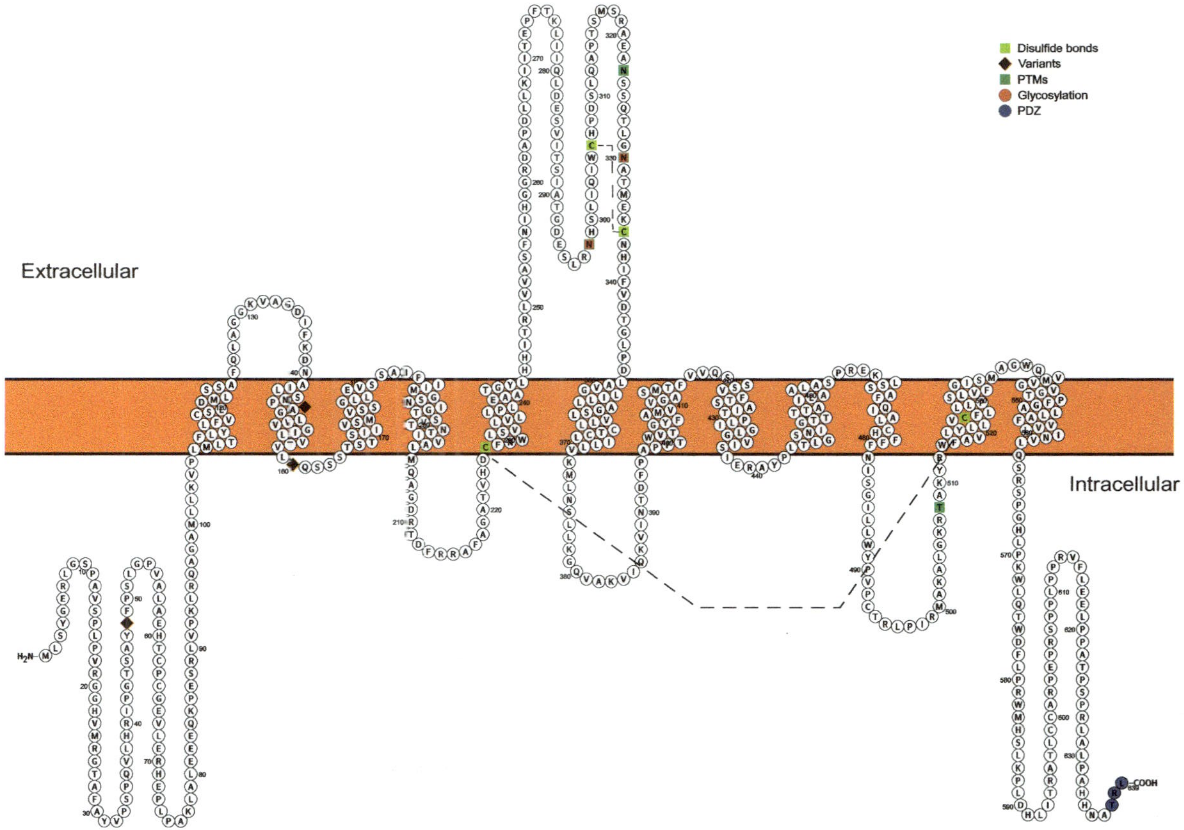

Jyothsna Gattineni and Peter A. Friedman, Figure 1 Structural organization of human NPT2A Na–Pi cotransporter. Described mutations (A48F, V147M), disulfide bridges, and the carboxy-terminal PDZ-binding domain are shown. Putative N-glycosylation sites are based on those described for the rat Npt2a (Hayes et al., 1994). The model was constructed using a hidden Markov model for topology prediction and visualized with Protter http://wlab.ethz.ch/protter/start/ (Omasits, Ahrens, Muller, & Wollscheid, 2014).

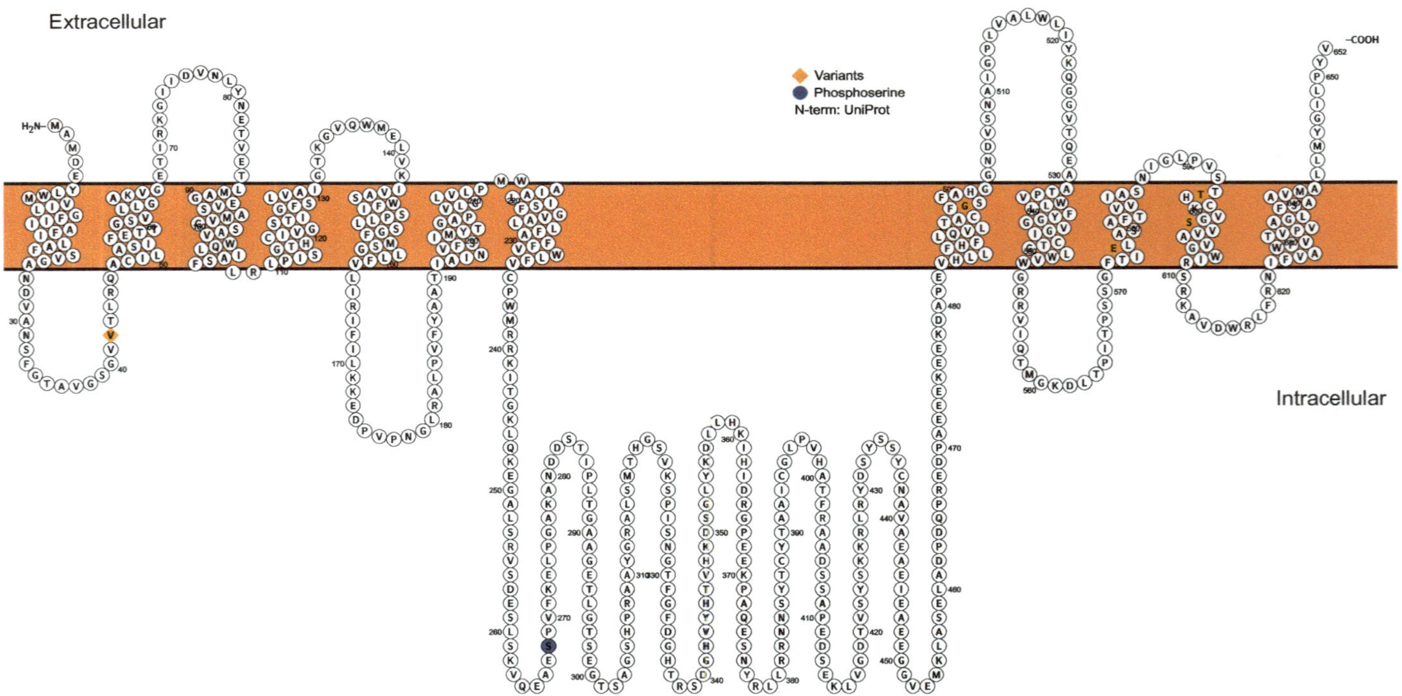

Jyothsna Gattineni and Peter A. Friedman, Figure 2 Human PIT-2 organization. The model was constructed as above and conforms to experimental findings (Bottger & Pedersen, 2011; Salaun, Rodrigues, & Heard, 2001). Naturally occurring variants are indicated in filled orange diamonds (Wang, Li, et al., 2012).

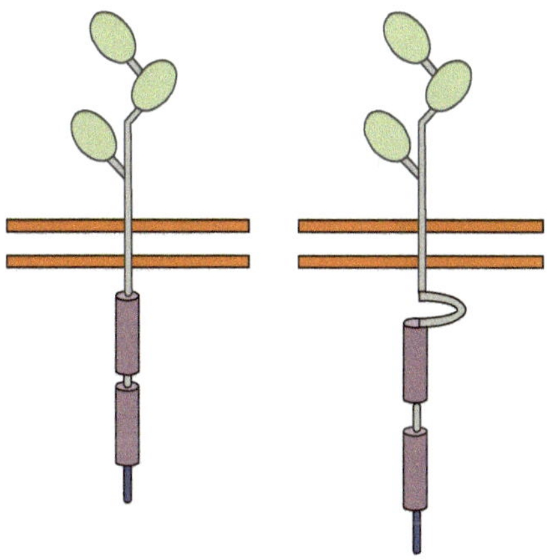

Jyothsna Gattineni and Peter A. Friedman, Figure 3 Features of FGFR. The extracellular domain contains 2 or 3 three immunoglobulin (IG)-like domains, a transmembrane domain, the absence or presence of a threonine–valine (TV) insert that defines a or b isoforms, and a split tyrosine kinase domain. Additionally, alternative splicing of the transcribed receptor genes results in a variety of receptor isoforms. Alternative splicing of the third IG-like domain occurs for FGFR1, FGFR2, and FGFR3.

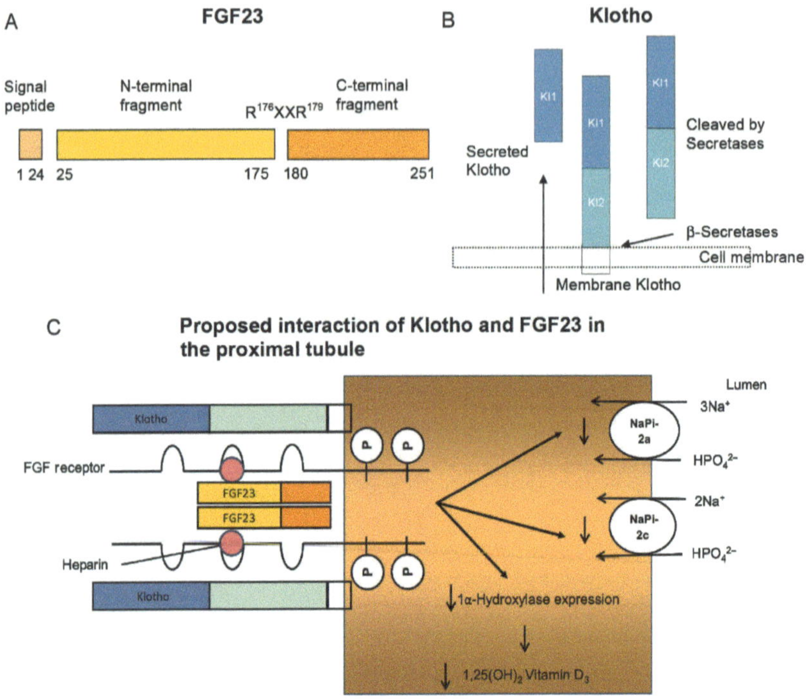

Jyothsna Gattineni and Peter A. Friedman, Figure 4 Structure of FGF23, Klotho, and the proposed interaction of FGF23, Klotho and FGFRs. (A) FGF23 structure depicting the signal sequence, N-terminal fragment that has homology to other FGFs, and the C-terminal which is divergent from other FGFs and has a high affinity to bind Klotho. (B) Structure of membrane Klotho with its two homologous Kl domains and short intracellular domain. The secreted form can be due to alternate splicing of the *Klotho* gene, alternatively, membrane Klotho can be cleaved by secretases. (C) The proposed interaction of 2FGF23:2FGFR:2Klotho with the site of heparin shown even though FGF23 has low affinity for heparin. Dimerization of the FGFRs with its ligands results in *trans*-phosphorylation of its tyrosine kinase domains which initiates intracellular signaling. Activation of the intracellular MAPK signaling ultimately decreases the expression of Npt2a and Npt2c with decreased phosphate reabsorption. Additionally, the expression of 1α-hydroxylase is decreased with decreased serum 1,25[OH]$_2$D$_3$ levels. *Adapted with permission from Gattineni and Baum (2012).*

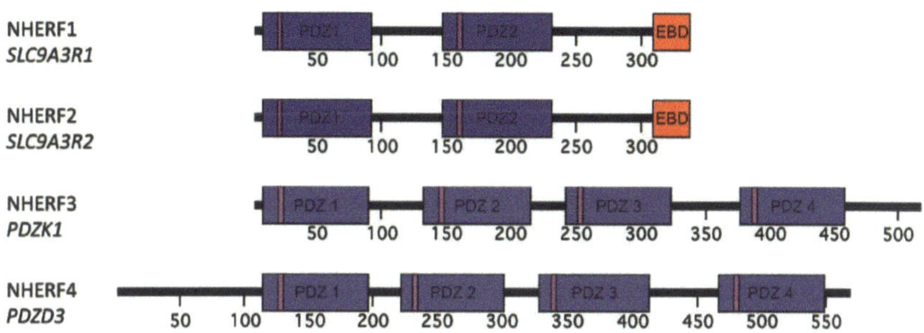

Jyothsna Gattineni and Peter A. Friedman, Figure 5 Schematic structural representation of the four described members of the human Na^+/H^+ exchanger (NHERF) family. Common and gene names are shown on the left. PDZ domains are shaded in blue and, where present, ezrin-binding domains are represented in red. Each PDZ domain possesses a single core-binding motif (pink band). The four-residue sequence may be the same (GYGF) as in the two PDZ domains of NHERF1 and NHERF2, or may be different as in the four PDZ domains of NHERF3 (GFFL, GFSL, GYGF, GYGF) or NHERF4 (GFHL, GFGF, GFGF, GFRL). Notably, NHERF1 and NHERF 2 terminate with a PDZ ligand (-SNL, -SNF, respectively) allowing them to form a closed confirmation (Wang, Means, et al., 2012) or functional heterodimers (Yang et al., 2014).

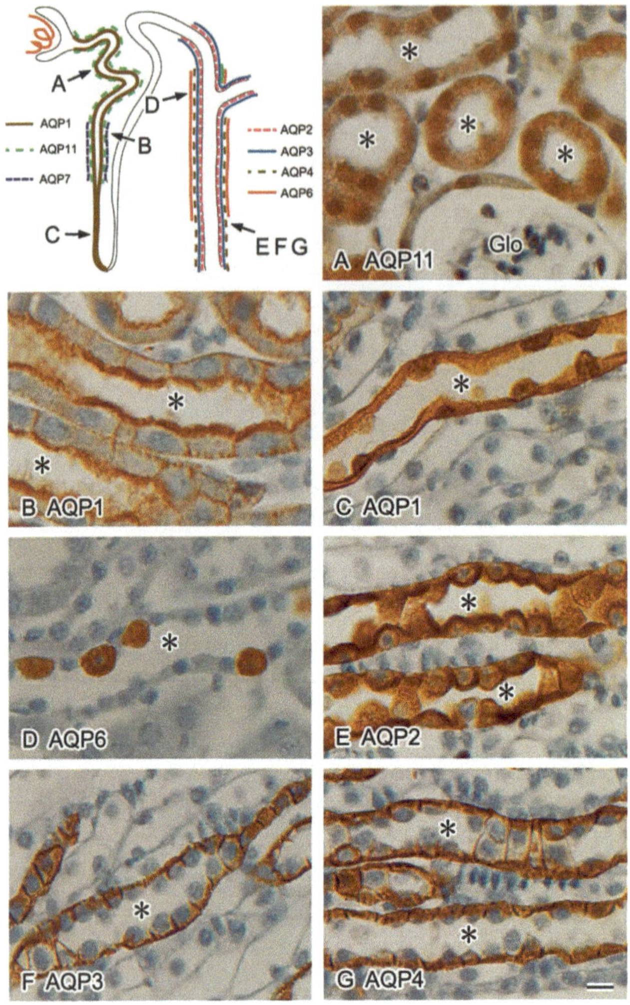

Masahiro Ikeda and Toshiyuki Matsuzaki, Figure 2 Aquaporins in the kidney. A schematic drawing of the distribution of AQP1, AQP2, AQP3, AQP4, AQP6, AQP7, and AQP11 in the urinary tubule and collecting duct system is shown at the upper left. Micrographs of each region indicated by arrows A–G in mouse kidney immunolabeled with isoform-specific antibodies by the immunoperoxidase method are shown in panels (A)–(G), respectively. Cell nuclei were counterstained with hematoxylin. (A) AQP11 is localized intracellularly in proximal tubular cells. Glo, glomerulus. (B) AQP1 is localized in both the apical and basolateral membranes of proximal tubular cells. (C) AQP1 in the descending limb of the loop of Henle. (D) AQP6 in the cortical collecting duct. AQP6 is localized intracellularly in intercalated cells. (E) AQP2 in the inner medullary collecting duct. Strong labeling is seen in the apical region of principal cells. AQP3 (F) and AQP4 (G) in the inner medullary collecting duct. Both are localized to the basolateral membrane of principal cells. Asterisks indicate the lumen of the immunopositive tubules and ducts. Bar: 10 μm in (A)–(G).

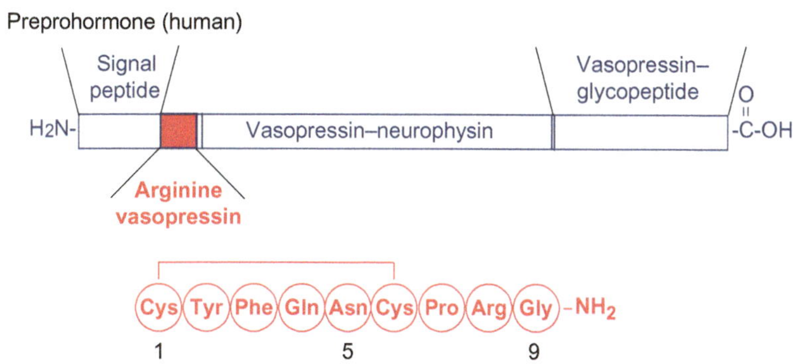

Masahiro Ikeda and Toshiyuki Matsuzaki, Figure 3 Schematic drawings of human arginine vasopressin and its preprohormone. The preprohormone includes four domains: an N-terminal signal peptide, vasopressin, vasopressin–neurophysin, and vasopressin–glycopeptide. During synthesis, the signal peptide is cleaved from the preprohormone to produce prohormone. The prohormone is then processed by sequential enzymatic cleavages to yield vasopressin.

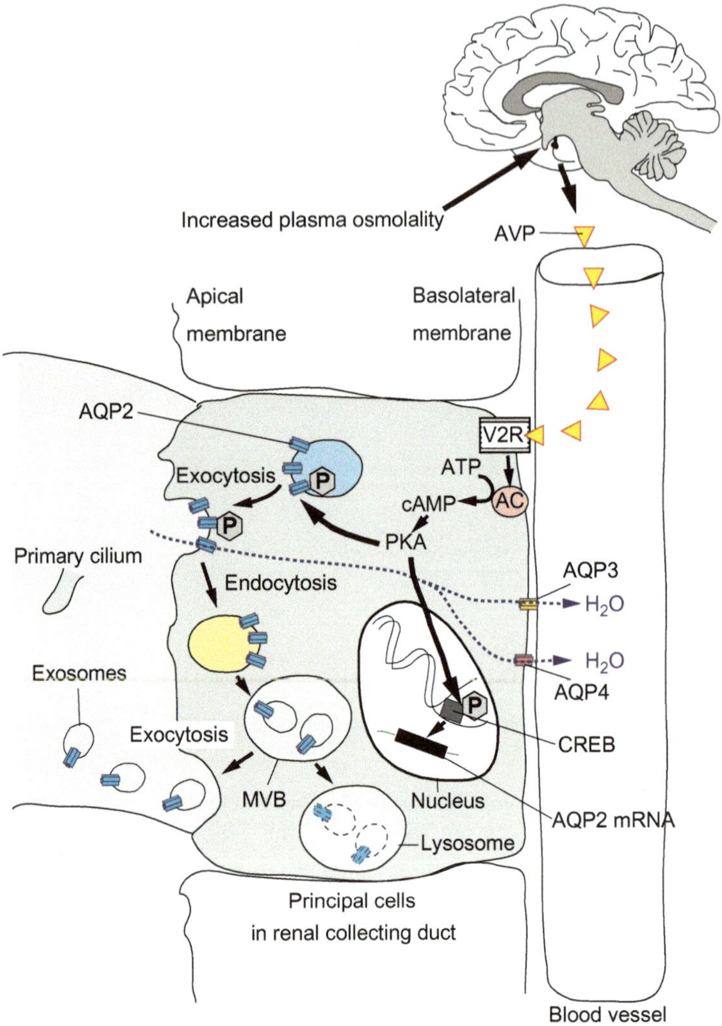

Masahiro Ikeda and Toshiyuki Matsuzaki, Figure 4 An illustration of the mechanism underlying the short- and long-term regulations of AQP2. Vasopressin released from the posterior pituitary binds to the V2 receptor (V2R) present on the basolateral membrane of principal cells in the renal collecting duct. Activation of the receptor stimulates adenylate cyclase (AC) activity, resulting in the production of cAMP, which in turn activates protein kinase A (PKA), and this protein kinase phosphorylates AQP2. Subsequently, AQP2 accumulates on the apical surface membrane through vesicular trafficking from intracellular storage vesicles expressing AQP2. Thereafter, water molecules flowing into the collecting duct cells through AQP2 are able to flow out to the interstitium via the basolateral water channels AQP3 and/or AQP4. The PKA also phosphorylates CRE-binding protein (CREB), and the phosphorylated CREB then stimulates transcription of the AQP2 gene. Vasopressin also regulates the abundance of renal AQP2 via excretion of exosomal AQP2. When the multivesicular bodies (MVBs) fuse with the plasma membrane in response to vasopressin, intraluminal vesicles, termed exosomes, are released into the urinary space.

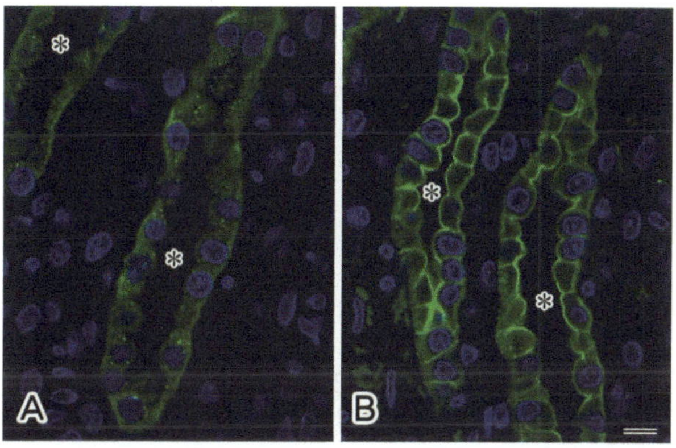

Masahiro Ikeda and Toshiyuki Matsuzaki, Figure 5 Short-term regulation of the subcellular localization of AQP2. Immunofluorescence microscopy of AQP2-specific antibody staining in kidneys of a water-loaded rat (A) and a vasopressin-administered rat (B). Inner medullary collecting ducts were observed with a laser confocal microscope. AQP2 is localized intracellularly in the kidney of a water-loaded rat (A). AQP2 has markedly accumulated in the apical membrane and slightly in the basolateral membrane at 30 min after vasopressin administration (B). Asterisks indicate the lumen of collecting ducts. Bar: 10 μm.

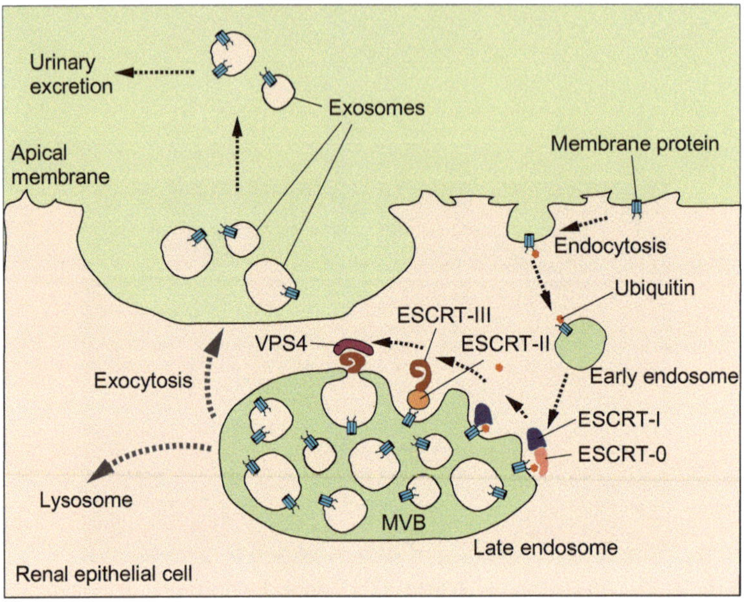

Masahiro Ikeda and Toshiyuki Matsuzaki, Figure 6 Biogenesis of exosomes in renal epithelial cells. The endosome containing nanosized intraluminal vesicles is known as the multivesicular body (MVB). The formation of this MVB is mediated by the endosomal-sorting complex required for transport (ESCRT) machinery, including the ESCRT-0, -I, -II, and -III protein complexes. Ubiquitination of cargo proteins such as membrane proteins causes endocytosis. Then, ESCRT-0 captures the ubiquitinated proteins at the endosome membrane, and this interaction then initiates the invagination of a portion of the endosome membrane. Thereafter, the cargo proteins are sequentially influenced by ESCRT-I, ESCRT-II, and ESCRT-III and then finally excised from the endosome membrane by the action of vacuolar protein sorting 4 (VPS4), resulting in the formation of intraluminal vesicles (exosomes). Once MVBs have fused with lysosomes, the cargo proteins are degraded, and when the MVBs have fused with the plasma membrane, exosomes are released into the extracellular space.

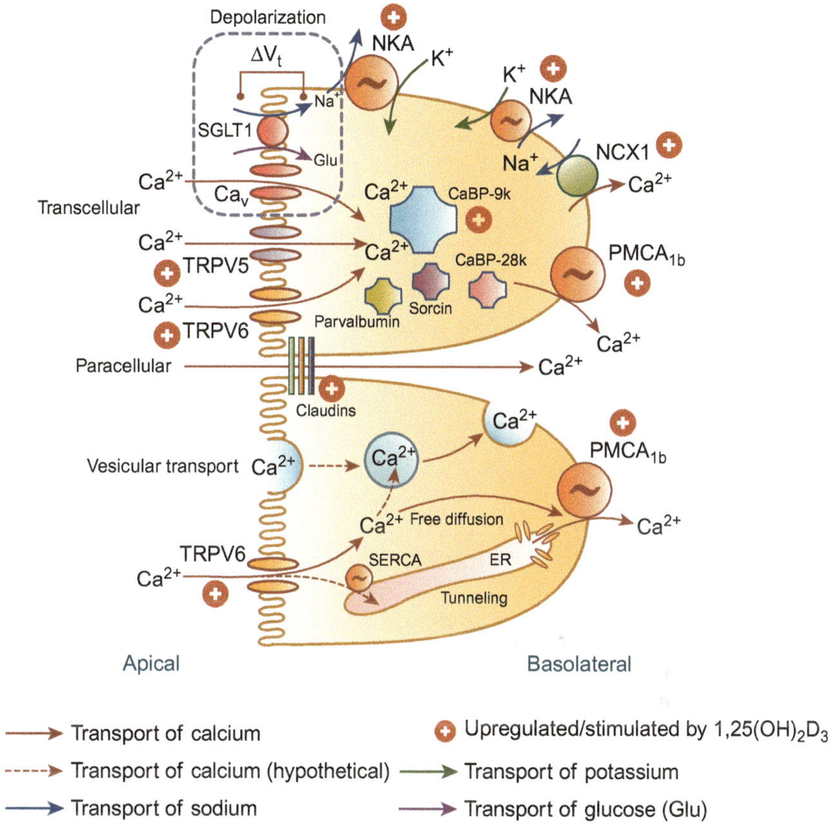

Kannikar Wongdee and Narattaphol Charoenphandhu, Figure 1 A schematic diagram shows possible mechanisms of transcellular and paracellular calcium transport in the rat duodenum. 1,25(OH)$_2$D$_3$ is capable of stimulating both paracellular and transcellular calcium transport [TRPV5/6-mediated calcium entry, calcium-binding protein (CaBP)-facilitated calcium diffusion, and basolateral active calcium extrusion], as indicated by red plus signs. ΔV_t, change in membrane potential. Please see text for details. Glu, glucose.

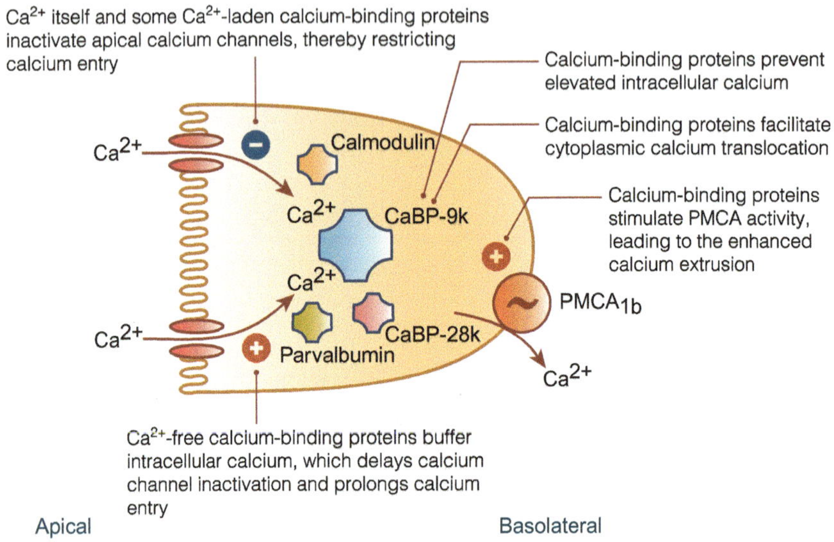

Kannikar Wongdee and Narattaphol Charoenphandhu, Figure 2 Cytoplasmic calcium-binding proteins (CaBP) in the duodenal enterocytes show a high degree of redundancy and have a number of functions. Calcium-free calcium-binding proteins (e.g., calbindin-D_{9k} and -D_{28k}) buffer the intracellular calcium, which can, in turn, delay calcium channel inactivation and prolong calcium entry. On the other hand, some calcium-laden calcium-binding proteins (e.g., calmodulin and calbindin-D_{28k}) as well as intracellular calcium itself lead to calcium-dependent inactivation of apical calcium channels, thereby restricting calcium entry. Several calcium-binding proteins, especially calbindin-D_{9k}, ferry the intracellular calcium from apical-to-basolateral compartment by a process known as translational diffusion rather than bucket brigade. Finally, they potentiate PMCA activity to accelerate the basolateral calcium extrusion. Ca^{2+}, free-ionized calcium.

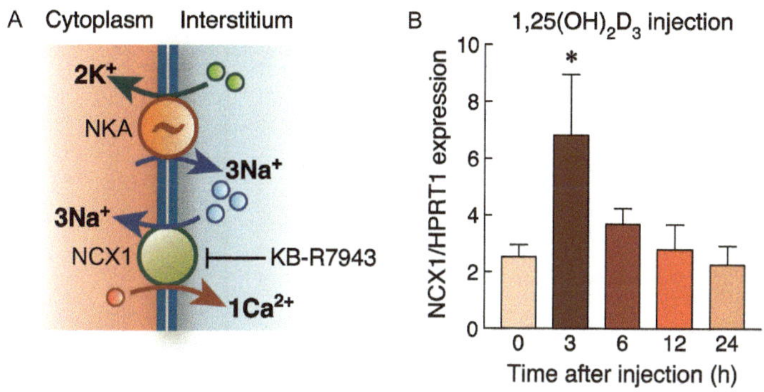

Kannikar Wongdee and Narattaphol Charoenphandhu, Figure 3 (A) Basolateral calcium extrusion through Na^+/Ca^{2+}-exchanger (NCX)-1. The NCX1 is coupled with Na^+/K^+-ATPase (NKA). NKA first moves three sodium ions out of the cell and moves two potassium ions into the cell, thereby generating sodium gradient for NCX1 function. NCX1 then allows influx of three sodium ions and efflux of one calcium ion. NCX1 activity can be inhibited by KB-R7943. (B) Time-dependent expression of NCX1 in duodenal epithelial cells of mice subcutaneously injected with a single-dose 2 μg/kg $1,25(OH)_2D_3$, as determined by quantitative real-time PCR. The duodenal epithelial cells were collected at 0, 3, 6, 12, and 24 h after $1,25(OH)_2D_3$ injection. Values are means ± SE of NCX1/HPRT1 expression ratios. The results show that, at 3 h postinjection, $1,25(OH)_2D_3$ significantly increases the mRNA level of NCX1 by 2.69-fold ($*P < 0.05$ vs. 0 h; one-way analysis of variance with Dunnett's posttest) before returning to the baseline thereafter. HPRT1, hypoxanthine phosphoribosyltransferase-1 (a housekeeping gene).

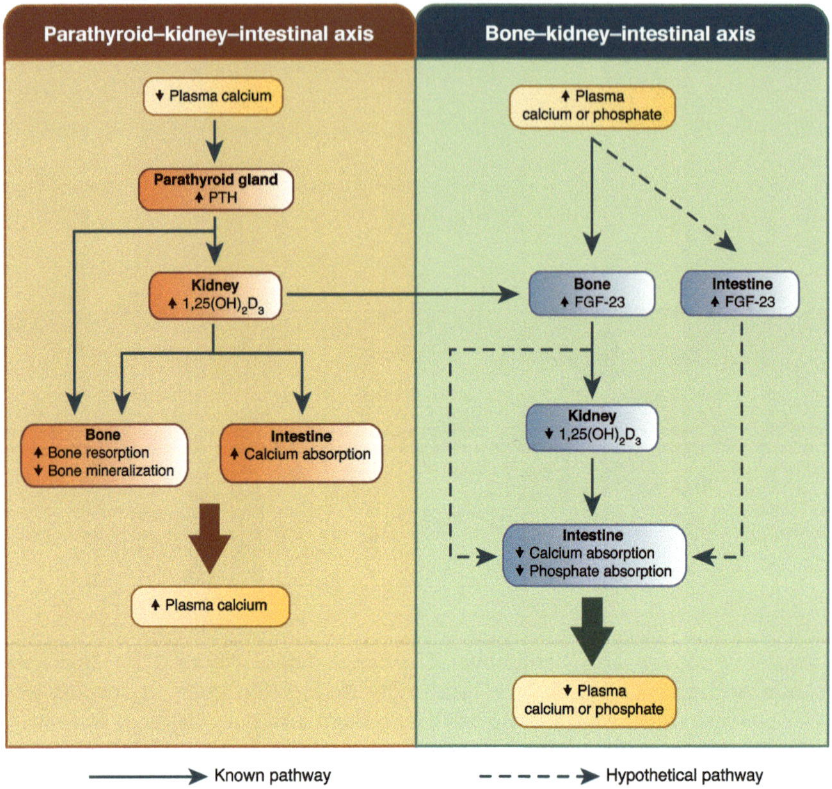

Kannikar Wongdee and Narattaphol Charoenphandhu, Figure 4 Interrelationship between the parathyroid–kidney–intestinal (left panel) and bone–kidney–intestinal axes (right panel) in regulating calcium homeostasis. In the parathyroid–kidney–intestinal axis, low plasma calcium level is detected by parathyroid glands, thereby stimulating PTH release. PTH stimulates 1α-hydroxylase in the kidney to convert $25(OH)D_3$ into $1,25(OH)_2D_3$. An increase in $1,25(OH)_2D_3$ level further enhances bone resorption, inhibits bone mineralization, and promotes intestinal calcium absorption, thus increasing plasma calcium level. In the bone–kidney–intestinal axis, high plasma calcium leads to an increase in circulating level of fibroblast growth factor (FGF)-23 produced by bone and perhaps by the intestine. FGF-23 suppresses renal $1,25(OH)_2D_3$ production by inhibiting 1α-hydroxylase, thereby decreasing intestinal calcium and phosphate absorption, which, in turn, lower plasma calcium and phosphate levels. An increase in circulating $1,25(OH)_2D_3$ level in the parathyroid–kidney–intestinal axis can also increase bone FGF-23 production, which links the two axes together. Furthermore, FGF-23 is hypothesized to exert a direct inhibitory effect on the intestine. Circulating FGF-23 might directly suppress intestinal calcium and phosphate absorption, whereas local FGF-23 production from the enterocytes might serve to prevent excessive intestinal calcium/phosphate absorption.

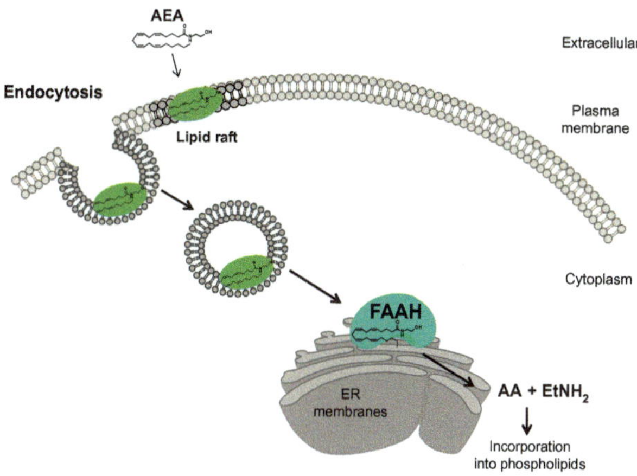

Simon Nicolussi and Jürg Gertsch, Figure 2 Model of endocytosis-mediated AEA uptake and intracellular sequestration.

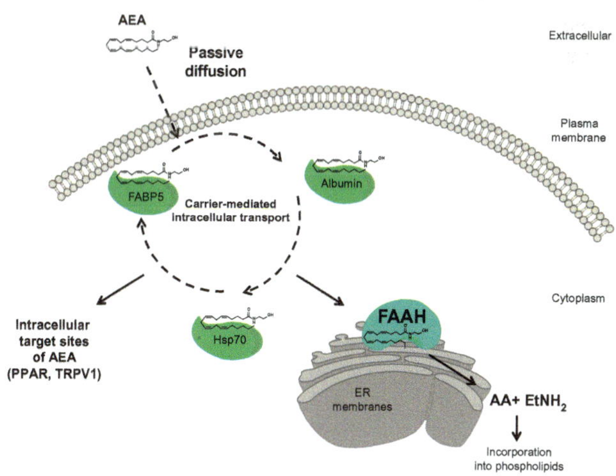

Simon Nicolussi and Jürg Gertsch, Figure 3 Model of passive diffusion of AEA across the plasma membrane following intracellular sequestration and/or carrier-mediated transport.

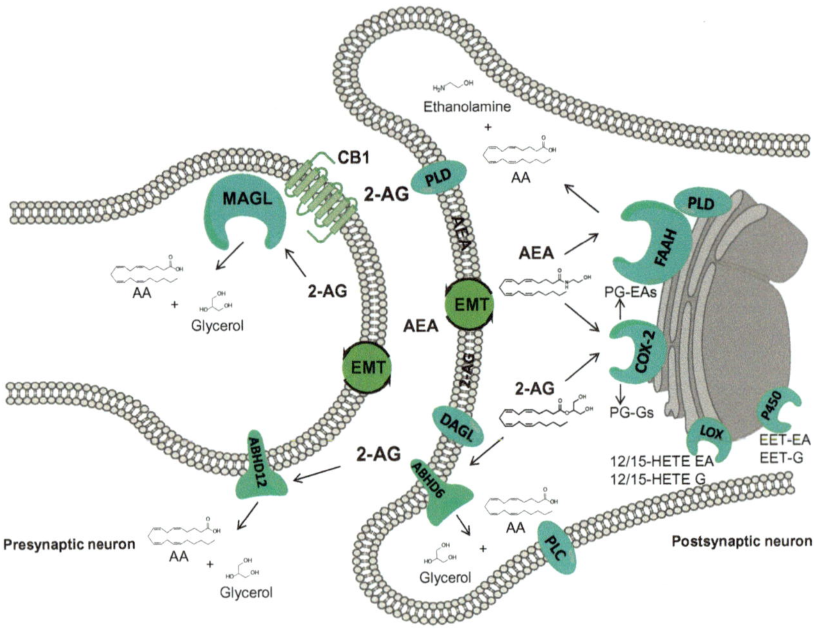

Simon Nicolussi and Jürg Gertsch, Figure 5 Transport, inactivation, and transformation of the endocannabinoids, 2-AG and AEA, at the synapse. AA = arachidonic acid, AEA = anandamide, 2-AG = 2-arachidonoyl glycerol, ABHD6/12 = α/β hydrolase 6/12, CB1 = cannabinoid receptor type 1, COX-2 = cyclooxygenase 2, DAGL = diacylglycerol lipase, EET-EA/-G = epoxyeicosatrienoic acid ethanolamide/glycerol, EMT = putative endocannabinoid membrane transporter, FAAH = fatty acid amide hydrolase, 12/15 HETE EA/G = 12/15-hydroxyeicosatetraenoic acid ethanolamide/glycerol, LOX = lipoxygenase, P450 = cytochrome enzymes, PLC = phospholipase C, PLD = phospholipase D, and PG-EAs/Gs = prostaglandin ethanolamides/glycerol esters.

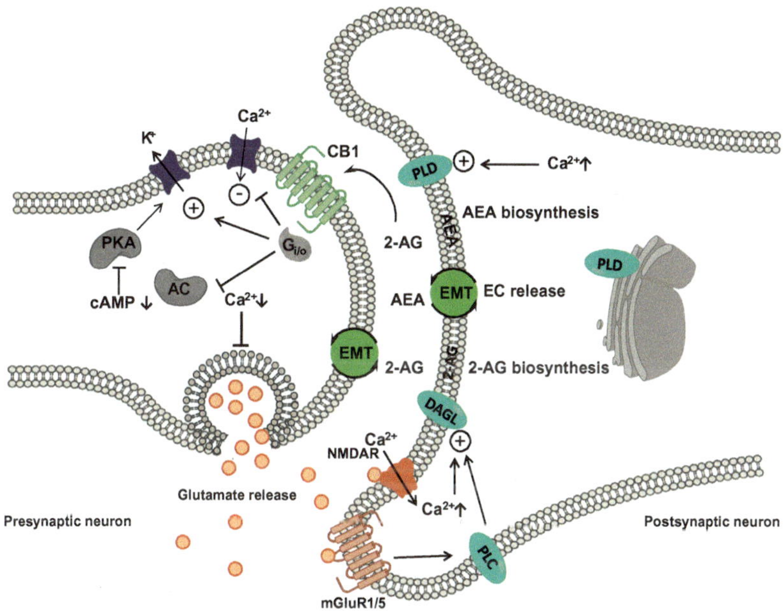

Simon Nicolussi and Jürg Gertsch, Figure 6 Schematic representation of the ECS in pre- and postsynaptic neurons mediating retrograde signaling at the glutamatergic synapse.

FSC
www.fsc.org
MIX
Papier | Fördert gute Waldnutzung
FSC® C083411

Kolibri 360 GmbH
Ferdinand-Jühlke-Straße 7
99095 Erfurt, Deutschland
produktsicherheit@kolibri360.de